Alfredo Rizzi · Maurizio Vichi
Hans-Hermann Bock (Eds.)

Advances in Data Science and Classification

Proceedings of the 6th Conference of the
International Federation of Classification Societies (IFCS-98)
Università "La Sapienza", Rome, 21–24 July, 1998

With 141 Figures
and 75 Tables

Springer

Professor Alfredo Rizzi
Università di Roma "La Sapienza"
Dipartimento di Statisticà, Probabilità e
Statistiche Applicate
Piazzale Aldo Moro 5
I-00185 Roma, Italia

Professor Maurizio Vichi
Università "G. D'Annunzio" di Chieti
Dipartimento di Metodi Quantitativi e
Teoria Economica
Viale Pindaro 42
I-65127 Pescara, Italia

Professor Dr. Hans-Hermann Bock
Rheinisch-Westfälische Technische Hochschule (RWTH) Aachen
Institut für Statistik und Wirtschaftsmathematik
Wüllnerstraße 3
D-52056 Aachen, Germany

ISBN 3-540-64641-8 Springer-Verlag Berlin Heidelberg New York

Cataloging-in-Publication Data applied for
Die Deutsche Bibliothek – CIP-Einheitsaufnahme
Advances in data science and classification:
Università "La Sapienza", Rome, 21–24 July, 1998; with 75 tables / Alfredo Rizzi ... (ed.). –
Berlin; Heidelberg; New York; Barcelona; Budapest; Hong Kong; London; Milan; Paris; Singapore;
Tokyo: Springer, 1998
(Proceedings of the ... conference of the International Federation of Classification Societies; 6)
(Studies in classification, data analysis, and knowledge organization)
ISBN 3-540-64641-8
International Federation of Classification Societies:
Proceedings of the ... conference of the International Federation of Classification Societies. –
Berlin; Heidelberg; New York; Barcelona; Budapest; Hong Kong; London; Milan; Paris; Singapore;
Tokyo: Springer
6. Advances in data science and classification. – 1998

Printed in Germany

Coverdesign: Erich Kirchner, Heidelberg

SPIN 10684513 43/2202-5 4 3 2 1 0 – Printed on acid-free paper

Studies in Classification, Data Analysis, and Knowledge Organization

Springer
Berlin
Heidelberg
New York
Barcelona
Budapest
Hong Kong
London
Milan
Paris
Singapore
Tokyo

Titles in the Series

H.-H. Bock and P. Ihm (Eds.)
Classification, Data Analysis, and Knowledge Organization. 1991
(out of print)

M. Schader (Ed.)
Analyzing and Modeling Data and Knowledge. 1992

O. Opitz, B. Lausen, and R. Klar (Eds.)
Information and Classification. 1993
(out of print)

H.-H. Bock, W. Lenski, and M.M. Richter (Eds.)
Information Systems and Data Analysis. 1994
(out of print)

E. Diday, Y. Lechevallier, M. Schader, P. Bertrand,
and B. Burtschy (Eds.)
New Approaches in Classification and Data Analysis. 1994
(out of print)

W. Gaul and D. Pfeifer (Eds.)
From Data to Knowledge. 1995

H.-H. Bock and W. Polasek (Eds.)
Data Analysis and Information Systems. 1996

E. Diday, Y. Lechevallier and O. Opitz (Eds.)
Ordinal and Symbolic Data Analysis. 1996

R. Klar and O. Opitz (Eds.)
Classification and Knowledge Organization. 1997

C. Hayashi, N. Ohsumi, K. Yajima, Y. Tanaka, H.-H. Bock,
and Y. Baba (Eds.)
Data Science, Classification, and Related Methods. 1998

I. Balderjahn, R. Mathar, and M. Schader (Eds.)
Classification, Data Analysis, and Data Highways. 1998

PREFACE

International Federation of Classification Societies

The International Federation of Classification Societies (IFCS) is an agency for the dissemination of technical and scientific information concerning classification and multivariate data analysis in the broad sense and in as wide a range of applications as possible; founded in 1985 in Cambridge (UK) by the following Scientific Societies and Groups:

- British Classification Society - BCS
- Classification Society of North America - CSNA
- Gesellschaft für Klassification - GfKl
- Japanese Classification Society - JCS
- Classification Group of Italian Statistical Society - CGSIS
- Société Francophone de Classification - SFC

Now the IFCS includes also the following Societies:

- Dutch-Belgian Classification Society - VOC
- Polish Classification Section - SKAD
- Portuguese Classification Association - CLAD
- Group at Large
- Korean Classification Society - KCS

IFCS-98, the Sixth Conference of the International Federation of Classification Societies, was held in Rome, from July 21 to 24, 1998.
Five preceding conferences were held in Aachen (Germany), Charlottesville (USA), Edinburgh (UK), Paris (France), Kobe (Japan).

Conference Organization

The scientific program of the Conference included 67 invited papers and 129 contributed papers, that discussed recent developments in the following topics:

- *Classification Theory*
- *Multivariate Data Analysis*
- *Multiway Data Analysis*
- *Proximity Structure Analysis*
- *Software Developments for Classification*
- *Applied Classification and Data Analysis in Social, Economic, Medical and other Sciences*

The novelty of IFCS-98 is the publication of the papers before the Conference since new electronic media allow a quick diffusion of the most recent developments of the topics discussed.
This volume, "*Advances in Data Science and Classification*", contains 39 invited and 52 contributed referred papers.
They are presented in six chapters as follows:

- ***Methodologies in Classification***
  - *Clustering and Classification*
  - *Comparison and Consensus of Classifications*
  - *Fuzzy Clustering and Fuzzy Methods*
  - *Optimization in Classification and Constrained Classification*
  - *Probabilistic Modelling Methods in Classification and Pattern Recognition*
- ***Other Approaches for Classification: Discriminant, Neural Network, Regression Tree***
  - *Discrimination and Classification*
  - *Neural Network and Classification*
  - *Regression Tree*
- ***Factorial Methods***
  - *Factorial Methods for Multivariate Data*
- ***Symbolic and Textual Data Analysis***
  - *Symbolic Data Analysis*
  - *Textual Data Analysis*
- ***Multivariate and Multidimensional Data Analysis***
  - *Proximity Analysis and Multidimensional Scaling*
  - *Multivariate Data Analysis*
  - *Non-Linear Data Analysis*
  - *Multiway Data Analysis*
- ***Case Studies***
  - *Applied Classification and Data Analysis*

The volume provides new developments in classification and data analysis and presents new topics which are of central interest to modern statistics. For most topics considered, this book provides a systematic state-of-the-art account written by some of the top researchers in the world. We hope this volume will serve as a helpful introduction to the area of classification and data analysis and contribute significantly to the quick transfer of new advances in data science and classification.

Acknowledgments

The Organizers of the Conference and the editors of this volume wish first to extend their sincere thanks to the authors for their contributions and for their enthusiastic participation which made this Conference into an international meeting for exchanging ideas and information about new developments in classification data analysis and related methods and in a wide range of applications.

We appreciate the cooperation of the 114 colleagues who reviewed papers for this volume and thank them for their time spent in this professional capacity.
We are also grateful to the organizers of invited session and colleagues who chaired sessions during the Conference.
Special thanks are also due to those working behind the scenes to make arrangements and see that many details were attended to.

Future IFCS Conference

The IFCS-2000 is being planned. The Seventh IFCS Conference is scheduled for the year 2000 in Namur (Belgium) with program chairs: Henk Kiers and Jean Paul Rasson.

Finally, thanks are extended to Springer-Verlag, Heidelberg for the quick and excellent publication of this volume.

Rome, July 1998

Alfredo Rizzi
Maurizio Vichi

TABLE OF CONTENTS

Part II: Other Approaches for Classification: Discrimination, Neural Network, Regression Trees

Discrimination and Classification

Neural Network and Classification

Software Developments for Classification and Multivariate Data Analysis

Part III: Factorial Methods

Factorial Methods for Multivariate Data

Part IV: Symbolic and Textual Data Analysis

Symbolic Data Analysis

Non-Linear Data Analysis

Multiway Data Analysis

Part VI: Case Studies

Applied Classification and Data Analysis

PART I

Methodologies in Classification

- Clustering and Classification
- Comparison and Consensus of Classifications
- Fuzzy Clustering and Fuzzy Methods
- Optimization in Classification and Constrained Classification
- Probabilistic Modelling Methods in Classification and Pattern Recognition

K-midranges clustering

J. Douglas Carroll
Rutgers University
Faculty of Management
Management Education Center # 125
81 New Street
Newark NJ 07102

Anil Chaturvedi
AT&T Laboratories
Room 5c-134
600 Mountain Avenue
Murray Hill NJ 07974-0636

Abstract: We present a new clustering procedure called K-midranges clustering. K-midranges is analogous to the traditional K-Means procedure for clustering interval scale data. The K-midranges procedure explicitly optimizes a loss function based on the L_∞ norm (defined as the limit of an L_p norm as p approaches infinity).

Keywords: Continuous data, Cluster analysis, Groups, Midrange, K-means

1. Introduction

In this paper, we present a clustering procedure called K-midranges, which is analogous to K-means or K-medians clustering, but explicitly optimizes an L_∞-loss function. The K-midranges procedure is as fast as K-means clustering, can handle large data set sizes and is computationally tractable. We also find it yields interpretable clusters for a marketing data set.

In the following sections, we describe the general bilinear clustering model, and show that when model parameters are estimated using an L_∞-norm-based loss function with partitioning constraints on the clusters, this results in the K-midranges clustering procedure.

2. The General Bilinear Clustering Model

Assume that we have data on N continuous variables for M objects (or subjects). Let K be the number of clusters being sought. Then, the general bilinear model can be written as:

$$C_{MxN} = S_{MxK}W_{KxN} + \text{error} \qquad (1)$$

where

-**C** is a subjects x variables data matrix,
-**S** is a binary indicator matrix for membership of the M subjects in K (possibly overlapping) clusters. In the present case, we constrain the model to yield mutually exclusive, non-overlapping clusters however (so that each row of **S** has exactly one element equal to one and the remaining elements are equal to zero), and
-**W** is a matrix of "generalized centroids", defined in this case as mid-ranges of certain observations.

It should be noted that only **C**, the data matrix, is known in (1), while both **S** and **W** are unknown and need to be estimated. This general bilinear model has been used in the past for determining overlapping clusters for interval scale data (where $bold C$ is continuous) by Mirkin (1990) and Chaturvedi, Carroll, Green, and Rotondo (1997), and is a special case of the CANDCLUS approach developed by Carroll and Chaturvedi (1995). In this paper, we assume that the data matrix **C** consists of interval scale valued elements, while **S** will be constrained, as stated earlier, to define a partition.

3. Parameter Estimation via an L_∞ norm

As in Chaturvedi, Carroll, Green, and Rotondo (1997), we define the parameter estimation problem via minimizing an L_p-norm based loss function

$$L_p = \left(\sum_{m=1}^{M} \sum_{n=1}^{N} |c_{mn} - \hat{c}_{mn}|^p \right)^{1/p}$$

where $\hat{c}_{mn}$ is the $(m,n)^{th}$ element of $ bold C hat = **SW**, to estimate **S** and **W**, for positive values of $p \geq \infty$.

In this paper, we concentrate on estimating the model in equation (1) using the L_∞ loss function, while **S** is constrained to be a partitioning matrix. The L_∞ norm, defined as the limit of the L_p norm as p approaches infinity, can be shown to be: $L_p = \max(|c_{mn} - \hat{c}_{mn}|)$, m = 1, M, n = 1, N. It is often called the "max metric" or the "dominance metric".

4. Estimation Procedure

The matrices **S** and **W** are estimated in an iterative fashion (estimating **S** given estimates of **W**, then revising the estimates of **W** given the new estimates of **S**) until the L_0 loss function does not improve. The procedures for estimating **S** and **W** are given below:

To estimate **S**, the cluster membership, given estimates of **W**, consider the following illustrative case by first defining

$$\mathbf{C} = \begin{bmatrix} 1 & 5 & 0 & 3 \\ 2 & 6 & 1 & 3 \\ 3 & 6 & 0 & 3 \\ 0 & 6 & 1 & 2 \\ 1 & 5 & 0 & 1 \\ 2 & 4 & 1 & 2 \end{bmatrix} \quad \mathbf{S} = \begin{bmatrix} s_{11} & s_{12} \\ s_{21} & s_{22} \\ s_{31} & s_{32} \\ s_{41} & s_{42} \\ s_{51} & s_{52} \\ s_{61} & s_{62} \end{bmatrix} \text{ and } \hat{\mathbf{W}} = \begin{bmatrix} 2 & 6 & 1 & 3 \\ 1 & 5 & 0 & 1 \end{bmatrix}$$

The matrices **C** and $\hat{\mathbf{W}}$ are assumed known, while a current (conditional) estimate of **S** is sought. The $(i,j)^{th}$ entry of matrix **W** corresponds to the continuous value defining the centroid for the ith cluster and the jth continuous variable. In the general case, we wish to find the L_∞-norm based estimates of **S**, where $\mathbf{C} = \mathbf{S}\hat{\mathbf{W}} + \text{error}$ and elements of **S** are either 0 or 1. If we let

$$f_1 = \max(|1 - 2s_{11} - 1s_{12}| + |5 - 6s_{11} - 5s_{12}| + |0 - 1s_{11} - 0s_{12}| + |3 - 3s_{11} - 1s_{12}|),$$

$$f_2 = \max(|2 - 2s_{21} - 1s_{22}| + |6 - 6s_{21} - 5s_{22}| + |1 - 1s_{21} - 0s_{22}| + |3 - 3s_{21} - 1s_{22}|),$$

$$f_3 = \max(|3 - 2s_{31} - 1s_{32}| + |6 - 6s_{31} - 5s_{32}| + |0 - 1s_{31} - 0s_{32}| + |3 - 3s_{31} - 1s_{32}|),$$

$$f_4 = \max(|0 - 2s_{41} - 1s_{42}| + |6 - 6s_{41} - 5s_{42}| + |1 - 1s_{41} - 0s_{42}| + |2 - 3s_{41} - 1s_{42}|),$$

$$f_5 = \max(|1 - 2s_{51} - 1s_{52}| + |5 - 6s_{51} - 5s_{52}| + |0 - 1s_{51} - 0s_{52}| + |1 - 3s_{51} - 1s_{52}|), \text{ and}$$

$$f_6 = \max(|2 - 2s_{61} - 1s_{62}| + |4 - 6s_{61} - 5s_{62}| + |1 - 1s_{61} - 0s_{62}| + |2 - 3s_{61} - 1s_{62}|),$$

then the total mismatch (L_∞ loss function) is given by F = max(f_1, f_2, f_3, f_4, f_5, f_6). Note that f_1 is a function only of s_{11} and s_{12}; f_2 is a function only of s_{21} and s_{22}; f_3 is a function only of s_{31} and s_{32};and f_4 is a function only of s_{41} and s_{42}; Thus, F is *separable* with respect to parameters for each row of **S**. To minimize F, one can separately minimize f_1 w.r.t parameters for row 1, f_2 w.r.t parameters for row 2, etc. Certainly, if f_1 through f_M are minimized, then F = max(f_m), (m = 1, M), will be minimized.[1]

To minimize, say, f_1 w.r.t. parameters for row 1, [s_{11} s_{12}] with partitioning constraints, one evaluates f_1 explicitly at the two permissible values of row 1, [1 0] and [0 1], given the constraint to a partitioning solution. For the pattern [1 0], f_1 is obtained by computing the maximum of absolute differences between the data value and the predicted value, computed for each element in the row. Thus, for the pattern [1 0], the absolute differences for the four elements are 1, 1,1, and 0, resulting in a maximum absolute difference of 1. For the pattern [0,1], the element-wise differences are 0,0,0, and 2. This results in a maximum difference of 2 for the pattern [0 1]. The winning pattern is the pattern that results in the smallest maximum difference. Thus, in this case, the winning pattern is [1 0]. Thus,

$\hat{s}_{11} = 1$ and $\hat{s}_{12} = 0$ are the optimal estimates. The other rows of $\hat{\mathbf{S}}$, the current estimate of **S**, can be determined using a similar procedure.

We note that since **S** induces a partition on the rows of **C**, that rows of **W** correspond to "generalized centroids" of the associated clusters. Given that we are minimizing the L_∞ loss function, these "generalized centroids" will be vectors of midranges since it can easily be seen that the measure of central tendency minimizing the L_∞-norm based loss function is the midrange (mean of the two most extreme values). We can use this to advantage in estimating **W**, given **C** and conditional estimates of **S**. To illustrate, using the previous example, to estimate **W** (the cluster "generalized centroids"), given estimates of **S**, we first define:

$$\mathbf{C} = \begin{bmatrix} 1 & 5 & 0 & 3 \\ 2 & 6 & 1 & 3 \\ 3 & 6 & 0 & 3 \\ 0 & 6 & 1 & 2 \\ 1 & 5 & 0 & 1 \\ 2 & 4 & 1 & 2 \end{bmatrix} \quad \hat{\mathbf{S}} = \begin{bmatrix} 1 & 0 \\ 1 & 0 \\ 1 & 0 \\ 0 & 1 \\ 0 & 1 \\ 0 & 1 \end{bmatrix} \text{ and } \mathbf{W} = \begin{bmatrix} w_{11} & w_{12} & w_{13} & w_{14} \\ w_{21} & w_{22} & w_{23} & w_{24} \end{bmatrix}.$$

[1] While, given the definition of F as the max (f_m), (m =1, M), we should minimize F by simply minimizing the largest of the f_m's, we resolve the implied indeterminacy by choosing the solution that separately minimizes each of the f_m's.

The matrices **C** and $\hat{\mathbf{S}}$ are assumed known, while a current (conditional) estimate of $\hat{\mathbf{W}}$ is sought. In order to determine the L_∞ estimate of **W**, where $\mathbf{C} = \hat{\mathbf{S}}\mathbf{W} + \text{error}$ and **W** is continuous, we can show that the best estimate of w_{11} that optimizes the L_∞ loss function is the mid-range of (1,2,3) = 2, and hence, $\hat{w}_{11}$ = 2. Similarly, $\hat{w}_{12}$ will be the mid-range of (5,6,6) = 5.5, ..., and $\hat{w}_{24}$ will be the mid-range of (4,4,2) = 3. Thus, $\hat{\mathbf{W}}$, the current estimate of **W**, is a matrix of midranges of the variables for each cluster, and hence, we call this procedure a K-midranges clustering procedure. We break ties arbitrarily in the current version of the software. We repeat the estimation of $\hat{\mathbf{S}}$ given $\hat{\mathbf{W}}$, and $\hat{\mathbf{W}}$ given $\hat{\mathbf{S}}$ until the L_∞ loss function does not improve. It should be noted that upon convergence, this procedure can yield *locally optimal* solutions (as can K-means, K-medians, K-modes, and other related methods). The best strategy when using K-midranges with real data is to use multiple random starting seeds, and choose the solution with the lowest L_∞-norm measure.

Actually, in this case of non-overlapping clusters (i.e., a partition), the K-midranges procedure can be described in much simpler terms, closely analogous to the standard description of K-means, K-medians, or K-modes. Namely, we alternate between defining cluster "generalized centroids" (in this case midranges) for an existing partition, into K non-overlapping clusters, by assigning each object (m) to that cluster for which the maximum deviation over n between c_{mn} and the midrange for that cluster is minimized; i.e., we assign each object to the cluster whose "generalized centroid" it is closest to in the sense of minimizing the L_∞ metric. This is equivalent to the K-means, K-medians, or K-modes algorithms, but with L_2, L_1, or L_0 norms, respectively, replaced by the L_∞ norm, and means, medians, or modes by midranges.

As with K-means, K-medians, and K-modes, the (non-overlapping) K-midranges approach to clustering can be generalized to overlapping K-midranges - but we shall not consider that further in this paper (except to say that this relies on the same "one cluster at a time" type of estimation procedure discussed by Chaturvedi, Carroll, Green, and Rotondo, 1997, in their paper describing overlapping K-means, and K-medians approaches, and discussing other "overlapping K-centroids" approaches based on optimizing different Lp-norm-based loss functions).

5. Application of K-midranges to Market Segmentation

The K-midranges procedure was applied to segment the mail order insurance market based on data obtained from a conjoint analysis study. Six hundred potential buyers participated in a study designed to evaluate new financial service concepts for the mail order insurance marketin the southwest U.S. Nine attributes – price of the basic service, credit management, credit recovery, assisted savings and investment, budgeting and control of discretionary spending, basic insurance services, general advice/information, bill payment, and service provision method – were used in a conjoint design. Data were collected from 600 potential purchasers.

Conjoint analysis was applied to derive importances of the nine financial services attributes for each of the 600 subjects. The derived attribute importances were then used in a subsequent market segmentation analysis. The basic data consisted of a 600 x 9 (subjects by variables) matrix of derived importances. The derived importances were normalized across the nine attributes for each subject such that the row sum for each subject was 1.0.

K-means clustering

K-means cluster analysis was used to find clusters in the marketplace. Different initial seeds were used for the K-means algorithm. Solutions from Ward's minimum-variance algorithm were also used as starting seeds for the K-means algorithm. A 5-cluster K-means solution, obtained from random starting seeds, was chosen on the grounds of highest VAF (Variance-Accounted-For) and interpretability. The details of the 5-cluster K-means solution are given in Table 1.

The 5-cluster, traditional K-means solution yielded a VAF value of 50.6%. The five segments were interpreted as:

1. **Price only**: The mean importance of price was the highest for this segment (0.60). Price is the only major benefit that is sought by this segment.
2. **Credit management & control:** This segment has the highest mean importances for credit management, credit recovery, control of spending, and provision method.

Table 1: *5-cluster K-means solution: Entries are cluster means for variables*

Attributes	Price	Credit Mgt. And Control	General Services	Savings	Price and Convenience
Basic Price	0.60	0.09	0.09	0.07	0.28
Credit Mgt	0.07	0.25	0.14	0.15	0.15
Credit Recovery	0.03	0.12	0.03	0.02	0.03
S&I	0.02	0.06	0.07	0.31	0.06
Control Spending	0.04	0.13	0.10	0.08	0.08
Insurance	0.02	0.05	0.11	0.06	0.06
General Advice	0.03	0.06	0.12	0.11	0.06
Bill Payment	0.07	0.10	0.18	0.09	0.10
Provision Method	0.12	0.13	0.18	0.12	0.18
Segment Size	32	181	187	62	138

Table 2: *Goodness of Fit for the K-midranges procedure*

# Clusters	L_∞⁺ Measure	Normalized L_∞* Measure
1	0.48	0.50
2	0.47	0.49
3	0.35	0.37
4	0.35	0.36
5	0.34	0.36
6	0.31	0.33
7	0.29	0.30
8	0.29	0.30
9	0.27	0.28
10	0.24	0.25
11	0.23	0.24
12	0.20	0.21
13	0.18	0.19
14	0.18	0.19

+ The L_∞ measure reported here is the maximum of the absolute differences between each element of the data matrix and the corresponding value predicted by the K-midranges procedure.

* The normalized L_∞ measure is obtained by dividing the L_∞ measure by the range of the input data.

3. **General services:** The differentially most important attributes characterizing this segment are insurance, general advice, bill payment, and provision method.

4. **Savings & investment oriented**: This segment has the highest importance for the savings and investment (S&I) attribute.
5. **Price and convenience:** This segment places a high importance on price and provision method.

K-midranges clustering

We also applied the K-midranges procedure on the same data set. Table 2 presents the goodness-of-fit measure (the L_∞ and the normalized L_∞ norms) for one through fourteen clusters. Table 2 suggests that we extract 3 clusters, since the normalized L_∞ measure has the steepest fall (about 12%) in going from 2 to 3 clusters. The 3-cluster solution is reported in Table 3.

The sizes of the three clusters have a large variance (proportion of observations ranging from 80% to 2%), and the largest cluster is not quite interpretable. Cluster 1 corresponds to the "Price only" cluster, where the midrange for the price attribute has a very high value of 0.60, and the other attributes have a very low value. This cluster contains roughly 18% of the observations. Cluster 2, the largest cluster, has relatively medium midranges on most attributes, except for price and advice/information. Cluster 3 corresponds to financial advice seekers, (roughly 2% of the sample), and has a very high midrange value of 0.61for the Advice/information attribute. Thus, the 3-cluster solution is not very useful from a marketing perspective. Apart from the "Price only" cluster, and the very small "Advice seekers" cluster, the rest of the observations were grouped into one large cluster.

We decided to pick the 13-cluster solution, for reasons of interpretability.
The thirteen cluster solution is presented in Table 4. Clusters 1 through 9 have small cluster sizes, and correspond to observations that have a large value on just one attribute, and very small values on the other attributes. Thus, Clusters 1 through 9 can be called the "Advice extremes", "Credit recovery only", "Extreme Price only", "Insurance only", "Provision method only", "Advice only", "Savings only", "Control only", and "Credit Management only" clusters.

Cluster 10 corresponds to the "Price only" cluster. It is similar to cluster 3, but the midranges are not that extreme for this cluster. Cluster 11 can be termed as the "Price and Convenience" cluster, since the largest two midranges correspond to the price and bill provision attributes. It should be noted that a cluster with a very

Table 3: *3-cluster K-midrange solution*

Segment	Size	Pr	CM	CR	SI	Ctrl	Ins	Adv	BP	PM
Price only	105	0.60	0.27	0.12	0.26	0.22	0.17	0.08	0.14	0.20
Majority	479	0.15	0.31	0.34	0.35	0.24	0.30	0.14	0.24	0.35
Advice seekers	16	0.07	0.10	0.06	0.18	0.10	0.06	0.61	0.08	0.12

Table 4: *13-cluster K-midrange solution*

Segment	Size	Pr	CM	CR	SI	Ctrl	Ins	Adv	BP	PM
1	2	0.01	0.01	0.01	0.01	0.01	0.01	0.93	0.01	0.01
2	8	0.08	0.09	0.51	0.09	0.04	0.06	0.05	0.10	0.12
3	10	0.79	0.06	0.02	0.01	0.04	0.01	0.02	0.07	0.07
4	11	0.14	0.13	0.03	0.07	0.08	0.44	0.05	0.08	0.18
5	11	0.12	0.06	0.05	0.10	0.04	0.09	0.06	0.13	0.55
6	12	0.07	0.13	0.03	0.18	0.10	0.04	0.41	0.08	0.12
7	13	0.17	0.11	0.00	0.53	0.08	0.04	0.13	0.07	0.10
8	37	0.19	0.18	0.08	0.10	0.32	0.08	0.11	0.11	0.11
9	42	0.18	0.45	0.15	0.13	0.13	0.10	0.11	0.12	0.15
10	67	0.43	0.14	0.12	0.09	0.08	0.10	0.08	0.11	0.14
11	108	0.23	0.17	0.08	0.17	0.17	0.12	0.12	0.11	0.23
12	120	0.14	0.14	0.11	0.15	0.10	0.12	0.13	0.29	0.22
13	159	0.13	0.16	0.16	0.17	0.16	0.16	0.15	0.17	0.16

Notation for Tables 3 and 4:
Pr –Basic Price
CM – Credit Management
CR – Credit Recovery
SI – Savings and Investment
Ctrl –Control over Discretionary spending
Ins – Insurance
Adv – Advice/Information
BP – Bill Payment;
PM – Provision Method
Entries are segment centroids for variables

similar interpretation had been extracted using the K-means procedure (see Table 1). In fact, 48 observations from the "Price and Convenience" cluster from the K-means solution lie in cluster 11. Cluster 12 corresponds to the "Convenience" cluster, since the largest midrange values occur for the Bill payment and Provision method attributes, both of which are aimed at enhancing consumer convenience.

Cluster 13 can be termed the general services cluster, where-in most attributes are equally important to the consumers in that cluster.
Thus, of the thirteen clusters extracted, we see that the nine smallest clusters are based on large values on just one of the nine variables, while the four largest clusters are based on high attribute importance weights for multiple attributes. From a marketing standpoint, the four largest clusters would be the most attractive and possibly profitable not only because of their larger size, but also because they would enable the management of the concerned financial services firm to market a mixture of financial services (e.g., mixture of bill payment service and telephone banking service for cluster 11, since cluster 11 has the highest midranges for these two variables, thereby providing the benefit of convenience to the subjects in that cluster) to consumers in these clusters, thereby potentially increasing their revenues, and hence profits, from these clusters.

It would appear that the midranges may be particularly good for isolating outliers; the first nine clusters in this case could be viewed as outlying clusters, each based on extreme values on a single attribute (or variable). Whether the outlying clusters are substantively interesting or not is debatable. Each of these nine clusters appear to be associated uniquely with one of the nine variables, and so could be viewed as analogous to "specific factors" in factor analysis. The four larger clusters, in this view, can be interpreted as representing "common structure" remaining after these outliers are removed from the data. In fact - without pressing the analogy too far - this "common structure" is strongly reminiscent of a "simple structure" configuration in factor analysis!

In terms of cluster membership, the degree of correspondence between the 13-cluster K-midrange solution and the 5-cluster K-means solution, computed using the modified RAND index measure of Hubert and Arabie (1985) is 0.154. We also compared the 5-cluster K-means solution with the four largest clusters derived from the 13-cluster K-midranges clustering procedure. The modified RAND index was only 0.170, indicating a very low correspondence between the 5-cluster K-means solution and the four largest clusters derived from the 13-cluster K-midranges solution. Thus, the K-midranges procedure and the K-means procedures offer very different solutions in terms of segment membership for this data set.

6. Conclusions

We have presented an approach to cluster analysis via K-midranges clustering. It is analogous to the K-means (MacQueen 1967), K-medians (Vinod 1969;

Rousseeuw 1987), and K-modes (Chaturvedi, Green, and Carroll 1996) clustering procedures. The proposed approach can handle data of relatively large size (e.g.,

up to 5,000 consumers or other entities) and, hence, can be used fruitfully in social and behavioral sciences. Though we presented an empirical illustration of the K-midranges procedure, we feel more empirical research is needed to assess the longer term potential of K-midranges clustering.

References

Carroll, J. D., & Chaturvedi, A. D. (1995). A General Approach to Clustering and Multidimensional Scaling of Two-way, Three-way, or Higher-way Data, in *Geometric Representations of Perceptual Phenomena: Papers in Honor of Tarow Indow's 70th Birthday.* Luce, R. D., D'Zmura, M., Hoffman, D. D., Iverson, G., & Romney, A. K. (Eds.), Lawrence Erlbaum, 295-318.

Chaturvedi, A. D., Carroll, J. D., Green, P., and Rotondo, J. A. (1997). A Feature based Approach to Market Segmentation via OverlappingK-centroids Clustering. *Journal of Marketing Research*, 34, 370-377

Chaturvedi, A. D., Paul E. Green, and J. Douglas Carroll (1996). Market Segmentation via K-modes clustering. Invited paper presented at the American Statistical Association conference held in Chicago.

Hartigan, J. A. (1975). *Clustering Algorithms*, Wiley, New York.

Hubert, L., & Phipps Arabie (1985). Comparing Partitions. *Journal of Classification*, 2, 193-218.

Jain, A. K. & Richard C. Dubes (1988), *Algorithms for Clustering Data*, Prentice-Hall, New Jersey.

Kaufman, L. & Peter J. Rousseeuw (1989). *Finding Groups in Data: An Introduction to Cluster Analysis*, Wiley, New York.

MacQueen, J. B. (1967). Some Methods for Classification and Analysis of Multivariate Observations. in *Proceedings of the Fifth Berkeley Symposium on Mathematical Statistics and Probability*. LeCam, L. M., & Neyman, J. (Eds.), 1, 281-297.

Mirkin, J. B. (1990). A Sequential Fitting Procedure for Linear Data Analysis Models. *Journal of Classification*, 7, 167-195.

Vinod, H. D. (1969). Integer Programming and Theory of Grouping. *Journal of the American Statistical Association*, 64, 506-519.

A New Method for Detecting Influential Observations in Nonhierarchical Cluster Analysis

Andrea Cerioli
Istituto di Statistica, Università di Parma
Via Kennedy 6, 43100 Parma, Italy (email: statecl@ipruniv.cce.unipr.it)

Abstract: In this paper we propose a new approach to the exploratory analysis of multivariate clustered data. Our technique is based on a fast forward search algorithm which orders multivariate observations from those most in agreement with a specified clustering structure to those least in agreement with it. Simple graphical displays of a variety of statistics involved in the forward search lead to the identification of multiple outliers and influential observations in nonhierarchical cluster analysis, without being affected by masking and swamping problems. The suggested approach is applied to the convergent K-means method in two examples, both with real and simulated data.

Keywords: cluster validity; forward search; K-means; masking; multivariate outliers.

1. Introduction

The aim of influence detection is to identify those data points that have a large impact on the partitions obtained from a clustering algorithm. This purpose is especially relevant to the areas of clustering validation and stability (see, e.g., Milligan, 1996, and Gordon, 1996). Recent contributions in the field of influence detection include Jolliffe *et al.* (1995) and Cheng and Milligan (1996), which explicitly consider only hierarchical methods. An extension to nonhierarchical clustering, for variable number of clusters, has been suggested by Cerioli (1997). These techniques are based on the comparison of a reference partition, computed on the complete data set through a specified clustering algorithm, and a modified partition, obtained from a reduced data set using the same algorithm. The reduced data set is produced by simply deleting a single case from the complete one. However, it is well known that single-case deletion diagnostics can suffer from the problems of *masking* and *swamping*, when multiple outliers are present in the data (Barnett and Lewis, 1994).

At the opposite end of the spectrum from the "traditional plus diagnostics" approach, Cuesta-Albertos, Gordaliza and Matrán (1997) have proposed a high-breakdown procedure with the aim of robustifying the K-means clustering method. Despite its mathematical elegance, a disadvantage of this procedure is that it can prove difficult to apply in practical situations with a large number of variables. Furthermore very precise estimates of cluster means may not be necessary when interest lies in identifying multivariate outliers. The purpose of the present paper is to propose a simple and fast technique which leads to the detection of multiple outliers and influential observations in nonhierarchical clustering, without being affected by masking and swamping problems. Specifically, our approach is based on a forward search algorithm which orders the observations from those most in agreement with a specified clustering structure to those least in agreement with it.

2. The forward search for clustered data

Forward search procedures for the identification of multiple outliers have been recently developed to deal with unstructured multivariate data (Hadi, 1992; Atkinson, 1994) and grouped data with known group structure (Atkinson and Riani, 1997). The present proposal extends this field of research to the case of multivariate clustered data with unknown clusters. The focus here is on the K-means method, although the suggested technique can be conveniently applied to any nonhierarchical clustering algorithm.

Suppose that the data matrix consists of an $n \times p$ set of multivariate observations. Let $\mathbf{x}_i'$ be the observation corresponding to unit i. The main steps of our algorithm can be summarized as follows.

Step 1. Given a fixed number of clusters, say K, find K "robust" centroids. These are intended to be the centroids of clusters computed from outlier-free observations. Our method for performing this step is explained below, after the algorithm description. Take the unit which is closest to the k-th robust centroid as the initial seed of cluster k ($k = 1, \ldots, K$).

Step 2. Define the initial centroid of cluster k as its initial seed. Denote initial centroids by $\mathbf{c}_1^0, \ldots, \mathbf{c}_K^0$. Compute nearest-centroid distances as

$$d_i = \min_k \; [d(\mathbf{x}_i, \mathbf{c}_k^0)] \qquad\qquad i = 1, \ldots, n,$$

where $d(\mathbf{x}, \mathbf{y})$ is the (euclidean) distance between p-dimensional vectors $\mathbf{x}$ and $\mathbf{y}$. Set $s = 1$.

Step 3. Put $m = K + s$. Arrange the observations in ascending order according to d_i. Let $\mathcal{C}_m$ be the "clean subset" of cardinality m, i.e. the set of indexes of the m observations with the smallest d_i.

Step 4. Perform the K-means algorithm on observations indexed in $\mathcal{C}_m$, using $\mathbf{c}_1^{(s-1)}, \ldots, \mathbf{c}_K^{(s-1)}$ as initial centroids. Denote by $\mathbf{c}_k^s$ the final centroid of cluster k after performing the clustering algorithm ($k = 1, \ldots, K$). Compute nearest-centroid distances as

$$d_i = \min_k \; [d(\mathbf{x}_i, \mathbf{c}_k^s)] \qquad\qquad i = 1, \ldots, n.$$

Step 5. Stop if $s = n - K$. Otherwise increase s by 1 and go to Step 3.

Our method for performing Step 1 of the forward search is to compute a preliminary clustering with $K' \gg K$ groups, and take as robust centroids those of the K largest clusters. This is because (multiple) outliers, being unusual, should stand alone or cluster by themselves at a large distance from the bulk of the data (see SAS, 1990).

However, it must be remarked that results from our algorithm are largely unaffected by different choices of $\mathbf{c}_1^0, \ldots, \mathbf{c}_K^0$, if very precise estimates of cluster means are not required. When outlier detection and ordering of multivariate data are the focus, the crucial requirement is that the set of initial centroids be outlier free. The preliminary use of cluster analysis for defining a clean subset of multivariate observations has been suggested also by Hadi and Simonoff (1993) in the context of multiple regression models.

In most moves from cardinality m to $m + 1$, just one new observation joins the clean subset. This provides an ordering of the data according to the given K-clusters structure, with observations furthest to it joining the subset at last stages of the forward search. Furthermore, multivariate outliers and other influential observations can be detected by simple graphical displays of a variety of measures of cluster cohesion computed at successive iterations of Step 4 of our procedure. In the examples that follow we consider the largest of maximum within cluster distances

$$D(s) = \max_k \left\{ \max_{i,i' \in k} \left[d(\mathbf{x}_i, \mathbf{x}_{i'}) \right] \right\}, \tag{1}$$

where, at step s, units i and i' both belong to cluster k. The rationale behind (1) is that, if the true structure of the data consists of K clusters plus a number of atypical observations, $D(s)$ stays small as long as $\mathcal{C}_m$ remains outlier free. In addition, the plot of $D(s)$ shows a dramatic change when the first outlier joins the clean subset.

3. Examples

The effectiveness of the suggested method in detecting masked multiple outliers, and more generally in ordering multivariate clustered observations, is shown by means of two examples. In the first instance, we simulate a data set where a large number of outliers are known to exist. Secondly, we apply our algorithm to nonhierarchical clustering of European Regions.

3.1. Simulated data

In our first example we generate 90 realizations from a eight-variate normal distribution with constant variance, through a slightly modified version of the algorithm in Jolliffe *et al.* (1995). Observations are generated to fall into six distinct clusters of equal size, whose centroids all lie within the unit hypercube. Then we add 10 contaminated observations falling near (or slightly outside) the upper boundaries of the hypercube.

A K-clusters partition is obtained through the convergent K-means method, with careful selection of seed points. In particular, a number of distance tests among potential seeds are performed before their selection, as in the FASTCLUS procedure of SAS (1990). Although contaminated values are clearly outliers with respect to the bulk of the data, they are not revealed by single-case deletion statistics, due to the masking effect, even if we correctly set $K = 6$.

Then we apply the forward search algorithm described in section 2

Figure 1: *Simulated data. Plot of D(s) against m for several values of K.*

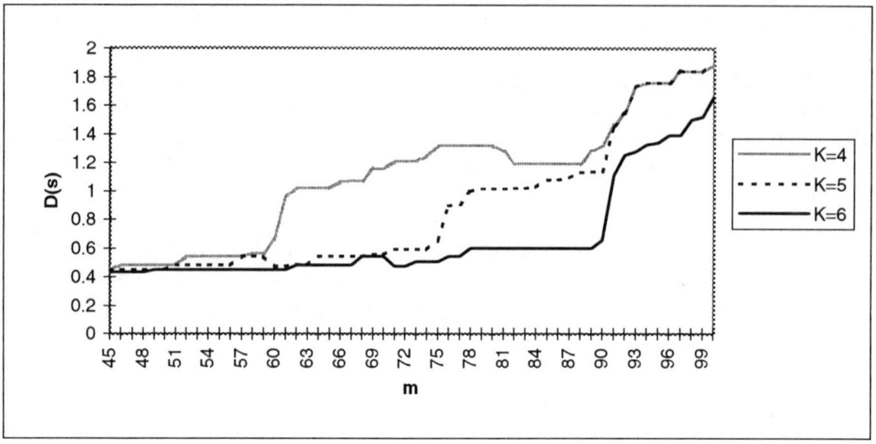

for different values of K. Robust centroids are computed from a preliminary clustering with 20 clusters. In all searches, contaminated observations join the clean subset at the last 10 steps of the procedure. However, the behaviour of (1) is markedly different according to what value of K is actually assumed. Figure 1 shows the plot of $D(s)$ against m for $K = 4$, 5 and 6. For simplicity we only display steps such that $m \geq 45$. For $K < 6$, it is apparent that many well-behaved observations produce high values of $D(s)$ when included in $\mathcal{C}_m$, due to misspefication of the true clustering structure. On the contrary, the plot for $K = 6$ reveals a close agreement between the bulk of the data and the underlying (true) structure, while contaminated values are clearly declared as outliers. Plots of $D(s)$ for $K > 6$ also produce similar displays, as far as contaminated observations are not included in the set of initial centroids.

3.2. Clustering of European Regions

In order to identify similar areas with respect to wealth and economic development, we classify the Regions of the European Union (EU) according to the following (standardized) variables:

X_1 = activity rate;

X_2 = dependence rate (i.e. ratio between people not aged in the range 14-65 and people aged in that range);

X_3 = unemployment rate;

X_4 = young unemployment rate;

X_5 = % of people employed in Industry;

X_6 = % of people employed in Trade and Services;

X_7 = index number of Gross National Product per inhabitant (measured in Purchasing Power Standards).

Data are taken from database REGIO (Eurostat, 1997) and refer to 1995, with the exception of X_7 which refers to 1993. European Regions are defined according to the NUTS2 aggregation level, except for Sverige and some areas in U. K. and Germany, where a coarser aggregation

(NUTS1) is adopted due to lack of available information at the required level. For the same reason Austria has been excluded from the analysis. As a result, we have $n = 160$ data units in this application. The corresponding data set is available from the author upon request.

As in the example of section 3.1, K-clusters partitions are obtained through the convergent K-means method, for a number of values of K. However, here we adopt an improved version of the FASTCLUS seed selection technique, in order to reduce the possible effect of the order of the observations on clustering results. The adopted seed selection procedure is described in Cerioli (1997).

The forward search algorithm is applied to the computed partitions, for $8 \leq K \leq 20$. Robust centroids are obtained from a preliminary clustering with $K' = 3K$ clusters, if $K \leq 16$; otherwise we set $K' = 50$. Figure 2 shows the plot of $D(s)$ against m for selected values of K in the range of interest. For simplicity we only display steps such that $m > 100$. Although the behaviour of $D(s)$ is not as clearcut as in Figure 1, there is some evidence that small values of K do not provide satifactory classifications, as a large number of Regions would then disagree with the supposed group structure.

Furthermore, all the reported curves of $D(s)$ show a sharp increase in the last stages of the forward search. This means that the last Regions in the ordering provided by the forward algorithm can be considered as potentially anomalous given the chosen number of clusters. Specifically, the Regions joining the clean subset at the last four steps are reported in Table 1 for selected values of K. Note that Voreio Aigaio (Greece), Brussels (Belgium) and Ceuta y Melilla (Spain) are always the last to come in, although in a different order, irrespective of the value of K.

References

Atkinson A. C. (1994). Fast Very Robust Methods for the Detection of Multiple Outliers, *Journal of the American Statistical Association*, 89, 1329-1339.

Atkinson A. C. and Riani M. (1997). Bivariate Boxplots, Multiple Outliers, Multivariate Transformations and Discriminant Analysis: the 1997 Hunter Lecture, *Environmetrics*, 8, 583-602.

Barnett V. and Lewis T. (1994). *Outliers in Statistical Data. 3rd Edition*, Wiley, Chichester.

Cerioli A. (1997). Measuring the Influence of Individual Observations and Variables in Cluster Analysis, submitted for publication in the series: *Classification, Data Analysis and Knowledge Organization*, Springer-Verlag, Berlin.

Cheng R. and Milligan G. W. (1996). Measuring the Influence of Individual Data Points in a Cluster Analysis, *Journal of Classification*, 13, 315-335.

Cuesta-Albertos J. A., Gordaliza A. and Matrán C. (1997). Trimmed k-Means: an Attempt to Robustify Quantizers, *The Annals of Statistics*, 25, 553-576.

Eurostat (1997). *Regionen. Statistisches Jahrbuch. 1996*, Luxembourg.

Gordon A. D. (1996). Hierarchical Classification, in: *Clustering and Classification*, P. Arabie, L. J. Hubert and G. De Soete (eds.), World

Figure 2: *Clustering of European Regions. Plot of D(s) against m for selected values of K.*

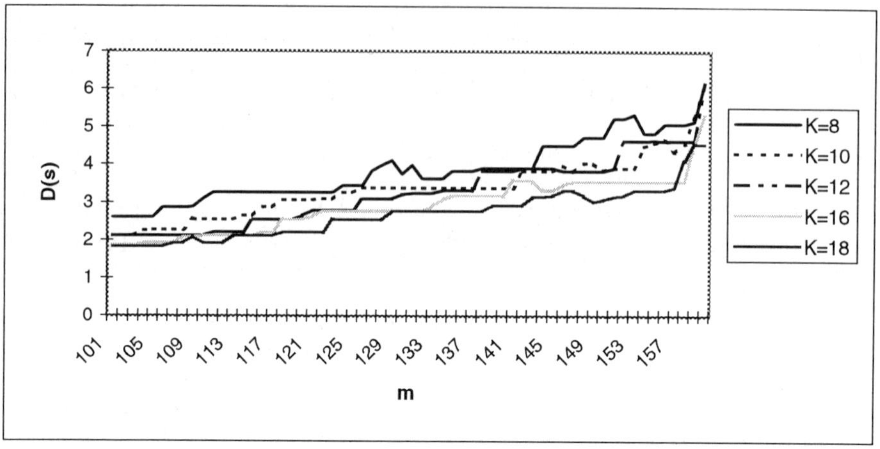

Table 1: *Regions of the EU joining the clean subset at the last four steps of the forward search for selected values of K.*

	$K=8$	$K=10$	$K=12$	$K=16$	$K=18$
$m=157$	Hamburg	Sicilia	Sicilia	Corse	Corse
$m=158$	Voreio Aigaio	Voreio Aigaio	Voreio Aigaio	Brussels	Ceuta y Melilla
$m=159$	Brussels	Brussels	Brussels	Voreio Aigaio	Voreio Aigaio
$m=160$	Ceuta y Melilla	Ceuta y Melilla	Ceuta y Melilla	Ceuta y Melilla	Brussels

Scientific, Singapore, 65-121.

Hadi A. S. (1992). Identifying Multiple Outliers in Multivariate Data, *Journal of the Royal Statistical Society*, B, 54, 761-771.

Hadi A. S. and Simonoff J. S. (1993). Procedures for the Identification of Multiple Outliers in Linear Models, *Journal of the American Statistical Association*, 88, 1264-1272.

Jolliffe I. T., Jones B. and Morgan B. J. T. (1995). Identifying Influential Observations in Hierarchical Cluster Analysis, *Journal of Applied Statistics*, 22, 61-80.

Milligan G. W. (1996). Clustering Validation: Results and Implications for Applied Analyses, in: *Clustering and Classification*, P. Arabie, L. J. Hubert and G. De Soete (eds.), World Scientific, Singapore, 341-375.

SAS (1990). *SAS/STAT User's Guide. Ver. 6. 4th Edition*, SAS Institute, Cary, NC.

Non-probabilistic Classification

John C. Gower and Gavin J.S. Ross

Department of Statisitics,
The Open University,
Walton Hall,
Milton Keynes MK7 6AA,
U. K.
e-mail: j.c.gower@open.ac.uk

Statisitics Department,
IACR Rothamsted,
Harpenden,
Herts AL5 2JQ,
U. K.
e-mail: gavin.ross@bbscr.ac.uk

Abstract: Much of the classification literature ignores notions of probability. In our view, this is due in part to a dominant tendency in the early days of computers for developing heuristic clustering algorithms and in part due to long traditions in classification outside the statistical/probabilistic orbit, of which biological taxonomy and book classification are primary examples. Statisticians have rightly stressed the role of probabilistic concepts in formulating classification problems and in interpreting classifications but we believe that they are wrong in suggesting, as they sometimes seem to, that other approaches are unsatisfactory. Probability has its proper place in classification but it is neither an essential nor always an appropriate tool. We discuss circumstances where non-probabilistically-based classifications are fully justified.

Considerations influencing the differences between the two approaches include: 1) Irrespective of whether things are to be assembled into classes (arranged hierarchically or not) or assigned to previously recognised classes, methodology depends on whether the things may be regarded as representing groups or as samples from groups; 2) Models are basic to the formulation of statistically based classifications, but they may also underpin nonprobabilistic classifications; overt models are not a characteristic of heuristic classification algorithms; 3) In principle, probabilistic models allow the significance and number of clusters justified by data to be assessed. In non-probabilistic classifications (probabilistic too), the eighteenth century concept of *approximation* offers a good basis for assessing the adequacy and stability of clusters.

Keywords: Probabilistic Classification, Non-probabilistic Classification, Classes, Groups, Assignment, Class Construction.

1. Introduction

The art of classifying knowledge has a long history reaching back into classical times. Book classification has long been essential for librarians and scholars and in the eighteenth century Linnaeus standardised the taxonomy of life-forms. The organisation of government and of large business enterprises is essential for efficiency. Preliminary classification is essential for, and may be designed to simplify, information retrieval. Just by naming things we recognise the existence of the class of things that bear that name. When the number of associated named things grows large, some form of classifying them into manageable, named, groups is essential in order to organise and describe what would otherwise be chaotic. Thus, the notion of classification is a fundamental human activity.

In the above examples, probability and models have little if any role to play. Although evolutionary concepts, which offer the possibility of modelling, are of current interest, they played no part in Linnaeus' foundations of taxonomy. Indeed, one of the reasons why biological taxonomy is so fascinating is because of the interplay between the cladists who seek classifications based on evolution and the pheneticists who are content with utilitarian classifications. The two approaches are not independent because species which are close in evolution will tend to share many common features and hence are likely also to be phenetically close. This argument which dominated the early days of numerical classification is of importance to biologists but has little relevance to classifications in most other fields of application, except, perhaps, to highlight what should be obvious, that there is no uniquely correct classification - it depends why one is classifying.

When classifying books or major species-groups, notions of variability are of little relevance. Statistics is concerned with handling variability. Discriminant analysis and the multivariate mixture problem are two classical classification problems handled by statistical methods. Both are concerned with two or more populations with several variables whose values are characterised by probability distributions within the populations. To force other situations into a framework based on populations modelled by probability distributions can be misguided.

Below, we first focus on how the status of the things being classified interacts with admissible criteria for defining classes and then we identify a few problems where nonprobabilistic classification is appropriate. Finally, we comment on how in non-probabilistic classifications the concepts of statistical inference may be replaced, and perhaps bettered, by the older concept of approximation.

2. Types of variable and the things being classified

Three main ingredients of all classification problems are (i) the things being classified, (ii) the variables used to describe these things and (iii) the purposes of the proposed classification. In this section we mainly discuss (i) ending with a brief reference to (ii); some purposes of classification are discussed in section 3. Customarily, data for classification are presented in a matrix $\boldsymbol{X}$ with N rows pertaining to the objects and p columns pertaining to the variables; we shall see that this can be a simplistic view of the structure of the data.

Table 1: *Types of classification problem depending on whether (i) the problem is probabilistic or non-probabilistic, (ii) is for assignment to classes or construction of classes, or (iii) is concerned with structured or unstructured objects. (Simplification of a similar table in Gower, 1998)*

	Objects	*Assignment*	*Construction*
Non-probabilistic	*Structured*	Matching Diagnostic keys	Maximal predictive classes Cluster analysis: k-groups, hierarchic, other
Probabilistic	*Unstructured*	(Null)	Mixture problems
	Structured	Discrimination	Undeveloped

Table 1 is a three-way table giving a breakdown of some well-known classification problems. As well as distinguishing between (i) probabilistic and non-probabilistic classifications and (ii) between assignment/identification and class-construction problems, Table 1 distinguishes (iii) between the objects being classified as *structured* or *unstructured.* We take the meanings of (i) and (ii) to need no further explanation; what we mean by (iii) is best understood by considering the usual between and within groups sample-structure that is familiar in simple analysis of variance and in canonical variate analysis. Thus, the N rows of our data matrix $\boldsymbol{X}$ refer to objects drawn from g groups the kth of which has n_k members and $N = \Sigma n_k$. The g groups may have further *a priori* structure but, at least for the present, we shall assume that this is not so.

We begin by considering two important special cases of this general type of data-structure, Type A: $g = 1$ so that $N = n_1$ and Type B: $n_k = 1$ for $k = 1,2,...,g$. Type A is the classical multivariate data-matrix of statistical multivariate analysis. Typically, its rows represent a random sample of size $N = n_1$. The samples have no *a priori* structure, so are said to be *unstructured.* Note that it is the *data* that are unstructured, but nevertheless they might be modelled by a highly structured model. Indeed, that assumption is the starting point of the multivariate mixture problem which seeks a representation of the underlying probability distribution as a mixture of g more simple distributions. After analysis, the previously unstructured samples can be assigned to the g newly-found distribution-groups and then $\boldsymbol{X}$ becomes structured. With Type B we have g groups at the outset where each group is represented by *one* object, or row, of $\boldsymbol{X}$. This is quite different from the familiar statistical data-matrix and would be an inadequate representation of reality when variability within groups is substantial. However, there are many classification problems where within-group variability may be ignored. For example, the gth group may have a unique member (there is only one Rome) or we may be able to choose variables which do not vary within groups (all cats have claws) but do between groups. The latter is a very common situation and one may note that those interested in classifying groups will seek to describe the groups by variables which are constant; indeed, they would be foolish not to do this when it is a possibility. When such variables are unavailable, the only remaining possibility is to base classifications on variables which do vary within groups, in which case probabilistic methods become relevant and the choice $n_k = 1$ becomes untenable; then replication of samples within groups is essential to capture the within-group variability. Within groups variation is a familiar component of the assignment problem of classical discriminant analysis but at the end of section 4 we sketch how it can also be important when constructing classes.

With groups, even when $n_k = 1$, $\boldsymbol{X}$ is said to be structured. Normally, the groups will bear names indicating an initial classification. Botanists recognise species like daisies and dandelions, linguists recognise languages, geographers recognise countries, librarians recognise book-titles and so on. This initial classification does not inhibit a desire for further classification; rather it encourages classification to organise what can be very large bodies of information.

To recapitulate, the objects to be classified are said to be structured when they fall into named groups, which implies that a preliminary classification is recognised. Otherwise they are said to be unstructured, which usually means that the rows of $\boldsymbol{X}$ are undifferentiated as with a random sample assumed to be drawn from some notional population or mixture of populations. In table 1 all non-probabilistic classification problems are concerned with objects of Type B. In probabilistic classification problems we meet with both structured (e.g. canonical variate analysis) and unstructured (e.g. multivariate mixtures) sets of objects.

With replication within groups, whether or not the n_k replicates are chosen by some random process is important. It is not obvious that simple random samples are necessarily a desirable basis for forming classifications. For example, the number of speakers of a language is unlikely to be relevant for classifying languages thus excluding simple random sampling. Indeed, it is the *distribution* of variables within the groups which are important, not the relative frequencies of speakers; each language might be represented by a single set of characteristics as in taxonomy where within-group variation is often handled by representing each group by a single invented object with *typical* values; this is acceptable when within group variation is negligible. Type B data are an extreme form of non-random sampling but are often the proper object structure.

We have considered the most simple between-within structure where the *g* groups represent the totality of objects to be classified. This does not preclude the possibility of additional groups being added at a later stage. Also the groups might be regarded as a random sample of some larger set, and then probabilistic methods might have a role to play; to us this seems an artificial set-up. The groups might have an elaborate *a priori* imposed structure of the crossed and nested kinds, indicating an equally elaborate *a priori* classification. Then it seems unlikely that one would wish to use this classification as the starting point for further classification. Rather, one would begin again, subsuming the complex structure into a simple group structure and comparing any new classification with the old one. No further account is taken of these possibilities here but they point the way to some areas of future research in classification theory.

The above has said little about the types of variable used in classification. Variables may be numerical or categorical. If numerical, they may be continuous on ratio or interval scales, a distinction which rarely affects classification. Categorical variables may be nominal, ordinal or dichotomous, a special binary categorical variable with one category merely not the other; all these types contain different kinds of information which may be exploited when classifying things. Numerical variables are less important for non-probabilistic classification than are categorical variables. This is because numerical variables are likely to vary within groups, so when non-varying categorical variables can be found they are to be preferred. Just as objects may be structured, so may variables. Structure in variables has been long-recognised in survey design and Gower (1971) took it into account when designing a general similarity coefficient in which primary variables could be associated with sets of secondary variables, in turn associated with tertiary variables, and so on.

3. Non probabilistic classification problems

Table 1 lists some non-probabilistic classification problems, both for assignment to classes and for the construction of classes; the starting point is data of Type B

Identification keys constructed by *ad hoc* methods have been known to botanists for several centuries. To use a key one has a single object which has to be identified, that is it has to be named. The most simple thing to do is to compare the object with each row of $\boldsymbol{X}$ until a match is found, when identification is achieved. This is inefficient because an enormous number of comparisons may have to be made, and it may be impracticable because variables included in $\boldsymbol{X}$ may be unavailable (e.g. a flowering plant may give no information on the characteristics of its seeds). Diagnostic keys overcome these difficulties. A key is essentially a tree with one binary variable at each node. The value of the binary variable for the object to be identified determines which of the two possible branches of the tree one traverses to reach the next node. In this way one answers a series of questions until one reaches an end-point of the tree, where the identification is given. Many interesting problems in constructing keys have been reviewed by Payne and Preece (1980). We may require the tree with fewest nodes or using fewest different among the binary variables or with minimum average numbers of steps to achieve identification. Costs may be associated with ascertaining the values of the binary variables, in which case we may require the key that is cheapest to use. Probability concepts are irrelevant for these interesting problems but may enter if we recognise that the objects have different frequencies of occurrence, so affecting average numbers of steps to identification and average costs. However, any probabilistic distribution associated with the binary variables themselves is not required nor is it material.

When it comes to constructing classifications with the same form of $\boldsymbol{X}$ (i.e. with binary variables) we can classify the g objects into $k<g$ groups in such a way that on being told that an object belongs to one of these groups, more correct statements about the likely value of its binary variables can be made than for any other classification. This is maximal predictive classification (Gower, 1975) which models the dictum of a distinguished botanical taxonomist, Gilmour (1937), that *a system of classification is the more natural the more propositions there are that can be made regarding its constituent classes*. No probability distribution is associated with the binary variables, although again the relative frequencies of the objects may be accommodated.

We may apply the *k*-means algorithm to numerical Type B data, thus forming homogeneous groups among the *N* objects without any appeal to probability. This is an interesting example, because if ***X*** is regarded as of Type A then the algorithm finds the maximum likelihood solution to the mixture problem modelled as a combination of multinormal populations. Thus, the same algorithm may be used to compute two different classification problems, one probabilistic and the other not. Of course, different mixture models have different maximum likelihood solutions while the *k*-means algorithm is only valid for multinormal mixtures.

Many, mostly heuristic, algorithms applied to Type B data give a hierarchical classifications of the *g* objects. Heuristic algorithms are acceptable when there is no practicable way of optimising an objective criterion (e.g. NP complete problems). Ultrametrics and additive trees give objective criteria for fitting trees by least squares, thus providing well-defined models. The use of least-squares does not necessarily imply an appeal to probability. One may note the eighteenth century work of mathematicians like Legendre and Laguerre who used L_1, L_2, or L_∞ norms to approximate complicated functions (e.g. Bessel functions) by polynomials. Polynomials give an acceptable approximation to the function - probability is irrelevant. Analogously, we may regard nonprobabilistic classifications as giving similar approximations to ***X*** where the goodness of fit may be used to assess the adequacy of the tree approximation or of the class-predictors fitted to maximise prediction. Such measures of approximation are at least as useful as significance tests associated with probabilistic models. Remember, *significant* is not synonymous with *important* and *not significant* is not synonymous with *unimportant*, a point we return to below.

Other nonprobabilistic objectives for classification concern mixtures of hierarchical and non hierarchical organisation. We may wish to classify the *g* groups into $k<g$ classes arranged hierarchically with each class containing one or more undifferentiated members. Indeed, it seems to us that this is more frequently required than full hierarchical classification but it rarely gets mentioned, except when pointing out that branches of a full tree may be amalgamated, perhaps governed by specifying some threshold level in a dendrogram. Again with Type A data we could formulate a variant of the mixture problem in which the *k* classes are to be arranged hierarchically. Gower (1975) gives a method for constraining maximal predictive classes to have hierarchical organisation which immediately extends to other classification criteria; unfortunately the computational problems are formidable.

4. Conclusion

We hope we have shown that nonprobabilistic classification criteria are often relevant and lead to interesting and useful methods. Although we distinguish between assignment to and construction of classes, a perfectly good criterion for constructing classes is so that they can be assigned to optimally; nonprobabilistic maximal predictive classes and maximum likelihood mixture classes both have optimal assignment properties. We have highlighted two extreme forms of $\boldsymbol{X}$, Type B ($g=N$) for which nonprobabilistic methods are often applicable and Type A ($g=1$) where probabilistic methods are often applicable. That computers cannot distinguish between Types A and B has encouraged some misuse of classification software, a situation made the more confusing when the same algorithm may be used on either form of $\boldsymbol{X}$ to solve different problems.

Important as it is, not all useful classificatory information need be probabilistic. Suppose we wish to classify g normal populations into $k<g$ groups. Following discrimination ideas, we could seek boundaries which minimise the overlap between the k groups, thus minimising future errors of misclassification. With limiting point-densities *any* set of boundaries give no overlap and no possibility of misclassification. Nevertheless, it remains reasonable to require a grouping of the populations into k classes, possibly nested, by grouping together pairs of point-populations that are closer than others, as judged by distances based on the nonprobabilistic information contained in the values of the variables. When the point densities expand to conventional distributions, it seems that a combination of probabilistic and nonprobabilistic information should be used for constructing classifications; assignment to these classes might be entirely probabilistic.

References

Gilmour, J. S. L. (1937). A taxonomic problem. *Nature*, **134**, 1040 - 1042.

Gower, J. C. (1971). A general coefficient of similarity and some of its properties. *Biometrics*, **27**, 857-871.

Gower, J. C. (1975). Maximal predictive classification. *Biometrics*, **30**, 643-654.

Gower, J. C. (1998). Classification in: *Encyclopaedia of Biostatistics*, Armitage, P. and Colton, T. (Eds.), Wiley, Chichester. (in press).

Payne, R. W. and Preece, D. A. (1980). Identification keys and diagnostic tables: a review (with discussion). *J. R. Statist. Soc. A.*, **143**, 253-292.

An investigation of nine procedures for detecting the structure in a data set

André HARDY, Paul ANDRE
Unité de Statistique, Département de Mathématique,
Facultés Universitaires N.-D. de la Paix,
8 Rempart de la Vierge, B-5000 Namur, Belgium
e-mail: aha@math.fundp.ac.be

Abstract: A problem common to all clustering techniques is the difficulty of deciding the number of clusters present in the data. The aim of this paper is to assess the performance of the best stopping rules from the Milligan and Cooper's (1985) study, on specific artificial data sets containing a particular cluster structure. To provide a variety of solutions the data sets are analysed by four clustering procedures. We compare also these results with those obtained by three methods based on the hypervolume clustering criterion.

Keywords: clustering, stopping rule, number of clusters, hypervolume criterion.

1. Introduction

Most clustering algorithms can partition a data set into any specified number of clusters even if the data set contains no structure. So one of the most important problems when validating the results of a cluster analysis is: how many clusters are in the data? Some studies have been proposed, to compare procedures for determining the number of clusters. For example, Milligan and Cooper (1985) conducted a Monte Carlo evaluation of thirty indices for determining the number of clusters. Hardy (1994) compared three methods based on the hypervolume clustering criterion with four other methods available in the Clustan (1978) software. More recently, Gordon (1997) modified the five stopping rules whose performance was best in Milligan and Cooper's (1985) study in order to detect when several different, widely-separated values of c, the number of clusters, would be appropriate, that is, when a structure is detectable at several different scales.

This paper addresses the problem of assessing the performance of the best procedures from the Milligan and Cooper's (1985) study, on specific artificial two-dimensional data sets showing a particular cluster structure considered neither in Milligan and Cooper's (1985) study nor in Gordon's (1997) paper. We also compare these results with those obtained by three methods based on the hypervolume clustering criterion.

2. The hypervolume clustering criterion

We assume a clustering model when the observed points are a realization of a Poisson process in a set D of R^m, where D is the union of k disjoint convex domains $D_1, D_2, ..., D_k$; $C_i \subset \{x_1, x_2, ..., x_n\}$ is the subset of observations belonging to D_i $(1 \leq i \leq k)$. The problem is to estimate the unknown domains D_i in which the points are distributed. The maximum likelihood estimation of the k subsets $D_1, D_2, ..., D_k$ is constituted by the k subgroups C_i of points such that the sum of the Lebesgue measures of their disjoint convex hulls $H(C_i)$ is minimum $(1 \leq i \leq k)$ (Hardy and Rasson (1982); Hardy (1983)). The hypervolume clustering criterion is thus defined as

$$W(P, k) = \sum_{i=1}^{k} m(H(C_i))$$

where P is a partition of the observed points into k clusters, $H(C_i)$ is the convex hull of the points belonging to C_i and $m(H(C_i))$ is the m-dimensional Lebesgue measure of that convex hull.

3. Methods to determine the number of clusters

The first six indices - those whose performance was best in Milligan and Cooper's (1985) study - are the the Caliński and Harabasz (1974) method (M_1), the J index (Duda and Hart (1973)) (M_2), the C index (Hubert and Levin (1975)) (M_3), the γ index (Goodman and Kruskal (1954)) (M_4), the Beale (1969) test (M_5), and the Cubic Clustering Criterion (M_6) (Sarle, 1983).

The last three methods are based on the hypervolume criterion.

3.1 A classical geometric method (M7)

This well-known method consists in plotting a criterion value W against k, the number of clusters. With every increase in k there will be a decrease in W. A discontinuity in slope should correspond to the true number of "natural" clusters. W is here the hypervolume criterion.

3.2 Method based on the estimation of a convex set (M8)

If we consider a realization of a homogeneous planar Poisson process of unknown intensity within a compact convex set D, the best estimate of D is given by $D' = g(H(D)) + c \cdot s(H(D))$ where $H(D)$ is the convex hull of the points belonging to D, $g(H(D))$ is the centroid of $H(D)$ and $s(H(D)) = H(D) - g(H(D))$. So D' is a dilation of the convex hull about its centroid (Ripley, Rasson (1977)). The realization of a Poisson process within the union D of k subsets $D_1, D_2, ..., D_k$ can be considered as the realization of k Poisson processes of the same intensity within the k subsets $D_1, D_2, ..., D_k$. We apply that result to the hypervolume clustering problem. For k = 2, 3, ..., we consider the dilated convex hulls of the clusters of the best partition of the set of objects into k clusters. If k_0 is the first value of k for which at least one pair of dilated convex hulls do intersect, we then consider k_0 - 1 as the optimal number of clusters (Hardy (1994)).

3.3 A likelihood ratio test (M9)

Let x_1, x_2, ,..., x_n be a random sample from a Poisson process on k disjoint convex sets D_1, D_2, ..., D_k. For a given integer $k \geq 2$, we test whether a subdivision into k clusters is significantly better than a subdivision into $k-1$ clusters, i.e. the hypothesis $H_0 : t = k$ against the alternative $H_1 : t = k-1$, where t represents the number of clusters.
The statistic of the test is given by (Hardy, 1994): $S(x) = \frac{W(P,k)}{W(P,k-1)}$.
We have $S(x) \in [0,1]$. Thus we will reject H_0 iff $S(x) > C$, where C is a constant. Unfortunately we do not know the H_0 distribution of the statistic S. Nevertheless in practice, we can use the following rule: reject H_0 if S takes large values i.e. if S is close to 1. So we will apply the test in a sequential way: if k_0 is the first value of k for which we reject H_0, we shall consider $k_0 - 1$ as the appropriate number of natural clusters.

4. Results

In order to make the assessment of the six best stopping rules from the Milligan and Cooper's study, we have considered four well-known hierarchical clustering procedures (nearest neighbour, furthest neighbour, group average and Ward) and six artificial data sets containing a known structure (one single homogeneous group, two or three elongated clusters, linearly nonseparable clusters, unequal-size hyperspherical-shaped clusters, nonconvex clusters). We have then applied the six indices to the results obtained by the application of the clustering methods to the data sets test.

4.1 First set of data: absence of any group structure

Here we have simulated a Poisson process into a single rectangular set in the plane; so the 150 points are independently and uniformly distributed in this set. The performance of the six indices are tabulated in Table 1.

	M1	M2	M3	M4	M5	M6
nearest neighbour	4	1	1	1	1	1
furthest neighbour	4	1 or 3	1	1	1	1
group average	4	1	1	1	1	1
Ward	4	1 or 3	1	1	1	1

Table 1: One homogeneous group.

The first column of a table lists the names of the clustering procedures. The other columns show the results given by the best six stopping rules from the Milligan and Cooper's study.
Methods M_3, M_4, M_5, M_6 detect one cluster. The Duda and Hart's rule (M_2) requires the computation of a critical value. One of the parameters used in the computation is z, a standard normal score. Milligan and Cooper chose z to be 3.2 because for the examples considered in their paper, this value of z produces the best recovery. In his study Gordon specifies z to be 4.0. If z is equal to 4, M_2 gives in this example the correct number of clusters. If we choose 3.2 for z, M_2 detects three clusters in two cases (see Table 1). The Caliński and Harabasz's method (M_1) doesn't recover the true structure in this example.

4.2 Second set of data: two elongated clusters

In this example, the natural structure consists of two elongated parallel clusters. The results appear in Table 2.

		M1	M2	M3	M4	M5	M6
nearest neighbour	+	6	1	1	1	1	1
furthest neighbour	-	2	2	1	1	1	1
group average	-	2	2	1	1	1	1
Ward	-	2	2	1	1	1	1

Table 2: Two elongated clusters.

A "+" in the second column of Table 2 indicates that if we apply a clustering method in order to get two clusters, we obtain the two "natural" elongated clusters; dashed entries correspond to situations where the classification obtained is not the expected one. Only the nearest neighbour method reproduces the "natural" classification of the data, and in that case, the six best rules from the Milligan and Cooper's study are ineffective in predicting the correct number of clusters. Milligan and Cooper noted that the two clusters case was the most difficult structure for the stopping rules to detect. Let us remark that some rules determine the correct number of clusters, for a partition into two clusters that is not the natural one.

4.3 Third set of data: three elongated clusters

In this example, artificial data were generated to contain three elongated parallel clusters. The results obtained are given in Table 3.

		M1	M2	M3	M4	M5	M6
nearest neighbour	+	2	1	1	1	1	1
furthest neighbour	-	2	2	1	1	1	1
group average	-	2	2	1	1	1	1
Ward	-	2	2	1	1	1	1

Table 3: Three elongated parallel clusters.

The results, here, are similar to those of the second data set. Only the nearest neighbour method reproduces the "natural" classification when we fix the number of clusters to three. The furthest neighbour, group average and Ward procedures tend to produce hyperspherical-shaped clusters. So it is not surprising to see that the results obtained here are similar to those of the previous data set; the Caliński and Harabasz's method and the J index detect two clusters in almost all cases in both examples; the other rules dectect only one cluster.

4.4 Fourth set of data: three elongated nonseparable clusters

In this example, we have chosen three elongated clusters such that no one is linearly separable from the two others. Many clustering methods fail to detect the natural partition into three groups. The results appear in Table 4.

		M1	M2	M3	M4	M5	M6
nearest neighbour	+	3	1	1	1	1	1
furthest neighbour	-	1	1 or 3	1	1	1	1
group average	-	1	1	1	1	1	1
Ward	-	1	1 or 3	1	1	1	1

Table 4: Three elongated nonseparable clusters.

The nearest neighbour method is efficient in recovering the underlying structure into three clusters. The Caliński and Harabasz's method (M_1) applied to the results of the nearest neighbour procedure gives the correct number of clusters. If we choose 3.2 for z, the Duda and Hart's method (M_2) detects three clusters in two cases. If z is equal to 4, M_2 gives only one cluster. The other rules detect only one cluster.

4.5 Fifth set of data: unequal-size hyperspherical-shaped clusters

This set of data consists of two groups; the first one contains eighty points, and the second one twelve points. The distance between the two groups is larger than the diameter of each of the groups. So the two natural clusters are well separated. The results obtained are tabulated in Table 5.

		M1	M2	M3	M4	M5	M6
nearest neighbour	+	2	2	2	2	1	2
furthest neighbour	+	2	2	2	2	1	3
group average	+	2	2	2	2	1	6
Ward	+	2	2	2	2	1	6

Table 5: Unequal-size hyperspherical-shaped clusters.

The Beale test (M_5) and the cubic clustering criterion (M_6) perform poorly; they fail in recognizing the true structure of the data, except when M_6 is applied to the results given by the nearest neighbour method. The first four rules predict the correct number of clusters.

4.6 Sixth set of data: nonconvex linearly nonseparable clusters

This set of data contains two clusters. Both are nonconvex. They are not linearly separable. The results obtained are given in Table 6.

		M1	M2	M3	M4	M5	M6
nearest neighbour	+	5	1	1	1	1	1
furthest neighbour	-	1	1	1	1	1	1
group average	-	1	1	1	1	1	1
Ward	-	1	1	1	1	1	1

Table 6: Nonconvex linearly nonseparable clusters.

Here also, only the nearest neighbour method retrieves the natural structure of the data when we fix the number of clusters to two. The six best methods from the Milligan and Cooper's study are unable to determine the correct number of clusters.

Results associated with the hypervolume criterion

We've applied the hypervolume clustering procedure to the same data sets and we've recorded the number of clusters given by the three methods M_7, M_8 and M_9, based on the hypervolume criterion.

In the first five examples the hypervolume method retrieves the natural structure of the data when we fix the number of clusters to its correct value. The hypervolume method is based on a hypothesis of convexity for the clusters.

In the sixth example, the clusters are nonconvex linearly nonseparable. So in that case it is impossible for the hypervolume method to detect the two natural nonconvex clusters.
The three methods for the determination of the number of clusters based on the hypervolume criterion (M_7, M_8, M_9) always correctly estimate the number of clusters in the first five examples. In the sixth data set the three methods detect only one group.

5. Conclusions

In the Milligan and Cooper's study, the clusters were internally cohesive and well separated in the variable space. So the aim of this paper was first to investigate the performance of the six best procedures from the Milligan and Cooper's study on specific data structures not considered by these authors. They themselves pointed out that the results that they have obtained are likely to be data dependant, and that it would not be surprising that the ordering of the indices would change if different data structures were used.
From the examples considered in this paper, we can conclude that the six best methods of the Milligan and Cooper's investigation are usually unable to detect elongated clusters, linearly nonseparable clusters and nonconvex clusters. That's not very surprising because some of these rules, being based on between-groups and within-groups sum-of-squares, are better able to detect hyperspherical-shaped clusters. In addition, the Caliński and Harabasz method (M_1), and the J index if z is equal to 3.2, have difficulties to detect the presence of one single homogeneous group. The Beale test (M_3) and the Cubic Clustering Criterion (M_6) perform very poorly in presence of unequal cluster sizes.
Concerning the methods based on the hypervolume criterion, they performed at a competitive rate on the examples considered here, except in the case of nonconvex linearly nonseparable clusters.
Nevertheless we can recommend the collective use of the procedures considered here, being aware of the fact that the data may contain a particular cluster structure.
Overall, the problem of estimating the number of clusters in multidimensional data remains a difficult and challenging problem. No completely satisfactory solution is available. Maybe the question is incapable of any formal or complete solution simply because there is nowadays no general agreement on an universally acceptable definition of the term cluster.

References:

BEALE, E. M. L. (1969): Euclidean cluster analysis. *Bulletin of the International Statistical Institute*, 43, 2, 92-94.

CALINSKI, T., and HARABASZ, J. (1974): A dendrite method for cluster analysis. *Communications in Statistics*, 3, 1-27.

DUDA, R.O., and HART, P.E. (1973): *Pattern Classification and Scene Analysis.* Wiley, New York.

GOODMAN, L.A. and KRUSKAL, W.H. (1954): Measures of association for cross-classifications. *Journal of the American Statistical Association*, 49, 732-764.

GORDON, A.D. (1997): How many clusters? An investigation of five procedures for detecting nested cluster structure, in *Proceedings of the IFCS-96 Conference*, Kobe (in print).

HARDY, A., and RASSON, J.P. (1982): Une nouvelle approche des problèmes de classification automatique. *Statistique et Analyse des données*, 7, 41-56.

HARDY, A. (1983): Statistique et classification automatique : Un modèle - Un nouveau critère - Des algorithmes - Des applications. *Ph.D Thesis*, F.U.N.D.P., Namur, Belgium.

HARDY, A. (1994): An examination of procedures for determining the number of clusters in a data set, in *New Approches in Classification and Data Analysis*, E. Diday et al. (Editors), Springer-Verlag, Paris, 178-185.

MILLIGAN, G.W. and COOPER, M.C. (1985): An examination of procedures for determining the number of clusters in a data set. *Psychometrika*, 50, 159-179.

RIPLEY, B.D., and RASSON, J.P. (1977): Finding the edge of a Poisson Forest. *Journal of Applied Probability*, 14, 483-491.

SARLE, W.S. (1983): Cubic Clustering Criterion. *Technical Report: A-108*, SAS Institute Inc., Cary, NC, USA.

WISHART, D. (1978): *CLUSTAN User Manual*, 3rd edition, Program Library Unit, University of Edinburgh.

Clustering and Outlier Identification: Fixed Point Cluster Analysis

C. Hennig
Institut für Mathematische Stochastik,
Universität Hamburg, D-20146 Hamburg, Germany

Abstract: Fixed Point Cluster Analysis (FPCA) is introduced in this paper. FPCA is a new method for non-hierarchical cluster analysis. It is related to outlier identification. Its aim is to find groups of points generated by a common stochastic model without assuming a global model for the whole dataset. FPCA allows for points not belonging to any cluster, for the existence of clusters with a different shape, and for overlapping clusters. FPCA is applicated to the clustering of p-dimensional metrical data, 0-1-vectors, and linear regression data.

Keywords: Stochastical clustering, overlapping clusters, mixture model, outlier identification, linear regression

1. The cluster concept of Fixed Point Clusters

Sometimes, a dataset does not consist of a partition into some "natural clusters", but nevertheless there seem to be clusters, perhaps of different types. The following dataset is taken from Hand et al. (1994), p. 58, and gives the rates of child mortality (y-axis) and adult female literacy (x-axis) of 29 countries.

Figure 1: *UNICEF-data*

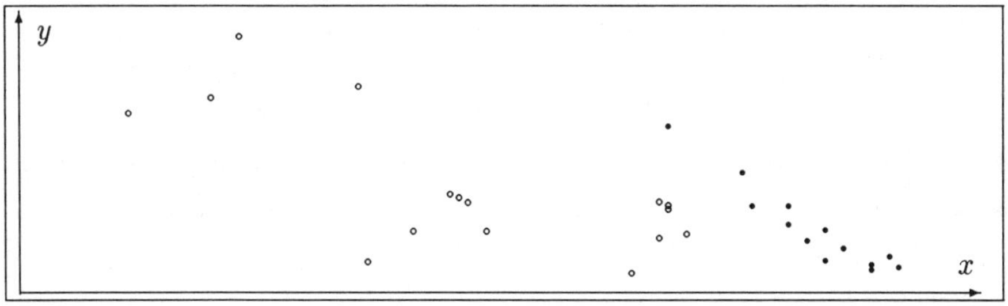

There seems to be some "grouping" in the data. A linear regression model could be adequate for all or a large part of the filled points, but hardly for considerable parts of the rest.

The aim of Fixed Point Clustering is to find "natural clusters" in a stochastic sense. For a given dataset, a "cluster" in a stochastic sense is a subset whose points are described adequately by a common distribution. This distribution comes from a parametric family of "cluster reference distributions" which model the form of the clusters of interest. For example, it could be Normal or other suitable unimodal distributions, while bimodal distributions heuristically mostly generate two clusters. Frequently, such stochastic cluster analysis problems are treated by means of a mixture model

$$P = \sum_{i=1}^{s} \epsilon_i P_i, \quad \sum_{i=1}^{s} \epsilon_i = 1, \quad \epsilon_i > 0, \ i = 1, \ldots, s, \tag{1}$$

where the $P_i, i = 1, \ldots, s$, are cluster reference distributions with different parameters and the $\epsilon_i, i = 1, \ldots, s$, are the proportions of the s clusters. While the estimation of s is difficult, the parameters of the P_i can often be estimated consistently by Maximum Likelihood estimators if s is known.

In difference to that, FPCA is a local cluster concept, i.e. the cluster property of a data subset depends only on the particular set and its relation to the whole dataset, but not on a global model. FPCA bases on "contamination models" of the form

$$P = (1 - \epsilon)P_0 + \epsilon P^*, \ 0 < \epsilon \leq 1, \tag{2}$$

where P_0 is a cluster reference distribution. P^* can be arbitrary. In particular, it can be a mixture of other cluster reference distributions so that this model includes mixture distributions of the form (1). Additionally, P^* has to be "well seperated" from P_0 in order to guarantee that the points generated by P_0 form a cluster with respect to the whole dataset. An ideal FPC should consist of the points generated by P_0. Contamination models are often used in robust statistics and outlier identification, and a cluster could be seen as a data subset which belongs together, i.e. it contains no outliers, and which is separated from the remaining data, i.e. all other data points are outliers w.r.t. the cluster. FPCA provides a formalization of this imagination.

This concept can cope with more complex data situations:

- If a dataset contains outliers, they need not to be included in any cluster.
- Sometimes a dataset contains clusters of various shapes. For example, there are growth curves in biology, where linear growth is followed by logarithmic growth.
- Clusters can overlap.
- No assumption on the number of clusters is needed.

2. Definition of fixed point clusters

An FPC is a subset of the data, which does not contain any outlier. All other points of the dataset have to be outliers with respect to the FPC.

More formally: Let M be the space of the data points and $\mathbf{Z_0} \in M^n = (z_1, \ldots, z_n)'$ be the dataset. For $m \in I\!N$ let

$$I_m: \; M^m \mapsto \{0,1\}^M \quad (\mathbf{Z} \in M^m \text{ is mapped on a mapping.}) \tag{3}$$

an "outlier identifier" with the interpretation that a point $z \in M$ is outlier with respect to $\mathbf{Z} \in M^m$, if $I_m[\mathbf{Z}](z) = 1$. The construction of reasonable outlier identifiers will be discussed in later sections. For an indicator vector $g = (g_1, \ldots, g_n)' \in \{0,1\}^n$, let $\mathbf{Z}_0(g)$ be the matrix of the data points z_i, for which $g_i = 1$, and $n(g)$ be the number of these points.

Definition. *With outlier identifiers $I_1, \ldots, I_n$ define*

$$f_{\mathbf{Z}_0}: \; \{0,1\}^n \mapsto \{0,1\}^n, \; g \mapsto (1 - I_{n(g)}[\mathbf{Z}_0(g)](z_i))_{i=1,\ldots,n}, \tag{4}$$

i.e. $f_{\mathbf{Z}_0}$ maps g on the indicator vector of the $\mathbf{Z}_0(g)-$non-outliers. Then $\mathbf{Z}_0(g)$ is an FPC if g is a fixed point of $f_{\mathbf{Z}_0}$. Thus, the non-outliers w.r.t. $\mathbf{Z}_0(g)$ have to be precisely the points of $\mathbf{Z}_0(g)$.

This definition can be related to stochastic modeling as follows: Let M be a space with $\sigma-$Algebra $\mathcal{B}$. Let $\mathcal{P}$ be the space of distributions on $(M, \mathcal{B})$, and $\mathcal{P}_0 \subset \mathcal{P}$ the set of cluster reference distributions.

Definition. *(Davies and Gather 1993) $A(P)$ is called $\alpha-$outlier region w.r.t. P under*

$$A: \; \mathcal{P} \mapsto \mathcal{B} \text{ with } \forall P \in \mathcal{P}_0: \; P(A(P)) \leq \alpha. \tag{5}$$

$0 < \alpha < 1$ should be small so that the points in $A(P)$ "lie really outside". Now, the set of points $A_m(\mathbf{Z}) := \{z: \; I_m[\mathbf{Z}](z) = 1\}$ is an estimator of $A(P)$, if $\mathbf{Z}$ is a dataset of i.i.d. points distributed by P. If $\mathbf{Z} = \mathbf{Z}_0(g)$ forms an FPC w.r.t. $\mathbf{Z}_0, A_m(\mathbf{Z})^c$, i.e. the set of $\mathbf{Z}$-non-outliers, is an estimator (whose quality is to be assessed) for an FPC-set of a distribution:

Definition. *For a given distribution $P \in \mathcal{P}$ and $B \in \mathcal{B}$ let $P_B := P(\bullet|B)$ the distribution P restricted to the points of B. Let A be an outlier region. Define*

$$f_P: \; \mathcal{B} \mapsto \mathcal{B}, \quad B \mapsto A(P_B)^c. \tag{6}$$

B is an FPC-set w.r.t. P if B is a fixed point of f_P.

With help of this definition, the quality of the results of an FPCA with respect to distributions P of the form (2) can be investigated by comparing the FPC-sets B of P (the distributions P_B, respectively) to P_0, e.g. the high density areas of P_0.

The computation of all FPCs of a dataset requires the evaluation of (4) for every subset of data. This is impossible, except for very small datasets. Instead, one can apply the usual fixed point algorithm

$$g^{i+1} = f(g^i) \tag{7}$$

to various starting vectors g^0.

Now, the application of this concept to various clustering problems will be demonstrated.

3. Multidimensional Normal clusters

The estimation of Normal mixture parameters is discussed broadly in the literature, see e.g. Titterington et al. (1985). In this setup, a cluster of points from $I\!R^p$ is considered as a group of points generated by a common $\mathcal{N}_p(\mu, \mathbf{\Sigma})$-distribution. The definition of FPCs is given by the definition of the cluster reference distributions and the corresponding outlier identifier. Here:

$$\mathcal{P}_0 := \{\mathcal{N}_p(\mu, \mathbf{\Sigma}) : \ \mu \in I\!R^p, \mathbf{\Sigma} \text{ pos. semidef. } p \times p\}. \tag{8}$$

An easy $\alpha-$outlier region for the distributions of $\mathcal{P}_0$ has the form

$$A(\mathcal{N}_p(\mu, \mathbf{\Sigma})) := \{x \in I\!R^p : (x - \mu)'\mathbf{\Sigma}^{-1}(x - \mu) > \chi^2_{p;1-\alpha}\}. \tag{9}$$

Such outlier regions can be estimated by replacing μ and $\mathbf{\Sigma}$ by estimators $\hat{\mu}_m$ and $\hat{\mathbf{\Sigma}}_m$. This leads to the following outlier identifier: For $\mathbf{X} \in (I\!R^p)^m$ and $x \in I\!R^p$:

$$I_m[\mathbf{X}](x) := 1\left((x - \hat{\mu}_m(\mathbf{X}))'\hat{\mathbf{\Sigma}}_m(\mathbf{X})^{-1}(x - \hat{\mu}_m(\mathbf{X})) > \chi^2_{p;1-\alpha}\right). \tag{10}$$

Chosing $\hat{\mu}_m$ and $\hat{\mathbf{\Sigma}}_m$ as mean vector and sample covariance matrix leads to an outlier identification by means of the Mahalanobis distance. This approach is computationally simple and is discussed e.g. in Rousseeuw and Leroy (1988). They show that a better identification procedure can be obtained if $\hat{\mu}_m$ and $\hat{\mathbf{\Sigma}}_m$ are replaced by more robust estimates.

4. Clustering 0-1-vectors

Let $\mathbf{Z} \in (\{0,1\}^k)^n$ a dataset of $k-$dimensional 0-1-vectors. Then data points can be considered as belonging to a common cluster if they seem to be generated by the same independent product of Bernoulli distributions. That is,

$$P[k, p_1, \ldots, p_k] := \bigotimes_{j=1}^{k} B[1, p_j], \quad \mathcal{P}_0 := \left\{P[k, p_1, \ldots, p_k] : (p_1, \ldots, p_k) \in [0,1]^k\right\},$$

where $B[k,p]$ denotes the Binomial distribution with parameters k,p. An α-outlier region for this kind of data can be defined by counting the components for which a data point does not belong to the expected majority in the cluster:

$$\begin{aligned} &\forall P[k,p_1,\ldots,p_k] \in \mathcal{P}_0 : \\ A(P[k,p_1,\ldots,p_k]) &:= \left\{z \in \{0,1\}^k : v[p_1,\ldots,p_k](z) \geq c(\alpha)\right\}, \qquad (11) \\ &\text{where } c(\alpha) := \min\left\{c : 1 - B[k,\tfrac{1}{2}](c) \leq \alpha\right\}, \\ v[p_1,\ldots,p_k](z) &:= \sum_{j=1}^{k} \left[z_j 1\left(p_j \leq \frac{1}{2}\right) + (1-z_j) 1\left(p_j > \frac{1}{2}\right)\right]. \end{aligned}$$

Here v counts the "minority components" of z and $c(\alpha)$ is chosen such that (5) holds for all $P \in \mathcal{P}_0$, because $B[k,\frac{1}{2}]$ is the distribution of $v[\frac{1}{2},\ldots,\frac{1}{2}](z)$. This is stochastically largest among the $v[p_1,\ldots,p_k](z)$, where $\mathcal{L}(z) \in \mathcal{P}_0$.

Again, a proper outlier identifier can be defined by replacing $p_1,\ldots,p_k$ by estimators in the definition of v, e.g.

$$\hat{p}_{jm}(\mathbf{X}) := \frac{1}{m}\sum_{i=1}^{m} x_{ij}, \text{ where } \mathbf{X} := (x_1,\ldots,x_m) \in (\{0,1\}^k)^m. \qquad (12)$$

5. Linear regression clusters

For analyzing the UNICEF-Data, the female literacy can be considered as explanatory variable for the child mortality. While there is no clear structure for the whole dataset, the relation for certain subsets of the data seems to be approximatively linear. Cluster reference distributions for linear regression can be defined as follows:

$$\mathcal{P}_0 := \{P_{\beta,\sigma^2,G} : \ \beta \in I\!R^{p+1}, \sigma^2 \in I\!R^+, G \in \mathcal{G}\} \qquad (13)$$

where $P_{\beta,\sigma^2,G} \in I\!B^{p+1}$ is the distribution of $(x,y) \in I\!R^{p+1} \times I\!R$ defined by

$$y = \beta' x + u, \quad u \sim \mathcal{N}_{0,\sigma^2}, \qquad (14)$$

$$(x_1,\ldots,x_p)' \sim G \text{ independently of } u, \ x_{p+1} = 1. \qquad (15)$$

$\mathcal{G}$ is some suitable space of distributions on $(I\!R^p, I\!B^p)$, and a simple α-outlier region can be constructed:

$$A(P_{\beta,\sigma^2,G}) := \{(x,y) \in I\!R^p \times \{1\} \times I\!R : (y - x'\beta)^2 > c(\alpha)\sigma^2\}, \qquad (16)$$

where $c(\alpha)$ is the $1-\alpha$-quantile of the χ^2-distribution.

If β and σ^2 are replaced by estimators $\hat{\beta}_m$ and $\hat{\sigma}^2_m$, the indicator function of A is again a reasonable outlier identifier. The best choices for $\hat{\beta}_m$ and $\hat{\sigma}^2_m$

would be robust estimators, but they impose formal and numerical problems. If one uses Least Squares and

$$\hat{\sigma}_m^2(\mathbf{X}) := \frac{1}{m-1}\sum_{i=1}^n (y_i - x_i'\hat{\beta}_m(\mathbf{X}))^2, \quad (17)$$
$$\text{where } \mathbf{X} := ((x_1, y_1), \ldots, (x_m, y_m)) \in (\mathbb{R}^p \times \{1\} \times \mathbb{R})^m,$$

then the subset marked by full points in figure 1 turns out as an FPC with respect to the UNICEF-data. This subset contains all south american countries, Oman and Czechoslovakia, the only european country in the sample.

6. Conclusion

FPCA needs less restrictive assumptions than usual cluster analysis methods based on stochastic models. It enables the analysis of more complex data situations. The idea applies to various problems of cluster analysis. Theoretical background is presented in Hennig (1997), especially for the linear regression setup.

References

Davies, P. L. and Gather, U. (1993). The identification of multiple outliers, *Journal of the American Statistical Association* 88, 782-801.

Hand, D. J., Daly, F., Lunn, A. D., McConway, K. J. and Ostrowski, E. (1994). *A Handbook of Small Datasets*, Chapman & Hill, London.

Hennig, C. (1997). *Datenanalyse mit Modellen für Cluster linearer Regression*, dissertation, Universität Hamburg.

Rousseeuw, P. J. and Leroy, A. M. (1987). *Robust Regression and Outlier Detection*, Wiley, New York.

Titterington, D. M., Smith, A. F. M. and Makov, U. E. (1985). *Statistical Analysis of Finite Mixture Distributions*, Wiley, New York.

Clustering Large Datasets of Mixed Units

Simona Korenjak-Černe, Vladimir Batagelj
University of Ljubljana, Faculty of Mathematics and Physics, and
Institute of Mathematics, Physics and Mechanics, Dept. of TCS
Jadranska 19, 1 000 Ljubljana, Slovenia

Abstract: In this paper we propose an approach for clustering large datasets of mixed units based on representation of clusters by distributions of values of variables over a cluster – histograms, that are compatible with merging of clusters. The proposed representation can be used also for clustering symbolic data. On the basis of this representation the adapted versions of leaders method and adding method were implemented. The proposed approach was successfully applied to several large datasets.

Keywords: large datasets, clustering, mixed units, distribution description compatible with merging of clusters, leaders method, adding method.

1. Introduction

In this paper we propose an approach for clustering large datasets of *mixed* (nonhomogenous) units – units described by variables measured in different types of scales (numerical, ordinal, nominal).

Let E be a finite *set of units*. A nonempty subset $C \subseteq E$ is called a *cluster*. A set of clusters $\mathbf{C} = \{C_i\}$ forms a *clustering*. In this paper we shall require that every clustering $\mathbf{C}$ is a partition of E.

The clustering problem can be formulated as an optimization problem:

Determine the clustering $\mathbf{C}^* \in \Phi$, for which

$$P(\mathbf{C}^*) = \min_{\mathbf{C}\in\Phi} P(\mathbf{C})$$

where Φ is a set of *feasible clusterings* and $P : \Phi \to \mathbb{R}_0^+$ is a *criterion function*.

Most approaches to the clustering problem are based explicitly or implicitly on some kind of criterion function that measures the deviation of units from a selected description or *representative* of its cluster. Usually $P(\mathbf{C})$ takes the form

$$P(\mathbf{C}) = \sum_{C\in\mathbf{C}} \sum_{X\in C} d(X, R_C)$$

where R_C is a representative of cluster C and d a dissimilarity (Brucker, 1978).

The cluster representatives usually consist of variable-wise summaries of variable values over the cluster, ranging from central values (mean, median, mode), (min, max) intervals (symbolic objects; Diday 1997), Tukey's (1977) box-and-whiskers plots, to detailed distribution (histogram or fitted curve).

In this paper we investigate a description satisfying two requirements:

- it should require a fixed space per variable;
- it should be *compatible* with merging of clusters – knowing the description of two clusters we can, without additional information, produce the description of their union.

Note that only some of the cluster descriptions are compatible with merging, for example mean (as sum and number of units) for numerical variables and (min, max) intervals for ordinal variables.

2. A cluster representative compatible with merging

In our approach a cluster representative is composed from representatives of each variable. They are formed, depending on the type of the scale in which a variable is measured, in the following way. Let $\{V_i, i = 1, \ldots, k\}$ be a partition of the range of values of variable V. Then we define for a cluster C the sets

$$Q(i, C; V) = \{X \in C : V(X) \in V_i\},\ i = 1, \ldots, k$$

where $V(X)$ denotes the value of variable V on unit X.

In the case of an ordinal variable V (numerical scales are a special case of ordinal scales) the partition $\{V_i, i = 1, \ldots, k\}$ usually consists of intervals determined by selected threshold values $t_0 < t_1 < t_2 < t_3 < \cdots < t_{k-1} < t_k$, $t_0 = \inf V$, $t_k = \sup V$.

For nominal variables we can obtain the partition, for example, by selecting $k-1$ values $t_1, t_2, t_3, \ldots, t_{k-1}$ from the range of variable V (usually the most frequent values on E) and setting $V_i = \{t_i\}$, $i = 1, \ldots, k-1$; and putting all the remaining values in class V_k.

Using these sets we can introduce the *concentrations*

$$q(i, C; V) = \operatorname{card} Q(i, C; V)$$

and *relative concentrations*

$$p(i, C; V) = \frac{q(i, C; V)}{\operatorname{card} C}$$

It holds

$$\sum_{i=1}^{k} p(i, C; V) = 1$$

The description of variable V over C is the vector of concentrations of V.

Note that in the special case $C = \{X\}$ we have

$$p(i, C; V) = \begin{cases} 1 & X \in Q(i, C; V) \\ 0 & \text{otherwise} \end{cases}$$

Similarly we can consider a missing value on V for unit X by setting $p(i, \{X\}; V) = \frac{1}{k}$, $i = 1, \ldots, k$. Using concentrations we can describe also symbolic data.

It is easy to see that for two clusters C_1 and C_2, $C_1 \cap C_2 = \emptyset$, we have

$$Q(i, C_1 \cup C_2; V) = Q(i, C_1; V) \cup Q(i, C_2; V)$$

$$q(i, C_1 \cup C_2; V) = q(i, C_1; V) + q(i, C_2; V)$$

The description is compatible with merging.

The threshold values are usually determined in such a way that, for the given set of units E (or the space of units $\mathcal{E}$), it holds that $p(i, E; V) \approx \frac{1}{k}$, $i = 1, \ldots, k$.

As a compatible description of nominal variable over a cluster C also its range $V(C)$ can be used, since we have $V(C_1 \cup C_2) = V(C_1) \cup V(C_2)$.

3. Dissimilarity

Most clustering methods are based on some dissimilarity between clusters or between a unit and a cluster . For our descriptions we define

$$d(C_1, C_2; V) = \frac{1}{2} \sum_{i=1}^{k} |p(i, C_1; V) - p(i, C_2; V)|$$

and in the case of nominal variable described by set of values

$$d(C_1, C_2; V) = \frac{\text{card}\,(V(C_1) \oplus V(C_2))}{\text{card}\,(V(C_1) \cup V(C_2))},$$

where $A \oplus B = A \cup B - A \cap B$.
We shall use the abbreviation $d(X, C; V) = d(\{X\}, C; V)$. In both cases it can be shown that

- $d(C_1, C_2; V)$ is a semidistance on clusters for variable V; i.e.
  1. $d(C_1, C_2; V) \geq 0$
  2. $d(C, C; V) = 0$
  3. $d(C_1, C_2; V) + d(C_2, C_3; V) \geq d(C_1, C_3; V)$
- $d(C_1, C_2; V) \in [0, 1]$

and for the vector representation also

$$X \in Q(i, E; V) \Rightarrow d(X, C; V) = 1 - p(i, C; V)$$

The semidistances on clusters for individual variable can be combined into a semidistance on clusters by

$$d(C_1, C_2) = \sum_{j=1}^{m} \alpha_j d(C_1, C_2; V_j), \qquad \sum_{j=1}^{m} \alpha_j = 1, \quad \alpha_j \geq 0$$

where m is the number of variables and α_j weights (Batagelj and Bren, 1995); often $\alpha_j = \frac{1}{m}$.

4. Clustering procedures

The proposed approach allows us to first recode the original nonhomogenous data to a uniform representation by integers – indices of intervals. For the recoded data efficient clustering procedures can be built by adapting leaders method (Hartigan, 1975) or adding clustering method (Batagelj and Mandelj, 1993). In this paper, because of limited space, we shall describe only the procedure based on the dynamic clusters method (a generalization of the leader method).

To describe the dynamic clusters method for solving the clustering problem let us denote (Diday, 1979; Batagelj, 1985): Λ a set of *representatives*; $\mathbf{L} \subseteq \Lambda$ a *representation*; Ψ a set of *feasible representations*; $W : \Phi \times \Psi \to \mathbb{R}_0^+$ *an extended criterion function*; $G : \Phi \times \Psi \to \Psi$ *a representation function*; $F : \Phi \times \Psi \to \Phi$ *a clustering function* and suppose that the following conditions are satisfied:

W0. $P(\mathbf{C}) = \min_{\mathbf{L}\in\Psi} W(\mathbf{C}, \mathbf{L})$

the functions G and F tend to improve (diminish) the value of the extended criterion function W:

W1. $W(\mathbf{C}, G(\mathbf{C}, \mathbf{L})) \leq W(\mathbf{C}, \mathbf{L})$
W2. $W(F(\mathbf{C}, \mathbf{L}), \mathbf{L}) \leq W(\mathbf{C}, \mathbf{L})$

then the dynamic clusters method can be described by the scheme:

$\mathbf{L} := \mathbf{L}_0; \mathbf{C} := \mathbf{C}_0;$
repeat
  $\mathbf{L} := G(\mathbf{C}, \mathbf{L})$
  $\mathbf{C} := F(\mathbf{C}, \mathbf{L});$
until the goal is attained

To this scheme corresponds the sequence $v_n = (\mathbf{C}_n, \mathbf{L}_n), n \in \mathbb{N}$ determined by relations

$$\mathbf{L}_{n+1} = G(\mathbf{C}_n, \mathbf{L}_n) \qquad \text{and} \qquad \mathbf{C}_{n+1} = F(\mathbf{C}_n, \mathbf{L}_{n+1})$$

and the sequence of values of the extended criterion function $u_n = W(\mathbf{C}_n, \mathbf{L}_n)$.

Let us assume the following model $\mathbf{C} = \{C_i\}_{i\in I}$, $\mathbf{L} = \{L_i\}_{i\in I}$, $\mathbf{L}(X) = L_i : X \in C_i$, and further $L = [L(V_1), \ldots, L(V_m)]$, $L(V) = [s(1, L; V), \ldots, s(k, L; V)]$, $\sum_{j=1}^{k} s(j, L; V) = 1$ and

$$d(C, L; V) = \frac{1}{2} \sum_{j=1}^{k} |p(j, C; V) - s(j, L; V)|$$

For

$$W(\mathbf{C}, \mathbf{L}) = \sum_{X\in E} d(X, \mathbf{L}(X)) = \sum_{i\in I} p(C_i, L_i)$$

where

$$p(C, L) = \sum_{X\in C} d(X, L)$$

we define $F(\mathbf{L}) = \{C'_i\}$ by

$$X \in C'_i : i = \min \operatorname*{Argmin}_j \{d(X, L_j) : L_j \in \mathbf{L}\}$$

each unit is assigned to the nearest leader; and we define $G(\mathbf{C}) = \{L'_i\}$ by

$$L'_i = \operatorname*{argmin}_{L \in \Psi} p(C, L)$$

To solve the last optimization problem we consider

$$p(C, L; V) = \sum_{X \in C} d(X, L; V) = \operatorname{card}(C) - \sum_{j=1}^{k} q(j, C; V)s(j, L; V)$$

This expression has a minimum value iff the sum has a maximum value. The unique symmetric optimal solution is

$$s(i, L'; V) = \begin{cases} \frac{1}{t} & \text{if } j \in M \\ 0 & \text{otherwise} \end{cases}$$

where $M = \{j : q(j, C; V) = \max_i q(i, C; V)\}$ and $t = \operatorname{card} M$.

Evidently W0, W1 and W2 hold. It holds also $P(\mathbf{C}) = W(\mathbf{C}, G(\mathbf{C}))$.

Usually $t = 1$ – we include in the representative of a cluster the most frequent range of values of variable on this cluster. This provides us with very simple interpretations of clustering results.

For example, in a clustering of types of cars we obtained among 14 clusters: (Remark: SIT is an abbreviation for a Slovenian money called tolar.)

Cluster 5: (42 units) coupes
doors = 2, height(mm) ∈ (– , 1388], power(KW) ∈ (125, –), acceleration time(s) ∈ (– , 9.1], max speed(km/h) ∈ (215 , –), price(1000 SIT) ∈ (6550, –).
Cluster 8: (57 units) minivans
doors = 5, passengers = 7, length(mm) ∈ (4555, 4761], height(mm) ∈ (1490, –), fuel tank capacity(l) ∈ (66, 76], weight(kg) ∈ (1540, –), wheelbase(mm) ∈ (2730, –).
Cluster 9: (193 units) small-cars
doors = 4, passengers = 5, length(mm) ∈ (– , 4010], width(mm) ∈ (– , 1680], luggage capacity(l) ∈ (–, 279], fuel tank capacity(l) ∈ (–, 48], weight(kg) ∈ (– , 940], power(KW) ∈ (– , 55], torque ∈ (– , 124], brakes = discs/drums, max speed(km/h) ∈ (– , 163], price(1000 SIT) ∈ (– , 2100].
Cluster 13: (62 units) sport-utility
doors = 5, height(mm) ∈ (1490, –), fuel tank capacity(l) ∈ (76, –), weight(kg) ∈ (1540, –), cargo capacity(l) ∈ (560, –), torque ∈ (270, –).

For another criterion function

$$W(\mathbf{C}, \mathbf{L}) = \alpha \sum_{i \in I} d(C_i, L_i) + \sum_{i \in I} p(C_i, L_i)$$

where α is a large constant, the first term can be set to 0 by setting $s(j, L_i; V) = p(j, C_i; V)$, $j = 1, \ldots, k$. If this defines the function G, and F is defined as in the previous case, we obtain a procedure which works very well on real data. Note also that for the obtained (local) minimum $(\mathbf{C}^*, \mathbf{L}^*)$ it holds

$$P(\mathbf{C}^*) = \min_{\mathbf{L} \in \Psi} W(\mathbf{C}^*, \mathbf{L}) = W(\mathbf{C}^*, G(\mathbf{C}^*)) = \sum_{C \in \mathbf{C}^*} p(C, G(C))$$

5. Conclusion

We successfully applied the proposed approach to the dataset of types of cars (1349 units) and also to some large datasets from AI collection

http://www.ics.uci.edu/~mlearn/MLRepository.html

An extended version of this paper is available at

http://vlado.fmf.uni-lj.si/pub/cluster/

Acknowledgment: This work was supported by the Ministry of Science and Technology of Slovenia, Project J1-8532.

References

Batagelj, V. (1985). Notes on the dynamic clusters method, in: *IV conference on applied mathematics*, Split, May 28-30, 1984. University of Split, Split, 139-146.

Batagelj, V. & Bren, M. (1995). Comparing Resemblance Measures, *Journal of Classification*, **12**, 1, 73-90.

Batagelj, V. & Mandelj, M. (1993). Adding Clustering Algorithm Based on L-W-J Formula, Paper presented at: *IFCS 93*, Paris, 31.aug-4.sep 1993.

Brucker, P. (1978). On the complexity of clustering problems, *Lecture Notes in Economics and Mathematical Systems 175*, in: Optimization and Operations Research, Proceedings, Bonn. Henn,R., Korte,B., Oettli,W. (Eds.), Springer-Verlag, Berlin 1978.

Diday, E. (1979). *Optimisation en classification automatique*, Tome 1.,2.. INRIA, Rocquencourt, (in French).

Diday, E. (1997). Extracting Information from Extensive Data sets by Symbolic Data Analysis, in: *Indo-French Workshop on Symbolic Data Analysis and its Applications*, Paris, 23-24. September 1997, Paris IX, Dauphine, 3-12.

Hartigan, J.A. (1975). *Clustering Algorithms*, Wiley, New York.

Tukey, J.W. (1977). *Exploratory Data Analysis*, Addison-Wesley, Reading, MA.

A Critical Approach to Non-Parametric Classification of Compositional Data

J. A. Martín-Fernández[1], C. Barceló-Vidal[1], V. Pawlowsky-Glahn[2]
[1] Dept. d'Informàtica i Matemàtica Aplicada, Escola Politècnica Superior, Universitat de Girona, Lluis Santaló, s/n, E-17071 Girona, Spain
[2] Dept. de Matemàtica Aplicada III, ETS de Eng. de Camins, Canals i Ports, Universitat Politècnica de Catalunya, Jordi Girona Salgado, 1 i 3, E-08034 Barcelona, Spain

Abstract: The application of hierarchic methods of classification needs to establish in advance some or all of the following measures: difference, central tendency and dispersion, in accordance with the nature of the data. In this work, we present the requirements for these measures when the data set to classify is a compositional data set. Specific measures of difference, central tendency and dispersion are defined to be used with the most usual non-parametric methods of classification.

Key words: Compositional Data, Cluster analysis, Classification.

1. Introduction

Any vector $\mathbf{x} = (x_1, \ldots, x_D)$ with non-negative elements $x_1, \ldots, x_D$ representing proportions of some whole is subject to the unit-sum-constraint $x_1 + \ldots + x_D = 1$. Compositional data, consisting of such vectors of proportions (compositions), play an important role in many disciplines. Frequently, some form of statistical analysis is essential for the adequate analysis and interpretation of the data. Nevertheless, all too often the unit-sum-constraint is either ignored or improperly incorporated into the statistical modelling giving rise to an erroneous or irrelevant analysis. The purpose of this paper is to revise the specific statistical requirements of standard hierarchic agglomerative classification methods when they are performed on compositional data.

In the next section we present the ideas proposed by Aitchison (1992) about the conditions that have to be satisfied by any distance between two compositions, and by any measure of central tendency and dispersion of a compositional data set. Next, we propose a modification of the most standard hierarchic agglomerative classification methods to make them suitable for the classification of a compositional data set. Finally, we present two examples where the proposed methodology is applied to simulated compositional data sets.

2. Statistical analysis of compositional data

If a vector $\mathbf{w}=(w_1,\ldots,w_D)\in\Re^D$ with non-negative components is compositional, we are implicitly recognising that the total size $w_1+\ldots+w_D$ of the composition is irrelevant. Therefore, a suitable sample space for compositional data is the unit simplex

$$\mathbf{S}^{D-1}=\left\{(x_1,\ldots,x_D):x_j>0\,(j=1,\ldots,D),\ x_1+\ldots+x_D=1\right\},$$

and any meaningful function f of a composition must be invariant under the group of scale transformations; i.e. $f(\lambda\mathbf{w})=f(\mathbf{w})$, for every $\lambda>0$. Note that only functions expressed in terms of ratios of the components of the composition satisfy this condition (Aitchison, 1992).

As an analogy to the role played by the group of translations when the sample space is the real space $\Re^D$, Aitchison (1986, Section 2.8) introduces the group of perturbations as a means to characterize the `difference' between two compositions. If we denote the perturbation operation by `$\circ$', then the perturbation $\mathbf{p}=(p_1,\ldots,p_D)\in\mathbf{S}^{D-1}$ applied to a composition $\mathbf{x}$ produces the new composition

$$\mathbf{p}\circ\mathbf{x}=(p_1x_1,\ldots,p_Dx_D)\Big/\sum_j p_j x_j .$$

If $\mathbf{x},\mathbf{x}^*\in\mathbf{S}^{D-1}$ are two compositions it is easy to prove that the perturbation

$$\mathbf{x}^*\circ\mathbf{x}^{-1}=(x^*_1/x_1,\ldots,\ x^*_D/x_D)\Big/\sum_j x^*_j/x_j ,$$

moves $\mathbf{x}$ to $\mathbf{x}^*$.

2.a Distance between two compositions

The requirements which any scalar measure of distance between two compositions should verify, according to the definitions given by Aitchison (1992), are scale invariance, permutation invariance, subcompositional dominance and perturbation invariance. These requirements are sensible, as they acknowledge the compositional nature of the data. A feasible distance between two compositions $\mathbf{x},\mathbf{x}^*\in\mathbf{S}^{D-1}$ is given by

$$\Delta(\mathbf{x},\mathbf{x}^*)=\left[D^{-1}\sum_{j<k}\left\{\log\left(\frac{x_j}{x_k}\right)-\log\left(\frac{x^*_j}{x^*_k}\right)\right\}^2\right]^{\frac{1}{2}}, \tag{2}$$

which is equivalent to the distance proposed by Aitchison (1992). Using the definition of *centred logratio transformation* (clr) from $\mathbf{S}^{D-1}$ to $\Re^D$ given by

$$\mathrm{clr}(\mathbf{x}) = \left(\log\left(\frac{x_1}{g(\mathbf{x})}\right), \ldots, \log\left(\frac{x_D}{g(\mathbf{x})}\right) \right), \tag{3}$$

where $g(\mathbf{x})$ is the geometric mean of the composition $\mathbf{x}$, it is easy to establish that

$$\Delta(\mathbf{x}, \mathbf{x}^*) = d_{eu}(\mathrm{clr}(\mathbf{x}), \mathrm{clr}(\mathbf{x}^*)), \tag{4}$$

where d_{eu} represents the euclidean distance. Moreover, since $\mathrm{clr}(\mathbf{p} \circ \mathbf{x}) = \mathrm{clr}(\mathbf{p}) + \mathrm{clr}(\mathbf{x})$, for any $\mathbf{p}, \mathbf{x} \in \mathbf{S}^{D-1}$, it is clear that the distance (4) is perturbation invariant.

2.b Measure of central tendency of a compositional data set

If $\mathbf{X} = \left\{ \mathbf{x}_i = (x_{i1}, \ldots, x_{iD}) \in \mathbf{S}^{D-1} : i = 1, \ldots, N \right\}$ represents a set of compositions, the arithmetic mean $\overline{\mathbf{X}}$ of the data set is usually not representative of the 'centre' of the set, and neither is compatible with the group of perturbations. Aitchison (1997) proposes the geometric mean $\mathrm{cen}(\mathbf{X})$ as more representative of the central tendency of a compositional data set. It is defined as

$$\mathrm{cen}(\mathbf{X}) = \frac{(g_1, \ldots, g_D)}{g_1 + \ldots + g_D}, \tag{5}$$

where $g_j = \left(\prod_{i=1}^{N} x_{ij} \right)^{\frac{1}{N}}$ is the geometric mean of the jth component of the data set. Figure 1 shows the ternary diagram of a simulated data set (adapted from Aitchison, 1986, data set 1) where it can be observed that the geometric mean lies inside the bulk of data, while the arithmetic mean lies slightly apart.

Figure 1: *Data set in the simplex (geometric mean = (0.6058, 0.2719, 0.1223) and arithmetic mean = (0.54, 0.2756, 0.1844))*

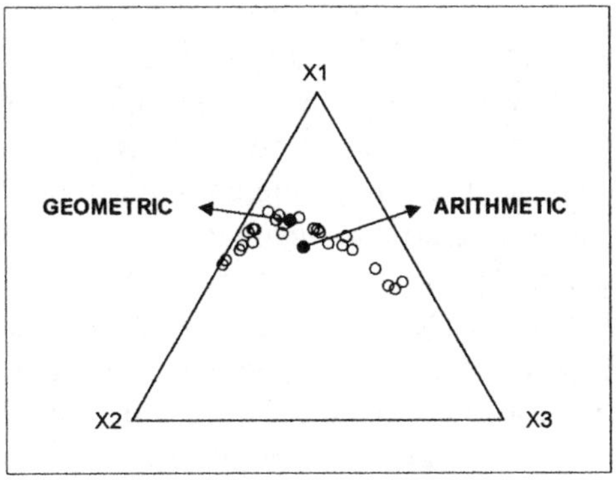

It is easy to prove that $\text{cen}(\mathbf{p} \circ \mathbf{X}) = \mathbf{p} \circ \text{cen}(\mathbf{X})$ for any perturbation $\mathbf{p} \in \mathbf{S}^{D-1}$, and that $\text{clr}(\text{cen}(\mathbf{X})) = \overline{\text{clr}(\mathbf{X})}$. Therefore, it will be true that

$$\Delta(\mathbf{x}, \text{cen}(\mathbf{X})) = d_{eu}(\text{clr}(\mathbf{x}), \overline{\text{clr}(\mathbf{X})}). \tag{6}$$

2.c Measure of dispersion of a compositional data set

It is sensible to assume that any measure of dispersion of a compositional data set should be invariant under the group of perturbations. The measure of dispersion defined by Aitchison (1992,1997) satisfies this condition; it is based on the trace of the covariance matrix of the centred logratio transformed compositions. In accordance with his definition, a measure of total variability of a compositional data set $\mathbf{X}$ can be defined as

$$\text{totvar}(\mathbf{X}) = \sum_{j=1}^{D} \sum_{i=1}^{N} \left\{ \log\left(\frac{x_{ij}}{g(\mathbf{x}_i)} \right) - m_j \right\}^2, \tag{7}$$

where $m_j = \frac{1}{N} \sum_i \log\left(\frac{x_{ij}}{g(\mathbf{x}_i)} \right)$ $(j = 1, \ldots, D)$. It is easy to prove that

$$\text{totvar}(\mathbf{X}) = \sum_{i=1}^{N} d_{eu}^2\left(\text{clr}(\mathbf{x}_i), \overline{\text{clr}(\mathbf{X})}\right) = \sum_{i=1}^{N} \Delta^2\left(\mathbf{x}_i, \text{cen}(\mathbf{X})\right), \tag{8}$$

which proves that the proposed measure of total variability (7) is compatible with the distance defined in (2). It can also be proved that $\text{totvar}(\mathbf{p} \circ \mathbf{X}) = \text{totvar}(\mathbf{X})$, for any perturbation $\mathbf{p} \in \mathbf{S}^{D-1}$, which proves that the measure of total variability (7) is invariant under perturbations.

3. Hierarchic cluster analysis of compositional data

Before applying any hierarchic method of classification to a data set, it is necessary to establish in advance some or all of the following measures: of difference, central tendency and dispersion, to be used in accordance with the nature of the data. Therefore, if we are using a hierarchic method to classify a compositional data set, we have to take into account that all these measures must be invariant under the group of scale transformations. It is clear that the definitions given in (2), (5) and (8) are scale-invariant, while the euclidean distance is not. Therefore, from this point of view, it is wrong to use the euclidean distance between two compositions to calculate the matrix of distances associated with hierarchic methods like single linkage, complete linkage and average linkage. We propose to use the distance defined in (2). By property (4), this distance is equivalent to the euclidean distance between the compositions transformed by the centred logratio transformation *clr* defined in (3).

Likewise, any method of classification which reduces the distance from a composition to a set of compositions to the distance between the composition and the `centre' of the group, has to take into account the considerations made in the previous section *2.b*. We propose to use (5) as a definition of the `centre' of a set of compositions, in addition to the distance defined in (2). Then, by property (6), it is easy to conclude that the centroid classification method can be used if it is applied to the data transformed by the centred logratio transformation.

On the other hand, the well-known method of Ward is a hierarchic method which uses the measure of dispersion to classify the data. In essence, this method is based on the concept of variability on a cluster C. This variability is defined by the sum $\sum_{\mathbf{x}\in C} d_{eu}^2(\mathbf{x},\overline{C})$ (see Everitt, 1993, Section 5.2), where $\overline{C}$ denotes the centre of the class. When the data set is compositional, we suggest replacing the squared Euclidean distance $d_{eu}^2(\mathbf{x},\overline{C})$, which appears in the previous sum, by $\Delta_{eu}^2(\mathbf{x},\mathrm{cen}(C))$ defined in (2). Then, by definition (7) and property (8), the variability of a cluster C of compositions will be equal to $\mathrm{totvar}(C)$. Thus, modifying the method of Ward to make it suitable for the classification of a compositional data set $\mathbf{X}$ is equivalent to applying the standard procedure to $\mathrm{clr}(\mathbf{X})$.

4. Two examples

Consider the 50 points plotted on a ternary diagram in Figure 2a, corresponding to a simulated compositional data set **X1** characterised by three components. Samples belong to two groups obtained one from the other by the application of a perturbation. Figure 3a shows the ternary diagram of a second simulated data set **X2** (adapted from Aitchison, 1986, data set 1) with 50 elements, which has been generated and labelled in a similar manner. Figures 2b-3b show the plots in $\Re^3$ of the *clr*- transformed data set clr(**X1**) and clr(**X2**), respectively. In each case, original groups are separated by a line and the groups resulting of a cluster method are distinguished by a different symbol.

As can be observed, original groups show no overlapping neither in $\mathbf{S}^2$ nor in $\Re^3$ but, while Figures 3a and 3b show also a clear visual separation of the two groups, this is not the case for the data represented in Figure 2a and 2b. For the sake of comparison, different standard classification methods have been applied to the four sets using the Euclidean distance. Misclassification rates are listed in Table 1.

Figure 2: *Example 1 (a) Plot in the simplex (groups from Ward's method) (b) clr-transformed set (groups obtained using single linkage)); classification results distinguished by symbols '+' and 'o'.*

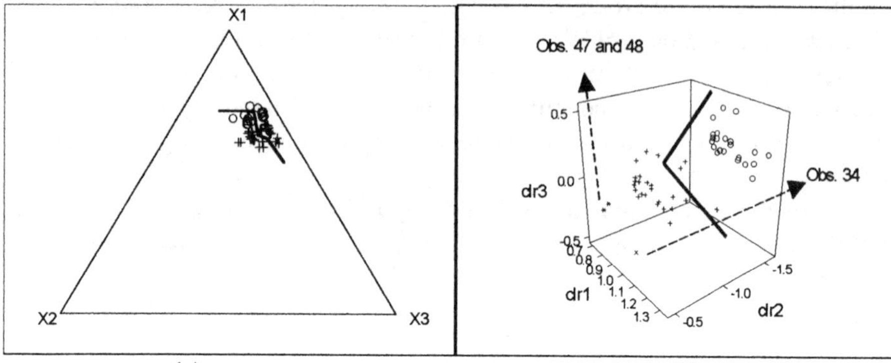

(a) (b)

Figure 3: *Example 2 (a) Plot in the simplex (groups obtained using single linkage) (b) clr-transformed set (groups from Ward's method); classification results distinguished by symbols '+' and 'o'.*

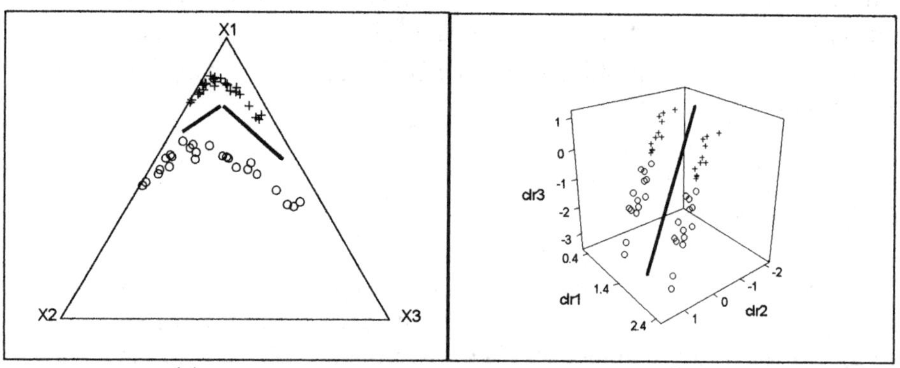

(a) (b)

Table 1 *Misclassification rates on the two examples*

	Example 1		Example 2	
Method	**X1**	*clr*(**X1**)	**X2**	*clr*(**X2**)
Single Linkage	48%	12%	0%	8%
Ward	50%	6%	8%	50%
Complete Linkage	22%	6%	42%	50%
Centroid	48%	6%	42%	50%
Average Linkage	50%	6%	28%	50%

As could be expected, a poor classification power is obtained for **X1**, because the two groups are *close* from an euclidean point of view (only the complete linkage method gives an acceptable classification). However the classification power seems to be reasonable for *clr*(**X1**), because only the three elements close to the border are misclassified. For *clr* **(X1)** the poorest result is obtained when the single linkage is used. Figure 4 shows the associated dendrogram. The samples of the first group are labelled from 1 to 25, while the others are labelled from 26 to 50. This dendrogram is similar to the dendrograms associated to the others methods whit only one difference: compositions labelled by 34, 47 and 48 are considered by the single linkage method as separated groups (see Figure 2(b)).

Figure 4: *Dendrogram of the single linkage method applied to clr* **(X1)**.

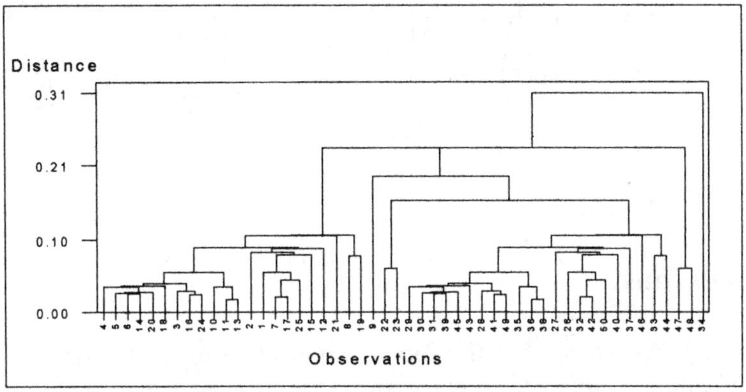

Results are more striking for **X2** and *clr*(**X2**), given the clear separation between groups both in the simplex and in real space: only single linkage and Ward's methods show a high classification power for **X2,** while complete linkage, centroid and average linkage methods have a poor classification power. They work still worse when applied to *clr*(**X2**) because the two groups are parallel and elongated. Figure 5(b) shows the associated dendrogram when the single linkage method is applied to the data set *clr*(**X2**)**.** Numbers from 1 to 25 correspond to the first group and 26 to 50 to the second. It can be observed that the almost all samples are well classified: only observations labelled 10 and 17, and observations 35 and 42 are misclassified as separated groups (see Figure 5(a)).

Figure 5: *Single linkage method applied to clr* **(X2)**: *(a) Plot (b) Dendrogram*

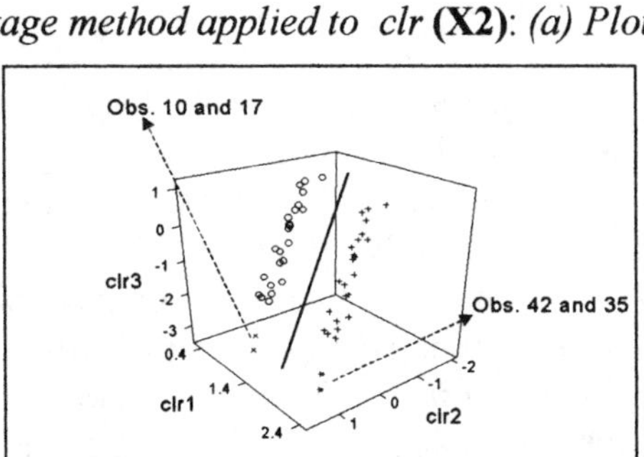

(a)

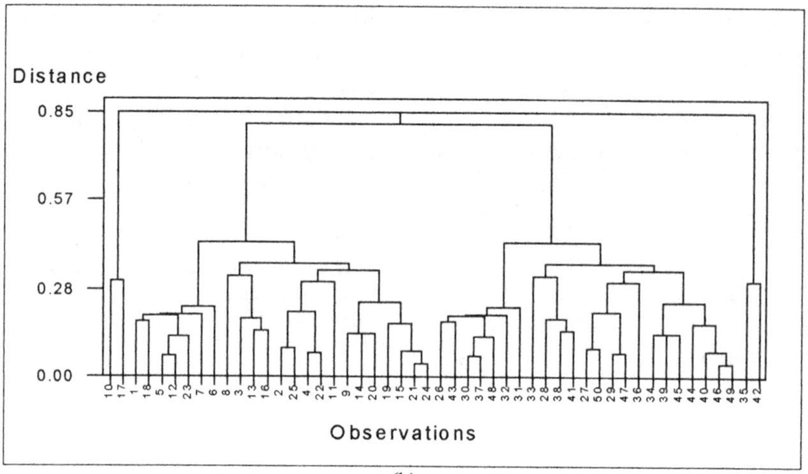

(b)

5. Conclusions

- There are theoretical objections to the application of the standard hierarchic classification methods to compositional data sets because they don't take into account the nature of this kind of data.
- To classify a compositional data set, we suggest adapting the usual hierarchic methods using the definitions of distance, centre and variability defined in (2), (5), and (7), which are compatible with the compositional nature of the data. This is equivalent to applying standard methods to the centred logratio transformed data set.
- Further research is needed to understand more thoroughly the performance of modified standard parametric classification methods when they are applied to compositional data.

References

Aitchison, J., (1986). *The Statistical Analysis of Compositional Data.* Chapman and Hall, New York (USA), 416 p.

Aitchison, J., (1992). On Criteria for Measures of Compositional Difference, *Mathematical Geology*, vol. 24, No. 4, pp. 365-379

Aitchison, J., (1997). The one-hour course in compositional data analysis or compositional data analysis is simple, in: *Proceedings of IAMG'97 - The 1997 Annual Conference of the International Association for Mathematical Geology*, Ed. Pawlowsky-Glahn, V., CIMNE, Barcelona (E), Part I, pp. 3-35.

Everitt, BS., (1993). *Cluster Analysis.* Edward Arnold, Cambridge (UK), 170 p.

Clustering Based on Wavelet Transform: Applications to Point Pattern Clustering and to High-Dimensional Data Analysis

F. Murtagh (1), J.L. Starck (2), M. Berry (3)
(1) Faculty of Informatics, University of Ulster
Northern Ireland BT48 7JL, email fd.murtagh@ulst.ac.uk
(2) CEA/DSM/DAPNIA, F-91191 Gif-sur-Yvette cedex, France
(3) Department of Computer Science, University of Tennessee,
107 Ayres Hall, Knoxville TN 37996-1301, USA

Abstract: We describe an effective approach to object or feature detection in point patterns via noise modeling. This is based on use of a redundant or non-pyramidal wavelet transform. Noise modeling is based on a Poisson process. We illustrate this new method with a range of examples. We use the close relationship between image (pixelated) and point representations to achieve the result of a clustering method with constant-time computational cost. We then proceed to generalize this method for high-dimensional data. Using a dataset of very well-known structure as a test case, we show proof of concept for this approach to analysis of high-dimensional boolean hyperlink datasets.

Keywords: Cluster Analysis, Wavelet Transform, Multiresolution Analysis.

1 Point Pattern Clustering

Point pattern clustering (Murtagh and Starck, 1998) has constituted one of major strands in cluster analysis. We will describe (i) a multiscale approach which is computationally very efficient, and (ii) a direct treatment of noise and clutter which leads to improved cluster detection.

The à trous wavelet transform (Shensa, 1992; Murtagh, 1998; Starck et al., 1998) is redundant and non-orthogonal. It is efficient, being of linear computation cost in terms of the size of the input data. Treating a point pattern set as a 2-dimensional image allows point occurrences to be accumulated, and allows an image of fixed dimensionality to represent the user field of view. Given such a two-dimensional image of fixed dimensionality, there is no computational dependence on the actual number of points. This analysis method is therefore of $O(1)$ computational cost in terms of the point patterns examined.

Although other analysis frameworks could be easily envisaged, we have applied this approach in the following way. We have visualized the data, examining each resolution level. We have retained some particular level, with greatest interpretational value. Contiguous regions of a robustly-thresholded version of such a resolution level are examined. Cluster properties are measured.

Data are usually noisy. The wavelet coefficients will therefore be noisy too. We have determined significance levels for detection of signal at each resolution level, on the assumption of low-count Poisson noise. This allows noise filtering and reconstruction of the data.

We will show examples of this approach, including the finding of scarcely perceptible point pattern clusters in a noisy background (e.g. 550 'signal' patterns with 40,000 'noise' points in the background). High-quality recovery of known clusters is shown.

2 Point Patterns and Wavelet Transforms

Given a planar point pattern, a 2-dimensional image is created by:

1. Considering a point at (x, y) as defining the value one at that point, yielding the tuple $(x, y, 1)$.

2. Projection onto a plane by (i) using a regular discrete grid (an image) and (ii) assigning the contribution of points to the image pixels by means of the interpolation function, ϕ, used by the chosen wavelet transform algorithm (in our case, the à trous algorithm with a B_3 spline).

3. The à trous algorithm is applied to the resulting image. Based on a noise model for the original image (i.e. tuples $(x, y, 1)$), significant structures are detected at each resolution level.

The wavelet transform used is the *à trous* ("with holes") method. It is a redundant (i.e. non-pyramidal) method and has computational cost which is linear as a function of the number of pixels in the input data.

In a wavelet transform, a series of transformations of an image is generated, providing a resolution-related set of "views" of the image. The properties satisfied by a wavelet transform, and in particular the à trous wavelet transform (so called because of the interlaced convolution used in successive levels: see step 2 of the algorithm below) are further discussed in Bijaoui et al. (1994).

A summary of the à trous wavelet transform is as follows. Index k ranges over all pixels.

1. Initialize i to 0, starting with an image $c_i(k)$.

2. Increment i, and carry out a discrete convolution of the data $c_{i-1}(k)$ using a filter h (see below). The distance between a central pixel and adjacent ones is 2^{i-1}.

3. From this smoothing we obtain the discrete wavelet transform, $w_i(k) = c_{i-1}(k) - c_i(k)$.

4. If i is less than the number p of resolution levels wanted, return to step 2.

The set $\mathcal{W} = \{w_0, w_1, ..., w_p, c_p\}$, where c_p is a residual, represents the wavelet transform of the data. The filter h is based on a B_3 spline: the 5×5 filter used and other implementation details such as treatment of boundaries are given in Starck and Murtagh (1994). We have the following additive decomposition of the input image:

$$c_0(k) = c_p + \sum_{i=1}^{p} w_i(k) \tag{1}$$

3 Poisson Noise with Few Counts

A large number of counts would permit us to model the Poisson characteristics of the noise with a Gaussian distribution, and for the latter, Murtagh et al. (1995) may be referred to. The small count situation is of interest to us here. It holds for all parts of point pattern images and there is no alternative for the sparsely-populated regions.

A wavelet coefficient at a given position and at a given scale j is

$$w_j(x, y) = \sum_{k \in K} n_k \ \psi(\frac{x_k - x}{2^j}, \frac{y_k - y}{2^j}) \tag{2}$$

where K is the support of the wavelet function ψ and n_k is the number of events which contribute to $w_j(x, y)$, i.e. the number of events included in the support of the dilated wavelet centred in (x,y).

If a wavelet coefficient $w_j(x, y)$ is due to noise, it can be considered as a realization of the sum $\sum_{k \in K} n_k$ of independent random variables with the same distribution as that of the wavelet function (n_k being the number of events used for the calculation of $w_j(x, y)$). This allows comparison of the wavelet coefficients of the data with the values which can be taken by the sum of n independent variables.

The distribution of one event in wavelet space is then directly given by the histogram H_1 of the wavelet ψ. As we consider independent events, the distribution of a coefficient w_n (note the changed subscripting for w, for convenience) related to n events is given by n autoconvolutions of H_1:

$$H_n = H_1 \otimes H_1 \otimes ... \otimes H_1 \tag{3}$$

For a large number of events, H_n converges to a Gaussian.

4 Multiresolution Support

The multiresolution support (Starck et al., 1995) will be obtained by detecting the significant coefficients at each scale. The multiresolution support is defined by:

$$M(j,x,y) = \begin{cases} 1 & \text{if } w_j(x,y) \text{ is significant} \\ 0 & \text{if } w_j(x,y) \text{ is not significant} \end{cases} \tag{4}$$

We will say that a multiresolution support of an image describes in a logical or boolean way if an image I contains information at a given scale j and at a given position (x,y). The algorithm to create the multiresolution support is as follows:

1. Compute the wavelet transform of the image.
2. Estimate the noise standard deviation at each scale. Deduce the statistically significant level at each scale.
3. Booleanization of each scale leads to the multiresolution support.
4. Modification using a priori knowledge is carried out if desired.

5 Example

Fig. 1 shows two Gaussian clusters designed with centers $(64,64)$ and $(190,190)$; and with standard deviations in x and y directions respectively $(10,20)$ and $(18,10)$. In the first (lower) of these clusters, there are 300 points, and there are 250 in the second. Background Poisson clutter was provided by 40,000 points. Even with this amount of noise, the recovery properties of these clusters, based on the multiresolution support, were very good. We found the centroid values of the "island" regions to be respectively $(61,70)$ and $(187,186)$ which are good fits to the design values. The standard deviations in x and y were found to be respectively $(6.4,8.1)$ and $(7.2,6.4)$, again reasonable fits to the input data given that there are around eighty times more "noise" points present compared to "signal points".

6 High-Dimensional Data Spaces

Information retrieval and related areas require solution methods for data in spaces of dimensionality of many tens of thousands. Matrix reordering techniques are closely related to eigen-reduction methods (Murtagh, 1985; Berry et al., 1996). Matrix reordering can be used to transform high-dimensional data arrays to a canonical form. On the basis of a data array in canonical

Figure 1: The image showing 550 "signal" points, i.e. two Gaussian-shaped clusters, with in addition 40,000 Poisson noise points added. Details of recovery of the cluster properties are discussed in the text.

form, we search for structure using an efficient multiscale approach – using the same wavelet transform approach as above, which handles sparse data effectively.

We explore how this works in practice, using a dataset with very well-known structure. This is booleanized to emulate a document/keyword or other high-dimensional dependency array in the following way. We take Fisher's (and Anderson's) iris data which consists of a 150 $\times$ 4 real-valued array, with 3 well-known classes. We booleanize this to create a binary dataset. This is done by roughly multiplying each value by 100, truncating and rounding each variable to 1 or 0. A 4-dimensional dataset is recoded in this manner to a 147-dimensional dataset. Each observation has exactly four 1-values. The total number of 1-values is 600. The total number of values in the array is 22050. We note again that the use of such an array is purely to provide a test framework of known properties. To quickly check on the change of information content in moving from a 4- to a 147-dimensional embedding of the 150 observations, we can use principal components analysis and correspondence analysis, respectively.

With such a boolean array, we seek a canonical ordering of the rows and

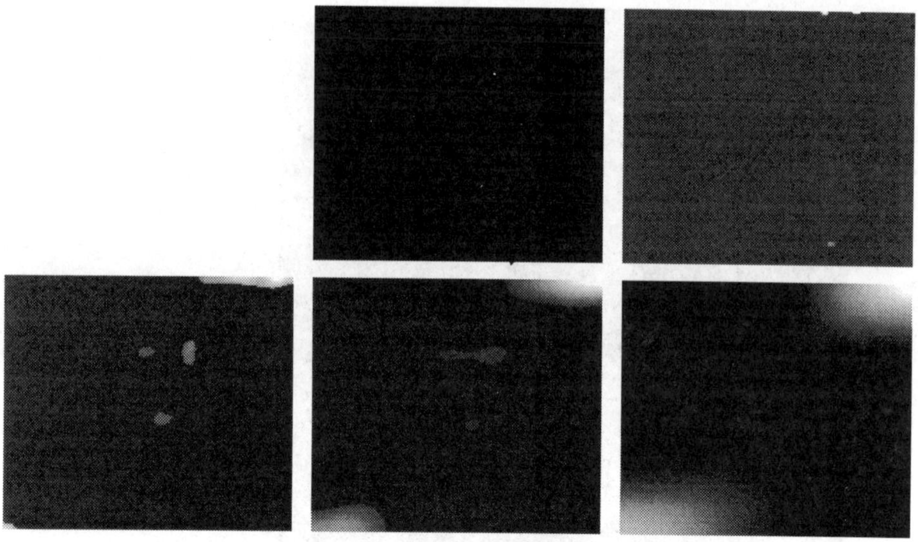

Figure 2: Wavelet transform of Fisher data. Left to right, from upper left: resolution levels 1, 2, 3, 4, 5 and the residual.

columns. For the small dataset considered here, a convenient ordering is that which reflects the ordering of the first factor in correspondence analysis.

We wavelet-transform this row- and column-permuted array. The à trous method is used. The wavelet transform is shown in Fig. 2. Resolution level 5 is the most interesting output scale relative to the clustering – contiguous regions – indicated. We note that of necessity such a resolution level is of zero mean, which facilitates thresholding.

We label the contiguous regions of the multiresolution support at resolution level 5. The corresponding observation sequence numbers are read off, taking account of the permuting previously carried out. The findings are that Fisher's class 1 is nearly completely recovered (this corresponds to the large positive region in the upper right-hand corner). This is of course in line with the known fact that this class is isolated from the others. Fisher classes 2 and 3 are less well resolved, in line with the correspondence analysis principal plane obtained from analysis of the 147-dimensional boolean data.

We conclude that this novel approach to clustering is reliable. For n rows and m columns, it has linear computational cost in terms of these dimensions, and hence is $O(nm)$. The constant of proportionality is not important. As noted before, the computational complexity in terms of the actual array contents (one-values or non-boolean values) is constant, $O(1)$. These computational results may be contrasted with other $O(n^2)$ methods, or methods which are of better computational complexity but which lack robustness and stability.

We are currently applying this new clustering method to the analysis of a $12,025 \times 9,778$ article-by-link matrix, relating to articles by cross-references from the *Condensed Columbia Encyclopedia* (1989, 2nd ed., online version). This boolean array has 33,396 links (nonzeros) = 0.0284% density. Having effective methods for the analysis of such data is important in practice since voluminous data of this kind arise naturally from communications networks, and from the entertainment and communications media. With the ability to determine clusters quickly and effectively, subsequent phases of the analysis may, for example, entail finer-tuned inferential modeling of the individual clusters.

Acknowledgements

Partial support from Office of Naval Research Grant N-00014-91-J-1074 ("Model-based clustering for marked spatial point processes") and discussions within the associated Working Group are acknowledged by F. Murtagh.

References

1. Berry, M.W., Hendrickson, B. and Raghavan, P. (1996). Sparse matrix reordering schemes for browsing hypertext, in *Lectures in Applied Mathematics (LAM) Vol. 32: The Mathematics of Numerical Analysis*, J. Renegar, M. Shub, and S. Smale (Eds.), American Mathematical Society, 99–123.

2. Bijaoui, A., Starck, J.-L. and Murtagh, F. (1994). Restauration des images multi-échelles par l'algorithme à trous, *Traitement du Signal*, 11, 229–243.

3. Murtagh, F. (1985). *Multidimensional Clustering Algorithms*, Physica-Verlag, Würzburg.

4. Murtagh, F. (1998). Wedding the wavelet transform and multivariate data analysis, *Journal of Classification*, in press.

5. Murtagh, F., Starck, J.-L. and Bijaoui, A. (1995). Image restoration with noise suppression using a multiresolution support, *Astronomy and Astrophysics Supplement Series*, 112, 179–189.

6. Murtagh, F. and Starck, J.-L. (1998). Pattern clustering based on noise modeling in wavelet space, *Pattern Recognition*, in press.

7. Shensa, M.J. (1992). The discrete wavelet transform: wedding the à trous and Mallat algorithms, *IEEE Transactions on Signal Processing*, 40, 2464–2482.

8. Starck, J.-L. and Murtagh, F. (1994). Image restoration with noise suppression using the wavelet transform, *Astronomy and Astrophysics*, 288, 342–348.

9. Starck, J.-L., Bijaoui, A. and Murtagh, F. (1995). Multiresolution support applied to image filtering and deconvolution, *Graphical Models and Image Processing*, 57, 420–431.

10. Starck, J.-L., Murtagh, F. and Bijaoui, A. (1998). *Image and Data Analysis: the Multiscale Approach*, Cambridge University Press, in press.

Beyond Simpson's Paradox: One Problem in Data Science

Chikio Hayashi
Institute of Statistical Mathematics
and
Kazue Yamaoka
Department of Hygiene and Public Health, School of Medicine, Teikyo University
e-mail: kazue@med.teikyo-u.ac.jp.

Abstract: In the present paper, the conditions under which Simpson's paradox does not occur are discussed for various cases. These conditions are first obtained from the descriptive point of view and then on the assumption of prior probability distributions of parameters. The robustness of the results is discussed with respect to the prior probability distributions. Practically, the result is given as the magnitude of odds ratio (or relative risk), i.e., Simpson's paradox does not occur if the odds ratio is more or less than a certain values, depending on various cases.

Keywords: Simpson's paradox, conditions of non-paradox, descriptive approach, prior probability distribution of a parameter, robustness of solution, Monte Carlo solution.

1. Introduction

Simpson's or Yule's paradox (Simpson, 1951), treated in the present paper, is well known and elementary but nonetheless very important, even in modern data analysis. The paradox sometimes causes serious problems for interpretation of results. These misleading effects have been illustrated in many epidemiological publications (Miettinen, 1976; Kleinbaum, 1982). Simpson's paradox has also been discussed by other investigators; for example, Shapro (1982), Geng (1992), Hayashi (1993), Hintzman (1993), Hand (1994), and Vogt (1995).

Although many methods have been proposed for adjusting the effects of confounding factors in experimental design, most of these efforts center on statistical tests in the case of known confounding factors. However, we cannot ignore the effects of unknown confounding factors. What must be noticed is that the existence of latent classes (groups) having the reverse or independent association from a given contingency table yields Simpson's paradox. Therefore, examining the condition under which such latent groups exist offers a key to understanding the occurrence of Simpson's paradox.
In the present paper, the conditions under which Simpson's paradox does not occur are discussed. In order to examine these conditions, we tried to find conditions exhibiting the nonexistence of two latent groups with independent association, using two parameters relating to those latent groups, because independent association is regarded as the extreme case of inverse association.

2. Explanation by Simple Examples

Suppose that a contingency table (*Table 1a*) is given in a group where A and B are categories in a factor (we call this as I), and + and – are categories in an outcome (we call this as II).

Table 1a: *Example 1*

I \ II	+	-	Total
A	3000	2000	5000
B	2000	3000	5000
Total	5000	5000	10000

Table 1b: *Proportions for Example 1*

I \ II	+	-	Total
A	0.6	0.4	1
	V		
B	0.4	0.6	1

From this table, we calculated proportions of + and – by A or B (*Table 1b*). From this analysis, we usually consider that outcome + in A is larger than outcome + in B. Here we suppose that *Table 1a* is obtained by the sum of the contingency tables in two latent groups *(Table 2a)* which are derived by the division or classification of the original group. It can be said, too, that the original contingency table is given by the sum of the contingency tables in two latent groups, G_1 and G_2.

Table 2a: *Frequency in two latent groups*

G_1-group

I \ II	+	-	Total
A	2800	1200	4000
B	800	200	1000
Total	3000	1400	5000

G_2-group

I \ II	+	-	Total
A	200	800	1000
B	1200	2800	4000
Total	1400	3000	5000

From these tables (*Table 2a*), we calculated proportions of + and – by A or B in G_1 and G_2 (*Table 2b*).

Table 2b: *Proportions in each group*

G_1-group

I \ II	+	-	Total
A	0.7	0.3	1
	Λ		
B	0.8	0.2	1

G_2-group

I \ II	+	-	Total
A	0.2	0.8	1
	Λ		
B	0.3	0.7	1

In these two groups, we find that outcome + in A is smaller than outcome + in B. That is, the inverse result has been obtained in both groups. A discrepancy between the total and the parts has appeared.

Here we consider, by a simple example, the condition under which such inconsistency does not occur. Notwithstanding that the extreme case of inverse association, we take up the following example (*Table 3a*).

Table 3a: *Original contingency table*

I \ II	+	-	Total
A	20	4980	5000
B	10	4990	5000
Total	30	9970	10000

Table 3b: *Contingency table in G_1-group: no association*

I \ II	+	-	Total
A	4000S	4000(1-S)	4000
B	1000S	1000(1-S)	1000
Total	5000S	5000(1-S)	5000

Table 3c: *Contingency table in G_2-group: no association*

I \ II	+	-	Total
A	1000T	1000(1-T)	1000
B	4000T	4000(1-T)	4000
Total	5000T	5000(1-T)	5000

Note: S and T are the proportions of outcome + by A and B in each group.

From these tables, the following relations hold;

$$4000S + 1000\,T = 20,$$
$$1000S + 4000\,T = 10.$$

From these equations, we have

$$S = 7/1500,$$
$$T = 2/1500.$$

In the total (original group), we have the result that outcome + by A is larger than outcome + by B, while we have no association between I and II in both groups. For some generalizations, we consider $20k_0$ ($k_0 \geq 0$) instead of 10 in *Table 3a* in the original group. What are k_0's such that this inconsistency does not occur? If no association case dose not occur, much less inverse association.
First, we calculate the conditions under which such inconsistency does occur. From *Table 3b* and *Table 3c*, we have the simultaneous linear equations:

$$4000\ S + 1000\ T = 20$$
$$1000\ S + 4000\ T = 20\ k_0$$

and the solution $750S = 4 - k_0$.
From $1 \geqq S \geqq 0$ and $k_0 \geqq 0$, the inequalities

$$750 \geqq 4 - k_0 \geqq 0$$
$$\text{i.e.} \quad 4 \geqq k_0 \geqq 0 \qquad (1)$$

are obtained. We have another solution

$$750T = 4\ k_0 - 1.$$

From $1 \geqq T \geqq 0$ and $k_0 \geqq 0$,

$$750 \geqq 4k_0 - 1 \geqq 0$$
$$\text{i.e.} \quad 751/4 \geqq k_0 \geqq 1/4 \qquad (2)$$

From (1) and (2), the range of k_0 in which the inconsistency does occur is calculated as $4 \geqq k_0 \geqq 1/4$. Thus, if $k_0 > 4$ ($K_{max}=4$) or $k_0 < 1/4$ ($K_{min}=1/4$), no inconsistency occurs, i.e., Simpson's paradox does not occur. Thus some idea of our treatment of Simpson's paradox has been given.

3. Descriptive Approach

We generalize the example shown in Section 2. Before beginning the discussion, we would like to explain the notation.

(i) Original group and two latent groups with classification based on unknown criteria

The notations are described in *Figure 1* and *Figure 2*.

Figure 1. *Original table (sample size=N)*

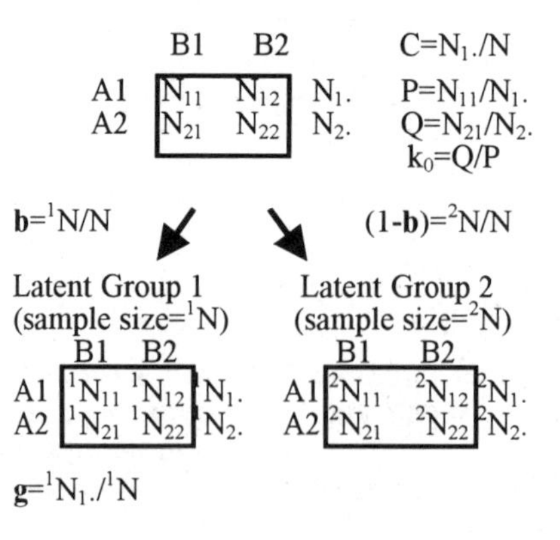

Figure 2. *Decomposition of original table.*

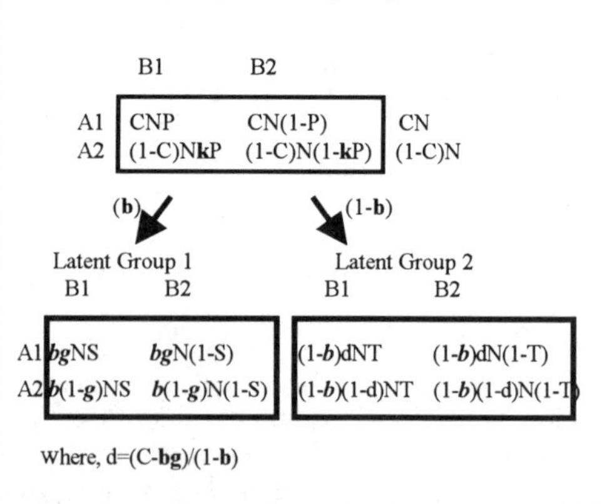

(ii) Restriction of Parameters

The conditions that these parameters must satisfy are shown below.

(a) Fundamental conditions

$\boldsymbol{b} \geq 0$, $\boldsymbol{g} \geq 0$, $k_0 \geq 0$,
$1 \geq P \geq 0$, $1 \geq S \geq 0$ and $1 \geq T \geq 0$.

(b) Restrictions among parameters from the fundamental conditions and *Figure 2.*

$1>d>0$, that is, $1>(C-\mathbf{bg})/(1-\mathbf{b})>0$
$\mathbf{b}>0$, $\mathbf{g}>0$, $k_0>0$
$1>P>0$, $1>S>0$, $1>T>0$.

(c) Restrictions on cell size

For practical use, the restriction $0<S<1$ and $0<T<1$ is assumed and we set the least cell frequency L (≥ 1) such that

$$^1N_{11},\ ^1N_{12},\ ^1N_{21},\ ^1N_{22},\ ^2N_{11},\ ^2N_{12},\ ^2N_{21},\ ^2N_{22} \geqq L.$$

(iii) Result

Under restrictions (a), (b), and (c), the values of k_0 were calculated for given values P, C, and N, and unknown parameters $\boldsymbol{b}$ and $\boldsymbol{g}$ with pre-assigned constant values. The results are shown in Yamaoka (1996).

4. Solution by Monte Carlo Method

Pre-assigned values for $\boldsymbol{b}$ and $\boldsymbol{g}$ were used in Section 3. That is, in the previous study, parameters $\boldsymbol{b}$ and $\boldsymbol{g}$ were moving with constant intervals. However, the values of $\boldsymbol{b}$ and $\boldsymbol{g}$ are essentially unknown. We therefore assume that these parameters have some prior probability distribution (hereinafter referred to as PPD) and examine how the solutions change depending on the PPD. When the solutions vary notably depending on the PPD, we must carefully discuss the solutions. On the other hand, if they prove to be robust to the PPD, this brings practical usefulness to the solutions. To examine the robustness of the solutions to several PPD's is the subject of the present study. For simplicity, restriction (c) is not used.

(i) Procedure

The value k_0 changes in accordance with changes in the parameters $\boldsymbol{b}$ and $\boldsymbol{g}$, which change in accordance with the decomposing condition. It should be noted that N_{11} and marginal distributions were fixed. To specify the meaning of the value k_0 calculated from the original table and that from the situation in *Figure 2*, we refer to the latter as **k**. The value **k** has the minimum and maximum values under the conditions of the parameters $\boldsymbol{b}$ and $\boldsymbol{g}$ with permission of some restrictions. Thus, the strength of the association can be examined by a comparison of the given value k_0 with the minimum value of the value **k** (K_{min}) or the maximum value of the value **k** (K_{max}). In such a situation, K_{min} is obtained when **k**<1, and K_{max} is obtained when **k**>1.

From (a) and (b) in Section 3 we have restrictions $\boldsymbol{bg} < C$, $C<1-\boldsymbol{b}+\boldsymbol{bg}$, and $0<\mathbf{k}P<1$ from nature of a contingency table. In this analysis, we have the following solutions from the formulas (1) and (2);

when $C>\boldsymbol{g}$:

$$K_{max} = \text{minimum}\left\{\frac{\boldsymbol{b}(C-\boldsymbol{g})+CP(1-\boldsymbol{b}+\boldsymbol{bg}-C)}{P(C-\boldsymbol{bg})(1-C)},\ \frac{C(1-\boldsymbol{g})}{\boldsymbol{g}(1-C)}\right\}$$

$$K_{min} = \text{maximum}\left\{\frac{C(1-\boldsymbol{b}+\boldsymbol{bg}-C)}{(C-\boldsymbol{bg})(1-C)},\ \frac{\boldsymbol{g}-C+CP(1-\boldsymbol{g})}{\boldsymbol{g}P(1-C)}\right\},$$

when $\boldsymbol{g}>C>\boldsymbol{bg}$:

$$K_{max} = \text{minimum}\left\{\frac{C(1-\boldsymbol{b}+\boldsymbol{bg}-C)}{(C-\boldsymbol{bg})(1-C)},\ \frac{\boldsymbol{g}-C+CP(1-\boldsymbol{g})}{\boldsymbol{g}P(1-C)}\right\},$$

$$K_{min} = \text{maximum}\left\{\frac{\boldsymbol{b}(C-\boldsymbol{g})+CP(1-\boldsymbol{b}+\boldsymbol{bg}-C)}{P(C-\boldsymbol{bg})(1-C)},\ \frac{C(1-\boldsymbol{g})}{\boldsymbol{g}(1-C)}\right\}. \quad (3)$$

In the case that $C \leqq \boldsymbol{bg}$, we have no solution.

In order to examine the robustness, Monte Carlo method is a very useful strategy. In the Monte Carlo method, we assume four types of PPD and generate random numbers. That is, instead of moving ***b*** and ***g*** with constant interval, we assume four types of distributions, U1, U2, U3 and U4, shown as below, on ***b*** and ***g***, and calculate medians and percentiles of $(K_{min})_{lo}$ and $(K_{max})_{up}$ (calculated as the 5th percentile for $(K_{min})_{lo}$ and the 95th percentile for $(K_{max})_{up}$).

Four types of PPD were considered corresponding to subsequent distributions:

U1: uniform distribution on [0,1]
U2-U4: triangle distribution on [0,1]
U2: isosceles triangle (△)
U3: right angle triangle with mode 1 (⊿)
U4: right angle triangle with mode 0 (◺). (4)

(ii) Simulation Results

(a) Robustness

We have sixteen combination for PPD's of the parameters ***b*** and ***g***. For each, the medians of K_{min} and K_{max} and the percentiles of $(K_{min})_{lo}$ and $(K_{max})_{up}$ were calculated. The random numbers were generated 20000 times for each.
Here, we concentrated our attention on the differences in the solutions among the PPD

types. As a whole, the differences between the medians among the PPD types (except four combination and those of $(K_{max})_{up}$) were not very large; i.e., robustness was generally confirmed. However, as $(K_{min})_{lo}$'s are scattered, this information must be treated carefully. Here, for example, solutions of the case for P=0.1 is shown in *Figure 3*.

Figure 3. *Medians and percentiles of $(K_{min})_{lo}$ and $(K_{max})_{up}$ with given values C as the horizontal axis and the four distribution types as the corresponding lines, for P= 0.1.*
Thin line = medians, dotted line = percentiles.

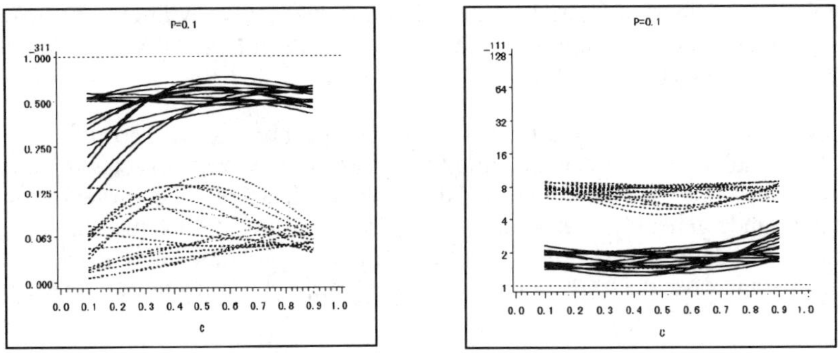

(b) Case of Mixed Distribution

Mixed distribution cases were calculated under the conditions of mutual independence between ***b*** and ***g*** with respect to the occurrence of PPD's with equal probability 1/4, respectively. That is, ***b*** and ***g*** were generated using random numbers from the mixed distribution of the four types of PPD with 1/4 probability for each. The random numbers were generated 50000 times.

Most of the medians for K_{min} varied around 0.5 (except for the case of small P value), and the percentiles of $(K_{min})_{lo}$ varied around 0.05. On the other hand, for K_{max}, the medians varied around 2 and $(K_{max})_{up}$ varied around 8. The behavior of solutions are being stable in this mixed case. For example, solutions of the case for P=0.1 is shown in *Figure 4*.

Figure 4. *Medians and percentiles of $(K_{min})_{lo}$ and $(K_{max})_{up}$ with given values C as the horizontal axis for mixed distribution, for P=0.1.*
Thin line = medians, broken line = percentiles.

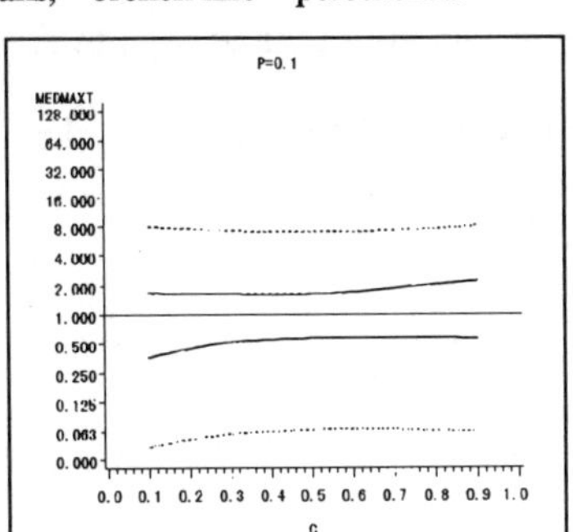

(c) Discussion

The analysis presented overall observation of the K_{min}, K_{max}, and C. These results suggested that the boundary line (for the median) of the occurrence of Simpson's paradox may exist around k_0=0.5 (or 2). However, in order to deny the occurrence of Simpson's paradox, without regard to any restrictions on the condition of the classified subgroups, k_0 should be generally smaller than approximately 0.05 or larger than 8.

5. Concluding Remarks

The importance of division (classification) of an original group by known or unknown criteria has been shown, through the treatment of Simpson's paradox, in data analysis from the exploratory point of view.

In the case of classification by known criteria, the problem is not too difficult, even though the results are very interesting, whereas in the case of classification by unknown criteria, i.e. the existence of latent groups (classes), the problems is very complicated and generally quite difficult to solve. Such problems have serious implications for data science. In the present paper, a fundamental problem called Simpson's paradox has been treated and discussed under the concept of data science.

References

Geng, Z.H.I. (1992). Collapsibility of relative risk in contingency tables with a response variable, *J R Statist Soc B*, 54, 585-593.

Hand, D.J. (1994). Deconstruction statistical questions, *J R Statist Soc,* A157, 317-356.

Hayashi, C. (1993). Treatise on behaviormetrics (in Japanese), *Asakura-Syoten*, 108-122.

Hintsman, D.L. (1993). On variability, Simpson's paradox, and the relation between recognition and recall: reply to Tulving and Flexser, *Psychol Rev*, 100, 143-148.

Miettinen, O.S. (1976). Stratification by multivariate confounder score, *Am J Epidemiol*, 104(6), 609-620.

Shapiro, S.H. (1982). Collapsing contingency tables: a geometric approach. *Am Statistn*, 36, 43-46.

Simpson, E.H. (1951). The interpretation of interaction in contingency tales, *J R Statist Soc*, B13, 238-241.

Vogt, A. (1995). Simpson's paradox revisited, *Student*, 1, 2, 99-108.

Weinberg, C.R. (1993). Toward a clear definition of confounding, *Am J Epidemiol*, 137, 1, 1-8.

Wermuth, N. (1989). Moderating effects of subgroups in linear models, *Biometrika*, 76, 81-92.

Yamaoka, K. (1996). Beyond Simpson's Paradox: A descriptive approach, *Data Analysis and Stochastic Models*, 12, 239-253.

For Consensus (With Branch Lengths)

François-Joseph Lapointe
Département de sciences biologiques, Université de Montréal,
C.P. 6128, Succursale Centre-ville, Montréal, Québec, H3C 3J7, Canada

Abstract: A debate over the use of consensus methods for combining trees, as opposed to combination of data sets, is currently growing in the field of phylogenetics. Resolution of this question is greatly influenced by the consensus method employed (i.e., for unweighted or weighted trees). In the present paper, my main objective is to review some of these methods that allow for the combination of weighted trees. These consensus with branch lengths will be compared to some of the commonly used consensus methods for *n*-trees. Finally, I will extend the results to supertrees.

Key words: Consensus Trees, Weighted Trees, Phylogenies, Supertrees.

1. Introduction

To combine, or not to combine? This is one of the hottest debate in phylogenetics today (de Queiroz et al., 1995; Huelsenbeck et al., 1996). One the one hand, proponents of *total evidence* (i.e., character congruence) claim that all available data must be combined prior to phylogenetic reconstruction (e.g., Kluge, 1989). On the other hand, proponents of *consensus* (i.e.,taxonomic congruence) prefer to combine the phylogenies derived from independent data sets (e.g., Miyamoto and Fitch, 1995). Resolution of this dilemma greatly depends on the consensus method selected, however, and how it fares against total evidence. In particular, it is striking that the advocates of character congruence have used conservative techniques to assess consensus (i.e., the strict consensus); when the consensus tree was less resolved than, or different from, the total-evidence phylogeny, the latter approach was declared superior (Barrett et al., 1991; de Queiroz, 1993). Thus, the important question of what consensus method to chose remains central to this debate.
Recently, it was shown that when branch lengths are accounted for when combining trees, total-evidence and consensus phylogenies do not differ much (Lapointe, 1997; Lapointe et al., submitted). In this paper, I will review the consensus methods currently available to combine trees while taking into account branch lengths. I will use examples to illustrate the differences between these methods and some other consensus techniques that ignore branch lengths. I will present methods

to combine ultrametric trees. Then, the case of additive trees will be considered, and a new consensus method based on spectral decomposition will be introduced.

2. Consensus methods for unweighted trees

Let us consider a simple example involving three dendrograms derived from different genes (Fig. 1). These phylogenies differ with respect to topological relationships and path-lengths among the species considered. In order to reconcile these results, two options are available: the first is to combine the data to obtain a "total-evidence" phylogeny (Fig. 1d), whereas the other is to combine the trees using some consensus method.

Fig. 1. Three ultrametric phylogenies derived from different genes: (a) 12s rRNA, (b) IRBP, (c) von Willebrand's factor (see Kirsch et al., submitted). The total-evidence tree (d) was obtained by combining all data in one phylogenetic analysis.

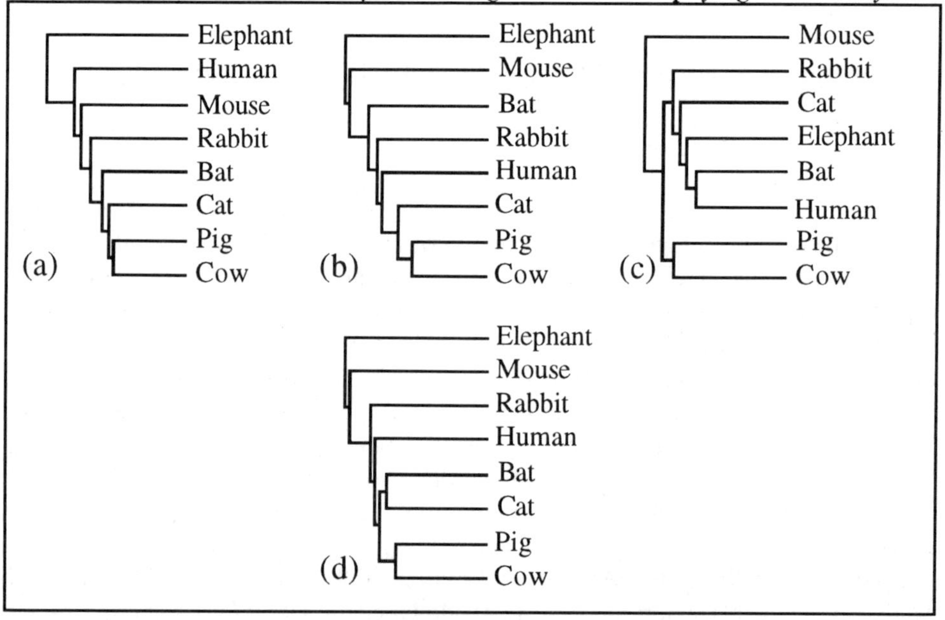

The choice of a consensus technique is again central to this problem as more conservative methods will usually produce less resolved trees. Indeed, a strict consensus (Sokal and Rohlf, 1981) of the three phylogenies is barely resolved (Fig. 2a), whereas the consensus obtained with the majority-rule method (Margush and McMorris, 1981) is somewhat better (Fig. 2b). The Adams (1972) consensus tree is even more resolved (Fig. 2c), but it contains clades (e.g., {Rabbit, Cat}) which were not observed in any of the original trees (see Fig. 1).

Fig. 2. (a) strict, (b), majority-rule, and (c) Adams consensus of the trees in Fig. 1.

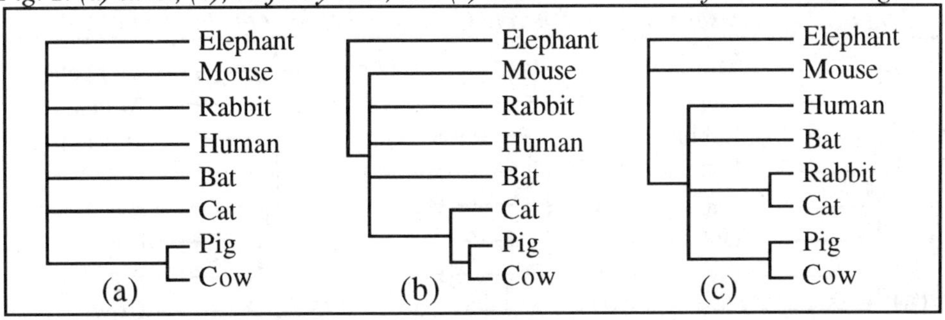

In this example, all of the consensus trees convey less phylogenetic information than the total-evidence tree (Fig. 1d); thus character congruence should be preferred here to reconcile the results of otherwise inconsistent phylogenies (Fig. 1). But this conclusion may not always be true. I will now show that consensus methods which take into account branch lengths often produce trees which are closer to total-evidence phylogenies than those produced by consensus methods that ignore these branch lengths.

2. Consensus methods for ultrametric trees

Some of the earlier methods proposed to compute consensus with branch lengths were simple extension of consensus methods for unweighted trees. These techniques were also restricted to ultrametric trees (i.e., dendrograms). Three distinct methods will be presented here: the intersection method of Stinebrickner (1984b), the Euclidean consensus method of Lefkovitch (1985), and the average procedure of Cucumel (1990), as modified by Lapointe and Cucumel (1997).

2.1. Intersection method

The method proposed by Stinebrickner (1984b) to combine dendrograms is a modification of intersection methods for *n*-trees (Adams, 1972; Neumann, 1983; Stinebrickner, 1984a). In short, the consensus tree is obtained by taking the maximum ultrametric distance observed over all dendrograms, for every pair of objects; the resulting pairwise distance matrix is also ultrametric and uniquely represents the consensus dendrogram. In the present example, the consensus obtained with Stinebrickner's method (Fig. 3a) is much more resolved than any of the consensus trees constructed with topological techniques (see Fig. 2). Nevertheless, that tree still differs quite a lot from the total-evidence phylogeny (Fig. 1d). Also, this method can produce clades which were not encountered in any of the individual phylogenies (e.g., {cat, human}).

Fig. 3. (a) Intersection (Stinebrickner, 1984b), (b), Euclidean (Lefkovitch, 1985), and (c) average consensus (Lapointe and Cucumel, 1997) of the trees in Fig. 1.

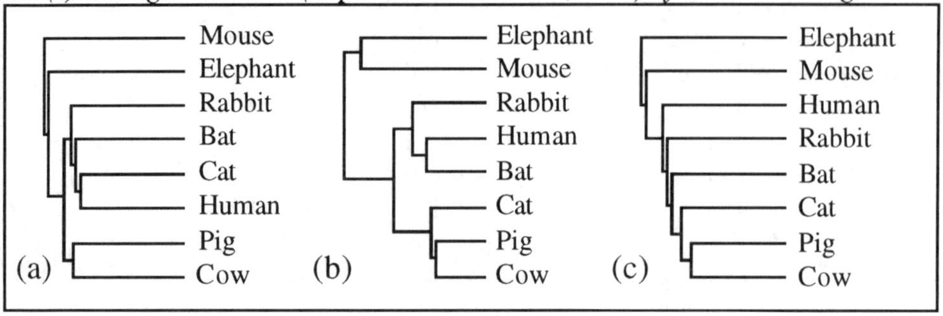

2.2. Euclidean method

The consensus method proposed by Lefkovitch (1985) is derived from his work on principal-coordinates representations of dendrograms (Lefkovitch, 1976; 1978). It uses the Euclidean representation of ultrametric distances, from which consensus coordinates can be computed. The consensus obtained in this fashion is the dendrogram that minimizes the sum of the squared distances from the objects in each of the separate representations to their positions in the consensus. Interestingly, one can obtain consensus trees containing overlapping groups with such an approach (see Lefkovitch, 1985). In the present example, the Euclidean consensus is a strictly bifurcating dendrograms (Fig. 3b). However, it is quite different from the consensus tree obtained with Stinebrickner's method (Fig. 3a).

2.3. Average method

This method initially introduced by Cucumel (1990; see also Vichi, 1993; Lapointe and Cucumel, 1997) is related to Lefkovitch's Euclidean consensus. In the case of dendrograms, the average consensus is the ultrametric tree that minimizes the sum of the squared distances from the trees in the input profile to the output consensus solution. That tree is obtained by applying a least-squares algorithm to a matrix of average path-length distances, computed over all dendrograms combined. In the present example, the average consensus tree (Fig. 3c) is the closest from the total-evidence tree (Fig. 1d), with only one different node (i.e., {Bat, Cat}).

3. Consensus methods for additive trees

When one is dealing with phylogenies, imposing the ultrametric constraint is to assume that every lineage is evolving at the same rate (i.e., the molecular clock hypothesis is satisfied, Fig. 1). In general, additive trees satisfying the four-point condition are preferred to obtain a better representation of the data matrix (see Fig.

4). Most of the consensus methods for weighted trees only apply to dendrograms, however. In the following sections, I will present two methods to obtain consensus phylogenies from additive trees in general.

3.1. Average method

As for ultrametric trees, the average procedure can be used to construct consensus additive trees (see Lapointe and Cucumel, 1997). The procedure is basically the same and consists of applying a least-squares algorithm for additive trees to a matrix of average path-length distances, computed over all trees combined. When such a method is used to combine our three additive phylogenies (Fig. 4), the average consensus obtained (Fig. 4e) is much closer to the corresponding total-evidence phylogeny (Fig. 4d) than what was obtained with the ultrametric procedure (Fig. 1d *vs.* Fig. 3c).

Fig. 4. (a-c) Three additive phylogenies derived from different genes (see Fig. 1). The total-evidence tree (d) was obtained by combing all data in one phylogenetic analysis. The average consensus tree (e) was obtained by combining all trees.

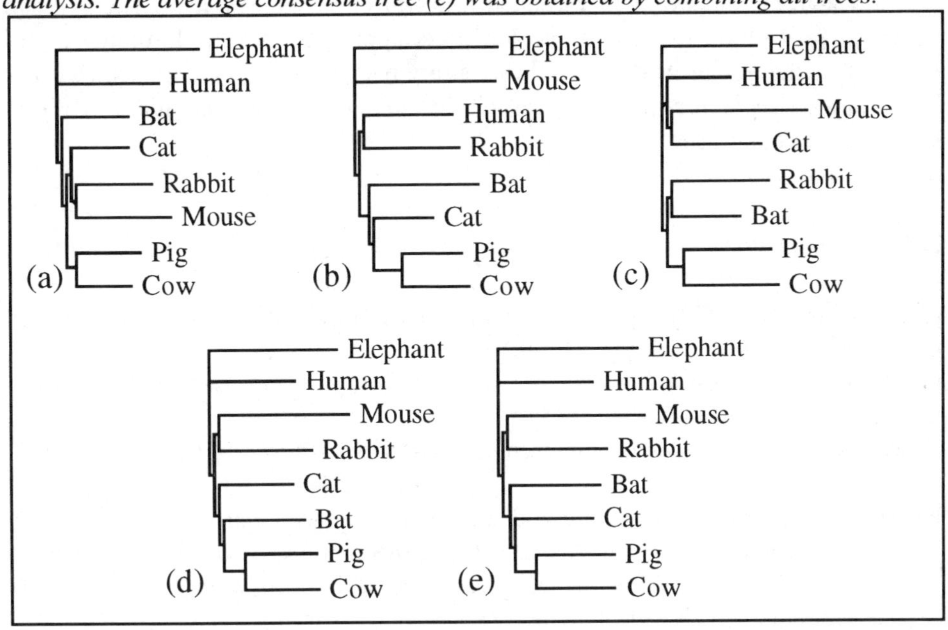

3.2. Spectral method

When a least-squares algorithm is used to a build a consensus from average path lengths, internal and terminal branches of the trees combined are not equally weighted (Lapointe et al., 1994). The spectral decomposition method (Hendy and Penny, 1993) can be applied to circumvent this problem, however. The procedure (see Hendy, 1991) is to find the tree spectrum that minimizes the sum of the squared distances from a given data spectrum. In a similar fashion, a spectral consensus can be derived by finding the closest tree from an average spectrum representing different tree spectra. In the present example, the consensus computed with the spectral decomposition method (not shown) is topologically identical to the average consensus tree (Fig. 4e), with small differences in branch lengths. This interesting result can be explained by the very short internodes of the trees considered (Fig. 4), thus reducing the gap between the spectral and average procedure.

4. Consensus supertrees, with branch lengths

While some supertree methods have been proposed to combine unweighted trees (Gordon, 1986), there have been less effort devoted to the combination of weighted trees containing overlapping sets of leaves. Brossier (1990) has been a pioneer in this field with his piecewise hierarchical procedure for dendrograms. The average consensus procedure was also applied to overlapping trees (e.g., Lapointe et al., 1994), using the tree properties to fill the empty part of the average matrix (Landry and Lapointe, 1997). It is still not clear yet that such a procedure could be extended to the spectral consensus.

References

Adams, E. N. III. (1972). Consensus techniques and the comparison of taxonomic trees, *Systematic Zoology,* 21, 390-397.

Barrett, M., Donoghue, M. J. & Sober, E. (1991). Against consensus, *Systematic Zoology,* 40, 486-493.

Brossier, G. (1990). Piecewise hierarchical clustering, *Journal of Classification,* 7, 197-216.

Cucumel, G. (1990). Construction d'une hiérarchie consensus à l'aide d'une ultramétrique centrale, in: *Recueil des Textes des Présentations du Colloque sur les Méthodes et Domaines d'Application de la Statistique 1990,* Bureau de la Statistique du Québec, Québec, 235-243.

de Queiroz, A. (1993). For consensus (sometimes), *Systematic Biology,* 42, 368-372.

de Queiroz, A., Donoghue, M. J. & Kim, J. (1995). Separate versus combined analysis of phylogenetic evidence, *Annual Review of Ecology and Systematics,* 26, 657-681.

Gordon, A. D. (1986). Consensus supertrees: The synthesis of rooted trees containing overlapping sets of labeled leaves, *Journal of Classification,* 3, 335-348.

Hendy, M. D. (1991). A combinatorial description of the closest tree algorithm for finding evolutionary trees, *Discrete Mathematics,* 96, 51-58.

Hendy, M. D. & Penny, D. (1993). Spectral analysis of phylogenetic data, *Journal of Classification,* 10, 5-24.

Huelsenbeck, J. P., Bull, J. J. & Cunningham, W. (1996). Combining data in phylogenetic analysis, *Trends in Ecology and Evolution,* 11, 152-158.

Kirsch, J. A. W., Springer, M. S. & Lapointe, F.-J. Out of Africa: a distance-based approach to character and taxonomic congruence in placental phylogeny, *submitted.*

Kluge, A. G. (1989). A concern for evidence and a phylogenetic hypothesis of relationships among *Epicrates* (Boidae, Serpentes), *Systematic Biology,* 38, 7-25.

Lapointe, F.-J. (1997). How to validate phylogenetic trees? A stepwise procedure, in: *Data Science, Classification, and Related Methods: Studies in Classification, Data Analysis, and Knowledge Optimization,* Hayashi, C., Bock, H. H., Yajima, K., Tanaka, Y., Oshumi, N. & Baba, Y. (Eds.), Springer, Tokyo, *in press.*

Lapointe, F.-J. & Cucumel, G. (1997). The average consensus procedure: Combination of weighted trees containing identical or overlapping sets of taxa, *Systematic Biology,* 46, 306-312.

Lapointe, F.-J., Kirsch, J. A. W. & Bleiweiss, R. (1994). Jackknifing of weighted trees: Validation of phylogenies reconstructed from distances matrices, *Molecular Phylogenetics and Evolution,* 3, 256-267.

Lapointe, F.-J., Kirsch, J. A. W. & Hutcheon, J. M. Total evidence, consensus, and bat phylogeny: a distance-based approach, *submitted.*

Landry, P.-A. & Lapointe, F.-J. (1997). Estimation of missing distances in path-length matrices: problems and solutions, in *Mathematical Hierarchies and Biology*, Mirkin, B., McMorris, F. R., Roberts, F. S. & Rzhetsky, A. (Eds.), DIMACS Series in Discrete Mathematics and Theoretical Computer Science, American Mathematical Society, Providence, 209-218.

Lefkovitch, L. P. (1976). Hierarchical clustering from principal coordinates: An efficient method from small to very large number of objects, *Mathematical Biosciences,* 31, 157-176.

Lefkovitch, L. P. (1978). Consensus coordinates from qualitative and quantitative attributes, *Biometrical Journal,* 20, 679-691.

Lefkovitch, L. P. (1985). Euclidean consensus dendrograms and other classification structures, *Mathematical Bioscience,* 74, 1-15.

Margush, T. & McMorris, F. R. (1981). Consensus *n*-trees. *Bulletin of*

Mathematical Biology, 43, 239-244.
Miyamoto, M. M. & Fitch, W. M. (1995). Testing species phylogenies and phylogenetic methods with congruence, *Systematic Biology,* 44, 64-76.
Neumann, D. A. (1983). Faithful consensus methods for n-trees, *Mathematical Bioscience,* 63, 271-287.
Sokal, R. R. & Rohlf, F. J. (1981). Taxonomic congruence in the Leptopodomorpha re-examined, *Systematic Zoology,* 30, 309-325.
Stinebrickner, R. (1984a). S-consensus trees and indices, *Bulletin of Mathematical Biology,* 46, 923-935.
Stinebrickner, R. (1984b). An extension of intersection methods from trees to dendrogram, *Systematic Zoology,* 33, 381-386.
Vichi, M. (1993). Un algoritmo dei minimi quadrati per interpolare un insieme di classificazioni gerarchiche con une classificazione consenso, *Metron,* 51, 139-163.

Consensus of classifications: the case of trees

Bruno Leclerc
Centre d'Analyse et de Mathématique Sociales,
École des Hautes Études en Sciences Sociales
F-75270 Paris cedex 06, France
E-mail: leclerc@ehess.fr

Summary: We present a survey of the literature on the consensus of classification trees, based on a corpus (in progress) of about ninety papers.

Key words: classification tree, hierarchy, dendrogram, ultrametric, consensus

1. Classification trees

1.1. Classification trees. Among the various types of trees (connected and acyclic graphs) of the literature, trees used in classification have two common properties. In a classification tree (CT), where, as usual, V is the finite *vertex* set of T and E, the *edge* set of T is a set of unordered pairs of V: (i) *no vertex, except possibly a unique distinguished one, not a leaf, has two neighbors;* and (ii) T *is leaf-labelled*, that is the labels of the set X of the leaves of T are the only relevant ones.

When it exists, the distinguished vertex of Condition (i) above is the *root* of T and T is said to be *rooted* (and *unrooted* otherwise). The tree T may be *valued*, that is endowed with an edge length (a positive real function on E); otherwise, it is *unvalued*. Hereunder, the length of the unique path between two vertices v and v' is denoted as $t(vv')$.

Each usual type of CTs admits several other equivalent definitions. Every consensus approach is related with a particular description of the involved CTs.

1.2. Hierarchies. A hierarchy H on X is a *rooted* and *unvalued* CT. It is well-known that, equivalently, H is a set of subsets (*clusters*) of X satisfying the conditions: $\varnothing \notin H$; $X \in H$; for all $x \in X$, $\{x\} \in H$; for $A, B \in H$; $A \cap B \in \{\varnothing, A, B\}$. Then, for any $A \subseteq X$, there is a minimum cluster A_H in H containing A.

Two other equivalent definitions of hierarchies appear in consensus studies:

A hierarchy may be defined as the set of the *triples* xy/z of elements of X such that $\{x,y\}_H \subset \{x,y,z\}_H$. Such sets of triples have been characterized by Colonius and Schultze (81). More generally, a *nesting* (Adams 86) is a pair (A,B) of subsets of X such that $A_H \subset B_H$ and a hierarchy is defined by the set of all its nestings.

1.3. Dendrograms. Dendrograms correspond to *rooted* and *valued* CTs, with the further assumption that the length of any path from the root to a leaf is constant. It is well-known (Johnson 67) that a dendrogram is equivalent to an *ultrametric* function, that is a real non-negative function u on X^2 satisfying $u(x,x) = 0$, $u(x,y) = u(y,x)$, and $u(x,z) \leq \max(u(x,y), u(y,z))$, for all $x, y, z \in X$. The correspondence between a dendrogram and its ultrametric u is given by $2u(x,y) = t(xy)$. Again, some other equivalent structures are useful:

- An *indexed hierarchy* (Benzécri 67) is a pair (H,ι), where H is a hierarchy on X and ι is the real *height* function on H defined by $\iota(A) = max\{u(x,y)$: $x, y \in A\}$.

- A *residual* mapping from $\mathbb{R}^+$ to the set of the partitions π of X: the converse image of the set of all partitions coarser than π has a minimum (Janowitz 78, Leclerc 91).

1.4. Ranked hierarchies. Ranked hierarchies are intermediate between hierarchies and indexed hierarchies. A ranked hierarchy is endowed with a complete, compatible with inclusion, preorder on its clusters. Equivalent structures are sets of pairwise nested partitions (including the coarsest $\{X\}$ and the finest $\{\{x\}: x \in X\}$ ones), and ultrametric dissimilarity preorders, that is preorders on X^2 which are complete and satisfy $xx < xy$, $xx \approx yy$ and $xz \leq xy$ or $xz \leq yz$, for all $x, y, z \in X$ (Barthélemy et al. 84, 86).

1.5. Phylogenetic trees. Now we consider *unrooted* CTs on X, where phylogenetic trees (abbreviated here as PTs) correspond to the *unvalued* case. Equivalent structures are:

- sets of pairwise compatible splits (bipartitions) (A,A') of X. Two splits (A,A') and (B,B') are compatible when one of the sets $A \cap B$, $A \cap B'$, $A' \cap B$, $A' \cap B'$ is empty (Buneman 71); these splits correspond with the edges of the tree;
- sets of quadruples xy/zw of elements of X satisfying some conditions given by Colonius and Schultze (81).

Consensus methods on hierarchies and PTs are frequently closely related by the following one-to-one correspondence (Day 85). Let $a \notin X$; a PT T on $X+\{a\}$ is obtained by adding the edge ar to a hierarchy H with root r. The split $(A,(X-A)+\{a\})$ of T comes from the cluster A of H, and the quadruple xy/za of T from the triple xy/z of H (so, the entire set of quadruples is redundant to define the tree T).

1.6. Tree metrics. An *unrooted* and *valued* CT is associated with a *tree metric*, that is a dissimilarity satisfying the *four-point condition*: for all $x, y, z, w \in X$, $t(x,y)+t(z,w) \leq \max(t(x,z)+t(y,w),t(x,w)+t(y,z))$. This is a well-known characteristic property of the path lengths functions t associated to these trees (Buneman 71). Equivalently, unrooted and valued CTs may be defined as weighted sets of pairwise compatible splits.

2. The consensus problem

2.1. Aims and approaches. The consensus problem consists of aggregating several objects into a unique object of the same type. In the case of classification trees, the most frequently mentioned situation is that of a clustering method applied to several data sets, all describing the same objects, and giving different CTs. In another situation, several clustering methods are applied to the same data set, each giving a CT. The purpose may be either to recover an hypothetical "true" tree (as in phylogenetic reconstruction), or to find a tree summarizing at best the given set of trees. Consider a set $\mathcal{T}$ of CTs. A *profile* of length k of $\mathcal{T}$ is a k-tuple $T^* = (T_1,\ldots,T_k)$ of elements of $\mathcal{T}$; we set $K = \{1,\ldots,k\}$. Let $\mathcal{T}^*$ be the set $\cup_k \mathcal{T}^k$ of all the profiles of finite length of $\mathcal{T}$. A *k-procedure* is a map from $\mathcal{T}^k$ to $\mathcal{T}$. A *complete procedure* is a map from $\mathcal{T}^*$ to $\mathcal{T}$. A *complete multiprocedure* is a map from $\mathcal{T}^*$ to $2^{\mathcal{T}}-\{\emptyset\}$.

Three overlapping approaches have been used to tackle the consensus problem:

- The *constructive approach*: a way to construct a consensus is explicitly given. Obvious constructions (as the unanimity, or *strict consensus* rule) are generally not adequate and more sophisticated constructive rules are needed.

- The *combinatorial optimization approach.* A criterion measuring the *remoteness* $R(T,T^*)$ of any CT T to the given profile T^* is defined; we search for the solutions of: Min $R(T,T^*)$, subject to $T \in \mathcal{T}$.
- The *Axiomatic approach*: "to sit in an armchair and think of desirable properties that a consensus method should possess, and then attempt to find the methods satisfying these properties" (McMorris 85). This approach leads frequently to problems of existence and uniqueness.

In the case of valued trees, e.g. dendrograms, two strategies are encountered. Instead of a direct study of the consensus of dendrograms, it is possible to consider the corresponding indexed hierarchies, then stripping heights to obtain hierarchies and use a consensus method defined on hierarchies. A final (possible) step may be to determine an height function for the obtained consensus hierarchie(s). Such an approach is more transparent when one focuses attention on clusters.

2.2. A corpus. Starting from 1972 (Adams), we have found more than 90 references, not all given here because lack of place, of papers partly or totally devoted to consensus tree methods. Among them are mathematical papers on consensus where cases of CTs are explicitely referred to (e.g., Barthélemy et al. 84, 86, Monjardet 90, Barthélemy and Janowitz 91) or surveys on topics including CTs where the consensus problem is explicitely referred to (Arabie and Hubert 96, Gordon 87, 96, Leclerc and Cucumel 87, Lapointe 97). Most of these papers consider hierarchies, whereas few works concern unrooted phylogenetic trees or tree metrics.
Papers dealing with related topics (consensus of other types of classifications as partitions or weak hierarchies, formal theory of consensus, comparison of two or more CTs and consensus indices) are quoted here only when needed.

3. Consensus methods and uses on classification trees

3.1. Quota rules and federation consensus functions. The q quota rule consists of retaining the features present in at least q elements of the profile. Quota rules are constructive k-procedures (complete procedures when replacing q with a percentage). For special values of q, one gets classical consensus rules: the *unanimity rule*, or *strict consensus* for $q = k$, and the *majority rule* when q is equal to p, the least integer greater than half the length k of the profile.
When trees are described as subsets (of clusters, of splits...), the q quota rule tree is equal to $T = \cup_{I \subseteq K,\ \mathrm{card}(I) \geq q} (\cap_{i \in I} T_i)$, provided this set corresponds to a tree. In the case of a set $\mathcal{T}$ of CTs endowed with a lattice, or a semilattice structure (with a meet $\wedge$ and a join $\vee$), *lattice quota rules* may be defined by the similar formula:

$$T = \vee_{I \subseteq K,\ \mathrm{card}(I) \geq q} (\wedge_{i \in I} T_i),$$

with, as above, algebraic unanimity ($q = k$) and majority ($q = p$) rules. A further generalization is constituted by the so-called *federation consensus functions* (FCFs). Let $\mathcal{F}$ be a family of subsets of K; define a consensus function $F_{\mathcal{F}}$ by $T = F_{\mathcal{F}}(T^*) = \vee_{I \in \mathcal{F}} (\wedge_{i \in I} T_i)$. Oligarchic consensus functions $T = \wedge_{i \in I} T_i$ correspond to $\mathcal{F} = \{I\}$, $I \subseteq K$, dual oligarchic $T = \vee_{i \in I} T_i$ to $\mathcal{F} = \{\{i\}: i \in I\}$ for a fixed subset I of K, and dictatorial consensus functions $T = T_i$ to $\mathcal{F} = \{\{i\}\}$ for a fixed $i \in K$.

These mathematical formalizations are very efficient in the search of optimization and axiomatic properties, especially for the majority rule. A basic result of formal

consensus theory is that, if the tree structure is a matter of pairwise compatibility, then the ordered set $(\mathcal{T},\subseteq)$ is a *median semilattice* and the set T corresponds to a tree as soon as $q \geq p$. Moreover, the quota rule is axiomatically characterized and the majority rule element is the solution of an optimization problem (Monjardet 80, Bandelt and Hedlikova 83, Bandelt and Barthélemy 84, Barthélemy and Janowitz 91, McMorris and Powers 95, McMorris et al. 96; see Sections 3.3 and 3.4). Formal axiomatic results about FCFs are given in Monjardet (90; see Section 3.4).
Their degree of abstraction may make such consensus rules appear somewhat formal. In fact, they are effectively computable in many cases (Day 83, 85, Day and Wells 84).

Strict consensus. Applied to hierarchies as sets of clusters, it is the early consensus method for CTs (as in other domains - see Day 86); it is still the most widely used one (Sokal and Rohlf 81, Rohlf 82, Wilkinson 94), sometimes with the addition of other selected compatible clusters (Nelson 79, Bremer 90).
The Adams consensus method may be also described as a unanimity rule on hierarchies as sets of nestings (Adams 86). Axiomatic results about this approach, and the unanimity rule on hierarchies as sets of triples are given in Vach (94).
Majority rule. It applies in the cases of median semilattice structures: hierarchies as sets of clusters (Margush and McMorris 81, Day 83, Barthélemy et al. 84, 86, Barthélemy and McMorris 86, Barthélemy and Janowitz 91, Leclerc 93, 94; see Sections 3.4 and 3.5), ranked hierarchies as sets of partitions (Barthélemy et al. 84, 86), phylogenetic trees as sets of splits (Barthélemy et al. 86, Barthélemy and Janowitz 91). Margush and McMorris introduced it as an alternative to the too strong strict consensus, which frequently gives an almost trivial hierarchy.
Other values of q have been considered for hierarchies as sets of clusters (McMorris et al. 83, Barthélemy 88 with an axiomatic characterization, Barthélemy and Janowitz 91) and phylogenetic trees as sets of splits (Barthélemy and Janowitz 91). A variant may be found in Wilkinson (96; see Section 3.5).
Federation consensus functions. They have been considered for hierarchies as sets of clusters (McMorris and Neumann 83). In this case, the elements of $\mathcal{F}$ must pairwise intersect to ensure that T is a hierarchy. Stinebrickner (84b) gives axiomatic properties of the oligarchic functions on dendrograms and, then, hierarchies by introduction of a height function (in relation with the intersection methods of the next section). In the study, in the same direction, of Neumann and Norton (86), the dual oligarchic rule is also considered. An axiomatization of the dual oligarchic rules (and of rules belonging to a wider class) on ultrametrics is obtained in Leclerc 84 (see also Barthélemy et al. 84, 86, Leclerc 91 and Leclerc and Monjardet 95).

3.2. Intersection methods. As observed by several authors, the description of hierarchies as sets of clusters may lose some common information, recoverable by methods able to produce clusters present in no element of the profile. Intersection methods are constructive complete procedures on hierarchies where any cluster $C \in H$ is obtained by intersection of clusters $C_i \in H_i$. The intersections should be made selectively, since the collection of all possible intersections does not generally correspond to a hierarchy.
In the celebrated Adams method (Adams 72), the C_i's are at the top of the not yet intersected clusters of each hierarchy H_i. Equivalent definitions and axiomatic characterizations are given in Adams (86), Vach and Degens (88) and Vach (94).

Neumann (83) defines generalized intersection rules, where the choice of the C_i's depends on the introduction of a height function r on the H_i's (then becoming dendrograms): the intersected C_i's are those of common height in each H_i (that is, an unanimity rule for dendrograms is applied). For instance, the Durschnitt or top-down height $r(C_i)$ is the number of edges of a path from the root to C_i, the bottom-up height is the maximum length of a path from C_i to a leaf $x \in C_i$, and the cardinality height is the number of elements of C_i. Axiomatic results are obtained in Neumann (83), Neumann and Norton (86) and Powers (95). Especially, an interesting betweenness axiom is satisfied by the Durschnitt intersection rule (Neumann 83): for each $C_i \in H_i$, there is $C \in H$ such that $\cap_i C_i \subseteq C \subseteq \cup_i C_i$. Such results may help to justify the choice, arbitrary at first sight, of the height function.
A variant (s-consensus methods) is proposed by Stinebrickner (84a): an index h measures the adequacy of each cluster obtained by intersection and the clusters with h greater than a fixed level s (for instance $h(C) = \text{card}C/\text{card}(\cup_i C_i)$), are selected. In the case of the cardinality height function, such a method is intermediate between pure intersection ($s = 0$) and strict consensus ($s = 1$). Further formalization and developments are found in Day and McMorris (85) and Stinebrickner (86).

3.3. The median procedure. In the case of CTs as elsewhere, the most frequently considered consensus method based on optimization is the median procedure. Consider a distance function d on $\mathcal{T}$. The problem is to minimize the remoteness function $\rho(T,T^*) = \sum_{i \in K} d(T,T_i)$, subject to $T \in \mathcal{T}$. This is a complete multiprocedure, since it may exist several solutions called *medians*. The median procedure applies to all metric spaces (Fréchet 49; see the survey of Barthélemy and Monjardet 81 with its 88 updating where cases of CTs are considered).
For CTs, the median procedure was introduced by Margush and McMorris (81), in the case of hierarchies as sets of clusters endowed with the symmetric difference metric. It is shown in this paper that the majority rule hierarchy is a median, a fact proved afterwards in all median semilattice structures (see Section 3.1). The algebraic characterization of median hierarchies is completed in Barthélemy et al. (84, 86) among similar cases; an axiomatic characterization of the median procedure is given in Barthélemy and McMorris (86), leading to the formal studies of Barthélemy and Janowitz (91) and McMorris and Powers (95). A probabilistic interpretation of the median procedure in the spirit of Condorcet, as read by Young (88), is given in McMorris (90). It is pointed out in Leclerc (93) that the medians remain the same for a wide variety of metrics. These results apply directly on phylogenetic trees as sets of splits, and also on ranked hierarchies as sets of partitions (Barthélemy et al. 84, 86).
When a median semilattice structure is no longer available, the median procedure generally leads to *NP*-hardness results, as for binary hierarchies (McMorris and Steel 94), or ultrametrics endowed with the classical L1 or L2 distances, by reduction to partitions (see, e.g., Barthélemy and Leclerc 95). Such situations lead to interesting algorithmic studies, as in the unusual approach of Philips and Warnow (96), replacing symmetric differences with set differences (on phylogenetic trees as sets of splits).
As a special instance of a family of solvable monotonic regression problems, Chepoi and Fichet (97) give the optimal solution for the consensus of ultrametrics with the criterion $\rho'(u,u^*) = \max_{i \in K} d(u,u_i)$, where d is the L_∞ distance.

3.4. Arrowian axiomatics. Many axiomatic studies are based on a main axiom (independance, stability,...) of the type introduced by Arrow (51) in the domain of social choice: if two profiles T^* and T'^* agree on their restrictions to a subset S, then the consensus trees T and T' also agree on S. A stronger neutral monotonicity axiom states that: if the profile T^* dominates (in some ordinal sense) the profile T'^* on S, then T dominates T' on S too. Such studies apply generally to k-procedures. A first remark is that there are generally various ways of setting an axiom of the above type; for instance, for hierarchies, one may consider restrictions of trees to a subset S of X (stability axioms), or restrictions of hierarchies as sets of clusters to a subset S of the power set of X (decisiveness axioms).
The results obtained in a formal theory of arrowian consensus in semilattices (Barthélemy et al. 86, Monjardet 90, Leclerc and Monjardet 95) apply to hierarchies as sets of clusters (characterization of federation consensus functions using neutral monotonicity axiom by McMorris and Neumann 83) and ultrametrics (characterization of generalized oligarchic consensus functions using a decisiveness axiom by Leclerc 84). Otherwise, the results concern often (but not always) dictatorial consensus functions:
Hierarchies. Characterization of dictatorial consensus functions using various stability axioms (Neumann 83, Barthélemy and McMorris 89, Barthélemy et al. 91); axiomatic studies on dictatorial consensus functions (Barthélemy et al. 92); study of various stability conditions and the related consensus functions (Barthélemy et al. 95).
Phylogenetic trees. Characterization of dictatorial consensus functions using stability axioms (McMorris 85, McMorris and Powers 93).

3.5. Pruning, reduction and supertrees. Many recent works concern pruning or supertree approaches. In pruning, one renounces to obtain a consensus tree on the entire set X and seeks a maximal subset Y of X such that all T_i's agree on Y. So, the consensus tree is a CT on Y. The idea, initiated by Gordon (79) for ranked hierarchies, has been mainly studied for hierarchies (Gordon 80, 81, Finden and Gordon 85 - with a regrafting procedure, thus leading to a hierarchy on X, Kubicka et al. 92, 95, Goddard et al. 95). In the "reduced" (Adams, majority rule) methods of Wilkinson (94, 96), maximum subsets corresponding with a consensus tree are also obtained.
A converse approach is the search of consensus supertrees for trees defined on different overlapping sets (Gordon 86, Steel 92, Lanyon 93, Constantinescu and Sankoff 95, Ronquist 96). As noted by Steel, this problem is related with the fundamental cladistic compatibility one. Considerations on the relations between genes and species lead to another kind of supertrees (Page and Charleston 97).

3.6. Other constructive approaches. Baum (92) proposes to use a binary encoding (Farris et al. 70) of hierarchies H_i and reach a consensus tree by addition of the encodings and parsimony techniques. Several authors have completed or improved this method (Ragan 92, Rodrigo 93, Baum and Ragan 93, Purvis 95a).
Lefkovitch (85) merges ultrametrics in an euclidean space and applies principal component analysis to find a "mean ultrametric".This approach is more developed and discussed in the work on average ultrametrics of Lapointe and Cucumel (97).

4. Final remarks

In the phylogeny reconstruction domain, there is a controversy about the relevance of consensus trees (Anderberg and Tehler 90, Barrett et al. 91, 93, de Queiroz 93, 95,

Nelson 93), the alternative being to combine first the data sets before the construction of a tree. From a formal point of view, the question is perhaps: what step of the process is the best to apply a consensus method? Indeed, several authors have made advances, frequently based on asymptotic statistical distributions, in the problem of the validation of consensus trees (Shao and Rohlf 83, Shao and Sokal 86, Steel 88, Hendy et al. 88, Sanderson 89, Lapointe and Legendre 90, 92a, b, 95, Purvis 95b, Cucumel and Lapointe 97, Lapointe 97).

In the domain addressed here, the increase of knowledge has provided tools for good choices among the many available methods. Besides statistical validation, the most interesting results in the consensus problem concern relations between the three approaches described in Section 2.1. Here, interactions between consensus of classifications and formal consensus theory have been important and fruitful. Especially, consensus of trees have prompted significant theoretical progresses. Conversely, results concerning trees have been obtained (or now appear as) special cases of mathematical results (Margush and McMorris 81). They also lead to extensions to other classification models (McMorris and Powers 97).

References

E.N. Adams III (1972), Consensus Techniques and the Comparison of Taxonomic Trees, *Syst. Zool.* 21, 390-397.

E.N. Adams III (1986), *N*-Trees as Nestings: Complexity, Similarity and Consensus, *J. of Classification* 3 (2), 299-317.

A. Anderberg, A. Tehler (1990), Consensus trees, a necessity in taxonomic practice, *Cladistics* 6, 399-402.

P. Arabie, L.J. Hubert (1996), An overview of combinatorial data analysis, in: P. Arabie, L.J. Hubert, G. De Soete, eds., *Clustering and Classification*, World Scientific Publ., River Edge, NJ, pp. 5-63.

K.J. Arrow (1951), *Social Choice and Individual Values*, New-York, Wiley.

H.J. Bandelt, J. Hedlikova (1983), Median algebras, *Discrete Math.* 45, 1-30.

H.J. Bandelt, J.P. Barthélemy (1984), Medians in Median Graphs, *Discrete Applied Math.* 8, 131-142.

M. Barrett, M.J. Donoghue, E. Sober (1991), Against consensus, *Syst. Zool.* 40, 486-493.

M. Barrett, M.J. Donoghue, E. Sober (1991), Crusade? A response to Nelson, *Syst. Zool.* 42, 216-217.

J.P. Barthélemy (1988), Thresholded consensus for *n*-trees, *J. of Classification* 5 (2), 229-236.

J.P. Barthélemy, M.F. Janowitz (1991), A formal theory of consensus, *SIAM J. Discr. Math.* 4, 305-322.

J.P. Barthélemy, B. Leclerc (1995), The median procedure for partitions, *in* I.J. Cox, P. Hansen, and B. Julesz, eds., *Partitioning data sets*, DIMACS Series in Discrete Mathematics and Theoretical Computer Science 19, Amer. Math. Soc., Providence, RI, pp. 3-34.

J.P. Barthélemy, B. Leclerc, B. Monjardet (1984), Ensembles ordonnés et taxonomie mathématique, in *Orders : Descriptions and Roles*, eds M. Pouzet and D. Richard, Annals of Discrete Mathematics 23, Amsterdam, North-Holland, 523-548.

J.P. Barthélemy, B. Leclerc, B. Monjardet (1986), On the use of ordered sets in problems of comparison and consensus of classifications, *J. of Classification* 3, 187-224.

J.P. Barthélemy, F.R. McMorris (1986), The Median Procedure for *n*-trees, *J. of Classification* 3 (2), 329-334.

J.P. Barthélemy, F.R. McMorris (1989), On an independance condition for consensus *n*-trees, *Appl. Math. Lett.* 2 (1), 75-78.

J.P. Barthélemy, F.R. McMorris, R.C. Powers (1991), Independance conditions for consensus *n*–trees revisited, *Appl. Math. Lett.* 4 (5), 43-46.

J.P. Barthélemy, F.R. McMorris, R.C. Powers (1992), Dictatorial consensus functions on *n*-trees, *Math. Soc. Sci.* 25, 59-64.

J.P. Barthélemy, F.R. McMorris, R.C. Powers (1995), Stability conditions for consensus functions defined on n-trees, *Mathl. Comput. Modelling* 22 (1), 79-87.

J.P. Barthélemy, B. Monjardet (1981), The median procedure in cluster analysis and social choice theory, *Math. Soc. Sci.* 1 (3), 235-238.

J.P. Barthélemy, B. Monjardet (1988), The median procedure in data analysis: new results and open problems, in *Classification and related methods in data analysis* (H.H.Bock, ed.), North-Holland, Amsterdam, 309-316.

B.R. Baum (1992), Combining trees as a way of combining data sets for phylogenetic inference, and the desirability of combining gene trees, *Taxon* 41, 3-10.

B.R. Baum, M.A. Ragan (1993), Reply to A.R. Rodrigo's "A comment on Baum's method for combining phylogenetic trees, *Taxon* 42, 637-640.

J.P. Benzécri (1967), Description mathématique des classifications, in *L'analyse des données 1. La taxinomie*, Paris, Dunod 1973.

K. Bremer (1990), Combinable component consensus, *Cladistics* 6, 369-372.

P. Buneman (1971), The Recovery of Trees from Measures of Dissimilarity, in *Mathematics in Archaeological and Historical Sciences*, eds. F.R. Hodson, D.G. Kendall and P. Tautu, Edinburgh, Edinburgh University Press, 387-395.

V. Chepoi, B. Fichet (1997), l_∞-approximations via subdominants, *Math. Social Sci.*, to appear.

H. Colonius, H.H. Schulze (1981), Tree structure for proximity data, *British J. Math. Statist. Psychol.* 34, 167-180.

M. Constantinescu, D. Sankoff (1995), An efficient algorithm for supertrees, *J. of Classification* 12, 101-112.

G. Cucumel, F.J. Lapointe (1997), Un test de la pertinence du consensus par une méthode de permutations, *Actes des XXIXèmes journées de statistique de Carcassonne*, 299-300.

W.H.E. Day (1983), The role of complexity in comparing classifications, *Math. Biosciences* 66, 97-114.

W.H.E. Day (1985), Optimal Algorithms for Comparing Trees with Labelled Leaves, *J. of Classification* 2, 7-28

W.H.E. Day (1986), Foreword: Comparison and Consensus of Classifications, *J. of Classification* 3, n°2, 183-185.

W.H.E. Day, F.R. McMorris (1985), A Formalization of Consensus Index Methods, *Bull. of Math. Biol.* 47, 215-229.

W.H.E. Day, R.S. Wells (1984) Extremes in the complexity of computing metric distances between partitions, *IEEE Trans. Pattern Anal. and Mach. Intel.* PAMI-6, 69-73.

A. de Queiroz (1993), For consensus (sometimes), *Syst. Biol.* 42, 368-372.

A. de Queiroz, M.J. Donoghue, J. Kim (1995), Separate versus combined analysis of phylogenetic evidence, *Annual Review of Ecology and Systematics* 26, 657-681.

J.S. Farris, A.G. Kluge, M.J. Eckart (1970), A numerical approach to phylogenetic systematics, *Syst. Zool., 19*, 172-189.

C.R. Finden, A.D. Gordon (1985), Obtaining Common Pruned Trees, *J. of Classification* 2, 255-276.

M. Fréchet (1949), *Réhabilitation de la notion statistique de l'homme moyen,* Les Conférences du Palais de la Découverte, Paris.

W. Goddard, E. Kubicka, G. Kubicki, F.R. McMorris (1995), Agreement subtrees, metrics and consensus for labeled binary trees, in *Partitioning Data Sets*, I. Cox, P. Hansen, B. Julesz, eds., DIMACS Series, vol. 19, AMS, Providence, R.I., 1995, pp. 97-104.

A.D. Gordon (1979), A measure of the agreement between rankings, *Biometrika* 66, 7-15.

A.D. Gordon (1980), On the assessment and comparison of classifications, in : R. Tomassone, ed., *Analyse de données et informatique*, Le Chesnay, INRIA, pp. 149-160.

A.D. Gordon (1981), *Classification: methods for the exploratory analysis of multivariate data*, Chapman and Hall, London.

A.D. Gordon (1986), Consensus Supertrees: the synthesis of rooted trees containing overlapping sets of labeled leaves, *J. of Classification* 3, n°2, 335-348.

A.D. Gordon (1987), A review of Hierarchical Classification, *J. of the Royal Statist. Soc. (series A)* 150, 119-137.

A.D. Gordon (1996), Hierarchical Classification, in: P. Arabie, L.J. Hubert, G. De Soete, eds., *Clustering and Classification*, World Scientific Publ., River Edge, NJ, pp. 65-121.

M. Hendy, M.A. Steel, D. Penny, I.M. Henderson (1988), Families of Trees and Consensus, in H.H.Bock, ed., *Classification and related methods in data analysis*, North-Holland, Amsterdam, pp. 355-362.

M.F. Janowitz (1978), An order theoretic model for cluster analysis, *SIAM J. Appl. Math.* 37, 55-72.

S. C. Johnson (1967), Hierarchical clustering schemes, *Psychometrika 32*, 241-254.

E. Kubicka, G. Kubicki, F.R. McMorris (1992), On agreement subtrees of two binary trees, *Congressus Numerantium* 88, 217-224.

E. Kubicka, G. Kubicki, F.R. McMorris (1995), An algorithm to find agreement subtrees, *J. of Classification* 12, 91-99.

S. Lanyon (1993), Phylogenetic frameworks: towards a firmer foundation for the comparative approach, *Biological J. of the Linnean Society* 49, 45-61.

F.J. Lapointe (1997), How to validate phylogenetic trees? A stepwise procedure, preprint.

F.J. Lapointe, G. Cucumel (1997), The average consensus procedure: combination of weighted trees containing identical or overlapping sets of objects, *Syst. Biol.* 46, 306-312.

F.J. Lapointe, P. Legendre (1990), A statistical framework to test the consensus of two nested classifications, *Syst. Zool.* 39, 1-13.

F.J. Lapointe, P. Legendre (1992a), A statistical framework to test the consensus among additive trees (cladograms), *Syst. Biol.* 41, 158-171.

F.J. Lapointe, P. Legendre (1992b), Statistical significance of the matrix correlation coefficient for comparing independent phylogenetic trees, *Syst. Biol.* 41, 378-384.

F.J. Lapointe, P. Legendre (1995), Comparison tests for dendrograms: a comparative evaluation, *J. of Classification* 12, 265-282.

B. Leclerc (1984), Efficient and binary consensus functions on transitively valued relations, *Math. Social Sci.* 8, 45-61.

B. Leclerc (1991), Agregation of fuzzy preferences: a theoretic Arrow-like approach, *Fuzzy Sets and Systems* 43 (3), (291-309).

B. Leclerc (1993), Lattice valuations, medians and majorities, *Discrete Math.* 111, 345-356.

B. Leclerc, G. Cucumel (1987), Consensus en classification : une revue bibliographique, *Math. Sci. hum.* 100, pp. 109-128.

B. Leclerc, B. Monjardet (1995), Latticial theory of consensus, in *Social choice, Welfare and Ethics*, V. Barnett, H. Moulin, M. Salles & N. Schofield, eds, Cambridge University Press, 145-159.

L.P. Lefkovitch (1985), Euclidean consensus dendrograms and other classification structures, *Math. Biosciences* 74, 1-15.

T. Margush, F.R. Mc Morris (1981), Consensus *n*-trees, *Bull. Math. Biology* 43, 239-244.

F.R. McMorris (1985), Axioms for consensus functions on undirected phylogenetic trees, *Math. Biosciences* 74, 17-21.

F.R. McMorris (1990), The median procedure for *n*-Trees as a Maximum Likelihood Method, *J. of Classification* 7, 77-80.

F.R. McMorris, D.B. Meronk, D.A. Neumann (1983), A view of some consensus methods for trees, *in* J. Felsenstein, ed., *Numerical Taxonomy*, Berlin, Springer-Verlag, pp. 122-126.

F.R. McMorris, H.M. Mulder, F.S. Roberts (1996), The median function on median graphs and median semilattices, *Discrete Applied Math.*, to appear.

F.R. McMorris, D.A. Neumann (1983), Consensus functions defined on trees, *Math. Soc. Sci.* 4, 131-136.

F.R. McMorris, R.C. Powers (1993), Consensus functions on trees that satisfy an independance axiom, *Discrete Applied Math.* 47, 47-55.

F.R. McMorris, R.C. Powers (1996), Intersection rules for consensus hierarchies and pyramids, in *Ordinal and Symbolic Data Analysis* (E. Diday et al., eds.), Berlin, Springer-Verlag, pp. 301-308.

F.R. McMorris, R.C. Powers (1995), The median procedure in a formal theory of consensus, *SIAM J. Disc. Math.* 14, 507-516.

F.R. McMorris, R.C. Powers (1997), The median function on weak hierarchies, in *Mathematical hierarchies and Biology* (B. Mirkin, F.R. McMorris, F. Roberts, A. Rzhetsky, eds.), DIMACS Series, vol. 37, AMS, Providence, RI, 1997, pp. 265-269.

F.R. McMorris, M.A. Steel (1994), The complexity of the median procedure for binary trees, in *New Approaches in Classification and Data Analysis* (E. Diday *et al.*, eds), *Studies in Classification, Data Analysis and Knowledge Organization*, Berlin, Springer-Verlag (1994), 136-140.

B. Monjardet (1980), Théorie et application de la médiane dans les treillis distributifs finis, *Annals of Discrete Math.* 9, 87-91.

B. Monjardet (1990), Arrowian characterizations of latticial federation consensus functions, *Math. Soc. Sci.* 20, 51-71.

G. Nelson (1979), Cladistic Analysis and Synthesis: principles and definitions, with a historical note on Adanson's *Familles des plantes* (1763-1764), *Syst. Zool.* 28, 1-21.

G. Nelson (1993), Why crusade against consensus? A reply to Barrett, Donoghue and Sober, *Syst. Biol.* 42, 215-216.

D.A. Neumann (1983), Faithful Consensus Methods for *n*-Trees, *Math. Biosci.* 63, 271-287.

D.A. Neumann, V.T. Norton Jr. (1986), On lattice consensus methods, *J. of Classification* 3, 225-255.

R.D.M. Page (1989), Comments on component-compatibility in historical biogeography, *Cladistics* 5, 167-182.

R.D.M. Page, M.A. Charleston (1997), Reconciled trees and incongruent gene and species trees, in *Mathematical hierarchies and Biology* (B. Mirkin, F.R. McMorris, F. Roberts, A. Rzhetsky, eds.), DIMACS Series, vol. 37, AMS, Providence, RI, 1997, pp. 57-70.

C. Philips, T.J. Warnow (1996), The asymmetric median tree - A new model for building consensus trees, *Discrete Applied Math.* 71, 311-335.

R.C. Powers (1995), Intersection rules for consensus *n*-trees, *Appl. Math. Lett.* 8, 51-55.

A. Purvis (1995a), A modification to Baum and Ragan's method for combining phylogenetic trees, *Syst. Biol.* 44, 251-255.

A. Purvis (1995b), A composite estimate of primate phylogeny, *Phil. Trans. of the Royal Soc. of London (B)* 348, 405-421.

M.A. Ragan (1992), Phylogenetic inference based on matrix representation of trees, *Molecular Phylogenetics and Evolution* 1, 53-58.

A.G. Rodrigo (1993), A comments on Baum's method for combining phylogenetic trees, *Taxon* 42, 631-636.

F.J. Rohlf (1982), Consensus indices for comparing classifications, *Math. Biosci.* 59, 131-144.

F. Ronquist (1996), Matrix representation of trees, redundancy and weighting, *Syst. Biol.* 45, 247-253.

M.J. Sanderson (1989), Confidence limits on phylogenies: the bootstrap revisited, *Cladistics* 5, 113-129.

M. Schader (1981), *Scharfe und unscharfe klassification qualitative daten*, Athenäum, Königstein.

K. Shao, F.J. Rohlf (1983), Sampling distribution of consensus indices when all bifurcation trees are equally likely, *in* J. Felsenstein, ed., *Numerical Taxonomy*, Berlin, Springer-Verlag, pp. 132-136.

K. Shao, R.R. Sokal (1986), Significance tests of consensus indices, *Syst. Zool.* 35, 582-590.

R.R. Sokal, F.J. Rohlf (1981), Taxonomic Congruence in the *Leptopodomorpha* Reexamined, *Syst. Zool.* 30, 309-325.

M.A. Steel (1988), Distribution of the symmetric difference metric on phylogenetic trees, *SIAM J. Discrete Math.* 9, 91-116.

M.A. Steel (1992), The complexity of reconstructing trees from qualitative characters and subtrees, *J. of Classification* 9, 91-116.

R. Stinebrickner (1984a), *s*-Consensus Trees and Indices, *Bull. of Math. Biology* 46, 923-935.

R. Stinebrickner (1984b), An extension of intersection methods from trees to dendrograms, *Syst. Zool.* 33 (4), 381-386.

R. Stinebrickner (1986) *s*-Consensus Index Method: an additional axiom, *J. of Classification* 3, n°2, 319-327.

W. Vach (1994), Preserving consensus hierarchies, *J. of Classification* 11, 59-77.

W. Vach, P.O. Degens (1988), The system of common lower neighbors of a hierarchy, *in* H.H. Bock, ed., *Classification and related methods of data analysis*, Amsterdam, North Holland, pp. 165-172.

M. Wilkinson (1994), Common cladistic information and its consensus representation: reduced Adams and reduced cladistic consensus trees and profiles, *Syst. Biol.* 43, 343-368.

M. Wilkinson (1996), Majority-rule reduced consensus trees and their use in bootstrapping, *Molecular Biology and Evolution* 13, 437-444.

H.P. Young (1988), Condorcet Theory of Voting, *Amer. Political Science Rev.* 82, 1231-1244.

The Information Content of Consensus Trees

Joseph L. Thorley[1], Mark Wilkinson[1,2] and Mike Charleston[3]
1. School of Biological Sciences, University of Bristol, Bristol BS8 1UG, UK
2. Department of Zoology, Natural History Museum, London SW7 5BD, UK
3. Department of Zoology, University of Oxford, Oxford 0X1 3PS, UK

Abstract: Phylogenetic Information Content, a class of measures of the information provided by consensus trees based on the number of permitted resolutions of the consensus, is introduced. A formula for the number of permitted resolutions of Adams consensus trees is derived and a proof given. We argue that maximising PIC measures provides a sensible criterion for choosing among alternative consensus trees and we illustrate this for consensus trees of cladograms.

Keywords: Consensus, Phylogeny, Adams resolutions, Information Content.

1. Introduction

Consensus trees (CTs) are used by systematic biologists, in many different contexts, to summarise graphically the agreement between multiple fundamental phylogenetic trees. The development of many alternative consensus methods raises the issue of choice between methods and/or tree(s). In this paper we derive a class of measures of the information provided by phylogenetic trees, argue that in certain contexts the optimal CTs are those which convey the most information, and illustrate the approach.

2. Phylogenetic Information Content

In order to convey information a CT must prohibit a subset of the possible phylogenetic relationships (Mickevich and Platnick, 1981; Wilkinson, 1994). Given that the information (I) conveyed by an event is

$$I = -\log \frac{\text{probability of the event}}{\sum \text{probabilities of all possible events}} \qquad (1)$$

where the base of the logarithm determines the unit of information (e.g. $\log_2$- *bits*; ln - *nats*), then, under the assumption that all possible phylogenies are bifurcating and equiprobable, the Phylogenetic Information Content (PIC) of a tree is

$$\mathrm{PIC} = -\log \frac{\text{number of permitted bifurcating trees } (n_\mathrm{R})}{\text{number of possible bifurcating trees } (n_\mathrm{T})}. \tag{2}$$

Because the concept of a phylogenetic tree is a general one, PIC defines a class of tree information measures with specific measures existing for each type of phylogenetic tree. In this paper, we focus exclusively on cladograms, which are the *n*-trees of Bobisud and Bobisud (1972), and the corresponding PIC measure which we refer to as Cladistic Information Content (CIC).

For a CT to provide phylogenetic information, n_R must be less than n_T, i.e., it must be possible to deduce from the CT alone which of the possible trees could not have been represented among the fundamentals. We refer to CTs which, unless totally unresolved, always fulfil this condition as *prohibitive*. *Permissive* CTs are those which permit all possible trees (i.e. $n_\mathrm{R} = n_\mathrm{T}$) irrespective of their resolution. In order to be prohibitive, CTs must be strict *sensu* Wilkinson (1994), i.e., their groupings must represent a particular type of phylogenetic relationship (components, *n*-taxon statements or nestings) that occurs in all the fundamentals.

3. Calculating CIC

The number of possible bifurcating rooted cladograms (n_T) is a function of the number of leaves *n* and is given by RB(*n*)

$$RB(n) = (2n-3)!!. \tag{3}$$

Rohlf (1982) provided equation (4) for calculating the number of permitted bifurcating trees (n_R) for strict CTs of rooted cladograms whose groupings represent components or other *n*-taxon statements, i.e., the CTs produced by the strict component (Sokal and Rohlf, 1981), strict Reduced Cladistic (RCC), Reduced Adams (RAC), Disqualifier-Faithful (DF) (Wilkinson, 1994), and Largest Common Pruned (LCP) (Gordon, 1980) consensus methods.

$$n_R = \prod_{i \in V(T)} RB(d_i) \tag{4}$$

where d_i is the number of vertices immediately descendant from vertex *i* in the set of vertices *V(T)* of tree *T*.

The number of permitted resolutions (n_R) of an Adams CT (Adams, 1972), the only strict CT to capture nestings (Adams, 1986), is

$$n_R = \mu_{root} \tag{5}$$

where

$$\mu_i = \sum_{(A,B)\text{ a split of } D_i} \frac{\left(\prod_{j \in D_i} \mu_j\right) RB(g_A) RB(g_B)}{\prod_{j \in D_i} RB(h_j)} \tag{6}$$

where D_i is the set of vertices immediately descendant from vertex i, a *split* of a set D, is in this case, a partition of D into two non-empty, mutually exclusive subsets whose union is D, g_A is the number of leaves descendant from subset A of $V(T)$, and h_j is the number of leaves in the jth subtree descendant from vertex i. The proof of this is given in the appendix.

4. Choosing a CT

The production of a consensus profile, a set of CTs, by several consensus methods means that each method is not synonymous with a single CT. Furthermore, the information content of each CT depends upon both the properties of the method and the characteristics of the particular fundamentals under consideration. Hence our emphasis is on the choice of CT as opposed to method, although the question of which method(s) will produce the most informative CT and sets of CT(s) is also discussed.

In Figure 1, the strict component CT is completely unresolved, and therefore provides no information, despite the occurrence of considerable agreement among the fundamentals. Insensitivity of the strict component CT has been well documented (e.g. Adams, 1986; Swofford, 1991). The Adams CT, in contrast, is well resolved and provides CIC = 5.159 nats of information. However, the groupings in the Adams CT represent nestings and thus the polytomies of this tree are more permissive than those in CTs whose groupings represent n-taxon statements. The strict RCC 1 and strict RCC 2 CTs provide more information than the Adams despite being less resolved and/or including fewer taxa.

The LCP, RAC and strict RCC consensus methods overcome the problems of insensitivity and ambiguity that affect the strict component and the Adams respectively, by the exclusion from CTs of taxa whose variable position among the fundamentals prevents unambiguous summary of the relationships common to the remaining taxa. Multiple CTs are produced when, as is usually the case, there is more than one way to exclude taxa to achieve unambiguity.

Figure 1: Two fundamental cladograms (a,b), and their strict CTs [Modified and extended from Adams, 1972; Wilkinson, 1994]

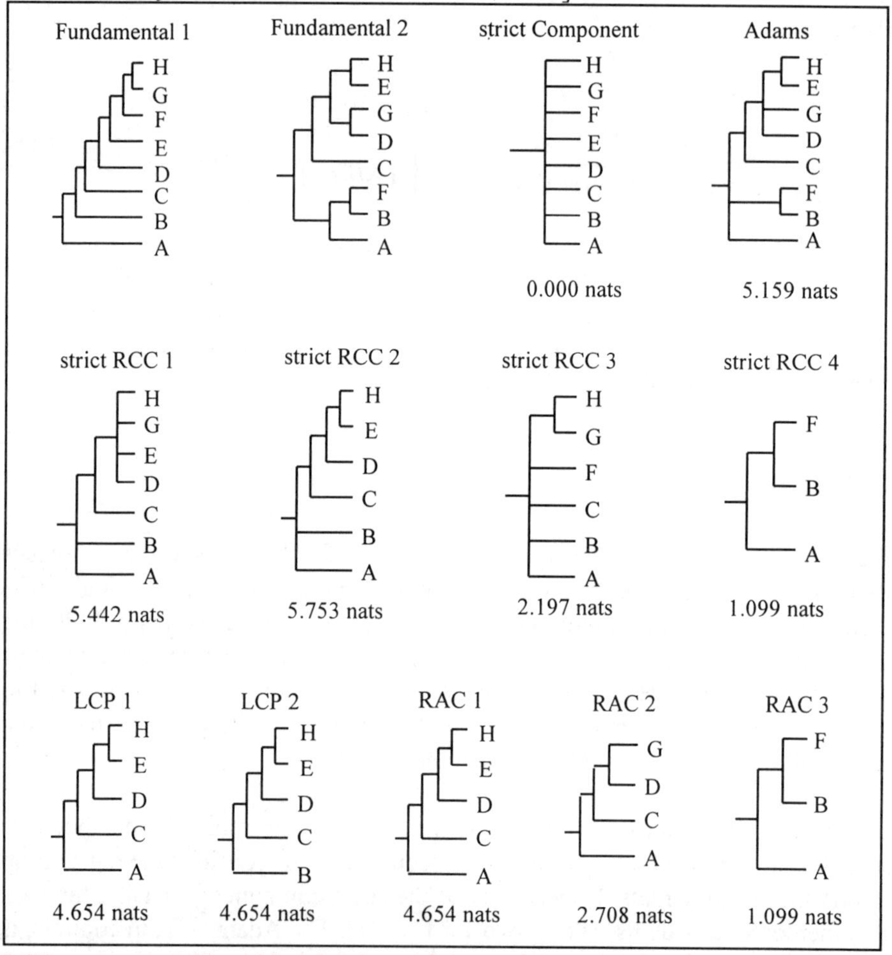

LCP trees are produced by pruning single taxa from the fundamental trees until they are rendered identical. RAC trees are produced by pruning branches from the polytomies of Adams CTs until the polytomies can be correctly interpreted as *n*-taxon statements. The strict RCC method first identifies the complete set of non-redundant *n*-taxon statements common to all the fundamentals. These *n*-taxon statements are then represented graphically through the production of a consensus profile.

Because of the way the RCC profile is constructed, no LCP or RAC tree can ever provide more information than the most informative RCC tree (e.g. Figure 1). Given that the strict component CT is included in the strict RCC profile unless it is completely unresolved, then choice of the most informative CT must be between the Adams CT or a tree from the RCC profile. In this example, strict RCC 2 maximises CIC. An example in which the Adams CT is the most informative CT is provided by Adams (1986; Figure 2).

5. Choosing a set of CTs

CIC can also be used to select the set of CTs which provide the most cladistic information. However, the occurrence of the same information in more than one tree means that the combined CIC of a set of CTs can be less than the sum of their individual CICs. If two strict CTs, both representing *n*-taxon statements, have the relationship that all taxa present in one of the trees also occur in the other, then the information common to the two CTs is the cladistic information content of the single CT in their strict RCC profile. However, if this relationship does not hold, then the only currently available way to determine combined CIC is with a brute force approach (i.e. compare every single possible tree with the set of CTs to determine whether it is permitted). The problem of selecting a set of CTs from a consensus profile so as to maximise combined information content appears to be computationally complex.

Applying the brute force approach to the strict RCC profile in figure 1, we find that the pair of CTs which maximise combined cladistic information is {RCC 2, RCC 3} (CIC = 6.954 nats). RCC 1 provides no information that is not already conveyed by RCC 2 and RCC 3, whilst conversely, all the information provided by RCC 4 is non-redundant with respect to the other three trees in the profile. Thus the combined cladistic information content of the entire strict RCC profile is 8.053 (6.954 + 1.099) nats. Because the strict RCC trees graphically represent all *n*-taxon statements common to all the fundamentals, the LCP (combined CIC = 5.753 nats) and RAC (combined CIC = 5.589 nats) profiles will never provide more information than the strict RCC profile. Furthermore, all the information provided by these two alternative profiles will be represented in the RCC profile. In fact, only the Adams CT can convey cladistic information that is non-redundant with respect to the strict RCC profile.

6. Discussion

Measures of information content provide a basis for choosing amongst alternative strict CTs and methods. Due to the probabilistic nature of information, it may also be possible to calculate the phylogenetic information content of majority-rule and other CTs with groupings that represent relationships occurring in less than 100% of the fundamentals.

Although this paper focuses solely on cladistic information, dendrograms, which are equivalent to cladograms with internally ranked nodes, are another very important type of phylogenetic tree. Whilst consensus methods which take into account the rank of internal nodes have been developed (Neumann, 1983; Stinebrickner, 1984) they suffer from problems of ambiguity or are permissive. Prohibitive dendritic consensus methods are currently under development by the authors.

The utility of our measure of phylogenetic tree information content extends far beyond the selection of CT(s). For example, a natural measure of tree similarity is the information common to the set of trees (Mickevich, 1978). Conversely, a measure of tree dissimilarity could be based on the symmetric difference of the information provided by each tree. Thus, PIC defines a class of tree similarity (and dissimilarity) measures for studies of congruence that can also be normalised to produce Consensus Indices (Mickevich, 1978).

Appendix

Proof of equation 6.

We need the following simple lemma:

Suppose we have a forest of k rooted binary trees $T_1,\ldots,T_k$ with leaf sets $L_1,\ldots,L_k$ respectively, such that $\bigcup_{i=1}^{k} L_i = \{1,\ldots,n\}$, and $L_i \cap L_j = \varnothing$ for all $1 \le i < j \le k$. Let the number of leaves in each tree T_i be a_i. The number of rooted binary trees T^* with leaf set $\{1,\ldots,n\}$ which contain as subtrees each of the T_i is given by

$$N\left(n;\{a_i\}_{i=1}^{k}\right) = \frac{RB(n)}{\prod_{i=1}^{k} RB(a_i)}.$$

Proof: It is well known that the number of rooted binary trees with n leaves is $(2n-3)!! = (2n-3)(2n-5)\ldots(3)(1) = (2n-3)!2^{2-n}/(n-2)!$. One way in which we may generate all the possible rooted binary trees is by successively adding leaves $1,\ldots,n$ to each of the edges of the tree. By adding a_2 leaves in a similar way to tree T_1 we can create all the possible trees with (a_1+a_2) leaves, containing T_1 as a subtree, in $(2a_1-1)(2a_1+1)\ldots(2(a_1+a_2)-3) = RB(a_1+a_2)/RB(a_1)$ ways. Similarly, the number of trees with (a_1+a_2) leaves containing T_2 as a subtree is $RB(a_1+a_2)/RB(a_2)$. The above argument is independent of the underlying trees T_1 and T_2: ergo the number of trees on $n > (a_1+a_2)$ leaves containing T_1 and T_2 is $RB(n)/(RB(a_1)RB(a_2))$. The argument is easily extended to $T_1,\ldots,T_k$.

Let an *Adams resolution* of a vertex v in a tree T be a resolution of v which is consistent with treating T as an Adams consensus tree. Now let μ_i be the number of resolutions of the subtree rooted at vertex i of a given Adams consensus tree T. Then μ_{root} is the number of Adams resolutions of the complete tree T. We proceed from the tips of T to the root, finding the number of Adams resolutions of each vertex j before we continue to the ancestor of j. Let D_j be the set of vertices which are immediately descendant from vertex j and let $|D_j| = d_j$. Let L_j be the set of leaf vertices in the subtree T_j rooted at j, and $|L_j| = l_j$. Thus $n = l_{root}$ and the degree of vertex j is d_j+1. Lastly let $G(A)$ be the set of leaves descendant from the vertices in the *set* $A \subseteq V(T)$, with cardinality g_A.

Consider vertex $i \in V(T)$, the immediately descendant vertices $v_j \in D_i$ of which have been resolved into trees $T_1,\ldots,T_{d_i}$. According to Nelson and Platnick's Interpretation 2b (Nelson & Platnick, 1980; Wilkinson, 1994), any Adams resolution of vertex i must maintain the branching order of the vertices in $T_1,\ldots,T_{d_i}$, but this resolution has no other constraints. Thus we may take any split (A,B) of D_i and resolve the vertices descendant from A to one side of i and the vertices descendant from $B = D_i \setminus A$ to the other side, while maintaining the branching order within each of the $T_1,\ldots,T_{d_i}$. The number of Adams resolutions at vertex i is thus the sum over all possible splits (A,B) of D_i, of the number of rooted binary trees on leaf set $\bigcup_{j \in D_i} L_j$, containing each of the resolved subtrees immediately descendant from i. The result follows.

References

Adams, E.N. (1972). Consensus techniques and the comparison of taxonomic trees, *Syst. Zool.*, 21, 390-397.

Adams, E.N. (1986). *N*-trees as nestings: complexity, similarity and consensus, *J.Classif.*, 3, 299-317.

Bobisud, H.M., and L.E. Bobisud. (1972). A metric for classification, *Taxon*, 21, 607-613.

Gordon, A.D. (1980). On the assessment and comparison of classifications, in: *Analyse de Donnees et Informatique*, Tomassone, R. (Ed.), Le Chesnay: INRIA, 149-160.

Mickevich, M.F. (1978). Taxonomic congruence, *Syst. Zool.*, 27, 143-158.

Mickevich, M.F., and N.I. Platnick. (1981). On the information content of classifications, *Cladistics*, 5, 33-47.

Nelson, G., and N.I. Platnick. (1980). Multiple branching in cladograms: two interpretations, *Syst. Zool.*, 29, 86-91.

Neumann, D.A. (1983). Faithful consensus methods for *n*-trees, *Mathematical Biosciences*, 63, 271-287.

Rohlf, F.J. (1982). Consensus indices for comparing classifications, *Mathematical Biosciences*, 59, 131-144.

Sokal, R.R., and F.J. Rohlf. (1981). Taxonomic congruence in the Leptopodomorpha re-examined, *Syst. Zool.*, 30, 309-325.

Stinebrickner, R. (1984). An extension of intersection methods from trees to dendrograms, *Syst. Zool.*, 33, 381-386.

Swofford, D.L. (1991). When are phylogeny estimates from molecular and morphological data incongruent?, in: *Phylogenetic analysis of DNA sequences*, Miyamoto, M.M. and Cracraft, J. (Eds.), Oxford Univ. Press, 295-333.

Wilkinson, M. (1994). Common cladistic information and its consensus representation: reduced Adams and reduced cladistic consensus trees and profiles, *Syst. Biol.*, 43, 343-368.

Compatible and Complementary Fuzzy Partitions

Slavka Bodjanova
Department of Mathematics, Texas A&M University-Kingsville,
Kingsville, TX 78363, U.S.A.

Abstract: Fuzzy partitions with large (small) coefficients at the same locations are considered to be compatible. We investigate to what extent two or more fuzzy partitions are compatible and to what extent two fuzzy partitions are complementary. Linguistic expressions *large* and *small* are defined by comparison with a threshold $\alpha \in (0, 1)$. We discuss how to find α at which two or more fuzzy partitions are compatible and how to find α at which two fuzzy partitions are complementary. Then we explain how α-compatibility and α-complementarity can be used in decision making.

Keywords: Compatibility, Complementarity, Decision Making, Fuzzy Partitions, α-Cuts, Aggregation.

1. Introduction

Fuzzy set theory introduced by Lotfi Zadeh (1965) has been an inspiration for many investigators in the theory and practice of classification aimed at development of methods which deal with vague description of categories and with classes with non precise boundaries. During the last two decades fuzzy clustering techniques and methods of fuzzy pattern recognition have received a great deal of attention (Bezdek (1981), Miyamoto (1990), Klir (1995), and others). Bezdek (1981) introduced the following partition spaces associated with a finite set of objects $X = \{x_1, x_2, \dots, x_n\}$:
Hard k–partition space ($k \geq 2$):

$$P_k = \left\{ U \in V_{kn} : u_{ij} \in \{0,1\}; \sum_i u_{ij} = 1 \; for\; all\; j; \sum_j u_{ij} > 0 \; for\; all\; i \right\}, \tag{1}$$

where V_{kn} denotes the set of real $k \times n$ matrices. Fuzzy k-partition space ($k \geq 2$):

$$P_{fk} = \left\{ U \in V_{kn} : u_{ij} \in [0,1]; \sum_i u_{ij} = 1 \; for\; all\; j; \sum_j u_{ij} > 0 \; for\; all\; i \right\}. \tag{2}$$

It is obvious that $P_k \subset P_{fk}$. Bezdek has also defined the degenerate fuzzy and hard k-partition spaces, denoted by P_{fko} and P_{ko} respectively, as the superset of P_{fk} and P_k obtained by the condition $0 \le \sum_j u_{ij}$ for all i. The coefficient u_{ij} in (1) and (2) is the grade of membership of object x_j, $j \in \{1,\ldots,n\} = N_n$ in fuzzy cluster u_i, $i \in \{1,\ldots,k\} = N_k$.

Suppose that we have fuzzy partitions $U, V \in P_{fko}$. If U and V have their large (small) coefficients at the same locations, we say that U and V are compatible. If U has its large (small) coefficients at the locations where V has its small (large) coefficients, then U and V are complementary. When we want to exhibit an element $x_j \in X$ that typically belongs to a fuzzy cluster u_i from U, we may demand its membership value to be large, i.e., greater that some threshold $\alpha \in [0,1]$. For each $\alpha \in [0,1]$ the α-cut of $U \in P_{fko}$ is the matrix $U^\alpha \in V_{kn}$, where $u^\alpha_{ij} = 1$ if $u_{ij} \ge \alpha$ and $u^\alpha_{ij} = 0$ otherwise. Then for each $(i,j) \in N_k \times N_n$: $u_{ij} = \sup_{ij} \min\{\alpha, u^\alpha_{ij}\}$.

We say that fuzzy partitions $U, V \in P_{fko}$ are α-compatible if

$$\frac{\sum_i \sum_j |u^\alpha_{ij} - v^\alpha_{ij}|}{nk} = 0. \tag{3}$$

The partitions $U, V \in P_{fko}$ are α-complementary if $D(U^\alpha, V^\alpha) = 1$. The value of $D(U^\alpha, V^\alpha)$ indicates to what degree U and V are α-complementary. The value of $1 - D(U^\alpha, V^\alpha)$ indicates to what degree U and V are α-compatible. We address the following problems: How to find $\alpha \in [0,1]$ at which given fuzzy partitions are compatible? Does such an α always exist? How to find $\alpha \in [0,1]$ at which two fuzzy partitions are complementary? Does such an α always exist?

The existence of two or more fuzzy partitions of the same set of objects creates a problem: which one to choose for further analysis or application? The choice may depend on several criteria. The last section of our paper describes an approach where the decision is based on information about α-compatibility and α-complementarity of the fuzzy partitions under consideration.

2. Alpha – Compatible Fuzzy Partitions

***Definition** 1 Let $\alpha \in [0,1]$ and $U, V \in P_{fko}$. We say that U and V are α-compatible if and only if $u^\alpha_{ij} = v^\alpha_{ij}$ for all $(i,j) \in N_k \times N_n$.*

Suppose that we have a collection $\Omega = \{U_1,\ldots,U_m\}$ of fuzzy partitions $U_t \in P_{fko}$, $t \in N_m$. We want to find whether there is an $\alpha \in [0,1]$ at which $U_1,\ldots,U_m$ are compatible.

Theorem 1 *Let* $\alpha \in [0,1]$. *Then* $U_1^\alpha = U_2^\alpha = \ldots = U_m^\alpha$ *if and only if there exists a coefficient* $u_{(r)pq}, r \in N_m, (p,q) \in N_k \times N_n$ *such that for all* $(i,j) \in N_k \times N_n$ *hold: for all* $t \in N_m$: $u_{(t)ij} < u_{(r)pq}$ *or for all* $t \in N_m$: $u_{(t)ij} \geq u_{(r)pq}$.

A question arises: when does the coefficient $u_{(r)pq}$ described in Theorem 1 exist? There are two trivial cases:

1. Let $\delta_1 = \min_{ijt}\{u_{(t)ij}\}$. Put $\delta_1 = u_{(r)pq}$. Then for all $(i,j) \in N_k \times N_n$ hold: for all $t \in N_m$ we have that $u_{(t)ij} \geq u_{(r)pq}$. Therefore for $\alpha \in [0,\delta_1]$ we get that $U_1^\alpha = U_2^\alpha = \ldots = U_m^\alpha = [1]$. In this case all coefficients are classified as large.
2. Let $\delta_2 = \max_{ijt}\{u_{(t)ij}\} \neq 1$. Put $\delta_2 = u_{(r)pq}$. Then for all $(i,j) \in N_k \times N_n$ hold: for all $t \in N_m$ we have $u_{(t)ij} < u_{(r)pq}$. Therefore for $\alpha \in (\delta_2, 1]$ we get $U_1^\alpha = U_2^\alpha = \ldots = U_m^\alpha = [0]$. In this case all coefficients are classified as small.

If $\delta_2 = 1$ then $U_1^\alpha = U_2^\alpha = \ldots = U_m^\alpha$ at $\alpha = 1$ if and only if the following hold: If $u_{(s)ij} = 1$ then for all $t \in N_m$: $u_{(t)ij} = 1$.

We say that U^α represents α - complete partition U if and only if for each $j \in N_n$ there exists $i \in N_k$ such that $u_{ij}^\alpha = 1$. This means that each object $x_j \in X$ belongs with a large coefficient of membership (compared with the threshold α) to at least one class $u_i \in U$. Our goal is to find the largest nontrivial $\alpha \in (\delta_1, \delta_2)$, if it exists, at which the fuzzy partitions $U_1,\ldots,U_m$ are α-compatible and $U_t^\alpha, t \in N_m$, represents an α- complete partition. We can accomplish this goal using the method described below.

Method 1

Let $U_1,\ldots,U_m$ be partitions from P_{fko} of a set $X = \{x_1, x_2, \ldots, x_n\}$.
Let $\delta_1 = \min_{ijt}\{u_{(t)ij}\}$ and $\delta_2 = \max_{ijt}\{u_{(t)ij}\}$.

Step 1

Put $\alpha_0 = \min_{jt}\{\max_i u_{(t)ij}\}$. If $\alpha_0 = \delta_1$ or $\alpha_0 = \delta_2$ then stop, because there is only a trivial solution. Else go to *Step 2*.

Step 2

Put $\alpha = \alpha_0$ and construct U_t^α for all $t \in N_m$.

Step 3

Find set *S* of elements $u_{(r)ij}$ from matrices $U_1,\ldots,U_m$ such that $u_{(r)ij}^\alpha = 0$ and there exists $s \in N_m$ such that $u_{(s)ij}^\alpha = 1$. If $S = \varnothing$ then stop, because $U_1^\alpha = U_2^\alpha = \ldots = U_m^\alpha$. Else go to *Step 4*.

Step 4

Put $\alpha_0 =$ minimal element from *S*. If $\alpha_0 = \alpha$ or $\alpha_0 \leq \delta_1$ then stop, because there is only a trivial solution. Else go to *Step 2.*

3. Alpha – Complementary Fuzzy Partitions

***Definition** 2 Let $\alpha \in [0,1]$ and $U,V \in P_{fko}$. We say that U and V are α-complementary if and only if $u_{ij}^{\alpha} + v_{ij}^{\alpha} = 1$ for all $(i,j) \in N_k \times N_n$.*

Suppose that we have fuzzy partitions $U_1, U_2 \in P_{fko}$. We want to find whether there is an $\alpha \in [0,1]$ at which U_1 and U_2 are complementary.

***Theorem** 2 Let $\alpha \in [0,1]$. Then U_1 and U_2 are α-complementary if and only if there exists a coefficient $u_{(r)pq}, r \in N_2, (p,q) \in N_k \times N_n$ such that for all $(i,j) \in N_k \times N_n$ hold: $\min\{u_{(1)ij}, u_{(2)ij}\} < u_{(r)pq} \le \max\{u_{(1)ij}, u_{(2)ij}\}$.*

When does the coefficient $u_{(r)pq}$ described in Theorem 2 exist and how to find it? The answer is given in Method 2 below.

Method 2

Let $U_1, U_2 \in P_{fko}$ be fuzzy partitions of a set $X = \{x_1, x_2, \ldots, x_n\}$.

Step 1

Put $j = 1$.

Step 2

Put $\gamma_{2j} = \min_i \{\max\{u_{(1)ij}, u_{(2)ij}\}\}$. If there is a class $r \in N_k$ such that $\min\{u_{(1)rj}, u_{(2)rj}\} \ge \gamma_{2j}$ then stop, because there is no $\alpha \in [0,1]$ such that U_1 and U_2 are complementary. Else go to *Step 3.*

Step 3

Let $\gamma_{1j} = \max_i \{\min\{u_{(1)ij}, u_{(2)ij}\}\}$. Then for $\alpha \in (\gamma_{1j}, \gamma_{2j}]$ we get $u_{(1)ij}^{\alpha} + u_{(2)ij}^{\alpha} = 1$. Increase *j* to $j+1$. If $j \le n$ then go to *Step 2.* Else go to *Step 4.*

Step 4

Calculate $\cap_j (\gamma_{1j}, \gamma_{2j}]$. If $\cap_j (\gamma_{1j}, \gamma_{2j}] = \varnothing$ then there is no $\alpha \in [0,1]$ such that U_1 and U_2 are complementary. Otherwise for all $\alpha \in \cap_j (\gamma_{1j}, \gamma_{2j}]$ we get $u_{(1)ij}^{\alpha} + u_{(2)ij}^{\alpha} = 1$ for all $(i,j) \in N_k \times N_n$.

4. Application

Suppose that we have a collection of fuzzy partitions $\Omega = \{U_1, \ldots, U_m\}, U_t \in P_{fko}$, $t \in N_m$ of the same set of objects. Different partitions arise because of different classification procedures or because of different opinions of several experts. If we want to find a representative of Ω, the following approach might be useful:

Step 1

We will find all nontrivial $\alpha \in [0,1]$ at which fuzzy partitions from Ω are compatible. We will choose α closest to *1/k* or α closest to the user's recommendation. Because of the compatibility of fuzzy partitions, it make sense to create an aggregate fuzzy partition U_{ag} from $U_1,\ldots,U_m$. Several methods of aggregation are available (Bodjanova (1997)). Then the partition U_{ag} or a partition $U_t \in \Omega$ which is the most similar to U_{ag}, will be chosen as the representative of the collection Ω.

Step 2

If there is no acceptable $\alpha \in [0,1]$ at which the fuzzy partitions from Ω are compatible, we will investigate whether Ω has some pairs of α-complementary fuzzy partitions. Let, for example, U_r and U_s be complementary at an acceptable level of. α. Then we will repeat *Step 1* using sets $\Omega_1 = \Omega - \{U_r\}, \Omega_2 = \Omega - \{U_s\}$ and $\Omega_3 = \Omega - \{U_s, U_r\}$, provided that *card* $\Omega_q \geq 2$ for $q = 1, 2, 3$.

Step 3

If *card* $\Omega_q \leq 1$, or the fuzzy partitions from Ω_q are not compatible at an acceptable level of α, we should not try to find a single representative of Ω. The appropriate characterization of Ω is simply a table of values $D(U_r^\alpha, U_s^\alpha)$ at $\alpha =$ *1/k* or α specified by the user for all $(r, s) \in N_m \times N_m$.

Example

Let *U*, *V*, *W*, *Z* be fuzzy partitions of a set $X = \{x_1, x_2, x_3, x_4, x_5\}$ into 3 classes characterized by matrices

$$U = \begin{pmatrix} 1.00 & 0.80 & 0.20 & 0.20 & 0.10 \\ 0.00 & 0.00 & 0.20 & 0.50 & 0.50 \\ 0.00 & 0.20 & 0.60 & 0.30 & 0.40 \end{pmatrix}, V = \begin{pmatrix} 0.00 & 0.00 & 0.40 & 0.75 & 0.90 \\ 0.30 & 0.50 & 0.40 & 0.10 & 0.10 \\ 0.70 & 0.50 & 0.20 & 0.15 & 0.00 \end{pmatrix},$$

$$W = \begin{pmatrix} 0.70 & 0.70 & 0.22 & 0.15 & 0.20 \\ 0.15 & 0.10 & 0.18 & 0.60 & 0.40 \\ 0.15 & 0.20 & 0.60 & 0.25 & 0.40 \end{pmatrix}, Z = \begin{pmatrix} 0.35 & 0.46 & 0.10 & 0.34 & 0.00 \\ 0.32 & 0.20 & 0.34 & 0.66 & 0.50 \\ 0.33 & 0.34 & 0.56 & 0.00 & 0.50 \end{pmatrix}.$$

It is easy to check that there is no nontrivial α such that *U*, *V*, *W*, and *Z* are compatible. Method 2 reveals that fuzzy partitions *U* and *V* are complementary at $\alpha \in (0.2, 0.3]$ and *V* is α-complementary to partition *W* at $\alpha \in (0.22, 0.25]$. Fuzzy partitions *Z* and *V* are neither α-complementary nor α-compatible at nontrivial α. Therefore we will use the subset $\Omega_3 = \{U, W, Z\}$ and check whether it contains α- compatible fuzzy partitions. Method 1 leads to the conclusion that $U^\alpha = W^\alpha = Z^\alpha$ at $\alpha = 0.35$. Then we can represent Ω_3 by a single representative $U^* \in P_{f3o}$ created by an aggregation of partitions *U*, *W*, *Z*.

For example, the simplest representative U_1^* of Ω_3 is created as the arithmetic mean of U, W and Z, i.e., $u^*_{(1)ij} = (1/3)(u_{ij} + w_{ij} + z_{ij})$. Representative U_2^* is created as the lower bound of Ω_3 with respect to the relation α-sharpness, α=0.35. This means that for small elements we have $u^*_{(2)ij} \leq \min\{u_{ij}, w_{ij}, z_{ij}\}$ and for large elements we have $u^*_{(2)ij} \geq \max\{u_{ij}, w_{ij}, z_{ij}\}$. The matrices of U_1^* and U_2^* are

$$U_1^* = \begin{pmatrix} 0.683 & 0.653 & 0.173 & 0.230 & 0.100 \\ 0.157 & 0.100 & 0.240 & 0.587 & 0.467 \\ 0.160 & 0.247 & 0.587 & 0.183 & 0.433 \end{pmatrix}$$

and

$$U_2^* = \begin{pmatrix} 1.000 & 0.800 & 0.100 & 0.150 & 0.000 \\ 0.000 & 0.000 & 0.1800 & 0.850 & 0.500 \\ 0.000 & 0.200 & 0.720 & 0.000 & 0.500 \end{pmatrix}.$$

It is easy to check that $U_1^{*\alpha} = U_2^{*\alpha} = U^\alpha = W^\alpha = Z^\alpha$ at α=0.35.

References

Bezdek, J. C. (1981). *Pattern Recognition with Fuzzy Objective Function Algorithms,* Plenum Press, New York.

Bodjanova, S. (1997). Aggregation Operations on Fuzzy Partitions, in: *IFSA '97 Proceedings,* Mares et al. (Eds.), Academia, Prague, 181-186.

Klir, J. G. & Youan, B. (1995). *Fuzzy Sets and Fuzzy Logic: Theory and Applications*, Prentice Hall, Englewood Cliffs.

Miyamoto, S. (1990). *Fuzzy Set in Information Retrieval and Cluster Analysis*, Kluwer, Dordrecht.

Zadeh, L. A. (1965). Fuzzy Sets, *Information and Control*, 8, 3, 338-353.

Two Methods of Fuzzy c-Means and Classification Functions

Sadaaki Miyamoto and Kazutaka Umayahara
Institute of Information Sciences and Electronics
University of Tsukuba, Ibaraki 305, Japan
email: miyamoto@esys.tsukuba.ac.jp

Abstract: A regularization method using an entropy function is studied and contrasted with the ordinary fuzzy c-means. The way in which two algorithms lead to similar formulas is discussed. Classification functions derived from the two methods, which are naturally obtained when the algorithm of clustering is convergent, are compared. Theoretical properties of the two classification functions are studied.

Keywords: Fuzzy c-means, Regularization, Entropy, Classification Rule

1. Introduction

Recently, a new method of fuzzy c-means has been remarked that is comparable to the ordinary method of Dunn (1974) and Bezdek (1981). The new method may be called an entropy method, as it uses an entropy criterion. Several authors have studied the entropy method from different viewpoints. Rose *et. al* (1990) has studied this method from the statistical mechanics viewpoint. Li and Mukaidono (1995) have considered the same method as a method of fuzzy c-means using the maximum entropy approach. Massulli *et al.* (1997) have applied this method to analysis of medical images.

In contrast with the foregoing formulations, the same entropy method is formulated within the framework of the two-stage optimization of the fuzzy c-means in this paper. Only the objective function to be minimized is different from the standard one. The objective function is *regularized* by adding an entropy-type function. Thus, relationships between the standard and present methods of fuzzy c-means can be disclosed.

Fuzzy classification rules are naturally derived from the clustering algorithms, of which the property is the second subject of the present study. It will be shown that the fuzzy classification functions by the two methods behave differently, and their behaviors are compared at the cluster centers and when the point moves toward the infinity. This result shows remarkable difference between the two methods of clustering and thus is useful in selecting an algorithm appropriate for a particular application.

2. Fuzzy c-Means and Regularization by Entropy

The problem herein is that n objects, each of which is represented by an

h-dimensional real vector $x_k = (x_{k1}, ..., x_{kh}) \in \mathbf{R}^h$, $k = 1, ..., n$, should be divided into c fuzzy clusters. Namely, the grade u_{ik}, $1 \le i \le c$, $1 \le k \le n$, by which the object k belongs to the cluster i should be determined. For each object k, the grade of membership should satisfy the condition of the fuzzy partition:

$$M = \{ (u_{ik}) \ : \ \sum_{i=1}^{c} u_{ik} = 1, \ 1 \le k \le n; \\ 0 \le u_{ik} \le 1, \ 1 \le i \le c, \ 1 \le k \le n \}.$$

The formulation by Bezdek (1981) uses the optimization of the objective function

$$J(U, v) = \sum_{i=1}^{c} \sum_{k=1}^{n} (u_{ik})^m d(x_k, v_i) \tag{1}$$

in which $d(x_k, v_i) = \|x_k - v_i\|_2^2$ is the square of the Euclidean norm between x and v, m is a real parameter such that $m > 1$, v_i is called the center of the fuzzy cluster i, and $U = (u_{ik})$ and $v = (v_1, ..., v_c) \in \mathbf{R}^{ch}$.
No useful procedure for the direct optimization of J by (U, v): $\min_{U \in M, v \in \mathbf{R}^{ch}} J(U, v)$ has been reported, and a two-stage iteration algorithm is used.

A General Algorithm of Fuzzy c-Means (Bezdek, 1981):

(a) Initialize $U^{(0)}$; Set $s = 0$.

(b) Calculate cluster centers $v^{(s)} = (v_1^{(s)}, ..., v_c^{(s)})$:

$$J(U^{(s)}, v^{(s)}) = \min_{v \in \mathbf{R}^{ch}} J(U^{(s)}, v).$$

(c) Calculate $U^{(s+1)}$:

$$J(U^{(s+1)}, v^{(s)}) = \min_{U \in M} J(U, v^{(s)}).$$

(d) If the convergence criterion $\|U^{(s+1)} - U^{(s)}\| < \epsilon$ where $\epsilon > 0$ is a given parameter is satisfied, stop; otherwise $s = s + 1$ and go to (b).

It is well-known (e.g., Bezdek(1981)) that the optimal solutions in (b) and (c) are obtained using the Lagrange multiplier:

$$v_i = \frac{\sum_{k=1}^{n} (u_{ik})^m x_k}{\sum_{k=1}^{n} (u_{ik})^m}, \qquad u_{ik} = \left[\sum_{j=1}^{c} \left(\frac{d(x_k, v_i)}{d(x_k, v_j)} \right)^{\frac{1}{m-1}} \right]^{-1} \tag{2}$$

(Remark that when $x_k = v_i$, put $u_{ik} = 1$).

In contrast to the above standard method, the present method uses the following objective function.

$$J(U,v)=\sum_{i=1}^{c}\sum_{k=1}^{n}u_{ik}d(x_k,v_i)+\lambda^{-1}\sum_{k=1}^{n}\sum_{i=1}^{c}u_{ik}\log u_{ik} \tag{3}$$

in which the second term $\sum_{k=1}^{n}\sum_{i=1}^{c}u_{ik}\log u_{ik}$ is the regularizing functional that fuzzifies the membership matrix.

The two-stage algorithm is applied for the new objective function. The optimal solutions in (b) and (c) are derived by using the Lagrange multiplier:

$$\hat{v}_i=\frac{\sum_{k=1}^{n}u_{ik}x_k}{\sum_{k=1}^{n}u_{ik}},\qquad \hat{u}_{ik}=\frac{e^{-\lambda d(x_k,v_i)}}{\sum_{j=1}^{c}e^{-\lambda d(x_k,v_j)}}. \tag{4}$$

where the optimal solutions in the entropy method are written as $\hat{v}_i$ and $\hat{u}_{ik}$ to distinguish them with v_i and u_{ik} in the standard method.

Although the two membership matrices appear to be quite different, they have close relationships. To see this, Let $\bar{d}_{ik}=d(x_k,v_i)$ and $\beta=\frac{1}{m-1}>0$. Since $a^z=e^{z\log a}$,

$$u_{ik}=[\sum_{j=1}^{c}(\frac{\bar{d}_{ik}}{\bar{d}_{jk}})^{\beta}]^{-1}=[\sum_{j=1}^{c}e^{\beta(\log\bar{d}_{ik}-\log\bar{d}_{jk})}]^{-1}=\frac{e^{-\beta\log\bar{d}_{ik}}}{\sum_{j=1}^{c}e^{-\beta\log\bar{d}_{jk}}}.$$

Thus, membership matrices in (2) and (4) appear the same by putting $\lambda=\beta$ and $d_{jk}=\log\bar{d}_{jk}$. In particular, the logarithmic transformation $d_{jk}=\log\bar{d}_{jk}$ should be remarked.

3. Classification Functions

It is strange that only a small number of studies mention fuzzy classification functions derived from fuzzy c-means.

Keller *et al.* (1985) discuss fuzzy k-nearest neighbor classification rule and fuzzy prototype classification rule. The latter has a most natural classification function that has been derived from the clustering algorithm. Namely, the function for fuzzy classification rule F_i for the class i is

$$F_i(x)=\left[\sum_{j=1}^{c}\left(\frac{d(x,v_i)}{d(x,v_j)}\right)^{\frac{1}{m-1}}\right]^{-1} \tag{5}$$

for the observation x, where v_i is the prototype for the class i. (Remark: when $x=v_i$, we put $F_i(x)=1$ and $F_j(x)=0$ for $j\neq i$.) It is obvious that $\sum_j F_j(x)=1$ and this function has the same form as the optimal membership in the clustering algorithm, since $u_{ik} = F_i(x_k)$ in (2) and (5). The above function is thus obtained from the standard fuzzy c-means algorithm. In other words, when the standard algorithm has converged, the rule of determining optimal membership functions provides the classification rule (5), which interpolates the membership values of the clustered objects: $F_i(x_k) = u_{ij}$.

The above argument means that a new classification rule is derived from the entropy method. Thus we obtain

$$\hat{F}_i(x) = \frac{e^{-\lambda d(x,v_i)}}{\sum_{j=1}^{c} e^{-\lambda d(x,v_j)}}. \tag{6}$$

It is obvious to see that $\hat{F}_1(x), \ldots, \hat{F}_c(x)$ form a fuzzy partition:

$$\sum_{j=1}^{c} \hat{F}_j(x) = 1.$$

Moreover this function interpolates the membership values in (4):

$$\hat{F}_i(x_k) = \hat{u}_{ik}. \tag{7}$$

The proof of (7) is trivial and omitted.

Comparison between these two functions leads us to a number of observations. They are given here without the proofs to save the space, since most of the proofs are straightforward.

First, notice that the maximum value of $F_i(x)$ is at the center v_i by the definition, whereas the maximum value of $\hat{F}_i(x)$ is not necessarily at v_i.

Define regions in $\mathbf{R}^h$ in which the membership for class i is greatest among $j = 1, ..., c$:

$$R_i = \{y \in \mathbf{R}^h : F_i(y) > F_j(y),\ j \neq i\}, \tag{8}$$

$$\hat{R}_i = \{y \in \mathbf{R}^h : \hat{F}_i(y) > \hat{F}_j(y),\ j \neq i\}. \tag{9}$$

Moreover notice that the crisp c-means is based on the nearest prototype allocation. Namely, when the crisp c-means algorithm has converged, the object x_k is allocated to the cluster i of the nearest center v_i: this means that the region in which an object represented by a point y is allocated to the cluster i is given by

$$K_i = \{y \in \mathbf{R}^h : \|y - v_i\| > \|y - v_j\|,\ j \neq i\}.$$

Notice moreover that K_i is a region determined by the Voronoi diagram in $\mathbf{R}^h$ (Preparata and Shamos, 1985). Then we have

Proposition 1.

$$R_i = \hat{R}_i = K_i.$$

We note a remarkable difference between the behaviors of the two functions as $\|x\| \to \infty$.

Proposition 2. Assume that R_i $(= \hat{R}_i)$ is unbounded. Then we can move x toward infinity ($\|x\| \to \infty$) in R_i. Then,

$$\lim_{\|x\|\to\infty} F_i(x) = \frac{1}{c}, \qquad \lim_{\|x\|\to\infty} \hat{F}_i(x) = 1.$$

Another notable property in the entropy regularization is that the classification function has convex level sets.

Proposition 3. $\hat{F}_i(x)$ is a quasi-convex function, in other words, an arbitrary α-cut for $\hat{F}_i(x)$

$$\{\hat{F}_i(\cdot)\}_\alpha = \{x \in \mathbf{R}^h \mid \hat{F}_i(x) \geq \alpha\}$$

is a convex set.

Note the following equation (where $\langle \cdot, \cdot \rangle$ is the scalar product), from which it becomes easier to see that Propositions 2 and 3 hold.

$$\{\hat{F}_i(x)\}^{-1} = \sum_{j=1}^{c} e^{\lambda\{d(x,v_i)-d(x,v_j)\}} = 1 + Const \cdot \sum_{j\neq i} e^{2\lambda\langle x, v_j - v_i\rangle} \tag{10}$$

The following observation is similar to the discussion in Rose *et al.* (1990), but the approach is different from ours.

Let $O(x)$ be a neighborhood of x (e.g., sphere or cube) and assume that $P(O(x)|C_j)$ which is the probability that the observation is in $O(x)$ with the condition C_i is given by a Gaussian distribution, and moreover the prior distribution is assumed to be the equal probability $P(C_j) = c^{-1}$. Then from the Bayes rule, we have

$$\begin{aligned} P(C_i|O(x)) &= \frac{\displaystyle\int_{O(x)} \exp(-\lambda\|y - z_i\|^2)dy}{\displaystyle\sum_{j=1}^{c}\int_{O(x)} \exp(-\lambda\|y - z_j\|^2)dy} \\ &= \frac{\displaystyle\frac{1}{|O(x)|}\int_{O(x)} \exp(-\lambda\|y - z_i\|^2)dy}{\displaystyle\frac{1}{|O(x)|}\sum_{j=1}^{c}\int_{O(x)} \exp(-\lambda\|y - z_j\|^2)dy}. \end{aligned}$$

Proposition 4. As $O(x) \to \{x\}$,

$$P(C_i|O(x)) \to \frac{\exp(-\lambda\|x - z_i\|^2)}{\displaystyle\sum_{j=1}^{c} \exp(-\lambda\|x - z_j\|^2)} = \hat{F}_i(x).$$

This proposition implies that the entropy method allows an interpretation of Bayesian rule of classification.

4. Conclusions

A new concept of regularization in fuzzy clustering has been discussed and a method of regularization by entropy has been introduced. Fuzzy classification functions that interpolate the membership values for the objects have been discussed. The contrast between the two classification functions by the standard fuzzy c-means and the entropy method reveals characteristics of these clustering algorithms. Moreover the entropy method has the Bayesian interpretation.

Future studies include investigations of geometric and analytical properties of the classification functions. Moreover, study for new methods of fuzzy c-means on the basis of the concept of regularization is promising, as the concept of regularization is general and has been found to be useful in different fields of applications.

Acknowledgment

This research has partly been supported by TARA (Tsukuba Advanced Research Alliance), University of Tsukuba.

References

Bezdek, J.C. (1981). *Pattern Recognition with Fuzzy Objective Function Algorithms.* Plenum.

Dunn, J.C. (1974). A fuzzy relative of the ISODATA process and its use in detecting compact well-separated clusters, *J. of Cybernetics*, 3, 32–57.

Li, R.-P. and Mukaidono, M. (1995). A maximum entropy approach to fuzzy clustering, *Proc. of the 4th IEEE Intern. Conf. on Fuzzy Systems (FUZZ-IEEE/IFES'95)*, Yokohama, Japan, March 20-24, 1995, 2227–2232.

Keller, J.M., Gray, M.R., and Givens, J.A. (1985). A fuzzy k - nearest neighbor algorithm, *IEEE Trans., on Syst., Man, and Cybern.*, 15, 580–585.

Masulli, F., Artuso, M., Bogus, P., and Schenone, A. (1997). Fuzzy clustering methods for the segmentation of multivariate medical images, *Proc. of the 7th International Fuzzy Systems Association World Congress (IFSA'97)*, June 25-30, 1997, Prague, Chech, 123–128.

Preparata, F.P. and Shamos, M.I. (1985). *Computational Geometry: An Introduction*, Springer, New York.

Rose, K., Gurewitz, E., and Fox, G. (1990). A deterministic annealing approach to clustering, Pattern Recognition Letters, 11, 589–594.

Dynamic Determination of Mixing Parameters in Fuzzy Clustering

Mina Ryoke[1], Yoshiteru Nakamori[1] and Hiroyuki Tamura[2]

[1]Japan Advanced Institute of Science and Technology
1-1 Asahidai, Tatsunokuchi, Ishikawa 923-1291, Japan
[2]Osaka University, 1-3 Machikaneyama, Toyonaka 560-8531, Japan

Abstract: This paper considers the simultaneous determination of data classification and linear regression models. The clustering criterion introduced in this paper includes two types of mixing parameters to make a balance between two objectives of clustering: the minimization of variances within clusters and the minimization of regression errors. The paper proposes an idea on dynamic determination of those parameters in the clustering process.

Key words: Fuzzy clustering, fuzzy modeling, mixing parameters.

1. Introduction

This paper considers the simultaneous determination of data classification and regression models in the context of fuzzy modeling. This problem was first considered in Hathaway and Bezdek (1993), in which the Fuzzy c-Regression Models (FRCM) were proposed. One of the problems is how to obtain a balance between the data scattering and regression errors within each cluster.

The idea proposed in Dave (1990) can be applied to this problem, where Dave considered an adaptive determination of weights between Fuzzy c-Means (FCM) and Fuzzy c-Lines (FCL). Extending this approach, Ryoke et al. (1996) developed the Adaptive Fuzzy c-Regression Models (AFCR) that determines the mixing parameters dynamically in order to make a balance between the data scattering and regression errors within each cluster.

The purpose of this paper is to explore a little further into the dynamic determination of parameters toward the discovery of interesting relationship and structures in a data set describing various objects and their properties. The paper proposes a new idea on the determination of mixing parameters in the context of fuzzy modeling.

2. Fuzzy Model

The fuzzy model proposed in Takagi and Sugeno (1985) is a nonlinear model that consists of a number of fuzzy rules such as

$$\text{Rule } R_i: \quad \text{if } \boldsymbol{z} \text{ is } F_i, \quad \text{then} \quad \boldsymbol{y} = \boldsymbol{g}_i(\boldsymbol{x}). \tag{1}$$

Here, $\boldsymbol{x} = (x_1, x_2, \cdots, x_s)^\top$ is the vector of consequence variables, $\boldsymbol{z}=(z_1, z_2, \cdots, z_t)^\top$ is the vector of premise variables, and $\boldsymbol{y} = (y_1, y_2, \cdots, y_r)^\top$ is the vector of response variables. Often, there is an intersection between two variable sets $\{x_1, x_2, \cdots, x_s\}$ and $\{z_1, z_2, \cdots, z_t\}$. F_i denotes a fuzzy subset with the membership function $f_i(\boldsymbol{z})$ which has premise parameters to be identified. The regression model $\boldsymbol{g}_i(\boldsymbol{x})$ is linear in regression parameters which should be identified as well.

The degree of confidence of each fuzzy rule is defined by

$$f_i(\boldsymbol{z}) = \prod_{j=1}^{t} \mu_{ij}(z_j;\ p_{ij1}, p_{ij2}, p_{ij3};\ t_{ij1}, t_{ij2}) \tag{2}$$

where μ_{ij} is a membership function defined in Ryoke et al. (1996). The prediction of $\boldsymbol{y}$ is given by

$$\hat{\boldsymbol{y}} = \frac{\sum_{i=1}^{c} f_i(\boldsymbol{z}^*) \cdot \boldsymbol{g}_i(\boldsymbol{x}^*)}{\sum_{i=1}^{c} f_i(\boldsymbol{z}^*)}, \tag{3}$$

where $\boldsymbol{x}^*$ and $\boldsymbol{z}^*$ denote actual inputs, and c is the number of rules.

To develop a fuzzy model mentioned above, the simultaneous determination of data classification and regression models is required. Recently, Phansalkar and Dave (1997) proposed an idea to make a balance between the data scattering and the distances of data from a hyperplane in each cluster. The objective of this paper is slightly different from theirs. This paper considers the balance between data scattering and regression errors.

3. Adaptive Fuzzy c-Regression Models

The AFCR proposed in Ryoke et al. (1996) has two mixing parameters which will be explained later in this section. The original proposal is an extension of the work in Dave (1990) in which a balance between the data scattering and distances of data from a line in each cluster was considered.

This paper modifies the original idea in Ryoke et al. (1996) so that the balance between the data scattering and the errors of any dimensional regression hyperplane can be treated well.

Let U be the fuzzy partition matrix with u_{ik} for (i, k)-entry. The membership grades u_{ik} satisfy the following conditions:

$$u_{ik} \in [0, 1], \quad \sum_{k=1}^{n} u_{ik} > 0, \quad \sum_{i=1}^{c} u_{ik} = 1. \quad 1 \le i \le c,\ 1 \le k \le n \tag{4}$$

where c is the number of clusters and n is the number of data points. Let V be the set of cluster centers: $V = \{\bar{z}_1, \bar{z}_2, \cdots, \bar{z}_c\}$, and let Ω be the set of regression parameters.

The clustering objective function $J(U, V, \Omega)$ is defined by

$$J(U, V, \Omega) = \sum_{k=1}^{n} \sum_{i=1}^{c} (u_{ik})^m L_{ik}(V, \Omega, \alpha_i, \eta_i) \tag{5}$$

where m (> 1) is the fuzzifier and $L_{ik}(\Omega, V, \alpha_i, \eta_i)$ is defined by

$$L_{ik}(\Omega, V, \alpha_i, \eta_i) = \alpha_i E_{ik}(\Omega) + (1 - \alpha_i)\eta_i D_{ik}(V). \tag{6}$$

Here, $E_{ik}(\Omega)$ and $D_{ik}(V)$ are defined as follows:

$$E_{ik}(\Omega) = ||\boldsymbol{y}_k - \boldsymbol{g}_i(\boldsymbol{x}_k; \Omega)||^2, \tag{7}$$

$$D_{ik}(V) = ||\boldsymbol{z}_k - \bar{\boldsymbol{z}}_i||^2. \tag{8}$$

By minimizing the objective function $J(U, V, \Omega)$, the fuzzy partition matrix U, the cluster centers V, and the regression parameters Ω are determined. The first term in the function $L_{ik}(\Omega, V, \alpha_i, \eta_i)$ is related to the square error of the regression model, on the other hand, the second term is related to the square distance of a data point from the cluster center. Here, α_i and η_i are parameters which are changed dynamically in the clustering process. The parameter α_i is introduced to make a balance between the distances of data of the objective variable from the regression hyperplane and the distances of data of the conditional variables from the cluster center within each cluster. The parameter η_i is necessary because the spaces considered in the first and second terms in the function are different.

Let us consider how to determine these clustering parameters. Using the data of the objective variable y_p and all explanatory variables, we define data vectors $\boldsymbol{w}_{kp}$:

$$\boldsymbol{w}_{kp} = (y_{kp}, \boldsymbol{x}_k^\top)^\top, \qquad p = 1, 2, \cdots, r. \tag{9}$$

Let λ_{i1p}, $\cdots$, $\lambda_{i(s+1)p}$ be eigenvalues of the scatter matrix:

$$S_{ip} = \sum_{k=1}^{n} (u_{ik})^m (\boldsymbol{w}_{kp} - \bar{\boldsymbol{w}}_{ip})(\boldsymbol{w}_{kp} - \bar{\boldsymbol{w}}_{ip})^\top, \tag{10}$$

and let ν_{i1}, $\cdots$, ν_{it} be eigenvalues of another scatter matrix:

$$S_i' = \sum_{k=1}^{n} (u_{ik})^m (\boldsymbol{z}_k - \bar{\boldsymbol{z}}_i)(\boldsymbol{z}_k - \bar{\boldsymbol{z}}_i)^\top. \tag{11}$$

The idea in this paper is to determine the parameters α_i and η_i as follows:

$$\alpha_i = \min\{\alpha_{i1}, \alpha_{i2}, \cdots, \alpha_{ir}\}, \quad \alpha_{ip} = \frac{Var[\lambda_{ijp}]}{Var[\lambda_{ijp}] + Var[\nu_{ij}]}, \tag{12}$$

$$\eta_i = \max\{\eta_{i1}, \eta_{i2}, \cdots, \eta_{ip}\}, \quad \eta_{ip} = \frac{\sum_{j=1}^{s+1} \lambda_{ijp}}{\sum_{j=1}^{t} \nu_{ij}}. \tag{13}$$

If the variance of eigenvalues in the space where we are building a linear model is large, there is a chance to develop a good linear model. In this case, we give added weight to the first term in the function. In order to obtain better models for all objective variables, the minimum one is selected for α_i. The sum of eigenvalues is one of the indicators to show how data are scattered. The reason of using the maximum one for η_i is that we want to develop individual linear model in a smaller region.

In this paper, we omit to describe the clustering algorithm which is similar to the ordinary fuzzy clustering algorithm.

4. Numerical Examples

A numerical example is shown in the following.

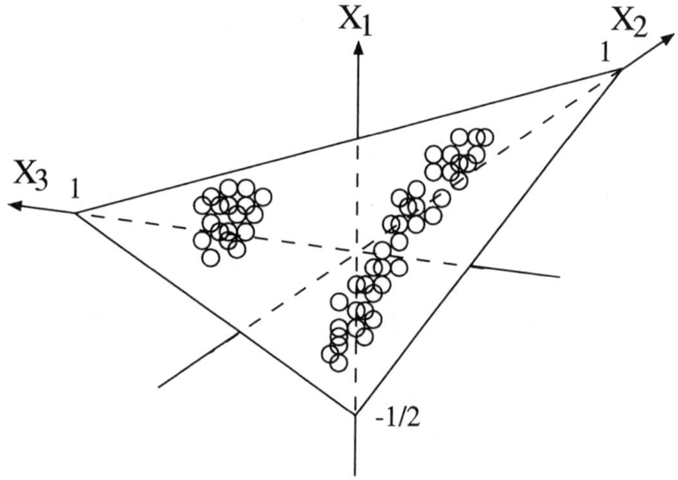

Figure 1: Data Distribution.

In Figure 1, the data are located close to a plane $2X_1 - X_2 - X_3 + 1 = 0$ in the three dimensional space. The data have an error value $\delta \in [-1, +1]$ produced by randomization respectively. Therefore the data are on a plane $2X_1 - X_2 - X_3 + 1 = \delta$ actually.

For such a case, it is difficult to detect two separate clusters because two linear equations expressing two planes are similar. Suppose X_1 is the response variable and X_2 and X_3 are the explanatory variables and they also are variables for which distances between data points are considered. The total number of data points is 64. The desirable number of the clusters is 2.

We can obtain a successful clustering result presented by two kinds of points in Figure 2 by the above mentioned method. The two kinds of points are determined by the element of the fuzzy partition matrix U, and the detail value $\{u_{ik}\} \in U$ are plotted in Figure 3, Let the white points denote cluster 1 and let the black points denote cluster 2. The detail of the result is shown as follows:

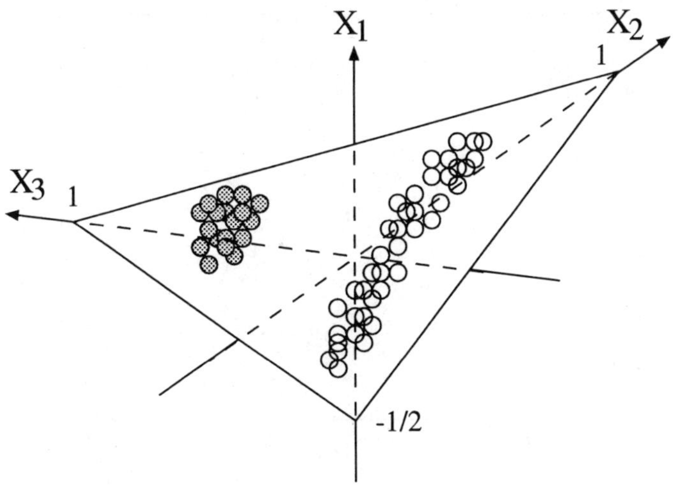

Figure 2: Result of Data Partition.

Result of Numerical Example

Cluster 1 (White Points)

$$X_1 = -0.8408 + 0.6357X_2 + 0.3971X_3$$

$$Final\ Parameters : \alpha_1 = 0.7972, \qquad \eta_1 = 1.538$$
$$Errors : \sum_{k=1}^{n} (u_{1k})^m E_{1k} = 0.6624$$
$$Distances : \sum_{k=1}^{n} (u_{1k})^m D_{1k} = 39.46$$

Cluster 2 (Black Points)

$$X_1 = -0.3613 + 0.4316X_2 + 0.4348X_3$$

$$Final\ Parameters : \alpha_2 = 0.7067, \qquad \eta_2 = 1.055$$
$$Errors : \sum_{k=1}^{n} (u_{2k})^m E_{2k} = 0.8235$$
$$Distances : \sum_{k=1}^{n} (u_{2k})^m D_{2k} = 15.46$$

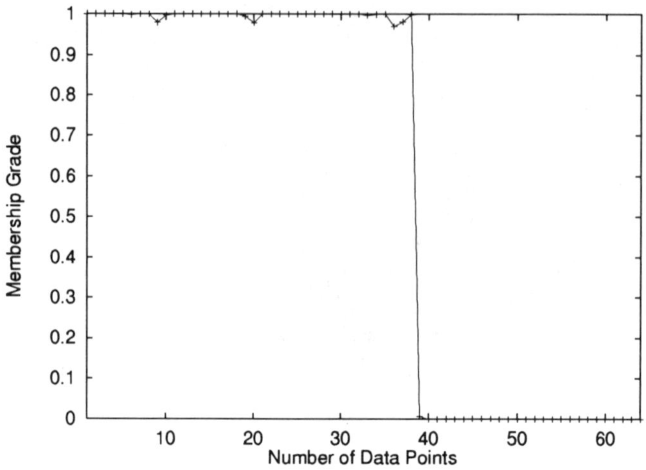

Figure 3: Grade of Membership Function for Cluster 1.

5. Conclusion

In this paper, a dynamic determination procedure of mixing parameters in fuzzy clustering is proposed. These parameters have a role to make a balance between the data scattering and the prediction errors. In building a fuzzy model, we can use the clustering information to identify membership functions of premise variables.

References

Dave, R. N. (1990). An adaptive fuzzy c-elliptotype clustering algorithm, *Proceedings of NAFIPS 90: Quarter Century of Fuzziness*, Vol. I, pp. 9–12.

Hathaway, R. J. & Bezdek, J. C. (1993). Switching regression models and fuzzy clustering, *IEEE Trans. on Fuzzy Systems*, Vol. 1, No. 3, pp. 195–204.

Phansalkar, G. & Dave, R. N. (1997). On generating solid models of mechanical parts through fuzzy clustering, *Proceedings of FUZZ-IEEE'97*, Vol. I, pp. 225–230.

Ryoke, M. et al. (1996). Fuzzy modeling though adaptive fuzzy clustering, *Proceedings of 4th conference of the European Congress on Fuzzy and Intelligent Technologies*, Vol. 2, pp. 1416–1419.

Takagi, T. & Sugeno, M. (1985). Fuzzy identification of systems and its applications to modeling and control, *IEEE Trans. on Systems, Man and Cybernetics*, Vol. SMC–15, No. 1, pp. 116–132.

A Dynamic Additive Fuzzy Clustering Model

Mika Sato-Ilic[1] and Yoshiharu Sato[2]

[1] University of Tsukuba
Tenodai 1-1-1, Tsukuba 305, Japan
e-mail : mika@sk.tsukuba.ac.jp
[2] Hokkaido University
Kita 13, Nishi 8, Kita-ku, Sapporo 060, Japan
e-mail : ysato@main.eng.hokudai.ac.jp

Abstract: This paper presents a dynamic clustering model in which clusters are constructed in order to find the features of the dynamical change.

If the similarity between the objects is observed depending on time or parameters which are satisfying the total order relation, then it is important to capture the change in the results of clustering according to the change in time. In this paper, we construct a model which can represent dynamically changing clusters by introducing the concepts of conventional dynamic MDS (Ambrosi, K. and Hansohm, J., 1987) or dynamic PCA (Baba, Y. and Nakamura, Y., 1997) into the additive clustering model (Sato, M. and Sato, Y., 1995).

Keywords: 3-Way Data, Dynamic MDS, Clustering Model.

1 Introduction

INDSCAL (Carroll, J.D. and Chang, J.J., 1970) or INDCLUS (Carroll, J.D. and Arabie, P., 1983) are typical methods for data with similarities between the objects through time.

Dynamic MDS has been discussed for the dissimilarities which are measured at each of T time periods. The essential meaning of the Dynamic MDS is to capture the insightful, changing nature of the relationship among the objects with respect to time by plotting their path over time. Suppose D as a super-dissimilarity matrix, that is,

$$D = \begin{bmatrix} D^{(11)} & D^{(12)} & D^{(13)} & \cdots & D^{(1T)} \\ D^{(21)} & D^{(22)} & D^{(23)} & \cdots & D^{(2T)} \\ \vdots & \vdots & \vdots & \vdots & \vdots \\ D^{(T1)} & D^{(T2)} & D^{(T3)} & \cdots & D^{(TT)} \end{bmatrix},$$

where $D^{(tt)} = (\delta^t_{rs})$ is the dissimilarity matrix at tth time. The element $\delta^{t,t'}_{rs}$ of the matrix $D^{(tt')}, (t \neq t')$ is the dissimilarity object r at the tth time with object s at the t'th time. The purpose of this method is to analyze the super-dissimilarity matrix in the usual manner of metric or nonmetric multidimensional scaling. Dynamical PCA is also applied to the super-matrix formed by the correlation matrices.

By introducing the idea of the construction specified by a super-dissimilarity matrix into the additive fuzzy clustering model, we propose a dynamical additive fuzzy clustering model.

2 Dynamic Additive Fuzzy Clustering

2.1 Additive Fuzzy Clustering Model

An additive fuzzy clustering model is a model which introduces the concepts of fuzzy clusters and aggregation functions into ADCLUS. This model is defined in (1) and (2) according to the division between symmetric and asymmetric similarities.

(1) Symmetric similarity ($s_{ij} = s_{ji}$): $s_{ij} = \sum_{k=1}^{K} \rho(u_{ik}, u_{jk}) + \varepsilon_{ij}$

(2) Asymmetric similarity ($s_{ij} \neq s_{ji}$):

$$\text{(2-1)} s_{ij} = \frac{1}{K} \sum_{k=1}^{K} \sum_{\ell=1}^{K} w_{k\ell} \rho(u_{ik}, u_{jl}) + \varepsilon_{ij} \qquad \text{(2-2)} s_{ij} = \sum_{k=1}^{K} \gamma(u_{ik}, u_{jk}) + \varepsilon_{ij}.$$

In these models, K denotes the number of clusters and it is assumed that u_{ik} represents the fuzzy grade of which object i belongs to cluster k. Usually u_{ik} belongs to an unit interval, that is $u_{ik} \in [0,1]$. $\rho(u_{ik}, u_{jk})$ is a symmetric aggregation function which satisfies the following conditions:

1. $0 \leq \rho(u_{ik}, u_{jl}) \leq 1, \quad \rho(u_{ik}, 0) = 0, \quad \rho(u_{ik}, 1) = u_{ik}$
2. $\rho(u_{ik}, u_{jl}) \leq \rho(u_{sk}, u_{tl})$ whenever $u_{ik} \leq u_{sk}, u_{jl} \leq u_{tl}$
3. $\rho(u_{ik}, u_{jl}) = \rho(u_{jl}, u_{ik})$

In particular, t-norm (Menger, K., 1942) is well-known as a concrete example which satisfies the above conditions. Originally, t-norm has been defined by K. Menger 1942 (Ref. B. Schweizer and A. Sklar 1983) as a function satisfying a triangle-like inequality in a statistical metric space. Supposing $F(x)$ to be a probability distribution function, Menger has proposed the distance between two points p and q in a set S is defined as a probability, $F_{pq}(x)$, that is, for any real number x, $F_{pq}(x)$ is interpreted as the probability that the distance between p and q is less than x. To introduce the metric property, he defined the inequality $F_{pr}(x+y) \geq t(F_{pq}(x), F_{qr}(y))$ for all $p, q, r \in S$ and all real numbers x, y. In this inequality, the function $t(\cdot,\cdot) : [0,1] \times [0,1] \to [0,1]$ is referred to as a t-norm satisfying the above conditions 1, 2, 3 and:

4. $t(t(x,y),z) = t(x,t(y,z))$. (Associativity)

Most of the t-norms are defined by the use of a generator function as follows: Suppose $f(x)$ is a generator function of t-norms, which is a continuous monotone decreasing function under these conditions:

$$f : [0,1] \to [0,\infty], \quad f(1) = 0.$$

Then $t(x,y)$ is defined by

$$t(x,y) = f^{[-1]}(f(x) + f(y)),$$

where

$$f^{[-1]}(z) = \begin{cases} f^{-1}(z), & z \in [0, f(0)) \\ 0, & z \in [f(0), \infty] \end{cases},$$

i.e. $f^{[-1]}$ is the pseudo inverse of f, and f^{-1} is the usual inverse of f.

$\gamma(u_{ik}, u_{jk})$ is an asymmetric aggregation function and is defined as follows:

$$\gamma(x,y) = f^{[-1]}(f(x) + \phi(x)f(y)).$$

Suppose $f(x)$ is a generating function of t-norm and $\phi(x)$ is a continuous monotone decreasing function satisfying

$$\phi : [0,1] \to [1,\infty], \quad \phi(1) = 1.$$

2.2 Dynamic Additive Fuzzy Clustering Model

The data is observed by the values of similarity with respect to n objects for T times, and the similarity matrix of tth time is shown by $S^{(t)} = \{s_{ij}^{(t)}\}$. Then a $Tn \times Tn$ matrix $\tilde{S}$ is denoted as follows:

$$\tilde{S} = \begin{bmatrix} S^{(1)} & S^{(12)} & S^{(13)} & \cdots & S^{(1T)} \\ S^{(21)} & S^{(2)} & S^{(23)} & \cdots & S^{(2T)} \\ \vdots & \vdots & \vdots & \vdots & \vdots \\ S^{(T1)} & S^{(T2)} & S^{(T3)} & \cdots & S^{(T)} \end{bmatrix}, \tag{2.1}$$

where the diagonal matrix is the $n \times n$ matrix $S^{(t)}$. $S^{(rt)}$ is an $n \times n$ matrix and the element is defined as

$$s_{ij}^{(rt)} \equiv m(s_{ij}^{(r)}, s_{ij}^{(t)}),$$

where $s_{ij}^{(t)}$ is (i,j)th element of the matrix $S^{(t)}$. $m(x,y)$ is an average function from the product space $[0,1] \times [0,1]$ to $[0,1]$ and satisfies the following conditions:

(1) $\min\{x,y\} \le m(x,y) \le \max\{x,y\}$
(2) $m(x,y) = m(y,x)$:(symmetry)
(3) $m(x,y)$ is increasing and continuous

Moreover, from (1), $m(x,y)$ satisfies the following condition:

(4) $m(x,x) = x$:(idempotency)

The examples of the average function are shown in Table 2.1.

Table 2.1. Example of the Average functions

harmonic mean	$\frac{2xy}{x+y}$
geometric mean	$\sqrt{xy}$
arithmetic mean	$\frac{x+y}{2}$
dual of geometric mean	$1-\sqrt{(1-x)(1-y)}$
dual of harmonic mean	$\frac{x+y-2xy}{2-x-y}$

For the element $\tilde{s}_{ij}$ of $\tilde{S}$, we propose a new fuzzy clustering model which is named the dynamic additive fuzzy clustering model and is as follows:

$$\tilde{s}_{ij} = \sum_{k=1}^{K} \rho(u_{i^{(t)}k}, u_{j^{(t)}k}), \quad 1 \leq i, j \leq Tn, \quad i = n(t-1) + i^{(t)}, \tag{2.2}$$

where K is a number of clusters and $u_{i^{(t)}k}$ shows a degree of belongingness of an object i to a cluster k for time t, and $1 \leq i^{(t)} \leq n, \quad 1 \leq t \leq T$.

3 Dynamic Additive Fuzzy Clustering Model for Asymmetric Similarity Data

If the observed similarity is asymmetric, then the model (2.2) cannot be used. In order to apply the asymmetric data to the model (2.2), we have to reconstruct the super-matrix (2.1). Suppose the upper triangular matrix of $S^{(t)}$ is denoted as $S^{(tU)}$ and the lower triangular matrix is $S^{(tL)}$. Then we assume the super-matrix as:

$$\tilde{\tilde{S}} = \begin{bmatrix} S^{(1U)} & S^{(1U)(1L)} & S^{(1U)(2U)} & \cdots & S^{(1U)(TU)} & S^{(1U)(TL)} \\ S^{(1L)(1U)} & S^{(1L)} & S^{(1L)(2U)} & \cdots & S^{(1L)(TU)} & S^{(1L)(TL)} \\ S^{(2U)(1U)} & S^{(2U)(1L)} & S^{(2U)} & \cdots & S^{(2U)(TU)} & S^{(2U)(TL)} \\ \vdots & \vdots & \vdots & \vdots & \vdots & \vdots \\ S^{(TL)(1U)} & S^{(TL)(1L)} & S^{(TL)(2U)} & \cdots & S^{(TL)(TU)} & S^{(TL)} \end{bmatrix},$$

where $s_{ij}^{(rU)(tL)} \equiv m(s_{ij}^{(rU)}, s_{ij}^{(tL)})$, and $s_{ij}^{(rU)}$ is (i,j)th element of the matrix $S^{(rU)}$. Then model (2.2) is rewritten as

$$\tilde{\tilde{s}}_{ij} = \sum_{k=1}^{K} \rho(u_{i^{(c)d}k}, u_{j^{(c)d}k}), \tag{2.3}$$

$$1 \leq i, j \leq 2Tn, \quad i = n(c-1) + i^{(c)}, \quad 1 \leq i^{(c)} \leq n, \quad 1 \leq c \leq 2T.$$

$\tilde{\tilde{s}}_{ij}$ is (i,j)th element of $\tilde{\tilde{S}}$ and $u_{i^{(c)d}k}$ is a degree of belongingness of an object i to a cluster k, and c and d are suffixes for representing time and the direction of the asymmetric proximity data respectively. The relationship is shown as follows:

$$d = \begin{cases} c \equiv 0 \pmod{2} & , \quad d = L \text{ and } t = \frac{c}{2} \\ c \equiv 1 \pmod{2} & , \quad d = U \text{ and } t = \frac{c+1}{2} \end{cases},$$

where $1 \leq t \leq T$. L and U show the lower and upper triangular matrixes, respectively, that is, asymmetric relationship in the same time. t shows tth time, and $c \equiv 0 \pmod{2}$ means that c is congruent to 0 modulo 2.

4 Numerical Example

To demonstrate the applications of model (2.3), we will use data related to a number of high school students who moved from one prefecture to another for matriculation (School Basic Survey, Ministry of Education in Japan, 1992). We define the number between two prefectures i and j as s_{ij}. Usually, the number in which the students moved from prefecture i to j is different from the number from prefecture j to i, so we can write $s_{ij} \neq s_{ji}$, $(i \neq j)$. For considering the total number of students for each prefectures, we used the following transformation:

$$s_{ij} = s_{ij} / \sum_{j=1}^{n} s_{ij}. \quad (i \neq j) \tag{4.1}$$

Self-similarity s_{ii} was avoided in this numerical examination. Using (4.1), we can regard this as an asymmetric similarity data. In the optimization algorithm used in this example, 20 sets of initial values are given by using uniform pseudorandom numbers in the interval $[0, \frac{\pi}{2}]$, and in the end we select the best result. The number of clusters is determined to be 3, which is based on the value of fitness.

We have proposed a clustering method for 3-way data (Sato M. and Sato Y., 1997). This method is to get the optimum clusters which minimizes extended within-class dispersion and shows the difference on times as weights of clusters for every time. The criterion of the method is defined as

$$F(U, \boldsymbol{w}) = \sum_{t=1}^{T} F^{(t)}(U, \boldsymbol{w}^{(t)}),$$

where $F^{(t)}(U, \boldsymbol{w}^{(t)})$ is the goodness of clustering in the t-th time which is given by the least square error,

$$F^{(t)}(U, \boldsymbol{w}^{(t)}) = \sum_{i=1}^{n}\sum_{j=1}^{n}(s_{ij}^{(t)} - \sum_{k=1}^{K} w_k^{(t)} \gamma(u_{ik}, u_{jk}))^2.$$

$s_{ij}^{(t)}$ is an asymmetric similarity between objects i and j at t-th time. The value of the membership functions of K fuzzy subsets, namely, the degree of belonging of each object to the cluster are denoted by

$$U = (u_{ik}), \quad u_{ik} \geq 0, \quad \sum_{k=1}^{K} u_{ik} = 1,$$
$$i = 1, 2, \cdots, n, \quad k = 1, 2, \cdots, K.$$

And we denote the weight of the cluster at t-th time as $\boldsymbol{w}_k^{(t)}, (t = 1, 2, \cdots, T;\ k = 1, 2, \cdots, K)$. The result of weights is shown in Table 4.1.

Table 4.1: Weights of Clusters for Times

	Time		
Cluster	T_1	T_2	T_3
C_1	0.98	0.98	0.98
C_2	0.99	0.99	0.99
C_3	0.67	0.59	0.25

According to the result, we can find that cluster 3 will move through times. Then we used the prefectures which are consisted of cluster 3 and again classified them to three clusters using model (2.3), in order to find the exact movement of these prefectures for the determined clusters. The results are shown in figures 4.1 to 4.6. In these figures, the abscissa shows the prefectures and the ordinate shows the grade to each cluster. From figure 4.3, cluster 3 is consisted of prefectures which are located in the central part of Japan and the central prefecture is Osaka. However, in time 2, Osaka and Okayama prefectures which have large capacity of students moved to cluster 2. In time 3, Okayama prefecture is still in cluster 2, but Osaka prefecture moved again to cluster 3. From this consideration, the large two prefecture's movement during the period might be caused by some influences which determine the unique clusters through the times.

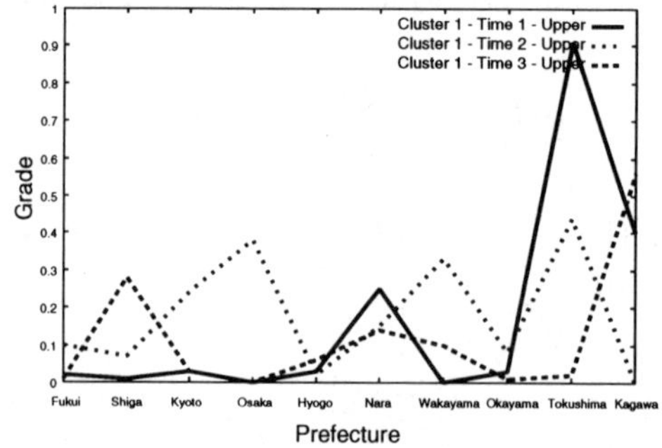

Figure 4.1: Dynamic Changes of Cluster 1 on the Upper Matrix

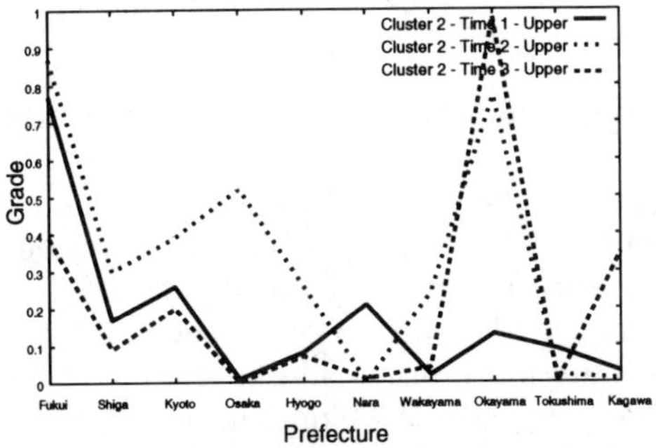

Figure 4.2: Dynamic Changes of Cluster 2 on the Upper Matrix

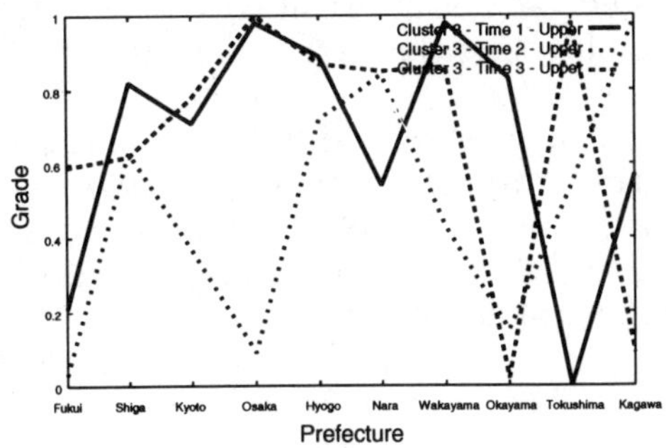

Figure 4.3: Dynamic Changes of Cluster 3 on the Upper Matrix

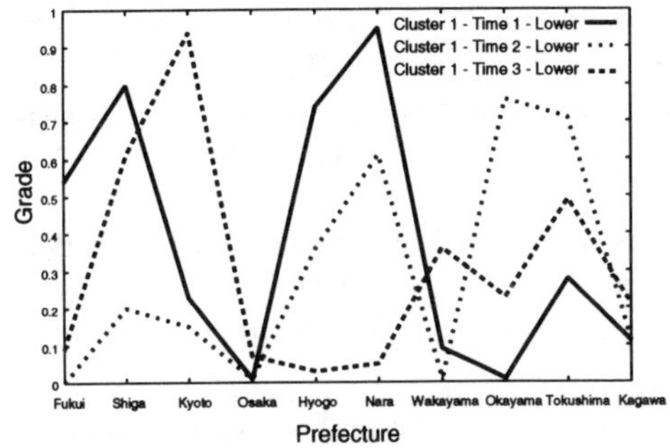

Figure 4.4: Dynamic Changes of Cluster 1 on the Lower Matrix

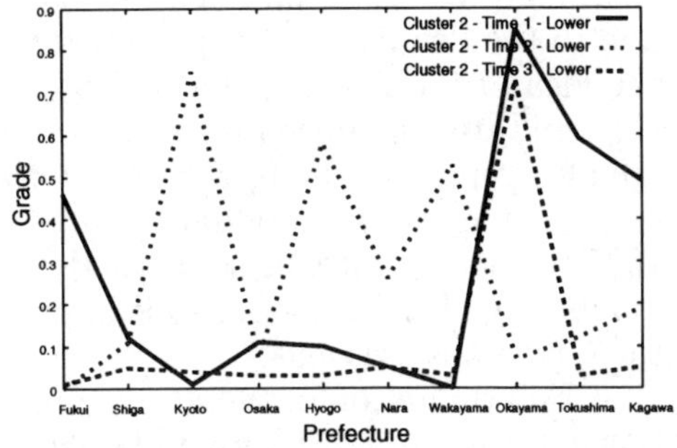

Figure 4.5: Dynamic Changes of Cluster 2 on the Lower Matrix

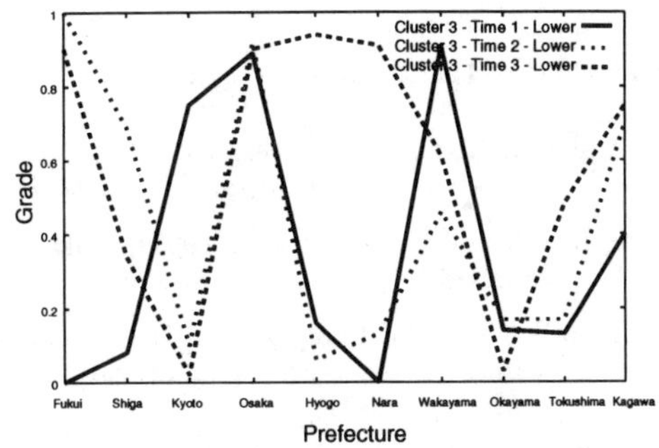

Figure 4.6: Dynamic Changes of Cluster 3 on the Lower Matrix

5 Conclusion

The dynamic fuzzy clustering model is an efficient technique of structural analysis for time dependent multivariate data. This method is effective for describing the dynamic structural change of clusters. Technically, this model amounts to dynamic MDS which is obtained by the super-matrix construction.

However, in the case of the dynamic additive fuzzy clustering model, the coordinate of the result of the fuzzy grade makes no sense mathematically, due to the definition of the aggregation function family. The merit for using the idea of the super-matrix is to obtain the grade of belongingness to fixed clusters through time. In the super-matrix, if cross time period dissimilarities are all assumed to be 0, then the usual evaluation function for the classification techniques of 3-way data will be obtained. (Sato, M. and Sato, Y., 1994, 1997) However, with these methods, clusters are not invariant with respect to numbering of the clusters, so it is impossible to get the latent phenomenon of the mixed observations.

Determination of the average functions may be related to the representation of the asymmetric proximity between objects or the factors of investigation over time which are dependent on the properties of the observations. Choosing the adaptable function corresponding to the relationship among objects or times will be discussed through some numerical investigations.

References

Ambrosi, K. and Hansohm, J. (1987), "Ein dynamischer Ansatz zur Repräsentation von Objekten", *In Operations Research Proceedings 1986*, Berlin: Springer-Verlag.

Baba, Y. and Nakamura, Y.(1997), "Jikan Henka wo tomonau Syuseibun Bunsekihou (in Japanese)", *11th Japanese Society of Computational Statistics*, 82-85.

Carroll, J.D. and Chang, J.J. (1970), "Analysis of individual differences in multidimensional scaling via an N-way generalization of "Eckart-Young" decomposition", *Psychometrika, 35*, 283-319.

Carroll, J.D. and Arabie, P. (1983), "INDCLUS: An individual differences generalization of the ADCLUS model and MAPCLUS algorithm", *Psychometrika, 48*, 157-169.

Menger, K.(1942), "Statistical Metrics", *Proc. Nat. Acad. Sci. U.S.A., 28*, 535-537.

Sato, M. and Sato, Y.(1994), "On a Multicriteria Fuzzy Clustering Method for 3-Way Data ", *International Journal of Uncertainty, Fuzziness and Knowledge-Based Systems, 2*, 127-142.

Sato, M. and Sato, Y.(1995), "On a General Fuzzy Additive Clustering Model", *International Journal of Intelligent Automation and Soft Computing, 1*, No.4, 439-448.

Sato M. and Sato, Y.(1997), "Generalized Fuzzy Clustering Model for 3-way Data", *International Conference on Fuzzy Logic and its Applications*, 132-137.

Schweizer, B. and Sklar, A.(1983), *Probabilistic Metric Space* , North-Holland, New York.

School Basic Survey, Ministry of Education in Japan, (1992).

Application of fuzzy mathematical morphology for pattern classification

Sandrine Turpin-Dhilly and Claudine Botte-Lecocq
Laboratoire d'Automatique I^{3}D
Cité Scientifique - Bâtiment P2 59655 Villeneuve d'Ascq
Tél. :03.20.43.48.76 E-Mail : Sandrine.Turpin@univ-lille1.fr

Abstract : In this paper, we present a new approach to unsupervised pattern classification, based on fuzzy morphology. The different modes associated to each cluster are detected by means of a fuzzy morphological transformation applied to a membership function defined from the mode concavity properties. The results of this clustering method are shown using artificially generated data sets.

Key Words : Fuzzy Morphology, Fuzzy structuring function, Mode detection, Pattern recognition, Probability density function.

1. Introduction

Determining the structure of multidimensional data is an important problem in exploratory data analysis. Most of the clustering procedures, developed in the sixties, were based on some similarity or dissimilarity measures [1]. More recently, some degree of fuzziness have been introduced in the metric criterions in order to cope with the irregularities of the data distribution [2].
Another way to classify the observations is to analyse the local density function of the pattern distribution in the data space [3]. Note that many statistical procedures aim at the detection of the modes of the probability density function underlying the distribution of the data [4]. These modes can be considered as regions of the data space with high local concentration of observations, separated by valleys, which are regions characterised by low local concentration of observations. Mathematical morphological tools have proved to be well-suited for discriminating between these two kinds of regions [5][6]. In this paper, we improve such morphological clustering techniques by introducing fuzzy concepts in the basic erosion and dilation operations [7]. As fuzzy mathematical morphology has been built on fuzzy sets, we first show how a multidimensional data set can be transformed as a fuzzy set, with its associated membership function. Then, we present a fuzzy morphological transformation, which directly detects the different modes associated to each cluster. Finally, the whole clustering method is illustrated using artificially generated data sets.

2. Evaluation of the membership function to a mode : the fuzzification phase

In order to detect the different modes associated to each cluster, we first evaluate, from the data themselves, a membership function to a mode of each observation, or point, of the analysed set.
As many other mode detection procedures working in an unsupervised context, we have first to estimate the underlying probability density function (pdf) related to the distribution of the observations. In the k nearest neighbours (knn) environment, the pdf is estimated at each point of the data space, by considering the smallest domain, centred at this point, which contains a fixed number k_1 of neighbours. To be more specific, the k_1 nearest neighbours estimator, denoted by $\hat{p}(X_m)$, is computed at each N-dimensional observation X_m of the analysed set **X** of size M such as : $\mathbf{X} = \{X_1, X_2, ..., X_m, ..., X_M\}$. At the end of the estimation step, which takes advantage of the fast algorithm presented in [8], the estimator $\hat{p}(X_m)$ is evaluated of each point X_m by :

$$\hat{p}(X_m) = \frac{k_1 / M}{V[D(X_m)]} \tag{1}$$

where $V[D(X_m)]$ denotes the volume of the smallest domain in which fall the k_1 nearest neighbours of the multidimensional observation X_m.

The membership function to a mode can then be evaluated by considering each mode as a region of the data space where the pdf is concave. This approach is based on a test which determines locally the convexity of the multivariate pdf by assigning the "concave" or "convex" label to each point of the analysed set [9]. In a knn environment, the local convexity of the pdf at a point X_m is determinated by analysing the variation of the knn estimator when the domain grows around X_m. More precisely, we fixe a number k_2 of neighbours ($k_2 > k_1$), and we grows the domain $V_1[D(X_m)]$ until it includes the number k_2 of neighbours in order to yield the domain $V_2[D(X_m)]$. Then, the two estimators $\hat{p}_1(X_m)$ and $\hat{p}_2(X_m)$ are evaluated with the equation (1).
If $V_2[D(X_m)] > V_1[D(X_m)]$ implies that $\hat{p}_2(X_m) > \hat{p}_1(X_m)$, then, the point X_m will have the **convex** label. Otherwise, X_m will have the **concave** label.

As a mode can be defined by a set of points characterised by the concave label, we introduce, here, a confidence value on the event : "X_m is a modal observation", from a Pi-fuzzification function (figure 1), where the belief degree on this event depends on the number of the "concave" observations in the neighbourhood of the studied point. So, the more numerous the nearest neighbours of the studied point with a concave label are, the more important the belief degree on the event "X_m is a modal observation" is.

Figure 1 : *The membership function*

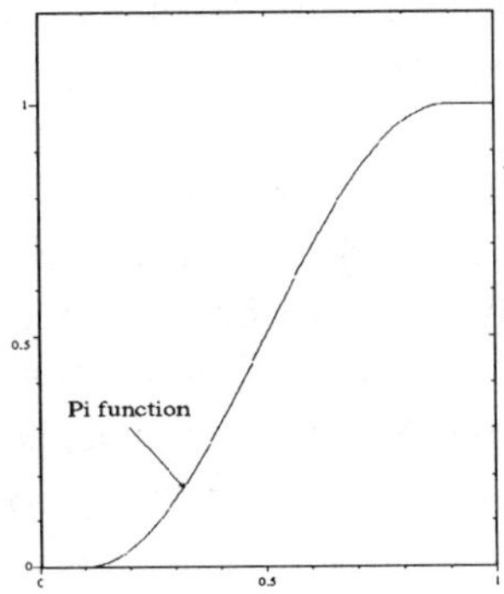

with

$$x = \frac{\text{Nb of neighbours } k_i \text{ (concave(ki)=1)}}{k_2}$$

At the end of this fuzzification step, the fuzzy set is composed of the initial set **X** with its associated mode membership function μ_M.

3. Mode detection by a new fuzzy morphological transformation

In classical morphology, the basic operations of erosion and dilation are often combined in pairs ; the resulting operation, namely an opening, is well-known for its filtering properties. Indeed, the principal effect of the erosion is to enlarge the valleys, in eliminating the irregularities of the distribution, but it also tends to shrink the modes. On the other hand, dilation is used to enlarge the modes, but it also tends to enhance the valleys. So, it would be more interesting to erode, or to dilate, more or less, depending on the localisation of the datas in the representation space by means of fuzzy operators. Indeed, in the definitions of the fuzzy erosion and the fuzzy dilation presented respectively in (2) and (3), the membership function μ_M is compared to some structuring function ν, which is defined on a domain S. Note that S is, here, an hypersphere including a number k_f of neighbours. The definition of the function ν depends on the problem to solve.

$$E_\nu(\mu_M)(X_m) = \inf_{X_{ki} \in S} \max\left[\mu_M(X_{ki}),\ 1 - \nu(X_{ki} - X_m)\right] \qquad (2)$$

$$D_\nu(\mu_M)(X_m) = \sup_{X_{ki} \in S} \min\left[\mu_M(X_{ki}),\ \nu(X_{ki} - X_m)\right] \qquad (3)$$

As our problem is to enhance the modes while deepening the valleys, we first used binary structuring functions depending on the mode membership function μ_M, only evaluated at the studied point X_m [10]. The promising results of such a procedure can be improved in taking into account the likeness between the studied point X_m and its nearest neighbours X_{ki}. Indeed, the confidence value on the event "X_m is a modal observation" should be higher when a great number of

its neighbours have themselves high confidence values on this event. As well, if the nearest neighbours have low confidence values, the confidence value of the studied point must be lower. So, in this paper, the binary structuring functions used for erosion and dilation will depend on both $\mu_M(X_m)$ and $\mu_M(X_{ki})$.
To be more specific, the structuring function υ_e, used in the fuzzy erosion of the equation (2), must be defined such as this operation will enlarge the valleys while not deepening the modes. This function $\upsilon_e(X_m-X_{ki})$, evaluated at each neighbour X_{ki} of the studied point X_m, will affect different confidence values to the neighbours of X_m, according to a binary criterion depending on the product $\mu_M(X_m).\mu_M(X_{ki})$ (figure 2). These confidence values must be as more important than X_{ki} and X_m are both probably located at the valleys, so than the product $\mu_M(X_m).\mu_M(X_{ki})$ is low.
As well, the structuring function υ_d, used in the fuzzy dilation of the equation (3), must enlarge the modes while not enhancing the valleys. So, we will use a binary criterion which will affect an important confidence value to each neighbour X_{ki} of the studied point X_m when they are both probably located at the modes, i.e. when the product $\mu_M(X_m).\mu_M(X_{ki})$ is high.

Figure 2 : *Representation of the binary criterion for the erosion and dilation*

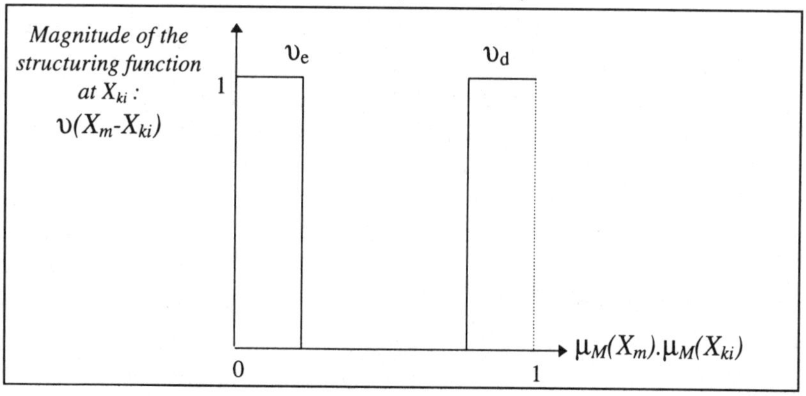

In that way, the mode detection algorithm consists in performing a fuzzy morphological transformation, which is defined by a fuzzy erosion using the specific structuring function ν_e followed by a fuzzy dilation using the specific structuring function ν_d. The different modes are directly detected by iteratively applying this transformation until stabilisation of the resulting function.

4. Example of this clustering method

The important difficulty in cluster analysis are non-spherical clusters. The bivariate data set of figure 3, has been generated keeping these well-know difficulty in mind. Indeed, a gaussian cluster is flanked by two non-spherical

clusters. Figure 4 presents the resulting mode membership function evaluated after the convexity test, performed on the estimated pdf. After the fuzzy morphological transformation presented in this paper, the 3 detected modes are well-separated (figure 5).
When the observations constituting the different modal regions are identified (figure 6), they can be considered as the prototypes of the clusters. Each remaining observation is then assigned to the cluster attached to its nearest neighbours among these prototypes. The result of such a classification technique (figure 7) presents an error rate equal to 8.3 %, which can be compared to the Isodata one which is equal to 31.3%.

Figure 3 : *Raw data set*

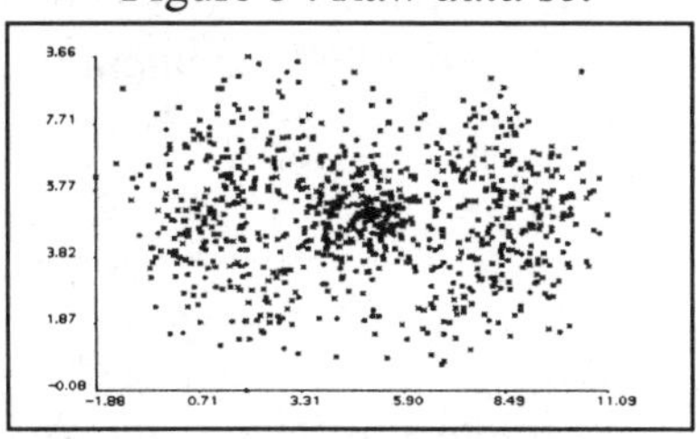

Figure 4 : *Membership function*

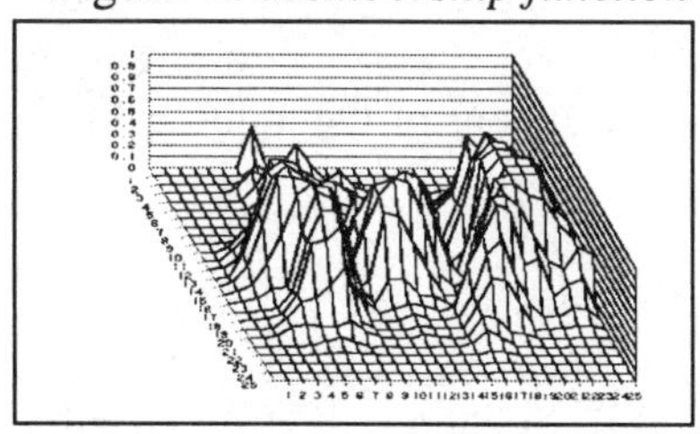

Figure 5 : *Result of the fuzzy morphological transformation*

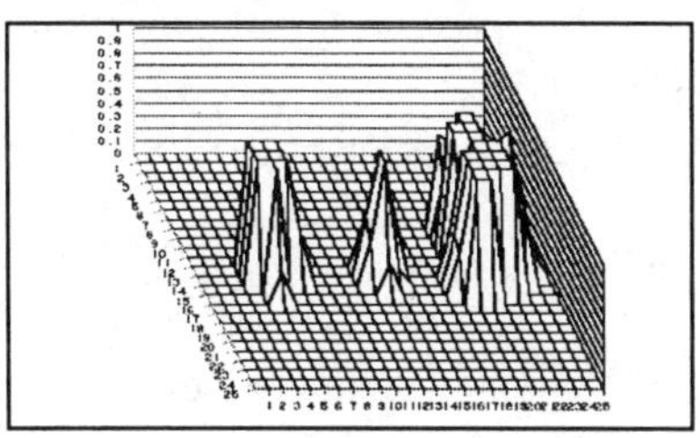

Figure 6 : *The different cores*

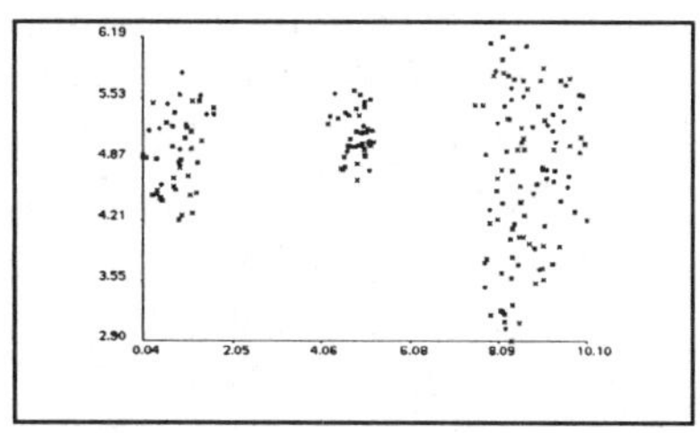

Figure 7 : *Result of the classification*

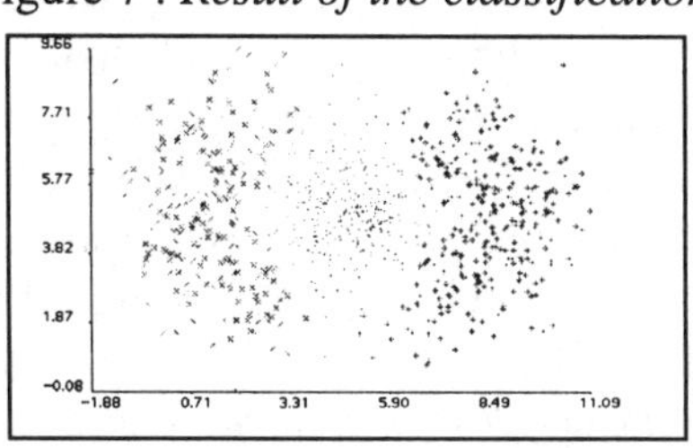

6. Conclusion

We have presented a classification procedure based on the fuzzy morphological mode detection of the pdf underlying the distribution of the data. As fuzzy morphological operators have to be performed on fuzzy sets, we have first defined a mode membership function, using the concavity properties of the modes. These different modes are then directly detected by iteratively applying a fuzzy morphological transformation to this fuzzy set, which combines fuzzy erosions and fuzzy dilations. Finally the classification phase consists in assigning each observation to its nearest mode.

The fuzziness notion has been introduced in the morphological operators in order to take into account the local structure of the data. With such fuzzy operators, defined using specific fuzzy structuring functions adapted to our mode detection problem, we keep advantages of the classical morphological operators, without the withdraws. Indeed, the fuzzy transformation, presented here, yields directly well-separated modes, with respect to the original shape of the data.

Face to the promising results to such fuzzy techniques, authors are now working in the development of different fuzzy morphological procedures including the local variations of the mode membership function.

References :

[1] R.R. Sokal and P.H.A. Sneath 1963. *Principles of numeral taxonomy*, W.H. Freeman Edition, San Francisco.

[2] Cannon R.L., Dave J.L., Bezdek J.C. 1986. Efficient implementation of the fuzzy c-means clustering algorithms, *I.E.E.E. Transactions on Pattern Analysis and Machine Intelligence,* 8, 2, 248-255.

[3]Tremolieres R., 1979. The percolation Method of an efficient grouping of data, *Pattern Recognition,* 11, 4.

[4]Asselin de Beauville J.C. 1989. Panorama de l'utilisation du mode en classification automatique, *A.P.I.I*, 23, 113-137.

[5] Botte-Lecocq C., Zhang R.D., Postaire J.-G. 1993. Cluster analysis by binary morphology, *I.E.E.E Transactions on Pattern Analysis and Machine Intelligence*, 2, 15, 170-180.

[6] Zhang R.D., Postaire J.-G. 1993. Convexity dependent morphological transformations for mode detection in cluster analysis, *Pattern Recognition*, 1-14

[7] Bloch I., Maitre H. 1995. Fuzzy Mathematical Morphologies : A Comparative Study, *Pattern Recognition*, 28, 9, 1341-1387.

[8] Czesnalowicz E., Postaire J.-G. 1990. Adaptive median filtering for mode detection with application to pattern classification, *9th COMPSTAT Symposium on Computational Statistics, Dubrovnik, Yugoslavia.*

[9] Postaire J.-G., Vasseur C. 1980. A convexity testing method for cluster analysis, *I.E.E.E Transactions on Systems and Man Cybernetics,* 10, 145-149

[10]Turpin S., Botte-lecocq C., Postaire J.-G. 1997. Mode extraction for pattern classification by fuzzy mathematical morphology, *EUFIT'97, Aachen, Germany*, 1689-1693.

Statistical Conditions for a Linear Complexity for an Algorithm of Hierarchical Classification Under Constraint of Contiguity

Kaddour BACHAR [1, 2] and Israël-César LERMAN [3]

[1] ESSCA- 1, rue Lakanal, B.P. 348 - 49003 - Angers - France
[2] LTSI -Université de Rennes 1, Campus de Beaulieu - Rennes - France
[3] IRISA-Université de Rennes 1, Campus de Beaulieu - Rennes - France

Abstract: In the best conditions, the time complexity of classical algorithms of hierarchical ascendent classification is for a number n of elements to be classified of order $O(n^2)$. In this paper the proposed CAHCVR algorithm is based on the principle of the "reciprocal neighbourhood" algorithm (CAHVR). It takes into account a contiguity constraint defined by a given connected graph. We study the properties of its results (absorbing class, inversion,...) and demonstrate that the average time complexity is in O(n) (linear complexity). The proposed CAHCVR algorithm is applied to image analysis in order to solve the image segmentation problem.

Keywords : hierarchical ascendent classification, connected graph of contiguity, reciprocal neighbourhood, image segmentation.

1. The CAHCVR algorithm

The proposed CAHCVR algorithm includes several steps, each of which constituted of two distinct phases. The first phase consists in finding all the couples of adjacent elements, according to a given connected graph, such that each element of the couple is the closest (in the sense of inertia) element to the other. These couples are called contiguous reciprocal neighbours (RN). The second phase **(i)** aggregates all the couples of contiguous RN and **(ii)** forms a new class for each of them. Both phases are performed alternatively until all the elements are grouped into a same class. The update of the contiguity condition is performed only at the end of the two previous phases ; in other words, at the end of each step in the CAHCVR algorithm. The new graph will be formed by, first, new vertices corresponding to the elements which have not been incorporated. The edges of this new graph are defined in the following way : if vertex x is adjacent to vertex y (in the sense of the graph) all vertices which contain y are also adjacent to all vertices adjacent to either one of the two incorporated vertices. Such an update conserves the connexity nature of the graph ; the connexity insures, in particular, the convergence of the algorithm, in other words it will always be possible to gather all the elements into one class (see proposition 1). The CAHC algorithm **[Lebart, 1978]** performs, at each step, one aggregation only. This aggregation corresponds to the pair of elements which are adjacent according to the graph and minimizes the distance (based on an inertia criterion, for example). The update of the contiguity is similar to that of the CAHCVR algorithm. The two algorithms, CAHC and CAHCVR, do not produce equivalent results [**Bachar, 1994**].

2. Theoretical properties of CAHCVR algorithm

Proposition 1 *A necessary and sufficient condition for CAHC and CAHCVR to end in a connected classification tree, is that the initial graph of contiguity is connected.*

Proof : obvious using the following update of contiguity.■

Theorem *Let G_0 be any initial connected graph of constant maximal degree and independent of n (denoted v_0). Suppose that the "distances" D between an element and its contiguous neighbours (in the sense of G_0) are independent non-negative random variables of same continuous probability law. For the CAHCVR algorithm , the average proportion of contiguous reciprocal neighbours (RN) at step k varies in the interval]1/4 , 3/8] , except in the last step where this proportion is equal to the maximum value 1/2, and assuming that the graph can be divided into $n_k / 2$ classes of contiguous couples, where n_k is the number of elements at the step k. If the initial graph G_0 is complete, the previous interval is]1/4 , 1/3].*

Proof : We have to calculate the probability p that two contiguous elements (x,y) will form an RN couple ; we suppose that $n > 2$ ($p = 1$ if $n = 2$). We denote by P the probability function (using the notations and the hypothesis of the theorem), and then $P(D < \rho) = F(\rho)$ is the cumulative distribution function, $f(\rho)$ will designate the probability density function ($\rho \geq 0$). First, we specify the law $G(\rho)$ ($g(\rho)$ is the corresponding probability density) of the distance between x and its closest contiguous element y among $v(x)$ elements contiguous to x ($x_1, x_2, \ldots x_{v(x)}$ are the contiguous elements to x). Next we specify the conditionnal probability $H_y(\rho)$ that x is the closest element to y (with the condition that y is the closest element to x ; ($y_1, y_2, \ldots y_{v(y)}$ are the contiguous elements to y).

$$P(D(x,y) > \rho) = P(\bigcap_i \{D(x,x_i) > \rho\}) = (1 - F(\rho))^{v(x)} \Rightarrow G(\rho) = 1 - (1 - F(\rho))^{v(x)}$$

$H_y(\rho) = P(\bigcap_{y_i \neq x} \{D(y, y_i) > \rho\}) = (1 - F(\rho))^{v(y)-1}$. Then the probability $p_x(y)$ that (x,y) is a couple of contiguous RN starting from x in the graph G_0 is $p_x(y) = \int_0^\infty H_y(\rho) g(\rho) d\rho = v(x)/(v(x)+v(y)-1)$. The probability p verifies : $2p = p_x(y) + p_y(x)$ and then $p = (v(x)+v(y))/2(v(x)+v(y)-1)$. As $v(x)+v(y) \geq 3$ (the graph is connected and $n > 2$) we have the result $1/2 < p \leq 3/4$. Then the average number of contiguous RN couple belongs to interval $]n_k/4, 3n_k/8\,]$ where n_k is the number of elements at step k ; in the last step the number of elements is 2 and $p = 1$. The case of a complete initial graph G_0 corresponds to the CAHVR algorithm (without contiguous constraint) and then $v(x)+v(y) \geq 4$ (for $n > 2$) : therefore $1/2 < p \leq 2/3$.■

The result **[Benzécri, 1982]** appears to be a particular case of our theorem

Proposition 2 *According to the hypothesis of the previous theorem, the average time complexity of the CAHCVR algorithm is in O(n).*

Proof : We start **(i)** by estimation the average proportion of contiguous elements and **(ii)** by computing the size of calculations in CAHCVR. Afterwards we estimate the average number of distances to compute and the average number of comparisons necessary in order to identify the contiguous RN. We denote by n the large number of elements to be classified and by θ_i the proportion of contiguous RN couples at step i ($1 \leq i \leq K$ where K is the (finite) number of the last step ; $0 < \theta_i \leq 1/2$). We have $n(1-\theta_1)(1-\theta_2)...(1-\theta_K) = 1$. If θ denotes the average proportion of contiguous RN couples, we have $n(1-\theta_1)(1-\theta_2)...(1-\theta_K) = 1 = n(1-\theta)^K$ $\quad (K = -\log(n)/\log(1-\theta) \geq \log_2(n))$
Subsequently, we use this average proportion θ .

The average number of distances to compute. At the begining of step 0, there are n elements and we perform less than $nv_0/2$ distances (v_0 is the constant maximal number of the contiguous elements to any initial element). At the end of step 0, an average number $n - n\theta = n(1-\theta)$ of elements remain. At the start of step 1 we perform less than $n(1-\theta)v_0/2$ distances. At the end of step k, the average number of remaining elements is $n(1-\theta)^k$ and we perform less than $n(1-\theta)^k v_0/2$ distances. Then we perform a maximal number of distances equal to $D_{\max}(\theta) = \sum_{k=0}^{K} n(1-\theta)^k v_0/2$. $D_{\max}(\theta) \approx nv_0/(2\theta)$ is decreasing with θ.

The average number of comparisons necessary in order to identify the contiguous RN couples. At the begining of first step we perform less than nv_0 comparisons to identify the closest elements to any element. In the start of the number k ($k \geq 1$) we perform a number of comparisons less than equal to $C_1(\theta) = \sum_{k \geq 1} n(1-\theta)^{k-1} v_0$. To identify the whole set of RN couples, we need to perform less than $C_2(\theta)=\sum_{k\geq 1} n(1-\theta)^{k-1}v_0^2$ comparisons (it is easy to prove that the average number of contiguous elements of any element is low or equal v_0 at step k). The CAHCVR algorithm requires, in average, less than $C_1(\theta)+C_2(\theta)\approx \theta v_0(1+v_0)n$ comparisons and $D_{\max}(\theta)\approx nv_0/(2\theta)$ distances. $C_1(\theta)+C_2(\theta)+D_{\max}(\theta)$ is a linear function in n, because θ is independent of n (following theorem).■

Proposition 3 *According to the hypothesis of the previous theorem, the CAHCVR algorithm facilitates (in probability) the formation of contiguous classes whose neighbourhoods are the smallest as possible. All possible incorporations of couples of contiguous elements are equally probable in the case of the CAHC algorithm.*

Proof : The following expression $p=(v(x)+v(y))/2(v(x)+v(y)-1)$ (following the theorem) is decreasing with respect to $v(x)+v(y)$: therefore, the CAHCVR algorithm facilitates (in probability) the formation of contiguous classes whose neighbourhoods are the smallest possible. This is not the case in the CAHC algorithm : indeed, if m_k is the number of edges at step k, the probability q that two contiguous elements x et y form a class at the kth step is $q=\int_0^\infty (1-F(\rho))^{m_k} f(\rho)d\rho = 1/m_k$: then following the same hypothesis, all possible incorporations of couples of contiguous elements are equally probable in the case of the CAHC algorithm.■

An absorbing class is a class that is created by successive aggregation of lower size classes. However, a lack of balance between the resulting aggregated classes is often observed (for a given partition). In practice, to avoid this lack of balance, an arbitrary constraint on class sizes is introduced (it is the case in the CAHC algorithm). In [**Bachar, 1994**] we demonstrate that the absorbing class increases (in probability) the presence of inversion. Proposition 3 highlights the inconvenience of absorbing classes ; it should be recalled that inversion phenomenon complicates the automatic choice of an "optimal" partition and deteriorates the final "quality" of the obtained result.

3. The likelihood of the maximal link criterion

This criterion of similarity between classes refers to a probability scale. This scale can be directly associated with a dissimilarity measure between classes in the form of a quantity of information. Following a very general construction, such a criterion can be applied both to cluster the set of variables and the described objects set ; and this is the case whatever what the mathematico-logical nature of the data. The established inequalities permit us to understand that with the criterion of the likelihood of the maximal link [See **Lerman, 1991**] , it is much more difficult for inversion to take place than in the case of explained inertia (criterion of inertia). In addition, the phenomenon of the absorbing classes decreases. These properties will be studied in a future paper.

4. Applications and conclusion

The aim of segmentation of a plane image is the identification of bounderies between the objects that form the image. The notion of contiguity is then natural here : the pixels are on a plane. The segmentation of an image is a partition which can be obtained after classification under contiguity constraint. This constraint can be done directly in a classification algorithm, in addition to using the attributes which characterize the measuring points. It can also intervene in a previous factor analysis, or in both cases. The choice of the variables which characterize a point of the image defines the segmentation criteria. The treatment proposed is applied to several real medical images of radiology. We use the original image ($n = 512 \times 512$ pixels) describing an abdomen in **fig.1**. The image segmentation in **fig.2** (partition in 70 classes using the CAHCVR algorithm with 3 statistical attributes characterizing the measuring pixels and with $v_0 = 8$) is obtained in 129 seconds (HP9000 Computer, HPUX V.10). The time evaluation is displayed in **fig.3** (for different values of the number n of pixels : $50 \times 50,\ 100 \times 100,\ 150 \times 150, ...,\ 500 \times 500$). The linearity of the time complexity and the quality of results are verified...

Acknoledgments : We thank Johanne Bezy-Wendling and Fabrice Wendling (LTSI) for the acquisition and evaluation of the images.

References

Bachar, K [1994] *Contributions en analyse factorielle et en classification ascendante hiérarchique sous contrainte de contiguïté. Applications à la segmentation d'images.* Thèse de doctorat, Université de Rennes 1 (France).

Benzécri, J-P [1982] *Sur la proportion des paires de voisins réciproques pour une distribution uniforme dans un espace euclidien.* Les cahiers de l'analyse des données, Vol.7, p.185-188.

Lebart, L [1978] *Programme d'agrégation avec contraintes.* Les cahiers de l'analyse des données, Vol.3, p.275-287.

Lerman, I-C [1991] *Foundations of the Likelihood Linkage Analysis(LLA) Classification method.* Applied Stochastic Models and Data Analysis, Vol.7, #1, march 1991, p.63-76, John Wiley.

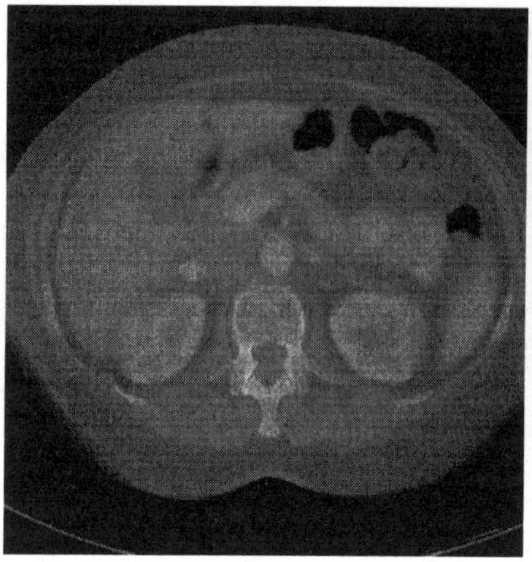

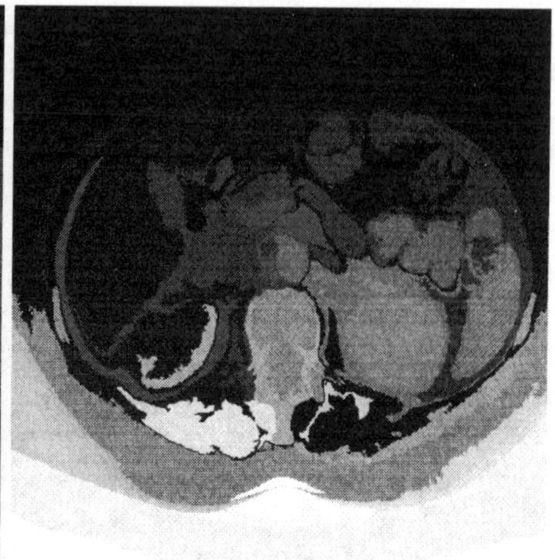

Figure 1 : *Original image of the abdomen, acquired with an helical CT (Siemens Somatom Plus 4, South Hospital, Rennes, France)*

Figure 2 : *Result of the CAHCVR algorithm applied to the image of Figure 1. Several organs and bone structures are quite well identified (vertebra, liver, kidneys, fatty tissue...)*

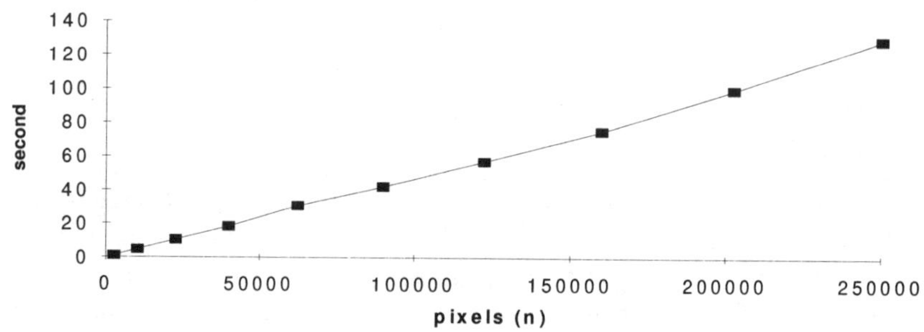

Figure 3 : *Time evaluation (seconds) for different sizes of image*

Constrained Clustering Problems

Vladimir Batagelj[1], Anuška Ferligoj[2]

[1] University of Ljubljana, Faculty of Mathematics and Physics, and Institute of Mathematics, Physics and Mechanics, Dept. of TCS, Jadranska 19, 1 000 Ljubljana, Slovenia

[2] University of Ljubljana, Faculty of Social Sciences, P.O. Box 47, 1 109 Ljubljana, Slovenia

Abstract: In the paper a view, based on the optimization approach, is given on different types of constrained clustering problems and methods for their solution.

Keywords: clustering methods, constraints, blockmodeling, optimization.

1. Introduction

For constrained clustering, grouping similar units into clusters has to satisfy some additional conditions. This class of problems is relatively old. One of the most frequently treated problems in this field is regionalization: clusters of similar geographical regions have to be found, according to some chosen characteristics, where the regions included in the cluster have to be also geographically connected. A number of approaches to this problem have been taken. The majority of authors (e.g., Lebart, 1978; Lefkovitch, 1980; Ferligoj and Batagelj, 1982; Perruchet, 1983; Gordon, 1973, 1980, 1987; Legendre, 1987) solve this problem by adapting standard clustering procedures, especially agglomerative hierarchical algorithms, and local optimization clustering procedures. The geographic contiguity is a special case of *relational constraint*. Ferligoj and Batagelj (1982, 1983) first treated this clustering problem for general symmetric relations and then for nonsymmetric relations. It is possible to work also with other, non-relational conditions. Murtagh (1985) provides a review of clustering with symmetric relational constraints. A more recent survey of constrained clustering was given by Gordon (1996).

2. Formalization

The (constrained) clustering problem can be posed as an optimization problem (Ferligoj and Batagelj, 1982, 1983):

Let E be a finite *set of units*. Its nonempty subset $C \subseteq E$ is called a *cluster*. A set of clusters $\mathbf{C} = \{C_i\}$ forms a *clustering*.

The clustering problem $(\mathbf{\Phi}, P, \min)$ can be expressed as: Determine the clustering $\mathbf{C}^* \in \mathbf{\Phi}$, for which

$$P(\mathbf{C}^*) = \min_{\mathbf{C} \in \mathbf{\Phi}} P(\mathbf{C})$$

where $\mathbf{\Phi}$ is a *set of feasible clusterings* and $P : \mathbf{\Phi} \to \mathbb{R}_0^+$ is a *clustering criterion function*. We denote the *set of minimal solutions* by $\mathrm{Min}(\mathbf{\Phi}, P)$.

Let us introduce some notions which we shall need in the following.

The clustering $\mathbf{C}$ is a *complete* clustering if it is a partition of the set of units E. We shall denote by $\mathbf{\Pi}(E)$ the set of all complete clusterings of E. Two among them $\mathbf{O} \equiv \{\{X\} : X \in E\}$ and $\mathbf{I} \equiv \{E\}$ deserve to be denoted by special symbols. The set of feasible clusterings $\mathbf{\Phi}$ can be decomposed into "strata" (layers)

$$\mathbf{\Phi}_k = \{\mathbf{C} \in \mathbf{\Phi} : \mathrm{card}(\mathbf{C}) = k\}$$

Let $(\mathbb{R}_0^+, \oplus, 0, \leq)$ be an ordered abelian monoid. A *simple* criterion function P has the form:

$$P(\mathbf{C}) = \bigoplus_{C \in \mathbf{C}} p(C), \quad p(C) \geq 0 \quad \text{and} \quad \forall X \in E : p(\{X\}) = 0$$

For almost all criterion functions used in applications, it holds also $p(C_1 \cup C_2) \geq p(C_1) \oplus p(C_2)$. For a simple criterion function satisfying this condition, it holds for $k < \mathrm{card}\, E$: $\forall \mathbf{C} \in \mathbf{\Pi}_k \exists \mathbf{C}' \in \mathbf{\Pi}_{k+1} : P(\mathbf{C}') \leq P(\mathbf{C})$. Since $P(\mathbf{C}) \geq 0$ and $P(\mathbf{O}) = 0$, it holds that $\mathbf{O} \in \mathrm{Min}(\mathbf{\Pi}, P)$. To avoid this trivial problem we usually introduce the obvious constraint – we restrict the problem to $\mathbf{\Pi}_k$, where k is a given number of clusters.

Not all clustering problems can be expressed by a simple criterion function. In some applications a *general* criterion function of the form

$$P(\mathbf{C}) = \bigoplus_{(C_1, C_2) \in \mathbf{C} \times \mathbf{C}} q(C_1, C_2), \quad q(C_1, C_2) \geq 0$$

is needed. An example of general criterion functions can be found in blockmodeling (Batagelj, Ferligoj, Doreian, 1992; Batagelj, 1997)

$$P(\mathbf{C}; \mathcal{T}) = \sum_{(C_1, C_2) \in \mathbf{C} \times \mathbf{C}} \min_{T \in \mathcal{T}} w(T) \delta(C_1, C_2; T)$$

where $\mathcal{T}$ is a set of feasible types, and δ measures the deviation of blocks, induced by a clustering, from the ideal block structure. The blockmodeling methods considering also (dis)similarities between units have still to be developed. The proposed optimization approach essentially expresses the constraints with a penalty function.

Another such example is a problem of partitioning of a generation of pupils into a given number of classes so that the classes will consist of (almost) the same number of pupils and that they will have a structure as similar as possible. An appropriate criterion function is

$$P(\mathbf{C}) = \max_{\substack{\{C_1, C_2\} \in \mathbf{C} \times \mathbf{C} \\ \mathrm{card}\, C_1 \geq \mathrm{card}\, C_2}} \; \min_{\substack{f : C_1 \to C_2 \\ f \text{ is surjective}}} \; \max_{X \in C_1} d(X, f(X))$$

where $d(X, Y)$ is a measure of dissimilarity between pupils X and Y.

3. Types of Constrained Clusterings

Various types of the constraints are discussed below.

3.1. Relational Constraints

Generally, the set of feasible clusterings for this type of constraint can be defined as:

$$\mathbf{\Phi}(R) = \{\mathbf{C} \in \mathbf{\Pi} : \text{each cluster } C \in \mathbf{C} \text{ is a subgraph } (C, R \cap C \times C) \text{ in the graph } (E, R) \text{ with the required type of connectedness}\}$$

We can define different types of sets of feasible clusterings for the same relation R if it is nonsymmetric (Ferligoj and Batagelj, 1983). Some examples of clusterings with (nonsymmetrical) relational constraint $\Phi^i(R)$ are

type of clusterings	type of connectedness
$\Phi^1(R)$	weakly connected units
$\Phi^2(R)$	weakly connected units that contain at most one center
$\Phi^3(R)$	strongly connected units
$\Phi^4(R)$	clique
$\Phi^5(R)$	the existence of a trail containing all the units of the cluster

A center of a cluster C in the clustering type $\Phi^2(R)$ is the set of units $L \subseteq C$ iff the subgraph induced by L is strongly connected and $R(L) \cap (C \setminus L) = 0$ where $R(L) = \{y : \exists x \in L : xRy\}$.

3.2. Constraining Variables

The set of feasible clusterings for this particular type of constraint is defined as follows (Ferligoj, 1986):

$$\mathbf{\Phi}[a, b] = \{\mathbf{C} \in \mathbf{\Pi} : \text{for each cluster } C \in \mathbf{C} \text{ holds: } v_C \in [a, b]\}$$

where v_C is a function of values of the constraining variable, V, for the units in the cluster C.

3.3. Optimizational Constraint

The set of feasible clusterings for an optimizational constraint is defined as:

$$\mathbf{\Phi}(F) = \{\mathbf{C} \in \mathbf{\Pi} : \text{for second criterion } F \text{ the condition } F(\mathbf{C}) \leq f \text{ holds }\}$$

where f is a given threshold value for the second criterion.

A combination of all three types of constraints (relational, constraining variable and optimizational) can be considered simultaneously.

3.4. Pre-specified Blockmodels

Given a network $\mathcal{N}$, set of types $\mathcal{T}$, and a model $\mathcal{M}$ (constraints!), determine μ (clustering) which minimizes the criterion function (for details see Batagelj, Ferligoj, Doreian, 1998).

4. Solving Constrained Clustering Problems

With few exceptions the clustering problem is too hard to be exactly solved efficiently. Therefore, approximative/heuristic methods have to be used. Among these, agglomerative (hierarchical) and local optimization (relocation) methods are the most popular.

4.1. Hierarchical Algorithm

The set of feasible clusterings $\mathbf{\Phi}$ determines the *feasibility predicate* $\Phi(\mathbf{C}) \equiv \mathbf{C} \in \mathbf{\Phi}$ defined on $\mathcal{P}(\mathcal{P}(E) \setminus \{\emptyset\})$; and conversely $\mathbf{\Phi} \equiv \{\mathbf{C} \in \mathcal{P}(\mathcal{P}(E) \setminus \{\emptyset\}) : \Phi(\mathbf{C})\}$.

In the set of all clusterings $\mathbf{\Phi}$ the relation of *clustering inclusion* $\sqsubseteq$ can be introduced by

$$\mathbf{C}_1 \sqsubseteq \mathbf{C}_2 \equiv \forall C_1 \in \mathbf{C}_1, C_2 \in \mathbf{C}_2 : C_1 \cap C_2 \in \{\emptyset, C_1\}$$

we say also that the clustering $\mathbf{C}_1$ is a *refinement* of the clustering $\mathbf{C}_2$.

It is well known that $(\mathbf{\Pi}, \sqsubseteq)$ is a partially ordered set (even more, semimodular lattice). Because any subset of partially ordered set is also partially ordered, we have: Let $\mathbf{\Phi} \subseteq \mathbf{\Pi}$ then $(\mathbf{\Phi}, \sqsubseteq)$ is a partially ordered set.

The clustering inclusion determines two related relations (on $\mathbf{\Phi}$):
$\mathbf{C}_1 \sqsubset \mathbf{C}_2 \equiv \mathbf{C}_1 \sqsubseteq \mathbf{C}_2 \wedge \mathbf{C}_1 \neq \mathbf{C}_2$ and
$\mathbf{C}_1 \sqsubset\!\!\cdot\, \mathbf{C}_2 \equiv \mathbf{C}_1 \sqsubset \mathbf{C}_2 \wedge \neg\exists \mathbf{C} \in \mathbf{\Phi} : (\mathbf{C}_1 \sqsubset \mathbf{C} \wedge \mathbf{C} \sqsubset \mathbf{C}_2)$.

4.1.1. Conditions on the Structure of the Set of Feasible Clusterings.

In the following we shall assume that the set of feasible clusterings $\mathbf{\Phi} \subseteq \mathbf{\Pi}$ satisfies the following conditions (Batagelj, 1984):

F1. The feasibility predicate Φ is *local* – it has the form $\Phi(\mathbf{C}) = \bigwedge_{C \in \mathbf{C}} \varphi(C)$ where $\varphi(C)$ is a predicate defined on $\mathcal{P}(E) \setminus \{\emptyset\}$ (clusters).

The intuitive meaning of $\varphi(C)$ is: $\varphi(C) \equiv$ the cluster C is "good". Therefore the locality condition can be read: a "good" clustering $\mathbf{C} \in \mathbf{\Phi}$ consists of "good" clusters.

F2. $\mathbf{O} \in \mathbf{\Phi}$

F3. The predicate Φ has the property of *binary heredity* with respect to the *fusibility* predicate $\psi(C_1, C_2)$, i.e.,

$$C_1, C_2 \neq \emptyset \wedge C_1 \cap C_2 = \emptyset \wedge \varphi(C_1) \wedge \varphi(C_2) \wedge \psi(C_1, C_2) \Rightarrow \varphi(C_1 \cup C_2)$$

This condition means: in a "good" clustering, a fusion of two "related" clusters produces a "good" clustering.

F4. The predicate ψ is *compatible* with clustering inclusion $\sqsubseteq$, i.e.,

$$\forall \mathbf{C}_1, \mathbf{C}_2 \in \mathbf{\Phi} : (\mathbf{C}_1 \sqsubseteq \mathbf{C}_2 \wedge \mathbf{C}_1 \setminus \mathbf{C}_2 = \{C_1, C_2\} \Rightarrow \psi(C_1, C_2) \vee \psi(C_2, C_1))$$

F5. The "*interpolation*" property holds in $\mathbf{\Phi}$, i.e., $\forall \mathbf{C}_1, \mathbf{C}_2 \in \mathbf{\Phi}$:

$$(\mathbf{C}_1 \sqsubset \mathbf{C}_2 \wedge \mathrm{card}(\mathbf{C}_1) > \mathrm{card}(\mathbf{C}_2) + 1 \Rightarrow \exists \mathbf{C} \in \mathbf{\Phi} : (\mathbf{C}_1 \sqsubset \mathbf{C} \wedge \mathbf{C} \sqsubset \mathbf{C}_2))$$

It is easy to verify that the sets of feasible clusterings $\mathbf{\Phi}^i(R)$, $i = 1, 2, 4, 5$ from Ferligoj and Batagelj (1983) satisfy the conditions F1 – F5. But, in the case of $\mathbf{\Phi}^3(R)$, the property F5 fails (in general). The counterexample is given in Figure 5 in Ferligoj and Batagelj (1983), for which we have $\{\{1,2\},\{3,4\},\{5,6\}\} \sqsubset \{\{1,2,3,4,5,6,\}\}$.

4.1.2. Criterion Function and Agglomerative Clustering. A *dissimilarity between clusters* is a function $d : (C_1, C_2) \rightarrow \mathbb{R}_0^+$ which is symmetric, i.e., $d(C_1, C_2) = d(C_2, C_1)$. Let $(\mathbb{R}_0^+, \oplus, 0, \leq)$ be an ordered abelian monoid. Then the criterion function P is *compatible* with dissimilarity d over $\mathbf{\Phi}$ iff:

$$\forall C \subseteq E : (\varphi(C) \wedge \mathrm{card}(C) > 1 \Rightarrow p(C) = \min_{(C_1, C_2) \in \Psi(C)} (p(C_1) \oplus p(C_2) \oplus d(C_1, C_2))$$

Now we can state our main theorem:

Theorem 1 *Let P be a simple criterion function compatible with d over $\mathbf{\Phi}$, $\oplus$ distributes over* min, *and F1 – F5 hold, then*

$$P(\mathbf{C}_k^*) = \min_{\mathbf{C} \in \mathbf{\Phi}_k} P(\mathbf{C}) = \min_{\substack{C_1, C_2 \in \mathbf{C} \in \mathbf{\Phi}_{k+1} \\ \psi(C_1, C_2)}} (P(\mathbf{C}) \oplus d(C_1, C_2))$$

From this theorem the following "greedy" approximation can be seen:

$$P(\mathbf{C}_k^*) \approx P(\mathbf{C}_{k+1}^*) \oplus \min_{\substack{C_1, C_2 \in \mathbf{C}_{k+1}^* \\ \psi(C_1, C_2)}} d(C_1, C_2)$$

which is the basis for the following agglomerative (binary) procedure for solving the clustering problem :

1. $k := n (= \mathrm{card}\, E)$; $\mathbf{C}(k) := \mathbf{O}$;
2. **while** $\exists C_i, C_j \in \mathbf{C}(k)$: $(i \neq j \wedge \psi(C_i, C_j))$ **repeat**
   - 2.1. $(C_p, C_q) := \mathrm{argmin}\{d(C_i, C_j) : i \neq j \wedge \psi(C_i, C_j)\}$;
   - 2.2. $C := C_p \cup C_q$; $k := k - 1$;
   - 2.3. $\mathbf{C}(k) := \mathbf{C}(k+1) \setminus \{C_p, C_q\} \cup \{C\}$;
   - 2.4. determine $d(C, C_s)$ for all $C_s \in \mathbf{C}(k)$
3. $m := k$

Note that this procedure is not an exact procedure for solving the clustering problem. But, because of the nature of clustering problem (Garey and Johnson, 1979; Shamos, 1976; Brucker, 1978), it seems that we are forced to search for and to use such procedures. This procedure has also some nice properties:

Theorem 2 *All clusterings* $\mathbf{C}(k), k = n, n-1, \ldots, m$ *obtained by the described procedure are feasible. It holds* $\mathbf{C}(k) \in \mathbf{\Phi}_k,\ k = n, n-1, \ldots, m$ *and* $\mathbf{C}(m) \in \text{Max}\,\mathbf{\Phi}$.

An agglomerative procedure is said to be *compatible* with $\mathbf{\Phi}$ iff: every clustering obtained by the procedure is feasible, and every feasible clustering can be obtained by the procedure if we can at each step fuse any pair of "related" clusters. For our procedure it can be shown:

Theorem 3 *If* $\mathbf{\Phi}$ *satisfies the conditions F1 – F5, then the described procedure is compatible with* $\mathbf{\Phi}$.

4.2. The Relocation Algorithm

The basic scheme for an adapted relocation algorithm is: Suppose that a reflexive and symmetric *neighborhood* relation $N \subseteq \mathbf{\Phi} \times \mathbf{\Phi}$ is given between feasible clusterings. Usually, for clustering problems, N is determined by the following two transformations: *moving* a unit X from cluster C_p to cluster C_q (*transition*); and *interchanging* units X and Y from different clusters C_p and C_q (*transposition*).

1. determine the initial feasible clustering $\mathbf{C}$;
2. **while** there exist $\mathbf{C}' \in N(\mathbf{C})$ such that $P(\mathbf{C}') \leq P(\mathbf{C})$ **repeat**:

2.1. move to $\mathbf{C}'$: $\mathbf{C} := \mathbf{C}'$.

The following two features are crucial in developing of an algorithm of this type: a method for randomly generating initial feasible clusterings; and an efficient procedure to scan the neighbourhood $N(\mathbf{C})$.

If the constraints are not too stringent, the relocation method can be applied directly on $\mathbf{\Phi}$; otherwise, we can transform (penalty function method) the problem to an equivalent nonconstrained problem $(\mathbf{\Pi}_k, Q, \min)$ with $Q(\mathbf{C}) = P(\mathbf{C}) + \alpha K(\mathbf{C})$ where $\alpha > 0$ is a large constant and

$$K(\mathbf{C}) = \begin{cases} 0 & \Phi(\mathbf{C}) \\ > 0 & \text{otherwise} \end{cases}$$

There exist several improvements of the basic relocation algorithm: simulated annealing, tabu search, ... (Aarts and Lenstra, 1997).

4.3. Dynamic Programming

Suppose that $\text{Min}(\mathbf{\Phi}_k, P) \neq \emptyset$. Denoting $P^*(E,k) = P(\mathbf{C}_k^*(E))$ we can derive the generalized Jensen equality (Batagelj, Korenjak and Klavžar, 1994):

$$P^*(E,k) = \begin{cases} p(E) & \{E\} \in \mathbf{\Phi}_1 \\ \min\limits_{\substack{\emptyset \subset C \subset E \\ \exists \mathbf{C} \in \mathbf{\Phi}_{k-1}(E \setminus C): \mathbf{C} \cup \{C\} \in \mathbf{\Phi}_k(E)}} (P^*(E \setminus C, k-1) \oplus p(C)) & k > 1 \end{cases}$$

This is a *dynamic programming* (Bellman) equation which, for some special constrained problems, allows us to solve the clustering problem by the adapted Fisher's algorithm. For an application of dynamic programming to hierarchies see Lebbe and Vignes (1996).

4.4. Multicriteria Clustering and Constraints

In a *multicriteria clustering problem* $(\mathbf{\Phi}, P_1, P_2, \ldots, P_k, \min)$ we have several criterion functions $P_t, t = 1, \ldots, k$ over the same set of feasible clusterings $\mathbf{\Phi}$, and our aim is to determine the clustering $\mathbf{C} \in \mathbf{\Phi}$ in such a way that $P_t(\mathbf{C}) \to \min, t = 1, \ldots, k$.

In general, solutions minimal for distinct criteria will differ from each other. This creates the problem how to find the 'best' solution so as to satisfy as many of the criteria as possible. In this context, it is useful to define the set of *Pareto efficient* clusterings: a clustering is Pareto efficient if it cannot be improved on any criterion without sacrificing on some other criterion. A multicriteria clustering problem can be approached in different ways (Ferligoj and Batagelj, 1992). It can be solved also by using constrained clustering algorithms where a selected criterion is considered as the clustering criterion and all other criteria determine the (optimizational) constraints. And conversely: a constrained clustering problem can be transformed to a multiciteria clustering problem by expressing the deviations from constraints by penalty functions.

4.5. Other

Some other optimizational approaches for solving constrained clustering problems can be found in Klauer (1994) and Hansen, Jaumard, and Sanlaville (1994).

5. Conclusion

In the paper we presented an overview of constrained clustering problems viewed through 'optimizational' glasses. For details the reader should consult the references. Some related papers are available at

`http://vlado.fmf.uni-lj.si/pub/cluster/`

Acknowledgment: This work was supported by the Ministry of Science and Technology of Slovenia, Project J1-8532.

References

Aarts, E. & Lenstra, J.K. (Eds.)(1997). *Local Search in Combinatorial Optimization.* Wiley, Chichester.

Batagelj, V. (1984). Agglomerative methods in clustering with constraints. *Preprint Series Dept. Math. Univ. Ljubljana,* 22(1984)102, 5-19.

Batagelj, V. (1985). Algorithmic aspects of clustering problem. (invited talk) *Proceedings of the 7 th International Symposium Computer at the University,* Cavtat, 27-29.5.1985, SRCE, Zagreb.

Batagelj, V. (1997). Notes on Blockmodeling. *Social Networks* **19**, 143-155.

Batagelj, V., Doreian, P. & Ferligoj, A. (1992). An Optimizational Approach to Regular Equivalence. *Social Networks* **14**, 121-135.

Batagelj, V., Ferligoj, A. & Doreian, P. (1992). Direct and Indirect Methods for Structural Equivalence. *Social Networks* **14**, 63-90.

Batagelj, V., Ferligoj, A. & Doreian, P. (1998). Fitting Pre-specified Blockmodels. C. Hayashi, & (Eds.): *Data Science, Classification, and Related Methods / IFCS'96,* Springer, Tokyo, 199-206.

Batagelj, V., Korenjak-Černe, S. & Klavžar, S. (1994). Dynamic Programming and Convex Clustering. *Algorithmica* **11**, 93-103.

Brucker, P. (1978). On the complexity of clustering problems, in: R.Henn, B. Korte, W. Oettli (Eds.): *Optimization and Operations Research.* Lecture Notes in Economics and Mathematical Systems, Vol. 157, Springer-Verlag, Berlin.

Ferligoj, A. (1986). Clustering with constraining variable. *Journal of Mathematical Sociology,* **12**, 299-313.

Ferligoj, A. & Batagelj, V. (1982). Clustering with relational constraint. *Psychometrika,* **47**, 4, 413-426.

Ferligoj, A. & Batagelj, V. (1983). Some types of clustering with relational constraints. *Psychometrika,* **48**, 4, 541-552.

Ferligoj, A. & Batagelj, V. (1992). Direct Multicriteria Clustering Algorithms. *Journal of Classification* **9**, 1, 43-61.

Garey, M.R. & Johnson, D.S. (1979). *Computer and intractability.* Freeman, San Francisco.

Gordon, A.D. (1996). A survey of constrained classification. *Computational Statistics & Data Analysis* **21**, 17-29.

Hansen, P., Jaumard, B. & Sanlaville, E. (1994). A Review of Mathematical Programming Approaches, in: P. Bertrand, B. Burtschy (Eds.): *New Approaches in Classification and Data Analysis / IFCS'93,* Springer-Verlag, Berlin, 228-240.

Klauer, K.C. (1994). Representing proximities by network models, in: P. Bertrand, B. Burtschy (Eds.): *New Approaches in Classification and Data Analysis / IFCS'93,* Springer-Verlag, Berlin, 493-501.

Lebbe, J. & Vignes, R. (1996). Optimal Hierarchical Clustering with Order Constraint, in: E. Diday, Y. Lechevallier, O. Opitz (Eds.): *Ordinal and Symbolic Data Analysis / OSDA'95,* Springer-Verlag, Berlin, 265-276.

Murtagh, F. (1985). A survey of algorithms for contiguity-constrained clustering and related problems. *The Computer Journal* **28**, 82-88.

Shamos, M.I. (1976). Geometry and statistics: Problems at the interface, in: J. Traub (Ed.): *Algorithms and Complexity (New directions and recent results).* Academic Press, New York, 251-288.

A constrained clusterwise procedure for segmentation

Tommaso Gastaldi, Donatella Vicari
Dipartimento di Statistica, Probabilità e Statistiche Applicate
Università degli Studi di Roma "La Sapienza", 00185 Roma, Italia,
e-mail: vicari@pow2.sta.uniroma1.it

Abstract: A procedure for segmentation by a constrained hierarchical clustering algorithm is proposed, using a criterion (or response) variable X and k structural factors or predictors, which yields classes different mainly as to the (conditional) distributions of X, computed within each segment. Since the procedure works on combinations of factor levels (and only indirectly on individuals), the methodology can be employed even for very large populations, with no increase of computational complexity.

Key words: Segmentation, structural factors, adjacency, constrained classification.

1. Introduction

Several methods have been proposed (Morgan et al.,1963, Celeux et al.,1982, Breiman et al.,1984, Quinlan, 1986) for generating a segmentation of a population by growing classification trees. The goal of such methods is to investigate the underlying data structure and to explain the behavior of a response variable from a given set of explanatory variables. The search for an optimal tree leads to recursive algorithms, which can be very complex for large data sets and very dependent on data (Aluja et al.,1996).

In this paper, a procedure for segmentation is proposed, based on an approach which employs constrained agglomerative cluster analysis methods. The goal is to operate a partitioning of the cells of the k-dimensional contingency table of the k predictor variables (referred to as *structural factors*) in contiguous classes of similar cells, according to a possibly multiple *criterion* or *response variable*. Such segments are structurally well-characterized, because cells which belong to each segment are contiguous and determine a "natural structure" (if any) to explain the distribution of the criterion variable.

The algorithm proposed here uses the dissimilarity between the conditional distributions of the criterion variable computed within each cell of the k-way contingency table.

The procedure is summarized by a tree diagram, which displays the whole hierarchical agglomerative process.

2. Notations and definitions

Let $k+1$ variables (S_1,..., S_k, X) be observed in a population, where S_i is the i-th *structural factor* and X is the variable with respect to which the classes will be discriminated (*dependent or criterion variable*). All the variables S_i are supposed to be at least on ordinal scale.

In the k-dimensional contingency table of S_1,...,S_k, let the h-th *cell* be denoted by C_h, $h \in \{1, \dots, p\}$, where p is the cardinality of the Cartesian product of the factor levels. Note that C_h represents a combination of the levels of k factors.

Let $D(X \mid C_h)$ be the conditional distribution of the criterion variable within each cell.

Two cells C_h and $C_{h'}$ $\left(\forall\, h,h' \in \{1,\dots,p\},\, h \neq h'\right)$ are *adjacent* if they differ exactly by one level of any factor. From a geometrical point of view, the cells C_h and $C_{h'}$ may be seen as hypercubes in a k-dimensional space and they are adjacent if they have one face (hypercube of dimension k-1) in common.

Consequently, an adjacency matrix **A** of a graph with vertices $\{C_h$: h=1, … , $p\}$, can be computed as a pxp matrix with entries {1,0}, depending on whether cells of the contingency table are adjacent or not.

Notice that the above notion of adjacency might be modified in order to take into account additional statistical information on the structural factors, or different types of constraints on the cells. See Gordon (1996) on the construction of different adjacency matrices in classification.

Let **D** be the matrix, whose entries are the dissimilarities between all pairs of the conditional distributions $D(X \mid C_h)$, $\forall\, h = 1,\dots,p$.

The dissimilarity between C_h and $C_{h'}$ will be considered according to the level of measurement of X. When X is on a nominal scale, the chi-square distance or the following ${}_1d_{h,h'}$ are considered in this paper and when X is a categorized quantitative variable with r levels, a measure of dissimilarity between C_h and $C_{h'}$ is ${}_2d_{h,h'}$:

$${}_1d_{h,h'} = \sum_{i=1}^{r} |{}^hf_i - {}^{h'}f_i| \quad (1) \qquad {}_2d_{h,h'} = \sum_{i=1}^{r-1} (x_{i+1} - x_i)|\,{}^hF(x_i) - {}^{h'}F(x_i)| \quad (2)$$

where ${}^hf$, ${}^{h'}f$ and ${}^hF$, ${}^{h'}F$ denote the observed probability density function and the cumulative distribution function of $D(X \mid C_h)$ and $D(X \mid C_{h'})$, respectively, and x_i and x_{i+1}, ($x_i \leq x_{i+1}$) are the end-points of the i-th interval of X.

3. Methodology

3.1 The algorithm

The aim of the procedure is to determine, on the basis of the pxp dissimilarity matrix **D**, subject to adjacency constraints of cells, disjoint classes of the

combinations of factor levels, by applying a constrained aggregative cluster analysis algorithm.
In the first step of the algorithm, the two closest adjacent cells C_h and $C_{h'}$ are merged to form a new class.
In the next step, dissimilarities between merged cells and other cells are updated, by considering for the new class the conditional distribution $D(X \mid C_h \cup C_{h'})$ and obtaining the new $(p-1)\times(p-1)$ dissimilarity matrix.
The binary adjacency matrix is also updated, according to the adjacency between the just-formed class and any other cell. The updated $(p-1)\times(p-1)$ matrix **A** can be eventually weighted to take into account the *adjacency degree* of each class, i.e., from a geometrical point of view, the number of hypercubes of dimension k-1 that are adjacent to each of the other classes.
In the next steps, the process of merging the two closest classes is repeated, until only one cluster is left, according to the usual aggregative hierarchical clustering.
At the end of the process a segmentation of the population can be determined, by selecting one of the generating partitions.
The main steps of the algorithm are summarized in the scheme 1.

Scheme 1: *Steps of the constrained aggregative algorithm*

Step 0: (inizialization) Let $k+1$ variables (S_1,..., S_k, X) be given. Determine the k-dimensional contingency table C of $S_1\times S_2\times\ldots\times S_k$; let p be the number of the cells of C.
Step 1: Compute the matrix $\mathbf{D}=\{d_{hh'}\}$, where $d_{hh'}$ is the dissimilarity between $D(X|C_h)$ and $D(X|C_{h'})$.
Step 2: Compute the matrix $\mathbf{A}=\{a_{hh'}\}$, where $a_{hh'}$ is an indicator variable for adjacency or an adjacency level.
Step 3: Merge the two closest adjacent cells or set of cells to form a new class C_h.
Step 4: Within the set of cells C_h compute the conditional distribution $(X|C_h)$.
Step 5: Update the dissimilarity matrix **D**.
Step 6: Update the adjacency matrix **A**.
Repeat Step 3 to **Step** 6, p-1 times.

Obviously, it is possible to perform a non-hierarchical algorithm, if a hierarchical structure is not required and a partition in a prespecified number of classes is desired, or when the number of cells p is very large.

3.2 The choice of the partition

For any two following steps, which correspond to the j and $(j-1)$-segment solution, it is possible to consider the arithmetic means of the corresponding **D** matrix entries and the ratio between such means (*mean distance ratio*), which indicates how much the average dissimilarity between segments increases (decreases), passing from the j to $(j-1)$-segment solution. Global or local peaks of these

plotting values with respect to the number of segments, can give some indications on how to choose the segmentation, because they correspond to an increase of the dissimilarity between segments, which denotes a good separation between the elements of partition.
Obviously, the choice of the number of segments is the well-known problem of identifying the most appropriate partition, by searching for a "stopping rule" to detect nested cluster structure, possibly at different levels of the hierarchical clustering, (Milligan and Cooper, 1985)

4. An application

The data were collected by an analysis of the choice of high school type, for children in families with different socio-economic features. For the sake of clarity, we recall that in Italy students, after their compulsory education, may choose among "Upper Secondary Schools" with different specializations (scientific/classical, technical, artistic, teacher training, professional training).
A number (1227) of students attending the 1995/96 class of Psychometry at the University of Rome "La Sapienza", were asked some information about their families. Table 1 displays the 3 factors recorded for each family and their levels.

Table 1

	Structural factors	Levels	Labels
S_1	Family income	3	1=Low 2=Medium 3=High
S_2	Family size	4	1 = Up to 3 components 2 = 4 components 3 = 5 components 4 = More than 5 components
S_3	Parents' highest educational level	3	1 = Up to compulsory school 2 = Secondary school 3 = University

Table 2 shows the levels of the variable under study, that is referred to each child of the family. The structure of the data set is appropriate to the proposed procedure, because the structural factors are observed for each family, while the criterion variable is recorded for each child within the family.

Table 2

	Criterion variable	Categories	Labels
X	Type of Secondary School attended by each child of the family	9	x_1= Scientific Lyceum x_2= Classical Lyceum x_3= Foreign Languages Institute x_4= Artistic Institute x_5= Teacher Training Institute (for primary and nursery school-teachers) x_6= Industrial Technical Institute x_7= Professional Institute x_8= Commercial and Business Institute x_9= Agricultural and Other Technical Institute

For each combination of factor levels (i.e., each cell of the 3×4×3 contingency table of the 3 factors), the observed conditional distribution of the type of Secondary School attended by children is given as input to the procedure.
The technique described in the previous section is employed to determine socio-economic strata that can "explain" the choice of children's Secondary School.
Clearly, in order to investigate whether different socio-economic characteristics can be considered relevant to the choice of the children, the 3 structural factors considered in the example are not exhaustive and further analysis would be desirable, but this is beyond the purpose of this paper.
Figure 1 shows a tree diagram of the whole hierarchical agglomerative process, which uses as ordinates for the various branches the cumulative levels of fusion distances (which are not monotonic because of the adjacency constraints).

Figure 1: *Tree diagram*

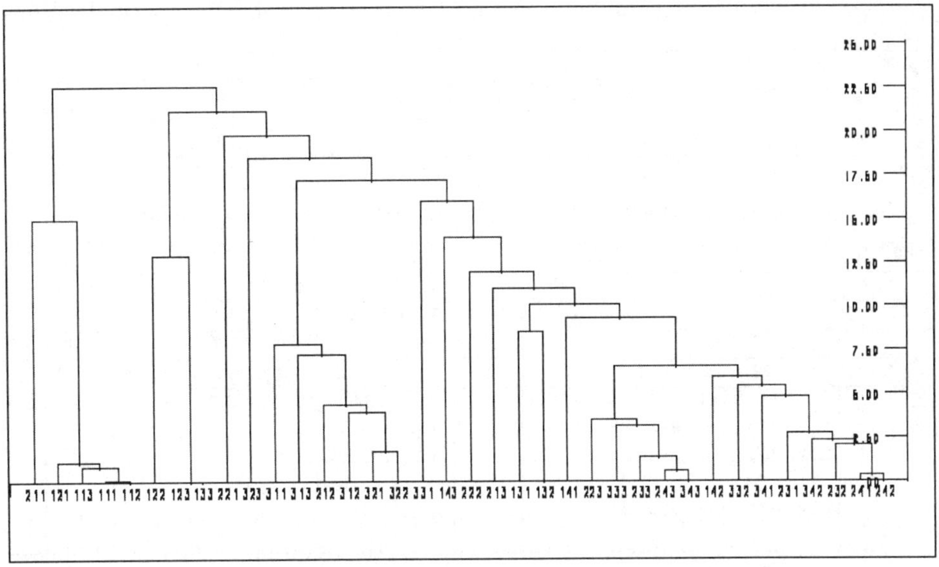

The distances between conditional distributions are computed according to formula (1).
Figure 2 shows the *mean distance ratio* with respect to the number of segments to give some indications on how to choose the segmentation.
The 9 and 5-segment solutions seem to be significant, because they correspond respectively to both a global and a local increase of the dissimilarity between segments, which denotes a good separation between the elements of partition.
The 5-segment solution is given in Table 3, that displays the factor combinations or cells in each segment and the corresponding conditional distributions of the variable X.
The analysis of the solution reveals that the segments are characterized as follows:

- SEGMENT 1: medium income, medium family size, parents' low educational level.

- SEGMENT 2: high income, medium family size, parents' high educational level.
- SEGMENT 3: low income, medium family size, parents' medium-high educational level.
- SEGMENT 4: low income, small family size, parents' low educational level.
- SEGMENT 5: medium-high income, medium-large family size (with some small families).

Figura 2: *Mean Distance Ratio Plot*

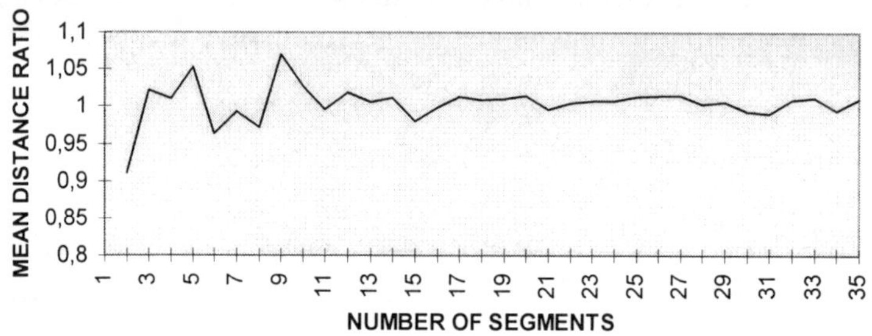

Table 3: *5-segment solution: composition and distribution of the variable X*

SEGMENT	FACTOR COMBINATIONS	X_1	X_2	X_3	X_4	X_5	X_6	X_7	X_8	X_9
1	221	0	0	0.33	0	0.33	0	0	0.33	0
2	323	0.11	0.44	0.11	0	0	0	0.33	0	0
3	122 123 133	0	0.33	0	0	0	0	0	0.67	0
4	111 112 113 121 211	0	0	0	0.02	0.08	0.35	0.45	0.10	0
5	131 132 141 142 143 212 213 222 223 231 232 233 241 242 243 311 312 313 321 322 331 332 333 341 342 343	0.23	0.12	0.04	0.13	0.09	0.04	0.09	0.21	0.05

The finest 9-segment solution reflects the same characterization because the only essential difference is in segment 5, which is separated into 4 subsegments: two formed only by one factor combination (331 and 143) and two other segments containing mostly small families with high income and medium-large families with medium income, respectively.
The procedure has also been carried out by using the chi-square distance between pairs of cells, with no significant differences in the results.

5. Conclusions

The proposed procedure determines a segmentation of a population which has well-defined structural characteristics, in the sense that different segments, each

characterized by similar combinations of factor levels, will be different mainly as to the conditional distributions of the criterion variable.
This kind of segmentation is especially useful in situations where there is a large number of statistical units. In fact, the computations involved depend on the numbers of combinations of factor levels and not on the size of the population. Therefore, the procedure is particularly appropriate for very large populations (virtually infinite) as, for example, in market segmentation, in socio-economic stratification problems, or in epidemiological surveys, where it can be employed both as a preliminary analysis or as an effective tool of investigation.

References

Aluja, T., Nafria, E. (1996). Robust impurity measures in decision trees, *Proceedings of IFCS-96: Data Science, Classification and Related Methods*, (Hayashi, C. et al., eds.), Springer Verlag, Tokio.

Breiman, L., Friedman, J.H., Olshen R.A., Stone C.J. (1984). *Classification and Regression Trees*, Wadsworth International Group, Belmont, California.

Celeux, G., Lechevallier, Y. (1982). Méthodes de Segmentation non Paramétriques, *Revue de Statistique Appliquée*, 4, 39-53.

Gordon, A. (1996). A survey of constrained classification, *Computational Statistics & Data Analysis*, 21, 17-29.

Milligan, G.W., Cooper, M.C. (1985). An examination of procedures for determining the number of clusters in a data set, *Psychometrika,* 50, 159-179.

Morgan, J.N., Sonquist, J.A. (1963). Problems in the analysis of survey data and a proposal, *Journal of American Statistical Association*, 58, 415-434.

Quinlan, J.R. (1986). Induction of Decision Trees, *Machine Learning*, 1, 81-106.

Maximal predictive clustering with order constraint: a linear and optimal algorithm

Laurent Guéguen, Régine Vignes and Jacques Lebbe
LIP6-Pôle IA UPMC-Boîte169 4, place Jussieu 75252 Paris Cedex 05
e-mail: gueguen@poleia.lip6.fr

Abstract: According to the idea of the *predictive power* of a class in a classification (Gilmour 1951), our aim is to build maximal predictive classifications (Gower 1974) of objects in a sequence, that respect the total order defined by this sequence. We propose a new dynamic programming algorithm able to discover a maximal predictive partition and which complexity is linear with the length of the sequence and with the number of possible predictors. This algorithm accepts vast range of predictor shapes and may be used for numerous possible applications. We present an experiment of this clustering algorithm on biological sequences.

Keywords: Optimal clustering, Maximal predictive classification, Maximal predictive partition, Constrained classification, Biological sequences.

1. Maximal predictive constrained classification

Since Gilmour (Gilmour 1951) many investigators have suggested that classifications need to be evaluated with respect to their *predictive power*. This means that a class is all the more useful when the knowledge *a priori* that an object belongs to that class allows for the most numerous *a posteriori* predictions on the properties of this object. Following this principle, it is better to classify flowering plants by the number of stamens than by the number of leaves because to know that a plant has three stamens permits, for example, to predict that its leaves are likely to have parallel venation. On the contrary few plant properties are correlated to the number of leaves. In many ways these ideas are largely promoted in biological applications of classification (Archie 1984) (Colless 1986) (Tassy 1991).

Gower (Gower 1974) was one of the first to give a precise operational meaning to these general guidelines. He proposes to measure the quality of a partition with the number of object properties correctly predicted by an optimal predictor assigned to each class of the partition. This is an additive criterion.

Maximal predictive classification shares with conceptual clustering (Michalsky & Stepp 1983) the advantages of (1) proposing an explanation of the creation of classes (the optimal predictors) which facilitates the output interpretation; (2) relying on the quality of these explanations for the construction of the clustering.

But unlike conceptual clustering this method is not limited to the construction of monothetic classes (i.e. able to be characterised by a conjunct of properties).

Sometimes in classification, the set of objects to be classified is structured *a priori* and this structure must be preserved by the analysis. We speak of constrained classification (Gordon 1981). We note here that the aim of the clustering procedure is neither to evaluate the structure given on the objects nor to propose another structure; the structure must be considered strictly as a constraint to be preserved. A usual structure is the total order structure: for example, if a geologist wants to study a drilling core, he has to usually take the structure of the successive layers into account. In genetics, the study of proteins or genetic sequences must consider the sequential structure of these objects.

We have tried to build maximal predictive partitions of sequences where each object (having a specific position in the sequence) is described by a letter of an alphabet and where predictors are patterns of letters in the classes. To preserve the order of the sequence, the admissible classes are not all the subsets of positions but are restricted to those including only successive positions.

2. The optimization problem

If the sequence is made of letters of an alphabet A, our problem is, for a fixed k, to build a partition of the sequence in k classes (intervals) in such a way that each class of this partition is best predicted by a pattern of letters of A. We consider that a predictor predicts correctly at a position of the sequence if it generates the right letter at this position. We define a quality for classes and for partitions as the total number of good predictions.

2.1 For sake of simplification, we first consider the case where only letters of the alphabet A are the possible predictors of classes.

If our sequence is $S = o_1 \ldots o_l \; (o_r \in \mathrm{A})_{r \in \{1,\ldots,l\}}$, for each class $c = o_i \ldots o_j$ we define the quality $\pi(\alpha, c)$ of the prediction of $\alpha \in \mathrm{A}$ on c as the number of letters of c correctly predicted by α:

$$\pi(\alpha,c) = \#\{r \in \{i,\ldots,j\}; o_r = \alpha\} \quad \text{where \# is the cardinal}$$

$$= \sum_{r=i}^{j} 1_{o_r = \alpha} \qquad \text{where } 1_{\text{true}} = 1 \text{ and } 1_{\text{false}} = 0$$

The **optimal predictors** of c are the letters α that reach the maximal quality, i.e. the most frequent letters. Thus, we define the **quality** of c as:

$$\pi(c) = \max_{\alpha \in \mathrm{A}} \pi(\alpha,c)$$

This quality measures the adequacy between a class and its optimal predictors. A class **redescription** is constructed by copying an optimal predictor along the full length of that class. For example, with A = {A, C, G, T}, the quality of c = ATTGT is 3, its optimal predictor T and the corresponding redescription of c is TTTTT.

For a partition $p = (c_1, \ldots, c_n)$ with $(c_z)_{z\in\{1,\ldots,n\}}$ classes of S, the quality of p will be:

$$\pi(p) = \sum_{z=1}^{n} \pi(c_z)$$

that is, the total number of letters of the sequence S correctly predicted when each class has a letter allocated as optimal predictor. This is according to the measure proposed by Gower. The concatenation of the redescriptions of all the classes of a partition gives a redescription of this partition. For example, if p = (ATTGT, ACC) with the same alphabet A as before, its quality is 3+2=5 and its corresponding redescription is TTTTTCCC.

2.2 If we use words as patterns of prediction, the quality π of the prediction of a word on a class will be the number of correct letters in the redescription of the class by this word. For example, if the word is AT and the class is ATTGAT, the redescription is ATATAT and the quality is 4. Using words permits to obtain more precise redescriptions.

3. A linear algorithm

For a fixed k, we want to optimize this quality function π on the set of partitions in k classes of the sequence.

3.1 First, we present our algorithm in the simplest case where predictors are single letters.

We use two properties of maximal predictive clustering :

- In a partition, if two letters of the sequence are in the same class, they are compared to the same predictor;
- The quality of a partition is the sum of the qualities of its classes, therefore any subset of the set of the classes of a maximal partition is a maximal partition of the concatenation of the elements of this subset.

In a sequence $S = o_1 \ldots o_l$, we define for any $m < l$, $d_m = o_1 \ldots o_m$.

To get the maximal predictive partitions in k classes of S, we will successively compute all the maximal predictive partitions in 1, 2, 3, ..., k classes of d_m for all m from 1 to l.

We show the recurrent steps of our dynamic programming algorithm. Let's call $P^n[m]$ the quality of the maximal predictive partitions in n classes of d_m. $P^k[l]$ is the final result we want to reach.

- $\left(P^1[m]\right)_{m=1}^{m=l}$ are the qualities of the corresponding classes $(d_m)_{1\leq m\leq l}$, easily computable.
- Recursively, having the series $\left(P^{n-1}[m]\right)_{m=n-1}^{m=l}$, we try to get $\left(P^n[m]\right)_{m=n}^{m=l}$.
  For all the letters α in A, we note for a partition $p = (c_1, ..., c_n)$:

$$\pi_\alpha(p) = \sum_{i=1}^{n-1} \pi(c_i) + \pi(\alpha, c_n)$$

and for all $m \leq l$, $P_\alpha^n[m]$ is the maximal quality $\pi_\alpha(p)$ among the set of the partitions p in n classes of d_m. These partitions reaching $P_\alpha^n[m]$ are called α-restrained maximal predictive partitions. We have:

$$\boxed{P^n[m] = \max_{\alpha \in A} P_\alpha^n[m]}$$

Now, we must get $P_\alpha^n[m]$. In a α-restrained maximal predictive partition R, the predictor of the class owning o_m is α and:

- either o_m is the only element of its class and so this partition R is a maximal predictive partition in n-1 classes of $o_1 \ldots o_{m-1}$ plus o_m alone predicted by α; so its quality is:

$$1_{\alpha=o_m} + P^{n-1}[m-1]$$

- or o_m is in the same class as o_{m-1} and, α being the predictor of this class, that partition R is a α-restrained maximal predictive partition in n classes of $o_1 \ldots o_{m-1}$ with the class owning o_{m-1} extended to own o_m. So its quality is:

$$1_{\alpha=o_m} + P_\alpha^n[m-1]$$

And, finally, $P_\alpha^n[m]$ being the best of the two possibilities (see figure 1):

$$\boxed{P_\alpha^n[m] = 1_{\alpha=o_m} + \max\left(P^{n-1}[m-1], P_\alpha^n[m-1]\right)}$$

Then, having $(P^{n-1}[m])_m$, we successively compute $(P_\alpha^n[m])_\alpha$ for all m, and at last $(P^n[m])_m$. At the end of this process, we get $P^k[l]$.

It is clear that this algorithm allows to have all the maximal predictive partitions in n classes for all d_m if we have all the maximal predictive partitions in n-1 classes for all d_m. So we get all the optimal partitions, i.e. maximally predictive in our case, of S in n classes for all n from 1 to k successively.

For each $\alpha \in$A, it takes in time $O(l)$ to compute from $(P^{n-1}[m])_m$ to $(P^n_\alpha[m])_m$ and, as we need all the #A $P^n_\alpha[m]$ to compute $P^n[m]$, the computation time from $(P^{n-1}[m])_m$ to $(P^n[m])_m$ is $O(l.\#A)$. So this algorithm has a linear time-complexity linear with the length of the sequence, the size of the alphabet and the desired number of classes : $O(k.l.\#A)$.

Figure 1: *working of the algorithm*

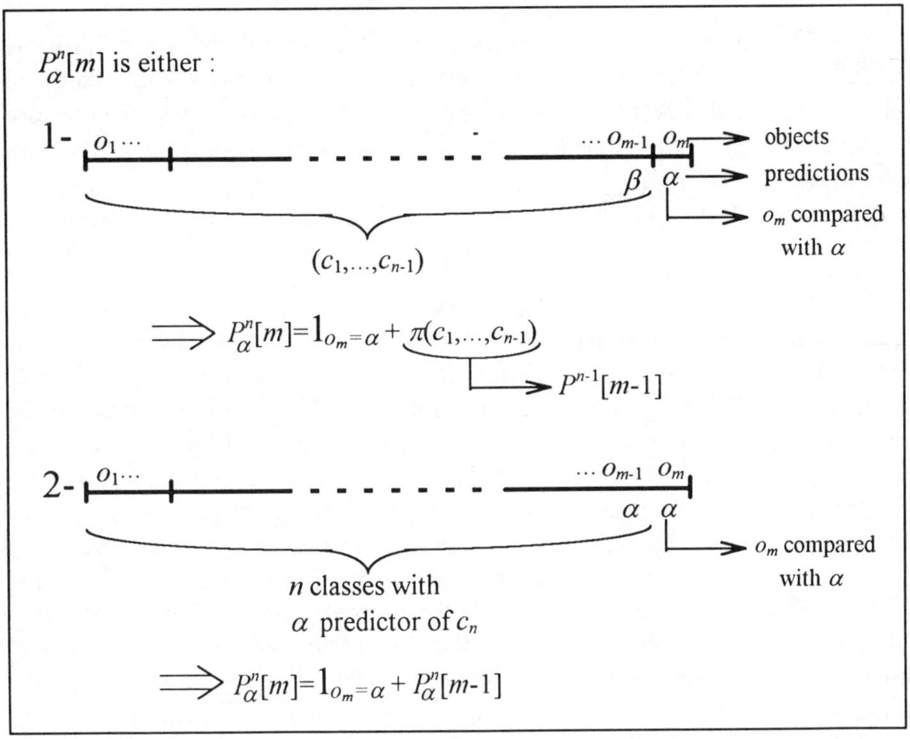

An essential feature of this algorithm is that for each possible predictor it handles variables bound to this predictor (the P^n_α with α a predictor). So, the set of all the possible predictors must be defined from the start of the computation. Here, this set was the alphabet A.

3.2 With a small change, the same algorithm is able to handle words as patterns of prediction. On the class c = ATACGTA, with AT as predictor, there are two possible redescriptions, ATATATA and TATATAT, which have different qualities. So we must consider the phase of the repeated word in the class. To do that, we can fix the phase of a predictor in a class in such a way that the last object of the class is predicted by the last letter of the pattern. On the same example c, with the predictor ACT, the redescription of c is then TACTACT. Moreover, we fix the set of all the possible predictors as the set of the circular permutations of the words in consideration. For example, for the word ACT, the possible predictors taken into account will be: ACT, CTA and TAC.

Now, if $a_1...a_r$ is a possible predictor, we define $P^n_{a_1...a_r}[m]$ as before with $a_1...a_r$ used as the predictor of the last class. It is easy to check that the former algorithm still works with the new formula:

$$P^n_{a_1...a_r}[m] = 1_{a_r=o_m} + \max\left(P^{n-1}[m-1], P^n_{a_r a_1...a_{r-1}}[m-1]\right)$$

The time-complexity grows naturally with the size of the set of possible predictors. If we want to make maximal predictive partitions with words of length λ on alphabet A, then the complexity will be $O(k.l.\#A^\lambda)$. We can also use together words of length 2 and 3 as possible predictors, in which case the time-complexity is $O(k.l.\#A^3)$.

4. Discussion

An algorithm based on Fisher's work (Fisher 1958) has already been applied to constrained predictive maximal clustering (Lebbe & Vignes 1994a) (Lebbe & Vignes 1994b). This algorithm has a quadratic complexity with the length of the sequence, but, unlike ours, is independent of the cardinal of the predictors set, and so it can be used with more general measures of homogeneity of a class.

On qualitative sequences, prediction measurement is a valuable homogeneity criterion of a class. For instance, with predictors restricted to be letters, low prediction is observed when all the letters have approximately the same frequency and high prediction when a letter is strongly predominant. In the case of more complex predictors it can also be interpreted as a more general homogeneity criteria. Moreover, because, unlike χ^2, prediction is roughly proportional to the size of the class, maximal predictive clustering has the advantage of not constraining the size of the classes in an optimal partition. The main drawback of this criteria is its potential ability to generate several equivalent partitions but, in practice, this number is a pertinent indicator of a bad value of k.

Another interest of the maximal predictive classification on sequences is the ability to produce a redescription of the original sequence which is a new optimally smoothed sequence. Moreover, the smoothing is, among all the sequences with the same length and made of k runs generated from the allowed predictors, the most similar to the initial one according to Hamming's distance. For example, if our sequence on A is S = AGCGTAGTAGTTCCC, if we look for the maximal predictive classification in 3 classes of S, and the set of predictors is A, the optimal quality is 9 and a resulting redescription is GGGGTTTTTTTTCCC. This redescription is an optimal smoothing of S according to this class specification. If the set of predictors is A^2, say the 2 letters words on A, the optimal quality is

10 and it is reached with the smoothing: AGAGTTTTTTTTCCC. If the set of predictors is A^3, the optimal quality is 14 and it is reached with the smoothing: AGCGTAGTAGTCCCC.

The smoothing gives a vision of the sequences by discarding the events with high frequency which blur large structures. It is interesting because the scale factor is given by the number of classes in the partition and not by the size of a moving window, difficult to set in a non arbitrary way. Moreover the scale factor varies naturally from place to place according to the modification of the local frequency variation of the events, and it can deal with more intricate smoothing than those made only by repetitions of letters but also with those formed by repetitions of words. Last but not least, the method chooses among all the possible smoothing an optimal one with properties clearly defined.

In our implementation of the algorithm, without changing its principles, we can deal with ambiguous predictors like {A, T} where a letter is considered predicted if it is A or T. We are also able to specify different successive sets of predictors along the successive classes of the partitions. Moreover, as in practice predictions errors have not necessarily the same consequences, we can treat different functions of letter prediction by replacing the function **1** by an other one. Finally, we consider the combination of the maximal predictive classification with a translation in a new alphabet to obtain both a partition and a characterization of the classes not only in terms of their objects but in terms of properties of their objects. In order to cover all these possibilities we have conceived a simple language for specifying the clustering problem to be solved ; each problem being specified by its set of acceptable predictors the complexity of the algorithm increases only linearly with the size of this set of predictors.

5. Applications

Because the complexity of our algorithm is linear with the length of the sequence, it is far easier to study clustering on large sequences than former algorithms. In particular, nowadays there are more and more large genetic sequences (> 10^6 letters) that are found and to be studied. Our algorithm may then be used to find structuration of complete genomes.

We demonstrated our algorithm on the complete genome of bacteriophage lambda, which is 48,502 bp length. The possible predictors are the 4 letters A, C, G and T. We stop the partitioning when we have 5 classes (Figure 2). Each class is represented by an arch marked above with its optimal predictor. In this application we notice that the five partitions which are constructed together but independently optimized define a clear hierarchy. This behavior which is quite common in practice, even with more complex predictors, will be the object of further studies.

In figure 2, the sequence is pictured by the lower horizontal line and the arrows represent the zones in which the genes are read in the direction shown by the arrows. When there is no arrow, it means that there are no genes in this part. We notice that the five classes found by our program match these different zones and their limits are rediscovered with an accuracy better than 250 bp, confirming its practical interest.

Figure 2: *Maximal predictive classification on bacteriophage lambda.*

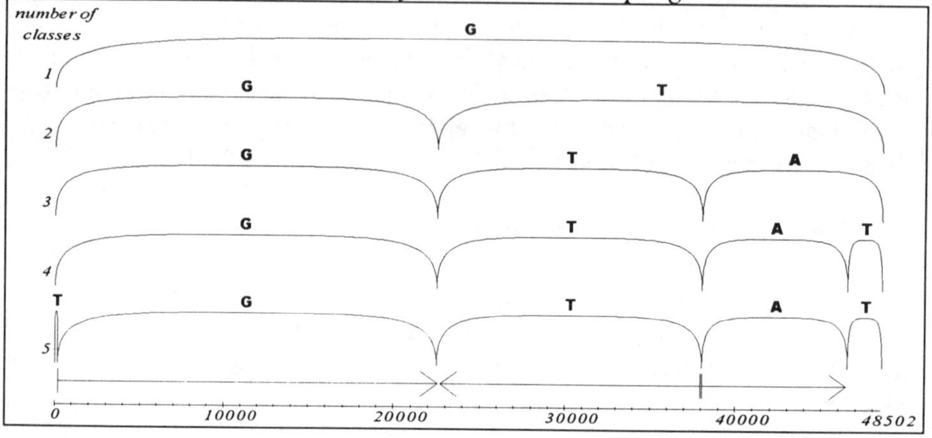

References

Archie, J. (1984). 'A new look at the predictive value of numerical classifications'. *Systematic Zoology* **33**, 30-51.

Colless, D. (1986). Classification et information. In P. Tassy (Ed.). 'L'ordre et la diversité du vivant'. Fayard. pp. 121-142.

Fisher, W. (1958). 'On grouping for maximal homogeneity'. *Journal of the American Statistical Association* **53,** 789-798.

Gilmour, J. (1951). 'The development of taxonomic theory since 1851'. *Nature* **168**, 400-402.

Gordon, A. (1981). *Classification*. Chapman and Hall.

Gower, J. (1974). 'Maximal predictive classification'. *Biometrics* **30**, 643-654.

Lebbe, J. & R. Vignes (1994a). Classification maximalement prédictive sous contrainte d'ordre total. In: *Actes de la société française de classification*. Tours. pp.127-128.

Lebbe, J. & R. Vignes (1994b). Optimal maximal predictive constrained classification. In: *Actes des septièmes entretiens du Centre Jacques Cartier*. Lyon. pp. 81-83.

Michalsky, R. & R. E. Stepp (1983). Automated construction of classifications: Conceptual clustering versus numerical taxonomy. In: *IEEE Transactions on Pattern Analysis and Machine Intelligence*. Vol PAMI-5 pp.396-410.

Tassy, P. (1991). *L'arbre à remonter le temps.* Christian Bourgois.

A linear programming based heuristic for a hard clustering problem on trees

Isabella Lari
Department of Statistics, "La Sapienza" University, P.le A. Moro 5, 00185, Rome, Italy, e-mail: lari@rosd.sta.uniroma1.it

Maurizio Maravalle
Department of Systems for Economics, University of L'Aquila, Via Assergi, 6 67100, L'Aquila, Italy, e-mail: maravalle@vaxaq.cc.univaq.it

Bruno Simeone
Department of Statistics, "La Sapienza" University, P.le A. Moro 5, 00185, Rome, Italy, e-mail: marsalis@rosd.sta.uniroma1.it

Abstract: Clustering problems with relational constraints in which the underlying graph is a tree arise in a variety of applications: hierarchical data base paging, communication and distribution network districting, biological taxonomy, and others. They are formulated here as optimal tree partitioning problems. In a previous paper, it was shown that their computational complexity strongly depends on the nature of the objective function and, in particular, that minimizing the total within-cluster dissimilarity or the diameter is computationally hard. We propose a heuristic which finds good partitions for the first problem within a reasonable time, even when its size is large. Such heuristic is based on the solution of a linear program and a maximal network flow one, and in any case it yields an explicit estimate of the relative approximation error. With minor variations a similar approach yields good solutions for the minimum diameter problem.

Key words: Contiguity-constrained clustering, trees, heuristics, computational complexity, linear programming, network flows.

1. Introduction

In many classification studies, given a set V of objects, one looks for clusters of V which are subject to constraints of different kinds. In particular, a special class of constraints has been studied by several authors, with slight differences from each other and under different names: *conditional clustering* (Lefkovitch, 1980), *relational constraints* (Ferligoj and Batagelj, 1983), *contiguity constraints* (Murtagh, 1985), *connectivity constraints* (Hansen *et al.*, 1993).
For example, if the objects to be classified are counties in a region, one usually requires that each cluster be formed by geographically contiguous counties. If

the objects are records in a relational data base, every record should be relationally accessible from any other record in the same cluster; and so on.
In this paper we shall consider a reflexive and symmetric binary relation R in V. Such relation can be effectively represented by an undirected graph G (for the graph-theoretic terminology, cf. Lovász, 1979). The node-set of G is the set V of all objects, which may always be taken as the standard set $\{1,...,n\}$; the set E of edges of G contains all pairs (2-element sets) (i,j) such that iRj. A partition $\pi = \{C_1,....,C_p\}$ of V is declared to be *feasible* if, for each $k = 1,2,...,p$, the subgraph $G(C_k)$ induced by cluster C_k is connected. The set of all feasible partitions of V into p cluster will be denoted by $\Pi_p(G)$.
In the present work we consider the particular case when the graph is a tree $T = (V,E)$. Clustering problems on trees arise in a variety of applications. Tree structures are featured in hierarchical data bases, in many communication or distribution networks (LANs, pipelines, etc.), in coastal highways, in phylogenetic trees and so on; and in many applications it makes sense to require that the clusters correspond to subtrees of the tree under consideration. In the case of a tree, by "cutting" (deleting) p-1 edges, one obtains p subtrees, whose node-sets are the p clusters of a feasible partition. Conversely, all feasible partitions into p clusters arise in this way.
As other Authors do (see, e.g., Rao, 1971; Mulvey and Crowder, 1979; Delattre and Hansen, 1980), we view clustering as a combinatorial optimization problem: with any partition π of V one can associate an indicator $f(\pi)$ of homogeneity within clusters or separation between clusters. For example, given a dissimilarity matrix $D = [d_{ij}]$, where $d_{ij} \geq 0$ for each pair (i,j) of objects, $d_{ii} = 0$ for all i and $d_{ij} = d_{ji}$ for all (i,j), two common indicators of homogeneity are the *inner dissimilarity* (or *total within-cluster dissimilarity*) and the *diameter*:

$$ind\ (\pi) = \sum_{k=1}^{p} \sum_{i,j \in C_k} d_{ij} \tag{1.1}$$

$$diam\ (\pi) = \max_{k=1,..,p} \ \max_{i,j \in C_k} d_{ij} \tag{1.2}$$

respectively. Notice that minimizing the inner dissimilarity is equivalent to maximizing the outer dissimilarity. Then the general clustering problem with relational constraints can be formulated as follows: find $\bar{\pi} \in \Pi_p(G)$ such that

$$f\ (\bar{\pi}) = \min\ (or\ \max) \left\{ f(\pi) : \pi \in \Pi_p(G) \right\} \tag{1.3}$$

In a previous paper, (Maravalle, Simeone and Naldini, 1997), the computational complexity of problem (1.3) for trees has been investigated*. It was shown that minimizing the inner dissimilarity or the diameter is NP-hard and thus, unless the widely supported $P \neq NP$ conjecture is false, problem (1.3) can be solved only by enumerative algorithms with exponential running time. This is unfortunately true even for trivial trees such as stars! On the other hand a statistician, who in real-life applications has to deal with trees having at least a few hundreds of nodes (objects), is probably interested in finding just a good feasible partition, perhaps not the best one, within a reasonable time. So he or she would be perfectly satisfied with a heuristic algorithm having the following two properties: i) it is fast; ii) an estimate of the percentage error is made available, either before or after the execution of the algorithm. In the former case one speaks of an *a priori* error bound, in the latter one of an *a posteriori* bound. Of course, for the heuristic to be of practical use, the percentage error should be relatively small.
As an example of an *a priori* error bound, in the above mentioned paper it was proved that the greedy algorithm outputs a final partition π whose outer dissimilarity, in the worst case, is within about 37% of the maximum one.
In Section 2 of the present paper, we formulate the problem of minimizing ind(π) over $\Pi_p(T)$ as an integer linear program, for whose solution we describe a heuristic consisting of at most three stages. Stage 1 consists in the (exact or approximate) solution of a (continuous) linear program, which is described in Section 3. Some optimality properties for this linear program are given in Section 4. Stage 2 and Stage 3, described in Section 5, consist in the solution of a maximal flow problem in a suitable network and in a "cut adjustment" procedure, respectively. Such a heuristic is fast enough as to solve practical problems with relatively large size. After the execution of heuristic, an upper bound on the relative approximation error is available. The heuristic has the interesting feature that, although it does not guarantee the optimality of the final partition, optimality does hold if it stops in Stage 1 or, in some cases, in Stage 2. While other heuristics (such as the greedy or the interchange ones) perform a *local search* and thus have a myopic nature, our heuristic takes a *global* approach, since it is based on the global solution of some optimization problem approximating the given one, $\min_{\pi \in \Pi_p(T)} ind(\pi)$.

On account of space, we will omit the proofs of all the theorems. For a full version of the results presented in this paper, including all the proofs, the reader is referred to Lari et al., 1998.

2. Formulation of the problem as an integer linear program

Given a tree $T = (V,E)$ with $V = \{1,..,n\}$, an integer p, $1 < p < n$ and, for each pair of nodes i and j in V, a dissimilarity index d_{ij} satisfying the above

---

* For a comprehensive exposition of complexity theory, the reader is referred to Garey and Johnson, 1979.

restrictions, we shall consider the tree clustering problem, where the objective function is (1.1). As mentioned above, the latter problem is NP-hard also for stars. In the following we will indicate with A the set of all unordered pairs (i,j) of elements of V. Moreover, for any $(i, j) \in A$, let P_{ij} be the unique path between i and j and let l_{ij} be the number of edges of P_{ij}. For any *ordered* pair of nodes i and j $(i \neq j)$, let $h(i,j)$ be the node adjacent to j in the path P_{ij}.
We define for each $(i,j) \in A$ the following variable:

$$x_{ij} = \begin{cases} 1 & \text{if } i \text{ and } j \text{ belong to the same cluster} \\ 0 & \text{otherwise} \end{cases}$$

Note that if $(i,j) \in E$, i. e. (i,j) is an edge, then one has $x_{ij} = 1$ if (i,j) is not cut and $x_{ij} = 0$ if (i,j) is cut. In order to find the constraints of the model we firstly impose the cardinality constraint on the number of edges that are cut:

$$\sum_{(i,j)\in E} x_{ij} = n - p \qquad \text{(cardinality)}$$

Now we have to impose that the clusters be connected. For each $(i,j) \in A\text{-}E$ the following statement must hold:

$$\exists\, (h,k) \in P_{ij} \text{ such that } x_{hk} = 0 \quad \Leftrightarrow \quad x_{ij} = 0.$$

The implications $\Rightarrow$ are equivalent to the following constraints:

$$x_{ij} \leq x_{hk}, \qquad (i,j) \in A\text{-}E\,,\ (h,k) \in P_{ij} \qquad \text{(order)}$$

The reverse implications $\Leftarrow$ are equivalent to the following constraints:

$$\sum_{(h,k)\in P_{ij}} x_{hk} - x_{ij} \leq l_{ij} - 1, \quad (i,j) \in A\text{-}E \qquad \text{(closure)}$$

Finally, by definition, the variables are subject to the integrality constraints and to the following constraints:

$$0 \leq x_{ij} \leq 1, \qquad (i,j) \in A \qquad \text{(unit-interval)}$$

The objective function is: $\sum_{(ij)\in A} d_{ij} x_{ij}$

We obtain the following formulation:

$$v^* = \min \sum_{(i,j)\in A} d_{ij}x_{ij} \quad \text{(Inner)}$$
$$\text{s.t. [cardinality, order, closure, unit-interval, integrality]}$$

This model has $\frac{n(n-1)}{2}$ variables and $O(n^3)$ constraints. In fact there is a variable for each pair (i,j) in A. Moreover, for each $(i,j) \in A$ we have $O(n)$ order constraints, hence the total number of such constraints is $O(n^3)$. Finally we have a closure constraint for each $(i,j) \in A\text{-}E$; then the total number of such constraints is $O(n^2)$.
The following theorems allow us to find a more concise formulation of Inner.
Theorem 1 - If all the variables x_{ij} are binary, then the closure and the order constraints imply the following $O(n^2)$ constraints:

$$x_{ij} \le x_{ih(i,j)}, \quad x_{ij} \le x_{h(i,j)j}, \qquad 1 \le i < j \le n,\ (i,j) \notin E \qquad \text{(s-order)}$$

$$x_{ih(i,j)} + x_{h(i,j)j} - x_{ij} \le 1, \qquad 1 \le i < j \le n,\ (i,j) \notin E \qquad \text{(transitivity)}$$

Theorem 2 - If $0 \le x_{ij} \le 1$ for all pairs (i,j), then the s-order constraints imply the order constraints and the transitivity constraints imply the closure constraints.
By theorems 1 and 2 the following model is equivalent to Inner.

$$v^* = \min \sum_{(i,j)\in A} d_{ij}x_{ij}$$
$$\text{s.t. [cardinality, s-order, transitivity, unit-interval, integrality]} \quad \text{(S-Inner)}$$

This model has the same variables as Inner, and $O(n^2)$ constraints. Moreover, all its constraints involve at most three variables.

3. The continuous relaxation

Consider the general integer programming problem:

$$\min\{\ f(x) : g(x) \le 0, x \in Z^n\ \}, \qquad \text{(P)}$$

where g: $\mathbb{R}^n \to \mathbb{R}^m$, f: $\mathbb{R}^n \to \mathbb{R}$, $\mathbb{R}$ and Z being the sets of reals and integers, respectively. The *continuous relaxation* of P (denoted by $\overline{\text{P}}$) is the (continuous) problem obtained by dropping the integrality constraints (this means that $x \in \mathbb{R}^n$).
Let us consider the linear program $\overline{\text{S-Inner}}$ (from now on denoted LP for short) given by the continuous relaxation of S-Inner. The optimal value v_{LP} of LP gives a lower bound of the minimum inner dissimilarity; moreover, if the optimal solution is integral, then it is the optimal solution to S-Inner. Notice

that, under the assumption that all dissimilarities are strictly positive, v_{LP} is also strictly positive in view of the cardinality constraint. If a linear program has only order constraints, then all its basic feasible solutions are integral (Picard, 1976); moreover Marcotorchino and Michaud, 1979, report that linear programming problems involving only transitivity constraints and where all variables take values in the interval [0,1], very frequently have integral solutions. Hence there is some hope that, in a non-negligible number of cases, the linear program LP admits an integral optimal solution, or, at least, that the percentage of fractional variables is small. In the former case, one has $v_{LP} = v^*$; in the latter one, it is expected that usually the lower bound on v^* given by v_{LP} is rather tight. In a very preliminary set of experiments, we solved LP for some small randomly generated test problems to see how many times non-integral solutions are obtained. In 450 examples with $n \leq 13$, we obtained non-integral solutions in about 30% of the cases and the average number of non-integral variables in the non-integral cases was about 15% of the total number of variables. Moreover, the percentage of problems with non-integral solutions increases when p approaches $n/2$. Of course, a much broader experimentation, involving larger test problems, is needed in order to reach some definitive conclusions.

Remark 1 - The optimal value of Inner is equal to the minimum inner dissimilarity of a partition $\pi \in \Pi_p(T)$, and hence it follows that the cardinality constraint in Inner may be replaced by the inequality $\sum_{(i,j)\in E} x_{ij} \geq n-p$.

A similar remark applies to S-Inner and to LP.

4. Some optimality properties

In the following we shall call *edge-variables* the variables x_{ij} such that $(i,j) \in E$ and *non-edge-variables* all other x_{ij}. We shall show that at the optimum of LP, every non-edge-variable x_{ij} can be expressed as a simple function of the edge-variables x_{hk} such that (h,k) is an edge of the path P_{ij}. Moreover, we shall exhibit a model equivalent to LP, involving only the n-1 edge-variables, the cardinality constraint and having a convex objective function.

Theorem 3 - Given an optimal solution x to Inner or S-Inner or LP, then, the following solution is optimal to Inner or S-Inner or LP, respectively:

$$\bar{x}_{ij} = \begin{cases} x_{ij}, & (i,j) \in E \\ \max\{0, \sum_{(h,k)\in P_{ij}} x_{hk} - l_{ij} + 1\}, & (i,j) \in A - E \end{cases} \tag{4.1}$$

Corollary 4 - If the dissimilarities are strictly positive, in any optimal solution to Inner or S-Inner or LP, the values of the edge-variables uniquely determine the values of the non-edge ones.

Corollary 5 - If LP has an optimal solution where all edge-variables are binary, then it has an optimal binary solution.

Remark 2 - When all edge-variables are binary, they define a set of cut edges, and hence a feasible partition of T. In this partition, nodes i and j belong to the same subtree iff x_{ij}, as given by (4.1), is equal to 1. This observation provides a direct, and more intuitive, way of computing the values of the non-edge-variables, when the edge-variables are binary.
By Theorem 3, we can consider a new model which gives the same optimal solutions as Inner and S-Inner:

$$v^* = \min \sum_{(i,j)\in A} d_{ij} \max\{\, 0, \;\; \sum_{(h,k)\in P_{ij}} x_{hk} - l_{ij} + 1 \,\} \qquad \text{(Convex)}$$
$$\text{s.t.} \;\; [\text{cardinality, unit-interval, integrality}]$$

Similarly we can consider the continuous relaxation $\overline{\text{Convex}}$ of Convex which gives the same optimal solutions as LP. In the latter models only the edge-variables appear and the objective function is convex.

5. A heuristic based on a lagrangian relaxation

In order to find good solutions to the min inner dissimilarity tree clustering problem, we consider the lagrangian relaxation of S-Inner, obtained by dropping the cardinality and the transitivity constraints:

$$v_{LR}(\lambda,\mu) = \min \sum_{(i,j)\in A} d_{ij}x_{ij} - \lambda \,(\sum_{(i,j)\in E} x_{ij} - (n-p)) + \sum_{1\le i<j\le n} \mu_{ij}(x_{ih(i,j)} + x_{h(i,j)j} - x_{ij} - 1) \qquad \text{(Inner-LR)}$$
$$\text{s.t.} \;\; [\text{s-order, unit-interval, integrality}]$$

where $\mu_{ij} \ge 0$ for $1 \le i < j \le n$, and we can consider $\lambda \ge 0$ because, by Remark 1, the cardinality constraint may be replaced by the constraint $\sum_{(i,j)\in E} x_{ij} \ge n-p$. This problem is equivalent to a maximal network flow problem (Picard, 1976), hence its linear relaxation has an optimal integral solution. Since Inner-LR has an optimal integral solution, it is equivalent to its continuous relaxation and so it is a relaxation of LP. Hence $v_{LR}(\lambda,\mu) \le v_{LP}$. Moreover, LP is a relaxation of S-Inner and then $v_{LP} \le v^*$. So the following Proposition holds.

Proposition 6 - $v_{LR}(\lambda,\mu) \le v_{LP} \le v^*$

The integrality of the solution to the continuous relaxation of Inner-LR suggests to solve it for fixed values of λ and μ, and to adjust its optimal (integral) solution if it does not satisfy the cardinality constraint. In order to obtain good values for λ and μ, we solve LP and we consider the optimal values of the dual variables associated with the cardinality and the transitivity constraints. For these particular values of λ and μ, we have $v_{LP} = v_{LR}(\lambda,\mu)$ (Nemhauser and Wolsey,

1988). Finally, in order to satisfy the cardinality constraint, we choose an integer k having a small value (1, 2 or 3) and we change the value of at most k edge-variables at a time in an optimal way.

Inner-heuristic

1. Solve LP. If the optimal solution x' is integral then STOP: x' is optimal for Inner; else let λ and μ be the optimal values of the dual variables associated with the cardinality and transitivity constraints, respectively.
2. Solve Inner-LR. Let x^* be the optimal solution. If x^* satisfies the cardinality and the transitivity constraints then STOP: x^* is optimal for Inner; else let $q = \sum_{(i,j)\in E} x^*_{ij}$.
3. If $q = n\text{-}p$ then STOP: the edge-variables induce a partition of T into p clusters.
4. (Cut adjustment procedure) If $|n\text{-}p\text{-}q| > 0$ then while $|n\text{-}p\text{-}q| > 0$
   Let $k' := \min\{k, |n\text{-}p\text{-}q|\}$
   Choose a set of k' edge-variables x_{ij} and change their values into $x^*_{ij} := 1 - x^*_{ij}$, so that $|n\text{-}p\text{-}q|$ decreases by k' units and the resulting inner dissimilarity is as small as possible.

Step 1 requires the solution of a linear program in which all coefficients are 0, 1 or -1; hence its solution time is bounded above by a polynomial in n and p (Tardos 1986). In Step 2 a maximal flow problem is solved and, also in this case, the solution time is bounded by a polynomial in n and p (Ahuja, Magnanti and Orlin, 1993). Finally the cut adjustment procedure given in Step 4 requires at most $\left\lfloor \frac{p}{k} \right\rfloor$ iterations, each of which requires the computation of ind(π) for at most $\binom{n}{k}$ feasible partitions π. Since $k \leq 3$, it follows that also the cut adjustment procedure runs in polynomial time. Therefore, the running time of the overall procedure Inner-heuristic is bounded by a polynomial in n and p.

The use of the polynomial algorithm of Tardos (1986) to solve LP is mainly of theoretical value. More practical solution approaches are:

i. use any standard linear programming package;
ii. use a primal-dual (simplex or interior point) algorithm;
iii. solve directly the non-differentiable optimization problem Convex using, e. g., one of the algorithms described in Chapter VII (by C. Lemaréchal) of Nemhauser et al., 1989.

Strategy ii. has the advantage that, if the algorithm stops before reaching an optimal solution, good dual values λ and μ_{ij} are still available. Strategy iii. is especially suited for large problems, since the number of variables involved is $O(n)$, rather than $O(n^2)$. Moreover, if LP is solved only approximately, in Step 1, in order to obtain a better solution from Inner-LR, one could apply some iterations of the subgradient method to the dual Lagrangian problem (Nemhauser and Wolsey, 1988).

Remark 3 - If v_h is the inner dissimilarity given by the solution of the Inner-heuristic and if $v_{LR}(\lambda, \mu) > 0$, then one can obtain an upper bound of the relative error by computing: $\dfrac{v_h - v_{LR}(\lambda, \mu)}{v_{LR}(\lambda, \mu)}$.

Remark 4 - The Inner-heuristic does not guarantee optimality but if it finds an optimal solution in Step 1 or in Step 2, then it recognizes it.

Example

Let us consider the following example with $n = 10$ and $p = 5$.

The following tree and dissimilarity matrix are given

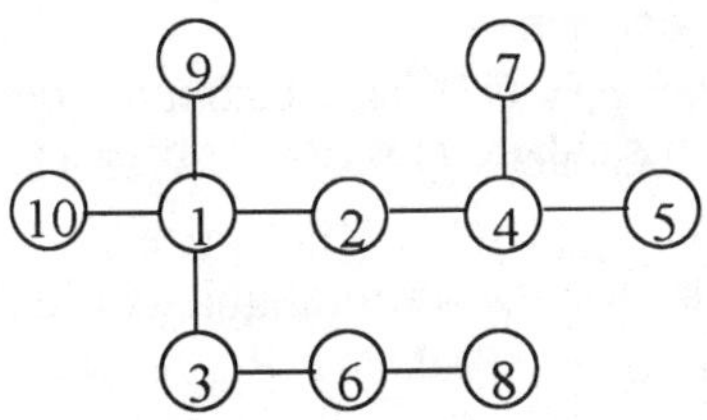

	2	**3**	**4**	**5**	**6**	**7**	**8**	**9**	**10**
1	20	20	34	9	11	22	26	5	11
2	.	8	3	33	6	26	26	13	22
3	.	.	26	28	27	11	7	22	21
4	.	.	.	27	4	7	31	32	34
5	.	.	.	.	6	39	25	11	34
6	.	.	.	.	.	36	16	3	39
7	.	.	.	.	.	.	40	10	18
8	.	.	.	.	.	.	.	38	39
9	.	.	.	.	.	.	.	.	11

The optimal solution to LP is: $x_{19} = x_{1\,10} = x_{68} = x_{9\,10} = 1$, $x_{24} = x_{36} = x_{38} = x_{45} = x_{47} = 0.5$, all other variables being equal to 0. The optimal value is $v_{LP} = 78.5$. In Figure 1 we show with a continuous line the optimal partition which gives $v^* = 79$, and by a dotted line the partition induced by the edge-variables of the solution to Inner-LR, where the inner dissimilarity is 43. By applying the cut adjustment procedure in order to reduce the number of cut edges, we obtain the optimal solution.

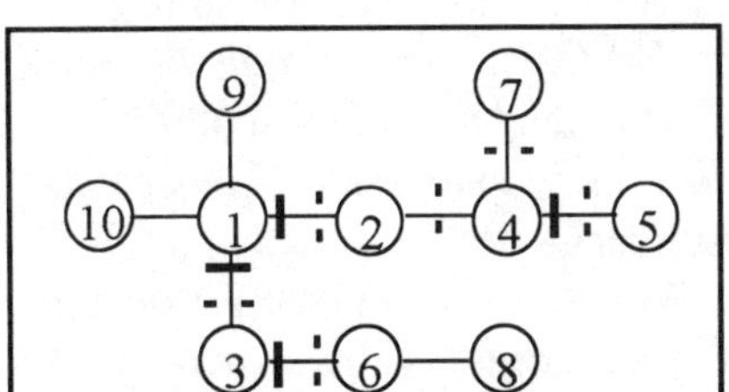

Figure 1

In Lari et al., it is shown that, by considering a formulation similar to S-Inner, it is possible to formulate as an integer linear program also the min diameter problem on trees. Moreover, good solutions to the latter problem can be obtained by applying a heuristic similar to Inner-heuristic. In this case the solution of a linear program is embedded in a binary search on the possible values of the diameter. Hence a sequence of linear programs, rather than a single one, must be solved. However, these linear programs are usually smaller, since many variables are forced to take the value zero.

References

Ahuja, R.K. & Magnanti, T.L. & Orlin, J.B. (1993), *Network flows,* Prentice Hall, New Jersey .

Delattre, M. & Hansen, P. (1980). Bicriterion cluster analysis, *IEEE Trans. on Pattern Analysis and Machine Intelligence*, 2, 277-291.

Ferligoj, A. & Batagelj, V. (1982). Clustering with relational constraints, *Psychometrika*, 47, 413-426.

Garey, M.R & Johnson, D.S.(1979). *Computers and intractability: a guide to the theory of NP-completeness*, Freeman, S.Francisco.

Hansen, P.& Jaumard, B. & Simeone, B. & Doring V.(1993). Maximum split clustering under connectivity constraints, Cahier G-93-06, GERAD, Montréal.

Lari, I. & Maravalle, M. & Simeone B. (1998). Linear Programming based heuristics for two hard clustering problems on trees, Technical Report, Dip. di Statistica, Probabilità e Statistiche Applicate, n. 2, 1998.

Lefkovitch, L.P. (1980). Conditional Clustering, *Biometrics,* 36, 43-58.

Lovász, L. (1979). *Combinatorial Problems and Exercises*, North Holland, Amsterdam.

Maravalle, M. & Simeone, B. & Naldini, R.(1997) Clustering on trees, *Computational Statistics and Data Analysis*, 24, 217-234.

Marcotorchino, J.F. & Michaud, P. (1979). *Optimisation en analyse des données*, Masson.

Mulvey, J.M. & Crowder, H.P. (1979). Cluster analysis: an application of lagrangian relaxation, *Management Science,* 25, 329-340.

Murtagh F. (1985) A survey of algorithms for contiguity-constrained clustering and related problems, *The Computer Journal*, 28, 82-88.

Nemhauser, G.L. & Rinnoy Kan, A.H.G. & Todd M.J. (1989). *Handbooks in Operations Research and Management Science*, vol. I , Elsevier Science Publishers B.V, Amsterdam.

Nemhauser, G.L. & Wolsey, L.A. (1988). *Integer and Combinatorial Optimization*, Wiley, New York.

Picard, J.C. (1976). Maximal closure of a graph and applications to combinatorial problems, *Management Science*, 22, 1268-1272.

Rao, M.R. (1971). Cluster analysis and mathematical programming, *Journal of the American Statistical Association,* 66, 622-626.

Tardos, E.(1986). A strongly polynomial algorithm to solve combinatorial linear programs, *Operations Research*, 34, 250-256.

Two Principal Points of Symmetric Distributions

Masahiro Mizuta
Division of Systems and Information Engineering, Hokkaido University, Japan,
e-mail: mizuta@main.eng.hokudai.ac.jp

Abstract: In the present article, we discuss some conditions on the existence of asymmetric 2-*principal points* of univariate symmetric distributions and show some examples of 2-principal points that are asymmetric.

k-Principal points of a distribution, proposed by Flury(1990), mean k points that minimize the expected squared distance from the nearest of the point with respect to the distribution. The criterion of principal points is the almost same as that of k-means clustering. The k-means clustering is a method for a *data set*, but the concept of principal points is a theory for *theoretical distributions*.

Many people may guess that principal points of symmetric distributions are symmetric but Flury shows counter examples. We study conditions for symmetric or asymmetric principal points. A characteristic function $p(c)$ for 2-principal points is defined in the paper. We investigate 2-principal points of mixed normal distributions with the characteristic function.

Keywords: Cluster Analysis, k-means, Optimal Allocation.

1. Introduction

Flury, Bernhard A.(1990) proposed principal points of distributions. The principal points is a set of points in p-dimensional space that approximate a given distribution. We will describe the definition of principal points.

Suppose $f(x)$ is the density function of a continuous random variable X and $F(x)$ is the distribution function. We assume that the distribution has finite second moments.

At first, we define the distance between a point $\boldsymbol{x} \in \mathrm{R}^p$ and a set of points $\{\boldsymbol{y}_j\}, \boldsymbol{y}_j \in \mathrm{R}^p$. The distance is defined by

$$d(\mathbf{x} \mid \mathbf{y}_1, \ldots, \mathbf{y}_k) = \min_{1 \le h \le k} \{(\mathbf{x} - \mathbf{y}_h)^T (\mathbf{x} - \mathbf{y}_h)\}^{1/2}. \quad (1)$$

Definition (Principal Points)

$\xi_j \in R^p$ $(1 \leqq j \leqq k)$ are called k-principal points of distribution F if

$$E_F\{d^2(X \mid \xi_1,\ldots,\xi_k)\} = \min_{y_j \in R^p} E_F\{d^2(X \mid y_1,\ldots,y_k)\}.$$

We put $P_F(k)=E_F\{d^2(X| \xi_1,\ldots,\xi_k)\}$. In the case of p=2 and k = 2, k-principal points are to find two points y_1, y_2 that minimize the function

$$M(y_1,y_2) = E(d^2(X \mid y_1,y_2)) = \int_{-\infty}^{\infty} \min\{(x-y_1)^2,(x-y_2)^2\} f(x)dx. \quad (2)$$

k-principal points $\xi_1,\ldots,$ ξ_k of random variable X minimize the expected squared distance of X from the nearest of the ξ_j. Two principal points of univariate normal distributions: $N(\mu, \sigma^2)$ are $\mu \pm \sigma(2/\pi)^{1/2}$ (Figure 1).

For the explanation of the concept of principal points, we give a concrete problem. When you would like to post a letter, you might go to the nearest letter box. What is the allocation of k letter boxes in a town relatively most convenient for citizens on the basis of the squared distances to the nearest letterbox (Figure 2)?

Figure 1: *Two principal points of the standard normal distribution*

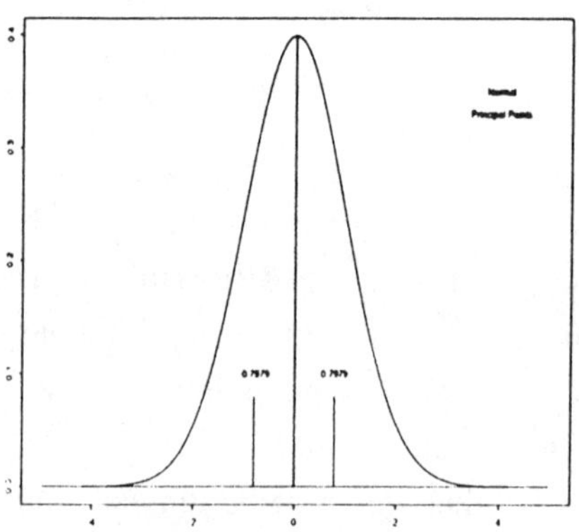

Figure 2: *A problem of letter boxes allocation*

2. Conditions on Symmetric 2-Principal Points

Flury (1990) gave a theorem for symmetric 2-principal points:

Theorem 1 (Flury 1990)

Let X denote a continuous random variable with mean $\mu = E(X)$, symmetric density $f(x)$, and finite second moments. Then the points

$$y_1 = \mu - E(|X-\mu|), \quad y_2 = \mu + E(|X-\mu|)$$

provide a local minimum of $E\{d^2(X|y_1, y_2)\}$ if and only if $f(\mu)E(|X-\mu|) < \frac{1}{2}$.

Strictly speaking, if $E\{d^2(X|y_1, y_2)\}$ is a local minimum, then $f(\mu)E(|X-\mu|) \leqq \frac{1}{2}$. If $f(\mu)E(|X-\mu|) < \frac{1}{2}$, then $E\{d^2(X|y_1, y_2)\}$ is a local minimum. We find the sufficient condition that y_1, y_2 provide a local minimum of the function $M(y_1, y_2)$.

Theorem 2 (Mizuta)

Suppose the same conditions as Theorem 1 and $\mu = 0$. Points y_1, y_2 provide a local minimum of $E\{d^2(X|y_1, y_2)\}$ if and only if

$$\begin{cases} p(c) = 0 \\ p'(c) < 0, \end{cases}$$

where $c = (y_1 + y_2)/2$, $G(c) = \int_{-\infty}^{c} x f(x)\,dx + c(1 - 2F(c))$, $p(c) = G(c)G'(c) - c$.

It can be proved with this result that the 2-principal points of *logistic distributions* and those of *Laplace distributions* are symmetric.

3. Two-Principal Points of Mixed Normal Distributions

We apply the above condition to some symmetric univariate distributions with S-Language (Becker *et al.* (1988)) to find out asymmetric 2-principal points. Specially, we deeply investigate of mixed normal distributions

$$F(x) = (1-\varepsilon)N(x;0,1) + \varepsilon N(x;0,\alpha^2) \qquad (3)$$

The distributions are classified into three types (Figure 3). Mixed normal distributions of Type 1 have unique symmetric 2-principal points. Mixed normal distributions of Type 2 have essentially unique asymmetric 2-principal points. Those of the Type 3 have essentially two candidates of 2-principal points: symmetric and asymmetric. Both candidates provide a local minimum of $M(y_1, y_2)$.

Figure 3: *Three types of 2-principal points of mixed normal distributions*

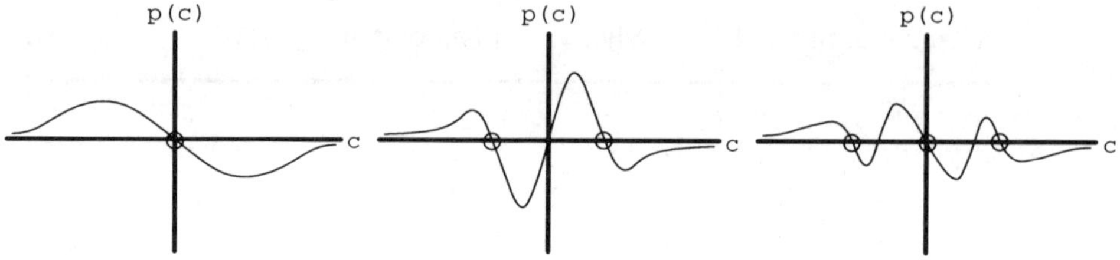

Type 1 Type 2 Type 3

Figure 4: *The relation between the three types and parameters* ε , α

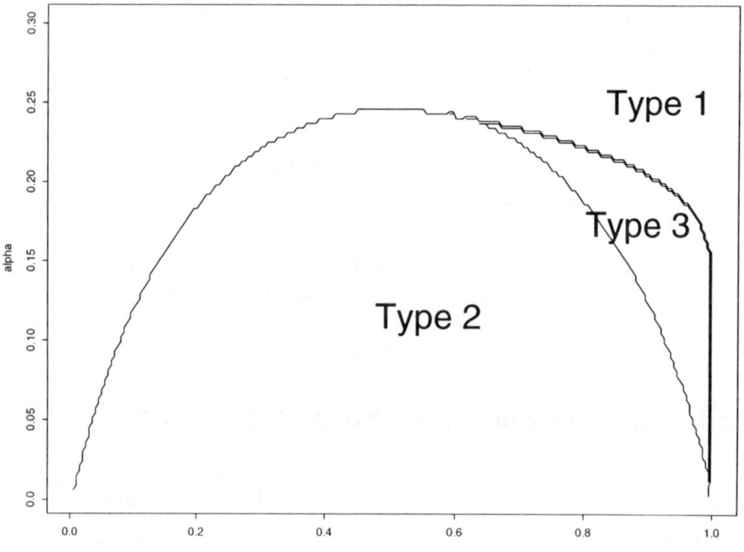

When ε =0.8 and α =0.2, we can find that the mixed normal distribution belongs to Type 3 with Figure 4. The figure 5 shows the graphs of $p(c)$, $M(y_1,y_2)$ and the range of c such that $p'(c)<0$. There are three points satisfied the conditions of the Theorem 2: c=−2.636, 0.0, 2.636. The function $M(y_1,y_2)$ has a local minimum when c=0, but not global minimum (Figure 6).

The 2-principal points of the mixed normal distribution is $(0.07623, -1.13063)$ and $(-0.07623, 1.13063)$ (Figure 7).

Figure 5: *The graphs of* $p(c)$, $M(y_1,y_2)$ *and the range of c such that* $p'(c)<0$

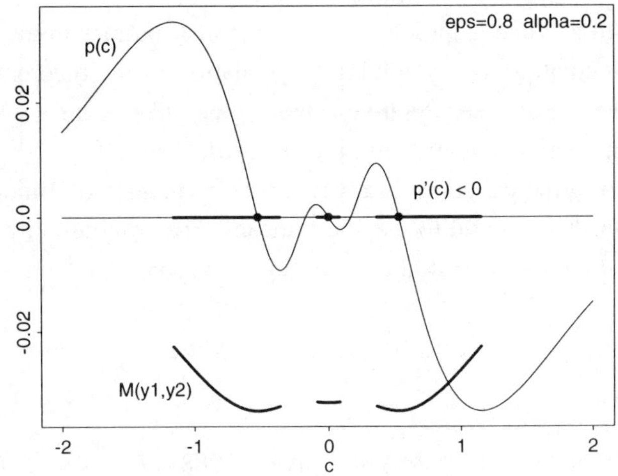

Figure 6: *Local minimum but not principal points* $(c = 0)$

Figure 7: *Global minimum i.e. principal points* $(c = 0.527)$

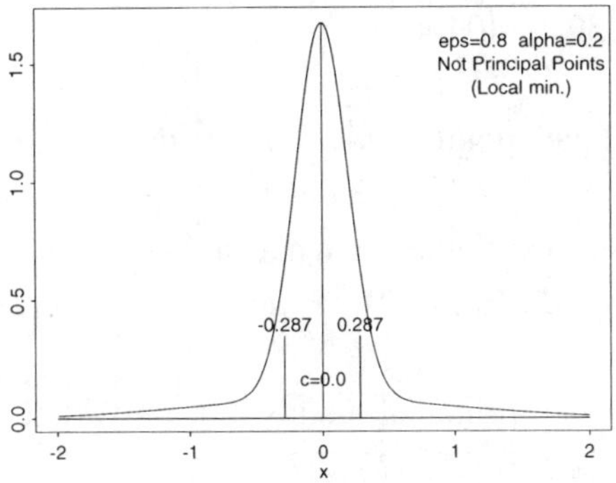

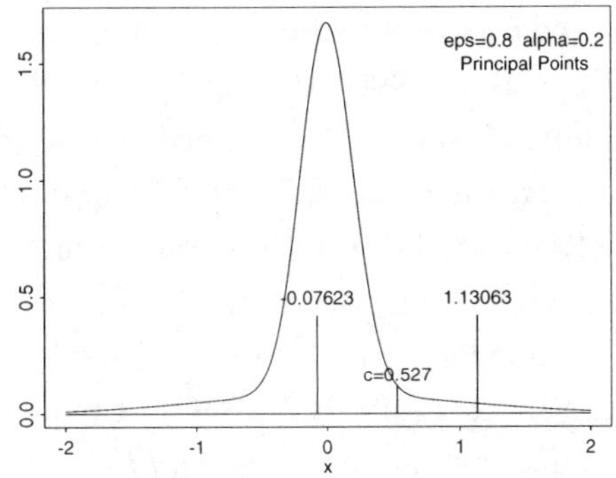

4. Concluding Remarks

In this article, we reveal sufficient conditions of 2-principal points and a family of symmetric distributions whose 2-principal points are asymmetric.

In $k > 2$, asymmetric k-principal points also exist. For example, 3-principal points of the mixed normal distribution (ε =0.8, α =0.3) are $y_1=-0.835$, $y_2=0.018$, $y_3=0.745$, which are asymmetric.

Tarpey(1994) and Li & Flury(1995) defined strongly unimodal and proved that if a univariate distribution is strongly unimodal, k-principal points are symmetric. But, with our Theorem 2, we find distributions that are not strongly unimodal and have symmetric 2-principal points.

We deal with 2-principal points of univariate distributions in the paper. So, we can find principal points by solving the equation of the theorem 2. In the other cases, we must use iterative algorithms (like k-means clustering algorithm) and numerical integration (Shimizu *et al.* (1997).)

A problem of principal points is a way for the estimation of principal points of sampling data. It is natural to use the k-means clustering algorithm, but we do not evaluate them. We would like to take up the problem in the near future.

References

Becker, R.A., Chambers, J.M. & Wilks, A.R (1988). *The New S Language*, Wadsworth & Brook/Cole Advanced Books & Software, Pacific Grove.

Chow, J.(1982). On the uniqueness of best $L_2[0,1]$ approximation by piecewise polynomials with variable breakpoints, *Mathematics of Computation*, 39, 160, 571-585.

Eubank, R.L.(1988). Optimal grouping, spacing, stratification, and piecewise constant approximation, *SIAM Review*, 30, 3, 404-420.

Flury, B.A.(1990). Principal points, *Biometrika*, 77, 1, 33-41.

Flury, B.A.(1993). Estimation of principal points, *Journal of the Royal Statistical Society, Series C, Applied statistics*, 42, 1, 139-151.

KPärna, K. (1991). Clustering in metric spaces: Some existence and continuity results for k-centers, *Analyzing and Modeling Data and Knowledge*, Springer-Verlag, 85-91.

Li, L. & Flury B.A.(1995). Uniqueness of principal points for univariate distributions, *Statistics and Probability Letters*, 25, 323-327.

Shimizu, N., Mizuta, M. & Sato, Y.(1997). Some properties of principal points in normal distribution, *Proceedings of The Ninth Korea and Japan Joint Conference of Statistics*,257-262.

Tarpey, T.(1994). Two principal points of symmetric, strongly unimodal distributions, *Statistics and Probability Letters*, 20, 253-257.

Pertinence for a Classification

N.Nicoloyannis, M.Terrenoire, D.Tounissoux
ESA 5047. Méthodes d'Analyse des Systèmes et des Structures
Université Claude-Bernard Lyon 1. Bât.101
43 Boulevard du 11 Novembre
69622 Villeurbanne Cedex. France

Abstract : The Clique Partitioning Problem constitutes a general framework for clustering, when similarities assume positive and negative values.
The effectiveness of various heuristic methods of solving the above problem has been proved by several authors.
In order to evaluate the validity of one optimal classification $\boldsymbol{P}^*$ obtained by this approach, we propose a hierarchical process, working on the clusters of $\boldsymbol{P}^*$. Thus, we obtain a family of classifications, and the concept of pertinence enables us to detect the interesting classifications.

Key words : *Clique Partitioning, Simulated annealing, Tabu search, Pertinence.*

1.Introduction

Cluster analysis sorts a set I of objects into "homogeneous" clusters. The usual approach [1,2,3,4,5] is to compute a symmetric measure of similarity s_{ij} for each pair (i,j) of objects.
In this paper, let us assume that the objects of I are evaluated with respect to p variables of the discrete type $V_1,V_2,\ldots,V_p$, and define [5] :

$$r_{ij} = \text{Card } [\{ k\in\{1,\ldots,p\} \,/\, V_k(i) = V_k(j) \}]$$

$$\bar{r}_{ij} = \text{Card } [\{ k\in\{1,\ldots,p\} \,/\, V_k(i) \neq V_k(j) \}]$$

then the quantity

$$s_{ij} = r_{ij} - \bar{r}_{ij}$$

is a similarity which takes values between $-p$ and p.

Let $\boldsymbol{P}$ be one partition of I into q clusters $\boldsymbol{P}=\{G_1,\ldots,G_q\}$, then the following quantity

$$val(\boldsymbol{P}) = \sum_{k=1}^{q} \sum_{\substack{i,j\in G_k \\ i<j}} s_{ij}$$

may be considered as a "good measure of quality" of $\boldsymbol{P}$.

Then the classification problem may be formulated as follows :

$$(\pi) \qquad \text{Max} \; [\; Val(\boldsymbol{P}) \; / \; \boldsymbol{P} \text{ partition of } I \;]$$

This problem, known as the "clique partitioning problem", is NP-hard. Exact algorithms based on branch-and-bound and polyhedral theory have been proposed by several authors [3]. But different heuristic approachs lead to good results, especially simulated annealing and tabu search [1].

In this paper we assume that problem (π) is resolved (exactly or not), and $\boldsymbol{P}^*$ will denote an optimal partition or a locally optimal partition obtained by heuristic techniques.

We propose a tool for evaluating the validity of the classification $\boldsymbol{P}^*$. This tool is based on one generalization of the similarity concept.

2. Consistency of a cluster, and similarity between two clusters

Definition.
Let $P=\{G_1,...,G_q\}$ be a partition, we define :

1- for each $k\in\{1,\ldots,q\}$

$$\begin{cases} \gamma(G_k) = \sum\limits_{\substack{i,j\in G_k \\ i<j}} s_{ij} & \text{if} \quad card(G_k) > 1 \\ \gamma(G_k) = 0 & \text{if} \quad card(G_k) = 1 \end{cases}$$

($\gamma(G_k)$ may be considered as a measure of consistency for the cluster G_k)

2- for each pair (k,l), $k,l \in\{1,\ldots,q\}$, $k\neq l$:

$$\delta(G_k, G_l) = \sum_{i\in G_k} \sum_{j\in G_l} s_{ij}$$

($\delta(G_k, G_l)$ may be considered as a measure of similarity between the clusters G_k, G_l).

As an immediate consequence of this definition, the following formula hold :

$$\delta(G_j \cup G_k, G_l) = \delta(G_j, G_l) + \delta(G_k, G_l) \quad (1)$$

$$\gamma(G_k \cup G_l) = \gamma(G_k) + \gamma(G_l) + \delta(G_k, G_l) \quad (2)$$

Then, with the partition $P=\{G_1,...,G_q\}$, we may associate the symmetric (q,q) matrix $S(P)$, defined as follows :

$$\begin{cases} s_{jk}(P)=\delta(G_j,G_k) \ \forall j,k\in\{1,...,q\} j\neq k \\ s_{kk}(P)=\gamma(G_k) \ \forall k\in\{1,...,q\} \end{cases}$$

Let P_n be the finest partition of I :

$$P_n=(\{1,\},\{2\},\ldots,\{n\}),$$

it is obvious that

$$s_{jk}(P_n)=s_{jk}$$

Moreover, we define :

$$\begin{cases} \gamma[P]=\sum_{k=1}^{q}\gamma(G_k) \\ \delta[P]=\sum_{j=1}^{q}\sum_{k=j+1}^{q}\delta(G_j,G_k) \end{cases}$$

$\gamma[P]$ measures the global consistency of the q clusters of P, and $\delta[P]$ measures the global similarity between clusters.
It may be noticed that

- $val(P) = \gamma[P]$

thus, the problem (π) consists of finding a partition which maximizes the global consistency.

- $\gamma[P] + \delta[P] = \sum_{i=1}^{n}\sum_{j=i+1}^{n} s_{ij}$

Then, (π) is equivalent to the following problem :

$$\text{Min } [\ \delta[P] \mid P \text{ partition of } I\ ]$$

- If P_1 denotes the partition of I in a unique cluster, $P_1=(\{1,2,\ldots,n\})$, then we get :

$$val(P_1)=\gamma[P_1]=\sum_{i=1}^{n}\sum_{j=i+1}^{n} s_{ij}$$

So, it appears, in this framework, that an element of the partition $P=(G_1,\ldots,G_q)$, for instance G_1 is pertinent if :

- $\gamma(G_1)\geq 0$ (the similarities prevail over the dissimilarities inside G_1)
- $\delta(G_1,G_k) \leq 0 \quad \forall k=2,...,q$ (the dissimilarities prevail over the dissimilarities between G_1 and any other cluster)

Thus we propose the following definition :

Definition.
The partition $\boldsymbol{P}=(G_1,\ldots,G_q)$ is *pertinent* if :

$$\begin{cases} \gamma(G_k) \geq 0 & k = 1,\ldots,q \\ \delta(G_j, G_k) \leq 0 & j \neq k \end{cases}$$

The interest of this definition is first illustrated by the following result.

Proposition 1.
i) One pertinent partition at least exists for any problem (π).
ii) Any optimal solution of (π) is a pertinent partition.

Proof.
i) First we make the following two remarks.
- Each diagonal element of the matrix $S(\boldsymbol{P}_n)$ is equal to zero.
- Let us assume that for a given partition $\boldsymbol{P}$, each diagonal element of $S(\boldsymbol{P})$ is positive or equal to zero, and consider the partition $\boldsymbol{P}'$, deriving from $\boldsymbol{P}$ by grouping together two clusters of $\boldsymbol{P}$, the similarity of which is positive (assuming that such a pair of clusters exists) . From (2) it follows that each diagonal element of $S(\boldsymbol{P}')$ is positive or equal to zero, moreover $\gamma[\boldsymbol{P}'] \geq \gamma[\boldsymbol{P}]$.

Thus the hierarchical process, working on the clusters of $\boldsymbol{P}_n$, grouping at each step together the two classes which have the greatest similarity, leads to a pertinent partition.

ii) Let $\boldsymbol{P}=(G_1,\ldots,G_q)$ be a non pertinent partition. Then either, there is a cluster G_k such that $\gamma(G_k)<0$, or there exists a pair of clusters (G_j,G_k) such that $\delta(G_j,G_k)>0$. Considering the first case, $\gamma(G_k)$ may be written as follows :

$$\gamma(G_k) = \frac{1}{2} \sum_{i \in G_k} \sum_{j \in G_k} s_{ij}$$

As $\gamma(G_k)<0$, there necessarily exists $i \in G_k$ such that $\sum_{j \in G_k} s_{ij} < 0$.

We define the partition $\boldsymbol{P}'$ deriving from $\boldsymbol{P}$ by splitting G_k in the two classes $\{i\}$and G_k-{i} , then we obtain :

$$\gamma[\boldsymbol{P}'] = \gamma[\boldsymbol{P}] - \sum_{j \in G_k} s_{ij} > \gamma[\boldsymbol{P}]$$

Considering the second case, we define the partition $\boldsymbol{P}'$ deriving from $\boldsymbol{P}$ by grouping $\boldsymbol{G}_j$ and $\boldsymbol{G}_k$, and we obtain :

$$\gamma[\boldsymbol{P}']) = \gamma[\boldsymbol{P}] + \delta(G_j,G_k) > \gamma[\boldsymbol{P}]$$

Thus for the two cases, $\boldsymbol{P}$ is not an optimal solution for (π).

3. Construction of a set of pertinent partitions

Let us consider a partition $\boldsymbol{P}^*=(G_1,\dots,G_q)$ which is locally optimal for (π).
First we have to be precise about what a locally optimal partition is :
For any cluster G_k of $\boldsymbol{P}^*$, for each element i of G_k. we consider the partition $\boldsymbol{P}'$, deriving from $\boldsymbol{P}^*$ by one of the two following ways :
- either transferring i from G_k to G_l ($l \neq k$),
- or defining a new cluster with i as the unique element.

Then we obtain : $val(\boldsymbol{P}') \leq val(\boldsymbol{P}^*)$
(assuming that $\gamma(\varnothing)=0$ in a logical way)
It is worth noticing that the heuristics solving (π) (especially simulated annealing and tabu search) always give a locally optimal solution.

The following result gives the method of generating pertinent partitions.

Proposition 2.
Let us assume that the locally optimal partition $\boldsymbol{P}^*$ is not pertinent. We consider the hierarchical process working on the clusters of $\boldsymbol{P}^*$, grouping at each step together, the two classes which have the greatest similarity. This process leads necessarily to a pertinent partition.

Proof.
Let us suppose that there exists a cluster G_k of $\boldsymbol{P}^*$ such that $\gamma(G_k)<0$. Then we could find (see the proof of proposition 1-ii) an element $i \in G_k$ such that $\sum_{j \in G_k} s_{ij} < 0$, and the partition $\boldsymbol{P}'$ deriving from $\boldsymbol{P}^*$ by splitting G_k in the two classes $\{i\}$and G_k-$\{i\}$ would verify :

$$\gamma[\boldsymbol{P}] > \gamma[\boldsymbol{P}']$$

That is in contradiction to the local optimality of $\boldsymbol{P}^*$ and then we get :

$$\gamma(G_k) \geq 0 \quad \forall k=1,\dots,q$$

As $\boldsymbol{P}^*$ is not pertinent, there exists a pair of clusters (G_j,G_k) such that $\delta(G_j,G_k)>0$. Then the arguments of the proof of the proposition 1-*i* allow us to show that the hierarchical process necessarily leads to a pertinent partition.

The essential interest of this approach may not be reduced to that of obtaining a single pertinent partition. For clustering applications we do not have to stop the hierarchical process when obtaining the first pertinent partition. Continuing the process, other pertinent partitions will often be obtained.
Thus the study of that set of pertinent partitions may be of great help for choosing a "good" classification, taking into account not only the value of the criterion, but also the number of clusters or any other particular aspect.

4. Examples

We have performed this method on several real-life problems.
To solve problem (π) we use a simulated annealing procedure as described in [1]. The computation time of the whole process is about 5 seconds (on a PC 200MgHz) for a problem size equal to 50 (the number of elements to classify).

Example 1.
It concerns the classification of 36 different types of cetacea which are described with respect to 15 characteristics. The detailed description and data are available from Grötschel and Wakabayashi [3].
The partition $\boldsymbol{P}^*$, obtained by the simulated annealing procedure, is the optimal partition given in [3] ; we set $\boldsymbol{P}^*=(G_1,G_2,G_3,G_4,G_5,G_6,G_7)$, using the same numbering than in [3]. In Table 1, we present the similarity matrix of $\boldsymbol{P}^*$.

Table 1 : *Similarity matrix of* $\boldsymbol{P}^*$ *(Example 1).*

	G_1	G_2	G_3	G_4	G_5	G_6	G_7
G_1	24						
G_2	-8	48					
G_3	-134	-202	102				
G_4	-280	-594	-200	704			
G_5	-32	-70	-44	-76	9		
G_6	-111	-176	-159	-304	-18	70	
G_7	-12	-46	-35	-158	-30	-66	10

According with the optimality of $\boldsymbol{P}^*$ and the proposition 1, it may be checked that $\boldsymbol{P}^*$ is pertinent (each diagonal element is positive and each non diagonal element is negative). In table 2, we present the partitions obtained at each step by the hierarchical process working on the clusters of $\boldsymbol{P}^*$. Looking at these results, it appears that Q_1,Q_2,Q_3 deriving from $\boldsymbol{P}^*$, are pertinent partitions, having less clusters than $\boldsymbol{P}^*$, for a slightly decrease of the criterion value.
Moreover the comparison of $val(\boldsymbol{P}^*)$ with $val(\boldsymbol{P}_1)$ (we note $\boldsymbol{P}_1$ the partition with a unique cluster) gives an interesting light on the significance of the partition $\boldsymbol{P}^*$.

Table 2 : *Hierarchical process (Example1).*

Partition	*Val*	Pertinence
$\boldsymbol{P}^*=(G_1,G_2,G_3,G_4,G_5,G_6,G_7)$	967	Yes (optimal)
$Q_1=(G_1 \cup G_2,G_3,G_4,G_5,G_6,G_7)$	959	Yes
$Q_2=(G_1 \cup G_2,G_3,G_4,G_5 \cup G_6,G_7)$	941	Yes
$Q_3=(G_1 \cup G_2,G_3 \cup G_7,G_4,G_5 \cup G_6)$	906	Yes
$Q_4=(G_1 \cup G_2,G_3 \cup G_5 \cup G_6 \cup G_7,G_4)$	607	Yes
$Q_5=(G_1 \cup G_2,G_3 \cup G_4 \cup G_5 \cup G_6 \cup G_7)$	-131	No
$\boldsymbol{P}_1=(G_1 \cup G_2 \cup G_3 \cup G_4 \cup G_5 \cup G_6 \cup G_7)$	-1780	No

Example 2.

It concerns the classification of 27 dogs which are described with respect to 7 characteristics. The detailed description and data are available from Saporta [6]. The partition $\boldsymbol{P^*}$, obtained by the simulated annealing algorithm, is made up of 6 clusters ; in table 3 we present its similarity matrix.

Table 3 : *Similarity matrix of* $\boldsymbol{P^*}$ *(Example 2).*

	G_1	G_2	G_3	G_4	G_5	G_6
G_1	51					
G_2	0	17				
G_3	-36	1	38			
G_4	-8	-13	-5	0		
G_5	-208	-101	-153	-5	67	
G_6	-36	-31	-129	-29	-31	26

Since $\delta(G_2,G_3)=1$, $\boldsymbol{P^*}$ is not pertinent, this is a case in which our simulated annealing algorithm does not give an optimal solution.
In table 4, we present the partitions obtained at each step by the hierarchical process working on the clusters of $\boldsymbol{P^*}$.

Table 4 : *Hierarchical process (Example 2).*

Partition	*Val*	Pertinence
$\boldsymbol{P^*}=(G_1,G_2,G_3,G_4,G_5,G_6)$	199	No
$(G_1,G_2\cup G_3,G_4,G_5,G_6)$	200	Yes (optimal)
$(G_1,G_2\cup G_3,G_4\cup G_5,G_6)$	195	Yes
$(G_1\cup G_2\cup G_3,G_4\cup G_5,G_6)$	159	Yes
$(G_1\cup G_2\cup G_3,G_4\cup G_5\cup G_6)$	99	Yes
$\boldsymbol{P_1}=(G_1\cup G_2\cup G_3\cup G_4\cup G_5\cup G_6)$	-585	No

References

[1] De Amorim S. & Barthélémy & J.P., Ribeiro C. (1992). Clustering and Clique Partitioning: Simulated Annealing and Tabu Search Approaches, in *Journal of Classification*, 9, 17-41.

[2] Bock H. (1994). Classification and Clustering : Problems for the Future, in : *E.Diday et al (eds). New Approaches in Classification and Data Analysis,* Springer Verlag, 3-24.

[3] Grötschel M.& Wakabayashi Y. (1989) A Cutting Plane Algorithm for a Clustering Problem, in : *Mathematical Programming*, 45, 59-96.

[4] Hansen P. & Jaumard B. & Sanlaville E. (1994). Partitionning Problems in Cluster Analysis : a Review of Mathematical Programming Approaches in : *New Approaches in Classification and Data Analysis,* E.Diday et al (eds), Springer Verlag, 228-240.

[5] Marcotorchino J.F. & Michaud P. (1981). Heuristic Approach to the Similarity Aggregation Problem, in : *Methods of Operations Research*, 43, 395-404.

[6] Saporta G. (1990). *Probabilités, Analyse des Données et Statistiques,* Technip, Paris,

Global stochastic optimization techniques applied to partitioning[1]

Javier Trejos Alex Murillo Eduardo Piza
PIMAD-CIMPA, School of Mathematics
University of Costa Rica
2060 San José, Costa Rica
⟨jtrejos,murillof,epiza⟩@cariari.ucr.ac.cr

Abstract: We have applied three global stochastic optimization techniques to the problem of partitioning: simulated annealing, genetic algorithms and tabu search. The criterion to be minimized is the within-variance. Results obtained are compared with those of classical algorithms and are shown to be better in nearly all cases.

Keywords: genetic algorithms, simulated annealing, tabu search, within-variance, numerical variables.

1 Partitioning

Given a set $\Omega = \{x_1, x_2, \dots, x_n\}$ in $\mathbb{R}^p$, with weights ω_i, we look for a partition $P = (C_1, \dots, C_k)$ of Ω in k classes, such that the within-variance

$$W(P) = \sum_{\ell=1}^{k} \sum_{x_i \in C_\ell} w_i d^2(x_i, g_\ell)$$

is a minimum, where g_ℓ is the centroid of C_ℓ and d is the Euclidean distance in $\mathbb{R}^p$. There exist many classical partionning methods, such as the Forgy k-means (1965), the McQueen k-means (1967), Diday's "nuées dynamiques" (1971), etc. All of these methods depend on an initial partition that is improved it by a gradient-descent type algorithm and find a local minimum of W. This characteristic of the classical methods led us to apply modern global stochastic optimization techniques to this problem. The techniques applied are simulated annealing, genetic algorithms and tabu search.

2 Global optimization for partitioning

2.1 Simulated annealing

Simulated annealing (SA) is a technique based on the Metropolis algorithm in Physical Statistics. It was proposed by Kirkpatrick et al. (1983) and by

[1]Research supported by the University of Costa Rica, project number 114-97-332.

Černy (1985), and a simplified version by Pincus (1968). Details on SA can be found in Aarts & Korst (1990). In a combinatorial optimization problem, SA improves the quality of a solution, but it has the special feature that it accepts some bad solutions. The probability of acceptance of bad solutions depends on a control parameter c that plays the role of the temperature in physical systems, and decreases as the system "gets cold", that is, as c tends to 0. Using Markov chain models, it has been proved that SA finds, asymptotically, the global optimum of the problem. Some conditions required for this convergence are *connectivity* of the states (any two states are connected by a finite sequence of states such that the probability of transition from one state to the following state is strictly positive) and *reversibility* (if the system passes from state a to state b with probability p, then the system may come back from state b to state a with the same probability p).

We have proposed (Piza & Trejos (1996)) a simulated annealing method based on the following definition of a new partition $\tilde{P}$ from partition P: (i) select at random an object $x_i \in \Omega$; (ii) select at random an index $\ell \in \{1, 2, \ldots, k\}$; and (iii) put x_i in class C_ℓ. This method is similar to one proposed in Klein & Dubes (1990). With this definition for the generation of a new state (that is, a partition), conditions of asymptotic convergence of SA are satisfied and we apply the usual SA algorithm (see Aarts & Korst (1990)). A simplification of the variation of the criterion ($\Delta W = W(\tilde{P}) - W(P)$) from one iteration to the next one makes the method converge quiet rapidly. Indeed, when object x moves from class C_j in partition P to class C_ℓ in partition $\tilde{P}$, it is only necessary to compute the difference

$$\Delta W = W(\tilde{P}) - W(P) = \frac{|C_j|}{n(|C_j| - 1)} ||g(C_j) - x||^2 - \frac{|C_\ell|}{n(|C_\ell| + 1)} ||g(C_\ell) - x||^2,$$

where $|C_j|$ is the cardinal of class C_j and we suppose that all the objects have the same weight (a similar formula can be found in the case of general weights). The centroids of the classes change in the following way:

$$g(C_j - \{x\}) = \frac{1}{|C_j| - 1}(|C_j| g(C_j) - x), g(C_\ell \cup \{x\}) = \frac{1}{|C_\ell| + 1}(|C_\ell| g(C_\ell) + x).$$

2.2 Genetic algorithm

A genetic algorithm (GA) is based on some features of the natural evolution of species. It handles a population of solutions of a combinatorial optimization problem, and combines the solutions with the so-called genetic operators, in order to avoid local minima. Classical genetic operators are selection, crossover and mutation. GA's were proposed by Holland (1976) and details can be found in Goldberg (1989). Modeling by Markov chains ensures the asymtotic convergence to the global optima (Rudolph (1994)).

For partitioning, we use a population (a sample) of partitions, described by "chromosomes" of length n and "alleles" in an alphabet of k letters. For example, let $P = (2311332321)$, where $n = 10$, $k = 3$ and x_1 is in class 2, x_2 in class 3, etc. The iterative method (Trejos (1996)) is such that in each iteration, the following genetic operators can be used: (i) Partitions are *selected* with a roulette-wheel procedure, such that the probability that a partition P will be selected is proportional to $W(P)$. (ii) A *mutation-type transfer*, by the selection of a single object, with probability p_m, and the selection at random of an index of a class where the object will be put. (iii) A *mutation-type exchange* between two objects selected at random with probability p_e. (iv) A *crossover*, such that a "cut" point is selected at random, for two partitions selected at random with probability p_c, and a combination of the partitions created as in the classical GA. (v) A *forced crossover*, such that two partitions P_1 and P_2 are selected at random with probability p_f, and a class selected at random on P_1 is copied into P_2; this procedure generates a new "son" S that will replace P_2. The last of the genetic operators is better than classical crossover, since it handles good partitioning information; indeed, objects that are together in P_1 will be together in S, a useful condition because P_1 has survived some time (that is, some iterations), thereby increasing its chances of being a good partition.

We have combined the above genetic operators with the classical Forgy k-means method: after a number of iterations, each partition in the population converges to a local optimum by the application of the k-means method. With each solution obtained, the GA continues until a stop criterion is satisfied: the maximum number of iterations is attained or the variance of the $W(P)$'s is less than a given threshold.

2.3 Tabu search

Tabu Search (TS) has been proposed by Glover et al. (1993), a combinatorial optimization technique that handles a tabu list, which forbids the transition to the states in this list. The technique is iterative and every state has a neighborhood of states that can be reached from it. The new state is selected as the best one in the neighborhood, unless it is in the tabu list.

In partitioning, we have applied TS (Murillo & Trejos (1996)) using a partition as the state of the problem and neighborhoods defined by the transfer of a single object to a new class. The tabu list is described by the class indicators of the objects that moved in the last m iterations; this list describes a set of partitions that cannot be reached unless they are the best partitions, in the sense of W, ever reached in the algorithm. The list must be long enough to avoid cycles, but not too long. Parameters such as m and the number of iterations need to be adjusted. The simplification of $\Delta(W)$, mentioned in 2.1, is also useful here, since it reduces considerably the time of execution.

3 Numerical results

We have applied the three methods to several data sets. Table 1 shows the values of W over the French Scholar Notes[2], Amiard's Fishes[3], the Thomas' Sociomatrix[2] and the Fisher's Iris[4]. Results for Forgy k-means and Ward's hierarchical classification (the tree was cut at the specified number of classes) are also presented for comparison.

Table 1: *Within-variance W using tabu search (TS), simulated annealing (SA), genetic algorithm (GA), Forgy k-means (kM) and Ward's method; % indicates the percentage of applications of the method that found the best solution reported.*

French Notes	TS W	TS %	SA W	SA %	GA W	GA %	kM W	kM %	Ward W
2 classes	28.2	100	28.2	100	28.2	100	28.2	12	28.8
3 classes	16.8	100	16.8	100	16.8	95	16.8	12	17.3
4 classes	10.5	100	10.5	100	10.5	97	10.5	5	10.5
Amiard's fishes	TS W	TS %	SA W	SA %	GA W	GA %	kM W	kM %	Ward W
2 classes	69849	96	69368	100	69368	52	69849	49	–
3 classes	32213	100	32213	100	32213	87	32213	8	33149
4 classes	18281	100	18281	100	22456	90	18281	9	19589
5 classes	14497	97	14497	100	20474	38	14497	1	14497
Thomas' sociom.	TS W	TS %	SA W	SA %	GA W	GA %	kM W	kM %	Ward W
3 classes	271.8	100	271.8	100	271.8	85	271.8	2	279.3
4 classes	235.0	100	235.0	100	235.0	24	235.0	0.15	239.4
5 classes	202.6	98	202.4	100	223.8	4	202.6	0.02	204.7
Fisher's Iris	TS W	TS %	SA W	SA %	GA W	GA %	kM W	kM %	Ward W
2 classes	0.999	100	0.999	100	0.999	100	0.999	100	–
3 classes	0.521	76	0.521	100	0.521	100	0.521	4	–
4 classes	0.378	60	0.378	55	0.378	82	0.378	1	–
5 classes	0.312	32	0.329	100	0.312	6	0.312	0.24	–

[2]Schektman, Y. (1978) "Estadística descriptiva: análisis lineal de datos multidimensionales", Ist. part. *Proc. Symp. Math. Meth. Appl. Sci.*, J. Badia et al. (eds.), Universidad de Costa Rica, San José: 9–67.

[3]Cailliez, F.; Pagès, J.P. (1976) *Introduction à l'Analyse des Données*. SMASH, Paris.

[4]Data can be found in Everitt, B.S. (1993) *Cluster Analysis*. 3rd ed. Edward Arnold, London.

4 Concluding remarks

Even though our methods depend on initial partitions, the optimization techniques used usually avoid local optima and search for a global optimum. The three methods proposed here require more time but yield better results than those obtained with classical methods. We believe the improvement in quality justifies the use of these new methods. It seems that results with simulated annealing are better than the rest, since in nearly all cases we found the global optimum in 100% of the experiments. However, simulated annealing did not find the optimum for the Fisher's Iris with 5 classes, and with 4 classes it only found it in 55% of the times. Tabu search is also a good technique for partitioning with medium-size data tables, and it is faster than simulated annealing (in another paper we will report a comparison of the time of each method). The genetic method is better than the k-means and the hierarchical, but it is worst than the simulated annealing and tabu search ones, and it is very slow in the implementation.

A deeper analysis of the parameters of the three methods is needed: length of the Markov chains, decrease of the control parameter, initial and final control parameters, in SA; stop criterion, size of the sample and probabilities for the genetic operators, in GA; and length of the tabu list and maximum number of iterations in TS. A combination of the techniques may give some insights into the handling of some of the parameters. For example, GA+SA can give a stop criterion when the control parameter of SA is close to zero; TS+SA gives also a stop criterion for TS. These ideas are under investigation at present. Some generalizations of the methods have also been undertaken: the use of distances other than the Euclidean one and the use of non-numeric data. At the present time, we are studying the application of these global optimization techniques to two-way and fuzzy partitioning, as well as to Multidimensional Scaling. Another area of current research is the application of a distributed search algorithm (Courrieu (1993)) to the partitioning problem.

References

Aarts, E.; Korst, J. (1990) *Simulated Annealing and Boltzmann Machines. A Stochastic Approach to Combinatorial Optimization and Neural Computing.* John Wiley & Sons, Chichester.

Courrieu, P. (1993) Distributed search algorithm for global optimization on numerical spaces, *Recherche Opérationnelle RAIRO*, 27(3), 281–292.

Diday, E. and coll. (1980) *Optimisation en Classification Automatique*, 2 vols. INRIA, Rocquencourt.

Klein, R. W.; Dubes, R.C. (1989) Experiments in projection and clustering by simulated annealing, *Pattern Recognition*, 22, 213–220.

Forgy, E. W. (1965) Cluster analysis of multivariate data: efficiency versus interpretability of classification, *Biometrics*, 21, 768–769.

Glover, F. et al. (1993) Tabu search: an introduction, *Annals of Opererations Research*, 41(1-4), 1–28.

Goldberg, D.E. (1989) *Genetic Algorithms in Search, Optimization and Machine Learning.* Addison-Wesley, Reading-Mass.

MacQueen, J. (1967) Some methods for classification and analysis of multivariate observations, *Proceedings 5th Berkeley Symposium*, 1, 281–297.

Murillo, A.; Trejos, J. (1996) Classification tabou basée en transferts, in *IV Journ. Soc. Franc. Classif.*, S. Joly & G. Le Calvé (eds.), Vannes, 26.1–26.4

Murillo, A.; Piza, E.; Trejos, J. (1997) A tabu search algorithm for partitioning, Preprint, CIMPA, Universidad Costa Rica.

Piza, E.; Trejos, J. (1996) Partitionnement par recuit simulé, in *IV Journ. Soc. Franc. Classif.*, S. Joly & G. Le Calvé (eds.), Vannes, 27.1–27.4

Piza, E.; Trejos, J.; Murillo, A. (1997) A genetic algorithm for clustering, Preprint, CIMPA, Universidad de Costa Rica.

Rudolph, G. (1994) Convergence of a genetic algorithm, *IEEE Transactions on Neural Networks*, 5(1), 96–101.

Trejos, J. (1996) Un algorithme génétique de partitionnement, in *IV Journ. Soc. Franc. Classif.*, S. Joly & G. Le Calvé (eds.), Vannes, 31.1–31.4

MCMC Inference for Model-based Cluster analysis

Halima Bensmail and Jacqueline J. Meulman

Department of Education, Data Theory Group, Leiden University, P.O. Box 9555, 2300 RB Leiden, The Netherlands (e-mail: bensmail@rulfsw.fsw.leidenuniv.nl).

Supported by The Netherlands Organization for Scientific Research (NWO) by grant nr. 575-67-053 for the 'PIONEER' project 'Subject Oriented Multivariate Analysis' to the first author.

Abstract: A new approach to cluster analysis has been introduced based on parsimonious geometric modelling of the within-group covariance matrices in a mixture of multivariate normal distributions, using Bayesian calculation and the Gibbs sampler. The approach answers many limitations of the Maximum likelihood approach. Here we propose to investigate more general models dealing with the shape and orientation of the clusters using the Gibbs Sampler.

Keywords: Bayes Factor, Gaussian Mixture, Gibbs Sampler.

1. Introduction

Cluster analysis has been developed mainly through the invention of empirical, and lately Bayesian investigation of ad hoc methods, in isolation from more formal statistical procedures. In recent years it has been found that basing cluster analysis on a probability model can be useful both for understanding when existing methods are likely to be successful and for suggesting new methods (Symons, 1981: McLachlan and Basford 1988; Banfield and Raftery 1993). One such probability model is that the population of interest consists of K different subpopulation, and that the density of a p-dimensional observation $\mathbf{x}$ from the kth subpopulation is $f_k(\mathbf{x}, \theta)$ for some unknown vector of parameters θ. Given observations $\mathbf{x} = (\mathbf{x_1}, \ldots, \mathbf{x_n})$, we let $\nu = (\nu_1, \ldots, \nu_n)^t$ denote the identifiying labels, where $\nu_i = k$ if $\mathbf{x}_i$ comes from the kth subpopulation. In the so-called classification maximum likelihood procedure, θ and ν are chosen to maximize the likelihood

$$\prod_{i=1}^{n} \sum_{k=1}^{K} \pi_{\nu_i} f_{\nu_i}(\mathbf{x}_i, \theta_k), \qquad (1)$$

In Bayesian classification, we assume that the data to be classified $\mathbf{x}_i$, ($i = 1, \ldots, n$, $\mathbf{x}_i \in \mathbf{R}^p$) arise from a random vector with density $p(\mathbf{x}, \theta)$ as in (1)

and that the corresponding classification variables ν_i are unobserved. We are concerned with Bayesian inference about the model parameter θ, π and the classification indicators ν_i.
Bensmail et al (1997) proposed a Bayesian approach to overcome the difficulties which arose in Banfield and Raftery's work (1993) –hereafter BR–. In particular, with the BR algorithm, the user must specify the shape matrix A when using the model $[D_k A D_k^t]$. Dasgupta and Raftery (1995) used the same model to detect features in spatial point process where the matrix shape was unknown but they restricted the diagonal terms of the matrix shape to be equal: $A = diag\{1, \alpha, \ldots, \alpha\}$ and to have a low value (α is the shape parameter). In our approach, the shape matrix is completely unknown and is estimated using a fully Bayesian calculation.
In Bensmail et al (1997), four parsimonious models were explicitely considered to classify the data. These are the spherical models $[\lambda I]$, $[\lambda_k I]$, the linear model $[\Sigma]$ and the proportional model $[\lambda_k \Sigma]$. We propose to extend this work to the family of models where the covariance matrix is represented by $[\lambda D_k A D_k^t]$ and $[\lambda_k D_k A D_k^t]$ using a fully Bayesian inference via Gibbs sampling. These overcomes all the limitations mentioned. We use the Laplace-Metropolis approximation to calculate Bayes factor (Lewis and Raftery 1997); this approximation is used to choose the model and to determine the number of groups simultaneously. In section 2, we outline Bayesian calculation of the models, and how Bayes factor is approximated from the MCMC output. In section 3 we show the methods at work on diabetes data (Reaven and Miller 1979).

2. Model and Estimation

We assume that data $\mathbf{x}_i, (i = 1, \ldots, n; \mathbf{x}_i \in \mathbf{R}^p)$ to be classified arise from a random vector with density

$$p(\mathbf{x}, \theta) = \prod_{i=1}^{n} \sum_{k=1}^{K} \pi_k f_k(\mathbf{x}_i, \mu_k, \Sigma_k)$$

where $f(., \mu_k, \Sigma_k)$ is the multivariate density function with mean μ_k and covariance matrix Σ_k, $\pi = (\pi_1, \ldots, \pi_K)$ is the mixing proportion ($\pi_k \geq 0, \sum_{k=1}^{K} \pi_k = 1$). The approach is based on a variant of the standard spectral decomposition of Σ_k namely

$$\Sigma_k = \lambda_k D_k A D_k^t,$$

where λ_k is a scalar, $A_k = diag\{1, a_{k2}, \ldots, a_{kp}\}$, $1 \geq a_{k2} \geq \ldots \geq a_{kp} > 0$ and D_k is an orthogonal matrix. We are concerned with Bayesian inference about the model parameters θ, π and the classification indicators ν_i.

MCMC methods provide an efficient and general recipe for Bayesian analysis of mixtures. For instance, many authors have used the Gibbs sampler or the Data Augmentation method of Tanner and Wong (1987) for estimating the parameters in univariate and multivariate Gaussian mixtures and proved that both algorithms converge to the true posterior distribution of the mixture parameter.
The models we are investigating are: $[D_k A D_k^t]$, $[\lambda_k D_k A D_k^t]$ and $[\Sigma_k]$. We use conjugate priors for the parameters π and θ of the mixture model. The prior distribution of the mixing proportions is a Dirichlet distribution

$$(\pi_1, \ldots, \pi_K) \sim D(\alpha_1, \ldots, \alpha_K)$$

and the prior distributions of the means μ_k of the mixture components conditionally on the variance matrices Σ_k are Gaussian:

$$(\mu_k | \Sigma_k) \sim \mathcal{N}_p(\xi_k, \Sigma_k / \tau_k)$$

The prior distributions of the shape matrix terms $a_1, \ldots, a_p$ and the volumes $\lambda_1, \ldots, \lambda_K$ are gamma inverse

$$a_t \sim \mathcal{IG}(r_t/2, \rho_t/2) \quad (t = 1, \ldots, p)$$

$$\lambda_k \sim \mathcal{IG}(l_k/2, s_k/2) \quad (k = 1, \ldots, K).$$

We estimate the models by simulating from the joint posterior distribution of π, θ and ν using the Gibbs sampler and following the same steps as in Bensmail et al (1997), so the Gibbs sampler steps go as follows:
1. Simulate the classification variables ν_i according to their posterior probabilities $(t_{ik}, k = 1, \ldots, K)$ conditional on π and θ,

$$t_{ik} = \frac{\pi_k f_k(\mathbf{x}_i | \mu_k, \Sigma_k)}{\sum_j \pi_j f_k(\mathbf{x}_i | \mu_j, \Sigma_j)} \qquad i = 1, \ldots, n.$$

2. Simulate the vector π of mixing proportions according to its posterior distribution conditional on the ν_i's.
3. Simulate the parameters θ of the model according to their posterior distributions conditional on the ν's.

For choosing both a "model M_2 against a model M_1" and the number of groups in one step, we compute the approximate Bayes factor from the Gibbs sampler output using the *Laplace-Metropolis estimator* (Bensmail et al (1997)) given by the formulas:

$$B_{12} = \frac{p(\mathbf{x}|M_2)}{p(\mathbf{x}|M_1)} = \frac{|\Psi_2|^{1/2} p(\mathbf{x}|\tilde{\theta}^{(2)}) p(\tilde{\theta}^{(2)})}{|\Psi_1|^{1/2} p(\mathbf{x}|\tilde{\theta}^{(1)}) p(\tilde{\theta}^{(1)})}$$

$\tilde{\theta}^{(i)}, (i = 1, 2)$ is the posterior mode of $\tilde{\theta}^{(i)}$ ($\theta^{(i)}$ is the parameter of the Model M_i), and $\Psi^{(i)}$ is minus the inverse Hessian of $h(\theta) = \log p(\mathbf{x}|\theta)p(\theta)$, evaluated at $\theta = \tilde{\theta}^{(i)}$.
The Laplace method requires the posterior mode, $\tilde{\theta}$, and $|\Psi|$. The Laplace-Metropolis estimator estimates these from the Gibbs sampler output. The Likelihood at the approximate posterior mode is

$$p(\mathbf{x}|\tilde{\theta}) = \prod_{i=1}^{n} \sum_{k=1}^{K} \tilde{\pi}_k f_k(\mathbf{x}_i|\tilde{\mu}_k, \tilde{\Sigma}_k)$$

which are then substituted into the previous equation to obtain the Bayes factor.

3. Diabetes example

Reaven and Miller (1979) described and analyzed data for 145 subjects, consisting of the area under a plasma glucose curve (glucose area), the area under a plasma insulin curve (insulin area) and steady-state plasma glucose reponse (SSPG). The subjects were clinically classified into three groups, chemical diabetes (Type 1), overt diabetes (Type 2), and normal (nondiabetic). Symons (1981) reanalyzed the data using seven different clustering criteria. The data have the three-dimensional shape of a boomerang with two wings and a fat middle as we can see in Figure 1. One of the wings corresponds to patients with overt diabetes, the other wing is composed primarily of patients with chemical diabetes, and the "fat middle" is composed of normal patients. Banfield and Raftery(1993) evaluated the criterion based on the model $[\lambda_k D_k A D_k^t]$, where $A = diag\{1, \alpha, \alpha\}$, which means that the clusters have different sizes and orientations but have the same "tubular" shape. The estimated values of α are 0.09, 0.19 and 0.34, and the results reported by BR algorithm is $\alpha = 0.2$. The model $[\lambda_k D_k A D_k^t]$ with three groups is favored quite strongly over the alternatives. The posterior means of $\lambda_1 A = diag\{0.09, 0.09, 0.43\}$, $\lambda_2 A = diag\{0.30, 0.36, 0.47\}$ and $\lambda_3 A = diag\{0.13, 0.16, 0.35\}$.
The optimal classification resulted in only 9% of the points being misclassified which are mostly situated at the border of the clusters. Ten percent error rate was obtained by the BR algorithm, but the misclassified points were dispersed over the three clusters. The present results compare our approach favorably to the procedures of Symons(1981) and BR(1993).

Figure 1: *Three two-dimensional projections of the three-dimensional diabetes data. The symbols indicate the clinical classification of subjects as having chemical diabetes, overt diabetes, or being normal. The Lower right panel shows the three clusters in the diabetes data found by our approach using the model* $[\lambda_k D_k A D_k^t]$.

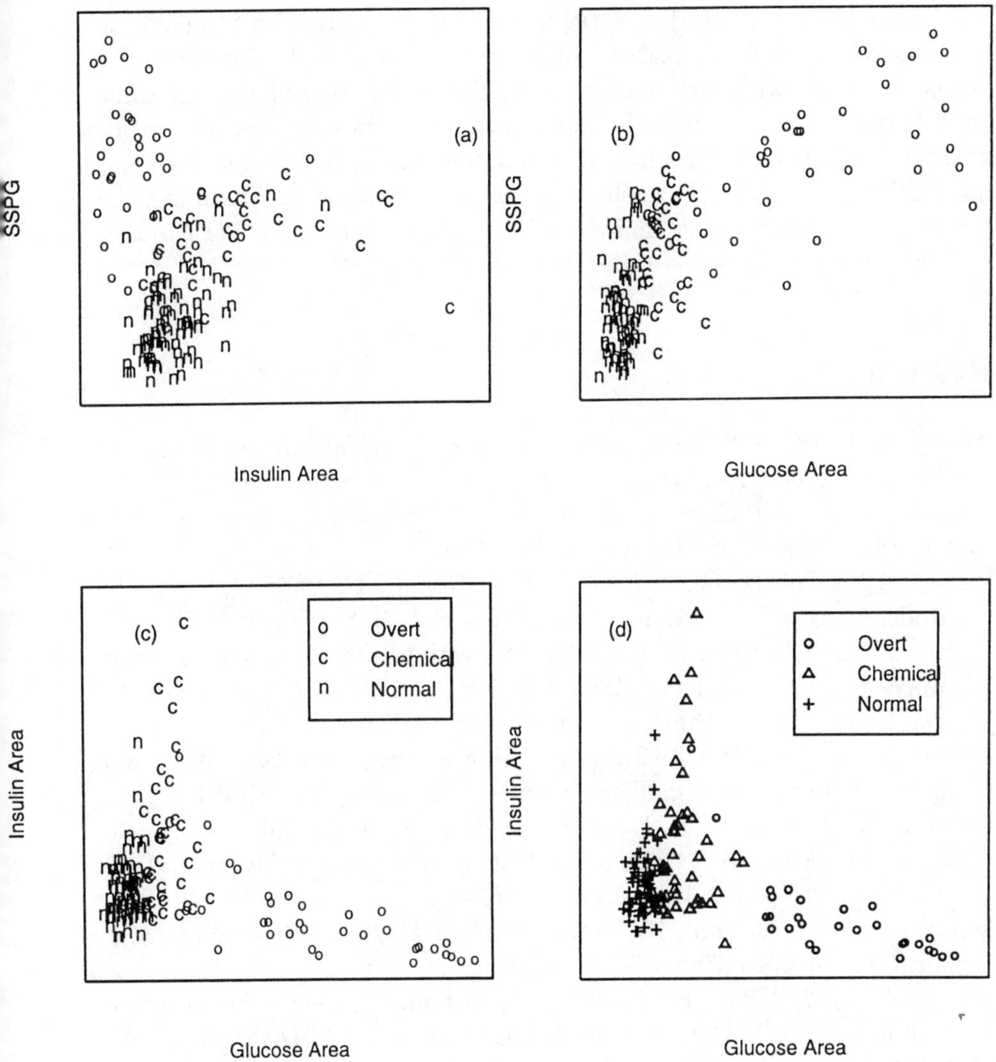

4. Discussion

We have presented a fully Bayesian analysis of a model-based clustering approach based on a mixture model in which the features of interest are presented my multivariate normal densities, specified by the shape being the same across clusters, while the features and orientations are different. Alternative frequentist approach, consist of maximizing the mixture likelihood using the EM algorithm. Dasgupta and Raftery (1995) considered the same model in which the features of interest are presented by multivariate normal densities with high linearity, specified by the *shape* being the same across features, which make the model restricted. In our case, we considered the case where shape, orientation and volume are totally unknown and no restriction is made on the shape parameter, thus our approach is more general and overcomes the limitations of Dasgupta and Raftery (1995) and is an extention of the work of Bensmail et al (1997).

References

Anderson, T. W. (1984,1958). *An Introduction to Multivariate Statistical Analysis.* Wiley, New York.

Banfield, J. D., & Raftery, A. E. (1993). Model-based Gaussian and non Gaussian Clustering, *Biometrics*, 49, 803-821.

Bensmail, H., Celeux, G., Raftery, A. E. & Robert, C.(1997). Inference in Model-Based Cluster Analysis, *Statistics and Computing*, 7, 1-10.

Dasgupta, A. & Raftery, A. E. (1995). Detecting features in spacial point processes with clutter via Model-based clustering, *Technical Report No 295*, Department of Statistics, University of Washington.

Lavine, M. & West, M. (1992). A Bayesian method for classification and discrimination, *The Canadian Journal of Statistics*, 20, 451-461.

Lewis, S. M. & Raftery, A. E. (1997). Estimating Bayes factors via posterior simulation with the Laplace-Metropolis estimator, *Journal of the American Statistical Association*, to appear.

McLachlan, G. & Basford, K. (1988). *Mixture Models: Inference and Applications to Clustering.* Marcel Dekker, New York.

Reaven, G. M. & Miller, R. G. (1979). An attempt to define the nature of chemical diabetes using a multidimensional analysis, *Diabetologia*, 16, 17-24.

Symons, M. (1981). Clustering criteria and multivariate normal mixtures, *Biometrics*, 37, 35-43.

Tanner, M. & Wong, W. (1987). The calculation of posterior distributions by data augmentation (with discussion), *Journal of the American Statistical Association*, 82, 528-550.

On the Asymptotic Normality of a Resubstitution Error Estimate [1]

László Györfi
Dept. of Computer Science and Information Theory
Technical University of Budapest
1521 Stoczek u. 2, Budapest, Hungary. e-mail:gyorfi@inf.bme.hu

Márta Horváth
Dept. of Computer Science and Information Theory
Technical University of Budapest
1521 Stoczek u. 2, Budapest, Hungary. e-mail:marti@inf.bme.hu

Abstract: The resubstitution error estimate for the partitioning classification rule from a sample $(X_1, Y_1), (X_2, Y_2), \ldots, (X_n, Y_n)$ is shown to be asymptotically normal under the condition that X has a density f, if the partition consists of rectangles.

Key Words: error estimation, central limit theorem, partitioning classification rule

1. Introduction

Let X be the d-dimensional feature vector with distribution μ, and let Y be the binary valued label. Denote the aposteriori probabilities by

$$P_i(x) = \mathbf{P}\{Y = i | X = x\},\ i = 0, 1$$

and the Bayes error probability by

$$L^* = \mathbf{E}\{\min(P_0(X), P_1(X))\}.$$

For the pattern recognition problem let $(X_1, Y_1), (X_2, Y_2), \ldots$ be i.i.d. random variables such that (X_i, Y_i) have the same distribution as (X, Y). Let $\mathcal{P}_n = \{A_{nj}, j = 1, 2, \ldots\}$ be a partition of $\mathcal{R}^d$, and let $A_n(x)$ denote the the subset in the partition that includes x. The resubstitution estimate $\hat{L}_n$ counts the number of errors commited on the training sequence $(X_1, Y_1), (X_2, Y_2), \ldots, (X_n, Y_n)$ by the classifier, i.e. for a classifier g_n it is defined as

$$\hat{L}_n = \frac{1}{n} \sum_{i=1}^{n} I_{\{g_n(X_i) \neq Y_i\}},$$

which for the partitioning classification rule

$$g_n(x) = \begin{cases} 0 & \text{if } \sum_{i=1}^{n} I_{\{Y_i=1\}} I_{\{X_i \in A_n(x)\}} \leq \sum_{i=1}^{n} I_{\{Y_i=0\}} I_{\{X_i \in A_n(x)\}} \\ 1 & \text{otherwise} \end{cases}$$

[1]The research was supported by the Computer and Automation Institute of the Hungarian Academy of Sciences (MTA SZTAKI).

can be written in the following form:

$$\hat{L}_n = \sum_{j=1}^{\infty} \min\{\nu_n(A_{nj}), \mu_n(A_{nj}) - \nu_n(A_{nj})\}, \tag{1}$$

where

$$\nu_n(A) = \frac{1}{n} \sum_{i=1}^{n} I_{\{X_i \in A\}} Y_i$$

and

$$\mu_n(A) = \frac{1}{n} \sum_{i=1}^{n} I_{\{X_i \in A\}}.$$

$\hat{L}_n$ is an estimate of the Bayes error R_n^* restricted to the partition $\mathcal{P}_n$:

$$R_n^* = \sum_{j=1}^{\infty} \min\{\nu(A_{nj}), \mu(A_{nj}) - \nu(A_{nj})\},$$

where

$$\nu(A) = \mathbf{E}\nu_n(A).$$

2. Asymptotic Normality

Concerning the resubstitution error estimate for partitioning rule the following inequalities are known (see Sec. 23.2 in Devroye, Györfi and Lugosi (1996)): for arbitrary partition

$$Var(\hat{L}_n) \le \frac{1}{n}$$

and

$$\mathbf{E}\hat{L}_n \le R_n^*.$$

For finite partition of size m_n

$$R_n^* - \mathbf{E}\hat{L}_n \le \sqrt{\frac{2m_n}{n}}.$$

Our aim is to derive the asymptotic normality of the distribution of $\hat{L}_n$.

Theorem 1 *Consider the partitions where A_{nj} are d-dimensional rectangles. Let $a_n^i(x)$, $i = 1, 2, ..., d$, denote the sidelengths of $A_n(x)$. Assume that for all x there exists a $K(x)$ so that $\frac{a_n^i(x)}{a_n^j(x)} \le K(x)$ for all $1 \le i, j \le d$ and n.*

Assume that for each sphere S centered at the origin

$$\lim_{n\to\infty} \sup_{A_{nj} \cap S \ne \emptyset} diam(A_{nj}) = 0 \tag{2}$$

and

$$\lim_{n\to\infty} \frac{|\{A_{nj} \cap S \neq \emptyset\}|^2}{n} = 0 \tag{3}$$

and

$$\lim_{n\to\infty} \frac{\log n}{n\lambda(A_n(x))} = 0, \tag{4}$$

where λ is the Lebesgue measure. If μ has a density f and

$$\mathbf{P}\{D(X) = 0\} = 0,$$

where $D(x) = P_0(x) - P_1(x)$, then there is a sequence C_n such that

$$n^{1/2}\left(\hat{L}_n - C_n\right)/\sqrt{L^*/2} \xrightarrow{\mathcal{D}} N(0,1).$$

Theorem 2 *If, in addition to the conditions of Theorem 1, for*

$$\tau(A) = \mu(A) - 2\nu(A) = \int_A D(x)\mu(dx)$$

there is a constant c such that

$$|\tau(A_{nj})| > c\mu(A_{nj}) \tag{5}$$

for all n and j then

$$n^{1/2}\left(\hat{L}_n - R_n^*\right)/\sqrt{L^*/2} \xrightarrow{\mathcal{D}} N(0,1). \tag{6}$$

Remark.
For cubic partitions with size h_n these conditions mean $h_n \to 0$, $nh_n^{2d} \to \infty$ and $nh_n^d/\log n \to \infty$ as $n \to \infty$.

3. Sketch of the proof of Theorem 1

The main difficulty in proving the theorem is that it states asymptotic normality of infinite sums of dependent random variables. Thus, we need a technique to extend asymptotic normality of the partial sums to the infinite sum. Assume that the $\{A_{nj}\}$ are ordered according to non-decreasing distances of their centers from the origin. For $\alpha \in (0,1)$ choose the integer m_n such that

$$1 - \alpha_n = \sum_{j=1}^{m_n} \mu(A_{nj}) \leq 1 - \alpha < \sum_{j=1}^{m_n+1} \mu(A_{nj}).$$

Roughly speaking, $S_{\alpha_n} = \bigcup_{j=1}^{m_n} A_{nj}$ is approximately a sphere centered at the origin with $1 - \alpha \sim \mu(S_{\alpha_n})$. Because of (2) and the absolute continuity of μ

$$0 \leq \alpha_n - \alpha \leq \mu(A_{n,m_n+1}) \leq \max_j \mu(A_{nj}) \to 0$$

as $n \to \infty$. Also define S_α as the sphere centered at the origin with the property $1-\alpha = \mu(S_\alpha)$.

The basics of our proof is the Poissonization. Introduce the notation N_n for a Poisson (n) random variable independent of $(X_1, Y_1), (X_2, Y_2), \ldots$, moreover put

$$n\tilde{\nu}_n(A) = \sum_{i=1}^{N_n} I_{\{X_i \in A\}} Y_i$$

and

$$n\tilde{\mu}_n(A) = \sum_{i=1}^{N_n} I_{\{X_i \in A\}}.$$

Proposition 1 *(Beirlant, Mason (1995)) Let g_{nj} be real measurable functions and let $M_{n,\alpha} = \sum_{j=1}^{m_n} g_{nj}\left(\tilde{\nu}_n(A_{nj}), \tilde{\mu}_n(A_{nj})\right)$. Assume that*

$$\Phi_{n,\alpha}(t,v) = \mathbf{E}\left(\exp\left(itM_{n,\alpha} + iv\frac{N_n - n}{\sqrt{n}}\right)\right) \to e^{-(t^2\rho_\alpha^2 + 2tv\gamma_\alpha + v^2)/2},$$

with $\rho_\alpha^2 = \int_{S_\alpha} h(x)dx$ and $\gamma_\alpha = \int_{S_\alpha} g(x)dx$, where $h(x)$ and $g(x)$ are some measurable functions such that $\gamma_0 = \int_{\mathcal{R}^d} g(x)dx < \rho_0^2 = \int_{\mathcal{R}^d} h(x)dx < \infty$. Then

$$\sum_{j=1}^{\infty} g_{nj}\left(\nu_n(A_{nj}), \mu_n(A_{nj})\right) \Big/ \sqrt{\rho_0^2 - \gamma_0} \xrightarrow{\mathcal{D}} N(0,1).$$

PROOF OF THEOREM 1. Because of $\min\{a,b\} = \frac{1}{2}(a+b-|a-b|)$

$$\hat{L}_n = \frac{1}{2}(1 - J_n), \tag{7}$$

where $J_n = \sum_{j=1}^{\infty} |\tau_n(A_{nj})|$ with $\tau_n(A) = \mu_n(A) - 2\nu_n(A)$, therefore the first statement of Theorem 1 is equivalent to

$$n^{1/2}\left(J_n - (1 - 2C_n)\right)/\sqrt{2L^*} \xrightarrow{\mathcal{D}} N(0,1). \tag{8}$$

In order to prove (8) let

$$1 - 2C_n = \sum_{j=1}^{\infty} \mathbf{E}|\tilde{\tau}_n(A_{nj})|,$$

where $\tilde{\tau}_n(A) = \tilde{\mu}_n(A) - 2\tilde{\nu}_n(A)$. So we need to prove that

$$n^{1/2}\left(\sum_{j=1}^{\infty}\left(|\tau_n(A_{nj})| - \mathbf{E}|\tilde{\tau}_n(A_{nj})|\right)\right)/\sqrt{2L^*} \xrightarrow{\mathcal{D}} N(0,1),$$

for which we will use the Proposition. We choose the functions g_{nj} as

$$g_{nj}(u,v) = \sqrt{n}\left(|v - 2u| - \mathbf{E}\left|\tilde{\mu}_n(A_{nj}) - 2\tilde{\nu}_n(A_{nj})\right|\right) \quad (j = 1, \ldots, m_n)$$

and let $h(x) = f(x)$ and $g(x) = |D(x)|f(x)$, so

$$\rho_0^2 - \gamma_0 = 1 - \int_{\mathcal{R}^d} |D(x)|f(x)dx = 2L^*.$$

Now, we check the conditions of the Proposition. Introduce

$$S_n = t\sqrt{n}\sum_{j=1}^{m_n} (|\tilde{\mu}_n(A_{nj}) - 2\tilde{\nu}_n(A_{nj})| - \mathbf{E}|\tilde{\mu}_n(A_{nj}) - 2\tilde{\nu}_n(A_{nj})|) + v\frac{N_n - n}{\sqrt{n}}$$

for which a central limit result is to be shown to hold. Remark that

$$Var(S_n) = n\sum_{j=1}^{m_n} \left\{t^2 Var(|\tilde{\tau}_n(A_{nj})|) + 2tv\mathbf{E}\left(|\tilde{\tau}_n(A_{nj})|(\tilde{\mu}_n(A_{nj}) - \mu(A_{nj}))\right)\right\} + v^2,$$

Apply Lemma 1 for $Z_i = I_{\{X_i \in A\}}(1 - 2Y_i)$ and $p = 2$ then

$$|nVar(\tilde{\tau}_n(A)) - \mu(A)| \leq \frac{A_2}{n^{1/2}}(\mu(A)^{1/2} + \mu(A)) \leq \frac{2A_2}{n^{1/2}}\mu(A)^{1/2}, \qquad (9)$$

therefore, on the one hand we get that

$$\limsup_n n\sum_{j=1}^{m_n} Var(|\tilde{\tau}_n(A_{nj})|) \leq \rho_\alpha^2.$$

Obviously

$$\liminf_n n\sum_{j=1}^{m_n} Var(|\tilde{\tau}_n(A_{nj})|)$$

$$\geq \liminf_n n\sum_{j=1}^{m_n} Var(\tilde{\tau}_n(A_{nj})) - \limsup_n n\sum_{j=1}^{m_n}((\mathbf{E}|\tilde{\tau}_n(A_{nj})|)^2 - (\mathbf{E}\tilde{\tau}_n(A_{nj}))^2) \geq \rho_\alpha^2,$$

if

$$\limsup_n n\sum_{j=1}^{m_n}((\mathbf{E}|\tilde{\tau}_n(A_{nj})|)^2 - \tau(A_{nj})^2) = 0.$$

It can be easily seen that by Lemma 2 for a set A

$$n((\mathbf{E}|\tilde{\tau}_n(A)|)^2 - \tau(A)^2) \leq 8n\mu(A)^2 e^{-n\tau(A)^2/(3\mu(A))}.$$

Because of (11)

$$n((\mathbf{E}|\tilde{\tau}_n(A)|)^2 - \tau(A)^2) \leq \frac{1}{n}Var(n\tilde{\tau}_n(A)) \leq \mu(A) + \frac{2A_2}{n^{1/2}}\mu(A)^{1/2}.$$

From these by the inequality

$$\min(a + b, c) \leq \min(a, c) + \min(b, c) \leq \min(a, c) + b^{2/3}c^{1/3}$$

we get that

$$n\sum_{j=1}^{m_n}((\mathbf{E}|\tilde{\tau}_n(A_{nj})|)^2 - \tau(A_{nj})^2)$$
$$\leq \sum_{j=1}^{m_n} \min\left(8n\mu(A_{nj})^2 e^{-n\tau(A_{nj})^2/(3\mu(A_{nj}))}, \mu(A_{nj})\right)$$
$$+8^{1/3}(2A_2)^{2/3}\sum_{j=1}^{m_n}\mu(A_{nj})e^{-n\tau(A_{nj})^2/(9\mu(A_{nj}))} \to 0 \tag{10}$$

if $\mathbf{P}(D(X) = 0) = 0$. To see this limit introduce the functions $\bar{f}_n(x) = \frac{\mu(A_n(x))}{\lambda(A_n(x))}$ and $D_n(x) = \frac{\tau(A_n(x))}{\mu(A_n(x))}$, then

$$\sum_{j=1}^{m_n} \min\left(8n\mu(A_{nj})^2 e^{-n\tau(A_{nj})^2/(3\mu(A_{nj}))}, \mu(A_{nj})\right)$$
$$= \int_{S_\alpha} \min\left(8\mu(A(_{nj})e^{-\frac{1}{3}n\lambda(A_n(x))\bar{f}_n(x)D_n(x)^2+\log n}, 1\right) f(x)dx$$
$$\leq \int_{S_\alpha} \min\left(8e^{-\frac{1}{3}n\lambda(A_n(x))\bar{f}_n(x)D_n(x)^2+\log n}, 1\right) f(x)dx$$

where the convergence is dominated, so we need to show that for $f(x) > 0$ the integrand tends to 0 almost everywhere w.r.t. λ, for which we can apply Theorem 7.16 in Wheeden, Zygmund (1977). The second term in (10) is similar. To complete the asymptotics for $Var(S_n)$ it remains to show that

$$n\sum_{j=1}^{m_n}\mathbf{E}\left(|\tilde{\tau}_n(A_{nj})|(\tilde{\mu}_n(A_{nj}) - \mu(A_{nj}))\right) \to \gamma_\alpha = \int_{S_\alpha}|D(x)|f(x)dx \text{ as } n\to\infty.$$

For a set A

$$n\mathbf{E}\left(|\tilde{\tau}_n(A)|(\tilde{\mu}_n(A) - \mu(A))\right) = n\mathbf{E}(|\tilde{\tau}_n(A)|\tilde{\mu}_n(A)) - n\mathbf{E}(|\tilde{\tau}_n(A)|)\mu(A)$$

For a set A let $Y(A)$ be a random variable with

$$\mathbf{P}(Y(A) \in B) = \mathbf{P}(Y \in B|X \in A)$$

then it can be easily seen that

$$n\mathbf{E}\left(|\tilde{\tau}_n(A)|(\tilde{\mu}_n(A) - \mu(A))\right)$$
$$= \mu(A)\sum_{k=0}^{\infty}\left(\mathbf{E}|\sum_{i=1}^{k+1}(1-2Y_i(A))| - \mathbf{E}|\sum_{i=1}^{k}(1-2Y_i(A))|\right)\frac{(n\mu(A))^k}{k!}e^{-n\mu(A)}.$$

Without loss of generality assume that $\tau(A) \geq 0$ then

$$\mu(A)\left(\mathbf{E}|\sum_{i=1}^{k+1}(1-2Y_i(A))| - \mathbf{E}|\sum_{i=1}^{k}(1-2Y_i(A))|\right)$$

$$= \mu(A)\left(\left(\mathbf{E}\sum_{i=1}^{k+1}(1-2Y_i(A)) + 2\mathbf{E}(-\sum_{i=1}^{k+1}(1-2Y_i(A)))^+\right)\right.$$
$$\left.-\left(\mathbf{E}\sum_{i=1}^{k}(1-2Y_i(A)) + 2\mathbf{E}(-\sum_{i=1}^{k}(1-2Y_i(A)))^+\right)\right)$$
$$= \mu(A)\left(\mathbf{E}(1-2Y_1(A)) + 2\mathbf{E}(-\sum_{i=1}^{k+1}(1-2Y_i(A)))^+ - 2\mathbf{E}(-\sum_{i=1}^{k}(1-2Y_i(A)))^+\right)$$

Because of $|2(-\sum_{i=1}^{k+1}(1-2Y_i(A)))^+ - 2(-\sum_{i=1}^{k}(1-2Y_i(A)))^+| \leq 2$ apply Hoeffdings inequality as in the proof of Lemma 2

$$|\mu(A)\left(\mathbf{E}|\sum_{i=1}^{k+1}(1-2Y_i(A))| - \mathbf{E}|\sum_{i=1}^{k}(1-2Y_i(A))|\right) - \mu(A)\mathbf{E}(1-2Y_1(A))|$$
$$= |\mu(A)\left(\mathbf{E}|\sum_{i=1}^{k+1}(1-2Y_i(A))| - \mathbf{E}|\sum_{i=1}^{k}(1-2Y_i(A))|\right) - |\tau(A)||$$
$$= |2\mathbf{E}(-\sum_{i=1}^{k+1}(1-2Y_i(A)))^- 2\mathbf{E}(-\sum_{i=1}^{k}(1-2Y_i(A)))^+|$$
$$\leq 2\mu(A)\left(\mathbf{P}\{\sum_{i=1}^{k+1}(1-2Y_i(A)) < 0\} + \mathbf{P}\{\sum_{i=1}^{k}(1-2Y_i(A)) < 0\}\right)$$
$$\leq 2\mu(A)\left(e^{-(k+1)\tau(A)^2/(2\mu(A)^2)} + e^{-k\tau(A)^2/(2\mu(A)^2)}\right)$$
$$\leq 4\mu(A)e^{-k\tau(A)^2/(2\mu(A)^2)}.$$

Summarizing the consequences of these inequlities we get that

$$|n\sum_{j=1}^{m_n}\mathbf{E}\left(|\tilde{\tau}_n(A_{nj})|(\tilde{\mu}_n(A_{nj}) - \mu(A_{nj}))\right) - \sum_{j=1}^{m_n}|\tau(A_{nj})||$$
$$\leq \sum_{j=1}^{m_n}4\mu(A_{nj})\sum_{k=0}^{\infty}e^{-k\tau(A_{nj})^2/(2\mu(A_{nj})^2)}\frac{(n\mu(A_{nj}))^k}{k!}e^{-n\mu(A_{nj})}$$
$$\leq \sum_{j=1}^{m_n}4\mu(A_{nj})e^{-n\tau(A_{nj})^2/(3\mu(A_{nj}))} \to 0,$$

therefore

$$n\sum_{j=1}^{m_n}\mathbf{E}\left(|\tilde{\tau}_n(A_{nj})|(\tilde{\mu}_n(A_{nj}) - \mu(A_{nj}))\right) = o(1) + \sum_{j=1}^{m_n}|\tau(A_{nj})| \to \int_{S_\alpha}|D(x)|f(x)dx = \gamma_\alpha.$$

To finish the proof of

$$S_n \xrightarrow{\mathcal{D}} N\left(0, t^2\rho_\alpha^2 + 2tv\gamma_\alpha + v^2\right) \text{ as } n \to \infty$$

by Lyapunov's central limit theorem, it suffices to prove that

$$n^{3/2}\left(\sum_{j=1}^{m_n}\mathbf{E}\left(\left|\,t\left(|\tilde{\tau}_n(A_{nj})|-\mathbf{E}|\tilde{\tau}_n(A_{nj})|\right)+v(\tilde{\mu}_n(A_{nj})-\mu(A_{nj}))\,\right|^3\right)\right.$$
$$\left.+\sum_{j=m_n+1}^{\infty}\mathbf{E}|v(\tilde{\mu}_n(A_{nj})-\mu(A_{nj}))|^3\right)\to 0,$$

or, by invoking the c_r inequality, that

$$n^{3/2}\sum_{j=1}^{m_n}\mathbf{E}\left(\left||\tilde{\tau}_n(A_{nj})|-\mathbf{E}|\tilde{\tau}_n(A_{nj})|\right|^3\right)\to 0,$$

and

$$n^{3/2}\sum_{j=1}^{\infty}\mathbf{E}\left(|\tilde{\mu}_n(A_{nj})-\mu(A_{nj})|^3\right)\to 0.$$

The second limit is shown in Beirlant, Györfi and Lugosi (1994). Concerning the first one apply Lemma 3 and Lemma 1 for $p=3$. □

The proof of Theorem 2 is similar, it is left to the reader.

Lemma 1 *(Beirlant, Mason (1995)) Let $\{Z_i\}$ i.i.d. random variables with $\mathbf{E}|Z_1|^p<\infty$, let N_n be a Poisson (n) random variable independent of $\{Z_i\}$, and N a standard normal random variable. Then there is a constant A_p (depending only on p) such that for $1\le p\le 3$*

$$\left|\mathbf{E}\left|\frac{\sum_{i=1}^{N_n}Z_i-n\mathbf{E}Z_1}{\sqrt{n\mathbf{E}Z_1^2}}\right|^p-\mathbf{E}|N|^p\right|\le\frac{A_p}{n^{1/2}}\left(\frac{\mathbf{E}|Z_1|^3}{(\mathbf{E}Z_1^2)^{3/2}}+1\right).$$

Lemma 2 *For a set A*

$$0\le\mathbf{E}|n\tilde{\tau}_n(A)|-|n\tau(A)|\le 2n\mu(A)e^{-\frac{n\tau(A)^2}{3\mu(A)}}.$$

Lemma 3 *If Z is a random variable with $\mathbf{E}|Z|^3<\infty$ then*

$$\mathbf{E}\left||Z|-\mathbf{E}|Z|\right|^3<27\mathbf{E}|Z-EZ|^3.$$

References

Beirlant, J., Györfi, L. and Lugosi, G. (1994). On the asymptotic normality of the L_1- and L_2- errors in histogram density estimation. *Canadian J. Statistics*, **22**, 309-318.

Beirlant, J. and Mason, D. (1995). On the asymptotic normality of L_p-norms of empirical functionals. *Mathematical Methods of Statistics*, **4**, 1-19.

Devroye, L., Györfi, L. and Lugosi, G. (1996). *Probabilistic Theory of Pattern Recognition.* Springer Verlag, New York.

Wheeden, R. L. and Zygmund, A. (1977). *Measure and Integral.* Marcel Dekker Inc., New York, Basel.

STOCHASTIC METHODS FOR GENERATIVE SYSTEMS ANALYSIS

Gabriel V. Orman
Department of Mathematics, "Transylvania" University of Brasov
50 Iuliu Maniu Street, 2200 Brasov, Romania,
e-mail: orman@lsbv.sfos.ro

Abstract: A very important part in the process of transmission of information is that of generation of the words. For some time past we tried to consider the process of generation of the words by generative systems, from a stochastic point of view involving Markov chains. Because the sequences of intermediate word (called *derivations*) by which a word can be generated are finite, it results that finite Markov chains will be connected to the process. In order to our discussion should be as general as possible we consider generative systems free of any restrictions. The model of such systems is offered by the most general class of formal grammars from the so-called *Chomsky hierarchy*, namely *the phrase-structure grammars*. The novelty that we have proposed consists in a new organization of the process of generation of the words by considering the set of all the derivations split into equivalence classes, each of them containing derivations of the same length. In this way our results are available up to an equivalence.

Keywords: generative systems, Markov chains, transition matrix.

1. The probability of generation of the words

We consider the process in which a word is obtained by a sequence of intermediate words starting from a fixed *initial symbol*: $\sigma=w_0, w_1, \ldots, w_j=w$. We refer to such a sequence to be a *derivation of length j*. In this way, in each word is stored the information obtained previously such that the process has a Markovian behavior. At the end w_j is just a word written as a sequence of symbols belonging to a known alphabet, called *terminal alphabet*.

We impose no restrictions on the passing from an intermediate word w_i to the next w_{i+1}, so that the generative system is free of any restrictions. Because the case of generation of a word directly from σ is of no particular interest, we suppose all the time that the length of a derivation is ≥ 2.

From now on we shall consider that the set of all the derivations is split into equivalence classes, each of them containing derivations of the same length. Thus, the results will be available up to an equivalence.

Let D_x, $x \geq 2$, be the equivalence class of the derivation $D(x)$ and let us denote by v_x the number of derivations into the class D_x by which a word w is generated. Obviously, it can or cannot be generated into the class D_x. Thus, if it is then, the probability that it will be generated again into D_{x+1} is denoted by γ ; but given that w is not generated into D_x, the probability that it should be generated into D_{x+1} is denoted by β. Hence we are in the case when the equivalence classes of the derivations are connected into a simple Markov chain. Furthermore, w can or cannot be generated directly from σ such that the probabilities that it will be or will not be in these cases are unknown to us. If we agree to denote by D_1 the class of derivations of length 1, it means that there is not an equivalence class preceding it. Then, let p_1 be the probability that w should be generated into D_1 and $q_1=1-p_1$ the contrary probability. Now let p_x be the probability that w should be generated into D_x and $q_x=1-p_x$ the contrary probability. The following result is obtained

Theorem 1. *A word w is generated into the class D_x, $x \geq 2$,* if

$$p_x \neq \frac{\beta}{1-\delta} \qquad \textit{where } \delta = \gamma - \beta. \tag{1}$$

But the constant to which p_x tends does not depend on the probability p_1. Because it plays the role of a limiting probability we introduce the notation

$$p = \frac{\beta}{1-\delta} \qquad \text{and} \qquad q = 1\text{-}p = \frac{1-\gamma}{1-\delta}$$

such that we have

$$p_x = (p_1 - p)\delta^{x-1} + p. \tag{2}$$

Now, let us propose to determine the probability that w should be generated into the class D_j if it was already generated into the class D_i, $i<j$. We denote by $p_j^{(i)}$ the probability that w should be generated into the class D_j if it was generated into the class D_i, $i<j$. In this case the following theorem is proved

Theorem 2. *The probability that a word should be generated into the class D_j if it was generated into the class D_i, $i<j$, $i \geq 2$, $j>2$, is given by the equality*

$$p_j^{(i)} = p + q\delta^{j-i} \tag{3}$$

2. The main characteristics of ν_x

It is obvious that ν_x defined before is a random variable that takes on values 0 and 1 with probabilities q_x=1-p_x and p_x respectively. Then, the number of derivations in n-1 equivalence classes, by which the word w is generated, is

$$\nu = \sum_{x=2}^{n} \nu_x$$

such that its expectation is

$$\mathbf{E}\nu = \sum_{x=2}^{n} \mathbf{E}\nu_x \qquad \text{while the variance is} \qquad \mathbf{D}\nu = \sum_{x=2}^{n} \mathbf{D}\nu_x \,.$$

The main result is the following

Theorem 4. *The expectation and the variance of the random variable giving the number of derivations by which a word is generated are as follows*

$$\mathbf{E}\nu = (n\text{-}1)p + a_n \qquad \text{and} \qquad \mathbf{D}\nu = pq[n\frac{1+\delta}{1-\delta}\ \text{-}1] + b_n$$

where a_n and b_n are some quantities that remain bounded as n increases.

3. The alternating generation procedure

Now we propose to study the case when a word w *is* generated by m equivalence classes from n on the following conditions:

1. It will be generated by the first class and the last and there is a direct rule (σ,w);
2. It will be generated by the first class but it will be not generated by the last and there is a direct rule (σ,w);
3. It will be not generated by the first class but it will be generated by the last and there is not a direct rule (σ,w);
4. It will be not generated both by the first class and the last and there is not a direct rule (σ,w).

We call such a way for generating words an *alternating generation procedure*. If $P_n(m)$ is the notation for the probability that we want to determine then, we have

$$P_n(m) = P_n(m,ww) + P_n(m,w\underline{w}) + P_n(m,\underline{w}w) + P_n(m,\underline{ww})$$

where $P_n(m,ww)$ is the notation for 1., and so on.

For example we obtain in the first case:

$$P_n(m,ww)=p_1\sum_{k=2}^{m}\binom{m-1}{k-1}\gamma^{m-k+1}(1-\gamma)^{k-1}\binom{n-m-1}{k-2}(1-\beta)^{n-m-k+1}\beta^{k-1}.$$

Now by computing these four probabilities we find the following main result which is of the form of a limit theorem

Theorem 5. *If the word w is generated by an alternating generation procedure, the derivations belonging to n equivalence classes then, the probability that w should be generated by m classes out of n is given by the relation*

$$P_n(m)=\frac{1}{\sqrt{2\pi npq}}\sqrt{\frac{1-\gamma+\beta}{1+\gamma-\beta}}\ell^{-\frac{z^2}{2}}(1+d_n)$$

where d_n vanish uniformly when z lies in some finite interval and z verifies the equality

$$m=np+z\sqrt{\frac{m\gamma(1-\gamma)+(n-m)\beta(1-\beta)}{(1-\gamma+\beta)^2}}$$

4. The fork-join generation procedure

Now we shall define a new procedure for generating words. We take into consideration only the case when a word cannot be generated by an equivalence class of derivations. Thus, if a word is not generated by a class D_x, $x\geq 2$, then it will be generated by the class D_{x-1} with probability q and by the class D_{x+1} with probability $p=1-q$. Relating to the first class and the last we suppose that a word can or cannot be generated by these two classes. But for the cases when it is not generated we put the following supplemental conditions:

1. If the word is not generated by the first class D_2 then, it will be certainly generated by the next;
2. If the word is not generated by the last class D_n then, it will be certainly generated by the last but one.

It is referred to as a *fork-join generation procedure*. For the other all classes D_x, $2<x<n$, we suppose that the word being in each of them, is subject to a fork-join generation procedure. The following cases can be distinguished:

1. The word will be generated by the first class and the last class;
2. It will be generated by the first class but it will be not generated by the last;
3. It will be not generated by the first class but it will be generated by the last;
4. It will be not generated both by the first class and the last class.

For each of them we have determined the two-step transition matrix and we have come to the following results:

- The rows of rank i=3,4, ..., n-3 contain, each of them, the triplet of elements q^2, $2pq$, p^2 disposed with q^2 and p^2 on two diagonals to the left and respective to the right of the main diagonal which contains the element $2pq$;
- The first two and the last two rows are different from a case to another. For these rows we have:

(i) the first case:

$$p_{11}=p_{n-1\ n-1}=1,\ p_{21}=q,\ p_{22}=p_{n-2\ n-2}=pq,\ p_{24}=p^2,\ p_{n-2\ n-4}=q^2 \text{ and } p_{n-2\ n-1}=p;$$

(ii) the second case

:

$$p_{11}=1,\ p_{21}=p_{n-1\ n-3}=q,\ p_{n-1\ n-1}=p,\ p_{22}=pq,\ p_{24}=p^2,\ p_{n-2\ n-4}=q^2$$
and $p_{n-2\ n-2}=p+qp$;

(iii) the third case:

$$p_{11}=q,\ p_{13}=p_{n-2\ n-1}=p,\ p_{n-1\ n-1}=1,\ p_{22}=q+pq,\ p_{24}=p^2,\ p_{n-2\ n-4}=q^2$$
and $p_{n-2\ n-2}=qp$;

(iv) the fourth case:

$$p_{11}=p_{n-1\ n-3}=q,\ p_{13}=p_{n-1\ n-1}=p,\ p_{22}=q+pq,\ p_{24}=p^2,\ p_{n-2\ n-4}=q^2$$
and $p_{n-2\ n-2}=p+qp$.

The main result is a *specific property of symmetry* of these four case that can be stated as follows

Theorem 6. *If a word is in a random process of generation by a fork-join generation procedure then, in all the cases of generation, the two-step transition matrix has n-5 successive rows, each of them containing the triplet of elements* q^2, $2pq$, p^2, *symmetrically disposed as against the first two and the last two rows. Furthermore,* q^2 *and* p^2 *are elements of two distinct diagonals symmetrically disposed as against the main diagonal which contains the element* $2pq$.

As regards the first and the fourth cases, they have conducted us to a surprising result. Let us denote by A_1 the event consisting in the word being generated by the class D_2, by A_2 it being generated by the class D_3, ..., by A_{n-1} it being generated by the class D_n.

Now in the first case, the two-step transition matrix is the same with the two-step transition matrix for a particle in a random walk between two absorbing barriers known in the theory of Markov chains. For example let us consider that a particle located on a straight line moves along the line via random impacts occuring at times t_1, t_2, t_3, ... The particle can be at points with integral coordinates a, $a+1$, $a+2$, ..., b. At points a and b there are absorbing barriers. Each impact displaces the particle to the right with probability p and to the left with probability $q=1-p$ so long as the particle is not located at a barrier. If the particle is at a barrier then, it remains in the states A_1 and A_{n-1} with probability 1.

As regards the fourth case of generation of a word by a fork-join generation procedure, the two-step transition matrix is the same with the two-step transition matrix for a particle in a random walk between two reflecting barriers also known in the theory of Markov chains. The conditions remain the same as in the former case, the only difference being that if the particle is at a barrier then, any impact will transfer it one unit inside the gap between the barriers.

These cases of generation of the word by a fork-join generation procedure become of a special interest. We also emphasize their practical caracter and we believe that they will be very useful in some studies on generative systems.

References

Ginsburg, S. (1975), *Algebraic and Automata-Theoretic Properties of Formal Languages*, North Holland Pub. Co., Amsterdam.

Orman, G.V., (1982), *Limbaje formale (Formal languages)*, Ed. Tehnica, Bucharest.

Orman, G.V., (1993), On some numerical functions utilized in the study of a class of discrete systems, *J. of Systems Science*, Vol 19, No.1, pp. 28-37.

Orman, G.V., (1994), On the Characterization of some Generative Grammars by Numerical Functions, in: *Current Issues in Mathematical Linguistics*, ed. by C. Martin-Vide, Elsevier Science B.V., North Holland, Amsterdam, pp. 151-160.

Orman, G.V., (1994), Markov chains connected to the process of generation of the words by a generative systems, *Yugoslav Journal of Operations Research*, Vol. 4, No. 2, pp. 167-178.

Orman, G.V., (1996), Random phenomena and some systems generating words, *Yugoslav Journal of Operations Research*, Vol. 6, No. 2, pp. 245-256.

Orman, G.V., (1996), On the symmetry phenomenon in generative systems, *Proceedings of the International Conference on Symmetry and Antisymmetry in Mathematics, Formal Languages and Computer Science*, Brasov, 18-19 July 1996, ed. by G.V. Orman, University of Brasov, pp.87-90.

Orman, G.V., (1997), New characterizations of generative systems, to be published in: *Proceedings of the 4th Balkan Conference on Operational Research*, Thessaloniki Greece, October 20-23.

Compensatory Rules for Optimal Classification with Mastery Scores

Hans J. Vos
Faculty of Educational Science and Technology, University of Twente
P.O. Box 217, 7500 AE Enschede, The Netherlands,
E-mail: vos@edte.utwente.nl

Abstract: A model for simultaneous optimization of combinations of test-based decisions in education and psychology is proposed using Bayesian decision theory. To illustrate the approach, one classification decision with two treatments each followed by a mastery decision are combined into a decision network. An important decision is made between weak and strong decision rules. As opposed to strong rules, weak rules are allowed to be a function of prior test scores in the series. Conditions under which optimal rules take weak monotone forms are derived. Results from a well-known problem in The Netherlands of selecting optimal continuation schools on the basis of achievement test scores are presented.

Keywords: Bayesian Decision Theory, Classification Decisions, Mastery Testing, Simultaneous Optimization, Compensatory Rules.

1. Introduction

Statistical decision problems arise when a decision-maker is faced with the need to choose an action that is optimal in some sense. In practice, one decision problem may often lead to another, which, in turn, may lead to a third one, and so on. A well-known example of combinations of test-based decisions is an Individualized Study System (ISS) in education. Such systems can be conceived of as instructional networks consisting of various types of decisions as nodes, with individual routes for the students (e.g., van der Linden & Vos, 1996).

An appropriate framework for dealing with instructional decision making in ISSs is Bayesian decision theory (e.g., Lindgren, 1976, Chapter 8). The application of Bayesian methods to, for instance, instructional decision making in ISSs, consists of two basic elements: A probability model relating test and criterion scores to each other, and a utility structure evaluating the total costs and benefits of all possible decision outcomes. To optimize series of decision problems within a Bayesian decision-theoretic framework, it is common to maximize the expected utility for each decision separately.

The purpose of this paper is to show that multiple decisions in networks can also be optimized simultaneously by maximizing the expected utility over all possible combinations of decision outcomes. Since more information usually results in better decision procedures, it is expected that a simultaneous approach has two main advantages compared with optimizing each decision separately. First, it is expected that efficiency of decisions can be increased using data from earlier tests as collateral information in later decisions. Second, more realistic utility structures can be obtained defining utility functions for earlier decisions on the ultimate success criterion in the network.

To illustrate the simultaneous approach, one classification decision with two treatments each followed by a mastery decision are combined into a decision network. Examples of mastery decisions include pass-fail decisions in education, certification, and successfulness of therapies. Well-known test-based classification problems are educational or vocational guidance situations where most promising types of schools or careers must be identified, classification of students into ISSs with tracks at different levels, selecting one out of different available therapies in clinical settings, and testing for the military service.

2. The Classification-Mastery Problem

In classification decisions, the decision problem consists of a choice among several alternative treatments to which subjects have to be assigned on the basis of their test scores. Prior to the treatments, all subjects are administered the same classification test and the success of each treatment is measured by its own criterion. Therefore, completion of each treatment is followed by different mastery tests, which the subjects may pass or fail.

In the following, we shall suppose that the test scores observed prior to the treatment for an arbitrary individual are denoted by a random variable X. Each treatment j is followed by a mastery test, with scores denoted by a random variable Y_j (j = 0,1). It is assumed that the mastery tests Y_j are unreliable representations of the criteria measuring the success of the treatments. The criteria to be considered are the classical test theory true scores (e.g., Lord & Novick, 1968) underlying the mastery tests Y_j, and will be denoted by T_j. The variables X, Y_j, and T_j will be considered to be continuous, with realizations x, y_j, and t_j. Furthermore, it is assumed that the classification of subjects into j treatments yields a joint distribution of scores X, Y_j, and T_j with density function $f_j(x,y_j,t_j)$.

The mastery score variables Y_0 and Y_1 will be rescaled such that they both take values on the same domain. As a result, for the realization y_0 and y_1 of the random variables Y_0 and Y_1 the indices 0 and 1 can be dropped in the remainder of this paper. Of course, this does not mean that a subject does receive the same value for y if s(he) follows different treatments. Similarly, since T_j is defined as the expectation of Y_j according to classical test theory, the indices 0 and 1 will be dropped for the realization t_0 and t_1 of T_0 and T_1.

3. Weak Monotone and Strong Monotone Rules

It is assumed that we can restrict the class of all possible simultaneous rules to monotone rules, that is, rules taking the form of cutting scores. Let x_c, y_{cj}, and t_{cj} denote the cutting scores on the random variables X, Y_j, and T_j, respectively, where t_{cj} is set in advance by the decision maker (j = 0,1). Clearly, the classification-mastery problem now consists of simultaneously setting cutting scores x_c and y_{cj} such that, given the values of t_{cj}, the expected utility over all possible combinations of decision outcomes is maximized. Note that the first advantage of the simultaneous approach (i.e., making more efficient use of the data) is nicely demonstrated by the fact that in optimizing mastery decisions, prior data on the classification test is used as collateral information.

In general, the observed scores on the classification test may or may not be taken explicitly into account in setting cutting scores on the mastery test score variables Y_j. As a case for the former, it seems reasonable that students who have been assigned to Treatment 1 with observed classification scores far above x_c are sooner granted mastery status than students with classification scores equal to or just above x_c. To distinguish between cases where cutting scores on the mastery tests do or do not depend on x, those rules will be denoted by weak and strong monotone rules, respectively. For each x, weak cutting scores on the mastery tests Y_0 and Y_1 have to be computed as functions $y_{co}(x)$ and $y_{c1}(x)$, respectively.

Let a_{jh} stand for the action either to retain (h = 0) or advance (h = 1) a subject who is classified into treatment j, then a weak simultaneous rule δ (i.e., a mapping from the sample space X x Y_j onto the action space) can be defined for our classification-mastery problem as follows:

$$\begin{aligned}
\{(x,y):\ \delta(x,y) = a_{00}\} &= A \times B_0(x) \\
\{(x,y):\ \delta(x,y) = a_{01}\} &= A \times B_0^C(x) \\
\{(x,y):\ \delta(x,y) = a_{10}\} &= A^C \times B_1(x) \\
\{(x,y):\ \delta(x,y) = a_{11}\} &= A^C \times B_1^C(x),
\end{aligned} \tag{1}$$

where A, A^C, $B_j(x)$, and $B_j^C(x)$ stand, respectively, for the sets of x and y values for which a subject is classified into Treatment 0, into Treatment 1, retained in treatment j, and advanced in treatment j. Note that at the left-hand side as well as at the right-hand side of (1) we have subsets in the X x Y_j-space, since both subsets A and A^C are sets of x values while both subsets $B_j(x)$ and $B_j^C(x)$ represent sets of y values depending on the variable x. It follows from (1) that a weak monotone rule δ (X,Y_j) can be defined for our example as:

$$\delta(X,Y_j)=\begin{cases} a_{00} & \text{for } X < x_c, Y_0 < y_{c0}(x) \\ a_{01} & \text{for } X < x_c, Y_0 \geq y_{c0}(x) \\ a_{10} & \text{for } X \geq x_c, Y_1 < y_{c1}(x) \\ a_{11} & \text{for } X \geq x_c, Y_1 \geq y_{c1}(x). \end{cases} \quad (2)$$

Conditions under which we can confine ourselves to the subclass of monotone simultaneous rules in looking for optimal rules will be considered below.

4. An Additive Utility Structure for the Combined Problem

A utility function $u_{jh}(t)$ evaluates the consequences of taking action a_{jh} while the true score of the subject is t. In the present paper, it is assumed that the utility structure of the classification-mastery problem can be represented as an additive function of the following form:

$$u_{0h}(t) = w_1 u_0^{(c)}(t) + w_2 u_{0h}^{(m)}(t)$$
$$u_{1h}(t) = w_1 u_1^{(c)}(t) + w_3 u_{1h}^{(m)}(t), \quad (3)$$

where $u_j^{(c)}(t)$, $u_{jh}^{(m)}(t)$ represent the utility functions for the separate classification and mastery decisions under treatment j, and w_1, w_2, and w_3 represent nonnegative weights. A common weight w_1 has been used in both functions because utility scales have an arbitrary zero (Luce & Raiffa, 1957).

Since utility functions are also measured with an arbitrary unit, and assuming $w_2 = w_3$ (i.e., the relative influence of the utility functions for both mastery decisions are equally weighted), the weights in (3) can always be rescaled, leading to the following result:

$$u_{jh}(t) = wu_j^{(c)}(t) + [(1-w)/2]u_{jh}^{(m)}(t), \quad (4)$$

where $0 \leq w \leq 1$.

It should be emphasized that both terms in the right-hand side of (4) are a function of the same variable. As indicated by van der Linden and Vos (1996), therefore, the utility structure assumed is not an instance of a utility function in a multiple-criteria decision problem. The usual assumptions needed to motivate an additive structure for multiple-criteria functions in the Luce-Raiffa sense to provide a utility function are thus not needed here.

Note that the second advantage of the simultaneous approach (i.e., more realistic utility structures) is effectively demonstrated by the additive utility structure of (4), in which utility functions defined on the ultimate criteria variables T_0 and T_1 are also used in previous decision problems, namely, the problem of classifying subjects into Treatment 0 and Treatment 1. This choice is in line with the philosophy underlying most ISSs.

5. Sufficient Conditions for Monotone Simultaneous Rules

For every integrable function f(x) and any set S of x values, it holds that

$$\int_S f(x)\,dx \le \int_{S_0} f(x)\,dx \text{ with } S_0 = \{x: f(x) \ge 0\}. \quad (5)$$

The theorem in (5) is implied by the obvious inequality $f(x) \le |f(x)|$. Applying this theorem to the expected utility in the simultaneous approach, the sets A^C and $B_j^C(x)$ can be derived for which an upper bound is obtained for this quantity. Next, it can be examined under which conditions the sets A^C and $B_j^C(x)$ take a weak monotone form.

Let $E[U_{sim}(A^C, B_0^C(x), B_1^C(x))]$ denote the expected utility in the simultaneous approach, then this quantity is calculated over all possible combinations of decision outcomes by adding the expected utilities associated with the four possible actions a_{00}, a_{01}, a_{10}, and a_{11}. It follows that $E[U_{sim}(A^C, B_0^C(x), B_1^C(x))]$ can be written as:

$$E[U_{sim}(A^C, B_0^C(x), B_1^C(x))] \equiv$$

$$\int_A \int_{B_0(x)} \int_R u_{00}(t) f_0(x,y,t)\,dt\,dy\,dx + \int_A \int_{B_0^C(x)} \int_R u_{01}(t) f_0(x,y,t)\,dt\,dy\,dx +$$

$$\int_{A^C} \int_{B_1(x)} \int_R u_{10}(t) f_1(x,y,t)\,dt\,dy\,dx + \int_{A^C} \int_{B_1^C(x)} \int_R u_{11}(t) f_1(x,y,t)\,dt\,dy\,dx, \quad (6)$$

where R denotes the set of real numbers. Taking expectations, completing integrals w.r.t. y, and rearranging terms, (6) can be written in its posterior form:

$$E[U_{sim}(A^C, B_0^C(x), B_1^C(x))] \equiv$$

$$\int_A E_0[u_{00}(T_0) \mid x] q(x)\,dx + \int_{A^C} E_1[u_{10}(T_1) \mid x] q(x)\,dx +$$

$$\int_A \int_{B_0^C(x)} E_0[u_{01}(T_0) - u_{00}(T_0) \mid x,y] h_0(x,y)\,dy\,dx +$$

$$\int_{A^C} \int_{B_1^C(x)} E_1[u_{01}(T_1) - u_{00}(T_1) \mid x,y] h_1(x,y)\,dy\,dx, \quad (7)$$

with $h_j(x,y)$ denoting the p.d.f. of (X,Y_j), and where $E_j[.]$ indicates that the expectation has been taken over a distribution indexed by j. Applying the theorem in (5) first to the inner integrals w.r.t. y, and using $h_j(x,y) \geq 0$, it can easily be verified that for each value x of X an upper bound to (7) is obtained for:

$$B^{*}{}_{0}{}^{C}(x) \equiv \{y: E_0[u_{01}(T_0)-u_{00}(T_0) \mid x,y] \geq 0\}, \tag{8}$$

$$B^{*}{}_{1}{}^{C}(x) \equiv \{y: E_1[u_{11}(T_1)-u_{10}(T_1) \mid x,y] \geq 0\}. \tag{9}$$

Next, inserting the sets $B^{*}{}_{0}{}^{C}(x)$ and $B^{*}{}_{1}{}^{C}(x)$ into (7), completing integrals w.r.t. x, and rearranging terms, (7) becomes

$$E[U_{sim}(A^C, B_0{}^C(x), B_1{}^C(x))] \equiv$$

$$E_0[u_{00}(T_0)] + \int_R \int_{B^{*}{}_{0}{}^{C}(x)} E_0[u_{01}(T_0)-u_{00}(T_0) \mid y]k_0(y \mid x)dy\}q(x)dydx +$$

$$\int_{A^C} \{E_1[u_{10}(T_1) \mid x] - E_0[u_{00}(T_0) \mid x] +$$

$$\sum_{j=0}^{1} (2j-1) \int_{B^{*}{}_{j}{}^{C}(x)} E_j[u_{j1}(T_j)-u_{j0}(T_j) \mid x,y]k_j(y \mid x)dy\}q(x)dx, \tag{10}$$

where $k_j(y \mid x)$ and $q(x)$ denote the p.d.f.'s of Y_j given $X = x$ and X, respectively. Finally, applying the theorem in (5) to the outer integral w.r.t. x, and using $q(x) \geq 0$, it can easily be verified that an upper bound to (10) is obtained for:

$$A^{*C} \equiv \{x: E_1[u_{10}(T_1) \mid x] - E_0[u_{00}(T_0) \mid x] + \sum_{j=0}^{1} (2j-1) \int_{B^{*}{}_{j}{}^{C}(x)} E_j[u_{j1}(T_j)-u_{j0}(T_j) \mid x,y]k_j(y \mid x)dy \geq 0\}. \tag{11}$$

Conditions for weak monotonicity can now be specified. For weak monotone rules, the sets $B^{*}{}_{j}{}^{C}(x)$ and A^{*C} take the form $[y_{cj}(x),\infty)$ and $[x_c,\infty)$, respectively. Hence, optimal rules take a weak monotone form if the left-hand sides of the inequalities in (8)-(9) and (11) are increasing functions in y for all x and in x, respectively. In the empirical example below it will be examined if these conditions hold.

6. Calculation of Optimal Weak Cutting Scores

Assuming the conditions for weak monotonicity are satisfied, optimal weak cutting scores can now be obtained for those values of $y_{cj}(x)$ and x_c for which the inequalities in (8)-(9) and (11) turn into equalities. Since (8)-(9) hold for all x, and thus for x_c, first the optimal weak cutting score on the classification test can be found by equating the left-hand sides of (8)-(9) and (11) to zero and solving the system of equations simultaneously for x_c, $y_{c0}(x_c)$, and $y_{c1}(x_c)$. Then for each $x < x_c$ and $x \geq x_c$, optimal weak cutting scores on the mastery tests Y_0 and Y_1 can be obtained by putting the left-hand sides of (8) and (9) equal to zero and solving for $y_{c0}(x)$ and $y_{c1}(x)$, respectively.

Since no analytical solutions for the system of equations could be found, Newton's iterative algorithm for solving nonlinear equations was used. Assuming a trivariate normal distribution for $f_j(x,y,t)$, a computer program called NEWTON was written to calculate the optimal weak cutting scores.

7. Empirical Example and Discussion

The empirical example concerns the assignment of students to appropriate continuation schools at the end of the elementary school (i.e., at grade 8), a problem well-known in the Dutch educational system. The utility functions $u_j^{(c)}(t)$ and $u_{jh}^{(m)}(t)$ involved for the separate classification and mastery decisions, respectively, were assumed to be the well-known threshold function (e.g., Huynh, 1976). The choice of this function implies that the costs and benefits of each possible decision outcome can be summarized by a constant.

It appeared that the left-hand sides of the inequalities in (8)-(9) were increasing functions in y for all values of x. Hence, the sets $B^{*C}_j(x)$ took the required form $[y_{cj}(x),\infty)$ for all x. Inserting this result into the left-hand side of the inequality in (11), it appeared that this expression was increasing in x. Thus, the set A^{*C} also took the required form $[x_c,\infty)$ implying that the monotonicity conditions were satisfied.

It turned out that, under the natural condition of the distributions of classification and mastery scores being stochastically increasing in the true mastery score (e.g., a trivariate normal distribution for $f_j(x,y,t)$), optimal weak cutting scores on the mastery tests were decreasing functions of the classification scores. Hence, optimal weak monotone rules were necessarily compensatory by nature. In other words, for instance, students with observed scores on the classification test X equal to or just above the cutting score x_c must compensate their relatively low classification scores with higher scores on the mastery test Y_1.

As a result of making more efficient use of the test data in a simultaneous approach, one might expect a gain in expected utility compared with optimizing each decision separately. To see whether this expectation could be confirmed, the expected utilities for the optimal weak monotone rules were compared with the weighted sum of the expected utilities for the optimal separate rules. The optimal separate cutting scores could easily be derived by imposing certain restrictions on the expected utility for the simultaneous approach. The results indicated that the expected utilities for the simultaneous approach were substantially larger than for the separate approach.

Two final remarks are appropriate. First, the procedures advocated in this paper have a larger scope of application than assigning students to optimal types of comprehensive education. For instance, the classification-mastery problem may be important in the area of psychotherapy in which patients have to be classified into the most appropriate therapy followed by a test, which has to be passed before they can be dismissed from the therapy. Generally, any situation in which subjects have to be assigned to one out of several possible treatments, training programs, or therapies on the basis of their scores on a test and in which each treatment is followed by a different mastery test, the optimal rules presented here have the potential to improve the quality of the decisions. Beside the areas of selecting optimal continuation schools and psychotherapy, the optimal decision procedures may be important in management sciences, agricultural sciences, medical decision making, and so forth.

Second, the problem of simultaneously optimize classification decisions with two treatments each followed by a mastery decision have been dealt here by applying Bayesian decision theory. However, other general classification approaches may also be appropriate, such as discriminant analysis (e.g., Tatsuoka, 1971, Chapter 8; McLachlan, 1992) or the use of dissimilarity measures techniques (e.g., Mirkin, 1996).

8. References

Huynh, H. (1976). Statistical considerations of mastery scores. *Psychometrika*, *41*, 65-79.

Lindgren, B.W. (1976). *Statistical Theory*, Macmillan, New York.

Lord, F.M. & Novick, M.R. (1968). *Statistical Theories of Mental Test Scores*, Addison-Wesley, Reading, Mass.

Luce, R.D. & Raiffa, H. (1957). *Games and Decisions,* Wiley, New York.

McLachlan, G.J. (1992), *Discriminant Analysis and Statistical Pattern Recognition*, Wiley, New York.

Mirkin, B. (1996), *Mathematical Classification and Clustering*, Kluwer, Dordrecht.

Tatsuoka, M.M. (1971). *Multivariate Analysis: Techniques for Educational and Psychological Research*, Wiley, New York.

van der Linden, W.J. & Vos, H.J. (1996). A compensatory approach to optimal selection with mastery scores. *Psychometrika*, *61*, 155-172.

PART II

Other Approaches for Classification: Discrimination, Neural Network, Regression Tree

- **Discrimination and Classification**
- **Neural Network and Classification**
- **Regression Tree**

About the Automatic Detection of Training Sets for Multispectral Images Classification

V. Bertholet, J.P. Rasson, S. Lissoir

Département de Mathématique
Facultés Universitaires Notre-Dame de la Paix.
Rempart de la Vierge, 8, B5000 Namur, Belgique
E-mail <Vincent.Bertholet@fundp.ac.be>
http : //www.fundp.ac.be/sciences/math/geosatel

Abstract : In images' discriminant analysis, training sets are usually given by experts of the field of interest. A procedure is proposed in this paper to avoid such a resort to experts. Two steps are required to achieve this scheme. Firstly, thanks to a multivariate and nonparametric approach of supports comparison, some homogeneous areas are detected on the image. Secondly, building a similarity measure based on the same criterion, those so found areas are merged into a small number of classes. Afterwards, these classes can be used as training sets for any discriminant analysis procedure.

Key words : training sets, images, support comparison, discriminant analysis, homogeneous areas, experts.

1. Introduction

Using an interdisciplinary approach inside the GEOSATEL Laboratory, mathematicians and geographers are working together with a view to joining the results of a long statistical research with a geographical application on environmental planning.

For several years, we have been interested in discriminant analysis applied to multispectral images at satellite scale. A supervised classification scheme has been elaborated which appears highly efficient on remote sensing data. Recently, we applied our procedure on airborne photographs and the classification results looked also very good.

As well at satellite as at airborne scale, the training sets we used were given by geographers. The use of such experts to get training sets can be expensive and their choice of training data can be subjective.

Therefore we wanted to pass to an unsupervised method to avoid those drawbacks. Some classical segmentation algorithms like edge detection, splitting and merging, were first tested but the provided results let us unsatisfied.

As a clustering of the entire image would lead to useless results, we thought then to automate the training sets selection procedure. This automation is discussed hereafter. We detail firstly how homogeneous areas are detected, and secondly how they are merged into training sets. The last paragraph is devoted to applications on satellite and airborne images.

2. Detection of homogeneous areas in the image

Due to mixels - pixels at the intersection of several regions in an image - we have to reduce the available information to the useful information. The philosophy consists of detecting homogeneous areas on the image, including not only the spectral intensities of the pixels but also the texture of the regions.

To detect changepoints –dual of finding homogeneous areas- Stephens and Smith (1993) propose a parametric way: use on each row and column of the image a test of hypothesis to detect if there is a changepoint or not.

In the same time, we pay attention to another approach. Using also parametric models, Wichern, Miller and Hsu (1976) propose a method to find sets of contiguous pixels containing almost surely no changepoint (up to a α-level of error) .

Based on those two points of view, we examine if a pixel is a changepoint - on a row for example - by testing the equality of the densities of the samples of size n on the left hand side and on the right hand side of this pixel. A pixel is a changepoint if it is a changepoint either on a row or on a column.

Due to the poor results we obtain we decide to modify our strategy : the test is no more applied to detect change of parameters but to detect change of distribution, whatever the direction may be. In order to perform this, classical non parametric tests such as Kolmogorov-Smirnov one are used.

In fact, as we apply this method on multispectral images, each coordinate of the distribution is treated separately so the complete distribution is not taken into account. This can explain the weak results we obtain.

We finally found an approach able to consider the full space in the paper of Devroye and Wise (1980). This approach, justified by asymptotic theorems, consists of a densities supports comparison.

So we deduce our procedure as follows :

- first we estimate the supports on a range of the same size on the left hand part and right hand part of each pixel (for the horizontal step), and we do the same vertically and along the two diagonals ;

- then we look at the number of points on one side being in the support of the points of the other side and inversely ; if the total number is sufficiently great in the four directions, the homogeneity is not rejected.

The estimation of the density support is made via union of discs centered on the pixels in the spectral space. The choice of the radius generates obviously a difficulty in this procedure.

3. Combining the homogeneous areas

Having at our disposal a high number of homogeneous parcels, we merge them into a small number of classes. For this, we compute a similarity measure between zones thanks to the same strategy as for testing the homogeneity. So for each pair of zones, we calculate the proportion of points being in the support of the other class. Therefore our similarity is a measure between 0 and 1.

With this similarity matrix, we can now group the areas by a hierarchical classification method like single linkage or complete linkage for example. We stop this process when the similarity between the groups go beyond a fixed level. A this time, we keep only the groups formed by the merging of several initial zones. We have now homogeneous groups at our disposal which are representative of different populations of the image. Those groups will be used as training sets.

4. Applications

We apply this scheme on satellite and on airborne scale images where the spectral space is composed of 3 channels. The satellite image is a scene of Marigny (France) and has been taken by SPOT satellite in early April season. The 3 SPOT channels were used (XS1, XS2, XS3). The airborne image is a picture of the region of Arlon (Belgium) and is composed by the classical 3 channels (Red, Green, Blue). The images are provided by IGN France and by the Walphot society (Belgium) respectively.

For each image, we detect automatically the training sets using the single linkage as clustering method. Afterwards, we classify the image using the behavioral method elaborated in our laboratory (Rasson and Granville 1996). This procedure is based on the non homogeneous Poisson Process. For this, we estimate the density in each group by the kernel method.

On the Marigny image, we succeed to detect the main classes. Nevertheless, the urban class is not recognized due to the small number of pixels in this group. The most interesting point is that several areas of the image are unclassified by our method. This point indicates that classes may be missing. With a view to reducing a too high number of unclassified pixels, the same scheme can be applied again on the unclassified areas only. By this way, some missing classes may be recovered.

On the Arlon image, we detect more classes than the expert. In fact, one class of the expert representing fields is split in two classes of fields characterized by their drainage's nuance. Thanks to that procedure we are also able to discriminate roofs and walls of the houses. That information is interesting for geographers and can be included in GIS (Geographic Information System) technology.

5. Numerical comparison

We compare some automatic training sets with the experts' ones for the satellite picture. For this, we compare the mean and standard deviation for the classes which appear on both sets (table 1).

Table 1 : *comparison of mean and standard deviation of channels XS1, XS2 and XS3 for training sets given by experts and automatic ones.*

	Experts' training sets			Automatic training sets		
	Number of pixels	Mean	Std. Dev.	Number of pixels	Mean	Std. Dev.
Deciduous	5400	77,24,22	20,13,12	1883	69,21,19	13,5,5
Grassland	3850	170,25,30	35,18,15	131	203,18,30	6,3,3
Fields	3800	94,126,111	27,44,41	884	78,136,108	7,9,9
Winter Cultures	5600	159,32,35	29,22,21	552	158,24,27	10,5,7

From table 1, we see that means are really different from one group of training set to the other. But the most interesting fact is that the standard deviation of the experts' sets are always much greater than those of the automatic sets. So our algorithm can discover training sets much more homogeneous than those given by experts.

Moreover, with the experts' sets, it is difficult to make the distinction between grassland and winter cultures. In the automatic sets, there is a clear cut between the two classes as we can see on figure 1.

Figure 1 : *Comparison of distribution for grassland and winter culture training sets in the XS1-XS3 plane. Figure 1.a : Grassland set given by experts. Figure 1.b : Winter cultures set given by experts. Figure 1.c : Automatic grassland set. Figure 1.d : Automatic winter cultures set.*

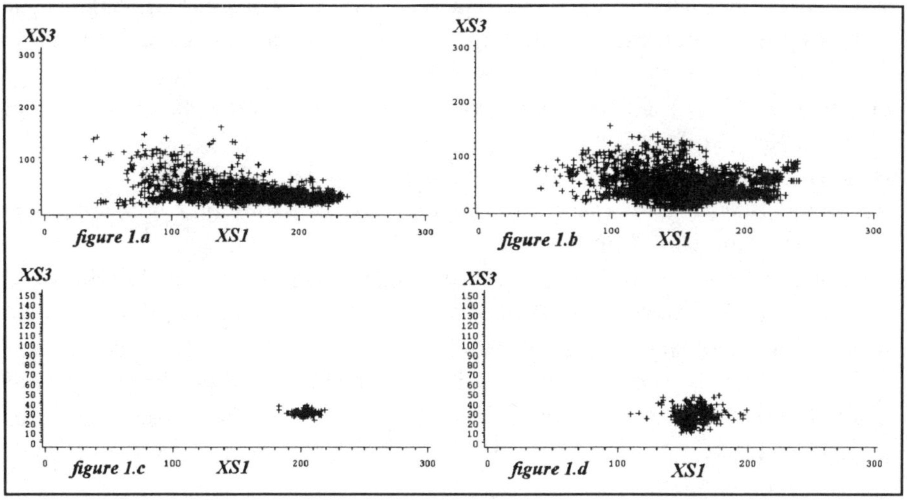

6. Conclusion

Using statistical tests, we are able to detect homogeneous areas inside an image up to a level of error. By tuning parameters, clustering methods provide automatic training sets which have several qualities : they are more homogeneous than those given by experts ; they are less subjective ; they can discover classes difficult to detect for the experts.

Segmentation of images with such training sets gives information to geographers and can help them to solve environmental problems and to establish land planning. The provided results may be used not only to grasp the evolution of the land cover, but also to cross the land cover inventory with other information useful for land planning through the GIS (Geographic Information System) technology.

References :

Mac Lachlan, G.J. (1992). *Discriminant analysis and statistical pattern recognition,* Wiley, New York.

Devroye L. and Wise G.L. (1980). Detection of abnormal behavior via non-parametric estimation of support, *SIAM Journal of Applied Mathematics,* volume 28, n°3, pp.480-488.

Stephens D.A. and Smith A.F.M. (1993). Bayesian edge detection in images via changepoint methods , in : *Computer intensive methods in statistics*, Härdle And Simar Edition, Physica-Verlag-Heidelberg, pp.1-18.

Wichern D. W., Miller R.B. and Hsu D.A. (1976). Changes of variance in first-order autoregressive time series models, *Applied Statistics,* volume 25, n°3, pp248-256.

Fukunaga K. (1972). *Introduction to Statistical Pattern Recognition,*Academic Press, New York.

Hand D.J. (1982). *Kernel discriminant analysis*, Wiley, Chichester.

Silverman B.W. (1986). *Density estimation for statistics and data analysis*, Chapman and Hall, London.

Rasson J.-P. and Granville V. (1996). Geometrical tools in classification, *Computational Statistics & Data Analysis*, volume 23, pp105-123.

Rasson J.-P., Montaigne F. and Orban-Ferauge F. (1995). From the supervised to the unsupervised behavioral classification, *Procedings of the International Geoscience and Remote Sensing Symposium 1995*, volume 2, pp1159-1161.

A "Leaps and Bounds" Algorithm for Variable Selection in Two-Group Discriminant Analysis

António Pedro Duarte Silva
Faculdade de Ciências Económicas e Empresariais
Universidade Católica Portuguesa at Porto, email: psilva@porto.ucp.pt
Rua Diogo Botelho 1327, 4150 Porto Portugal

Abstract:. An algorithm that identifies the variable subsets with most discriminatory power (in a predictive sense) is proposed. This algorithm minimizes parametric estimates of the error rate among all the possible variable subsets, evaluating only a fraction of the total number of subsets. The computational feasibility is illustrated by simulation experiments.

Keywords: Discriminant Analysis; Variable Selection Techniques; Optimization Techniques.

1. Introduction

In many applied discriminant analysis studies data is collected on a large number of variables, that latter are to be reduced to a small subset. In the early stages of the analysis the researcher may want not to disregard any potentially interesting variable. However, discriminant functions with too many variables are usually difficult to estimate and interpret and it is well known that the inclusion of irrelevant or redundant variables in data-based classification rules tends to weaken its prediction power (Jain and Waller 1978, McLachlan 1992, pp 391-392).
Most major statistical software packages implement stepwise selection methods that sequentially try to improve global measures of group discrimination. These methods tend to relay on statistical tests to measure the potential contribution of additional variables to the "global separation" between the groups (McKay and Campbell 1982a, Huberty and Wisenbaker 1992, Huberty 1994 pp 231-233, McLachlan 1992, pp 392-400). In spite of their popularity these procedures have important shortcomings:

(i) The criterion used to select variables should take into consideration the nature of the problem and the objectives of the analysis. In particular, when the primary objective of the analysis is prediction, the variable selection process should attempt to maximize prediction power rather than group separation or discrimination (McKay and Campbell 1982b).

(ii) Stepwise selection methods are suboptimal strategies that were developed to surpass the computational difficulties associated "all possible subsets" selection methods. However, when computationally feasible, "all subsets" strategies are preferable. As stepwise methods analyze one variable at the time, they tend to ignore the combined effect of several variables.

2. Choice of Selection Criterion

In descriptive DA studies variables are usually chosen in order to maximize some index of group separation. The most common index used for this purpose is $\eta^2 = 1 - \Lambda$ (Λ being the well known Wilk's statistic).
In predictive DA studies variables should be chosen in order to optimize some estimate of prediction ability with good statistical properties.
In two-group problems there are several parametric estimators of the conditional error rate that, under the usual assumptions of multivariate normality and equality of variance-covariance matrices across groups, tend to be superior to its nonparametric counterparts (McLachlan 1992, pp 369-370). For instance, Lachenbruch (1968) and McLachlan (1974) and have proposed respectively $\hat{e}_i^{(L)}$ and $\hat{e}_i^{(M)}$ as estimators for the conditional error rate in group i.

$$\hat{e}_i^{(L)} = \Phi[\, - M^{-1/2} * \{ \frac{1}{2}D - \frac{pN}{Dn_i(N-p-1)} \} \,] \quad (M = \frac{N(N-1)}{(N-p)(N-p-3)}) \qquad (1)$$

$$\hat{e}_i^{(M)} = \Phi(D/2) + \phi(D/2) *$$

$$\left\{ \left[\frac{p-1}{Dn_i} + \frac{D}{32m} * 4(4p-1) - D^2 \right] + \frac{(p-1)(p-2)}{4Dn_i^2} + \frac{p-1}{64mn_i}\left[-D^3 + 8D(2p+1) + {}^{16}\!/_{D} \right] + \right.$$

$$\left. \frac{D}{12288m^2}\left[3D^6 - 4D^4(24p+7) + 16D^2(48p^2 - 48p - 53) + 192(-8p+15) \right] \right\} \qquad (2)$$

where $\Phi(\bullet)$, $\phi(\bullet)$ are the cumulative probability, and the probability density of a standard normal variate, D is the sample Mahalanobis distance between group centroids, n_i is the number of training sample observations in group i, $N = n_1 + n_2$, $m = N - 2$, and p is the number of variables used in the allocation rule.
Estimators of the global conditional error rate can be defined as $\hat{e}^{(L)} = \pi_1\, \hat{e}_1^{(L)} + \pi_2\, \hat{e}_2^{(L)}$ or $\hat{e}^{(M)} = \pi_1\, \hat{e}_1^{(M)} + \pi_2\, \hat{e}_2^{(M)}$ where π_1 and π_2 are prior probabilities of group membership.
Once n_1, n_2 and p are fixed, $\hat{e}_i^{(L)}$ and $\hat{e}_i^{(M)}$ are simple functions of D. Thus, for a given subset size, minimizing $\hat{e}^{(L)}$ or $\hat{e}^{(M)}$ is equivalent to maximizing D. Furthermore, it is well known that D and η^2 are related by (3).

$$D^2 = \frac{m * N}{n_1 * n_2} \frac{\eta^2}{1-\eta^2} \qquad (3)$$

Therefore, for two-group problems, the use of η^2 (or equivalently D) as variable selection criterion can also be justified from a predictive perspective. Furthermore, although D and η^2 never decrease with the addition of new variables, $\hat{e}^{(L)}$ and $\hat{e}^{(M)}$ provide two alternative ways for comparing subsets of different dimensions.

3. Stepwise and All-Subsets Variable Selection Algorithms

The most important methods for variable selection in DA can be broadly divided in two major classes: Stepwise and All-Subsets methods. Stepwise methods can still be divided into forward (methods that start with an empty subset) and backward (methods that start with the full set of potential variables). Both these approaches have some limitations. The most important problem with forward stepwise methods, is that they may fail to recognize the combined discriminatory power provided by groups of variables which individually look uninteresting. Backward stepwise approaches are not so seriously affected by this problem. However, since backward strategies typically start with a much larger number of variables than necessary, they usually introduce a considerable amount of noise that may seriously affect the selection process.

The preferred strategy for variable selection, is to look at all the possible variable subsets, and choose the one who looks more "interesting". However, in a problem with P potential variables, the number of different subsets equals 2^P-1 which can easily be a formidable number. This problem also occurs in regression analysis, where a considerable research effort has been made in development of efficient algorithms for All-Subsets variable selection. For instance Furnival (1971) has showed, that performing successive Gaussian elimination operations on the Sum of Squares and Cross Products (SSCP) matrix (4), in an appropriate order, the Residual Sum of Squares (RSS) for the

$$\text{SSCP} = \begin{bmatrix} X'X \mid & X'Y \\ ---+ & --- \\ Y'X \mid & 0 \end{bmatrix} \tag{4}$$

regressions of Y on all the subsets of X can be evaluated with an effort of about six floating point operations per regression. McCabe (1975) has adapted Furnival algorithm to the DA problem, showing that the value of Wilk's Λ, for all possible variable subsets, can be evaluated with an effort of about twelve floating point operations per subset. Furnival and Wilson (1974) have refined Furnival algorithm, showing that the best variable subsets for each subset size can be found without computing all the 2^P-1 RSS. Furnival and Wilson algorithm combines the original Furnival strategy that moves from SSCP, working in a forward fashion (adding variables to previous subsets) with a backward strategy that works from SSCP^{-1}, deleting variables at each step. Working simultaneously their way up from SSCP and their way down from SSCP^{-1}, Furnival and Wilson build a search tree in a similar manner to branch and bound algorithms of integer programming. Using RSS from previous subsets as bounds, and the fact that the deletion of variables for a given subset can never reduce its RSS, large parcels of the search tree can be pruned. The details of the algorithm can be found in Furnival and Wilson (1974) and will not be repeated here.

Furnival and Wilson algorithm can be readily adapted to the search for the best subsets (in the sense of D) in two-group DA. For that purpose, the SSCP matrix (4), should be replaced by matrix M (5).

$$\mathrm{M} = \left[\begin{array}{c|c} S & \bar{X}_1 - \bar{X}_2 \\ \hline (\bar{X}_1 - \bar{X}_2)' & 0 \end{array}\right] \tag{5}$$

where S represents the sample variance-covariance matrix, and $\bar{X}_1$ and $\bar{X}_2$ the sample mean vectors for groups 1 and 2.

Matrix M is associated with an empty subset and matrix M^{-1} with the set that includes all variables. It can be shown that the lower-right corner element of M^{-1} equals $-D^2$. Using Gaussian elimination operations to move, "by all possible paths", from M in the direction of the identity matrix, the value of $-D^2$ is constantly updated from one subset to another with one additional variable. In a similar manner, moving from M^{-1} in the direction of the identity matrix, the value of $-D^2$ is constantly updated from one subset to another with one less variable. Choosing $\hat{e}^{(L)}$ or $\hat{e}^{(M)}$ as evaluation criterion, it is possible to compare subsets of different dimensions, leading to sharper bounds and further pruning of the search tree.

4. Simulation Experiments

The algorithm described in the previous section, was implemented in the C^{++} programming language. The evaluation criterion chosen was $\hat{e}^{(M)}$. In order to compare its computational effort for different values of P, the following simulation experiment was performed.

For a number of potential variables (P), ranging from 10 to 30, and population Mahalanobis distances (Δ) of 1, 2 and 3, balanced training samples of 50 observations per group, were generated from two multivariate normal populations $N(0,\Sigma)$, $N(\mu_2,\Sigma)$. The common matrix of variances-covariances, Σ, was built assuming that all correlations between $X_1, X_2, \ldots, X_P$ could be explained by a three-factor model with total communalities of 0.80 for all variables. The signs of the correlations between the variables and the factors and the relative importance of each factor in the total communalities were assigned at random. The mean vector in group 2 (μ_2), was controlled, in order to achieve the desired level of Δ, assuming

equal discriminatory power for all variables. Equal priors were assumed through the whole experiment.
This design tries to replicate a reasonably realistic, non-trivial correlation structure, while keeping everything else as simple as possible. The choice of variables with equal discriminatory power was done in order to illustrate a worse-case scenario (in terms of computational effort).
In each replication of the experiment, after searching for the best 20 variable subsets according to $\hat{e}^{(M)}$, statistics were collected on: a) The proportion of subsets that had to be evaluated (P.E.S.); b) The number of floating point operations required for the successive computations of D^2 (N.F.P.O)
For each combination of P and Δ, the experiment was replicated five times. The minimum median and maximum values of these statistics, for P=10, P=20 and P=30 are shown in Table 1.

Table 1: *Computational Statistics for selected values of P*

		P.E.S.			N.F.P.O.		
P	**Δ**	min	me	max	min	me	max
	1.0	29.03%	31.18%	41.35%	3,619	3,896	4,494
10	2.0	22.97%	31.57%	38.22%	3,536	3,945	4,398
	3.0	21.02%	25.51%	39.20%	3,374	3,676	4,484
	1.0	0.52%	0.82%	1.29%	117,143	223,077	373,393
20	2.0	0.78%	1.21%	1.69%	241,200	297,740	444,723
	3.0	0.88%	1.59%	2.35%	265,433	352,001	557,660
	1.0	0.01%	0.02%	0.05%	7,462,258	11,794,926	29,135,755
30	2.0	0.01%	0.03%	0.10%	7,921,745	18,956,392	45,606,974
	3.0	0.02%	0.05%	0.14%	8,636,676	26,129,459	57,380,001

This limited evidence, shows that the proportion of subsets that have to be evaluated, has a strong tendency to decrease as P grows. Therefore, although the number of floating point operations still increases exponentially with P, it no longer doubles at each additional variable.
All the problems considered in this experiment were analyzed with a 486DX2 Personal Computer with 16M of RAM, and no problem required more than 20 minutes of CPU time.

5. Conclusions

The variable selection process in predictive Discriminant Analysis studies should be based on estimators of prediction power. For two-group problems, when the classical assumptions hold, parametric estimators of error rates are known to have good statistical properties and can be used as criteria for variable selection. With these criteria, all-subsets analysis can and should be conducted as long as the number of potential variables is not “very large” . When the classical assumptions do not hold, non-parametric estimators of the error rate might have better properties. However, the computational effort required for an all-subsets analysis based on non-parametric estimators, may be prohibitive. In that case, the choice of selection criterion should balance the statistical advantages of non-parametric estimators against the dangers of stepwise analysis.

References

Furnival, G.M. 1971. All Possible Regressions with Less Computation. *Technometrics*, 13: 403-408.

Furnival, G.M. & Wilson, R.W. 1974. Regressions by Leaps and Bounds. *Technometrics*, 16: 499-511.

Huberty, C.J. 1994. *Applied Discriminant Analysis*, New York, NY: Wiley.

Huberty, C.J. & Wisenbaker, J.M. 1992. Variable Importance in Multivariate Group Comparisons. *Journal of Educational Statistics*, 17: 75-91.

Jain, A.K. & Waller, W.G. 1978. On the Optimal Number of Features in the Classification of Multivariate Gaussian Data. *Pattern Recognition*, 10: 365-374.

Lachenbruch, P.A. 1968. On Expected Probabilities of Misclassification in Discriminant Analysis, Necessary Sample Size, and a Relation with the Multiple Correlation Coefficient. *Biometrics*, 24: 823-834.

McCabe, G.P. 1975. Computations for Variable Selection in Discriminant Analysis. *Technometrics*, 17: 103-109.

McKay, R.J. & Campbell, N.A. 1982a. Variable Selection Techniques in Discriminant Analysis I. Description. *British Journal of Mathematical and Statistical Psychology*, 35: 1-29.

McKay, R.J. & Campbell, N.A. 1982b. Variable Selection Techniques in Discriminant Analysis II. Allocation. *British Journal of Mathematical and Statistical Psychology*, 35: 30-41.

McLachlan, G. J. 1974. An Asymptotic Unbiased Technique for Estimating the Error Rates in Discriminant Analysis. *Biometrics*, 30: 239-249.

McLachlan, G. J. 1992. *Discriminant Analysis and Statistical Pattern Recognition*, New York, NY: Wiley.

Canonical Discriminant Analysis of Multinomial Samples with Applications to Textual Data

Myung-Hoe Huh and Kyung-Sook Yang

Dept. of Statistics, Korea University
Anam-dong 5-1, Seoul 136-701, Korea

Abstract: We develop the canonical discriminant analysis of the G groups data consisting of $n_1,\cdots, n_G$ multinomial samples within each group, on scaled Euclidean space with chi-square distance. Our discriminant analysis produces quantification plots showing both q observation categories and $N\,(= n_1 +\cdots+ n_G)$ multinomial sample units, as well as G group centroids. We apply the proposed method to Korean text analysis to extract statistical characteristics of Korean language by genres.

Keywords: Canonical Discriminant Analysis, Multinomial Samples.

1. Aim and Background

For the G groups data of continuous q-variate observations, canonical discriminant analysis (CDA) is well developed and used frequently in practice: see Johnson and Wichern (1992) for instance. But, when all the observation units are multinomial samples with q observation categories, we need to develop a new version of CDA respecting the categorical nature of measured data. That situation arises, for example, when we study the statistical characteristics of texts by writers or by genres.

In the past, such data have been analyzed either by correspondence analysis or by Hayashi's quantification method III, that respects the categorical nature of obtained data but ignores the group membership of sample units, in achieving dimensional reduction. For instance, see Jin (1994) who studied on the positioning of commas in Japanese literary texts written by three famous novelists.

2. Geometry and Algebra of Multinomial Samples in R^q

Denote the j-th multinomial sample in the group i by

$$f_{ij} = (f_{ij1}, \cdots, f_{ijq})^t, \quad i=1, \cdots, G; \quad j=1, \cdots, n_i; \quad \sum_{i=1}^{G} n_i = N,$$

where f_{ijk} is the observed frequency of category k contained in that sample($k=1,\cdots,q$). Then, by stacking $\boldsymbol{f}_{ij}'$, $i=1,\cdots,G$; $j=1,\cdots,n_i$ in lexicographic order of i and j, we have the $N\times q$ frequency data matrix F. Transform $q\times 1$ frequency vectors $\boldsymbol{f}_{ij}$ to probability vectors $\boldsymbol{p}_{ij}$ by

$$\boldsymbol{p}_{ij}=(f_{ij1},\cdots,f_{ijq})'/f_{ij+},\quad f_{ij+}=\sum_{k=1}^{q} f_{ijk}\ ;\ \boldsymbol{p}_{ij}'\boldsymbol{1}_q=1\ .$$

So, $\boldsymbol{p}_{ij}$ $(i=1,\cdots,G;\ j=1,\cdots,n_i)$ are N points on (q-1)-dimensional simplex in R^q. Then, as in Greenacre and Hastie (1984), we may impose the geometry on the Euclidean space containing N point elements $\boldsymbol{p}_{ij}$ with weights w_{ij} proportional to f_{ij+}, i.e.

$$w_{ij}=f_{ij+}/\sum_{i=1}^{G}\sum_{j=1}^{n_i} f_{ij+}\ ,$$

and the inter-points distances in chi-square sense, defined by

$$d^2(\boldsymbol{p}_l,\ \boldsymbol{p}_m)=(\boldsymbol{p}_l-\boldsymbol{p}_m)'\,D_c^{-1}(\boldsymbol{p}_l-\boldsymbol{p}_m)\ ,$$

where $\boldsymbol{c}$ is the aggregated multinomial probability vector, i.e.

$$\boldsymbol{c}=\sum_{i=1}^{G}\sum_{j=1}^{n_i}\boldsymbol{f}_{ij}\Big/\sum_{i=1}^{G}\sum_{j=1}^{n_i} f_{ij+}=\sum_{i=1}^{G}\sum_{j=1}^{n_i} w_{ij}\,\boldsymbol{p}_{ij}$$

and D_c is the diagonal matrix with $\boldsymbol{c}$ as diagonal.

To locate the q-dimensional points $\boldsymbol{p}_{ij}$ in smaller dimensional subspace, consider the projection on $q\times 1$ vector $\boldsymbol{a}\neq\boldsymbol{0}$. That is

$$(\boldsymbol{p}_{ij}'\,D_c^{-1}\boldsymbol{a}\ /\boldsymbol{a}'\,D_c^{-1}\boldsymbol{a})\ \boldsymbol{a}\equiv x_{ij}\,\boldsymbol{a}\ ,\quad i=1,\cdots,G\ ;\ j=1,\cdots,n_i\ .$$

Thus, the objective is maximizing the ratio of weighted sum of squares between groups (SSB) to weighted sum of squares within groups (SSW):

$$\max\ \sum_{i=1}^{G}\sum_{j=1}^{n_i} w_{ij}\,(\bar{x}_{i.}-\bar{x}_{..})^2\ \Big/\ \sum_{i=1}^{G}\sum_{j=1}^{n_i} w_{ij}\,(x_{ij}-\bar{x}_{i.})^2.$$

It is not difficult to verify that

$$\begin{aligned} SSB &=\sum_{i=1}^{G}\sum_{j=1}^{n_i} w_{ij}\,\{(\bar{\boldsymbol{p}}_{i.}-\bar{\boldsymbol{p}}_{..})'D_c^{-1}\boldsymbol{a}\}^2/\,(\boldsymbol{a}'\,D_c^{-1}\boldsymbol{a})^2\\ &=\boldsymbol{a}'\,D_c^{-1}(P_G-P)'\,D_w\,(P_G-P)\,D_c^{-1}\,\boldsymbol{a}\ /\ (\boldsymbol{a}'\,D_c^{-1}\boldsymbol{a})^2,\\ SSW &=\sum_{i=1}^{G}\sum_{j=1}^{n_i} w_{ij}\,\{(\boldsymbol{p}_{ij}-\bar{\boldsymbol{p}}_{i.})'D_c^{-1}\boldsymbol{a}\}^2/\,(\boldsymbol{a}'\,D_c^{-1}\boldsymbol{a})^2\\ &=\boldsymbol{a}'\,D_c^{-1}(P-P_G)'\,D_w\,(P-P_G)\,D_c^{-1}\,\boldsymbol{a}\ /\ (\boldsymbol{a}'\,D_c^{-1}\boldsymbol{a})^2, \end{aligned}$$

where

$$P = \begin{pmatrix} \boldsymbol{p}_{11}{}^t \\ \vdots \\ \boldsymbol{p}_{ij}{}^t \\ \vdots \\ \boldsymbol{p}_{Gn_i}{}^t \end{pmatrix}, \qquad P_G = \begin{pmatrix} \bar{\boldsymbol{p}}_{1.}{}^t \\ \vdots \\ \bar{\boldsymbol{p}}_{i.}{}^t \\ \vdots \\ \bar{\boldsymbol{p}}_{G.}{}^t \end{pmatrix}, \qquad \underset{.}{P} = \begin{pmatrix} \bar{\boldsymbol{p}}_{..}{}^t \\ \vdots \\ \bar{\boldsymbol{p}}_{..}{}^t \\ \vdots \\ \bar{\boldsymbol{p}}_{..}{}^t \end{pmatrix}.$$

Therefore, the objective can be formulated as

$$\max \ \boldsymbol{a}^t B \boldsymbol{a} \qquad \text{subject to} \qquad \boldsymbol{a}^t W \boldsymbol{a} = \text{constant}, \tag{1}$$

where

$$B = D_c^{-1} (P_G - \underset{.}{P})^t D_w (P_G - \underset{.}{P}) D_c^{-1},$$
$$W = D_c^{-1} (P - P_G)^t D_w (P - P_G) D_c^{-1}.$$

Using Lagrange multiplier method, (1) leads to solve

$$B\boldsymbol{a} = \lambda \, W \boldsymbol{a}.$$

Since rank $(W) \equiv r < q$, decompose W into $V_1 \Lambda_1 V_1^t$, for $V_1 : q \times r$ matrix with orthonormal columns and $\Lambda_1 : r \times r$ diagonal matrix. Then

$$\Lambda_1^{-1/2} V_1^t B \boldsymbol{a} = \lambda \, \Lambda_1^{-1/2} V_1^t V_1 \Lambda_1 V_1^t \boldsymbol{a} = \lambda \, \Lambda_1^{1/2} V_1^t \boldsymbol{a}.$$

By letting $\boldsymbol{u} = \Lambda_1^{1/2} V_1^t \boldsymbol{a}$ and imposing the condition $\boldsymbol{0} = V_2^t \boldsymbol{a}$, where $V_2 : q \times (q - r)$ matrix with complementary orthonormal columns, we have

$$\boldsymbol{a} = V_1 \Lambda_1^{-1/2} \boldsymbol{u}. \tag{2}$$

Hence

$$\Lambda_1^{-1/2} V_1^t B V_1 \Lambda_1^{-1/2} \boldsymbol{u} = \lambda \, \boldsymbol{u}.$$

Therefore, λ and $\boldsymbol{u}$ are eigenvalues and eigenvectors, respectively, of $\Lambda_1^{-1/2} V_1^t B V_1 \Lambda_1^{-1/2}$. And, $\boldsymbol{a}$ can be obtained via (2).

3. Biplot Representation

We may apply the biplot representation, originally due to Gabriel (1971), in the following way ($i = 1, \cdots, G, \ \ j = 1, \cdots, n_i, \ \ k = 1, \cdots, q$).

Sample Units: $(\boldsymbol{p}_{ij} - \bar{\boldsymbol{p}}_{..})^t D_c^{-1} \boldsymbol{a} \ / \ \boldsymbol{a}^t D_c^{-1} \boldsymbol{a}$,

Group Centroids: $(\bar{\boldsymbol{p}}_{i.} - \bar{\boldsymbol{p}}_{..})^t D_c^{-1} \boldsymbol{a} \ / \ \boldsymbol{a}^t D_c^{-1} \boldsymbol{a}$,

Obs. Categories: $(\boldsymbol{e}_k - \bar{\boldsymbol{p}}_{..})^t D_c^{-1} \boldsymbol{a} \ / \ \boldsymbol{a}^t D_c^{-1} \boldsymbol{a}$.

Where $\boldsymbol{e}_k = (0, \cdots, 1, \cdots, 0)^t$, of which only the k-th position is filled by 1.

The reduced dimensional quantification plots will be helpful in the visual cognition of the multinomial samples, in reference to observation categories. Different scales for dimensional axes can be useful for proportionate expression, by re-scaling $\boldsymbol{a}$ requiring $\boldsymbol{a}^t T \boldsymbol{a} = 1$, where

$$T = B + W = D_c^{-1} (P - P_.)' D_w (P - P_.) D_c^{-1} .$$

4. An Example : Korean Text Data Analysis

We have collected textual data on 36 genres and around ten texts per genre. A subset data in Table 1 is used for methodological demonstration: three genres (G=3) of Korean texts (autobiography A, popular magazine P, and newspaper article N), three sample texts (n_1=n_2=n_3=3, N=9) from each. Measured variables are tallies of seven parts of speech such as adverb, noun, suffix, verb and so on (q=7). The two-dimensional quantification plots are shown in Figures 1 and 2, of which the first eigenvalue is 58.2 and the second one is 2.9.

Table 1: *Korean Texts Classified by Genres and Parts of Speech*

		Adverb	"Asc" (Mark)	"Eomi" (Ending)	"Josa" (Aux.)	Noun	Suffix	Verb
Auto-Biography	Sample 1	136	160	530	493	776	130	339
	Sample 2	181	122	546	393	664	104	390
	Sample 3	120	271	491	458	841	160	330
Popular Magazine	Sample 1	91	247	571	490	840	186	362
	Sample 2	112	237	564	436	725	171	387
	Sample 3	120	286	477	492	853	232	293
Newspaper Articles	Sample 1	54	174	242	297	853	109	117
	Sample 2	66	206	406	446	908	191	211
	Sample 3	51	270	314	351	904	128	162

From Figure 1, we see that Genre P (Popular magazine) is positioned between Genre N (Newspaper article) and Genre A (Autobiography) on the first axis. By superimposing Figure 2 over Figure 1, we also see that Genre A can be typified by the frequent use of verbs and Genre N is by "Eomi" (endings).

References

Gabriel, K. R. (1971). The biplot graphical display of matrices with applications to principal component analysis. *Biometrika*, 58. 453-467.

Greenacre, M. J. and Hastie, T. (1984). The geometric interpretation of correspondence analysis, *Journal of the American Statistical Association*, 82. 437-447.

Jin, M.-Z. (1994). Positioning of commas in sentences and classification of texts, *Mathematical Linguistics*, 19. 317-363. (written in Japanese)
Johnson, R. A. and Wichern, D. W. (1992). *Applied Multivariate Statistical Analysis*, 3rd Edition. Prentice-Hall, Inc.

Figure 1: *Sample Units*

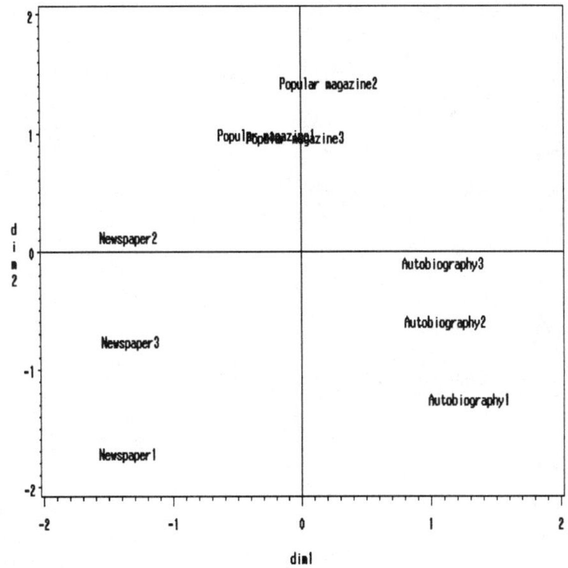

Figure 2: *Observation Categories*

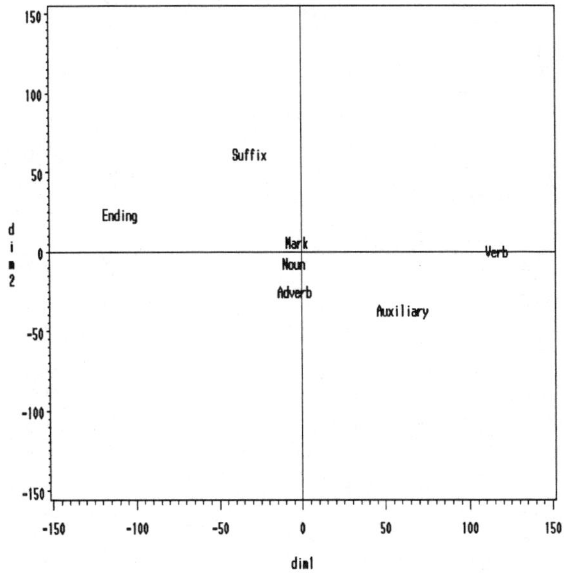

How to extract predictive binary attributes from a categorical one

I. C. Lerman
IRISA-INRIA
Rennes, France
lerman@irisa.fr

J. F. Pinto da Costa
Dep. Matemática Aplicada & LIACC
Universidade do Porto, Portugal
jpcosta@ncc.up.pt

Abstract: In this work, we present new ways of dealing with categorical attributes; in particular, the methodology introduced here concerns the use of these attributes in binary decision trees. We consider essentially two main operations; the first one consists in using the joint distribution of two or more categorical attributes in order to increase the final performance of the decision tree; the second - and the most important - operation, concerns the extraction of relatively few predictive binary attributes from a categorical attribute; specially, when the latter has a large number of values. With more than two classes to predict, most of the present binary decision tree software needs to test an exponential number of binary attributes for each categorical attribute; which can be prohibitive. Our method, ARCADE, is independent of the number of classes to be predicted, and it starts by reducing significantly the number of values of the initial categorical attribute. This is done by clustering the initial values, using a hierarchical classification method. Each cluster of values will then represent a new value of a new categorical attribute. This new attribute will then be used in the decision tree, instead of the initial one. Nevertheless, not all of the binary attributes associated with this new categorical attribute will be used; only those which are predictive. The reduction in the complexity of the search for the best binary split is therefore enormous, as will be seen in the application that we consider; that is, the old and still lively protein secondary structure prediction problem.

Keywords: Decision trees, categorical attributes, binarization, hierarchical clustering

1. Introduction

In our work, the construction of binary decision trees is based on a learning set O_l and a test set O_t . O_l is used to construct a very large tree, to find a sequence of pruned trees, and also to choose a tree from this sequence. The test set O_t is used to estimate the true performance of the chosen decision tree. The methodological developments that we introduce in this paper concern the construction phase; more precisely, the splitting of a categorical attribute. The splitting criterion used by us was the Gini coefficient. The pruning phase is the same as in CART [Breiman & al., 1984], although we do not use the test set to choose the final decision tree; instead, we split the learning set into two disjoint parts.

Let us now suppose that we have a categorical (nominal) attribute v with L values and that we have K classes to be predicted. In the construction of a binary decision tree, each attribute must be binarized; that is, we must consider all of the binary attributes associated with a given predictive attribute and, at each node of the decision tree, choose the best binary attribute. The main problem with categorical attributes is that they originate an exponential number of binary attributes ($2^{(L-1)}$-1), corresponding to all two-class partitions of the L values. If K=2, the authors of the celebrated method CART show how to find the best binary split originated by v in O(L.log(L)). Unfortunately, if the number of classes is greater than two, the complexity of the present version of CART is always exponential in L. However, it is possible to select the same binary attribute with a complexity O(2^K.L.log(L))., which can be very interesting if K is smaller than L. This can be done by using the super-class procedure discussed in CART book [Breiman & al., 1984], and then, for each division into two super-classes, choose the best binary split in O(L.log(L)) with the Gini criterion. [Asseraf, 1996], by generalizing the Kolmogorov-Smirnov distance to the case of categorical variables, has shown how to choose the optimum split in O(L.K^2).

In our application, the initial predictive attributes were nominal (categorical), each with 20 values, and there were three classes to be predicted (K=3). In order to clarify the exposition, we will now briefly describe the application.

2. Protein Secondary Structure Prediction Problem

The primary structure of a protein is formally a word taken from an alphabet of 20 letters (corresponding to the 20 amino acids A,T,P,...,L,M) and which length can reach many hundreds. The secondary structure of a protein can also be formally defined as a word of the same length, but the corresponding alphabet contains three letters, ***E***, ***H*** and ***X***, which are the names of the three concepts to be discriminated; ***H*** corresponds to an helix α, ***E*** to a strand β and ***L*** stands for loop (see figure 1).

Figure 1: *Example of the primary and secondary structures of a protein*

...	T	T	C	C	P	S	I	V	A	R	S	...
...	*E*	*E*	*E*	*E*	*X*	*X*	*H*	*H*	*H*	*H*	*H*	...

Although the number of known primary sequences increases extremely fast, the same is not true for the secondary structure, and so there is a widening number of protein sequences whose secondary structure needs to be predicted. Let us now suppose that we have a primary sequence like the one in figure 1, and that we want to predict the class (***E***, ***H*** or ***X***) corresponding to the letter (amino acid) S in the middle. Nevertheless, the prediction does not only depend on the letter S, but also on its neighborhood; and so, we have considered a window of size eleven (which we have found to be optimal), centered on the position to be predicted. To take account of the information present on that window, we have started to define 11 categorical attributes, v_1, v_2, ...,v_{11} ,each with 20 values, because there are 20 different amino acids. For example, if we did not know the secondary structure in figure 1 and we wanted to predict, for instance, the class (***E***, ***H*** or ***X***) corresponding to the letter S, the attribute v_1 would take the value T, and the attribute v_6 the value S (see figure 2).

Figure 2: *The initial predictive attributes* $v_1, v_2, \ldots, v_{11}$

...	T	T	C	C	P	S	I	V	A	R	S	...
						?						
	v_1	v_2	v_3	v_4	v_5	v_6	v_7	v_8	v_9	v_{10}	v_{11}	

Our data file is constituted by 151 non-homologous globular protein sequences [Colloc'h et al. 1993], which sizes range from a dozen residuals to almost nine hundred. The total number of residuals, that is, of amino acids to be predicted, is about 30,000 (46.6% ***X***, 29% ***H*** and 24.4% ***E***). We have therefore a data file with about 30,000 objects described by 11+1 categorical variables, and we want to learn from it how to predict the last variable.

3. The Results Obtained With CART and C4.5

We have started by considering the application of CART [Breiman & al., 1984] to this data. We must remember that at each node of the decision tree, and for each one of the eleven categorical attributes with 20 values, CART will test $2^{(20-1)}$-1 binary attributes in order to choose the best one. Although this is computationally expensive, the results obtained by us were 54.3% of correct predictions (using 63% of the data to learn and 37% to test). We have also considered the use of C4.5 [Quinlan, 1993] and the results were 52.5% of correct predictions (10-fold-cross-validation). Unlike CART, which always finds the best partition of the values into two classes (the best binary attribute), C4.5 has an option, -s, which also groups the values into classes; but not necessarily into two classes. Using this option, which is computationally expensive, the results of C4.5 were 47.3% of correct predictions (10-fold-cross-validation).

The attributes $v_1, v_2, \ldots, v_{11}$, although natural to consider, do not take into account the order between the different amino acids inside the window; nevertheless, this order is most important for prediction purposes. To take into account this order, we have decided to start by taking two adjacent amino acids at a time, like for instance the pair (4,5), which, in figure 2, is (C, P). This corresponds to use the joint distribution of the variables v_4 and v_5. As there are 400 (20x20) different pairs of amino acids, we were obliged to consider new categorical attributes, each one with 400 values. Using CART on these attributes is obviously prohibitive ($2^{(400-1)}$-1 binary attributes for each). Without using the option -s, we have applied C4.5 to these new attributes, which took very long time to run; the results were 52.6% of correct predictions (10-fold-cross-validation). The next step consisted in using the amino acids three at a time, four at a time, and so on. Obviously, if we take s amino acids at a time, the corresponding attribute will have 20^s values. The best results were obtained for the case $s = 4$. In this case, our predictive attributes have 20^4 = 160.000 values each, and therefore neither CART nor C4.5 are applicable. This was our motivation to develop a new method, ARCADE, which builds binary decision trees, and is capable of treating categorical attributes with a huge number of values.

4. The ARCADE Method

The ARCADE method [Pinto da Costa, 1996], [Lerman & Pinto da Costa, 1996], by grouping the values of each categorical (nominal) attribute v into clusters, allows the definition of another qualitative attribute with much fewer values. For example, let us consider an attribute w with 8 values m_1, m_2, …,m_8. By grouping these values into three clusters, for example $\{m_1, m_5, m_3\}$, $\{m_2, m_8\}$, $\{m_4, m_6, m_7\}$, we can define a new nominal attribute, $\overline{w}$, with three values. In order to make significant this new attribute, the values in the same cluster have to be similar, with respect to their statistical behavior on the classes to be predicted. In these conditions, a contingency table crossing the attribute w with the attribute c to be predicted, is established. The application of the hierarchical classification method AVL [Lerman, 1993] to the rows of this contingency table allowed us to find a significant partition of the values into clusters. For example, let us consider the contingency table crossing our attribute v (160.000 values) with c (3 values).

Table 1: *contingency table crossing v with c*

	1	2	3
V			
1(AAAA)	n^{V}_{11}	n^{V}_{12}	n^{V}_{13}
2(AAAC)	n^{V}_{21}	n^{V}_{22}	n^{V}_{23}
…	…	…	…
20^4(YYYY)	…	…	…

This is therefore the main idea of our method: clustering the values of a predictive categorical attribute in order to reduce the complexity. ???? However, in case where the number of values is to large (160.000 in our case), the direct utilization of a clustering method on the row set of the contingency table becomes - for complexity and statistical reasons - neither tractable nor significant; on the other side, these attributes with 160.000 values correspond to the case $s = 4$ above. Each one of these attributes v is therefore the ``multiplication" of two attributes, v^1 and v^2, both with $s = 2$. Thus, we establish two contingency tables for these two latter attributes, each one with 400 rows.

Table 2: *contingency table crossing v^1 with c*

	1	2	3
v^1			
1(AA)			
2(AC)			
…			
400(YY)			

Table 3: *contingency table crossing v^2 with c*

	1	2	3
v^2			
1(AA)			
2(AC)			
…			
400(YY)			

We apply AVL to the rows of both contingency tables, in order to get two significant attributes $\overline{v}^1$ and $\overline{v}^2$, both with much fewer values (for instance, 31 values each). We then "multiply" (that is, consider the joint distribution) these two latter attributes, getting a new attribute $\overline{v}^{1,2}$ with 31x31=961 values. The contingency table crossing this latter attribute with the attribute to be predicted is established, and AVL is applied to its 961 rows. A significant partition with, for instance, 12 classes is retained, and the values of the final attribute correspond to the classes of this final partition. Our initial attribute, with 160.000 values, was therefore converted into an attribute with 12 values, by means of three hierarchical classifications. We could now consider constructing a decision tree on these final attributes. Nevertheless, the values of these final attributes are hierarchically structured, because we have applied an hierarchical clustering algorithm (see figure 3). We profit from this and, instead of considering all of the $2^{(12-1)}$-1 binary attributes, we consider only about twenty, corresponding to the leaves and nodes of the hierarchical tree (see figure 4).

Figure 3 : *Hierarchical clustering tree on the final 12 macro-values*

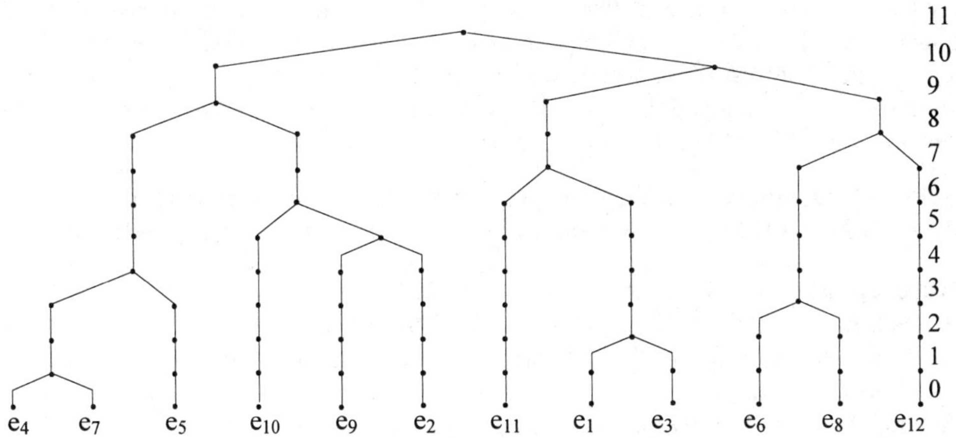

Figure 4 : *The binary attributes used by ARCADE*

a_1 : (a_1 = 1 in { e_1}) and (a_1 = 0 in { e_2, e_3, e_4, e_5, e_6, e_7, e_8, e_9, e_{10}, e_{11}, e_{12}}) (leaf)
a_2 : (a_2 = 1 in { e_2}) and (a_2 = 0 in { e_1, e_3, e_4, e_5, e_6, e_7, e_8, e_9, e_{10}, e_{11}, e_{12}}) (leaf)
...
a_{12}: (a_{12} = 1 in { e_{12}}) and (a_{12} = 0 in { e_1, e_2, e_3, e_4, e_5, e_6, e_7, e_8, e_9, e_{10}, e_{11}}) (leaf)
a_{13}: (a_{13} = 1 in { e_4, e_7}) and (a_{13} = 0 in { e_1, e_2, e_3, e_5, e_6, e_8, e_9, e_{10}, e_{11}, e_{12}}) (node 1)
a_{14}: (a_{14} = 1 in { e_1, e_3}) and (a_{14} = 0 in { e_2, e_4, e_5, e_6, e_7, e_8, e_9, e_{10}, e_{11}, e_{12}}) (node 2)
...
a_{20}: (a_{20} = 1 in { e_6, e_8, e_{12}}) and (a_{20} = 0 in {e_1, e_2, e_3, e_4, e_5, e_7, e_9, e_{10}, e_{11}}) (node 8)
a_{21}: (a_{21} = 1 in { e_4, e_7, e_5, e_{10}, e_9, e_2}) and (a_{21} = 0 in {e_{11}, e_1, e_3, e_6, e_8, e_{12}}) (node 9)

Thus the binary attributes that ARCADE chooses are a_1, a_2, ..., a_{21} (see figure 4). The reduction in complexity is therefore enormous: $2^{(160000-1)}-1 \rightarrow 2^{(961-1)}-1 \rightarrow 2^{(12-1)}-1 \rightarrow \approx 20$. This last part of our method is similar to the Bit-Per-Category Encoding scheme (see [Almuallim & al., 1995] for a description of this and other methods for dealing with tree-structured attributes). In our method, however and by difference, the main contribution is in the construction of a tree-structured attribute, from a nominal one; whereas in the work of [Almuallim & al., 1995], and other methods, the initial predictive attributes are already tree-structured.

5. Results and Conclusions

In this work we have set up a method, ARCADE, for extracting a few number of predictive binary attributes from a categorical one. This method is specially useful in the case of attributes with a large number of values. Nevertheless, even in the case of an attribute with a small number of values, the application of ARCADE can be interesting because it defines a tree-structure on these values. The aim of ARCADE is here situated on the context of the construction of a binary decision tree. However, our method is independent of the prediction method used and then it can be employed by any other prediction method based on binary attributes. The application of ARCADE to the protein secondary structure prediction problem, resulted in 65.1% of correct predictions, which is amongst the best results. The highest performance, for another data-set different from ours, was obtained for the method PHD [Rost & Sander, 1993], which is a combination of several neural networks. This method uses an additional information, which we do not use, given by the multiple sequence alignments of homologous proteins. If a new protein has no homologous proteins, the performance of PHD fells considerably.

Acknowledgements: Joaquim Costa´s work was partly supported by program PRAXIS XXI, FEDER, and program "Fin. Plurianual de Unid. de I & D da JNICT"

References

Almuallim, H., Akiba, Y. & Kaneda, S. (1995). On Handling Tree-Structured Attributes in Decision Tree Learning. *International Conference on Machine Learning*, 1995.

Asseraf, M. (1996). *Extension de la distance de Kolmogorov-Smirnov et stratégie de la coupure optimale*. IVème journées de la societé francophone de classification (SFC), Vannes, France.

Breiman, L., Friedman, J., Olshen, A. & Stone, C. (1984). *Classification and Regression Trees*. Wadsworth.

Colloc'h, N., Etchebest, C., Thoreau, E., Henrissat, B., & Mornon, J. (1993). *Protein Engineering*, 6, 377-382.

Lerman. I.C. (1993). *Likelihood linkage analysis (LLA) classification method: An example treated by hand.* Biochimie, Elsevier editions, volume 75, pp. 379-397.

Lerman, I.C. & Pinto da Costa, J. (1996). *Variables à trés grand nombre de catégories dans les arbres de décision. Application à l'identification de la structure secondaire d'une protéine.* Rapport de Recherche Inria, 2803, 46 pages.

Pinto da Costa, J. (1996). *Coefficients d'association et binarisation par la classification hiérarchique dans les arbres de décision. Application à l'identification de la structure secondaire des protéines*. Thèse de l'Université de Rennes I.

Quinlan, J.R. (1993). C4.5: Programs for Machine Learning. Morgan Kaufman, California.

Rost, B. & Sander, C. (1993). Prediction of protein secondary structures at better than 70%. J. Mol. Biology, 232, pp. 584-599.

A Density Distance Based Approach to Projection Pursuit Discriminant Analysis

Laura Lizzani
Dipartimento di Scienze Statistiche, Università di Bologna
Via Belle Arti 41, 40126 Bologna, Italia – e-mail: lizzani@stat.unibo.it

Abstract: In this paper a brief review of the main contributions to projection pursuit discriminant analysis is presented. A new procedure, based on distance measures between probability densities, is developed in order to obtain linear discriminant functions, which best separate different populations. In the two-group case, Matusita's distance between the projected population density functions is adopted as projection index; while, in the multi-group case, a monotone transformation of Matusita's affinity coefficient is employed to measure the separation among the marginal probability density functions of the different populations. Simulation studies stress the efficacy of the proposed method in comparison with classical parametric ones and with the projection pursuit based linear discriminant procedure developed by Posse.

Key words: Matusita's Affinity Coefficient, Projection Pursuit, Discriminant Analysis, Matusita's Distance Coefficient.

1. Introduction

Projection pursuit (Friedman and Tukey, 1974) is a multivariate statistical method whose goal is the search of the most interesting low-dimensional linear projections of a multidimensional data set, in order to detect the non-linear structure underlying the data.
The most important feature of projection pursuit is its ability to overcome the "curse of dimensionality" caused by the fact that high-dimensional space is mostly empty. Projection pursuit avoids this problem by working on low-dimensional linear projections. Furthermore, projection pursuit is able to detect and ignore irrelevant variables (Huber, 1985).
In this paper attention is focused on projection pursuit discriminant analysis. Different solutions to discrimination and classification problems based on projection pursuit methods have recently appeared in the statistical literature. The different proposals can be grouped into three main approaches. The first one consists in applying projection pursuit regression methods in order to model either the posterior probability distributions of group membership or the likelihood ratio (Henry, 1983; Friedman, 1985; Flick et al., 1990; Roosen and Hastie, 1993). The second one is based on projection pursuit density estimation

methods in order to approximate the probability density functions underlying the data in each population (Polzehl, 1995; Calò and Montanari, 1997). The third approach consists in constructing linear discriminant functions by optimising some suitably chosen projection index. Posse (1992) determines linear discriminant functions by minimising the total probability of misclassification, while Chen and Muirhead (1994) construct robust linear discriminant functions by inserting in Fisher's ratio of between class to within class variation different robust location and scale estimators.

The solution examined in this paper belongs to this last approach: we select the linear discriminant functions, which maximise an index measuring the separation between the groups. Separation is defined in terms of distance coefficients between the projected population density functions.

2. Projection pursuit discriminant analysis based on Matusita's distance coefficient

In the context of projection pursuit discriminant analysis, the most interesting linear projections of the multidimensional data set are those which permit to obtain the best separation among the groups.

In this paper we propose to measure the distance between two populations by Matusita's (1956) coefficient, also referred to as Hellinger's distance. Being X a p-dimensional random vector with probability density functions $f_1(x)$ and $f_2(x)$ in two distinct populations π_1 and π_2, respectively, Matusita's distance between the two multivariate populations π_1 and π_2, $D(\pi_1,\pi_2)$, is defined as

$$D(\pi_1,\pi_2)=\sqrt{\int\left[\sqrt{f_1(x)}-\sqrt{f_2(x)}\right]^2 dx}=\sqrt{2(1-\rho)} \tag{1}$$

where $\rho=\int\sqrt{f_1(x)f_2(x)}dx$ is the affinity coefficient, introduced by Bhattacharyya (1943), measuring the similarity between the two probability density functions $f_1(x)$ and $f_2(x)$, taking values in the [0,1] interval. Matusita's distance is a non-negative measure, it equals zero if and only if $f_1=f_2$; it is symmetric and it satisfies the triangular inequality.

Matusita's distance between two multivariate density functions has been exploited so far only in the context of parametric discriminant analysis (see Krzanowski 1988, for references).

In order to develop a projection pursuit discriminant analysis algorithm based on Matusita's distance, let $Y=\theta' X$ be the linear projection onto the unidimensional space spanned by the unit p-vector θ and let $f_{i\theta}(y)$ be the marginal probability density function of X in the direction θ if X comes from population π_i.

The linear discriminant function which shows the best separation between π_1 and π_2, in terms of Matusita's distance, can thus be defined as the direction θ maximising:

$$D_\theta(\pi_1,\pi_2)=\sqrt{\int\left[\sqrt{f_{1\theta}(y)}-\sqrt{f_{2\theta}(y)}\right]^2 dy}=\sqrt{2(1-\rho_\theta)} \tag{2}$$

where $\rho_\theta=\int\sqrt{f_{1\theta}(y)f_{2\theta}(y)}dy$.

The sample version of this projection index is obtained by replacing the unknown densities in (2) with their kernel estimates:

$$d_\theta(\pi_1,\pi_2)=\sqrt{\int\left[\sqrt{\hat{f}_{1\theta}(y)}-\sqrt{\hat{f}_{2\theta}(y)}\right]^2 dy} \tag{3}$$

Let $x_{i1},x_{i2},\ldots,x_{in_i}$ be a sample from the population π_i $(i=1,2)$. The univariate kernel estimate (Silverman, 1986) of the marginal density function of the *i-th* group is defined by:

$$\hat{f}_{i\theta}(y)=(n_i h_i)^{-1}\sum_{j=1}^{n_i}K\left(\left(y-\theta' x_{ij}\right)/h_i\right) \tag{4}$$

where $K(\cdot)$ is a kernel function, usually a symmetric probability density function, and h_i is the window width (one for each population), which controls the degree of smoothness of the estimate.

Once the unit vector θ_0, maximising the index (3), has been found, the allocation rule, proposed in order to classify a new observation, x, to one of the possible populations, consists in computing the linear combination $y=\theta_0'x$, and assigning x to π_1 if $\hat{f}_{1\theta_0}(y)\geq\hat{f}_{2\theta_0}(y)$, to π_2 otherwise.

Let $f_1(x),\ldots,f_G(x)$ be the probability density functions of the p-dimensional random vector X in G distinct populations $\pi_1,\ldots,\pi_G$ $(G\geq 2)$. Matusita (1967) developed an extension of the affinity coefficient ρ, measuring the similarity between $\pi_1,\ldots,\pi_G$, defined as:

$$\rho_G=\int[f_1(x)\cdots f_G(x)]^{1/G}dx \tag{5}$$

Glick (1973) proposed the following transformation of ρ_G as distance measure:

$$D_G(\pi_1,\dots,\pi_G)=1-\rho_G^{G/2} \tag{6}$$

It is symmetric, invariant under permutation of its arguments, non-negative and it equals zero if and only if $f_1=\dots=f_G$. The measure of total separation among the G populations is not less than the separation between any two of them.
The projection index based on this distance measure is defined as:

$$D_{G\theta}(\pi_1,\dots,\pi_G)=1-\left\{\int\left[f_{1\theta}(y)\cdots f_{G\theta}(y)\right]^{1/G}dy\right\}^{G/2} \tag{7}$$

The empirical version of (7) is obtained by estimating the unknown univariate density functions by means of kernel density estimators (4), on the basis of G samples $x_{i1},x_{i2},\dots,x_{in_i}$ $(i=1,\dots,G)$.
Let θ_0 be the unit vector which maximises this projection index. The allocation rule consists in assigning a new observation x to the i-th population if $\hat{f}_i(\theta_0{}'x)\geq\hat{f}_j(\theta_0{}'x)$ for each $j\neq i$ $(i,j=1,\dots,G)$.

3. Examples and conclusions

In order to test the performances of our projection pursuit discriminant analysis algorithm we have considered three situations described in Polzehl (1995).

$$S1: \quad \begin{cases}\pi_1: \frac{1}{3}N_2\left((0 \quad -2)',I_2\right)+\frac{1}{3}N_2\left((\sqrt{3} \quad 1)',I_2\right)+\frac{1}{3}N_2\left((-\sqrt{3} \quad 1)',I_2\right)\\ \pi_2: \frac{1}{3}N_2\left((0 \quad 2)',I_2\right)+\frac{1}{3}N_2\left((\sqrt{3} \quad -1)',I_2\right)+\frac{1}{3}N_2\left((-\sqrt{3} \quad -1)',I_2\right)\end{cases}$$

$$S2: \quad \begin{cases}\pi_1: LogN_2\left((0 \quad 0)',0.5I_2\right)\\ \pi_2: N_2\left((e^{1/4} \quad e^{1/4})',(e-e^{1/2})I_2\right)\end{cases}$$

Situation *S3* is obtained by adding two noisy variables to situation *S1*. It is used to study the behaviour of the procedure with an increasing number of dimensions. For each situation we have generated 100 training samples of size 150 and test samples of size 900 for each population. In our applications we have used a standard gaussian kernel whose window width has been chosen by Sheather and Jones's (1991) automatic selection method. Numerical optimisation has been carried out through the random search algorithm developed by Huber (1990). We have computed the mean percent error rates (percent error rates averaged over the 100 replications) both on the training samples and on the test samples and we have compared the results of our method (PPDA) with those obtained by linear discriminant analysis (LDA), quadratic discriminant analysis (QDA) and by Posse's algorithm (POSSE).

Table 1: *Mean percent error rates computed on the test (training) samples*

	LDA	QDA	POSSE	PPDA
S1	50.1 (48.8)	49.7 (46.6)	38.9 (31.3)	37.3 (33.9)
S2	50.9 (48.0)	48.7 (45.9)	37.0 (31.1)	36.4 (32.9)
S3	50.0 (46.6)	50.3 (43.1)	39.9 (30.6)	38.6 (33.8)

Our totally non-parametric procedure performs far better than the classical parametric ones in all of the situations considered (see Table 1) and seems to be a valid competitor to the projection pursuit based procedure developed by Posse (1992). Aimed at the minimisation of the total probability of misclassification Posse's algorithm is the best on the training samples, but, when evaluated on the test samples, it is outperformed by our procedure. In the case of normal populations with equal covariance matrices, the projection directions, optimising both indices, approximate the coefficient vector defining Fisher linear discriminant function. Otherwise, the different behaviour of the two indices is due to the different definition of separation: our index measures the distance between the projected probability density functions and takes into account their different shapes over the whole domain, while Posse's index depends only on the overlapping area between them.

It is worthwhile to note the ability of projection pursuit to ignore irrelevant variables. In all the replications of situation *S3*, the solution vector is about $\theta_0 = (0.866, -0.494, -0.009, -0.083)$. If we scale the absolute values of these coefficients with the standard deviations of the corresponding variables and normalise the resulting vector, we obtain the importance coefficient vector (Montanari, Lizzani 1996) $CI = (0.866, 0.497, 0.005, 0.047)$. By comparing each component of this vector, which measures the discrimination power of each variable, with the 10-th percentile of the uniform distribution on the unit *p*-sphere (0.078 in this case), we can deem irrelevant the contributions of the last two variables, which represent noisy terms. The 10-th percentile of the uniform distribution on the unit *p*-sphere is chosen as the cut-off value because, when $f_1 = f_2$, all the directions on the unit *p*-sphere are equally uninteresting and have the same probability to be chosen as solution vectors by the projection pursuit algorithm.

References

Bhattacharyya, A. (1943). On a measure of divergence between two statistical populations defined by their probability distributions, *Bulletin of the Calcutta Mathematical Society*, 35, 99-109.

Calò, D. & Montanari, A. (1997). An empirical discrimination algorithm based on projection pursuit density estimation, *Book of Short Papers, Classification and Data Analysis*, Meeting of the Classification Group of Società Italiana di Statistica, Pescara, 3-4 luglio, 65-68.

Chen, Z. & Muirhead, R.J. (1994). A comparison of robust linear discriminant procedures using projection pursuit methods, in *Multivariate Analysis and its Applications*, IMS Lecture Notes-Monograph Series, Hayward, 24, 163-176.

Flick, T.E. & Jones, L.K. & Priest, R.G. & Herman, N.C. (1990). Pattern classification using projection pursuit, *Pattern Recognition*, 12, 1367-1376.

Friedman, J.H. (1985). Classification and multiple regression through projection pursuit, Technical Report 12, Laboratory for Computational Statistics, Department of Statistics, Stanford University.

Friedman, J.H. & Tukey, J. (1974). A projection pursuit algorithm for exploratory data analysis, *IEEE Transactions on Computers*, 23, 881-889.

Glick, N. (1973). Separation and probability of correct classification among two or more distributions, *Annals of the Institute of Statistical Mathematics*, 25, 373-382.

Henry, D.H. (1983). Multiplicative models in projection pursuit, Ph. D. Thesis, Stanford Linear Accelerator Center, Stanford University.

Huber, P.J. (1985). Projection pursuit, *The Annals of Statistics*, 13, 435-475.

Huber, P.J. (1990). Algorithms for projection pursuit, Technical Report 3, Department of Mathematics, M.I.T. Cambridge.

Krzanowski, W.J. (1988). *Principles of multivariate analysis*, Oxford Science Publication, New York, 359-360.

Matusita, K. (1956). Decision rule, based on distance, for the classification problem, *Annals of the Institute of Statistical Mathematics*, 8, 67-77.

Matusita, K. (1967). On the notion of affinity of several distributions and some of its applications, *Annals of the Institute of Statistical Mathematics*, 19, 181-192.

Montanari, A. & Lizzani, L. (1996). Projection pursuit e scelta delle variabili, *Atti della XXXVIII Riunione Scientifica della Società Italiana di Statistica*, 2, 591-598.

Polzehl, J. (1995). Projection pursuit discriminant analysis, *Computational Statistics and Data Analysis*, 20, 141-157.

Posse, C. (1992). Projection pursuit discriminant analysis for two groups, *Communications in Statistics - Theory and Methods*, 21, 1-19.

Roosen, C.B. & Hastie, T.J. (1993). Logistic response projection pursuit regression, Statistics and Data Analysis Research Department, AT&T Bell Laboratories, Doc. BL011214-930806-09TM.

Sheather, S.J. & Jones, M.C. (1991). A reliable data-based bandwidth selection method for kernel density estimation, *Journal of the Royal Statistical Society, Series B*, 93, 683-690.

Silverman, B.W. (1986). *Density estimation for statistics and data analysis*, Chapman and Hall, London.

Two Group Linear Discrimination Based on Transvariation Measures

Angela Montanari Daniela G. Calò
Dipartimento di Scienze Statistiche, Università di Bologna,
Via Belle Arti, 41, 40126 Bologna, Italy – e mail: montanar@stat.unibo.it

Abstract: In this paper we derive a two group linear discriminant function (LDF) by minimizing Gini's transvariation probability. This solution is a special case of projection pursuit methods and improves the performances of Fisher's LDF when the conditions which guarantee its optimality do not hold.

Key words: Linear discriminant function, Transvariation, Projection Pursuit.

1. Introduction

In classification issues, the two group linear discriminant function (LDF) is commonly used mainly because of its simple form and interpretability.
Its origin goes back to Fisher (1936) who suggested to search directly the linear combination of the p measured characteristics which maximizes group separation, defined as the ratio of "between" to "within" group variance under homoscedasticity constraint. It apparently does not require any distributional assumption, but normality or at least symmetry is actually implicitly assumed. Fig. 1 clearly shows this point. For skewed distributions it may in fact happen that Fisher's criterion does not succeed in distinguishing between situation a) and b): they have the same "between" to "within" group variance ratio, while from a discrimination perspective case a) is clearly preferable.
Furthermore, under normality, and of course homoscedasticity, the allocation rule based on Fisher's LDF is also the one which minimizes the total probability of misclassification (or maximizes posterior probability of group membership).

Figure. 1 : *Two situations having the same between to within group variance ratio but different overlapping.*

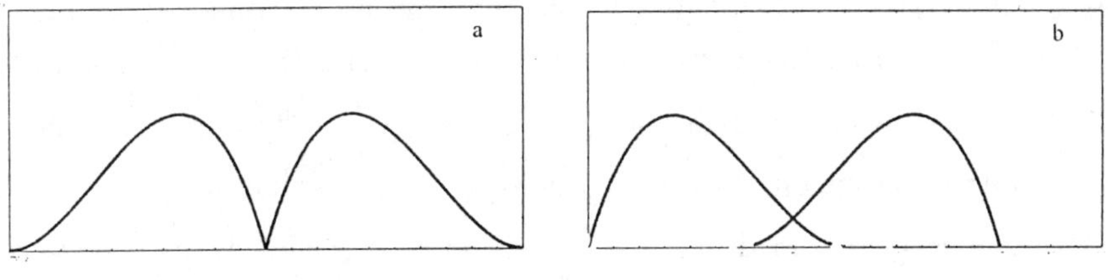

The performance of Fisher's LDF, when homoscedasticity and normality assumptions are violated, have largely been studied both theoretically and empirically (see Krzanowski and Marriott, 1995, for a detailed review). All the studies have shown that as covariance differences increase, the behavior of Fisher's function, i.e. its error rate, for moderate sample sizes deteriorates (for small samples Fisher's function seems to perform better than the quadratic allocation rule as it involves a smaller number of parameters, but its error rate may still be too high for practical use). The same happens for departures from normality whose effect varies according to the distribution shape.

To cope with heteroscedasticity in the normal case, while preserving the simple form of the linear function, Anderson and Bahadur (1962) have proposed an alternative they have called the "best" linear discriminant function, meaning that it's the linear function which minimizes the total probability of misclassification.

This idea has recently been generalized by Posse (1992), in a fully non-parametric context, exploiting the high potentiality of projection pursuit methods (Huber, 1985). In fact projection pursuit is concerned with "interesting" low dimensional projections of high dimensional data and if "interestingness" is equated to "showing the minimum total probability of misclassification" for any group density then Posse's method is obtained.

It obviously gives the best error rates as far as the classification of the units belonging to the training sample is concerned, but for small or moderate sample sizes nothing guarantees its good performance on new cases whose group membership is to be determined; in other words, it may be derailed by a sort of overfitting effect.

As an alternative distribution free solution we propose a LDF whose derivation is still based on the projection pursuit methodology but whose logic is closer to Fisher's than to Posse's one. Our linear discriminant function is obtained by maximizing group separation in terms of Gini's transvariation.

2. Transvariation and its measures

According to Gini (1916), two groups g_1 and g_2 are said to transvariate on a variable y, with respect to an average value m_y, if the sign of some of the $n_1 n_2$ differences $y_{i1} - y_{j2}$ $(i = 1,2,\ldots,n_1 \quad j = 1,2,\ldots,n_2)$ which can be defined between the y values belonging to the two groups is opposite to that of $m_{y1} - m_{y2}$. Any difference satisfying this condition is called "a transvariation" and $|y_{i1} - y_{j2}|$ is its intensity. Similarly the group g_k $(k = 1, 2)$ and a constant c are said to transvariate on y, with respect to an average value m_y, if the sign of some of the n_k $(k = 1, 2)$ differences $y_{ki} - c$ $(i = 1, 2, \ldots, n_k)$ is opposite to that of $m_{yk} - c$.

In order to measure the transvariation between a group and a constant or between two groups Gini first introduced, among the others, transvariation probability, transvariation intensity and transvariation area.
The discrete version of the transvariation probability with respect to an average value – which Gini assumed to be the median – is defined as the ratio of the number of transvariations (s_{kc} for transvariation with a constant and s_{12} for transvariation between two groups) to its maximum ($1/2\,n_k$ or $1/2n_1\,n_2$ respectively when the group character distribution is symmetric, or an higher value, whose maximum is $3/2n_1\,n_2$, for skewed distributions; see Gini for a thorough discussion).
Denoted by $f_k(y)$ $(k = 1, 2)$ the probability density function of y in the parent population of group g_k and by $F_k(y)$ its distribution function, and assuming that $m_{y1} > m_{y2}$, the continuous expression of the transvariation probability is given by $2\int_{-\infty}^{+\infty} F_1(y) f_2(y)\, dy$, and the transvariation area (which only admits a continuous formulation) is:

$$\int_{-\infty}^{+\infty} \Psi(y)\, dy \qquad \text{where } \Psi(y) = min(f_1(y), f_2(y)) \tag{1}$$

The transvariation probability equals zero when the sign of the difference between the two medians allows to predict with certainty the sign of the difference between the value of y in any two units belonging to the two different groups; it increases as the probability of such a forecast decreases, and reaches its maximum value, i.e. 1, when the relative positions of the medians of the two groups do not allow any prediction. Its complement can thus be assumed as a measure of the typicality of the medians which summarizes location, scale, and shape characteristics.
When the transvariation probability is zero, the two groups do not overlap and therefore the transvariation area is also zero, but the inverse is not always true (Gini, 1959 pag. 257). In fact for multimodal densities it may happen that the transvariation area is zero while the transvariation probability is not. This means that the two measures usually highlight different aspects of group transvariation. However, when y is normally distributed in the parent population of both g_1 and g_2 and homoscedasticity holds (σ_y denotes the common standard deviation), the transvariation probability and the transvariation area are monotonically linked, being

$$p_{tran} = 1 - \Theta\left(\frac{m_{y1} - m_{y2}}{2\sigma_y}\right) \quad \text{and} \quad a_{tran} = 1 - \Theta\left(\frac{m_{y1} - m_{y2}}{2\sqrt{2}\sigma_y}\right) \tag{2}$$

(where Θ is the Laplace integral $\Theta(x) = \frac{2}{\sqrt{\pi}}\int_0^x \exp(-z^2)\, dz$): they therefore give equivalent information.

While the previously described indices involve frequencies only, the transvariation intensity regards the character values too. It is defined with respect to the arithmetic mean and its discrete expression is the ratio of the sum of transvariation intensities (T_{12}) to its maximum. It too admits a continuous expression which allows to show the equivalence of transvariation intensity to transvariation probability and area in the normal homoscedastic case.

3. Transvariation based discrimination and classification

From the above discussion it emerges that transvariation measures can be profitably used to discriminate between two groups (having equal prior probabilities).

Let $\boldsymbol{x}$ be the p-dimensional random vector of the variables to be used in the discrimination process, $\boldsymbol{a}$ a p-dimensional unit vector defining a projection direction and $y = \boldsymbol{a}'\boldsymbol{x}$ the random variable that results from projecting $\boldsymbol{x}$ along $\boldsymbol{a}$. A LDF can then be derived, in a projection pursuit framework, as the linear combination ($y = \boldsymbol{a}'\boldsymbol{x}$) which minimizes transvariation probability, intensity or area. This possibility has already been indicated by Salvemini (1959) but, as far as we know, it hasn't yet been translated into an operative procedure. Projection pursuit offers a set up within which this solution may be pursued.

A closer look at equation (1) shows that (in the equal prior case) the transvariation area is but twice the total probability of misclassification i.e. the quantity Posse (1992) has suggested to minimize. In the multivariate normal homoscedastic case Posse has shown that his projection pursuit LDF coincides with Fisher's, but as transvariation probability, transvariation intensity and transvariation area (i.e. the total probability of misclassification) are monotonically linked (see (2)) minimization of the transvariation probability or

Figure. 2 : *Values of Fisher's criterion (dotted line), transvariation intensity (crosses), probability (solid line) and area (bold line) as a function of the rotation angle for two bivariate normal homoscedastic populations.*

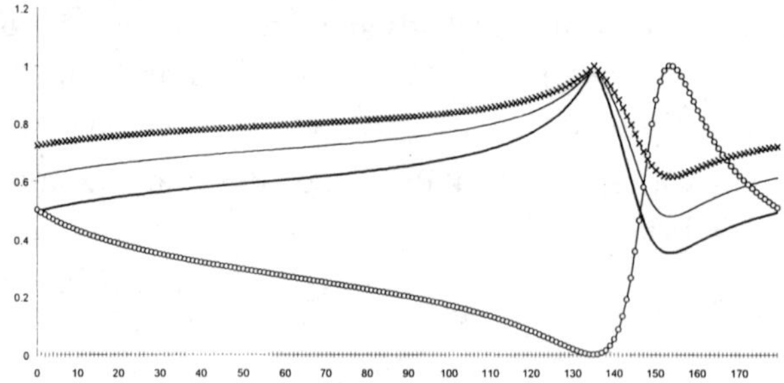

the transvariation intensity (with respect to $\boldsymbol{a}$) yields Fisher's function too. This is clearly represented in Fig. 2, where all the projections of two bivariate normal homoscedastic populations have been examined. Each discrimination measure has been evaluated as a function of the rotation angle, which ranged, at steps of one degree, from 0 to 180 degrees. Fisher's criterion attains its maximum exactly where all the transvariation measures are minimized.

In different distributional conditions, however, minimizing the transvariation probability, the transvariation intensity or the transvariation area may give profoundly different results in terms of misclassification rates. In particular, being not affected by the overfitting effect already discussed for Posse's procedure, allocation rules based on the transvariation probability and the transvariation intensity may perform better on a novel set of units. In the following applications, the discrete transvariation probability (the continuous one gives the same results but is computationally more time consuming) has been preferred due to its good robustness properties. In fact, while giving results similar to those obtained by transvariation probability in the no outlier case, transvariation intensity performances heavily deteriorate when extreme values are present in the data set (Montanari e Calò, 1998).

When the discriminant function is determined by the numerical minimization of transvariation probability a coherent classification rule is then given by projecting the unit c on the obtained discriminant direction and allocating it to the group with which it transvariates most, that is to the group for which p_{kc} is maximum.

4. Example and conclusions

The performances of our transvariation based discrimination procedure (resorting to simulated annealing as the optimization algorithm) have been tested on many simulated data sets and compared to those of Fisher's and Posse's methods (a GAUSS version of Posse's original FORTRAN algorithm has been employed). Here we present the results obtained on 50 replications of four simulation experiments involving two normal heteroscedastic six-variate populations (as described in Posse, 1992), and training sets of size $n_1 = n_2 = 150$ and $n_1 = 100$ $n_2 = 200$, and two lognormal four-variate populations (derived from gaussian densities centered in (0,0,0,0) and (2,0,0,0) and with identity covariance matrix) with the same training set sizes as the previous experiments.

In the lognormal case most projections show skewed training sample distributions. This, in principle, would require to determine, for each projection, the maximum attainable value for the number of transvariations. As no closed form exists for it, this value should be computed numerically, thus heavily slowing the procedure which involves a further numeric step in the minimization phase. Furthermore this additional computational load could not yield a remarkable gain in the determination of the best discriminant direction; as Gini himself adfirms "in practical cases, assuming the upper limit of

transvariation probability to be 1 is not far from truth". For this reason we have chosen to minimize the transvariation probability in its form suggested for symmetric distributions; this amounts to look for those linear projections along which the absolute number of transvariations is minimum.
Table 1 reports the results of the simulation experiments as the mean probability of misclassification, and the standard deviation of the probability estimates (in brackets), for each rule in the different cases, evaluated on a test set of 1000 individuals for each population.

Table 1: *Results of four simulation experiments*

LDF	heteroscedastic normal		lognormal	
	n_1=n_2=150	n_1=100 n_2=200	n_1= n_2=150	n_1=100 n_2=200
Fisher	0.149 (0.00966)	0.163 (0.01213)	0.254 (0.02016)	0.258 (0.02585)
Posse	0.143 (0.01643)	0.165 (0.01934)	0.243 (0.01787)	0.271 (0.01888)
Transvariation based	0.158 (0.01182)	0.161 (0.01152)	0.165 (0.01023)	0.165 (0.01074)

The simulation studies have shown that in the heteroscedastic normal case transvariation based discrimination, Fisher's function and Posse's one yield similar error rates (evaluated on a test set); while in the non normal case - that is when non parametric methods become necessary - transvariation based discrimination largely outperforms both of the other methods.

References

Anderson, T. W. & Bahadur, R. R. (1962). Classification into two multivariate normal populations with different covariance matrices, *Annals of Mathematical Statistics*, 33, 420-431.

Fisher, R. A. (1936). The use of multiple measurements in taxonomic problems, *Annals of Eugenics*, 7, 179-188.

Gini, C. (1916). Il concetto di transvariazione e le sue prime applicazioni, *Giornale degli economisti* and *Rivista di statistica*, and in Gini, C. (1959).

Gini, C. (1959). *Transvariazione* , Libreria Goliardica, Roma (Ottaviani G. ed.).

Huber, P. J. (1985). Projection Pursuit (with discussion), *The Annals of Statistics*, 13, 435-475.

Krzanowski, W. J. & Marriott, F. H. C. (1995). *Multivariate Analysis*, Part 2, Arnold, London

Montanari, A. & Calò, D. G. (1998). Una rilettura, in termini di tranvariazione, dell'analisi discriminante lineare, *XXXIX Riunione Scientifica della Società Italiana di Statistica*, in press.

Posse, C. (1992). Projection Pursuit Discriminant Analysis for two groups, *Communications in Statistics*, Theory and Methods, 21 (1), 1-19.

Salvemini, T. Transvariazione e analisi discriminatoria, in Gini, C. (1959).

A new geometrical hypothesis for clustering and discriminant analysis

Jean-Paul Rasson - Didier Jacquemin - Vincent Bertholet
Department of Mathematics
Facultés Universitaires Notre-Dame de la Paix
B-5000 Namur, Belgium
E-Mail: Didier.Jacquemin@fundp.ac.be

Abstract: In this paper, we propose a new clustering method and a new discriminant rule valid on the basic space $\Re^2$. These procedures make use of a new concept in clustering: the concept of closed and connex forms. This hypothesis generalizes the convex hypothesis and is very useful to find non convex natural clusters. Finally, we examine the admissibility of our procedure, in the sense of Fisher and Van Ness [4].

Keywords: Cluster Analysis, Discriminant Analysis, Lebesgue Measure, Stationary Poisson Point Process, Hypervolume Method, Replacing the convexity, Non Convex case, Connex and Closed domains, Optimal Closing, Admissibility Conditions.

1. Introduction

In the framework of the research of clustering by partition, none method can be defined as 'the best clustering method'. We can however easily see that many of them are based on a distance. To avoid the arbitrary choice of distances, similarities and dissimilarities always posed in classification problems, we use in this paper a statistical method based on the Poisson point process. So, the Lebesgue measure which is both the canonical measure of the classification space $\Re^d$ and the measure induced by the Stationary Poisson point process ("the natural model for points distributed at random "(Fisher (1972) [5], Cressie (1991) [3],...)) will be considered here.

The starting point of this article is the recall of a method based on the Stationary Poisson Point Process and on the Lebesgue measure: The Hypervolume method, proposed by Hardy and Rasson (1982) [6]. This method, assuming that the number of clusters (let us say k) is fixed beforehand, was based on the assumption of k CONVEX clusters. The concept of clustering models for convex clusters has been analysed by Bock (1997) [2]. By the definition of the Hypervolume method, such a method is not able to find natural non convex clusters. So, we judged the hypothesis of convexity too strong, and desired a hypothesis which would also allow us to find natural non convex clusters. We used then the notion of *connex and closed* forms to solve this problem. The method based on this new hypothesis is developped: As the hypervolume method, it is based on the Stationary Poisson Point Process and on the Lebesgue measure and requires to know the number of clusters (k) beforehand. We also expose the corresponding discriminant rule (after recalling the one corresponding to the convex case) and sift the procedure to the Fisher and Van Ness conditions (1971) [4] which "eliminate obviously bad clustering algorithms". We restrict ourselves in this paper in examining only the 2-dimensional case, but extension to $\Re^d$ is straightforward.

2. Replacing the hypothesis of convexity

2.1. The Hypervolume criterion

The reader is referred to Hardy and Rasson (1982) [6] who proposed the criterion. For details about the Point Processes theory, see Karr (1991) [7] or Krickeberg (1982) [8].

Suppose that we observe n data points $\{x_1, x_2, \ldots, x_n\}$ inside some unknown domain $D \subseteq \Re^d$. The goal is to find out the natural classes present in the data, assuming that their number is k (k is fixed beforehand). We consider that the n observations to classify $\{x_1, x_2, \ldots, x_n\}$ are a realization of a *stationary Poisson point process* in a domain D of $\Re^d$ which is the union of k disjoint CONVEX compact subdomains D_i (i=1,...,k) (with D_i unknown for all i). Applying the maximum likelihood method, the Hypervolume method consists in finding out the partition which minimizes the sum of the Lebesgue measure of the disjoint convex hulls of the points in the classes.

2.2. The closed hypothesis

The hypothesis of convexity used in the hypervolume method will be generalized here thanks to the notions of CLOSINGS and CLOSED sets, defined in Schmitt and Mattioli (1993) [11], Matheron (1975) [9] or Stoyan and Stoyan (1994) [14].

The (morphological) closing of a set X by a set B is noted X^B and is defined as the complement of the set of all points of X^c covered by the translates of B that are inside X^c. We restrict ourselves in considering only closings when B is a DISC of radius r>0. We will talk about *closing of radius r*.

We also include in the term *closing by a disc* the case of a closing of infinite radius which is the same as the convex hull.

A compact which is the same as its closing by a disc of some radius r is called *closed*. Let us now use the last definition in cluster analysis:

We still suppose that we observe n data points $\{x_1, x_2, \ldots, x_n\}$ inside $D \subseteq \Re^d$ (D unknown). We consider that the n observations are a realization of a stationary

Poisson point process in D which is the union of k disjoint CONNEX CLOSED compact domains D_i ($i=1,...,k$) (We recall here that k is still fixed beforehand).

The fact that the domains are chosen to be *connex and closed* can be explained : a domain must of course be connex to be considered as only one piece. However, the hypothesis of connexity is not sufficient : if we search for domains of minimal Lebesgue measure, we can indeed easily find connex domains which contain all the points and such that the sum of their measures is zero. Trivial cases like these must of course be avoided.
So, we add the closed hypothesis to the connex hypothesis. There are indeed closed forms which are not connex. As domains must be connex, we can not take only the hypothesis of closed domains and we have to ask domains to be connex *and* closed.

This hypothesis (connex + closed domains) generalizes the hypothesis of convexity because the convex are closed (a convex is equal to each of its closings (by discs)) (and a convex is of course connex) and because the set of closed forms is huge compared to the set of convex forms.
We have got therefore a hypothesis giving no triviality cases and less restrictive than the convexity.

Statistical solution :
The aim is to estimate the k subsets D_1, ..., D_k.
With our hypothesis, the points will be independently and uniformly distributed on D and the Likelihood function will take the form :

$$f_D((x_1,...,x_n)) = \prod_{i=1}^{n} \frac{1_D(x_i)}{(\lambda_d(D))} = \frac{1}{(\lambda_d(D))^n} \prod_{i=1}^{n} 1_D(x_i)$$

where $\lambda_D(D) = \sum_{i=1}^{k} \lambda_D(D_i)$ is the d-dimensional Lebesgue measure of D.
So, applying the maximum likelihood method, the domain D for which the likelihood is maximal is, among all those which contain all the points, the one whose Lebesgue measure is minimal.

Let us now define the *optimal closing of a set of points* as the connex closing (by a disc) whose Lebesgue measure is minimal.
Thanks to our closed and connex hypothesis, for each partition of the set of points into k subdomains having DISJOINT closings, the likelihood has a local maximum attained with the optimal closings of the k subsets.

The global maximum will then be attained with the partition which minimizes the sum of the Lebesgue measures of the **disjoint optimal closings** of the points in the k subgroups. This is the solution we seek.

We can see 2 examples of optimal partitions in figure 1 and figure 2, showing the ability of the method to find non convex natural clusters.

Figures 1-2: *Two examples of optimal partitions (which minimize the sum of the area of the disjoint optimal closings in the k clusters), with k=2.*

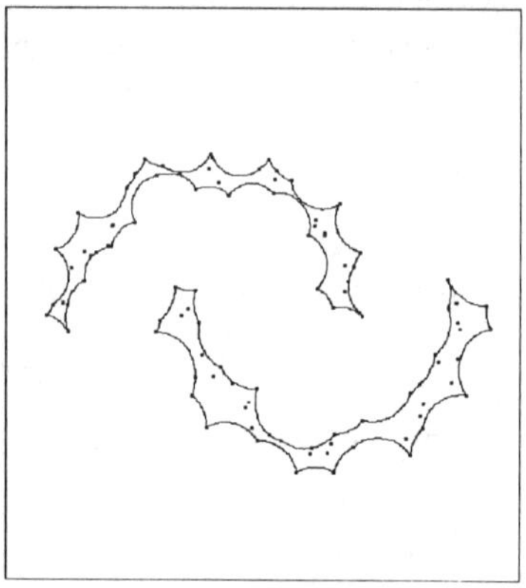

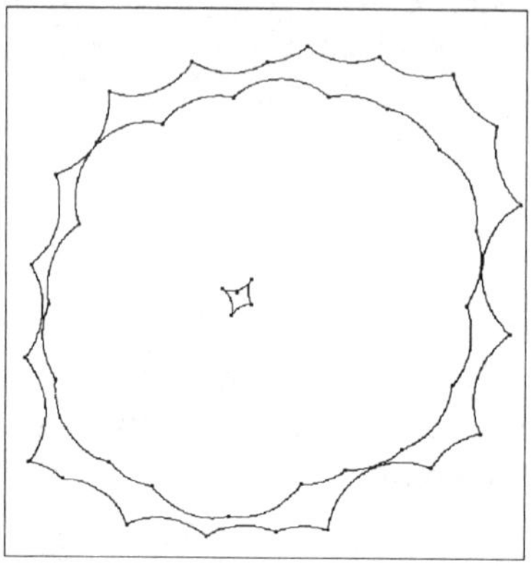

3 Corresponding discriminant rules

3.1. Based on the convexity

The details of the discriminant rule exposed here can be found in Beaufays and Rasson (1985) [1]. The allocation rule is :

Assign x to the ith population if and only if the difference between the area of the convex hulls of the labeled sample of the ith population with and without the individual to be assigned is minimal.

3.2. Based on the closings

We now base our discriminant rule on the closed and connex model.

So, we suppose that the n observations $\left\{x_1,\ldots,x_{n_1+1},\ldots,x_{n_1+n_2},\ldots,x_{n_1+\ldots+n_{k-1}+n_k}\right\}$ in the training sets are a realization of a **stationary Poisson point process** in a domain D of $\Re^2$ which is the union of k disjoint CLOSED and CONNEX compact subdomains D_i $(i=1,\ldots,k)$. We want to assign an individual x to one of the k populations D_i.

We follow here the same idea as the one exposed before : we judge the hypothesis of convexity too strong. So, we examine the case of closed and connex domains instead of convex domains as in **3.1**.

The conditional distribution for the ith population is assumed to be uniform in the closed and connex domain D_i, and the a priori probability p_i that an individual belongs to population i is proportional to the Lebesgue measure of D_i.

The density of population i, $f_i(x)$, and the unconditional density, $f(x)$, are respectively equal to :

$$f_i(x)=\frac{1_{D_i}(x)}{\lambda_2(D_i)}$$

and

$$f(x)=\sum_{i=1}^{k} p_i f_i(x)=\frac{\sum_{i=1}^{k} 1_{D_i}(x)}{\lambda_2(D)}$$

The decision rule is the one given by the Bayesian rule, with the unknown parameters -the convex sets D_i- replaced by their maximum likelihood estimations.

Let X_i be the labeled sample of population i, $F(X_i)$ its *optimal closing* and x the individual to be assigned to one of the k populations.

If x is allocated to the ith group, the maximum likelihood estimates of the domains D_j are, thanks to our closed and connex hypothesis :

$$\hat{D}_j=\begin{cases} F(X_j) & \text{if } j\neq i \\ F(X_j \cup \{x\}) & \text{if } j=i \end{cases}$$

The maximum likelihood estimate of $p_i f_i(x)$ is therefore :

$$\hat{p}_i \hat{f}_i(x)=\frac{1_{\hat{D}_i}(x)}{\sum_{j=1}^{k} \lambda_2(F(X_j))+S_i(x)}$$

where $S_i(x)\equiv\lambda_2(F(X_i \cup \{x\}))-\lambda_2(F(X_i))$.

As the closed and connex sets are assumed disjoint, x can not be allocated to the ith group, unless $F(X_i \cup \{x\})$) and $F(X_j)$ are disjoint for all $j \neq i$. Otherwise, we define $S_i(x) = +\infty$.

The allocation rule is :

Assign x to the ith population if and only if $\hat{p}_i \hat{f}_i(x) > \hat{p}_j \hat{f}_j(x) \quad \forall j \neq i$,

which comes to

Assign x to the ith population if and only if $S_i(x) < S_j(x) \quad \forall j \neq i$

namely

Assign x to the ith population if and only if the difference between the area of the OPTIMAL CLOSINGS *of the labeled sample of the ith population with and without the individual to be assigned is minimal.*

There are different possible cases about this allocation rule.
The optimal closings of the ith population with and without the individual to be assigned can have the same radius. But of course, the case of different radius is also possible.

Figure 3: *Illustration of the difference between the areas of optimal closings of the same radius used in the allocation rule based on the closings.*

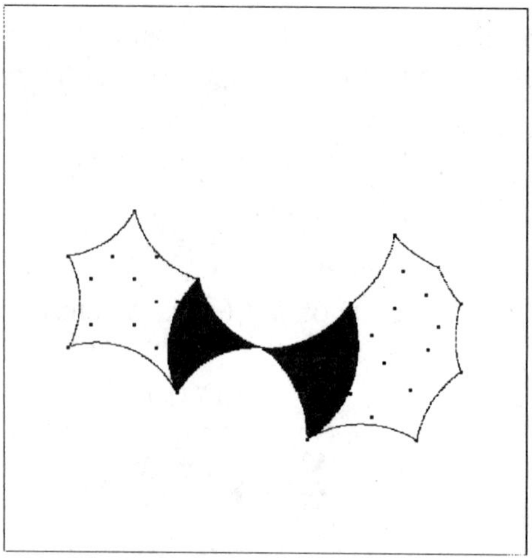

4. Admissibility of our procedure

The problem of choosing a clustering procedure among the myriad proposed is perplexing. The approach proposed here is the suggestion made by L. Fisher and J.W. Van Ness (1971) [4]. As our procedure is based on Lebesgue measure and not on a distance, some problems will arise when the notion of distance is used in their admissibility conditions. So, we will divide the 9 conditions into 3 groups.

Among the conditions directly applicable to $\Re^d$, we can prove that our procedure is *point proportion*, *cluster proportion* (these 2 conditions are trivially fulfilled by our procedure) and *not cluster omission admissible*. Moreover, it is not *convex admissible*... which is not surprising because our aim was precisely to be able to find non convex natural clusters.

For the conditions expressed in terms of distance, we can only test admissibility on $\Re$: our procedure is *image*, *k-groups* and *monotone admissible* on $\Re$. These conditions can however be generalized to $\Re^d$. The last condition (well-structured exact tree admissibility) is only applicable to hierarchical methods and thus cannot be fulfilled by our procedure.

Van Ness (1973) [15] added admissibility conditions to the precedent ones. Among them, we will point out one which is only applicable to clustering procedures with corresponding discriminant analysis procedures. Our procedure was proved to be *repeatable admissible*.

5. Conclusion

Using closings in both clustering analysis and discriminant analysis seems to be very interesting. Indeed, as the closed and connex hypothesis generalizes the convex hypothesis, we were able to create a clustering method and a discriminant analysis method generalizing the Hypervolume method and the corresponding discriminant rule. So, our aim to change the convexity in a less restrictive hypothesis was attained. Moreover, a lot of Fisher and Van Ness admissibility conditions were fulfilled by our procedure, which cannot then be considered as a bad clustering algorithm. Unfortunately, for the clustering method created, an algorithm seeking the best partition among all of these will not be suitable. So, researchs have to be made - and are made - to avoid the consideration of each different partition. As the domain of use of closed forms in clustering is quite unexplored, a lot of research domains are being examined. For example, the use of alpha-shapes in clustering instead of closings. Improvements of the method can also be made, such as the generalization to d dimensions, or the use of closings by squares or any convex instead of discs.

References:

[1] Beaufays, P. and Rasson, J.-P. (1985), A new geometric discriminant rule, in: *Computational Statistics Quarterly*, 2, 15-30, Elsevier Science B.V.

[2] Bock, H. H. (1997), Probability Models for Convex Clusters, in: *Classification and Knowledge Organization, Proceedings of the 20th Annual Conference of the Gesellschaft für Klassifikation e.V., University of Freiburg, March 6-8, 1996*, Klar-Opitz editors, Springer.

[3] Cressie, N. A.C. (1991), *Statistics for spatial data*, John Wiley & Sons, New York.

[4] Fisher, L., and Van Ness, J.W. (1971), Admissible clustering procedures, in: *Biometrika*, 58, 91-104.

[6] Hardy, A., and Rasson, J.-P. (1982), Une nouvelle approche des problèmes de classification automatique, in: *Statistiques et Analyse des Données*, 7, 41-56.

[7] Karr, A.F. (1991), *Point processes and their statistical inference*, Marcel Dekker, New-York.

[8] Krickeberg, K. (1982), Processus ponctuels en statistique, in: *Ecole d'été de Probabilités de Saint-Flour X-1980*, 205-313, Springer-verlag, Berlin.

[9] Matheron, G. (1975), *Random Sets and Integral Geometry*, John Wiley & sons, New-York.

[10] Rasson, J.-P., Hardy, A., Szary, L. and Schindler, P. (1982), An optimal clustering method based on the Rasson's criterion and resulting from a new approach, in: *Pattern Recognition Theory and Applications*, 63-71, Kitter, Fu and Pau editions.

[11] Schmitt, M. and Mattioli, J. (1993), *Morphologie mathématique*, Logiques Mathématiques Informatiques, Masson, Paris.

[12] Serra, J.P. (1982), *Image Analysis and Mathematical Morphology*, Academic Press, London.

[13] Spath, H. (1980), *Cluster analysis algorithms for data reduction and classification of objects*, Ellis Horwood Limited, Chichester.

[14] Stoyan, D., and Stoyan, H. (1994), *Fractals, Random Shapes and Point Fields, Methods of Geometrical Statistics*, John Wiley & Sons, Chichester.

[15] Van Ness, J.W. (1973), Admissible clustering procedures, in: *Biometrika*, 60, 422-424.

Clustering and Neural Networks

Hans-Hermann Bock
Institut für Statistik, Technische Hochschule Aachen,
D-52056 Aachen, Germany, *bock@stochastik.rwth-aachen.de*

Abstract: This paper considers the usage of neural networks for the construction of clusters and classifications from given data and discusses, conversely, the use of clustering methods in neural network algorithms. We survey related work in the fields of k-means clustering, stochastic approximation, Kohonen maps, Hopfield networks and multi-layer perceptrons. We propose various new approaches, reveal the asymptotic behaviour of Kohonen maps, and point to possible extensions.

Key words: Clustering Methods, Neural Networks, Asymptotics for Kohonen Maps, Hopfield networks, Multi-Layer-Perceptrons, Gravitational Clustering.

1. Introduction

The application of artificial neural networks (NNs) in pattern recognition refers often to problems of *discrimination*, i.e. the assignment of objects to more or less specified classes (*supervised learning*). In contrast, the present paper deals with *clustering problems* (*unsupervised learning*). We consider a set $\mathcal{O} = \{1, ..., n\}$ of n objects described by multidimensional data points $x_1, ..., x_n \in R^p$ or by an $n \times n$ matrix $D = (d_{kl})$ of pairwise dissimilarities. From these data we want to construct an appropriate partition $\mathcal{C} = (C_1, ..., C_m)$ of $\mathcal{O}$ with 'homogeneous' classes $C_1, ..., C_m$, suitable class representatives $z_1, ..., z_m \in R^p$ and, possibly, a corresponding subdivision $\mathcal{B} = (B_1, ..., B_m)$ of the whole feature space R^p (in order to classify further data points). We show how problems of this type can be solved with neural networks and, conversely, how classical clustering methods are incorporated into the neural network approach (see also Murtagh 1996).

Section 2 surveys the classical k-means clustering approach and describes a general stochastic approximation method for clustering problems. Section 3 deals with a generalized Kohonen algorithm for constructing 'self-organizing maps' (SOMs; Anouar et al. 1996,1997) and shows its relationship to the previous clustering methods. We propose a new 'continuous' clustering criterion (K-*criterion*) which reveals and characterizes the asymptotic performance of Kohonen maps. We suggest some generalized SOMs along the lines of classical cluster analysis. Section 4 is devoted to Hopfield networks, we discuss their use for constructing a fuzzy classification of objects and recall some related classical 'gravitational' clustering methods. In Section 5 we describe how multi-layer perceptrons (MLPs) can be used for clustering purposes.

2. k-means Clustering, Stochastic Approximation and Asymptotics

In this section we recall briefly the ideas of k-means clustering and stochastic approximation in order to combine them in Kohonen's algorithm in section 3. We start with the problem of partitioning the set $\mathcal{O}$ of objects with data points $x_1, ..., x_n \in R^p$ into a given number m of 'homogeneous' classes $C_1, ..., C_m$ by minimizing the well-known *variance* or *SSQ clustering criterion*

$$g_n(\mathcal{C}) \quad := \quad \frac{1}{n}\sum_{i=1}^{m}\sum_{k\in C_i} ||x_k - \overline{x}_{C_i}||^2 \to \min_{\mathcal{C}} \tag{2.1}$$

with respect to all m-partitions $\mathcal{C} = \{C_1, ..., C_m\}$. It is well-known (*cf.* Bock 1974) that this is equivalent to each of the two following problems:

$$g_n(\mathcal{C},\mathcal{Z}) \quad := \quad \frac{1}{n}\sum_{i=1}^{m}\sum_{k\in C_i} ||x_k - z_i||^2 \quad \to \min_{\mathcal{C},\mathcal{Z}} =: g_n^* \tag{2.2}$$

$$\gamma_n(\mathcal{Z}) \quad := \quad \frac{1}{n}\sum_{k=1}^{n} \min_{1\leq j\leq m} \{||x_k - z_j||^2 \quad \to \min_{\mathcal{Z}} \tag{2.3}$$

where minimization is over all m-partitions $\mathcal{C}$ and all m-tuples $\mathcal{Z} = (z_1, ..., z_m)$ of 'class centers' $z_1, ..., z_m \in R^p$. In fact, any solution $\mathcal{C}^*, \mathcal{Z}^*$ of one problem yields a solution of the other problem(s) due to the *stationarity equations* $z_i^* = \overline{x}_{C_i^*}$ (centroids) and $C_i^* = \{\ k \in \mathcal{O} \mid ||x_k - z_i^*|| = \min_{j=1,...,m}\{||x_k - z_j^*||\}\ \}$ $(i = 1, ..., m;$ *minimum-distance partition*, Voronoi classes).

There are essentially two types of iterative algorithms for solving (approximately) those optimization problems: (a) Methods which proceed *batch-wise* insofar as all n data points are processed or reclassified at the same time (see the k-means algorithm (i) below), and (b) methods which proceed *sequentially*, i.e., the data points $x_1, x_2, ...$ are presented one after the other, and the m-partition $\mathcal{C}^{(n)}$ and the center system $\mathcal{Z}^{(n)}$ obtained from the first data $x_1, ..., x_n$ are suitably updated after observing x_{n+1}. It is this second type which resembles the *learning algorithm paradigm* in the NN context and includes, in particular, the method (ii) of MacQueen (1967) and the stochastic approximation method (iii) by Bravermann (1966) and Tsypkin & Kelmans (1967). The mentioned clustering methods $(i), (ii), (iii)$ can be described as follows:

(i) k-means algorithm:
With a fixed number n of data points and an arbitrary initial m-partition $\mathcal{C}^{(0)}$, we minimize the criterion $g_n(\mathcal{C},\mathcal{Z})$, (2.2), with respect to $\mathcal{C}$ and $\mathcal{Z}$ in turn and obtain, for $t = 0, 1, ...$, a sequence of center systems $\mathcal{Z}^{(t)}$ (= centroids of $\mathcal{C}^{(t)}$) and m-partitions $\mathcal{C}^{(t)}$ (= minimum-distance partition to $\mathcal{Z}^{(t-1)}$) which improve steadily on the criteria (2.1) to (2.3):

$$g_n(\mathcal{C}^{(t)}) \equiv g_n(\mathcal{C}^{(t)},\mathcal{Z}^{(t)}) \geq g_n(\mathcal{C}^{(t+1)},\mathcal{Z}^{(t)}) \geq g_n(\mathcal{C}^{(t+1)},\mathcal{Z}^{(t+1)}) \equiv g_n(\mathcal{C}^{(t+1)}) \tag{2.4}$$

until we end with a stationary value of g_n which is (hopefully) close or identical to the global minimum g_n^*.

(ii) MacQueen's sequential procedure:
Begin with $n = m$ data points and m singleton classes $C_i^{(n)} := \{x_i\}$ for $i = 1, ..., m$. Then, for $n \geq m$, if $\mathcal{C}^{(n)}$ is the current m-partition of $x_1, ..., x_n$, the next observation x_{n+1} is assigned to the class $C_{i^*}^{(n)}$ with minimum distance $||x_{n+1} -$

$\overline{x}_{C_i^{(n)}}||$ ($i = 1, ..., m$; *competitive learning*), thus obtaining an updated partition $\mathcal{C}^{(n+1)}$ with classes $C_{i^*}^{(n+1)} = C_{i^*}^{(n)} \cup \{n+1\}$ for $i = i^*$ and $C_i^{(n+1)} = C_i^{(n)}$ for $i \neq i^*$. Similarly, the class centroids $z_i^{(n)} := \overline{x}_{C_i^{(n)}}$ are updated according to

$$z_i^{(n+1)} = \overline{x}_{C_i^{(n+1)}} = \begin{cases} z_i^{(n)} + \alpha_{in} \cdot (x_{n+1} - z_i^{(n)}) & \text{for} \quad i = i^* \\ z_i^{(n)} & \text{for} \quad i \neq i^* \end{cases}$$

where the 'learning rate' $\alpha_{in} := (|C_i^{(n)}| + 1)^{-1}$ is obviously data dependent.

(iii) Stochastic approximation:

This algorithm concentrates on the sequence $(\mathcal{Z}^{(n)})_{n \geq m}$ of class representatives. It begins, for $n = m$, with m distinct points $z_i^{(m)} := x_i$ ($i = 1, ..., m$) and updates, for $n \geq m$ and after observing x_{n+1}, the current system $\mathcal{Z}^{(n)} = (z_1^{(n)}, ..., z_m^{(n)})$ according to the recursive formula

$$z_i^{(n+1)} = \begin{cases} z_i^{(n)} + \alpha_{in} \cdot (x_{n+1} - z_i^{(n)}) & \text{for} \quad i = i^* \\ z_i^{(n)} & \text{for} \quad i \neq i^*, \end{cases} \tag{2.5}$$

now with m *predetermined, possibly class-specific* sequences $\{\alpha_{in}\}_{n \in N}$ of 'learning factors' with $\lim_{n\to\infty} \alpha_{in} = 0$, $\sum_{n=1}^{\infty} \alpha_{in} = \infty$ and $\sum_{n=1}^{\infty} \alpha_{in}^2 < \infty$ (such as, e.g., $\alpha_{in} = 1/n$). A suitable sequence of m-partitions results if we define $\mathcal{C}^{(n)}$ as the minimum-distance partition of $\{x_1, ..., x_n\}$ generated by the center system $\mathcal{Z}^{(n)}$.

The performance and asymptotic behaviour of all three methods has been investigated under the assumption that the data is a sequence $X_1, X_2, ...$ of i.i.d. random vectors with Lebesgue density $f(x)$ on R^p (e.g., a mixture of m p-dimensional normals). The results refer to the solution $\mathcal{B}^*, \mathcal{Z}^*$ of the following continuous analogue of the clustering criterion (2.2):

$$g(\mathcal{B}, \mathcal{Z}) := \sum_{i=1}^{m} \int_{B_i} ||x - z_i||^2 \cdot f(x)\, dx \quad \to \quad \min_{\mathcal{B},\mathcal{Z}} =: g^* \tag{2.6}$$

where minimization is over all m-partitions $\mathcal{B} = (B_1, ..., B_m)$ of R^p and all center systems $\mathcal{Z} = (z_1, ..., z_m)$. Note that optimum configurations $\mathcal{B}^*, \mathcal{Z}^*$ will typically be related by similar *stationarity conditions* as before (eventually after a relabeling of class indexes): $z_i^* = E_f[X|X \in B_i^*]$ is the conditional expectation of X in the domain B_i^* (under f), and $B_i^* = \{x \in R^p|\ ||x - z_i^*|| = \min_j\{||x - z_j^*||\}\ \}$ is the Voronoi region in R^p of the centroid z_i^*. With this notation we obtain, for a sufficiently regular density f:

Theorem 1:

(i) Denote by $\mathcal{C}^{*(n)}, \mathcal{Z}^{*(n)}$ an optimum configuration for the classical SSQ criteria (2.1), (2.2), (2.3) with a sample size n, and by $g_n^* = g_n(\mathcal{C}^{*(n)}, \mathcal{Z}^{*(n)}) = g_n(\mathcal{C}^{*(n)})$ the minimum criterion value. If the optimum m-partition $\mathcal{B}^*$ of the continuous

criterion (2.6) is unique, then for $n \to \infty$ and $i = 1, ..., m$:

$$z_i^{*(n)} := \overline{X}_{C_i^{*(n)}} \quad \to \quad z_i^* = E[X|X \in B_i^*] \tag{2.7}$$

$$g_n^* := \min_{\mathcal{C},\mathcal{Z}} g_n(\mathcal{C},\mathcal{Z}) \quad \to \quad g^* := \min_{\mathcal{B},\mathcal{Z}} g(\mathcal{B},\mathcal{Z}) = g(\mathcal{B}^*,\mathcal{Z}^*). \tag{2.8}$$

in probability and almost surely (Pollard 1981, 1982, see also Bock 1985, 1996a).
(ii), (iii): Under some regularity conditions for f analoguous limits are obtained for the centers $z_i^{(n)}$ resulting from the sequential algorithms (ii), (iii) when $n \to \infty$:

$$z_i^{(n)} \quad \to \quad z_i^* = E[X|X \in B_i^*] \tag{2.9}$$

$$g_n := g_n(\mathcal{C}^{(n)},\mathcal{Z}^{(n)}) \quad \to \quad g^* := g(\mathcal{B}^*,\mathcal{Z}^*). \tag{2.10}$$

in probability and almost surely, where $\mathcal{B}^*, \mathcal{Z}^*$ is a stationary pair (local minimum) for $g(\mathcal{B},\mathcal{Z})$ (and might sometimes be the unique global optimum).

This theorem underlines that all three algorithms are implicitly resolving the same continuous partitioning problem (2.6). Details for (*ii*) are given by MacQueen (1967), for (*iii*) see Braverman (1966), Tsypkin & Kelmans (1967) and Bock (1974, Chap. 29).

As a matter of fact, the results of Braverman pertain even to generalized partitioning criteria such as

$$G(\mathcal{B},\mathcal{Z}) \quad := \quad \sum_{i=1}^{m} \int_{B_i} \phi_i(x; z_1, ..., z_m) \cdot f(x)\, dx \quad \to \quad \min_{\mathcal{B},\mathcal{Z}} \tag{2.11}$$

where the squared Euclidean distance $||x - z_i||^2$ is replaced by a generalized distance function $\phi_i(x;\mathcal{Z}) \equiv \phi_i(x; z_1, ..., z_m)$, e.g., by the L_p-distance $||x - z_i||_p$ (for some $p > 1$) or even a class-specific distance which depends on all centers $z_1, ..., z_m$ as we will see in (3.2), (3.3), (3.4) below. Tsypkin & Kelmans (1967) derive stationarity conditions for an optimal pair $(\mathcal{B},\mathcal{Z})$: Whereas the optimum centers $z_i \in R^p$ must fulfill the 'normal equations':

$$\sum_{\nu=1}^{m} \int_{B_\nu} \nabla_{z_i} \phi_\nu(x; z_1, ..., z_m)\, f(x) dx \quad = \quad 0 \qquad i = 1, ..., m. \tag{2.12}$$

the optimum classes $B_i \subset R^p$ are the Voronoi regions belonging to the z_i (using the metric $\phi_i(x,\mathcal{Z})$.

In analogy to the SSQ case (2.2), we can formulate a related finite-sample clustering problem by:

$$G_n(\mathcal{C},\mathcal{Z}) \quad := \quad \frac{1}{n} \sum_{i=1}^{m} \sum_{k \in C_i} \phi_i(x_k; z_1, ..., z_m) \quad \to \quad \min_{\mathcal{C},\mathcal{Z}}. \tag{2.13}$$

Even if we can easily design a k-means-type algorithm for $G_n(\mathcal{B},\mathcal{Z})$, the minimization with respect to $\mathcal{Z}$ and the solution of (the empirical version of) (2.13) can be computationally difficult. On the other hand, we can easily formulate the stochastic approximation algorithm for minimizing G: After observing x_{n+1}, we update the current center system $\mathcal{Z}^{(n)} = (z_1^{(n)}, ..., z_m^{(n)})$ as follows:

$$z_i^{(n+1)} \quad = \quad z_i^{(n)} - \alpha_{in} \cdot \nabla_{z_i}\phi_i(x_{n+1}; z_1^{(n)}, ..., z_m^{(n)}) \tag{2.14}$$

with m class-specific learning sequences $(\alpha_{in})_{n \in N}$. For a dissimilarity of the type $\phi_i(x; \mathcal{Z}) = \phi(x - z_i)$ this reduces to:

$$z_i^{(n+1)} \quad = \quad \begin{cases} z_i^{(n)} - \alpha_{in} \cdot \nabla_x \phi(x_{n+1} - z_i^{(n)}) & \text{for } i = i^* \\ z_i^{(n)} & \text{for } i \neq i^* \end{cases} \tag{2.15}$$

where $i = i^*$ denotes the class with minimum distance $\phi(x_{n+1} - z_i^{(n)})$ (*cf.* (2.5) for $\phi(x) = ||x||^2/2$). Similarly as before we obtain (Tsypkin & Kelmans 1967):

Theorem 2:
If $X_1, X_2, ...$ are i.i.d.with a suitably regular density f the sequence of center systems $\mathcal{Z}^{(n)}$ obtained from stochastic approximation (2.14) together with the corresponding minimum-distance partitions $\mathcal{B}^{(n)}$ (using the class-specific distances $\phi_i(x, \mathcal{Z}^{(n)})$) converge in probability and almost surely to a stationary pair $(\mathcal{B}^*, \mathcal{Z}^*)$ of the criterion $G(\mathcal{B}, \mathcal{Z})$. Similarly, the corresponding criterion values $G^{(n)} := G(\mathcal{B}^{(n)}, \mathcal{Z}^{(n)})$ converge to $G^* := G(\mathcal{B}^*, \mathcal{Z}^*)$.

In the following section this theorem will be used for characterizing the asymptotic performance of Kohonen maps.

3. Clustering Methods for Classical and Generalized Kohonen Maps

Clustering methods are implicitly used in the construction of 'self-organizing maps' (SOM's; Kohonen 1982, 1995), a well-known method in the NN framework. We show how it is related to the previous clustering approaches and present various new or recent generalizations of Kohonen's basic proposal.

Kohonen's algorithm starts from a sequence of high-dimensional data vectors $x_1, x_2, ...$ in R^p (*learning sample*) and tries to find an illustrative display of these points in a (given) rectangular, hexagonal,... lattice $\mathcal{L}$ of points in R^2 or R^3 such that the high-dimensional structure of the data (neighbourhoods, topological ordering, clusters) can be easily detected in this line diagram. In statistics this problem has been tackled, e.g., by principal component analysis, projection pursuit clustering or multidimensional scaling (see Bock 1997). In contrast, Kohonen has proposed the following *sequential strategy* (where we consider w.l.o.g. a rectangular $a \times b$ lattice $\mathcal{L} = \{P_i = (r, s) | r = 1, ..., a; s = 1, ..., b\}$ of R^2):

(1) For each sample size n we look for a (large) number $m = a \cdot b$ of 'mini-classes' $C_1^{(n)}, ...C_m^{(n)}$ for $\{x_1, ..., x_n\}$ which are homogeneous in some SSQ sense, a system of suitable class representatives $z_1^{(n)}, ..., z_m^{(n)} \in R^p$ and the minimum-distance partition $\mathcal{B}^{(n)}$ of R^p generated by $\mathcal{Z}^{(n)}$.

(2) Each class representative $z_i = z_i^{(n)}$ is assigned to one of the vertices P_i of the rectangular lattice $\mathcal{L}$. (Note that neural network terminology calls P_i a

'neuron' and z_i a 'weight vector'). Ideally this assignment ψ should be such that close or neighbouring class centers z_i, z_j in R^p are mapped into vertices $P' := P_{\psi(i)}, P'' := P_{\psi(j)}$ which are close or neighbours in the lattice $\mathcal{L}$ as well, i.e. with a small *path distance* $\delta(P', P'')$. This is attained by the following construction:

(3) Similarly as in (2.5) or (2.14) we define a sequence of center systems $\mathcal{Z}^{(n)} = (z_1^{(n)}, ..., z_m^{(n)})$ in a sequential way, but with a modified updating formula (3.1) below: We start, for $n = 0$, with m random points $z_1^{(0)}, ..., z_m^{(0)}$ from R^p; they are arbitrarily assigned to the lattice points $P_i \in \mathcal{L}$ (e.g. by $\psi(i) \equiv P_i$; this assignment is never changed in the sequel). If now, for $n \geq 0$, a new data vector x_{n+1} is closest to the class center $z_{i^*}^{(n)}$ among all current centers $z_1^{(n)}, ..., z_m^{(n)}$, we update $z_{i^*}^{(n)}$ by moving it towards x_{n+1}, and *simultaneously* all other class centers $z_j^{(n)}$ whose counterparts P_j are close to, or neighbours of, P_{i^*} in $\mathcal{L}$. A quite general updating formula is given by:

$$z_j^{(n+1)} \quad = \quad z_j^{(n)} + \alpha_{jn} \cdot K(\delta(P_j, P_{i^*})) \cdot (x_{n+1} - z_j^{(n)}) \qquad j = 1, ..., m \quad (3.1)$$

where the transformation $K(\delta)$ is typically decreasing from $K(0) = 1$ to 0 and controls the influence of more or less neighbouring vertices P_j of $\mathcal{L}$. Kohonen's original methods are obtained for the threshold function $K(\delta) := 1$ or 0 if $\delta \leq$ or $> \epsilon$, respectively (with a given distance threshold $\epsilon \geq 0$) and a bathtub function K ('Mexican hat') where positive (negative) values of K yield attraction (repulsion) effects in (3.1).

In this context Anouar, Badran & Thiria (1997) have defined the following finite sample clustering criterion:

$$\tilde{g}_n(\mathcal{C}, \mathcal{Z}) \quad := \quad \frac{1}{2n} \sum_{i=1}^{m} \sum_{k \in C_i} [\sum_{j=1}^{m} K(\delta(P_i, P_j)) \cdot ||x_k - z_j||^2] \to \min_{\mathcal{C},\mathcal{Z}} =: \tilde{g}_n^* \quad (3.2)$$

where the bracket

$$\phi_i(x; \mathcal{Z}) \quad := \quad \sum_{j=1}^{m} K(\delta(P_i, P_j)) \cdot ||x - z_j||^2 \qquad (3.3)$$

defines a dissimilarity between a data point $x \in R^p$ and the vertex P_i from $\mathcal{L}$ (or the mini-class C_i) in which the term $||x_k - z_i||^2$ (i.e., with $j = i$) has maximum weight $K(0) = 1$.

As a continuous counterpart we define the *K-criterion*:

$$\tilde{g}(\mathcal{B}, \mathcal{Z}) \quad := \quad \frac{1}{2} \sum_{i=1}^{m} \int_{B_i} [\sum_{j=1}^{m} K(\delta(P_i, P_j)) ||x - z_j||^2] f(x) \, dx \;\to\; \min_{\mathcal{B},\mathcal{Z}} =: \tilde{g}^* \quad (3.4)$$

which is obviously a special case of the criterion (2.11). Similarly as in section 2 we can derive for the criteria (3.2) and (3.4) three 'self-organizing' clustering or mapping methods $(i^*), (ii^*), (iii^*)$:

(i*) A finite-sample k-means analogue of Kohonen's method:
We minimize $\tilde{g}_n(\mathcal{C},\mathcal{Z})$, (3.2), partially with respect to the center system $\mathcal{Z}$ and the m-partition $\mathcal{C}$ in turn: (a) From a current m-partition $\mathcal{C}^{(t)}$ ($t \geq m$) we obtain the optimum centers:

$$z_i^{(t)} = \sum_{j=1}^{m} w_{ij}^{(t)} \cdot \overline{x}_{C_j^{(t)}} \qquad i = 1,...,m \tag{3.5}$$

as a weighted mean of class centroids where the normalized weights

$$w_{ij}^{(t)} := |C_j^{(t)}| \cdot K(\delta(P_i, P_j))/w_i^{(t)} \qquad \text{with}$$
$$w_i^{(t)} := \sum_{j=1}^{m} |C_j^{(t)}| \cdot K(\delta(P_i, P_j))$$

depend on the current class sizes $|C_i^{(t)}|$. (b) Inversely, for a given center system $\mathcal{Z} = \mathcal{Z}^{(t)}$, (3.2) is minimized by the minimum-distance partition $\mathcal{C}^{(t+1)}$ with classes

$$C_i^{(t+1)} := \{ \, k \in \{1,...,n\} \mid i = \arg\min_j \phi_j(x_k, \mathcal{Z}^{(t)}) \, \} \tag{3.6}$$

with the class-specific dissimilarity measure $\phi_i(x;\mathcal{Z}^{(t)})$, (3.3).

(ii*) An analogue to MacQueen's sequential algorithm:
With the notation from (*ii*), a new data point x_{n+1} is assigned to the class $C_i^{(n)}$ with minimum dissimilarity $\phi_i(x_k, \mathcal{Z}^{(n)})$ from x_k (for $i = i^*$, say), and *all* centers $z_i^{(n)}$ are suitably updated:

$$z_i^{(n+1)} = \sum_{j=1}^{m} w_{ij}^{(n+1)} \cdot \overline{x}_{C_j^{(n+1)}} \qquad i = 1,...,m. \tag{3.7}$$

(iii*) The stochastic approximation algorithm:
Since the K-criterion (3.4) is a special case of the criterion $G(\mathcal{B},\mathcal{Z})$, (2.11), we can directly apply Tsypkin & Kelman's updating formula (2.14). Calculating the gradients in (2.14) yields the generalized Kohonen algorithm (3.1) which therefore proves to be the stochastic approximation device for the new K-criterion (see also Anouar & al. 1997).

The asymptotic behaviour of the three algorithms $(i^*), (ii^*), (iii^*)$ is characterized by Theorem 3 which assumes a sufficiently regular density f for the data $X_1, X_2, ...$:

Theorem 3:
(i^*) : *Assume that the optimum m-partition $\mathcal{B}^*$ for the K-criterion $\tilde{g}$, (3.4), is unique and denote by $\mathcal{C}^{*(n)}, \mathcal{Z}^{*(n)}$ the optimum configurations for Anouar's SSQ criterion $\tilde{g}_n$, (3.2), for a given sample size n, and by $\tilde{g}_n^* = \tilde{g}_n(\mathcal{C}^{*(n)}, \mathcal{Z}^{*(n)})$ the minimum criterion value. Then, for $n \to \infty$ and $i = 1,...,m$:*

$$z_i^{*(n)} := \sum_{j=1}^{m} w_{ij}^{(n)} \cdot \overline{X}_{C_i^{*(n)}} \;\to\; \zeta_i^* := \sum_{j=1}^{m} w_{ij}^* \cdot E[X|X \in B_i^*] \tag{3.8}$$

$$\tilde{g}_n^* := \min_{\mathcal{C},\mathcal{Z}} \tilde{g}_n(\mathcal{C},\mathcal{Z}) \;\to\; \tilde{g}^* := \min_{\mathcal{B},\mathcal{Z}} \tilde{g}(\mathcal{B},\mathcal{Z}). \tag{3.9}$$

in probability and a.s. *with normalized weights* $w_{ij}^* := P(X \in B_j^*) \cdot K(\delta(P_i, P_j))/w_i^*$ and $w_i^* := \sum_{\nu=1}^m P(X \in B_\nu^*) \cdot K(\delta(P_i, P_\nu))$.

(iii^*) : *Similar convergence properties hold for the sequence of center systems* $\mathcal{Z}^{(n)}$ *of the generalized Kohonen algorithm (3.1), i.e. the stochastic approximation method for the* K*-criterion (3.4):*

$$z_i^{(n)} \quad \rightarrow \quad \zeta_i^* \qquad\qquad i = 1, ..., m \tag{3.10}$$

$$\tilde{g}^{(n)} := \tilde{g}_n(\mathcal{C}^{(n)}, \mathcal{Z}^{(n)}) \quad \rightarrow \quad \tilde{g}^* := \tilde{g}(\mathcal{B}^*, \mathcal{Z}^*). \tag{3.11}$$

where $(\mathcal{B}^*, \mathcal{Z}^*)$ *is a stationary pair for (3.4) and* $\mathcal{C}^{(n)}$ *the minimum-distance partition of* $\{x_1, ..., x_n\}$ *corresponding to the center system* $\mathcal{Z}^{(n)}$ *(using the distances* ϕ_i*).*

The importance of Theorem 3 resides in the fact that it characterizes the performance of (original and generalized) 'self-organizing' Kohonen algorithms and can be a key for investigating the well-discussed question if Kohonen's algorithm produces 'topologically correct' results, e.g., in the case where the underlying density $f(x)$ is concentrated near a low-dimensioanl manifold of R^p. – Related problems are investigated in Cottrell & Fort (1989), Tolat (1990), Bouton & Pagès (1993), Fort & Pagès (1996) and Ambroise & Govaert (1996).

The criterion-based approach to Kohonen networks suggests various generalizations which are known from classical cluster analysis and may be appropriate in practical situations. Apart from the use of *Mahalanobis-type distances* and L_p-*distances* we mention models where each class is characterized (a) by a *class-specific hyperplane* H_i (instead of a single point z_i only) as in principal component clustering (Bock 1974, 1996b,c), see also the approaches by Ritter (1997), Moshou & Ramon (1997), Kohonen et al. (1997) and Bock (1998b); (b) by a *local non-linear* (e.g., quadratic) *surface* (*cf.* Ritter 1997); (c) by *regression planes* (as in Bock 1996b,c), see Badran et al. 1997) or (d) by a general class-specific distribution density $f(x, \vartheta_i)$ as in maximum-likelihood clustering (Bock 1996b), see also Bock (1998b).

4. Hopfield Networks for Clustering Purposes

Hopfield networks are designed to minimize a smooth loss function $L(v_1, ..., v_N) \geq 0$ with respect to N real-valued variables $v_1, ..., v_N$. In the clustering framework L denotes a clustering criterion and $v_1, ..., v_N$ describe a classification (e.g., class indicators, class centers, fuzzy memberships etc.). The method proceeds by analogy to the motion of N particles in time $t \geq 0$ which migrate under a system of forces (Hamiltonian differential equations) and approach finally a steady state $(v_1^*, ..., v_N^*)$ which provides a (global or local) minimum of the 'potential energy function' L.

More specifically, each variable v_i is modeled as a time function $v_i(t) := \sigma(a_i(t))$ (position of the particle i, output of the 'neuron' i) with an *activation function* $a_i(t)$ and a prespecified smooth sigmoid function σ, e.g., the logistic function $\sigma(a) = 1/(1 + e^a)$. The activations $a_i(t)$ evolve in time according to the system of differential equations (d.e.):

$$\dot{a}_i(t) \;=\; -\frac{\partial}{\partial v_i} L(v_1(t), ..., v_N(t)) \qquad t \geq 0,\ i = 1, ..., N \tag{4.1}$$

L is called a *Lyapunov function* for the system (4.1) which means that the loss function $K(t) := L(v_1(t), ..., v_N(t))$ is decreasing in $t \geq 0$ along any solution of (4.1); this follows easily from $\dot{K}(t) = \sum_{i=1}^{N} L_{v_i}(v_1(t), ..., v_N(t)) \cdot \dot{v}_i(t) \leq 0$. Therefore the limits $a_i^* := \lim_{t\to\infty} a_i(t)$ exist and the corresponding 'outputs' $v_i^* := \sigma(a_i^*)$ yield a (local or global) minimum $L(v_1^*, ..., v_N^*)$ of L. (Note that we did not mention a 'neural net' in this description.)

This classical optimization method can be applied to any clustering problem if we re-formulate it as a minimization problem $L \to min$, solve the corresponding system of differential equations (4.1) for $a_i(t)$ and determine the steady states a_i^* and v_i^* which are then re-interpreted in terms of an optimal clustering of data.

We exemplify this approach by two special cases taken from Adorf & Murtagh (1988) and Kamgar-Parsi & al. (1990). Both papers look for a *fuzzy classification* $\mathcal{U} = (u_{ik})_{m\times n}$ of n objects where $u_{ik} \in [0,1]$ is the degree of membership of the k-th object in the i-th fuzzy class U_i which leaves $N = m \cdot n$ variables $u_{ik} \equiv v_{ik} = \sigma(a_{ik})$ to be determined. Following the classical fuzzy clustering approach (Bock 1979, Bezdek 1981), Kamgar-Parsi & al. minimize a suitably modified SSQ criterion (4.2):

$$g_1(\mathcal{U}) \;:=\; \sum_{i=1}^{m}\sum_{k=1}^{n} u_{ik}^2 \cdot \Delta_{ki}(\mathcal{U}) = \sum_{i=1}^{m}\sum_{k=1}^{n} u_{ik}^2 \cdot ||x_k - \tilde{x}_{U_i}||^2 \tag{4.2}$$

with class centers $\tilde{x}_{U_i} := [\sum_{k=1}^{n} u_{ik}^2 x_k]/[\sum_{k=1}^{n} u_{ik}^2]$ and deviations $\Delta_{ki}(\mathcal{U}) := ||x_k - \tilde{x}_{U_i}||^2$. They add some weighted penalty terms:

- $g_2(\mathcal{U}) = \sum_{k=1}^{n}[\sum_{i=1}^{m} u_{ik} - 1]^2$ for forcing the norming conditions $\sum_{i=1}^{m} u_{ik} = 1$,
- $g_3(\mathcal{U}) := \sum_{k=1}^{n}[\sum_{i=1}^{m}\sum_{j\neq i}^{m} u_{ik}u_{jk}]$ for reducing the overlap between classes,
- $g_4(\mathcal{U}) := \sum_{k=1}^{n}\sum_{i=1}^{m}\int_0^{u_{ik}} \sigma^{-1}(y)dy$ for forcing the u_{ik} to the boundary of $[0,1]$.

This yields the clustering criterion:

$$L(\mathcal{U}) \;:=\; \frac{\alpha}{2}\cdot g_1(\mathcal{U}) + \frac{\beta}{2}\cdot g_2(\mathcal{U}) + \frac{\gamma}{2}\cdot g_3(\mathcal{U}) + g_4(\mathcal{U}) \to \min_{\mathcal{U}} \tag{4.3}$$

with positive weights α, β, γ and the following system of d.e. (4.1):

$$\begin{aligned}\dot{a}_{ik}(t) \;=\; & -a_{ik}(t) - \alpha\cdot v_{ik}(t)\cdot\Delta_{ik}(t) - \beta\cdot[\sum_{j=1}^{m} u_{ik} - 1] - \gamma\cdot\sum_{j\neq i}^{m} u_{ik}(t) \\ & -(\alpha/2)\cdot\sum_{l=1}^{n} u_{il}^2 \frac{\partial\Delta_{il}}{\partial u_{ik}}.\end{aligned} \tag{4.4}$$

It is solved by methods from numerical analysis, simulation or an analog device, eventually in discrete time. The limits for $t \to \infty$ yield a (sub-)optimum fuzzy classification $\mathcal{U}^* = (u^*_{ik})$. Note that Kamgar-Parsi et al. cancelled the last term in the system (4.4) (with $\partial\Delta_{il}/\partial u_{ik} = -(x_l - \tilde{x}_{U_i})'(x_k - \tilde{x}_{U_i})/(\sum_{s=1}^n u_{is})$) in order to get a linear system of d.e.'s. Moreover, they used a rescaled logistic function $u_{ik} = \tilde{\sigma}(a_{ik}) := \sigma((a_{ik} - \vartheta_{ik})/\tau)$ with various choices for the thresholds ϑ_{ik} and the scale $\tau > 0$ (where $\tau \to 0$ yields 'harder' classifications).

In contrast,, Adorf & Murtagh (1988) started from an $n \times n$ matrix $D = (d_{kl})$ of pairwise dissimilarities d_{kl} for n objects and looked for an optimum fuzzy m-partition $\mathcal{U}$ in the sense:

$$\begin{aligned} L(\mathcal{U}) &:= \frac{\alpha}{2}\sum_{i=1}^m\sum_{k=1}^n\sum_{l=1}^n d_{kl}u_{ik}u_{il} - \frac{\beta}{2}\sum_{i=1}^m\sum_{j\neq i}^m\sum_{k=1}^n\sum_{l=1}^n d_{kl}u_{ik}u_{il} + \frac{\gamma}{2}\sum_{i=1}^m\sum_{j\neq i}\sum_{k=1}^n b_k u_{ik}u_{jk} \\ &= -\frac{1}{2}\cdot\sum_{i=1}^m\sum_{j=1}^m\sum_{k=1}^n\sum_{l=1}^n W_{ik,jl}\cdot u_{ik}u_{jl} \quad\to\quad \min_{\mathcal{U}}. \end{aligned} \tag{4.5}$$

Here the first term supports a small heterogeneity in the classes, the second one a high separation among the classes, and the third term little overlap for all objects k (with constant weights α, β, γ and b_k (typically $b_k \equiv b$)). This is a *quadratic* function in $\mathcal{U}$ with coefficients $W_{ik,jl} := -\alpha\cdot d_{kl}\delta_{ij} + \beta\cdot d_{kl}(1-\delta_{ij}) - \gamma b_k\cdot(1-\delta_{ij})\delta_{kl}$ (and $\delta_{kl} \in \{0,1\}$ the Kronecker symbol) which yields a *linear* system of d.e.'s:

$$\dot{a}_{ik}(t) \;=\; \sum_{j=1}^m\sum_{l=1}^n W_{ik,jl}\cdot u_{jl}(t) \qquad t \geq 0 \tag{4.6}$$

for $i = 1, ..., m,\ k = 1, ..., n$. As before, the limiting values a^*_{ik} for $t \to \infty$ result in a stationary or optimal fuzzy classification $\mathcal{U}^* = (u^*_{ik}) = (\sigma(a^*_{ik}))$.

Clustering with Hopfield networks resembles very much the classical *gravitational clustering methods* proposed by Butler (1969), Coleman (1971) and Wright (1977) for n data vectors $x_1, ..., x_n \in R^p$. Essentially, these methods consider $N := m$ class centers $z_i = z_i(t) \in R^p$ as particles i moving in (discrete) time $t \geq 0$ from some arbitrary initial states $z_i(0) \in R^p$ along trajectories which are controlled by the 'gravitational' forces exercised by the data points $x_1, ..., x_n$ and derived from a suitable 'potential energy function' L. Centers which approach each other too closely, are aggregated. Typically, the centers converge, for $t \to \infty$, to some few locations $\zeta_1, ..., \zeta_m \in R^p$ which characterize the regions where the data $x_1, ..., x_n$ are most concentrated (modes). Insofar these locations can be considered as 'representatives' of classes C_i which collect all data points which are close to the same center ζ_i.

5. Multilayer Perceptrons for Additive Clustering Models

A *Multilayer Perceptron (MLP)* is built from several layers of elementary processing units i ('neurons') where each one realizes a nonlinear function of the type $v = \sigma(\sum_j w_j u_j)$ of input values $u_1, u_2, ...$ with a sigmoid function σ. An MLP is designed to solve optimization or approximation problems by an optimal choice of the (numerous) 'weights' w_j in the system (e.g., by using the backpropagation algorithm). Therefore an MLP can solve clustering problems provided they are formulated as a minimization problem (with continuous weight variables). As a typical example, we sketch the approach of Sato & Sato (1995, 1998) who start from an $n \times n$ matrix $S = (s_{kl})_{n \times n}$ of pairwise similarities $s_{kl} \geq 0$ and look for an optimal fuzzy classification $\mathcal{U} = (u_{ik})_{m \times n}$.

By generalizing the *classical additive clustering model (ADCLUS)* of Shepard & Arabie (1979), they assume that each pair of fuzzy classes U_i, U_j has an unknown underlying similarity W_{ij} such that the observed similarities s_{kl} are modeled, up to random errors, by the additive pseudo-similarity $\sigma_{kl} = \sum_{i=1}^{m} \sum_{j=1}^{m} W_{ij} \rho(u_{ik}, u_{jl})$ for any pair k, l of objects. This model uses a prespecified aggregation function $\rho(u_{ik}, u_{jl})$ that weights the *simultaneous* appearance of the events '$k \in U_i$ with a degree u_{ik}' and '$l \in U_j$ with a degree u_{jl}' (e.g., $\rho(u,v) = u \cdot v$, $\min\{u,v\}$, $uv/[1-(1-u)(1-v)]$ etc.). The model σ_{kl} is adapted to the data s_{kl} by minimizing the clustering criterion:

$$G(\mathcal{U}, W) \quad := \quad \sum_{k=1}^{n} \sum_{l=1}^{n} (s_{kl} - \sigma_{kl}(\mathcal{U}, W))^2 \ \rightarrow \ \min_{\mathcal{U}, W} \qquad (5.1)$$

with respect to all fuzzy partitions $\mathcal{U}$. This problem has been solved by Sato & Sato with the help of an MLP. Since the symmetry of s_{kl} or W_{ij} is not used in (5.1), their method applies to non-symmetric data as well, e.g., if s_{kl} is the amount of telephone traffic from one town k to another one l, with a similar interpretation for W_{ij} on the cluster level. In a similar way, other additive least-squares decompositions of data could be handled (e.g., overlapping clusterings or MUMCLUS models as in Carroll & Chaturvedi 1998).

6. Other Neural Network Approaches

The previous list of neural network clustering approaches is by far not exhaustive. As illustrated by the previous cases, many well-known classical clustering (minimization) problems could be reformulated in the NN framework even if, by definition, an NN can scarcely yield 'better' clusterings than classical minimization methods. The borderline between classical and NN approaches is a fuzzy one, as exemplified by the *associative resonance theory (ART)* (Grossberg 1987) where several clustering approaches combine aspects from classical ISODATA with neural network terminology (Postma & Hudson 1995). Firmin & Hamad (1996) consider the usual mixture approach in the NN context and Kovalenko (1996) designs a NN method for detecting high-density clusters from density estimates.

References:

Adorf, H.-M., Murtagh, F. (1988): Clustering based on neural network processing. In: D. Edwards, N.E. Raun (eds.): *COMPSTAT 1988*. Physica Verlag, Heidelberg, 1988, 239-244.

Anouar, F., Badran, F., Thiria, S. (1996): Topological maps for mixture density. Intern. Conf. on Artificial Neural Networks, Bochum, Germany, ICANN'96.

Anouar, F., Badran, F., Thiria, S. (1997): Cartes topologiques et nuées dynamiques. In: Thiria, S., et al. (eds.) (1997), 190-206.

Ambroise, Ch., Govaert, G. (1996): Constrained clustering and Kohonen self-organizing maps. *J. of Classification* 13, 299-313.

Badran, F., Daigremont, Ph., Thiria, S. (1997): Régression par carte topologique. In: Thiria, S., et al. (eds.) (1997), 207-222.

Bezdek, J.C. (1981): *Pattern recognition with fuzzy objective function algorithms.* Plenum Press, New York.

Bock, H.H (1974): *Automatische Klassifikation.* Vandenhoeck & Ruprecht, Göttingen, 1974, 480 pp.

Bock, H.H (1979): Fuzzy clustering procedures. In: R. Tommassonne (ed.): *Analyse de données et informatique.* INRIA, Le Chesnay, France, 1979, 205-218.

Bock, H.H. (1985): On some significance tests in cluster analysis. *J. of Classification* 2, 77-108.

Bock, H.H. (1996a): Probability models and hypotheses testing in partitioning cluster analysis. In: Ph. Arabie, L. Hubert, G. De Soete (eds.): *Clustering and classification.* World Scientific, River Edge, NJ, 1996, 377-453.

Bock, H.H. (1996b): Probabilistic models in partitional cluster analysis. In: A. Ferligoj, A. Kramberger (eds.): *Developments in data analysis.* FDV, Metodoloski zvezki, 12, Ljubljana, 1996, 3-25.

Bock, H.H. (1996c): Probabilistic models in cluster analysis. *Computational Statistics and Data Analysis* 23, 5-28.

Bock, H.H. (1997): Simultaneous visualization and clustering methods as an alternative to Kohonen maps. In: G. Della Riccia et al. (eds.), 1997, 67-85.

Bock, H.H. (1998a): Probabilistic aspects in classification. In: Hayashi, Ch., et al., Proc. IFCS-96, 1998, 3-21.

Bock, H.H. (1998b): Clustering and neural network approaches. In: H. Locarek-Junge et al. (eds.): Proc. 22th Annual Conference of the Gesellschaft für Klassifikation, University of Dresden, 4-6 March 1998. Springer-Verlag, Heidelberg (submitted).

Bouton, C., and G. Pagès (1993): Selforganization and convergence of the one-dimensional Kohonen algorithm with non-uniformly distributed stimuli. *Stochastic Processes and Applications* 47 (1993) 249-274.

Braverman, E.M. (1966): The method of potential functions in the problem of training machines to recognize patterns without a teacher. *Automation and Remote Control* 27, 1748-1771.

Butler, G.A. (1969): A vector field approach to cluster analysis. *Pattern Recognition* 1, 291-299.

Carroll, J.D., Chaturvedi, A. (1998): Fitting the CANDCLUS/MUMCLUS models with partitioning and other constraints. In: Hayashi, Ch., et al., Proc. IFCS-96, 1998, 496-505.

Coleman, J.S. (1971): Clustering in n dimensions by use of a system of forces. *Journal of Mathematical Sociology* 1, 1-47.

Cottrell, M., Fort, J.-C. (1989): Etude d'un processus d'auto-organisation. *Ann. Inst. Henri Poincaré Probab. Statist.* 23, 1-20.

Della Riccia, G., et al. (eds.) (1997): : *Learning, networks and statistics.* CISM Courses and Lectures no. 382. Springer, Wien - New York, 1997.

Firmin, Ch., and D. Hamad (1995): Gaussian neural networks applied to cluster analysis problem. In: W. Gaul and D. Pfeifer (eds.): *From information to knowledge.* Springer-Verlag, Heidelberg, 159-166.

Fort, J.-C., Pagès, G. (1996): About the Kohonen algorithm: strong or weak self-organisation? *Neural Networks* 9, 773-785.

Hayashi, Ch., Ohsumi, N., Yajima, K., Tanaka, Y., Bock, H.-H., Baba, Y. (eds.): *Data science, classification and related methods.* Proc. IFCS-96. Springer-Verlag, Tokyo, 1998.

Grossberg, S. (ed.) (1987): *The adaptive brain II: Vision, speech, language, and motor control.* Elsevier, Amsterdam.

Hopfield, J.J. (1982): Neural networks and physical systems with emergent collective computational capabilities. *Proc. Natl. Acad. Sci. USA 79*, 2554-2558.

Hopfield, J.J., Tank, D.W. (1985): Neural computation of decisions in optimization problems. *Biological Cybernetics 52*, 141-152.

Kamgar-Parsi, B., J.A. Gualtieri, J.E. Devaney and B. Kamgar-Parsi (1990): Clustering with neural networks. *Biological Cybernetics* 63, 201-208.

Kohonen, T. (1982): Self-organized formation of topologically correct feature maps. *Biological Cybernetics* 43, 59-69.

Kohonen, T. (1995): *Self-organizing maps.* Springer, New York.

Kohonen, T., S. Kaski, H. Lappalainen, J. Salojärvi (1997): The adaptive-subspace self-organizing map (ASSOM). Workshop on Neural Networks, Helsinki, Website.

Kovalenko, A. (1996): : A multi-layer neural network algorithm for high-density cluster analysis. Paper presented at IFCS-96, Kobe, Japan.

MacQueen, J. (1967): Some methods for classification and analysis of multivariate observations. In: L.M. LeCam et al. (eds): *Proc. 5th Berkely Symp. on Math. Stat. Probab.* Univ. of California Press, Los Angeles, vol. 1, 281-297.

Moshou, D., H. Ramon (1997): Extended self-organizing maps with local linear mappings for function approximation and system identification. Workshop on Neural Networks, Helsinki, Website.

Murtagh, F. (1995): The Kohonen self-organizing map method: an assessment. *J. of Classification* 12, 165-190.

Murtagh, F. (1996): Neural networks for clustering. In: Ph. Arabie, L. Hubert, G. De Soete (eds.): *Clustering and classification.* World Scientific, River Edge, NJ, 1996, 235-269.

Pollard, D. (1981): Strong consistency of k-means clustering. *Annals of Probab. 10*, 919-926.

Pollard, D. (1982): Quantization and the method of k-means. *IEEE Trans. Information Theory IT-28*, 199-205.

Postma, E.O., Hudson, P.T.W. (1995): Adaptive resonance theory. In: P.J. Braspenning, F. Thuijsman, A.J.M.M. Weijters (eds.): *Artificial neural networks.* Springer, Berlin, 1995.

Ritter, H. (1997): Neural networks for rapid learning in computer vision and robotics. In: G. Della Riccia et al. (eds.), 1997, 25-39.

Sato, M., Sato, Y. (1995): Neural clustering: Implementation of clustering model using neural networks. Proc. IEEE Conference, 1995, 3609-3614.

Sato, M., Sato, Y. (1998): Additive clustering model and its generalizations. In: Ch. Hayashi et al., Proc. IFCS-96, 1998, 312-319.

Shepard, R.N., Arabie, Ph. (1979): Additive clustering: Representation of similarities as combinations of discrete overlapping properties. *Psychological Reviews* 86, 87-123.

Thiria, S., Lechevallier, Y., Gascuel, O., Canu, S. (1997): *Statistique et méthodes neuronales.* Dunod, Paris, 311 pp.

Tolat, V.V. (1990): An analysis of Kohonen's self-organizing maps using a system of energy functions. *Biological Cybernetics* 64, 155-164.

Tsypkin, Y.Z., Kelmans, G.K. (1967): Recursive self-training algorithms. *Engineering Cybernetics USSR*, 1967, V, 70-79.

Wright, W.E. (1977): Gravitational clustering. *Pattern Recognition* 9, 151-166.

Application of Self-Organizing Maps to Outlier Identification and Estimation of Missing Data

Mariusz Grabowski,
Cracow University of Economics, Department of Computer Science,
Cracow, Poland.

Abstract: Kohonen's self-organizing maps (SOM) belong to the group of artificial neural network methods that are the most frequently applied to data analysis. The most common applications of SOM are multidimensional data visualization and huge data sets clustering. Some characteristics of SOM make this method interesting also in other aspects of data analysis. The following paper presents the possibility of SOM application to outliers identification and missing data estimation.

Keywords: Neural Networks, Self-organizing Maps, Outliers, Missing Data.

1. Self-Organizing Maps Method

SOM is an unsupervised artificial neural network, and is frequently used in many areas of research and in practice (3014 Studies ..., 1997). Usefulness of SOM is proved by its:

- Uniqueness, i.e. lack of direct similarity to other methods of data analysis (Sarle, 1994);
- Robustness;
- Rapid training;
- Possibility of analyzing huge data sets;
- Possibility of analyzing corrupted data (incomplete data vector) (Kohonen, 1995);
- Biological analogy.

The main area of SOM applications is data analysis – the discipline in which no additional information about data set (e.g. number of groups, group membership etc.) is provided. From this point of view, in many papers, SOM is presented as a visualization (Ultsh, 1993) and clustering method (Murtagh, 1995; Murtagh & Hernández-Pajeraz, 1995a, 1995b). SOM, as a clustering algorithm, is especially used in case of huge data sets.

In SOM all neurons (called nodes) are arranged in one- or two-dimensional array, however other dimensionalities are also possible.
In its basics, SOM algorithm consists of two phases which are repeated for every input object for a number of times; each step is called a presentation (Kangas,

1994; Kohonen, 1995; Kohonen et al., 1995). If the number of objects is small, e.g. less than 1000, objects are presented to the map recursively and consequently the total number of presentations would be in the range of 10,000 – 100,000 (Kohonen, 1995).

Phase 1. Finding the matching node.
For every pattern $\mathbf{x}(t)$ from input space a matching node on the map should be found: $\mathbf{w}_j$. In order to find the matching node, in majority of cases the Euclidean distance is used. Sometimes the dot product of $\mathbf{x}^T\mathbf{w}$ is used, but in this case the input vector normalization is necessary.

$$\left\|\mathbf{x}(t) - \mathbf{w}_m(t)\right\| = \min_j\{\left\|\mathbf{x}(t) - \mathbf{w}_j(t)\right\|\} \tag{1}$$

Where m is the index of matching node; t – number input object presentation; j – node index.

Phase 2. The adaptation of the matching node and its neighbors.
Nodes vector weights are being adopted. The goal of this phase is to lower the distance between the presented object $\mathbf{x}(t)$ and its matching node ($\mathbf{w}_j$) together with its neighbors. If the distance measure is Euclidean distance, the adaptation formula is as follows:

$$w_{ji}(t+1) = w_{ji}(t) + \eta(t)h_{mj}(t)[x_i(t) - w_{ji}(t)] \tag{2}$$

Where $w_{ji}(t)$ is i-th component of j-th node weight vector in t-th learning step; $\eta(t)$ – learning parameter, $0 < \eta(t) < 1$ (decreasing with t); $h_{mj}(t)$ – neighborhood function, m – index of matching node, j – updated node index, t – learning step, $0 < h_{mj}(t) < 1$.

The unique mechanism that distinguishes SOM from other competitive-learning networks is the neighborhood function – $h_{mj}(t)$. The main task of the learning function is to provide learning (weights updating) not only for best matching node (winning neuron), but also for its neighbors. Commonly used types of neighborhood functions are as follows:

$$h_{mj}(t) = \begin{cases} 1; & \left\|\mathbf{r}_m - \mathbf{r}_j\right\| < \sigma(t) \\ 0; & \textit{otherwise} \end{cases} \tag{3}$$

Where: $\mathbf{r}_m$ – coordinates of matching node; $\mathbf{r}_j$ – coordinates of updated node; $\sigma(t)$ – neighborhood radius – decreasing function depending on t.

$$h_{mj}(t) = e^{-\frac{\|\mathbf{r}_m - \mathbf{r}_j\|^2}{2\sigma^2(t)}} \qquad (4)$$

SOM can be perceived as:

- Nonlinear transformation of any multidimensional continuous space into one- or two-dimensional discrete space, which preserves order of objects from input space. In this sense SOM may be seen as similar to principal component analysis (Sarle 1994; Oja et al., 1995), multidimensional scaling (Jajuga, 1990, 1993) and clustering methods;
- Nonparametric, nonlinear regression (Cherkassky & Lari-Najafi, 1991; Kohonen 1995) which attempts to fit number of codebook vectors to the distribution of objects in the input space.

Interested reader may find a detailed description of SOM algorithm and related topics in (Kohonen, 1995; Kohonen et al. 1995).

2. Outlier Identification

Many of data analysis methods assume data set homogeneity, whereas in reality this assumption is not fulfilled. Such terms as elliptical homogeneity and the homogeneity in the sense of linear regression are defined in (Jajuga, 1993). The author points out the importance of the use of robust analytical methods and the necessity of outliers identification.

SOM can be helpful in dealing with the issue of data set homogeneity. As a visual-grouping method, SOM may be used in data set structure analysis. Additionally its nonparametric nonlinear regression capabilities may be used in visualizing data set homogeneity and identification of non-typical data, especially in case of multi-modal data sets.

Otliers are much different from the rest of data. In order to distinguish between typical and non-typical data, it is necessary to define a criterion of difference. This criterion can be defined as the distance (i.e. Euclidean distance) from a data object to the linear regression hyper-surface. The object located far (in terms of distance) from the hyper-surface can be perceived as non-typical.

However in multi-modal data sets the criterion of distance to linear regression hyper-surface cannot be accepted because the linear regression hyper-surface does not match the modal structure of the data set. In this case the use of a nonlinear, nonparametric regression method is necessary and SOM, being the one, can serve the purpose.

As it has been stated in then following paper, SOM can be perceived as a nonlinear nonparametric regression which is formed (in the learning process) in elastic lattice. Non-typical data, because of their uniqueness, are unable to form their "own" nodes – codebook vectors; this makes them distinct from the lattice. The measure of non-typicality is referred to as a quantization error in SOM literature. The quantization error connected with the *i*-th object of the input space, is defined as follows:

$$QE_i = \min_j \{\|\mathbf{x}_i - \mathbf{w}_j\|\} \qquad (5)$$

Where: $\mathbf{x}_i$ - i-th input object, i=1, ...,S; $\mathbf{w}_j$ - j-th codebook vector, j=1, ...,M.

The SOM algorithm is to find such a map in which an average quantization error:

$$AQE = \frac{1}{S}\sum_{k=1}^{S} QE_i \qquad (6)$$

reaches its minimum.

Such functions as Euclidean distance, Mahalanobis distance and others may be used as a distance measure.

The method for outliers identification can be summarized in the following steps:
1. Definition of the SOM with a satisfactory average quantization error – done in map design and learning process;
2. Data processing done by the defined SOM;
3. Outliers identification according to the height of quantization error.
4. Data non-typicality visualization through presentation of quantization error associated with every datum in a coordinate system.

The issue, which demands further studies, is to find the value of quantization error which would allow for classifying certain datum as typical or non-typical. Obviously it is necessary to analyze the quantization error distribution. In case when this distribution would be, for example normal, the demarcation criterion could be the three-sigma rule.

Taking into account the multimodality of the analyzed data sets, it seems that two cases are possible:
1. Assume that the distribution is based on all quantization errors, regardless of the group the quantization error is coming from. In this case finding the parameters of one distribution is necessary;

2. Assume that every group has its own quantization error distribution (in the sense of parameters). In this case it would be necessary to estimate parameters for each group in the data set.

3. Estimation of Missing Data

An important problem in data analysis often addressed by many authors, is their incompleteness (Ghahramani & Jordan 1994; Gałuszka 1994, 1995; Kordos, 1988) and probably a satisfactory solution for this problem does not exist at all. In certain disciplines, e.g. technical, missing data can be obtained by repetition of the experiment, in others, e.g. socioeconomic, it is not possible. In this case an analysis of incomplete data is necessary.

Considering missing data we usually think about multidimensional vectors, in which there are empty places in certain positions of the vector. There are following alternatives of dealing with incomplete data:

- Elimination of vectors or columns containing missing data from the data matrix;
- Estimation of missing data (using different methods);
- Application of robust data analysis methods.

SOM is a robust nonparametric data analysis method and can be used in order to estimate missing data. To fulfill this task certain modifications have to be done to classical SOM algorithm. These modifications are described in (Kohonen, 1995) and implemented in SOM_PAK software (Kohonen et al., 1995):

- In the first phase of SOM algorithm, the distance between codebook vector and analyzed object is calculated excluding missing data components. It means that the number of missing positions reduces the order of the vector.
- In the second phase of SOM algorithm, only these components of codebook vectors are updated in which there are no data losses.

This modified method was used by Kohonen in visualizing world's poverty map (Kohonen, 1995).

The algorithm mentioned above provides not only visualization but also estimation of all missing values. This is possible even when all data vectors are corrupted. Estimated values are represented by these output node weight vectors' components which are related to empty places in input objects. Naturally, SOM for missing values can be applied only in the case when missing data are not significantly different from the existing data, being the base for missing data estimation.

In order to test the ability of SOM to estimate the missing values, the experiments were made on Iris data file. Iris is a data set containing 150 observations concerning *Iris* flower characteristics, used for the first time by Fisher in 1936. The measured features are the following: length of the leaf, width of the leaf, length of the petal and width of the petal. The observations come from three *Iris* kinds: *Iris setosa* (observations 1 – 51), *Iris versicolour* (observations 51 – 100) and *Iris virginica* (observations 101 –150). This data set is used as a benchmark file in data analysis procedures testing.

The author has decided not to present Fisher's Iris file in detail assuming its popularity. Interested reader may find it in many publications, for example in (Jajuga, 1990).

The experiments were conducted using SOM_PACK software (Kohonen et al., 1995).

The test files were prepared in such a way that certain vector components were randomly eliminated. In the experiments five data files were used: 1-st original Fisher's file with no corrupted data, 2-nd with 90% of existing data, 3-rd with 80% of existing data, 4-th with 70% and 5-th with 60% of existing data. The SOM was compared with the most commonly used method of missing data estimation – substitution by means of the column.

Following measures of error were used:
Absolute error

$$E = |x_w - x| \tag{7}$$

Where: x_w – pattern value; x – estimated value.

Relative error:

$$Erel = \frac{E}{x_w} \tag{8}$$

Because the data objects in Iris data set are 4-component vectors, the above error measures were modified in the following way:

$$\mathbf{E} = \sqrt{\sum_{j=1}^{4} E_j^2} \tag{9}$$

$$\mathbf{Erel} = \sqrt{\sum_{j=1}^{4} Erel_j^2} \tag{10}$$

The above formula can be interpreted as the Euclidean distance of absolute and relative error vectors.

The results of experiments showed that the aggregated relative estimation error (10) in case of SOM method, tested in different variants of neighborhood and topology parameters, was approximately 4 – 5 times lower than in case of substitution-by-mean method.
These results prove the usefulness of SOM in the given application.

4. Conclusion

SOM besides its common application – visualization and clustering – may be used in other non-trivial data analysis tasks: outliers identification and estimation of missing data. This is possible due to SOM's robustness and nonparametric nonlinear regression characteristics. The features of SOM algorithm, mentioned above, demand further deep studies.

References

3014 Studies of Self-organizing Map (SOM) and Learning Vector Quantitization (LVQ), (1997), Compiled in the Laboratory of Computer and Information Science, Helsinki University of Technology, http://www.cis.hut.fi/nnrc/refs/references.ps,

Cherkassky V., Lari-Najafi H. (1991). Constrained Topological Mapping for Nonparametric Regression Analysis, *Neural Networks*, Vol. 4. pp. 27-40,

Gałuszka B. (1994), Szacowanie brakujących danych w szeregach czasowych (wyniki badań), in: *Przestrzenno-czasowe modelowanie i prognozowanie zjawisk gospodarczych*, Aleksander Zeliaś, Kraków, pp. 134-141,

Gałuszka B., (1995), Szacowanie brakujących informacji w szeregach przekrojowych (wyniki badań), in.: *Przestrzenno-czasowe modelowanie i prognozowanie zjawisk gospodarczych*, Aleksander Zeliaś, Kraków, pp. 199-204,

Ghahramani Z., Jordan M.I. (1994). *Learning From Incomplete Data*, MIT, ftp://publications.ai.mit.edu/ai-publications/1500-1999/AIM-1509.ps.Z,

Jajuga K. (1993). *Statystyczna analiza wielowymiarowa*, PWN, Warszawa,

Kohonen T. (1995). *Self-Organizing Maps*, Springer-Verlag Berlin,

Kohonen T., Hynninen J., Kangas J., Laaksonen J. (1995). *SOM_PAK The Self-Organizing Map Program Package Version 3.1*, Helsinki University of Technology, ftp://cochlea.hut.fi/pub/som_pak,

Kordos J. (1988). *Jakość danych statystycznych*, PWE, Warszawa,

Murtagh F. (1995). Interpreting Kohonen self-organizing map using contiguity-constrained clustering, *Pattern Recognition Letters*, 16, pp. 399-408,

Murtagh F. Hernández-Pajeraz (1995a). *The Kohonen self-organizing map method: an assessment, Journal of Classification*, 12, pp. 165-190,

Murtagh F., Hernández-Pajeraz (1995b). Clustering moderately-sized datasets using the Kohonen map approach, *Statistics in Transition - Journal of the Polish Statistical Association*, 2, pp. 151-162,

Oja E., Karhunen J., Wang L., Vigario R., (1995). *Principal and independent components in neural networks - recent developments*, Proc. VII Italian Workshop on Neural Nets WIRN'95, May 18 - 20, Vietri sul Mare, Italy,

Sarle W. S. (1994). Neural Networks and Statistical Models, *Proceedings of the Nineteenth Annual SAS Users Group International Conference*, April,

Ultsh A. (1993). Self organized feature maps monitoring and knowledge aquisition of a chemical process, in: *Procedings of the International Conference on Artificial Neural Networks (ICANN93),* S. Gielen, B. Kappen, pp. 864-867, Springer-Verlag, London.

Neuro-Fuzzy Classification

Detlef Nauck, Andreas Nürnberger and Rudolf Kruse
Faculty of Computer Science, University of Magdeburg
Universitaetsplatz, D-39106 Magdeburg, Germany
E-Mail: detlef.nauck@.cs.uni-magdeburg.de, WWW:fuzzy.cs.uni-magdeburg.de

Abstract. Neuro-fuzzy classification systems offer means to obtain fuzzy classification rules by a learning algorithm. It is usually possible to find a suitable fuzzy classifier by learning from data, but it can be hard to obtain a classifier that can be interpreted conveniently. However, the main reason for using fuzzy methods for classification is usually to obtain an interpretable classifier. In this paper we discuss the learning algorithms of NEFCLASS, a neuro-fuzzy approach for data analysis.

Keywords: fuzzy classification, neuro-fuzzy system

1 Introduction

Neuro-fuzzy systems are approaches to learning fuzzy systems from data by using learning algorithms derived from neural network theory. The learning capabilities of neural networks made them a prime target for a combination with fuzzy systems in order to automate or support the process of developing a fuzzy system for a given task. The first so-called neuro-fuzzy approaches were considered mainly in the domain of (neuro-) fuzzy control, but today the approach is more general. Neuro-fuzzy systems are applied in various domains, e.g. control, data analysis, decision support, etc.

Modern neuro-fuzzy-systems are usually represented as a multilayer feedforward neural network [Berenji and Khedkar, 1992; Buckley and Hayashi, 1995; Halgamuge and Glesner, 1994; Nauck et al., 1997; Tschichold-Gürman, 1997]. In neuro-fuzzy models, connection weights and propagation and activation functions differ from common neural networks. Although there are a lot of different approaches we want to restrict the term "neuro-fuzzy" to systems which display the following properties:

Acknowledgement: The research presented in this paper is partly funded by DFG contract KR 521/3-1

(i) A neuro-fuzzy system is a fuzzy system that is trained by a learning algorithm (usually) derived from neural network theory. The (heuristical) learning procedure operates on local information, and causes only local modifications in the underlying fuzzy system. The learning process is not knowledge based, but data driven.

(ii) A neuro-fuzzy system can be viewed as a special 3-layer feedforward neural network. The units in this network use t-norms or t-conorms instead of the activation functions common in neural networks. The first layer represents input variables, the middle (hidden) layer represents fuzzy rules and the third layer represents output variables. Fuzzy sets are encoded as (fuzzy) connection weights. This view of a fuzzy system illustrates the data flow within the system, and its parallel nature. However this neural network view is not a prerequisite for applying a learning procedure, it is merely a convenience.

(iii) A neuro-fuzzy system can always (i.e. before, during and after learning) be interpreted as a system of fuzzy rules. It is both possible to create the system out of training data from scratch, and it is possible to initialize it by prior knowledge in form of fuzzy rules.

(iv) The learning procedure of a neuro-fuzzy system takes the semantical properties of the underlying fuzzy system into account. This results in constraints on the possible modifications applicable to the system parameters.

(v) A neuro-fuzzy system approximates an n-dimensional (unknown) function that is partially given by the training data. The fuzzy rules encoded within the system represent vague samples, and can be viewed as vague prototypes of the training data. A neuro-fuzzy system should not be seen as a kind of (fuzzy) expert system, and it has nothing to do with fuzzy logic in the narrow sense [Kruse et al., 1994].

In this paper "neuro-fuzzy" has to be understood in the way given by the five points above. Therefore we consider "neuro-fuzzy" as a certain technique to derive a fuzzy system from data, or to enhance it by learning from examples. The exact implementation or "neuro-fuzzy model" does not matter.

In this paper we discuss neuro-fuzzy classification and use our NEFCLASS model as an example. In the following section we describe the model and its learning capabilities. After that we given an example where we apply NEFCLASS to a real-world data set.

2 Classification with NEFCLASS

Classification of data is an area of application where statistical methods, machine learning and neural networks are thoroughly examined and successfully used. It is also possible to use a fuzzy system for classification with rules like

if x_1 is μ_1 **and** x_2 is μ_2 **and** ... **and** x_n is μ_n
then pattern $(x_1, x_2, \ldots, x_n)$ belongs to class i,

where the $\mu_1, \ldots, \mu_n$ are fuzzy sets.

What is the advantage of having another method for classification? A fuzzy classifier is not a replacement for the aforementioned methods yielding better results, but a different way of achieving the same goal. If a decision is made for a fuzzy classifier usually the following advantages are considered:

- vague knowledge can be used,
- the classifier is interpretable in form of linguistic rules,
- from an applicational view the classifier is easy to implement, to use and to understand.

The rule base of a fuzzy classifier that uses rules like mentioned above, represents an approximation of an (unknown) function $\varphi : \mathbb{R}^n \to \{0,1\}^m$ that represents the classification task where $\varphi(\mathbf{x}) = (c_1, \ldots, c_m)$ such that $c_i = 1$ and $c_j = 0$ $(j \in \{1, \ldots, m\}, j \neq i)$, i.e. $\mathbf{x}$ belongs to class C_i. Because of the inferences process, the rule base actually does not approximate φ but the function $\varphi' : \mathbb{R}^n \to [0,1]^m$. We can obtain $\varphi(\mathbf{x})$ by $\varphi(\mathbf{x}) = \psi(\varphi'(\mathbf{x}))$, where ψ reflects the interpretation of the classification result obtained from the fuzzy classifier. Usually the class with the largest activation value is chosen.

Classifiers are usually derived from data and are not specified directly. In case of a fuzzy classifier there are two common methods: *fuzzy clustering*, and *neuro-fuzzy learning*.

In case of fuzzy clustering the input space is searched for clusters. The number of clusters is determined by an evaluation measure, and the size and shape of the clusters is given by the clustering algorithm. The obtained clusters can function as a fuzzy classifier, in fact, there is no need to express the classifier by rules. However, in this case the interpretability is lost, and therefore fuzzy rules are sometimes created by projection of clusters. This can cause a loss of information, because a fuzzy rule created by projection does not represent the cluster exactly, but only the smallest encompassing hyperbox. The performance of a fuzzy classifier obtained by clustering is therefore usually reduced, once it is expressed in form of fuzzy rules. In

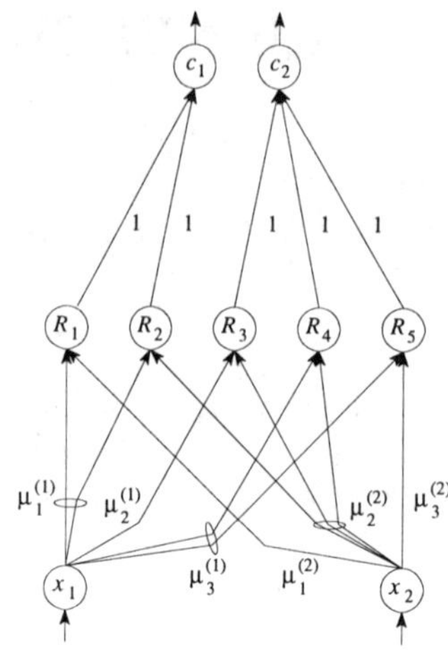

Figure 1: *A NEFCLASS system represented as a 3-layer neural network*

addition the rules are often hard to interpret, because the resulting fuzzy sets can have almost any shape.

Another method to obtain a fuzzy classifier from data is to use a neuro-fuzzy approach. This means the classifier is created from data by a heuristic learning procedure. If the neuro-fuzzy approach meets the five points listed in Section 1, then the interpretability of the resulting classifier, and a acceptable performance might be obtained. A neuro-fuzzy approach is also often less computationally expensive than a clustering approach, because of its simplicity.

A neuro-fuzzy classifier simply a fuzzy classifier obtained by a learning procedure, as described in Section 1. There are several known approaches to find a fuzzy classifier this way, e.g. FuNE [Halgamuge and Glesner, 1994], Fuzzy RuleNet [Tschichold-Gürman, 1997], or NEFCLASS [Nauck and Kruse, 1997a; Nauck and Kruse, 1997b]. After creation of the classifier there is usually no hint on how it was derived. Therefore the term "neuro-fuzzy" strictly only applies to the creation or training phase. Once the classifier is applied and not changed further, a simple fuzzy system for classification remains. However the term "neuro-fuzzy classifier" is usually kept, to stress the mode of obtaining the classifier. The interpretation of the classifier is also often illustrated by representing it in a neural network structure, like e.g. NEFCLASS in Fig. 1.

We have presented our NEFCLASS model (neuro-fuzzy classification) in [Nauck and Kruse, 1997a], and some refined learning algorithms in [Nauck and Kruse, 1997b]. Details about the model can also be found in [Nauck et al., 1997]. In this paper we only want to discuss the general idea of the learning process. For the following descriptions consider the illustration

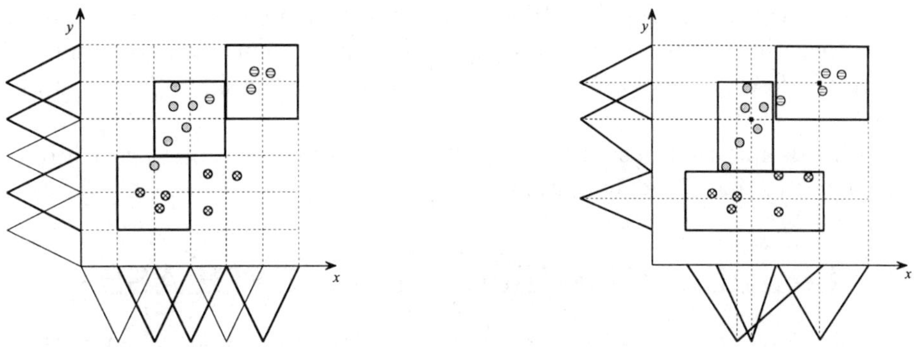

Figure 2: *Left: Situation after 3 fuzzy classification rules have been created using initial membership functions. Right: Situation after training the classifier, i.e. modifying the membership functions*

given in Figure 2

The rule learning algorithm needs an initial fuzzy partitioning for each variable. This is e.g. given by a fixed number of equally distributed triangular membership functions (see left part of Fig. 2). The combination of the fuzzy sets forms a "grid" in the data space, i.e. equally distributed overlapping rectangular clusters. Then the training data is processed, and those clusters that cover areas where data is located are added as rules into the rule base of the classifier. In a next step this rule base is reduced by just keeping the best performing rules [Nauck and Kruse, 1997a]. The result after this stage of training can e.g. look like the situation on the left hand side of Fig. 2.

After the rulebase has been created, the membership functions are tuned by a simple heuristic. For each rule a classification error is determined, and used to modify that membership function that is responsible for the rule activation (i.e. delivers the minimal membership degree of all fuzzy sets in the rules antecedent). The modification results in shifting the fuzzy set, and enlarging or reducing its support, such that a larger or smaller membership degree is obtained depending on the current error. The learning result might look like the situation on the right hand side of Fig. 2.

To obtain an interpretable classifier some restrictions can be specified by the user. The NEFCLASS software [Nauck et al., 1997] allows to impose the follwing restrictions on the learning algorithm:

- a membership function must not pass one of its neighbors,
- a membership function may be asymetrical,
- membership functions must intersect at 0.5.

After learning pruning methods [Nauck and Kruse, 1997b] can be used to reduce the number of rule and variables in the neuro-fuzzy system. A complete description of the learning algorithm can be found in [Nauck et al., 1997].

An example that describes the performance of the learning procedure is presented in the following section.

3 Learning Classifiers with NEFCLASS

As an example for the learning capabilities of NEFCLASS we use the "Wisconsin Breast Cancer" data set [Wolberg and Mangasarian, 1990]. This data set contains 699 cases distributed into two classes (benign and malign). We used only 683 cases (342 for training, 341 for testing), because 16 cases have missing values. Each pattern has nine features.

To show how NEFCLASS performs when prior knowledge is supplied, we used a fuzzy clustering method to obtain fuzzy rules. We used a modification of the algorithm by Gustafson and Kessel [Gustafson and Kessel, 1979] presented in [Klawonn and Kruse, 1997]. Fuzzy clustering discovered three clusters that were interpreted as fuzzy rules by projecting the clusters to each dimension and finding trapezoidal membership functions that closely matched the projections. The resulting three rules caused 94 classification errors on the whole data set.

Now we matched the membership functions obtained by projection with linguistic labels *small, medium* and *large*. By this we got the following rule base (s = small, m = medium, l = large):

R_1: **if** (s,s,s,s,s,s,s,s,s) **then** *benign*,
R_2: **if** (m,m,m,m,m,l,m,m,s) **then** *malignant*,
R_3: **if** (m,l,l,l,m,l,m,l,s) **then** *malignant*.

After we initialized a NEFCLASS system with these three rules, we at first got 240 classification errors. However, after 80 epochs of training, using the constraint, that fuzzy sets must not pass each other, we obtained a result of only 50 errors altogether (92.7% correct).

A neuro-fuzzy learning strategy is a tool to support the creation of a fuzzy classifier, but not to completely automate it. This means the user should supervise the learning process, and interpret the result. By analyzing the fuzzy sets obtained by training the three rules used as prior knowledge we found that the fuzzy set *medium* substantially overlapped with either the fuzzy set *small* or *large* for almost all variables. This can be seen as evidence that *medium* is superfluous.

Therefore we again trained a NEFCLASS system with "best per class" rule learning, allowing it to create four rules (two for each class). This time we only used two fuzzy sets to partition the domains of each variable. After 100 epochs of training NEFCLASS made only 24 errors on the complete set (96.5% correct). It has found the rules

R_1: **if** (s,s,s,s,s,s,s,s,s) **then** *benign*,
R_2: **if** (l,s,s,s,s,s,s,s,s) **then** *benign*,
R_3: **if** (l,l,l,l,l,l,l,l,s) **then** *malignant*,
R_4: **if** (l,l,l,l,s,l,l,l,s) **then** *malignant*.

If we additionally apply the constraint that the fuzzy sets must intersect at membership degree 0.5 we obtain 26 classification errors on the whole set.

With the pruning techniques of NEFCLASS [Nauck and Kruse, 1997b] this rule base can be further improved. After pruning we obtained a classification result of 6 errors on the training set and 20 errors on the test set. So we have again 26 classification errors, but now a much smaller rule base. The two remaining rules are:

R_1: If the variables 3,4,6,7 and 8 are all *large*, then class is *malignant*,
R_2: If the variables 3,4,6,7 and 8 are all *small*, then class is *benign*.

The linguistic terms *small* and *large* for each variable are represented by membership functions which overlap with their neighbors at membership degree 0.5, and so also the fuzzy sets are nicely interpretable.

4 Conclusions

We have discussed neuro-fuzzy classification using our NEFCLASS model as an example. We consider a neuro-fuzzy method to be a tool for creating fuzzy systems from data. It supports the user, but it cannot do all the work. It is difficult to find a good **and** interpretable fuzzy classifier by a completely automated learning process. The learning algorithm must take the semantics of the desired fuzzy system into account, and adhere to certain constraints. The learning result must also be interpreted, and the insights gained by this must be used to restart the learning procedure to obtain better results if necessary.

A fuzzy classifier, especially a neuro-fuzzy classifier, is only used, when interpretation, and the employment of (vague) prior knowledge is required. Fuzzy classification is not a replacement, but an addition to other methods like statistics or neural networks. The price for the interpretation of the classifier in form of simple fuzzy rules, and for the simple and computationally efficient learning algorithm might be paid by a classification result that is not as good as it could be if other methods are used.

The example with the Wisconsin Breast Cancer data shows that NEFCLASS can be used as an interactive data analysis method. It is useful to provide prior knowledge whenever possible, e.g. by a initializing NEFCLASS with rules found by fuzzy clustering. The learning result of NEFCLASS can be analyzed, and the obtained information can be used for another run that yields an even better result. The user is able to supervise and interpret the learning procedure in all its stages. The software tool and the data that was used by us to obtain the results presented in this paper, can be found on our WWW server fuzzy.cs.uni-magdeburg.de.

References

H. R. Berenji and P. Khedkar (1992). *Learning and Tuning Fuzzy Logic Controllers through Reinforcements.* IEEE Trans. Neural Networks, 3:724–740.

J. J. Buckley and Y. Hayashi (1995). *Neural Networks for Fuzzy Systems.* Fuzzy Sets and Systems, 71:265–276.

D. Gustafson and W. Kessel (1979). *Fuzzy Clustering with a Fuzzy Covariance Matrix.* In *Proc. IEEE CDC*, pages 761–766, San Diego, CA.

S. K. Halgamuge and M. Glesner (1994). *Neural Networks in Designing Fuzzy Systems for Real World Applications.* Fuzzy Sets and Systems, 65:1–12.

F. Klawonn and R. Kruse (1997). *Constructing a Fuzzy Controller from Data.* Fuzzy Sets and Systems, 85:177–193.

R. Kruse, J. Gebhardt and F. Klawonn (1994). *Foundations of Fuzzy Systems.* Wiley, Chichester.

D. Nauck, F. Klawonn and R. Kruse (1997). *Foundations of Neuro-Fuzzy Systems.* Wiley, Chichester.

D. Nauck and R. Kruse (1997a). *A Neuro-Fuzzy Method to Learn Fuzzy Classification Rules from Data.* Fuzzy Sets and Systems, 89:277–288.

D. Nauck and R. Kruse (1997b). *New Learning Strategies for NEFCLASS.* In *Proc. Seventh International Fuzzy Systems Association World Congress IFSA'97*, volume IV, pages 50–55, Prague.

N. Tschichold-Gürman (1997). *The Neural Network Model RuleNet and its Application to Mobile Robot Navigation.* Fuzzy Sets and Systems, 85:287–303.

W. Wolberg and O. Mangasarian (1990). *Multisurface Method of Pattern Separation for Medical Diagnosis Applied to Breast Cytology.* Proc. National Academy of Sciences, 87:9193–9196.

Computational Enhancements in Tree-Growing Methods

Jan Klaschka, Roberta Siciliano and Jaromír Antoch
ICS CAS, Pod vodárenskou věží 2, CZ – 182 07 Prague, Czech Rep.
Univ. of Naples, via Cinthia, I – 80 126 Naples, Italy
Charles Univ., Sokolovská 83, CZ – 186 75 Prague, Czech Rep.

Abstract: In this paper we show how to avoid unnecessary calculations and to save considerably the computational cost in a wide class of tree-based methods. So called auxiliary statistics, which enable to restrict handling the raw data, are introduced. Aside that, a fast splitting algorithm is outlined, which allows to recognize and avoid unnecessary split evaluations during the search of an optimal split. Relationships between the computational cost savings and properties of both a specific method and data are summarized.

Key Words: Classification and regression trees, computational cost, auxiliary statistics, fast splitting algorithm.

1. Introduction

The tree-growing methods of the data analysis (see a survey by Mola in this volume) constitute a family whose members have, besides methodological differences, a number of common features. Each of these methods fits to the data a kind of a "*piecewise model*" defined in the terms of the set of the terminal nodes of a tree. The trees are grown recursively. Each branching is a result of an extensive testing of a number of candidate partitions (splits). During the search of the optimal split, typically, analogous calculations are performed many times with one and the same portion of the data.

Our interest in the tree-based methods of data analysis, which dates back several years, was primarily methodological. However, the experience with computer implementation has attracted our attention also to the questions of computational effectivity and computational enhancements.

We believe that many interesting ideas of computational enhancements may have been implemented. Unfortunately, most of them are not described and documented (a rare exception being to find in Breiman at al., 1984 – see Section 4 of this paper).

In this contribution, we would like to advertise the following principle: *Do not compute what need not be computed*!

Section 2 recapitulates some basics of the tree-based methodologies. Main results are contained in Sections 3 and 4, where our proposals of computational enhancements are outlined. While in Section 3 we deal with so called *auxiliary statistics* – a tool that allows to restrict calculations from the raw data, Section 4 is devoted to the most straightforward demonstration of the above principle – the fast splitting algorithm. This algorithm enables to recognize that some computations are unnecessary and to avoid them. Computational cost savings resulting from the ideas outlined in Sections 3 and 4 will be discussed in Section 5, together with a concluding remark.

2. Basics and notation

One of the most important common building blocks of almost all of the tree-based methods is the task of finding the best split for a node of a tree over a set of candidate splits. Let us recall some basics and introduce necessary notation related to this procedure.

Throughout the paper, we shall confine our interest to the standard data structures and splits following Breiman et al. (1984). At the same time we shall explicitly deal only with *binary* splits, though much of the ideas presented in the paper can be easily adapted for the methods generating ternary or, more generally, n-ary splits with $n > 2$.

Each case l in the data set (learning sample) is assumed to have its values ${}^1x_l, \ldots, {}^kx_l$ of *predictors* ${}^1X, \ldots, {}^kX$ (left upper index will be left out whenever there is just one predictor of interest at the moment).

We shall use symbol $\mathcal{L}$ with possible arguments and/or subscripts for sets of cases. The size of set of cases $\mathcal{L}$ will be denoted $|\mathcal{L}|$. By $\mathcal{L}(t)$ we denote the set of those cases that belong to the node t of the tree. The left and right "*sons*", respectively, of t will be denoted t_L and t_R. Recall that $\mathcal{L}(t_L)$ and $\mathcal{L}(t_R)$ are disjoint, $\mathcal{L}(t)$ being their union.

Let $X(t)$ be the set of all values that predictor X takes on in node t, i.e. in $\mathcal{L}(t)$. Predictor X defines for each node t a set of splits $\mathcal{S}(X,t)$, each split s being given by a standard question. For a categorical predictor X standard questions are of the form "*Is the value of X in B?*", where B is any proper nonempty subset of $X(t)$. For a numerical predictor X, questions "$X < c$?" are used for a finite set of cutpoints c, e.g., the midpoints between the consecutive ordered values from $X(t)$. The positive and negative answers, respectively, define the left and right "*sons*" $t_L(s)$ and $t_R(s)$ corresponding to the given split s.

Search of the best split for node t is (or may be reduced to) maximization of a kind of statistic $\phi(s,t)$ – the *splitting criterion* – over the set $\bigcup_{i=1}^k \mathcal{S}({}^iX, t)$ of all splits based on all predictors.

3. Auxiliary statistics: Minimization of handling raw data

A naive way of computing the splitting criterion for candidate splits within a node t is to start calculation for each split "*from scratch*" using the raw data that "*belong*" to t. Many splitting criteria allow, nevertheless, utilization of partial results, so called *auxiliary statistics* (Klaschka and Antoch, 1997). By the auxiliary statistic for a set of cases $\mathcal{L}$ we mean such a structured set $\alpha(\mathcal{L})$ of numbers (a vector, a matrix, a list of vectors and/or matrices, etc.) that:

- it is computationally cheap, using the auxiliary statistics $\alpha(\mathcal{L}_1)$ and $\alpha(\mathcal{L}_2)$ available for disjoint sets of cases $\mathcal{L}_1$ and $\mathcal{L}_2$, to calculate $\alpha(\mathcal{L}_1 \cup \mathcal{L}_2)$;
- analogously, for $\mathcal{L}_2 \subset \mathcal{L}_1$, it is easy to calculate $\alpha(\mathcal{L}_1 \setminus \mathcal{L}_2)$, using $\alpha(\mathcal{L}_1)$ and $\alpha(\mathcal{L}_2)$;
- it is computationally cheap to calculate the splitting criterion $\phi(s,t)$ from the auxiliary statistics $\alpha\big(\mathcal{L}(t_L(s))\big)$ and $\alpha\big(\mathcal{L}(t_R(s))\big)$ for the left and right "*sons*" $t_L(s)$ and $t_R(s)$ of the node t.

Some simple examples of auxiliary statistics are as follows.

- In the classification trees (Breiman et al., 1984), the vector of frequencies of those cases from a set of cases $\mathcal{L}$ that are a priori classified into given classes $C_1, \ldots, C_m$, can be used as an auxiliary statistic assigned to $\mathcal{L}$.
- The splitting criterion $\phi(s,t)$ in the least squares regression trees (ibid.) with dependent variable Y can be expressed as

$$\phi(s,t) = \sum_{l \in \mathcal{L}(t_L(s))} \left(y_l - \overline{y}_L\right)^2 + \sum_{l \in \mathcal{L}(t_R(s))} \left(y_l - \overline{y}_R\right)^2,$$

  where $\overline{y}_L$ and $\overline{y}_R$ are means of Y within $\mathcal{L}(t_L(s))$ and $\mathcal{L}(t_R(s))$, respectively. Clearly, the triple consisting of $|\mathcal{L}|$, $\sum_{l \in \mathcal{L}} y_l$, and $\sum_{l \in \mathcal{L}} y_l^2$ may be used as an auxiliary statistic for $\mathcal{L}$.
- When the model of multiple linear regression $\boldsymbol{Y} = \boldsymbol{X\beta} + \boldsymbol{e}$ is fitted within nodes by the least squares method and the splitting criterion is defined as

$$\phi(s,t) = RSS\Big(\mathcal{L}\big(t_L(s)\big)\Big) + RSS\Big(\mathcal{L}\big(t_R(s)\big)\Big),$$

  $RSS(\mathcal{L})$ being the residual sum of squares in $\mathcal{L}$ (compare Ciampi, 1993), then the triple consisting of the matrix $\boldsymbol{X'X}$, the vector $\boldsymbol{X'Y}$, and the scalar $\boldsymbol{Y'Y}$ can be used as an auxiliary statistic. It follows from the fact that the residual sum of squares can be expressed as $RSS = \boldsymbol{Y'Y} - (\boldsymbol{Y'X})(\boldsymbol{X'X})^{-}(\boldsymbol{X'Y})$.

Analysis of more sophisticated auxiliary statistics utilizing QR-factorization, following up previous results of Antoch and Ekblom (1995), is in progress.

Usage of auxiliary statistics depends on the type of predictor:

- For a *categorical* predictor X, the set of cases $\mathcal{L}(t)$ is partitioned into subsets $\mathcal{L}_x = \{l \in \mathcal{L}(t); x_l = x\}, x \in X(t)$, and auxiliary statistics $\alpha(\mathcal{L}_x)$ are computed from the raw data for all these subsets. For any split s based on X, the auxiliary statistics $\alpha(\mathcal{L}(t_L(s)))$, $\alpha(\mathcal{L}(t_R(s)))$ are calculated by "*combining*" the pre-calculated auxiliary statistics $\alpha(\mathcal{L}_x)$, $x \in X(t)$. The splitting criterion $\phi(s,t)$ is then calculated, using $\alpha(\mathcal{L}(t_L(s)))$ and $\alpha(\mathcal{L}(t_R(s)))$.
- For a *numerical* predictor X, especially when many cutpoints are used within node t, we propose another procedure. Splits are processed in the natural order, so that cutpoints increase. For the leftmost cutpoint $c_{\min}$, auxiliary statistics $\alpha(\{l \in \mathcal{L}(t); X < c_{\min}\})$ and $\alpha(\{l \in \mathcal{L}(t); X \geq c_{\min}\})$ are computed from the raw data. For any other cutpoint c, the auxiliary statistics $\alpha(\{l \in \mathcal{L}(t); X < c\})$ and $\alpha(\{l \in \mathcal{L}(t); X \geq c\})$ are recalculated from:
  - the stored auxiliary statistics for the preceding cutpoint $c' < c$, namely $\alpha(\{l \in \mathcal{L}(t); X < c'\})$, $\alpha(\{l \in \mathcal{L}(t); X \geq c'\})$, and
  - $\alpha(\{l \in \mathcal{L}(t); c' \leq X < c'\})$, which is computed from scratch.

Auxiliary statistics are, of course, available only for some splitting criteria. We may also meet a situation "*somewhere in between*". For example, in the least absolute deviation regression trees (Breiman et al., 1984) median and the sum of absolute deviations from median "*have nearly*" the properties of an auxiliary statistic. To recalculate these statistics for union of two data subsets requires some handling the raw data, but much less than a complete calculation from scratch.

Finally, we should mention, that the auxiliary statistics represent rather a natural idea than a "*trick*". Therefore, it is quite possible that they are used in many existing tree-growing programs. However, we have not found any support of such a hypothesis in the literature.

4. Fast splitting algorithm: Omitting calculations for splits known to be nonoptimal

Undoubtedly, it is useful to recognize at time that some splits are nonoptimal, and the related calculations can be omitted.

Breiman et al. (1984) present a classical example of such a utilization of knowledge. In the two-class classification problem in CART, for a broad class of splitting criteria, the search of the best split among the $2^{k-1} - 1$ splits based on a k-valued categorical predictor X can be confined to a subset of $k-1$ splits only. The values $x_1, \ldots, x_k$ of X can be ordered into a sequence $x_{(1)}, \ldots, x_{(k)}$ so that the relative frequencies of cases classified into the first class (of the two) among the cases with $X = x_{(i)}$, $i = 1, \ldots, k$ are nondecreasing. Only those splits that for some j put cases with $X \in \{x_{(1)}, \ldots, x_{(j)}\}$ into one node (while the rest into the other node), are candidates of optimality (see Theorem 4.5 in the Breiman's book). A similar result concerning the least squares regression trees in CART can be found in Breiman (1984), Proposition 8.16.

Mola and Siciliano (1996, 1997) have suggested (and in several programs implemented) another result of that kind – so called *fast splitting algorithm*. Its idea is as follows. For various splitting criteria ϕ it is possible to define a statistic $\psi(X,t)$ that is a common upper bound for the values $\phi(s,t)$ for all splits s based on a single predictor X within a node t. When we already have such a split s_0 that the inequality $\phi(s_0,t) > \psi(X,t)$ holds for a predictor X, we know that none of the splits in $\mathcal{S}(X,t)$ can be the best one for the given node; therefore we can omit calculations of the splitting criterion for all these splits.

The algorithm first sorts the predictors into the sequence ${}^{(1)}X, \ldots, {}^{(k)}X$ so that the corresponding sequence $\psi\big({}^{(1)}X,t\big), \ldots, \psi\big({}^{(k)}X,t\big)$ is nonincreasing. Then all the splits based on ${}^{(1)}X, {}^{(2)}X$ etc. are evaluated in turn, until for some j the maximum ϕ-value found among the splits based on ${}^{(1)}X, \ldots, {}^{(j)}X$ is higher than $\psi\big({}^{(j+1)}X\big)$. Since it is clear that none of the yet unevaluated splits based on ${}^{(j+1)}X, \ldots, {}^{(k)}X$ can be optimal, the search can be stopped, and the best split found so far is the best one for the given node.

The algorithm should be applied to categorical predictors with at least three values, and to numerical (ordered) predictors with only a few (but more than two) values. (It is evident, that for a binary predictor X, which defines just one split, the possible computational cost savings related to $\phi(s,t)$ are lost in calculation of $\psi(X,t)$. For numerical predictors with many values the ψ-values become computationally too costly.)

The upper bound ψ mentioned above is very easy to construct for a broad class of splitting criteria (see Mola and Siciliano, 1998, and Klaschka and Antoch, 1997), covering, e.g. those of:

- the least squares regression and the least average distance regression in CART (Breiman et al., 1984);
- classification trees with the splitting criterion based on the Gini coefficient (ibid.);
- the methods fitting in each node of the tree a specific multivariate model, using the least squares method or the maximum likelihood method (see, e.g., Ciampi, 1993).

Maximization of $\phi(s,t)$ is often equivalent to maximization of criterion $\phi'(s,t) = G\big(\mathcal{L}(t_L(s))\big) + G\big(\mathcal{L}(t_R(s))\big)$, where G is a "*goodness of fit func-*

tion" such that $G(\mathcal{L}_1 \cup \mathcal{L}_2) \leq G(\mathcal{L}_1) + G(\mathcal{L}_2)$ for any pair of disjoint sets of cases $\mathcal{L}_1$ and $\mathcal{L}_2$. An upper bound of the splitting criterion $\phi(s,t)$ over $\mathcal{S}(X,t)$ is then obtained in the following way: The set of cases $\mathcal{L}(t)$ is partitioned into subsets $\mathcal{L}_x = \{l \in \mathcal{L}(t); x_l = x\}$, $x \in X(t)$. The sum $\sum_x G(\mathcal{L}_x)$ over $x \in X(t)$ is a common upper bound of values $\phi'(s,t)$ for $s \in \mathcal{S}(X,t)$, and it can usually be "*transformed back*" into an analogical bound for the original criterion ϕ.

It is often possible and, moreover, highly desirable to combine the use of auxiliary statistics and of the fast splitting algorithm, utilizing the same auxiliary statistics for calculation of both the upper bound ψ, and of the splitting criterion ϕ (see the next Section).

5. Computational cost savings

The effect of the enhancements described in Sections 3 and 4 may be estimated both theoretically and empirically. As regards the first approach, Siciliano and Mola (1996) described a part of their empirical experience with the performance of the fast splitting algorithm. As concerns the second approach, Klaschka and Antoch (1997) outlined some partial theoretical considerations regarding the effect of the fast splitting algorithm. In this paper, we present more recent and more complete theoretical results. They highlight the mutually complementary roles of the auxiliary statistics and the fast splitting algorithm, and show the virtues of the joint application of both enhancements.

Our analysis will deal with a single categorical predictor X. The overall computational cost savings during the whole process of tree-growing are, of course, composed of these partial savings. We shall utilize a number of simplifying assumptions which are listed below.

- We assume that X takes on the same number k of values in each of those nodes of the tree, where the search of the best split takes place. Thus, there are $2^{k-1} - 1$ candidate splits based on X in any of these nodes.
- We assume that within a node with n cases:
  - the cost of all the auxiliary statistics $\alpha(\mathcal{L}_x)$, $x \in X(t)$ computed from the raw data is $c\,n$, where c is a constant;
  - the cost of calculation of one value of the splitting criterion ϕ, and/or the cost of calculation of the upper bound ψ, from the auxiliary statistics is equal to $c\,b$;
  - the cost of calculation of one value of the splitting criterion ϕ, and/or the cost of calculation of the upper bound ψ, computed from scratch is $c\,(n+b)$.
- We neglect the cost of recalculation of auxiliary statistics for $\mathcal{L}(t_L)$ and $\mathcal{L}(t_R)$ from the auxiliary statistics $\alpha(\mathcal{L}_x)$, $x \in X(t)$.

- By ν we denote the mean number of cases in those nodes where the search of the best splits takes place. Notice that:
  - parameter ν grows with the total number of cases in the data set;
  - the larger number of nodes is allowed to exist by the stopping rule, the smaller is ν;
  - for a given total number of cases and a given number of nodes, it holds that the stronger is the tendency (of the method and/or of the data) to prefer extremely asymmetrical ("*end-cut*") splits the higher is the parameter ν.
- The fast splitting algorithm treats the splits based on X within a node in one of the two following manners.
  1. They are detected as nonoptimal by the fast algorithm, and the calculations of the splitting criterion for these splits are omitted.
  2. The splitting criterion for all these splits must be calculated.

  We assume that the choice between these two alternatives is random, and moreover, that the latter one has the same probability $\pi(X)$ in every node.

Under the above assumptions, the following formula for the relative computational cost savings (RS) attributable to the use of the auxiliary statistics (AUX) can be derived. (Subscript $PLAIN$ stands for unenhanced computation.)

$$RS_{AUX} = \frac{cost_{PLAIN} - cost_{AUX}}{cost_{PLAIN}} \approx \left(1 - \frac{1}{2^{k-1}-1}\right) \frac{1}{1 + b/\nu}\,.$$

Thus, as we can see, the effect of using auxiliary statistics is increased by:
- large number of values of categorical predictors;
- high overall number of cases in the data set;
- small number of terminal nodes (before possible pruning);
- tendency to very asymmetric ("*end-cut*") splits;
- simple calculation of the splitting criterion from the auxiliary statistics.

Under the same assumptions, the relative computational cost savings due to the application of the fast splitting algorithm ($FAST$) without using the auxiliary statistics can be expressed as

$$RS_{FAST} = \frac{cost_{PLAIN} - cost_{FAST}}{cost_{PLAIN}} \approx 1 - \pi(X) - \frac{1}{2^{k-1}-1}\,.$$

The effect of the fast splitting algorithm grows, as we can see, with:
- high number of values of categorical predictors;
- the probability that the fast splitting algorithm omits calculation of the splits related to X.

Notice, that the *global* savings (for *all* predictors) obviously depend on how the π-values are distributed in the set of all the predictors used. A big amount of calculations is saved when there are only a few splits close to optimum and many "*much worse*" splits.

Finally, the savings due to the combined use of the auxiliary statistics and the fast splitting algorithm $(AUX, FAST)$ can be evaluated as

$$RS_{AUX,FAST} = \frac{cost_{PLAIN} - cost_{AUX,FAST}}{cost_{PLAIN}}$$
$$\approx \left(1 - \frac{1}{2^{k-1}-1}\right)\frac{1}{1+b/\nu} + \frac{1-\pi(X)-\frac{1}{2^{k-1}-1}}{1+\nu/b} \,. \tag{1}$$

It is easy to see that $RS_{AUX,FAST} = RS_{AUX} + RS_{FAST}/(1+\nu/b)$. From (1) it is clear that the factors behind the parameters b and ν can eliminate the effects of either the auxiliary statistics or of the fast splitting algorithm, *but not both*.

As the conclusion, we would like to stress that the two enhancements described in Sections 3 and 4 cover together well the broad spectrum of situations where the computational cost of the tree-growing may become high. Therefore, their joint use is highly desirable whenever it is possible.

Acknowledgement: Research was supported by grants VaV/520/2/97 from the Czech Ministry of Environment, MURST 60%, and GAČR 1163/ 201/97.

References

Antoch J. and Ekblom H. (1995). Recursive robust regression, computational aspects and comparison, *CSDA*, 19, 115 – 128.

Breiman L., Friedman J.H., Olshen R.A. and Stone C.J. (1984). *Classification and Regression Trees*, Wadsworth, Belmont CA.

Ciampi A. (1993). Constructing prediction trees from data, the RECPAM approach, in: *Computational Aspects of Model Choice*, Antoch J. (Ed.), Physica – Verlag, Heidelberg, 105 – 156.

Klaschka J. and Antoch J. (1997). How to grow trees fast, in: *ROBUST'96*, Antoch J. and Dohnal G. (Eds.), JČMF, Prague, 91 – 106 (in Czech).

Mola F. and Siciliano R. (1997). A fast splitting procedure for classification trees, *Statistics and Computing*, 7, 209 – 216.

Mola F. and Siciliano R. (1998). A general splitting criterion for classification trees, *Metron*, to appear.

Siciliano R. and Mola F. (1996). A fast regression tree procedure, in: *Statistical Modelling, Proc. 11th Int. Workshop on Statistical Modelling*, Forcina A. et al. (Eds.), Graphels, Perugia, 332 – 340.

A New Way to Build Oblique Decision Trees Using Linear Programming

Guy Michel♣, Jean Luc Lambert♣
Bruno Cremilleux♣ & Michel Henry-Amar♠
GREYC♣, CNRS UPRESA 6072 - GRECAN♠, CNRS UPRES 1772
Esplanade de la paix - Université de Caen – F 14 032 Caen Cedex, France

Summary : Adding linear combination splits to decision trees allows multivariate relations to be expressed more accurately and succinctly than univariate splits alone. In order to determine an oblique hyperplane which distinguishes two sets, linear programming is proposed to be used. This formulation yields a straightforward way to treat missing values. Computational comparison of that linear programming approach algorithm with classical univariate split algorithms proofs the interest of this method.

Key words : Oblique decision tree, missing values, linear programming.

Introduction

Classification and decision trees already belong to the way of current research and many methods still have to be explored [5]. Unlike classical decision trees which produce univariate trees, linear programming is proposed to be used to generate oblique decision trees[1].

For many years, many methods have been shown to create ODT but finding best multivariate hyperplane is a NP complete problem. Murthy, Kasif and Salzberg [6] have proposed OC1, a solution based on impurity measures and perturbation algorithm. Other methods proposed by Mangasarian, Setiono and Wolberg [4] or Bennet [1] induce ODT using linear programming but these methods find the optimal linear discriminates by optimising specific goodness measures given by the authors.
Instead of having an analytic approach, geometrical results of linear programming is used and no special measures are needed. Only data set containing two classes can be treated for the moment but an original solution to solve the problem of missing values is proposed. That gives the possibility to work with a medical data set and to compare the performances of this algorithm to C4.5 [7], a standard univariate decision trees builder.

♣ {gmichel,lambert,cremilleux}@info.unicaen.fr

♠ amar@baclesse.fr

[1] Trees using oblique hyperplanes to partition data are called oblique decision trees and noted ODT.

1 Linear programming : geometrical approach

Two principal results of linear programming are essentially used in this work : the Simplex Algorithm and the Duality Theorem. Linear programming is an algebraic tool and so the data set has to be translated into an algebraic form.

Assume, for the moment, that data set contains two classes, that all data are numeric and that there is no missing value.

1.1 Global approach

The data set can be projected in an Euclidean space, noted E, of dimension n (n is the number of characteristics describing each data and one data becomes one point in E) and each class is represented by a cloud of points. The aim of learning, in this case, is to separate two classes (or two clouds of points) : in the case of the *Figure 1*, the hyperplane D divides x-class and y-class and allows to decide in which classes belong new points. As linear programming products linear borderline, dividing two clouds of points or their convex covers is equivalent.

Figure 1 : Duality theorem

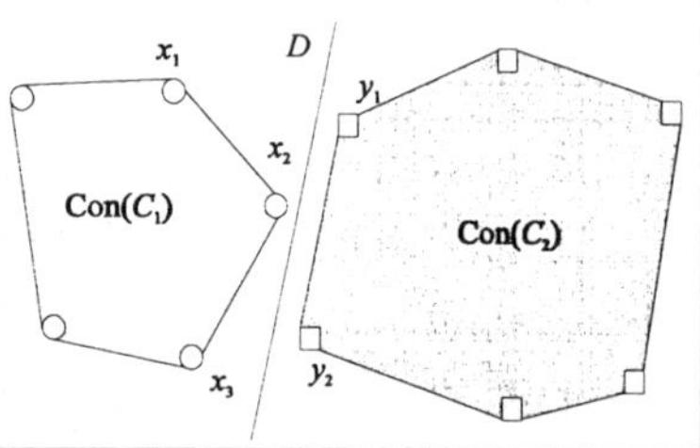

Let be $C_1 = \{ E_1,....E_p\}$ and $C_2 = \{ F_1,....F_q\}$ with E_i in $\mathfrak{R}^n$ for all i in $[1...p]$ Convex cover of C_1 and C_2 is obtained by introducing convex notions (They are noted $Con(C_1)$ and $Con(C_2)$). The intersection of $Con(C_1)$ and Con(C_2) is null (and a linear borderline can be found) if and only if *System 1* has no solution. In the other case, an other result of the linear programming gives a method for having good results.

$$\begin{cases} \sum_{i=1}^{p} \lambda_i E_i - \sum_{j=1}^{q} \mu_j F_j = 0 \\ \sum_{i=1}^{p} \lambda_i = 1 \\ \sum_{j=1}^{q} \mu_{ji} = 1 \end{cases} \quad \text{with} \quad \begin{matrix} \forall i \in [1..p] & \lambda_i \geq 0 \\ \forall j \in [1..q] & \mu_j \geq 0 \end{matrix}$$

System 1 : Intersection of convex covers.

1.2 Tools and methods

The Simplex Algorithm and the Duality Theorem [3] are mainly used in these twice situations.

1.2.1 A borderline exists

$$\exists a \in \mathfrak{R}^n, b \in \mathfrak{R} : \forall i \in [1..p], \forall j \in [1..q] \quad \begin{cases} a' E_i < b \\ a' F_j > b \end{cases}$$

System 2 : System 1 dual

Firstly, studying the case in which a borderline dividing convex covers exists (e.g. the situation look like at the *Figure 1*).

A linear borderline exists (e.g. a vector a) if and only if the *system 2* is verified. By noticing that the *System 2* is the dual of the *System 1* and using the Duality Theorem, following result can be proved : if the *System 1* has no solution, the base of the dual is solution of the *System 2*. This base, easily extractable from the *System 1*, gives, for each dimension, the coefficient of the vector a (and, of course, the equation of the hyperplane D).

1.2.2 No borderline exists

But, most of the time, it not exists hyperplanes dividing classes and the dual has no solution. See the *Figure 2* for having an example of this kind of situation.

Figure 2 : Simplex Algorithm

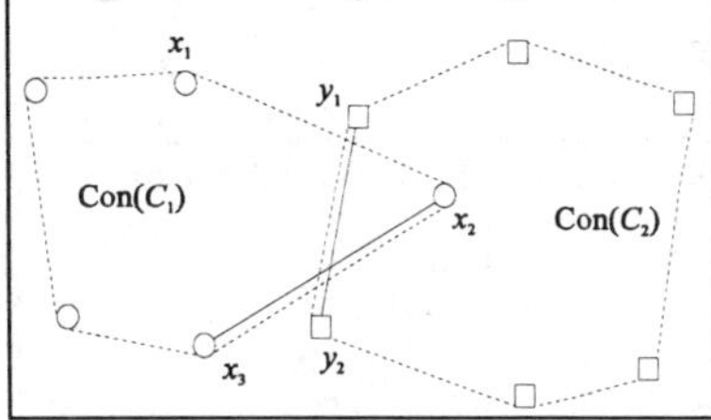

The Simplex Algorithm return a proof of the non existence of hyperplanes by giving a small set of points belonging to the both classes. The *Figure 2* shows easily that : $Con(C_1) \cap Con(C_2) \neq \varnothing$ and one proof can be given by the following points : $\{x_2,x_3,y_1,y_2\}$ (because $[x_2,x_3] \cap [y_1,y_2] \neq \varnothing$). In this case, the idea is to take out one point from this set and to consider new convex covers again.

Logically, after m iterations, two distinct convex covers have to be obtained and results of § 1.2.1 may be applied.

Those twice remarks allow to propose the following algorithm :

1.3 A trivial algorithm

Let C_1 and C_2, two sets of point in $\mathfrak{R}^n$. The procedure used to induce a split that divides two sets of points can be defined : *Algorithm 1*. The function that chooses one point in the intersection's set is very important and not easy to define. For example, in *Figure 2*, it is more interesting to extract x_2 than x_3, y_1 or y_2. For the moment, the choice function is trivial and chooses a point randomly. It is certainly easy to find a better solution and this point should be studied in the future.

Algorithm 1 : Application

```
Build system
While Indivisible do
        Find intersection
        Choose point
        Extract point from system
end while
Extract hyperplane from system
```

2 Missing values treatment

In this representation of data set, there is no place for missing value whereas it is very important to consider this kind of trouble highly present in real word data extracted.

2.1 Linear Programming interpretation of missing values

Algorithms usually propose [2][8] to replace missing values by values extracted from the data set (e.g. median value of the argument) or to have a probabilistic approach [7].

The linear programming approach allows us to apply an other treatment. Instead of *fixing* missing values, they are replaced by variables. Limits are given to those variables to be sure that they have *logical* values. That means an expert has to define a maximum value and a minimum value for all dimensions in which there exists missing values. Sometimes, those values could be extracted from the data base by finding real maximum and minimum if this data base is representative enough.

2.2 Geometric interpretation of missing values

For understanding the approach, it is interesting to have a geometrical interpretation of this operation. In fact, A data having p missing values is replaced by an hypercube of dimension p (e.g. an hypercube with 2^p vertices). Studying, in $\Re^2$, the following case : $A=(?,\alpha)$ and $B=(\beta,?)$ and have a look to the *Figure 3*. Constraints given by the linear programming approach are stronger than those given by the classical approaches and the choice for finding hyperplanes is more limited. It is even possible to have situations in which there exists hyperplanes in the first case whereas the intersection between the two hypercubes is not empty (it can not exist hyperplane in this condition). That proves those approaches are not equivalent. The algorithm given in § 1.3 finds an elegant solution for this trouble because *bad* points are eliminated.

Figure 3 : Missing values treatment

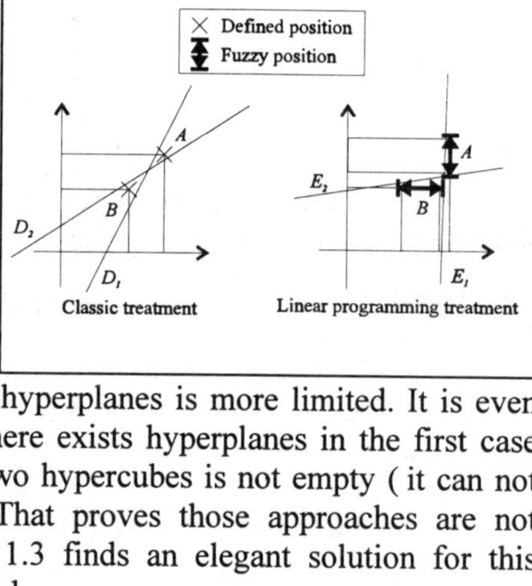

2.3 Generalisation

The algebraic form is generalised for C_1 and C_2, two sets of $\Re^n$. Let be $\Omega=(\Omega_1,...,\Omega_n)$ and $\Theta=(\Theta_1,...,\Theta_n)$, two sets of n sets. For all i in [1..n], let Ω_i (resp. Θ_i) the set of indices of variables of C_1 (resp. C_2) for which the ith component is unknown. For example, r is in Ω_i if and only if x_r is in C_1 and $x_{r,i}$ is a missing value. Let , for all i in [1..n], m_i and M_i as the limits for the ith component of the data base (read in the data base or given by an expert of the domain). The system obtained (*System 3*) is linear and linear programming algorithms may be applied to resolve it.

System 3 : Generalised system

$$\begin{cases} \forall h \in [1..n] \sum_{i=1, i \notin \Omega_h}^{p} \lambda_i E_{i,h} + \sum_{r \in \Omega_h} X_{r,h} - \sum_{k=1, k \notin \Theta_h}^{p} \mu_{kj} F_{k,h} - \sum_{r \in \Theta_K} Y_{r,k} = 0 \\ \forall j \in [1..n] \forall r \in \Omega_j \begin{cases} m_j \lambda_r - X_{r,j} \leq 0 \\ X_{r,j} - M_j \lambda_r \leq 0 \end{cases} \\ \forall i \in [1..p] \lambda_i \geq 0 ; \forall j \in [1..q] \mu_j \geq 0 \\ \forall k \in [1..n] \forall s \in \Theta_k \begin{cases} m_j \mu_s - Y_{s,k} \leq 0 \\ Y_{s,k} - M_j \mu_s \leq 0 \end{cases} \\ \sum_{i=1}^{p} \lambda_i = 1 ; \sum_{j=1}^{q} \mu_j = 1 \end{cases}$$

Notice that if all components are not numeric, it is always possible to transform them. For example, [Small,Medium,Big] becomes [1,2,3] and [Yellow,Blue,Red] is replaced with three binary components : [0,1] for Yellow, [0,1] for Blue and [0,1] for Red. This treatment is important and, at a rough estimate, can represent a negative point of this method because it could not be done automatically by algorithms and knowledge of experts is needed. But, with the view to have a semantic interpretation of attributes, only human users can be efficient.

3 Computational results

Before giving results of experiments that compare the performances of those algorithm to C4.5, the data set used has to be described.

3.1 The Hodgkin's disease

In this paper, reported results are issued from a data set collected by the Lymphoma Cooperative Group of the European Organisation for Research and Treatment of Cancer (EORTC) and provided by Dr. M. Henry-Amar

The data set describes more than 3000 patients treated with various protocols. After treatment[2], the data set currently has 824 entries for the learning data and 701 entries for the test data[3]. The patients, grouped as "Favourable" (369 cases) nor "Unfavourable" (455 cases), are described through 16 continuous attributes and three binaries attributes. The learning data set contains 330 missing values concentrated on five attributes.

3.2 Comments about results form

In this experience, extrema were extracted from the data base as described in § 2.1 and were supervised by Dr. M. Henry-Amar. Hyperplanes are given

---

[2] Different protocols products different descriptors for patients and choices have to be done.

[3] Those data sets are fixed like that for temporal reasons.

under the form explain in *Table 1* : for each attribute of the data base, the value obtained in the dual is the coefficient of the vector *a* described in § 1.2.1.

Table 1 : Results form

Coef. for age	: 0.0093584466	◊	Coef. for sexe	: -0.005338922
Coef. for cbdfus	: 0.0014359370	◊	Coef. for cbgfus	: 0.0007233347
Coef. for axdfus	: 0.0009392872	◊	Coef. for axgfus	: 0.0003019576
Coef. for medfus	: 0.1363806657	◊	Coef. for ext	: 0.0887066056
Coef. for sg	: 0.2534100762	◊	Coef. for vs	: 0.0048398000
Coef. for hb	: -0.001241328	◊	Coef. for gb	: 2.485281e-05
Coef. for polfus	: 0	◊	Coef. for lymfus	: 0
Coef. for monfus	: 0	◊	Coef. for plaq	: 0
Coef. for pa	: -0.000123759	◊	Coef. for ldh	: 7.475156 e-06
Coef. for histforus	: 0.0077649976			
Coef. for lambda_dua	: -0.411007586	◊	Coef. for mu_dua	: 0.4219074193

In the case of the *Table 1*, the vector *a* is :

$$a = (0.0093584466, -0.005338922, \ldots, 7.475156\,e-06, 0.0077649976)$$ and let be

$$b = \frac{0.0077649976 - (-0.4219074193)}{2}$$

A point *x* belongs to “Favourable” if $ax > b$; otherwise, it belongs to “Unfavourable”.

Results are given under a form that allowed a semantic interpretation. For example, the attribute polfus is unneeded to classify because its coefficient is null and the important value of the age’s coefficient means that the youngest are more concerned by the disease than the oldest. Those information are well known from medical experts but others hypotheses are confirmed by this way.

3.3 Results of the linear programming classifier

For having results, just three oblique hyperplanes are extracted (That means that the work space is devised in four regions with three linear borderline) whereas C4.5 use more than 20 rules.

Learning	Class 1	Class 2	Total
Linear Prog	88.1%	89.2%	88.7%
C4.5 (Rules)	90.8%	99.3%	94.3%

The results obtained are under those obtained by C4.5 but they are very interesting. With few hyperplanes (Three), results obtained on the test data base prove that the method described bellow is valid even on a real world extracted data base (C4.5 results show the nontrivial partition of the Hodgkin’s disease data base).

Test	Class 1	Class 2	Total
Linear Prog	85.3%	81%	82.8%
C4.5 (Rules)	87.4%	91.1%	89.6%

Notice that speed of linear programming algorithm, particularly adapted for this kind of situation, allows to obtain results very easily on common computer.

4 Conclusions

This paper has described a linear programming method to construct linear borderline speedily and easily. A new way to exploit linear programming in classification has been presented and an original treatment of missing values has been introduced. The implementation proposed was tested on a consistent data set extracted from the medical domain and interesting results were obtained : good hyperplanes that generalise very well were got.

But better results may be obtained by trying those following ideas : build a complete decision trees with linear borderline, define a clever choice function (§ 1.3), introduce the extracted elements in an other order, generate complete decision trees, use some pruning methods, test the algorithm with cross validation and other data sets, generalise to multi classes decision trees. These extensions are currently explored.

References

[1]. Bennet (1992). Decision tree construction via linear programming, *Computer Sciences Technical report 1067.*

[2]. Celeux (1988). *Le traitement des valeurs manquantes dans le logiciel SICLA.*

[3]. Chvátal (1993). *Linear Programming*, W.H. Freman and Compagny.

[4]. Mangasarian, Setiono & Wolberg (1990). Pattern recognition via linear programming : Theory and application to medical diagnosis, in : *S.I.A.M. Workshop on optimisation.*

[5]. Michie, Spiegelhalter & Taylor (1994). Machine learning, neural and statistical classification. *Ellis Herwood Series in Artificial Intelligence.*

[6]. Murthy, Kasif & Salzberg (1994). A system for induction of oblique decision trees in : *Journal of Artificial Intelligence Research 2,* 1-32.

[7]. Quinlan (1993). *C4.5 : Programs for machine learning.* Morgan Kaufmann.

[8]. Quinlan (1989). Unknown attribute values in induction in Segre. *Proceedings of the sixth International Workshop on Machine Learning,* 164-168.

Classification and Regression Trees Software and New Developments

Francesco Mola
Dip. Matematica e Statistica,
Università di Napoli Federico II
Via Cintia M.Sant'Angelo, I-80126
e-mail: mola@dms.unina.it

Abstract:.The recent interest for tree based methodologies by more and more researchers and users have stimulated an interesting debate about the implemented software and the desired software. A characterisation of the available software suitable for so called *classification and regression trees* methodology will be described. Furthermore, the general properties that an ideal programme in this domain should have, will be defined. This allows to emphasise the peculiar methodological aspects that a general segmentation procedure should achieve.

Keywords: Classification and Regression Trees, Segmentation, Discrimination, Splitting, Pruning, Amalgamation.

1. Introduction

During last ten years the interest about the tree based methodologies in statistics increased considerably. This is not surprising since a tree diagram, together with the information about its construction and node characteristics, provide a powerful summary of information contained in a data set. Typically, most of the known algorithms provide either the prediction of an *a priori* class (*classification trees*) or the expected value of a response variable (*regression trees*) and at the same time, they can also define a tree-structured predictor of the parameter(s) of a distribution, which can be incompletely specified and can contain nuisance parameters (*generalised trees*). For basic information and statistical background see referenced papers and monographs; this paper concentrates especially on the computational side of the problem. In the following we describe briefly some criteria which are, according to our point of view, the most important for the proper choice of the software. Considering these criteria, we analyse the most important programs/packages commercialised or free distributed. We concentrate (in concise manner) especially on the weak and strong sides of typical representatives of available software. In this paper, for the sake of brevity, we report only the main characteristics of each analysed software referring to the literature for methodological and technical details. In particular, we consider a general framework of the desired (or ideal) software. This is basically characterised by main phases that can be sorted in different ways depending on the nature of the variables, the adopted criteria, the desired output and the complementary use of other approaches such as factorial analysis and statistical modelling.

2. Criteria

In this section we present the main criteria for the proper choice of the software suitable for this type of statistical analysis. First group of parameters covers statistical properties, second one basic characteristics related to the programme control and the third one concentrates on characteristics related to its distribution and overall support both from the developers and both the distributors.

In the first group of characteristics, called *basic characteristics related to the methods implemented*, we consider the characteristics related to the statistical aspects of the methodologies and the quality of the outputs. Among the other parameters, we analyse which type of data one can deal with, the way how missing data are treated in the analysis, the types of methodologies implemented, the splitting procedure and splitting rules, the stopping rules and the pruning procedures implemented. Particular attention is paid to the possibility of automatically classify cases of unknown classification and the possibility to follow the same cases along the tree structure defined by the classification rule obtained. Concerning the results, we consider characteristics like the overall quality of the presentation of results, the numerical and graphical presentation of results, the presence or absence of interactive graphics, the possibility to export graphics and results in other environments.

A second important group is called *basic characteristics related to the programme control*. In this group we have enclosed all characteristics related to the informatics aspects, that is the way how to use the software. Here are considered characteristics as the input/output facilities; the ease of the use; the way of control; the error indication; the possibility of recovery after errors and the quality of the on-line help.

The third and last group of characteristics are *basic characteristics related to the distribution and overall support*. In this group we consider, among others, these characteristics:

- the type of operating system(s) the software is available for;
- whether the package is a stand alone package, part of the larger project or part of some general purpose software;
- whether it is a supported software;
- whether it is a software under continuous development, improvements and innovation;
- whether the possibility of user configurability is available.

3. Analysed software

Currently, more than ten commercial programs/statistical systems and several shareware programs offer (up to some extend) the methodology of classification and regression trees. Among them, the following are (from different reasons) of special interest, i.e. ID3, AnswerTree, C4.5, CART, CHAID, FIRM, RECPAM, SICLA, S-Plus, Spad.S, SYSTAT, TWO-STAGE and UNAIDED. Some of these packages (namely ID3, CART, S-Plus, SPAD.S, and Systat) refer to the well known and widely used binary segmentation methodologies, while other packages refer to some not so widely used binary segmentation procedures introduced in the literature by several authors during the last years.

ID3 software performs the AID methodology (Morgan and Sonquist, 1963). AID attempts to fill a gap overlooked by regression, discriminant analysis and logistic regression through identifying important interactions among the predictors. AID also provides a visual display of results in tree diagram form, where file segments are represented by the terminal tree branches and are accompanied by the associated response rates. While the AID Tree Analysis technique has been shown to generally discriminate between good and bad respondents and interactions, it has also been shown to contain a considerable amount of errors. Limited to dichotomous splits, the AID technique's primary inherent problem is a bias toward the use of variables containing many categories over variables with few, or just two, levels. AID does not properly adjust for the likelihood of finding a dichotomization by chance due to sampling variability and, therefore, it is biased in favour of variables having many categories because there are more ways to dichotomise such variables. AID segments tend not to validate well in rollouts-response rates, because the best responding segments tend to be lower (sometimes substantially) than predicted.

After the module **CHAID** implemented in SPSS since four years, **SPSS** released recently more algorithms, i.e. CHAID, Exhaustive CHAID, QUEST and CART (most of them are characterised elsewhere in this paper) and offers them under the name **AnswerTree**. Important feature of this module is that AnswerTree saves the segment identifier as a new variable for the use in subsequent analyses using SPSS. Moreover, you can easily extract data from a database for selected segments to generate a mailing list or new data set using the model's decision rule(s).

The program **C4.5** is one of a series of programs written by Quinlan and his colleagues in the field of decision trees. The basic idea is to generate all possible decision trees that correctly classify the training set and then to select the simplest of them. The program is designed for the situation when there are many attributes and the training set contains many objects, but where a reasonably good decision tree is required without much of computation. For the evaluation of trees and their validation, the authors do not use classical statistical approaches (as do most of the other programs and methodologies mentioned in this paper), but the measures from the field of information theory are adopted.

CART is a software that implements the binary segmentation procedure proposed by Breiman *et al.* (Breiman *et al*, 1984). As known, CART is one of the most important methodologies for binary segmentation. Two basic ideas characterise CART methodology: the definition of an impurity function at each node (and consequently the definition of the splitting criterion based on an impurity measure), and the pruning process to define honest size trees. Since this is probably the best known software in this domain, we leave place for less known packages.

Developed by Kass in late seventies, **CHAID** represents a statistically appropriate approach to the tree analysis. It is a chi-squared technique and requires all variables to be categorical. Like AID, CHAID is multivariable rather than a multivariate technique, and thus does not require that a full multiway table be formed during the analysis. CHAID is well suited as:

1. A replacement for AID in identifying those segments that differ in response.
2. A variable selector to screen out extraneous predictor variables.
3. A category collapsing technique that reduces the number of cells in a full multiway table.

Unlike AID, CHAID is not limited to dichotomous splits.

FIRM, developed by D. M. Hawkins in late eighties is parallel especially to the CART and the AID, however, it differs from them in several aspects, notably in varying the number of descendant nodes into which different nodes are split, and of using conservative formal (Neyman-Person) statistical inference for determining when to terminate analysis at each node. Further differences include:

1. A facility for handling *predictive missingness*, where the fact of a predictor being missing conveys some predictive information about the dependent variable.
2. The predictors are either on the nominal or the ordinal scale.
3. The dependent variable may be either on the categorical or on the interval scale of measurement.

RECPAM is essentially a method of growing generalized trees on the basis of generalized information measures, which add to the traditional RECursive partitioning steps of tree-growing an AMalgamation step; this step results in the identification of groups both homogeneous and distinct with respect to the parameters to predict. RECPAM is therefore a family of algorithms (and programs as well). Perhaps more precisely, it is a research programme which aims at applying the principles underlying tree-growing to a variety of situations occurring in data analysis, in which the information to be presented in a tree is not as simple as in the classical CART algorithm. Recent developments are RECPAM-GLIM and RECPAM-COX, which are the methods for treating, respectively, data with a response variables that can be modelled by a distribution of the exponential family, and data containing a censored survival times.

At the moment there does not exists a module suitable for the recursive modelling in **SAS**. However, recently SAS Inc. asked users of its system for the opinion whether and when to include such a module. Therefore, it seems highly probable that the methodology of *Classification and Regression Trees* type will appear soon within the SAS as well.

SICLA (Systeme Interactive pour CLAssification) is a programme oriented to the clustering and classification. In the SICLA software, the classification trees procedure **DNP** (Decoupage Non Parametrique) performs a classification tree analysis according to the methodology proposed by Celeux and Lechevallier in the 1982 and based on the Kolmogorov-Smirnov distance. Recent development in SICLA concerns the pruning process in binary segmentation; in fact a procedure (called *bonsai*) enabling to search for optimal trees was implemented.

SPAD.S (SPAD Segmentation) is an optional module of the package SPAD.N. The tree-based methodology implemented in SPAD.S is basically the CART methodology of Breiman *et al.* (1984) with several improvements suggested by J. P. Nakache. SPAD.N (and obviously also SPAD.S) is implemented in the MS Windows environment; this allows the users to use all the input/output facilities offered by the MS Windows. SPAD.S offers very detailed outputs as concern both the splitting and validation phases.

In the **S+** package, mixture of CART and RECPAM methodologies is implemented. Some interesting enhancements are relative to the possibility to choose among different pruning procedures. The procedure, however, is not well documented in either the manuals or books describing the use of S+.

The **TREE** module of **SYSTAT** v. 8.0 computes classification and regression trees according the approach suggested by Breiman *et al.* TREE procedure produces numerical and graphical (called mobiles) results. Algorithm used is that from Breiman *et al.* for the splitting computations. Unfortunately, neither pruning nor amalgamation step is implemented, so that the practical usage is of limited value. Missing data are eliminated from the calculation of the loss function for each split separately.

The binary segmentation methodology proposed by Mola and Siciliano in 1992 is implemented in the shareware software called **TWO-STAGE**. This software (running in MS Windows environment) performs the classical CART procedure with the fast splitting algorithm (Mola and Siciliano, 1997). As stopping rule the CATANOVA test is considered (see Mola and Siciliano, 1994).

The **UNAIDED** software developed by Capiluppi et al. (1997) is a software for binary and not binary segmentation according the approach suggested by Morgan and Sonquist (1963). All the facilities offered by the MS Windows environment are available. The possibility of considering the criterion variable on any measurement scale is available, too. Aside that, so called *look-ahead analysis* is available.

4. How should like look an ideal programme system for classification and regression trees methodology

Taking into account the above considerations about the existing software, let us try to imagine an ideal software for the classification and regression trees methodologies. We are involved in a project with the aim of realising such a software.

As first point let us take a look to some computational aspects of the classification and regression trees procedure. It is well known that many steps of the classification and regression trees programs are the same for all proposed procedures, independently of the splitting and pruning algorithm adopted. Let us consider at first the splitting algorithms. To split a node into the left and right offsprings, there are some computational efforts that cannot be "bypassed". We refer in particular to the seek, at each node, of the set of dichotomous (in case of binary segmentation) questions (splitting variables) generated by each predictor, the cross-classification of predictors and response variable, the class assignment, the description of the terminal and non terminal nodes, the graphical visualisation of results and so on. In other words, we can see it is possible to reduce the splitting algorithm into small phases, making it possible to offer the user with a general tool to grow binary and not binary trees by choosing not only different splitting rules (something already existing, see e.g. Steinberg and Colla, 1984, Capiluppi et al., 1997, AnswerTree, 1997) but also different splitting criteria. In other words, the researcher could have available some informatic tools and the possibility to paste them building up the desired classification and regression trees splitting procedure. We think that this is a very crucial point, since the possibility to analyse data sets with different splitting procedure is very important because the comparison of results of different splitting criteria is important for the researcher (some discrepancy in the results could reveal important aspects of the data). Another important aspect, related with this point, is to give the possibility for the researcher to implement a self-made splitting procedure suitable for particular data structure.

Concerning the choice of the best tree (either using the pruning approach or pruning and amalgamation approach, either using statistical tests or user defined thresholds), many phases are common for different methodologies.

Another important point is the possibility of analyse data sets with different statistical methodologies at each step of the growing tree process, i.e. descriptive statistics (some package do it), visualisation of dependency by factorial methods (Mola and Siciliano, 1997) and statistical modelling (Mola, Klaschka and Siciliano, 1996). This means to have the possibility to consider each node of the tree as a sample to be investigated. When we are dealing with exploratory trees, this point is very important.

5. Conclusions and remarks

Classification and regression trees is present as methodology in the most important general purpose statistical software. This is an important result if we think that the methodological developments of this kind of methodologies are recent. The efforts of many researchers in developing new and alternative algorithms give evidence of the strong interest about segmentation, either for splitting process (Steinberg, 1996; Capiluppi *et al.*, 1997; Klaschka and Mola, 1998; Klaschka *et al.*, 1998) either for pruning process (Cappelli *et al*, 1998).
On the other hand, we also think that tree based methods cannot strongly be enveloped in general purpose system; researchers who like investigate data needs for *flexible tools* and not *black boxes*. It is also true that only commercialised software, for economical reasons, can guarantee the development of software for this domain of research. At the end a question still remain opened: "To perform classification and regression trees, is it better a specialised software or a procedure inserted in a general purpose software?"

Acknowledgements: The work has been partially supported by MURST 40%. Author would like to thank to Prof. J. Antoch for thorough discussions which improved considerably this paper.

References

AnswerTree (1997). *SPSS-AnswerTree*, SPSS Inc., Chicago, IL,, USA.

Breiman, L., Friedman, J.H., Olshen, R.A: & Stone, C.J. (1984). Classification and Regression Trees, Wadsworth, Belmont, CA.

Capiluppi C., Fabbris L. & Scarabello M. (1997). UNAIDED: a PC system for Binary and Ternary Segmentation analysis, *Book of Short Papers of Classification and Data Analysis*, Pescara.

Cappelli, C., Mola, F. & Siciliano, R. (1998). An alternative pruning method based on impurity, In *Proceedings of COMPSTAT'98*, Bristol August (to appear)

CART (1984). *CART: A software for classification and regression trees*, California Statistical Software Inc., Yorkshire Ct. Lafayette, CA, USA.

CHAID (1996). *SPSS-CHAID*, SPSS Inc., Chicago, IL, USA.

Ciampi, A. (1994). Classification and discrimination: The RECPAM approach, In *COMPSTAT'94*, Dutter R. & Grossmann W. eds., 139-147, Physica Verlag, Heidelberg.

Clark, L. A. & Pregibon, D. (1992). Tree based models, In *Statistical Models in S*, Chambers J.M. & Hastie T.J. eds., Wadswoorth and Brooks/Cole.

FIRM (1990). *FIRM: Formal inference-based recursive modelling by D.M. Hawkins (for IBM PC)*, University of Minnesota, School of Statistics, Technical Report 546.

Hörmann, A. *et al.* (1990). Comparing statistical analysis systems, *Statistical Software Newsletter*, 16, 90-127.

Klaschka, J. & Mola, F. (1998). Minimization of computational cost in tree-based methods by a proper ordering of splits, In *Proceedings of COMPSTAT'98*, Bristol August (to appear).

Klaschka, J., Siciliano, R. & Antoch, J. (1998). Computational enhancements in Tree-Growing Methods, in *Proceedings of IFCS'98,* Rome (to appear).

Kass, G.V. (1980). An exploratory technique for investigating large quantities of categorical data, Applied Statistics, 29, 119-127.

Mola, F. & Siciliano, R. (1992). A two-stage predictive splitting algorithm in binary segmentation, In *COMPSTAT'92*, Y. Dodge & J. Whittaker eds., 179-184, Physica Verlag, Heidelberg.

Mola, F. & Siciliano, R. (1994). Alternative strategies and CATANOVA testing in TWO-STAGE binary segmentation, *New Approaches in Classification and Data Analysis*, Diday E. et al. eds., 316-323, Springer Verlag.

Mola, F. & Siciliano, R. (1997). A fast splitting procedure for classification trees, Statistics and Computing, **7**, 209-216.

Mola, F. & Siciliano, R. (1997). Visualizing data in tree-structured classification, In *Proceedings of the IFCS-96: Data Science, Classification and Related Methods,* Hayashi C. *et al.* eds., Springer Verlag, Tokyo.

Mola, F., Klaschka, J. & Siciliano, R. (1996). Logistic classification trees, In *COMPSTAT '96 Proceedings,* Prat A. ed., Physica Verlag, Heidelberg.

Morgan, J.N. & Sonquist, J.A. (1963). Problems in the analysis of survey data and a proposal, *JASA*, 58, 415-434.

Quinlan, J. R. (1992). *C4.5: Programs for machine learning*, Morgan Kaufmann, New York.

Safavian, S. R. & Landgrebe, D. (1991). A survey of decision tree classifier methodology. *IEEE Trans. on Systems, Man. and Cybernetics*, 21, 660-674.

Siciliano, R. & Mola, F. (1997). Multivariate data analysis and modeling through classification and regression trees, *Proceedings of the Second World Conference of IASC*, Wegman E. ed., Pasadena CA (in press).

Siciliano, R. & Mola, F. (1997). Ternary classification trees: a factorial approach, In *Visualization of Categorical Data*, Greenacre M. & Blasius J. Eds, Academic Press.

SICLA (1991), Manuel de l'utilisateur, INRIA, Le Chesnay Cedex, France.

S-PLUS (1997). *S-PLUS Computing Environment v.4.0*, StatSci, a division of MathSoft Inc., Seattle, WA, USA.

SPAD.S (1996). *Systeme Portable pour l'Analyse des Donnes-Segmentation, v.3.2*. CISIA, Saint Mande.

Steinberg, D., & Colla, P. (1995). CART. Salford System, San Diego, CA.

Steinberg,, D. (1996). New developments in CART (Classification & Regression Tree) software, In *Proceedings of the Fifth Conference of the IFCS*, Kobe, Japan.

Venables, W. N. & Ripley, B. D. (1994). *Modern Applied Statistics with S-Plus*, Springer-Verlag, Berlin.

STABLE: a visual environment for the study of multivariate data

Roger W. Payne and Simon A. Harding
Statistics Department, IACR-Rothamsted,
Harpenden, Herts AL5 2JQ, UK.

Abstract: STABLE is a European Union funded project that provides a completely new approach to statistical programming, analysis and visualisation. The system combines the statistical algorithms from the Genstat statistical system with the sophisticated visualisation and flexible visual programming environment of the IRIS Explorer system to provide a unique environment for exploring statistical data. The system will provide a stimulating environment for programming and investigating new multivariate techniques, and then powerful techniques for making them available in encapsulated IRIS Explorer *modules* with suitably customized interfaces.

Key words: data visualisation, distributed processing, interactive graphics, multivariate graphics, statistical analysis, visual programming.

1. Introduction

Visual inspection of data and results is an important part of any statistical analysis, particularly with multivariate data. Despite this need for sophisticated graphics, however, most statistical software offers little beyond static 2-dimensional displays. Furthermore, the statistical systems that do contain facilities for interactive or 3-dimensional graphics generally provide these in pre-programmed "canned" forms that cannot be modified by the end-user. (See Gower *et al.* 1997).

In contrast, systems are available for engineers and designers that offer considerable freedom in the ways in which displays can be assembled and viewed, and the way in which feedback can be incorporated from one display to another. Some also encourage flexibility by use of the visual programming paradigm, instead of the command language approach favoured by statistical software. One good example is the IRIS Explorer data visualisation and application building system, which is providing one of the basic components for STABLE.

The other main source of material for STABLE is the Genstat statistical system (Payne *et al.* 1993). This covers all the standard statistical techniques such as analysis of variance, regression and generalized linear models, curve fitting, multivariate and cluster analysis, estimation of variance components, analysis of unbalanced mixed models, tabulation and time series. It also offers many advanced techniques, including generalized additive models and the analysis of repeated measurements. Additional numerical and statistical components are being supplied from the NAG Fortran Library (NAG 1997).

Figure 1: *STABLE map for a principle coordinates analysis*

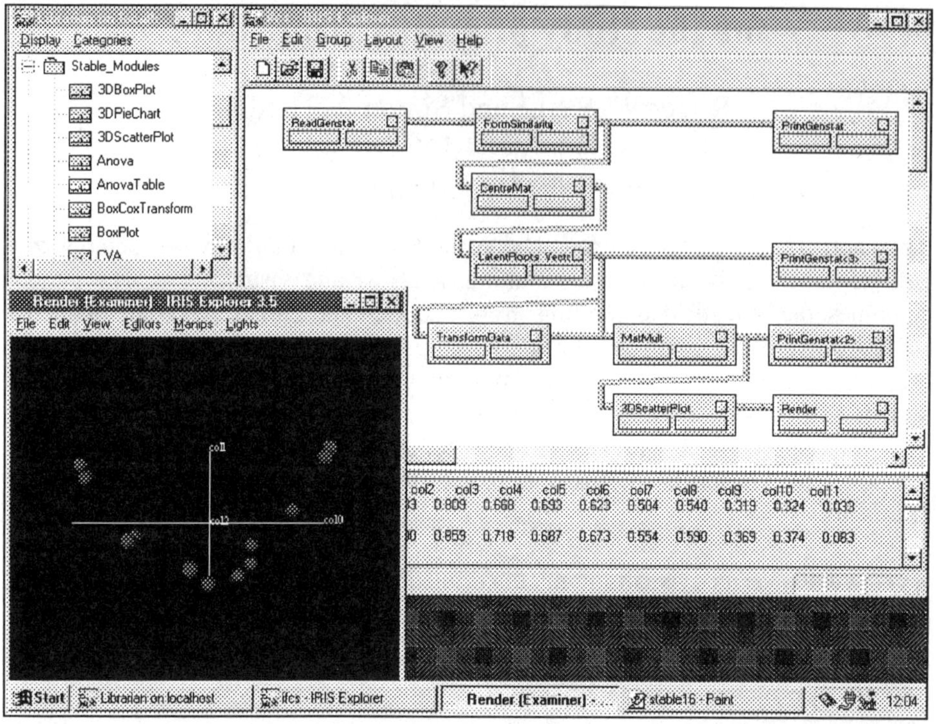

STABLE integrates the Genstat algorithms with the IRIS Explorer system to produce a "Statistical Explorer" system combining the statistical power and reliability of Genstat with the flexibility, power and extendability of IRIS Explorer. The visualisation facilities in IRIS Explorer will encourage the development of new visualisation methods in statistics. Its visual programming environment will enable non-programmers to develop their own statistical analyses in a user-friendly way. The ability of IRIS Explorer to define new interfaces, will allow specific applications to be tailored to meet the needs of non-specialist end-users. IRIS Explorer also facilitates links to other systems such as databases, graphical information systems, symbolic manipulation packages, wordprocessing and desk-top publishing systems, and to users' own source code. Furthermore it supports distributed heterogeneous operation, allowing graphical workstations, supercomputers and larger PCs to be used collectively in the solution of problems. Thus, it will provide a uniquely powerful environment, in particular, for research, development and promulgation of new ideas in multivariate graphics and display.

2. Example

Figure 1 illustrates visual programming by presenting the individual algorithmic components of the familiar Principal Coordinates Analysis. The right-hand side of the figure shows the *map*, or "visual programme", that does the analysis. The grey boxes, or *modules*, are the equivalent of commands in ordinary command-driven systems and the lines, or *pipes*, show the flow of data through the map.

The window in the top left of Figure 1 shows the IRIS Explorer *Librarian*, which contains the available modules, which can be dragged and dropped across to the right-hand window to form the map. The statistical modules in STABLE include all the (Fortran) statistical algorithms in Genstat. The full Genstat system can also be run, as a module that executes a script of Genstat commands. This guarantees upwards compatibility and, in particular, allows users to continue to use their existing libraries of Genstat procedures. Genstat has been re-engineered so that the STABLE modules and Genstat both use the same DLLs, to ensure compatibility of results and efficiency of operation. Furthermore, these DLLs will also be available for use in other systems. Each module has two rectangular buttons that provide access to menus that allow data to be input to the module (the left button) or taken as output to another module (the right button). Once an argument has been selected, buttons on other modules that can supply data of the required type (for input to the module) or receive data of the generated type (for output from the module) are highlighted in green, to facilitate correct inter-connection.

Figure 2: *Menu for FormSimilarity*

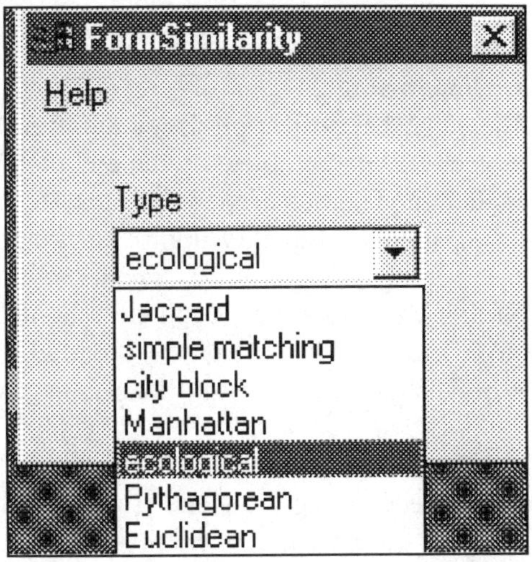

In Figure 1, the analysis starts with a ReadGenstat module which reads data from an ascii file. The top right-hand button on a module opens the control panel (or menu) associated with the module. For ReadGenstat this specifies the file to be read and provides various options controlling how its contents are interpreted, while for FormSimilarity it selects the way in which the similarity matrix is calculated (Figure 2). STABLE supports a wide range of widgets to construct the menus, including list boxes, sliders, radio buttons, check boxes, dials etc, allowing custom menus to be formed appropriate to the particular end-user. Next, the PCO is calculated. First of all the similarity matrix is centred (CentreMat), then an eigenvalue decomposition is done (LatentRoots). From the LatentRoots module, the eigenvalues are taken for printing by a PrintGenstat module and also to the TransformData module for a square-root transformation to be applied. The latent vectors are taken to the MatMult module to be transposed and then right-multiplied by the square root of the eigenvalues to form the scores. The scores are printed by a PrintGenstat module, and plotted by a 3DScatterPlot module which produces an IRIS Explorer *geometry* structure to be displayed by the Render module in the bottom right corner of the map.

The display generated by Render, shown in the bottom left-hand side of Figure 1, is an interactive 3-D plot which can be manipulated by changing viewpoints, lighting, colouring and so forth. Information can be extracted from the plot using what is known as a *pick* operation, and used as input to other modules, thus allowing the user's interpretation of the plot to modify other aspects of the analysis and display.

The system thus allows sophisticated systems to be built, at run time, for the interactive visual analysis of multivariate data. For example, in a biplot analysis, the user could interactively select variables to be displayed or highlighted on the graph, or to be used in back-predictions. (Gower *et al.* 1997). The system also facilitates intensive compuation. The pipes can go across the internet, so that different modules are running on different machines. IRIS Explorer runs on a wide range of machines, including UNIX Workstations, Crays and PCs running Windows NT. Once the map has been constructed, the details can be encapsulated in a *grouped module*. This appears like an ordinary module but in fact contains a map of individual modules (which may themselves be grouped modules). It is possible to specify which of the controls of the modules inside the group are made accessible to the outside user, allowing them appropriate control of the analysis while hiding unnecessary complications. Alternatively, the map can be stored to run again on a later occasion, by the developer or an end-user.

3. Conclusion

The STABLE project produced a first prototype at the end of 1997, and the whole project will be complete by the end of 1998. Also, during 1998, three diverse and challenging demonstration systems are being developed, for forecasting of time series, industrial quality improvement, and design and analysis of experiments.

Further information is available about STABLE from the project coordinators NAG Ltd, Wilkinson House, Jordan Hill Road, Oxford OX2 8DR, UK or on the World Wide Web at: http://extweb.nag.co.uk/projects/STABLE.html

Acknowledgements

IACR receives grant-aided support from the Biotechnology and Biological Sciences Research Council of the United Kingdom. STABLE is funded by EU ESPRIT4 Project 22832.

References

Gower, J.C., Hand, D.J. and Ngouenet, R.F. (1997). The development of interactive graphical software for a unified biplot methodology. Proceedings of the 2nd World Conference of the IASC, Pasadena.

NAG (1997) *NAG Fortran Library Mark 18 Manual*. NAG Ltd, Oxford.

Payne, R.W., Lane, P.W., Digby, P.G.N., Harding, S.A., Leech, P.K., Morgan, G.W., Todd, A.D., Thompson. R., Tunnicliffe Wilson, G., Welham, S.J. & White, R.P. (1993). *Genstat 5 Reference Manual, Release 3.* Oxford University Press, Oxford.

PART III

Factorial Methods

- **Factorial Methods for Multivariate Data**

Some recent Developments in Factor Analysis and the Search for proper Communalities

Jos M.F. ten Berge.
Heijmans Institute, University of Groningen, Grote Kruisstraat 2/1, 9712 TS Groningen, The Netherlands

The author is obliged to Henk Kiers, Anne Boomsma and Ivo Molenaar for commenting on a previous draft of this paper.

Abstract: This paper describes recent results on factor analytic theory which involve the so-called Ledermann bound, and discusses theoretical properties of Minimum Rank Factor Analysis as an alternative to Iterative Principal Factor Analysis and exploratory Maximum Likelihood Factor Analysis. In terms of the residual eigenvalues, the three methods have closely related object functions, and will often give highly similar solutions in practice. Nevertheless, there are important differences between the methods. The most notable points are that Maximum Likelihood Factor Analysis is a method of fitting a statistical model, and that Minimum Rank Factor Analysis yields communalities in the classical sense, thus showing how much of the common variance is explained with any given number of factors.

Key words: Factor analysis, proper solutions, Ledermann bound.

1. Introduction

Factor analysis is based on the notion that, given a set of standardized variables $z_1,\ldots,z_m$, each variable z_j can be decomposed into a common part c_j, giving rise to correlation between z_j and z_k, $j\neq k$, and a unique part u_j, assumed to be uncorrelated with any variable except z_j, $j=1,\ldots,m$. Upon writing

$$z_j = c_j + u_j, \tag{1}$$

$j=1,\ldots,m$, and using the assumption on u_j, we have

$$\Sigma=\Sigma_c+\Psi, \tag{2}$$

where Σ is the correlation matrix of the variables, Ψ is the diagonal matrix of unique variances, and Σ_c is the variance-covariance matrix of the common parts c_j, $j=1,\ldots,m$, of the variables. The variances of these common parts are in the diagonal of Σ_c. They are the so-called communalities of the variables.

The ideal of factor analysis is to find a decomposition (2) with Σ_c of low rank r, which can be factored as Σ_c=**FF**′, with **F** an $m \times r$ matrix. In practice, we have to settle for an approximation to the ideal situation, and decompose Σ, for some small value of r, as

$$\Sigma=\mathbf{FF}'+(\Sigma_c-\mathbf{FF}')+\Psi. \tag{3}$$

This means that the variances of the variables (diagonal elements of Σ) are decomposed in explained common variances (diagonal elements of **FF**′), unexplained common variances (diagonal elements of Σ_c-**FF**′), and unique variances (diagonal elements of Ψ). A proper solution for (3) requires that both Σ_c and Ψ are positive (semi)definite. Negative elements in Ψ, known as Heywood cases, have drawn a lot of attention. However, when Σ_c, the covariance matrix for the common parts of the variables, would appear to be indefinite, that would be no less embarrassing than having a negative unique variance in Ψ. The decomposition (3) is particularly relevant for "large sample methods" of factor analysis, where Σ is estimated by the sample correlation matrix. Distinguishing between the population matrix Σ and the sample correlation (or covariance) matrix **S**, say, will be deferred till Section 4.
The unexplained common variance in (3) will be zero only if r equals the rank of Σ_c. Accordingly, the (classical) question is to what extent the unique variances in Ψ, or, equivalently, the communalities in the diagonal of Σ-Ψ, can be chosen to reduce the rank of Σ_c. The answer has an intriguing history, revolving around Ledermann's bound, defined as

$$\varphi(m)=[2m+1-\sqrt{(8m+1)}]/2\ . \tag{4}$$

Ledermann (1937) proposed that the rank of (Σ-Ψ) could always be reduced to this bound. In a classic paper on the history of factor analysis, Mulaik (1986, p. 25) did mention Ledermann's bound, but left its alleged merits unquestioned. Some of these merits have not been known until recently. The first purpose of the present paper is to familiarize the reader with these results.
From electronic surveys of recent publications (e.g., see the PsychLITS 1991-1997 of the American Psychological Association), it is clear that, in practice, a vast majority of researchers adopt Principal Component Analysis when Exploratory Factor Analysis could also be used, and among those who do use Exploratory Factor Analysis, Iterative Principal Factor Analysis (Harman & Jones, 1966) and Exploratory Maximum Likelihood Factor Analysis (Jöreskog, 1967) are about equally popular. The second purpose of this paper is to discuss relationships between these two methods of factor analysis, and compare them to a third method, Minimum Rank Factor Analysis (Ten Berge & Kiers, 1991). The latter method has interesting properties that seem to have gone unnoticed. A broader exposition of these properties, in comparison with the two familiar

methods of factor analysis, seems warranted. Principal Components Analysis will not be considered here, because its purpose (condensation of variance) is not the purpose of factor analysis, see Widaman (1993).

2. The Ledermann bound

Ledermann (1937) compared numbers of equations and unknowns, and argued from this that the function $\varphi(m)$ of the number of variables, see (4), should be seen as an upper bound to r, the number of factors needed to solve (2). This optimistic view was shattered implicitly by counterexamples of Wilson and Worcester (1939), and explicitly by Guttman (1958). Guttman showed that the universal upper bound to r is as high as m-1, also see Bekker & De Leeuw (1987). Ledermann's bound seemed history, albeit that the bound still plays a role in the number of degrees of freedom for the chi-square test of the factor analysis model: That number is positive, if and only if the number of factors is below the bound (e.g., Jöreskog, 1967, [40]).

A revival of Ledermann's bound took place in the eighties. Shapiro (1982) showed that Ledermann's bound is almost surely a lower (rather than an upper) bound to the number of factors needed in factor analysis. Shapiro (1982) also proved that the minimum rank is unstable below the Ledermann bound. This means that, when the rank of Σ can be reduced to a value below the Ledermann bound, any small change of any off-diagonal element of Σ will increase the minimum rank of Σ.

Furthermore, Shapiro (1985) has shown that a Ψ that solves (2) properly, which means that Σ-Ψ is a positive semidefinite matrix of rank r, is almost surely non-unique (meaning that other proper choices of Ψ exist) when r is above the Ledermann bound, and almost surely locally unique at and below the Ledermann bound. Local uniqueness means that in a neighbourhood of any Ψ that satisfies (2) no other solutions exist. It is necessary but not sufficient for global uniqueness. Shapiro (1985) also conjectured, as a generalization of a result by Anderson and Rubin (1956, Theorem 5.1), also see Kano (1990, p. 281), that Ψ is almost surely globally unique when r is below the Ledermann bound. This conjecture has been proven correct (Bekker & Ten Berge, 1997).

The results on Ledermann's bound are important in that they provide answers to certain long-standing questions. However, from a practical point of view, the results are not what one might have hoped. Uniqueness of Ψ usually holds for cases of perfect fit, as defined by (2), below the Ledermann bound, but such cases do not arise in practice. We shall have to resort to solutions for (3), for which the uniqueness results mentioned above are not valid. A potentially useful method for obtaining a meaningful solution to (3) is MRFA.

3. Minimum Rank Factor Analysis

MRFA (Ten Berge & Kiers, 1991) is an iterative method which seeks those unique variances (diagonal elements of Ψ), that minimize the sum of the m-r smallest eigenvalues of Σ-Ψ, written as

$$f_1(\Psi)= \sum_{j=r+1,\dots,m} \lambda_j(\Sigma-\Psi), \tag{5}$$

subject to the constraint that Σ-Ψ has nonnegative eigenvalues $\lambda_1 \geq \lambda_2 \geq \dots \geq \lambda_m \geq 0$ and that Ψ has nonnegative diagonal elements. The smallest value of r for which f_1 attains the minimum of zero is the minimum rank of Σ, see Ten Berge & Kiers (1991). Incidentally, when the minimum rank of Σ should be less than $m/2$, the solution for Ψ is almost surely unique (because r is below the Ledermann bound), but, more importantly, it can be obtained in closed form, see Albert (1944) and Ihara & Kano (1986).

MRFA can be used to solve (2) subject to its constraints when the number of factors r is allowed to be large enough but such numbers of factors would be too large to be of practical interest. For smaller values of r, MRFA gives a rank r fit to the common parts $c_1,\dots,c_m$, see (1). This property can easily be derived from the well-known Eckart-Young theorem. Due to the nonnegativity constraints, MRFA is protected against improper solutions: The resulting Σ_c is positive semidefinite and Ψ is typically, but not always, positive definite (it may be singular).

The practical interpretation of MRFA is that it offers a proper solution to (3) by minimizing the common variance that is left unexplained when as few as r factors are used, see (5). This is precisely what factor analysis is about. However, MRFA does not just minimize the unexplained common variance: It also reveals how small it is. This is possible because MRFA preserves the distinction between communalities and explained common variances: The former are the diagonal elements of $\Sigma_c = \Sigma - \Psi$, and the latter are the row sums of squares of **F**. In terms of eigenvalues of Σ_c, the total common variance is $\lambda_{1+\dots+}\lambda_m$ and the unexplained part of this is given by (5), whence the percentage of explained common variance is $100\times(\lambda_1+\dots+\lambda_r)/(\lambda_1+\dots+\lambda_m)$.

To see how convenient MRFA can be, consider the analysis by Iterative Principal Factor Analysis (to be discussed below) of nine intelligence tests by Carroll (1993, Table 3.2). Carroll (1993, pp. 96-98) reports that a four factor solution is sufficient. When MRFA is applied with r=3 and r=4, we find apparently unique solutions for Ψ, with percentages of explained common variance of 93.16 and 99.10, respectively. This is very clear evidence that four factors are indeed enough. In terms of the resulting factor solutions, the MRFA loadings for r=4 are very similar to those reported by Carroll. The essential difference is that MRFA yields, as a bonus, the percentage of explained common variance associated with any number of factors.

The iterative algorithm for MRFA is started from $m+1$ different rational starts for Ψ. In practice, the algorithm has never failed to find the global minimum when that is known to be zero (Ten Berge & Kiers, 1991). A study on the sample size, needed to infer population factors from MRFA factors in a sample, is still in progress. Results so far indicate that samples of n=200 are more than enough.

4. Iterative principal factor and maximum likelihood factor analysis

The object function of MRFA is similar to that of Iterative Principal Factor Analysis (IPFA), first suggested by Thomson (1934), as implemented by Harman & Jones (1966). This method, also known as the Unweighted Least Squares option of LISREL (Jöreskog & Sörbom, 1996), is aimed at minimizing, for any given r, the function

$$f_2(\mathbf{F},\Psi)=SS\{\Sigma-\mathbf{FF}'-\Psi\} \tag{6}$$

where SS means sum of squares, **F** is an $m\times r$ matrix of factor loadings, and Ψ a diagonal matrix of unique variances. Harman and Jones (1966) suggested an iterative approach, updating **F** on the basis of the first r eigenvectors and eigenvalues of Σ-Ψ, and updating Ψ as Diag[Σ-$\mathbf{FF}'$]. The minimum obtained is also the minimum of the sum of squares of the last m-r eigenvalues of Σ-Ψ. This means that the covariance matrix Σ_c of the common parts of the variables is made as close as possible to a rank r matrix, rather than the matrix of these common parts themselves, as in MRFA. Still, the two minimization criteria of PFA and MRFA are quite similar. The most important difference is that MRFA minimizes the "size" of the smallest eigenvalues of Σ-Ψ subject to nonnegativity constraints, whereas IPFA minimizes their size without a constraint. Harman (1967, p. 195) has pointed to the important property that the last m-r eigenvalues in IPFA sum to zero, which means that they are either zero (perfect fit) or at least one of them is negative. It follows that IPFA yields an improper solution for (3) in every case of less than perfect fit.

As a computational alternative to IPFA, Harman and Jones (1966) have also proposed the MINRES method. This method is aimed at minimizing (6), expressed as a function of F alone, viz.

$$f_3(\mathbf{F})=SS\{\Sigma\text{-diag}(\Sigma)\text{-}\mathbf{FF}'\text{+diag}(\mathbf{FF}'\}. \tag{7}$$

To minimize f_3, they suggested updating the rows of **F** iteratively by multiple regression. When differences between IPFA and MINRES solutions arise, this is due to early termination of the iterative process, or to differences in the

suppression of Heywood cases. It is easy to adapt MINRES so as to prevent Heywood cases, but in IPFA it is not at all obvious how to do this without disturbing the monotonical convergence of the algorithm.
So far we have not distinguished between the correlation matrix in a sample and in the population. The reason is that both MRFA and IPFA are "large sample" methods, which require that the sample correlation matrix **S** provides an accurate estimate of the population correlation matrix Σ. When turning to MLFA, the distinction between **S** and Σ becomes essential.
The MLFA method (e.g., Jöreskog, 1967) determines the population parameters such that they maximize the likelihood function of the data, assuming multivariate normality. Interestingly, MLFA also has an interpretation in terms of the smallest eigenvalues of a certain matrix. Specifically, MLFA minimizes the variation around the value one of the smallest eigenvalues of $\Psi^{-1/2}\mathbf{S}\Psi^{-1/2}$, where **S** is the sample correlation matrix, in an "approximate least squares sense", as explained by Jöreskog (1967, p. 449). Roughly (when the approximation is close), this means that the sum of squares of the smallest eigenvalues of $\Psi^{-1/2}(\mathbf{S}-\Psi)\Psi^{-1/2}$ is minimized. This is very similar to the criterion of IPFA, and coincides with both IPFA and MRFA in case of perfect fit. Of course, the eigenvalue criterion for MLFA is not what MLFA is aimed at. However, it does explain the striking similarity that is often encountered between loadings obtained from MLFA and those obtained from IPFA.
Whereas MRFA and IPFA use the sample correlation matrix **S** as a direct estimate of the population correlation matrix Σ, MLFA estimates Σ indirectly as $\mathbf{FF}'+\Psi$, with **F** and Ψ based on fitting the factor analysis model by the ML-procedure. When oblique factors are allowed, MLFA estimates Σ as $\mathbf{F\Phi F}'+\Psi$, where Φ is the matrix of correlations between the factors. This may yield an indefinite Φ, in violation of (2). To avoid this kind of anomaly, a constrained method of MLFA, yielding Φ positive definite, has been proposed by Bentler & Ramshidian (1994).
In terms of decomposition of Σ, MLFA cannot yield an improper solution when Ψ is positive semidefinite, because the estimate of Σ satisfies (2) by construction. MLFA provides a test for the "ideal situation" represented in (2). It is not aimed at a decomposition as given in (3), which represents the situation of imperfect fit, to be faced in practice. An important implication of this will be discussed in the next section.

5. In search for proper communalities

In the classical view, factor analysis was conceived of as a two-step procedure. Assuming that Σ is well-estimated from **S**, unique variances are determined which reduce the rank of $\Sigma-\Psi$ as much as possible; next, $\Sigma-\Psi$ is to be factored as $\mathbf{FF}'$+residual. In modern approaches, the loading matrix is typically determined by iterative methods, and the communalities are obtained as by-products afterwards: Computer package implementations for IPFA and MLFA (using orthogonal factors) yield communalities as row sum of squares of the loading matrix **F**.

In case of perfect fit, there is nothing disturbing about this procedure. However, when the fit is less than perfect, neither IPFA nor MLFA yield proper communalities: Row sums of squares of **F** are explained common variances, to be distinguished from common variances to be explained. When the distinction is ignored, 100 % of the common variance would be explained by the *r* common factors, which would be incompatible with non-perfect fit. Because they fail to yield communalities that satisfy (3), it must be concluded that IPFA and MLFA do not indicate the percentage of common variance explained by the factors.

Although MRFA is fully protected against the improper solutions, and although the method restores the concept of communality in the traditional sense, the communalities of MRFA are not necessarily equal to the "true" communalities. By definition, communalities are a property of the set of variables under study, independent of the factors used to decompose the common variance. With MRFA they are not, as will be demonstrated in the next section: The communalities of MRFA do depend on the number of factors. Hence, MRFA cannot be claimed to reproduce the true communalities as long as the "true rank" is unknown. This embarrassing problem, however, seems to be inherent to any method of factor analysis.

Compared to MRFA, IPFA has a computational advantage. MRFA is quite laborious (it may take several hours on a 486PC) when the number of variables exceeds 20. For IPFA, this is not nearly as much of a problem. Compared to MLFA, MRFA lacks the statistical footing of estimating population parameters, and it lacks the flexibility in adjusting hypothesized loadings or correlations between factors: Confirmatory and exploratory MLFA can be carried out in a unified ML-framework, with exploratory MLFA as the special case where no restrictions are imposed other than those needed for identification of the loadings. On the other hand, MRFA is useful in evaluating, rather than testing, hypotheses about the number of factors needed for perfect fit: The percentage of unexplained common variance is a natural measure of departure from this hypothesis. Nowadays, MLFA is typically accompanied by measures of fit (Browne & Cudeck, 1992; Cudeck & Henly, 1991; Hu, Bentler & Kano, 1992), a number of which reflect the deviation of **S** from $\mathbf{FF}'+\Psi$. None of these, however, seem to be a direct measure of the percentage or proportion of explained common variance.

In deciding between MRFA and MLFA for Exploratory Factor Analysis, it is tempting to choose the best looking method according to one's taste. However, this attitude is incompatible with the view that applied statistics is an empirical science. One should simply adopt the method that works best., e.g., in Monte Carlo studies, where the truth is known by construction. MRFA and its fit measure, the percentage of explained common variance, will have to prove their potential merits in comparative simulation studies. Apart from that, MRFA may be helpful in theoretical work on factor analysis (e.g., Heywood cases), because it is the first method of factor analysis that yields a proper solution for (3) under a reasonable optimality criterion.

6. Lord's test data

A reanalysis was conducted of 15 psychological tests, selected by Jöreskog (1971, pp. 119-124) from a larger battery of Lord (1956), administered to a sample of 649 subjects. The tests are of three kinds: Vocabulary (1-5), Intersections (6-10) and Arithmetic Reasoning (11-15). For each kind of test, there are two speeded versions (1, 2, 6, 7, 11, and 12) and three level versions. Jöreskog applied confirmatory MLFA to these variables. From the high value for chi-square with the three factor hypothesis he inferred that more factors were needed. In fact, he adopted a five or six factor solution.

On the same data, MRFA was carried out, with r=3, r=4 and r=5. In each case, 16 independently started runs converged to a unique set of communalities. The percentages of explained common variance were 91.50, 94.67, and 96.52, respectively. The Varimax rotated loadings and communalities are reported in the left hand part (for r=3) and the right hand part (for r=4) of Table 1. Clearly, the first three (trait) factors for r=3 and r=4 are essentially the same. The additional factor IV when r=4 is a speed-level contrast factor: This factor has all correlations with speed tests positive and all correlations with level tests negative. It explains hardly any variance (2.4 % of the common variance) at all. There is no reason to adopt a fifth factor, let alone a sixth. Considering that factor analysis should strike a compromise between parsimony (small r) and meaningfulness, the MRFA results are quite satisfactory.

When IPFA is applied to the same data, the resulting loading matrices for r=3 and r=4, respectively, closely resemble those of Table 1. The key difference, however, is that IPFA does not show proper communalities, which means that (3) is not solved, and evaluating the percentage of explained common variance is not possible.

When exploratory MLFA (EQS: Bentler, 1995; LISREL: Jöreskog & Sörbom, 1996) is applied with r=3 and r=4, respectively, setting the upper triangular elements of **F** to zero for identification of the loadings, we find essentially the same loadings as with MRFA. However, the two models do not fit ($p \leq .01$), and the chi-square test would lead us to reject a perfectly satisfactory solution. Also, MLFA does not allow an evaluation of the percentage of explained common variance.

As was mentioned above, the MRFA communalities for r=3 and for r=4 appeared to be unique. In fact, multiple solutions were not encountered when r was stepped up to 5, 6,...,10 (values above 10, the Ledermann bound when m=15, were not considered). This means that uniqueness below and even at the

Ledermann bound, although not granted, appears to hold here. Nonetheless, Table 1 shows that the communalities do change when we go from r=3 to r=4 factors. This is the embarrassing problem of MRFA, alluded to above.

Table 1: *MRFA loadings and communalities h^2 for Lord's data (r=3; r=4)*

Test	I	II	III	h^2	I	II	III	IV	h^2
1	0.035	0.764	0.164	0.172	0.030	0.760	0.168	0.208	0.707
2	0.092	0.784	0.169	0.764	0.085	0.784	0.171	0.288	0.780
3	0.094	0.802	0.218	0.742	0.098	0.802	0.219	-.090	0.739
4	0.019	0.911	0.202	0.933	0.024	0.912	0.204	-.114	0.933
5	0.011	0.857	0.238	0.860	0.018	0.859	0.240	-.193	0.859
6	0.802	0.069	0.214	0.813	0.800	0.065	0.212	0.231	0.827
7	0.833	0.023	0.163	0.779	0.829	0.021	0.162	0.075	0.769
8	0.848	0.086	0.188	0.800	0.849	0.085	0.168	-.036	0.796
9	0.891	0.053	0.165	0.878	0.893	0.052	0.163	-.074	0.874
10	0.868	0.029	0.237	0.850	0.872	0.029	0.234	-.107	0.852
11	0.221	0.183	0.649	0.580	0.216	0.178	0.652	0.203	0.588
12	0.206	0.167	0.671	0.593	0.204	0.163	0.669	0.178	0.585
13	0.172	0.201	0.766	0.703	0.178	0.201	0.764	-.125	0.698
14	0.217	0.199	0.728	0.657	0.221	0.197	0.728	-.054	0.657
15	0.129	0.232	0.729	0.634	0.135	0.232	0.730	-.099	0.635
SS	3.806	3.618	2.913		3.813	3.610	2.911	0.363	

References

Albert, A.A. (1944). The minimum rank of a correlation matrix. *Proceedings of the National Academy of Sciences,* 30, 144-146.

Anderson, T.W. & Rubin, H. (1956). Statistical inference in factor analysis. In J. Neyman (Ed.): *Proceedings of the third Berkeley Symposium on Mathematical Statistics and Probability,* University of California Press, 5, 111-150.

Bekker, P.A. & De Leeuw, J. (1987). The rank of reduced dispersion matrices. *Psychometrika*, 52, 125-135.

Bekker, P.A. & Ten Berge, J.M.F. (1997). Generic global identification in factor analysis. *Linear Algebra and its Applications*, 264, 255-264.

Bentler, P.M. & Ramshidian, M. (1994). Gramian matrices in covariance structure analysis. *Applied Psychological Measurement*, 18, 79-94.

Bentler, P.M. (1995). *EQS Structural Equation Program Manual.* Multivariate Software, Encino (Cal.).

Browne, M.W. & Cudeck, R. (1992). Alternative ways of assessing model fit. *Sociological Methods & Research*, 21, 230-258.

Carroll, J.B. (1993). *Human cognitive abilities: A survey of factor analytic research.* New York: Cambridge University Press.

Cudeck, R. & Henly, S.J. (1991). Model selection in covariance structure analysis and the "problem" of sample size: A clarification. *Psychological Bulletin*, 109, 512-519.

Guttman, L. (1958). To what extent can communalities reduce rank ? *Psychometrika*, 23, 297-308.

Harman, H.H. (1967). *Modern factor analysis.* Chicago: The University of Chicago Press.

Harman, H.H. & Jones, W.H. (1966). Factor analysis by minimizing residuals (minres). *Psychometrika*, 31, 351-368.

Hu, L., Bentler, P.M. & Kano, Y. (1992). Can test statistics in covariance structure analysis be trusted ? *Psychological Bulletin*, 112, 351-362.

Ihara, M. & Kano, Y. (1986). A new estimator of the uniqueness in factor analysis. *Psychometrika*, 51, 563-566.

Jöreskog, K.G. (1967). Some contributions to maximum likelihood factor analysis. *Psychometrika*, 31, 443-482.

Jöreskog, K.G. (1971). Statistical analysis of a set of congeneric tests. *Psychometrika*, 36, 109-133.

Jöreskog, K.G. & Sörbom, D. (1996). *LISREL 8: User's reference guide.* Scientific Software Incorporation, Chicago.

Kano, Y. (1990). Noniterative estimation and the choice of the number of factors in exploratory factor analysis. *Psychometrika*, 55, 277-291.

Ledermann, W. (1937). On the rank of reduced correlation matrices in multiple factor analysis. *Psychometrika*, 2, 85-93.

Lord, F.M. (1956). A study of speed factors in tests and academic grades. *Psychometrika*, 21, 31-50.

Mulaik, S.A. (1986). Factor analysis and Psychometrika: Major developments. *Psychometrika*, 51, 23-33.

Shapiro, A. (1982). Rank reducibility of a symmetric matrix and sampling theory of minimum trace factor analysis. *Psychometrika*, 47, 187-199.

Shapiro, A. (1985). Identifiability of factor analysis: Some results and open problems. *Linear Algebra and its Applications*, 70, 1-7.

Ten Berge, J.M.F. & Kiers, H.A.L. (1991). A numerical approach to the exact and the approximate minimum rank of a covariance matrix. *Psychometrika*, 56, 309-315.

Thomson, G.H. (1934). Hotelling's method modified to give Spearman's g. *Journal of Educational Psychology*, 25, 366-374.

Widaman, K.F. (1993). Common factor analysis versus component analysis: Differential bias in representing model parameters ? *Multivariate Behavioral Research*, 28, 263-311.

Wilson, E. B. & Worcester, J. (1939). The resolution of six tests into three general factors. *Proc. National Academy of Sciences*, U.S.A., 25, 73-77.

Bayesian Factor Analysis Model and Choosing the Number of Factors Using a New Informational Complexity Criterion

Hamparsum Bozdogan
Department of Statistics
The University of Tennessee
Knoxville, TN 37996-0532 US
Bozdogan@utk.edu

Kazuo Shigemasu
Department of Life Sciences
The University of Tokyo
Tokyo 153, JAPAN
Kshige@bayes.c.u-tokyo.ac.jp

Abstract: This paper introduces two forms of informational complexity *ICOMP* criteria of Bozdogan (1988, 1990,1994) as a decision rule for model selection and evaluation in *Bayesian Confirmatory Factor Analysis (BAYCFA)* model due to Press and Shigemasu (1989) in contemporaneously *choosing the number of factors* and determining the *"best" approximating factor pattern structure*. A Monte Carlo simulation example with a known factor pattern structure and known actual number of factors is shown to demonstrate the utility and versatility of the new approach in recovering the true structure.

Key words: Bayesian Confirmatory Factor Analysis; Choosing the Number of Factors; and Informational Complexity.

1. Introduction

The main purpose of factor analysis as an important multivariate statistical technique, is to determine whether or not the correlations among a large number of observed variables can be explained in terms of a relatively small number of underlying factors and how many factors best fit the data (*the exploratory stage*). After choosing the best fitting number of factors, the question then becomes how the extracted factors depend upon the original observed variables, and how do we incorporate the prior information in the factor model (*the confirmatory stage*).

The standard *confirmatory factor analysis (CFA)* model currently practiced, does not incorporate the prior distributions of the parameters and the postulated prior information on the specific factor pattern structures within the model. In CFA, the prior information is fixed and it is used statically. Since information changes over time and prior information is often a random behavior, a new approach is needed to estimate *the number of factors* and to determine the *'best' approximating factor pattern structure* contemporaneously. This will help us to capture the expert judgement of the researchers to validate their substantive models as the prior information might change over time and expert to expert.

The purpose of this paper is to implement various information theoretic procedures as new post-data analytic procedures, that is, decision making in the light of the data in *BAYCFA* model which was developed by Press and Shigemasu (1989), where the number of factors and the postulated factor pattern structure were chosen arbitrarily. In this paper, two forms of *informational complexity (ICOMP) criteria* of Bozdogan (1988, 1990, 1994) are derived and estimated under the oblique Bayesian factor model using a generalized natural conjugate family of prior distributions. For comparative purposes the classic *Akaike's (1973) information criterion (AIC)*, and *Rissanen's (1978) minimum description length (MDL)* criteria are also developed. To eliminate the subjective specification of the hyperparameters, in this paper, we estimate the imperfectly known hyperparameters of the model using a clustering method. This will be based on the *sparse root algorithm* of Hartigan (1975) to obtain the "best" approximating factor pattern structure to be used in determining the initial prior hyperparameters for the *BAYCFA* model. Such an approach provides a valuable mechanism to achieve flexibility in *BAYCFA* modeling.

A simulation example with a known factor pattern structure and known actual number of factors is shown to demonstrate the utility and versatility of the new approach in recovering the true structure.

2. Bayesian Factor Analysis (BAYFAC) Model

Let **x** (px1) be an observable random vector with mean vector μ and the covariance matrix Σ. We write the usual factor model in the matrix form given by

$$x_j = \Lambda f_j + \mu + \varepsilon_j, j = 1,\ldots,n;\ \mathrm{m<p}, \tag{1}$$

where Λ is a (p× m) matrix called the *factor loading matrix*; $\boldsymbol{f}_j$ is an (m× 1) vector which represents the *common factors* or *factor scores* for individual j, $F' = (f_1,\ldots,f_n)$; and ε_j is a (p× 1) *vector of random errors* or *disturbances* and whose variances ψ_j represents the "uniqueness" of $\boldsymbol{x}_j$.

The basic assumptions for the model in (1) are:

$$\boldsymbol{f}_j \sim \boldsymbol{N}_m(0, \boldsymbol{I}_m);\ \text{and}\ \varepsilon_j \sim \boldsymbol{N}_p(0, \Psi), \tag{2}$$

where Ψ is a symmetric positive definite (p.d.) matrix of random errors, and that Ψ is not a diagonal matrix in our case as in the usual factor analysis model; and $\boldsymbol{f}_j$ and ε_j are independent.

We assume that $(\Lambda, \boldsymbol{F}, \Psi)$ are unobserved and fixed parameters. Further, we write the probability distribution of the model in (1) equivalently as

$$x_j \sim N_p(\Lambda f_j, \Psi). \tag{3}$$

For a random sample of n observations, the likelihood function is given by

$$L(\Lambda, F, \Psi | X) \propto |\Psi|^{-n/2} exp\left\{-\frac{1}{2} tr\Psi^{-1}(X - F\Lambda')'(X - F\Lambda')\right\}, \tag{4}$$

where "$\propto$" denotes the constant of proportionality which depends on (p, n), and not on (Λ, F, Ψ).

Following Press (1982), Press and Shigemasu (1989), we use a generalized natural conjugate family of prior distributions for (Λ, Ψ). We take as prior density for the parameters:

$$p(\Lambda, F, \Psi) = p(\Lambda|\Psi)p(\Psi)p(F). \tag{5}$$

We adopt the normal prior density

$$p(\Lambda|\Psi) \propto |\Psi|^{-\frac{m}{2}} exp\left\{-\frac{1}{2} tr(\Lambda - \Lambda_0)\mathbf{H}(\Lambda - \Lambda_0)'\Psi^{-1}\right\}, \tag{6}$$

where Λ_0 is a (p×m) prior factor loading matrix, and $\mathbf{H}$ is an (m×m) prior inter-factor correlation matrix. Both of these matrices are hyperparameters that need to be estimated in practice.

Next, we assume that the prior distribution for Ψ is:

$$p(\Psi) = c_0(\nu, p)|\Psi|^{-\frac{\nu}{2}} |\mathbf{B}|^{\frac{\nu - p - 1}{2}} exp\left\{-\frac{1}{2} tr\Psi^{-1}\mathbf{B}\right\} \tag{7}$$

an inverted Wishart distribution which is denoted by $W^{-1}(\mathbf{B}, p, \nu)$. In (7), $\mathbf{B}$ is a (p×p) scale matrix of hyperparameters, and $c_0(\nu, p)$ is a constant depending only upon (ν,p) and not on the unknown number of factors in the model, given by

$$c_0^{-1}(\nu, p) = 2^{(\nu-p-1)/2} \pi^{p(p-1)/4} \prod_{j=1}^{p} \Gamma\left(\frac{\nu - p - j}{2}\right), \tag{8}$$

with ν degrees of freedom, which is also a hyperparameter.

If we consider a vague (or uninformative) prior density for the factor scores F, i.e., if $p(F) \propto$ constant, then $F|X$ follows a matrix T-distribution. If we suppose f_j~N(0,1), and the f_j's are mutually independent, then

$$p(F) \propto exp\left\{-\frac{1}{2} trF'F\right\} \text{ (spherical prior density)} \tag{9}$$

Now, using the Bayes theorem, we obtain the joint posterior density of the parameters by combining the equations (4),(6),(7), and (9). This is given by

$$\begin{aligned} p(\Lambda, F, \Psi | X) &\propto L(\Lambda, F, \Psi | X) p(\Lambda|\Psi) p(\Psi) p(F) \\ &\propto p(F)|\Psi|^{-(n+m+\nu)/2} exp\left\{-\frac{1}{2} tr\Psi^{-1} G\right\}, \end{aligned} \tag{10}$$

where

$$G = (X - F\Lambda')'(X - F\Lambda') + (\Lambda - \Lambda_0)\mathbf{H}(\Lambda - \Lambda_0)' + \mathbf{B}. \tag{11}$$

As shown in Press and Shigemasu (1989), the respective marginal posterior distributions are obtained from (10) by integrating out the nuisance parameters sequentially. For more details, we refer the readers to Press and Shigemasu (1989). For brevity, here, it will suffice just to give the necessary Bayes estimates of factor scores, factor loading matrix, and the disturbance covariance matrix respectively, which we need to develop *ICOMP*, *AIC*, and *MDL* criteria. These Bayes estimates of the unknown parameters are as follows.

- First, the Bayes estimate of the factor scores is given by

$$\begin{aligned}\hat{\boldsymbol{F}} &= (\boldsymbol{I}_n - \boldsymbol{X}\boldsymbol{W}^{-1}\boldsymbol{X})^{-1}\boldsymbol{X}\boldsymbol{W}^{-1}\Lambda_0\mathbf{H} \\ &= \left[\boldsymbol{I}_n - \boldsymbol{X}(\boldsymbol{X}\boldsymbol{X}' - \boldsymbol{W})^{-1}\boldsymbol{X}'\right]\boldsymbol{X}\boldsymbol{W}^{-1}\Lambda_0\mathbf{H},\end{aligned} \tag{12}$$

where

$$\boldsymbol{W} = \boldsymbol{X}\boldsymbol{X}' + \mathbf{B} + \Lambda_0\mathbf{H}\Lambda_0 \,. \tag{13}$$

- The Bayes estimate of the factor loading matrix Λ, conditional on $\hat{\boldsymbol{F}}$ and $\boldsymbol{X}$ is given by

$$\hat{\Lambda} = (\boldsymbol{X}'\hat{\boldsymbol{F}} + \Lambda_0\mathbf{H})(\mathbf{H} + \hat{\boldsymbol{F}}'\hat{\boldsymbol{F}})^{-1}. \tag{14}$$

- The Bayes estimate of the disturbance matrix Ψ is given by

$$\hat{\Psi} = \hat{\boldsymbol{G}}/(n + m + \nu - 2p - 2),$$

where

$$\hat{\boldsymbol{G}} = (\boldsymbol{X} - \hat{\boldsymbol{F}}\hat{\Lambda}')'(\boldsymbol{X} - \hat{\boldsymbol{F}}\hat{\Lambda}') + (\hat{\Lambda} - \Lambda_0)\mathbf{H}(\hat{\Lambda} - \Lambda_0)' + \mathbf{B}, \tag{15}$$

and where ν is the degrees of freedom of the distribution of Ψ. Usually we do not have a substantial knowledge about Ψ, and so we probably want to make ν as minimal as possible. Rigorously speaking, one may want to change ν depending on the number of factors extracted, because our knowledge about Ψ may be different when we presuppose a different number of factors. However, in applications, we suppose this change is negligible. Certainly, it will be of interest to study the sensitivity of the number of factors as a function of ν. The results of this will be reported elsewhere.

3. ICOMP, AIC, and MDL Criteria for the BAYFAC Model

In this section, based on the Bayes estimates obtained in Section 2 above, we construct the covariance structure for the m-th factor model given by

- M_{OF}: $\hat{\Sigma}_m = \hat{\Lambda}_m\hat{\Lambda}_m{}' + \hat{\Psi}_m$ (the orthogonal factor model), and
- M_{OBF}: $\hat{\Sigma}_m = \hat{\Lambda}_m\hat{\mathbf{H}}_m\hat{\Lambda}_m{}' + \hat{\Psi}_m$ (the oblique factor model),

where m=1, 2,…, M is the number of factors, and $m \leq M \leq 1/2[2p+1-\sqrt{8p+1}]$. We call this bound *big-factor*. It is the largest number of factors we can allow and still have the number of free parameters $s=(mp+p)-1/2m(m-1)$ in the factor model be less than the number of parameters in the covariance matrix. Now using these covariance structures and treating the factor model as a Random Factor Model (RFM), we give the two closed form expressions of ICOMP criteria of Bozdogan (1988,1990, 1994), Akaike's (1973) AIC, and Rissanen's (1978) MDL criteria.

The first approach of ICOMP uses the covariance matrix properties of a model's parameter estimates starting from their finite sampling distributions. In this case, we have

$$ICOMP_{OBF}(BayFac) = \text{nplog}(2\pi) + n[log|\hat{\Sigma}_m| + tr\hat{\Sigma}_m^{-1}\mathbf{S}] + 2\left[(m+1)C_1(\hat{\Psi}_m) + pC_1(\hat{\mathbf{H}}_m^{-1})\right] \tag{16}$$

where $C_1(\mathbf{Q}) = {}^p\!/\!_2\, log\left[{}^{tr(Q)}\!/\!_p\right] - {}^1\!/\!_2\, log|\mathbf{Q}|$ is the maximal entropic measure of complexity of a (p.d.) covariance matrix **Q** of van Emden (1971) which was first introduced by Bozdogan (1988) in multivariate linear and/or nonlinear parametric modeling as a new penalty functional. Instead of penalizing a model based only on the number of parameters in the model, as in AIC, ICOMP penalizes based upon the covariance structure of the components of the model. In this manner, ICOMP incorporates the interactions among the components of the model and measures the inequalities among the variances and the contribution of the covariances. Note that if $\mathbf{H}=s\mathbf{I}$ and $\mathbf{H}^{-1}=1/s\mathbf{I}$, then $C_1(\mathbf{H}^{-1})=0$, where s>0, a scalar constant. Hence for the *orthogonal BAYFAC model* (16) reduces to

$$ICOMP_{OF}(BayFac) = \text{nplog}(2\pi) + n[log|\hat{\Sigma}_m| + tr\hat{\Sigma}_m^{-1}\mathbf{S}] + 2(m+1)C_1(\hat{\Psi}_m). \tag{17}$$

The first component of ICOMP in both (16) and (17) measures the lack of fit, the second component in (16) measures the complexity of the covariance matrix of the estimated disturbance which is <u>not a diagonal matrix</u> in BAYFAC model, and the third component measures the complexity of the estimated inter-factor correlation matrix.

The second approach to ICOMP uses the complexity of the estimated inverse-Fisher information matrix (IFIM) of the model with regards to the model's parameter space. We denote this by *ICOMP(IFIM)*. Similar to ICOMP we have

$$ICOMP_{OBF}(IFIM) = -2\log \boldsymbol{L}(\hat{\theta}_k) + 2C_1(\hat{\boldsymbol{F}}_m^{-1}), \tag{18}$$

where

$$\hat{\boldsymbol{F}}_m^{-1} = \begin{bmatrix} \hat{\Sigma}_m & 0 \\ 0 & \frac{2}{n}\mathbf{D}_p^{+}(\hat{\Sigma}_m \otimes \hat{\Sigma}_m)\mathbf{D}_p^{+\prime} \end{bmatrix} \tag{19}$$

is the *estimated IFIM,*and where ⊗ denotes the Kronecker product. $\mathbf{D}_p^+$ is the Moore-Penrose inverse of the duplication matrix $\mathbf{D}_p$. Operating the complexity measure C_1 on the *estimated IFIM*, and after some work, $ICOMP_{OBF}(IFIM)$ is given in an open form by

$$ICOMP_{OBF}\left(IFIM\right) = \text{nplog}(2\pi) + n[log\left|\hat{\Sigma}_m\right| + tr\hat{\Sigma}_m^{-1}\mathbf{S}]$$

$$+ \tfrac{p(p+3)}{2}\,\mathbf{log}\left[\frac{tr(\hat{\Sigma}_m) + \frac{1}{2n}[tr(\hat{\Sigma}_m^{\;2}) + (tr\hat{\Sigma}_m)^2 + 2\sum_j \left(\hat{\sigma}_{jjm}^{\;2}\right)^2]}{\frac{1}{2}p(p+3)}\right] \quad (20)$$

$$- plog(2) + p(p+1)/2log(n) - (p+2)log\left|\hat{\Sigma}_m\right|.$$

The first component of $ICOMP_{OBF}(IFIM)$ in (20) measures the lack of fit of the model, and the second component measures the complexity of the estimated inverse-Fisher information matrix (*IFIM*), which gives a scalar measure of the celebrated *Cramér-Rao lower bound matrix* which takes into account the accuracy of the estimated parameters and implicitly adjusts for the number of free parameters included in the model.

With both ICOMP criteria, complexity is viewed not as the number of parameters in the model, but as the *degree of interdependence*, (i.e., the *correlation structure* among the parameter estimates). By defining complexity in this way, *ICOMP* provides a more judicious penalty term than *AIC*, *MDL*, and other *AIC-type criteria.*

We derive Akaike's (1973) AIC given by

$$AIC(\,BayFac\,) = nplog(2\pi) + n[log\left|\hat{\Sigma}_m\right| + tr\hat{\Sigma}_m^{-1}\mathbf{S}]$$
$$+ 2[Number\ of\ nonzero\ loadings + 1/2m(m-1) + p] \quad (21)$$

Similarly, the expression of Rissanen's (1978) MDL, which is identical in form to the Schwarz (1978) Bayesian criterion (SBC), is given by

$$MDL/SBC(\,BayFac\,) = nplog(2\pi) + n[log\left|\hat{\Sigma}_m\right| + tr\hat{\Sigma}_m^{-1}\mathbf{S}]$$
$$+ [Number\ of\ nonzero\ loadings + 1/2m(m-1) + p]log(n) \quad (22)$$

Comparing ICOMP, AIC, MDL/SBC, we note that the difference between these criteria is in their crucial penalty term.

4. A Simulated Numerical Example

In this section, we give a simulated numerical example based on an oblique factor model structure provided by Mulaik (1989) with the population parameters for the actual number of factors m*=3, and p=9 variables given by

$$\Lambda' = \begin{bmatrix} 0.6 & 0.7 & 0.8 & 0.2 & 0 & 0 & 0.2 & 0 & 0 \\ 0.2 & 0 & 0 & 0.6 & 0.7 & 0.8 & 0 & 0.2 & 0 \\ 0 & 0.2 & 0 & 0 & 0.2 & 0 & 0.6 & 0.7 & 0.8 \end{bmatrix}, \mathbf{H} = \begin{bmatrix} 1.00 & 0.08 & 0.12 \\ 0.08 & 1.00 & 0.24 \\ 0.12 & 0.24 & 1.00 \end{bmatrix},$$

where Λ' is the transpose of (9×3) factor loading matrix Λ, and **H** is (3×3) the inter-factor correlation matrix. The unique variances are given by the diagonal matrix
Ψ=Diag(0.581,0.436,0.360,0.581,0.403,0.360,0.571,0.403,0.360).

In our study, first, we generated a random sample of size n=100 observations from a multivariate normal distribution with mean vector $\mu=\mathbf{0}$ and the covariance matrix $\Sigma=\Lambda\mathbf{H}\Lambda'+\Psi$, with Λ, **H**, and Ψ given above. Next, by pretending that we do not know the number of factors, we fit m=1, 2,..., 6 factors. We use the *maximum likelihood factor analysis (MLFA)* and obtain the varimax rotated (or unrotated) *ML factor loadings*, and the model correlation matrices for each of m=1, 2,..., 6 factors for this simulated data. Then, we apply the *sparse root algorithm* of Hartigan (1975, p.314) to obtain the *factor pattern structure* of Λ_0, the prior factor loading matrix data-adaptively for each of the models. This algorithm is an iterative clustering procedure, which works on a factor model correlation (or covariance) matrix with a sample correlation matrix of the data. This technique produces the loading matrix and its simple pattern structure as the root of the model correlation matrix. It seeks roots of the model correlation matrix with many zeros. The zeros correspond to entries (or variables) that are not members of a given cluster. A given column of the root matrix represents the cluster. The root matrix is determined iteratively by sweeping the model correlation matrix one column at a time. The approach is essentially a principle component type procedure which takes into account the linear combination of the original variables. The columns of the root matrix are chosen to be eigenvectors of the model correlation matrix for which the ratio, (eigenvalue)/(number of nonzero elements of the eigenvector), is a maximum. The model correlation matrix is modified in terms of a partial correlation matrix at each stage of the fitting process until the *root*; that is, the simple pattern structure of the loading matrix is reached. In this manner, we discover the simple pattern structure of a factor loading matrix data-adaptively, which permits an interpretable factor pattern of the final factors as clusters of variables. Now, we use these pattern structures corresponding to m=1, 2,..., 6 factors to initialize our prior factor loading matrix Λ_0, prior inter-factor correlation matrix $\mathbf{H}_0$, and the prior disturbance matrix Ψ_0, respectively in the *BAYFAC* algorithm. With this set up, we then ran 100 simulations and generated new data sets for each m=1, 2,..., 6 factors and scored the number of times the information criteria selected the true number of factors and the corresponding factor pattern structure for different sample sizes. The results are summarized in Table 1 below. These results show that *ICOMP* and *ICOMP(IFIM)* criteria outperform *AIC* and *MDL* in selecting the number of factors and the associated factor pattern structure in small samples which we like to see in Bayesian modeling. All the computations are carried out using new open architecture easy-to-use interactive computational modules called *BAYFAC,* which are developed by the first author in MATLAB®.

Table 1: The number of times information criteria selected the true m=3 factors.

Sample Sizes	**n=30**						**n=50**					
Criteria\m	**1**	**2**	**3**	**4**	**5**	**6**	**1**	**2**	**3**	**4**	**5**	**6**
AIC	8	20	**70**	0	0	0	**75**	25	0	0	0	0
MDL/SBC	**99**	1	0	0	0	0	**100**	0	0	0	0	0
ICOMP	1	19	**80**	0	0	0	**62**	**38**	0	0	0	0
ICOMP(IFIM)	0	0	**98**	2	0	0	0	8	**92**	0	0	0

References

Akaike, H. (1973). Information theory and an extension of the maximum likelihood principle, in: *Second International Symposium on Information Theory*, B.N. Petrov and F. Csaki (Eds.) Budapest: Academiai Kiado, 267-281.

Bozdogan, H. (1988). ICOMP: A new model selection criterion, in: *Classification and Related Methods of Data Analysis*, Hans H. Bock (Ed.), Amsterdam: North -Holland, 599-608.

Bozdogan, H. (1990). On the information-based measure of covariance complexity and its application to the evaluation of multivariate linear models, *Communications in Statistics, Theory and Methods*, **19**(1), 221-278.

Bozdogan, H. (1994). Mixture-model cluster analysis using a new informational complexity and model selection criteria, in: *Multivariate Statistical Modeling*, Volume 2, Proceedings of the First US/Japan Conference on the Frontiers of Statistical Modeling, H. Bozdogan (Ed.), Dordrecht: Kluwer Academic Publishers, the Netherlands, 69-113.

Hartigan, J. (1975). *Clustering algorithms*, John Wiley and Sons, New York.

Mulaik, S. A. (1989). Personal correspondence.

Press, S. J. (1982). *Applied multivariate analysis: using Bayesian and frequentist methods of inference*, Malabar, Florida: Robert E. Krieger Publishing Co., Inc.

Press, S. J. and Shigemasu, K. (1989). Bayesian inference in factor analysis, in: *Contributions to Probability and Statistics: Essays in Honor of Ingram Olkin*, L. Gleser, M. Perlman, S.J. Press, and A. Sampson (Eds.), New York: Springer-Verlag, 271-287.

Rissanen, J. (1978). Modeling by shortest data description, *Automatica*, **14**, 465-471.

Schwarz, G. (1978). Estimating the dimension of a model, *Annals of Statistics*, **6**, 461-464.

Van Emden, M.H. (1971). An analysis of complexity. Amsterdam: Math. Centre Tracts 35.

Nonlinear Biplots for Multivariate Normal Grouped Populations

Calvo M., Villarroya A. and Oller J.M.
Dept. Estadística. Universitat de Barcelona. 08028 Barcelona (Spain).

Abstract: In this paper a new method for the simultaneous representation of populations and variables associated with multivariate normal model (without covariance matrix restrictions) in a low dimensional space is developed. The method is based on non-linear biplot methodology over the Siegel metric, which allows us to define a distance between populations. Finally, the variables are represented by the directions of maximum variation of their means.

Keywords: Nonlinear biplots. Siegel distance. Multivariate normal distribution.

1. Introduction

Biplots are used to represent samples and variables jointly, in a low dimensional plot. Gower and Harding (1988) generalised the classical biplot method considering embeddable metrics in an Euclidean space, and proposed to extend this idea to any kind of metric. On the other hand, there is no doubt that one of the most used (continuous) models in multivariate data analysis is the multivariate normal distribution. But when we want to plot, in a low dimensional space, samples from populations associated with this model, we find that the existing methods are restricted to the assumption of a common variance matrix (as in the classical canonical discriminant analysis introduced by Rao, 1948, since the Mahalanobis distance is used) or, at least, to common eigen-vectors of their covariance matrices (Krzanowski, 1996). Since none of the previous assumptions holds in practice, it seems reasonable to look for a low dimensional representation without the previous assumptions that includes representation of the variables.

To reach this goal, we propose to follow the non-linear biplots approach, but we still need a metric for multivariate normal model without restrictions. Although the Hellinger or Bhattacharyya distances could have been chosen, in this paper we propose to use the *Siegel distance* for the following reasons (see Calvo and Oller, 1990, for more details): the Siegel distance is not upper bounded, as are the other two; it is *invariant* to affine transformations over the random variables and it is closely related to the Rao distance.

2. Representation of Populations

Let us consider the populations Ω_1, ..., Ω_p, where the probabilistic model of the $x_1, x_2, \cdots x_p$ vector samples is the multivariate normal model. For each population, Ω_i, its associated density function has parametric representation (μ_i, Σ_i). The multivariate normal embedding into the *Siegel group,* see Calvo and Oller (1990), is obtained by associating each density with the following symmetric definite-positive matrix:

$$(\mu_i, \Sigma_i) \rightarrow f(\mu_i, \Sigma_i) = S_i = \begin{pmatrix} \Sigma_i + \mu_i \mu_i^t & \mu_i \\ \mu_i^t & 1 \end{pmatrix}. \tag{1}$$

In practice, (μ_i, Σ_i) are unknown, so are substituted by their maximum likelihood estimators $\left(\bar{x}_i, \hat{S}_i^2 \right)$ to represent Ω_i, giving:

$$S_i = \begin{pmatrix} \hat{S}_i^2 + \bar{x}_i \bar{x}_i^t & \bar{x}_i \\ \bar{x}_i^t & 1 \end{pmatrix}. \tag{2}$$

The *Siegel distance* between the populations Ω_i and Ω_j is defined as the Riemannian distance between the two matrices into the Siegel group where the populations are embedded:

$$\mathrm{d}(S_i, S_j) = \frac{1}{\sqrt{2}} \left\| \ln \left(S_i^{-1/2} S_j S_i^{-1/2} \right) \right\| = \frac{1}{\sqrt{2}} \left(\sum \ln^2 (\lambda_i) \right)^{1/2}. \tag{3}$$

where $\|A\| = \sqrt{\mathrm{tr}(AA^t)}$, $\ln\left(S_i^{-1/2} S_j S_i^{-1/2}\right)$ stands for the ordinary matrix logarithm and λ_i are the Eigenvalues of $S_i^{-1/2} S_j S_i^{-1/2}$.
From (3) we compute the interdistance matrix D between the p populations:

$$D = \mathrm{d}(S_i, S_j) = (d_{ij}) i, j \in \{1, \cdots, p\}. \tag{4}$$

Now, we can use classical *Principal Coordinate Analysis* to reduce the dimension. The computing process starts by obtaining the matrix T:

$$T = (t_{ij}) \qquad \text{with} \quad t_{ij} = \frac{1}{2}\left(-d_{ij}^2 + \frac{1}{p}\sum_{h=1}^{p}\left(d_{ih}^2 - d_{hj}^2\right) - \frac{1}{p^2}\sum_{i,j=1}^{p} d_{ij}^2 \right) \tag{5}$$

and then, diagonalizing it

$$T = P \Lambda P^t = WW^t \text{ with } P P^t = I \text{ and } W = P\Lambda^{1/2}, \Lambda \text{diagonal}. \tag{6}$$

The principal coordinates are the columns of W. Usually, only the first and second principal coordinates are used in the final plot.

3. Representation of the Variables

To represent the *k-th* variable, the non-linear biplot defines a pseudo-sample $\alpha\mathbf{e}_k$ (where $\mathbf{e}_k$ is a unit vector of the original k-th axis). Then, the results of Gower (1968) for representing a new point in R^n, with known distances from p points, are used. However, in our case, this methodology can not be applied directly because there is no sense in defining a pseudo-sample, since our *points* are multivariate normal densities each one represented by a matrix S_i, depending on (μ_i, Σ_i), that is, $n + \frac{n(n+1)}{2}$ parameters (where n is the number of observed variables). To solve this problem, we propose to represent each variable X_i $(i = 1,\ldots,n)$ in the original space by the direction of the maximum variation of its mean $E(X_i)$. Notice that the above-mentioned pseudo-samples in a classical biplot also correspond to the directions of maximum variation of the observed variables.

In more technical terms, the gradient of $E(X_i)$ must be computed and, by integrating it, the directions of maximum variation of $E(X_i)$ are obtained as a bundle of curves. In the classical linear-biplot procedure, the concurrence point of the trajectories representing the variables coincides at the centroid of the sample-points. In our case, there is no such obvious point, but we propose as a candidate, S_0, the barycenter (centre of mass) of the populations, although other points could also be considered. The curve of maximum variation of the random variable X_i, which passes through S_0, is:

$$S(t) = S_0 + \begin{pmatrix} \Sigma_0 \mathrm{e}_i \mathrm{e}'_i \Sigma_0 \mathrm{t}^2 + \Sigma_0 \mathrm{e}_i \mu'_i t + \mu_0 \mathrm{e}'_i \Sigma_0 t & \Sigma_0 \mathrm{e}_i \mathrm{t} \\ \mathrm{e}'_i \Sigma_0 \mathrm{t} & 0 \end{pmatrix}, \tag{7}$$

where e_i denotes the i vector of the canonical basis, and t the curve parameter.

At this point, each variable is represented as a trajectory (collection of points) in the *Siegel* space and every point V_{ij} (the j-th point of the curve corresponding to the i-variable) will be mapped in the principal coordinate space following Gower and Harding (1988) by

$$V_{ij} \rightarrow W_{ij} = \frac{1}{2}\Lambda^{-1}P'd, \tag{8}$$

where Λ and P are the same as in expression (6), $d = (d_k)_{(p\times 1)}$ with

$$d_k = d_{kk}^2 - d^2(S_k, V_{ij}) \text{ and } d_{kk}^2 = \sum_{l=1}^{p}\frac{d^2(S_k, S_l)}{p} - \sum_{r=1}^{p}\sum_{s=1}^{p}\frac{d^2(S_r, S_s)}{2p^2}, \tag{9}$$

where $d(S_a, S_b)$ is computed as in (3).

4. Example

Thomson, A. and Randall-Maciver, R. (1905) reported four measurements of 150 male egyptian skulls from five different time periods. The variables measured were: MB (maximum breath of skull), BH (Basibregmatic height of skull), BL (basialveolar length of skull) and NH (nasal height of skull). And the populations to study are the five time periods: Early Predynastic (about 4000 years B.C), Late Predynastic (about 3300 B.C.), 12-13th dynasty (about 1850 B.C.), Ptolemaic (about 200 B.C.) and Roman (about 150 A.C.). A first analysis shows that the homogeneity of the variance-covariance matrix does not hold (Barlett's test: P=8.94×10^{-11}), so that classical techniques of graphical display that assume homogeneity of the covariance matrix should not be used. Applying the method described here, we compute the interdistance matrix according to (3):

$$D = \begin{pmatrix} 0.94308 & & & \\ 1.36731 & 1.22836 & & \\ 1.71955 & 1.59887 & 1.15124 & \\ 1.78993 & 1.59732 & 1.15836 & 1.18607 \end{pmatrix}$$

Then we compute the classical Principal Coordinate Analysis to reduce the

dimensionality. All the non-zero eigenvalues are positive and the first two represent 76.62 % of their sum.
Finally, we can represent the variables according to section 2. In figure 1 we have the joint plot of populations (points) as and variables (trajectories)

Figure 1: *Non-linear biplot of five egyptian periods. The barycenter of the populations is the common point of the trajectories. Equal intervals in the original scales of the variables are indicated.*

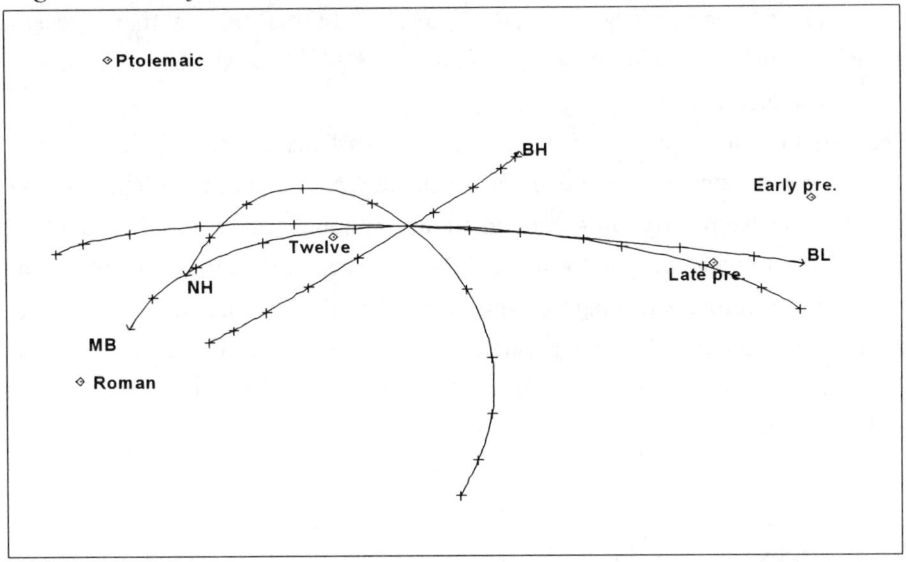

From the situation of populations in figure 1 it is clear that time progresses from left to right. Therefore *MB* and *NH* increase over time, while the other two (*BH* and *BL*) decrease, which agrees with the variable vs. time plots (not shown). It is also interesting to notice the irregular spacing of scale points along the trajectories suggests further distortions of the space (see the positive end of variable *BH* for instance).

5. Discussion

In this paper we have introduced a new method based on non-linear biplot techniques for simultaneous representation of populations and variables associated with multivariate normal models. Although simultaneous displays for special cases had been introduced (Ríos et al., 1991, for a common covariance matrix), the main advantage of our method is that there are no restrictions on the covariance matrix structure. Moreover, it is invariant to affine transformations over the random variables and is related to the Rao distance (the Siegel distance used here is a lower bound the Rao distance). One limitation of our method is the possibility of obtaining negative eigenvalues in (6), since the Siegel distance is not Euclidean. However, usually, the negative eigenvalues of the Λ matrix will be (in absolute value) quantitatively not very important relative to the positive ones and can therefore be ignored. If this is not the case, a more general technique such as multidimensional scaling could be used instead (see e.g. Gower and Hand, 1996).

The variables are represented by the directions of maximum variation of their means and a concurrence point for their trajectories must be chosen. We propose the barycentre of the populations, although others points could be taken. For instance, we can represent different trajectories from a variable in the same plot. Another advantage of this method is that we are not restricted to display only the first moment variations, but moments of any other order (including mixed moments) as well as other parametric functions of the random variables, too.

References

Calvo, M. and Oller, J.M. (1990). A distance between Multivariate Normal Distributions Based in Embedding into the Siegel Group. *Journal of Multivariate Analysis*, 35, 223-242.

Gower, J.C. (1968). Adding a point to a vector diagram in multivariate analysis. *Biometrika*, 55. 582-585.

Gower, J.C. and Hand, S. (1996). Biplots. London, Chapman & Hall.

Gower, J.C. and Harding, S. (1988). Non-linear Biplots. *Biometrika*, 73. 445-455.

Krzanowski, W.J. (1996). Rao's Distance Between Normal Populations that have Common Principal Components. *Biometrics*, 52. 1467-1471.

Ríos, M., Villarroya, A., Vegas, E. and Oller, J.M. (1991). Representación Canónica Simultánea de Poblaciones y Variables: Una aplicación al estudio de enfermedades renales. *Qüestiió*, 15. 279-298.

Thomson, A. and Randall-Maciver, R. (1905). *Ancient Races of the Thebaid.* Oxford: Oxford University Press.

Discretization as a tool in cluster analysis

Alicja Ciok
Institute of Computer Science, Polish Academy of Sciences
01-237 Warsaw, ul. Ordona 21, Poland

Abstract: The paper concerns clustering values of discrete-continuous variables X and Y when the numbers of clusters are given a priori. The clustering aims at minimal reduction of absolute dependence between X and Y, measured by the maximal value of Spearman's rho. The grade correspondence analysis (GCA) based on Spearman's rho is performed first and then followed by discretization of the pair of the GCA variables which is directly linked to discretization of X and Y. Discretization of only one variable is considered in Sec. 3. It is then used in Sec. 4 to construct an algorithm of simultaneous discretization of both GCA variables. Two examples are presented.

Keywords: Correspondence Analysis, Positive Dependence, Correlation Curve, Maximal Grade Correlation, Spearman's Rho, Mixed Variables, Clustering Algorithm

1. Introduction

Simultaneous aggregation of rows and columns in two-way contingency tables is considered in several papers since Hartigan (1975). In a slightly more general framework, pairs of categorical variables are replaced by pairs (X,Y) in which any variable may be discrete-continuous. An optimization criterion should be based on a measure of absolute dependence since dependence is inseparable from concentration and hence from clustering. We have chosen to maximize a generalized version of Spearman's rho (denoted ρ^*), applicable for any pair of mixed (discrete-continuous) variables. It is shown in Kowalczyk et al. (1996) that ρ^* can be represented as an index of concentration.

The present paper is a continuation and extension of Kowalczyk et al. (1996) and Ciok et al. (1994, 1995). In Ciok et al. (1995) the *grade correspondence analysis* (GCA) was introduced as an analogue of the classic correspondence analysis. GCA consists in maximizing ρ^* over the set of all possible pairs of permutations of the supports of X and of Y. The optimal pair is called the pair of GCA variables. As argued in Kowalczyk et al. (1996) and Ciok et al. (1995), the value of ρ^* for any pair of aggregations of the GCA variables can be *dominated* by the value of ρ^* for a pair of suitably chosen *discretizations* of these variables. This is why we can restrict to discretizations following GCA. But which scheme of discretization is then appropriate?

The algorithm proposed in Ciok et al. (1995) provides a *sequence* of discretizations such that the numbers (n,s) of clusters for X and Y derived at step i are changed at step $i+1$ into $(n-1,s)$ or $(n,s-1)$: the algorithm finds at each step a pair of *adjacent* rows or *adjacent* columns to be pooled which minimizes the loss of the value of ρ^*. But the authors observed that this may lead to serious drawbacks even for distributions with very regular positive dependence, because the discretizations in the sequence may be far from optimal ones with the same numbers of clusters.

The method proposed in the present paper tries to approximate optimal discretization when the cluster numbers (n,s) of the solution are given a priori. A suitable choice of (n,s) is beyond the scope of this paper.

Section 2 contains basic information on GCA. Discretization of just one variable is described in Sec. 3. Section 4 provides an iterative algorithm for two variables; it is based on alternating use of the solutions only for X and only for Y when in each case the other variable in the pair is previously discretized according to the scheme from the preceding iteration.

Applications of the proposed method are not restricted to clustering in contingency tables or, more generally, bivariate distributions. There are many situations in data analysis when it is desirable to treat a particular table with non-negative values as a contingency table. We can deal with sets of compositional data or objects-variables data. Analyses of real data based on GCA are given e.g. in Ciok et al. (1997), Ciok (1995). In both cases a proper clustering was crucial for investigation.

2. Grade parameters describing two-way probability tables

Let X and Y be categorical variables distributed according to $P=\left(p_{ij}\right), i=1,\ldots,m; j=1,\ldots,k$. Let (X^*,Y^*) be a pair of continuous variables distributed on $[0,1]^2$ with the constant density $p_{ij}/p_{i\bullet}p_{\bullet j}$ on rectangle $\{(u,v): S_{i-1}<u\le S_i, T_{j-1}<v\le T_j\}$ where $S_i=\sum_{l=1}^{i} p_{l\bullet}, i=1,\ldots,m, S_0=0$ and $T_j=\sum_{l=1}^{j} p_{\bullet l}, j=1,\ldots,k, T_0=0$. The density of (X^*,Y^*) is called the randomized grade density of (X,Y). The regression functions $E(Y^*|X^*=t)$ and $E(X^*|Y^*=t)$ are called the randomized grade regression functions of Y on X and of X on Y, denoted $r^*(Y{:}X)$ and $r^*(X{:}Y)$, respectively. Further, the correlation coefficient $cor(X^*,Y^*)$ is called the randomized grade correlation coefficient of (X,Y) and denoted $\rho^*(X,Y)$. It is equal to the Spearman's rho for pairs of continuous variables, to the Schriever's extension of Spearman's rho for pairs of categorical variables (Schriever (1985)), and

may be presented as:

$$\rho^*(X,Y) = 6\int_0^1 (u - C^*(Y:X)(u))du = 6\int_0^1 (u - C^*(X:Y)(u))du$$

where $C^*(Y:X)(t) = 2\int_0^t r^*(Y:X)(u)du$ is the randomized grade correlation curve while $C^*(X:Y)$ is defined analogously.

Parameters r^*, C^* are used to find the transformations of X and Y which maximize the value of ρ^* in the set of pairs $(f(X), g(Y))$, where f and g are 1-1 functions $f:S_X \to S_X$, $g:S_Y \to S_Y$, and S_X, S_Y are supports of X and Y, respectively. It is known (Kowalczyk et al. (1996)) that the maximal value of ρ^* is attained only if both regression functions $r^*(Y:X)$ and $r^*(X:Y)$ are non-decreasing or, equivalently, if curves $C^*(Y:X)$ and $C^*(X:Y)$ are convex. This property has been exploited for constructing a GCA algorithm described in Ciok et al. (1995), which approximates the maximal value of ρ^*, a pair of optimal transformations of X and Y, and the respective pairs of optimal grade regression functions as well as optimal grade correlation curves.

We will say that X and Y are GCA- *mutually positively dependent* if the value of $\rho^*(X,Y)$ cannot be increased for any pair of 1-1 transformations of X and Y. A well-known vague interpretation of mutual positive dependence between X and Y is that in each set of conditional distributions, $\{X|Y=y\}$ and $\{Y|X=x\}$, conditional distributions in the neighbourhood tend to be similar. This suggests an algorithm of mutual clustering of values of X and Y: firstly, X and Y are transformed to be GCA- mutually positively dependent; secondly, since pooling *adjacent* conditional distributions is then justified, we try to *discretize* both variables.

3. Discretization of only one variable

Throughout this section and the next one we assume that X and Y are GCA- mutually positively dependent. Discretization of only one variable in the pair is based on the following fact (the proof is omitted).

Let F be a convex function such that $F:[0,1] \to [0,1]$ and $F(0)=0, F(1)=1$. Numbers $w_0 = 0 < w_1 < \ldots < w_{n-1} < w_n = 1$ define a partition of interval $[0,1]$ into n segments. Let $\hat{F}$ denote a function which is linear inside every segment and such that $\hat{F}(w_i) = F(w_i), i = 0,\ldots,n$. If the partition defined by vector $w = (w_0, w_1, \ldots, w_{n-1}, w_n)$ minimizes the area between curves F and $\hat{F}$ then w fulfils the condition: for $i = 1,\ldots,n-1$

$$\frac{dF}{dx}\left(w_i - 0\right) \le \frac{F(w_{i+1}) - F(w_{i-1})}{w_{i+1} - w_{i-1}} \le \frac{dF}{dx}(w_i + 0) \tag{1}$$

If the function $f(x) = \frac{dF}{dx}(x)$ exists and is continuous then condition (1) reduces to the formula:

$$f(w_i) = \frac{F(w_{i+1}) - F(w_{i-1})}{w_{i+1} - w_{i-1}} \tag{2}$$

It follows that for any $i = 1,\ldots,n-1$ the minimal value of $\int_{w_{i-1}}^{w_{i+1}} (f(x) - A)^2 dx$ for $A \in (0,\infty)$ is attained at $f(w_i)$. Moreover, if f is increasing then $(w_1,\ldots,w_{n-1})$ satisfying (2) is unique and thus provides an optimal solution.

Generally, vectors w satisfying (1) form a set of *admissible* solutions in the problem of minimizing the area between F and $\hat{F}$. To obtain an admissible vector, we choose an initial vector $w^0 = (0, w_1^0, \ldots, w_{n-1}^0, 1)$, and modify it finding succesively: w_1^1 which fulfils condition (1) for $i = 1$ when w_2 is replaced by w_2^0; w_2^1 which fulfils (1) for $i = 2$ when (w_1, w_3) is replaced by (w_1^1, w_3^0), and so on until $w^1 = (w_0^1, w_1^1, \ldots, w_{n-1}^1, w_n^1)$ is obtained. In the next step, w^1 is treated as the initial vector and w^2 is the final one. This procedure is repeated until all thresholds satisfy condition (1). The above algorithm has been applied to many functions F and in each case the solution was easily obtained and found the same for any initial w^0.

Now, we put $C^*(Y{:}X)$ as F and $C^*(Y{:}\hat{X})$ as $\hat{F}$, where $\hat{X}$ refers to X aggregated according to discretization of X^* defined by vector w (discretization thresholds are quantiles of order w_i). The area between the two curves is equal to one-sixth of $\rho^*(X,Y) - \rho^*(\hat{X},Y)$, and in view of condition (1) $\hat{X}$ is an admissible discretization of X. It is approximately optimal when it is the best in a set of admissible ones obtained for various initial w^0.

If $r^*(Y{:}X)(u) = 1 - r^*(Y{:}X)(1-u)$ for $u \le 0.5$ then for any $n = 2k$, $w_k = 0.5$ and $w_i = 1 - w_{2k-i}$ for $i = 1,\ldots,k-1$. For binormal (X,Y) with the correlation coefficient ρ the regression function is: $r^*(Y:X)(u) = \Phi(\frac{\rho}{\sqrt{2-\rho^2}}\Phi^{-1}(u))$ (cf. Kowalczyk (1996)). In this case condition (2) has a unique solution. For instance, for $\rho = 0.6$ and $n = 16$, the respective thresholds w_i, $i = 1,\ldots,15$ are: 0.039, 0.093, 0.153, 0.218, 0.286, 0.357, 0.428, 0.5, 0.572, 0.643, 0.714, 0.782, 0.847, 0.907, 0.961. They define the optimal discretization of X according to $r^*(Y:X)$.

4. Simultaneous discretization of two variables

Let n and s be the numbers of categories settled a priori to discretize X and Y. It follows from Sec. 3 that we can find approximately optimal thresholds when we discretize only X or only Y. The resulting vectors can be taken as an initial pair (α_0, β_0) of thresholds vectors for *simultaneous* discretization. The proposed algorithm produces iteratively vectors (α_i^1, β_i^1) and (α_i^2, β_i^2) together with the corresponding values $\rho_i^{*l} = \rho^*(X(\alpha_i^l), Y(\beta_i^l)), l = 1,2$, where $X(\alpha_i^l)$ and $Y(\beta_i^l)$ denote respectively variable X discretized according to α_i^l and variable Y discretized according to β_i^l. For $i = 2j+1$, $\alpha_i^1 = \alpha_{i-1}^1$; for $i = 2j$ vector α_i^1 consists of n thresholds for X in the pair $(X, Y(\beta_{i-1}^1))$, derived from $r^*(Y(\beta_{i-1}^1):X)$ as described in Sec. 3, provided that $r^*(Y(\beta_{i-1}^1):X)$ is increasing; otherwise $(X, Y(\beta_{i-1}^1))$ is replaced by its *GCA* transformation. Similarly, $\beta_i^1 = \beta_{i-1}^1$ when $i = 2j$ and vector β_i^1 consists of s thresholds for Y derived from $r^*(X(\alpha_{i-1}^1):Y)$ if $i = 2j+1$. The sequence (α_i^2, β_i^2) is defined analogously; the difference is that the procedure starts from discretization of Y. It is convenient to calculate both pairs (α_i^1, β_i^1) and (α_i^2, β_i^2) paralelly. The iterations in "path" l stop in step N when (α_N^l, β_N^l) is equal to (α_i^1, β_i^1) or (α_i^2, β_i^2) for some $i < N$. Then we select j and l such that ρ_j^{*l} gains the maximum value and use (α_j^l, β_j^l) as the basis for the final discretization of X and Y. Examples show that this discretization well approximates an optimal one.

If positive dependence between X and Y is sufficiently regular, only one iteration is needed since its results are equal to (α_0, β_0). This happened in two examples mentioned in the sequel. In the first example (X,Y) is binormal with $\rho = 0.6$, and both variables are discretized according to the 15 thresholds calculated in Sec. 3 $(n = s = 16)$. The resulting discretized distribution with non-uniform marginals is graphically presented in Fig. 1a: the rectangles in the unit square are marked with four shades of grey according to the values of the respective grade densities for the pairs of discretized variables: black in the interval $(\frac{3}{2}, \infty)$, dark grey in $(1, \frac{3}{2}]$, light grey in $(\frac{2}{3}, 1]$, and white in $[0, \frac{2}{3}]$.

The second example concerns a data set taken from the files issued in 1992 by the Central Statistical Office of Poland (available at request from the author). A population of Polish farms in 1992 is taken into account, divided into subpopulations according to 12 geographical districts (X) and 6 categories of farm magnitude (Y). The data form a table $\{p_{ij}\}$, where p_{ij} is the fraction of farms in district i and magnitude category j. Clustering to four macroregions and four levels of magnitude is performed. The resulting 4×4 table is visualized in Fig. 1b. The clustering helped to extract macroregions with concisely described profiles of farm magnitude (cf Szczesny et al (1998)).

Figure 1: *Graphical presentation of the grade densities in the probability tables concerning: (a) the binormal discretized distribution with* $\rho = 0.6$ and $i, j = 1, \ldots, 16$, *(b) the fractions of farms in macroregion i and magnitude category* j, $i, j = 1, \ldots, 4$.

(a) (b)

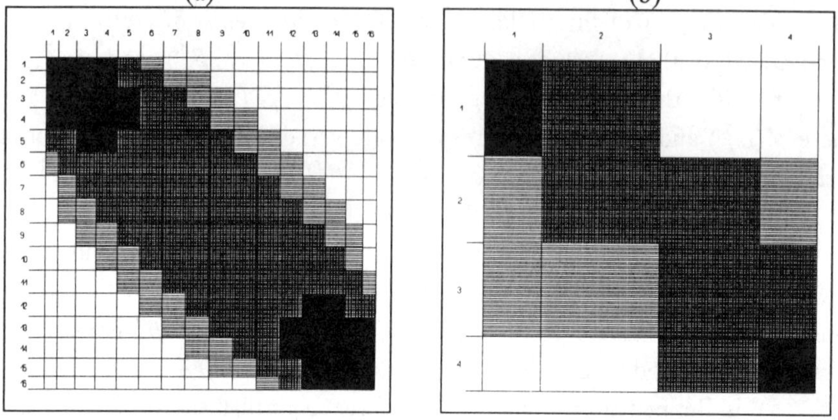

References

Ciok, A. & Bułhak, W. & Skoczylas, R. (1997). Exploration of control-experimental data by means of grade correspondence analysis. *Biocybernetics and Biomedical Engineering,* vol. 17, no 1-2 (in press).

Ciok, A. (1995). Establishing the pattern of attitudes among Polish businessmen, *Proceedings of 14th International Conference on Multivariate Statistical Analysis MSA '95, December 7-8, Warsaw.*

Ciok, A. & Kowalczyk, T. & Pleszczyńska, E. & Szczesny, W. (1995). Algorithms of grade correspondence-cluster analysis, *The Collected Papers on Theoretical and Applied Computer Science,* vol. 7, no 1-4, 5-22.

Ciok, A. & Kowalczyk, T. & Pleszczyńska, E. & Szczesny, W. (1994). On a system of grade dependence measures, *ICS PAS Reports*, No 741.

Hartigan, F. A. (1975). *Clustering algorithms*, New York, Wiley.

Kowalczyk, T. (1996): Link between grade measures of dependence and of separability in pairs of conditional distributions (submitted).

Kowalczyk, T. & Ciok, A. & Pleszczyńska, E. & Szczesny, W. (1996). Grade dependence in bivariate mixed data. A uniform approach based on concentration curves. *ICS PAS Reports*, No 815.

Schriever, B. F. (1985). *Order dependence*, Ph.D. Dissertation, Free University of Amsterdam.

Szczesny, W. & Ciok, A. & Pleszczyńska, E. (1998). Clustering land districts according to their farm magnitude repartition (submitted).

A Robust Biplot Representation of Two-way Tables

Christophe Croux and Peter Filzmoser

ECARE and Institut de Statistique, Université Libre de Bruxelles,
Av. F.D. Roosevelt 50, B-1050 Bruxelles, Belgium, e-mail: ccroux@ulb.ac.be

Department of Statistics and Prob. Theory, Vienna University of Technology,
Wiedner Hauptstraße 8-10, A-1040 Vienna, Austria,
e-mail: P.Filzmoser@tuwien.ac.at

Abstract: In this note we show how a two-way table can be fitted by an additive and multiplicative model in a robust way. This allows us to detect interactions between rows and columns, which can be displayed in a biplot. We show that the robust biplot we present here reflects the real interaction structure, and is not predetermined by atypical values.

Keywords: Biplot, Robustness, Two-way table, Alternating regression.

1. Introduction

Multivariate data are often represented in the form of a two-way table. We will denote the elements of this table by y_{ij}, where $1 \leq i \leq n$ indicates the row index, and $1 \leq j \leq p$ the column index. The classical model for such a two-way table is the Anova model

$$y_{ij} = \mu + a_i + b_j + \delta_{ij} \tag{1}$$

where μ is called the overall mean, a_i represent the row effect and b_j the column effect. The terms δ_{ij} can either be seen as residuals or as interaction terms between rows and columns. The above model (1) is called the additive model. It is however quite possible that the interaction terms δ_{ij} still contain some structure, and therefore one can model them by a factor model

$$\delta_{ij} = \sum_{l=1}^{k} f_{il}\sigma_l \lambda_{jl} + \varepsilon_{ij}. \tag{2}$$

To obtain symmetry between rows and columns, we assume that both $f_{i\cdot}$ and $\lambda_{j\cdot}$ are normed to one, so that σ_l needs to be considered as a scaling factor. As usual in factor analysis, the vectors of loadings $\lambda_{j\cdot}$ and scores $f_{i\cdot}$ are only determined up to an orthogonal transformation. The number of factors k is supposed to be chosen in such a way that the residuals ε_{ij} resemble the most a white noise. Taking $k = 2$ yields a biplot representation of the δ_{ij}. One can present now in the same two-dimensional plot the rows by (f_{i1}, f_{i2}) and the columns by $(\lambda_{j1}, \lambda_{j2})$.

The biplot was originally proposed by Gabriel (1971) and allows us to investigate the row and column interaction by visual inspection of a two-dimensional graphical display. More information about biplots can be found in Gower and Hand (1996).

Combining (1) and (2) yields an additive and multiplicative model for the two-way table:

$$y_{ij} = \mu + a_i + b_j + \sum_{l=1}^{k} f_{il}\sigma_l\lambda_{jl} + \varepsilon_{ij}. \tag{3}$$

In order to estimate the $(k+1)(n+p+1)$ unknown parameters in the above model from the $n \times p$ available data points, one usually performs a singular value decomposition (Hoaglin et al., 1983, p. 100-118). This classical approach corresponds essentially with a least squares fit of the model (3). It is however well known that an LS-based method is very vulnerable in the presence of outliers. In this paper, we will discuss a robust approach to fit the model (3).

Robust singular value decomposition for estimation of the parameters in the pure multiplicative model (2) was already considered by Daigle and Rivest (1992), who used an approach based on multivariate M-estimators. In the pure additive model (1) the Median Polish technique, described by Hoaglin et al. (1983) is quite popular and has been shown to be very robust. Median Polish is in fact an approximation to the L_1-fit of (1) and our proposed approach can be seen as a generalization of Median Polish to the more complicated model (3).

A first suggestion is to use the L_1-criterion to fit the model. In fact, the L_1 approach was already shown to be very robust for the additive fit of two-way tables (Hubert, 1997). If we denote by θ the vector of all unknown parameters in the model, and by $\hat{y}_{ij}(\theta)$ the corresponding fit, then this procedure tries to minimize the objective function

$$\sum_{i=1}^{n}\sum_{j=1}^{p} |y_{ij} - \hat{y}_{ij}(\theta)|. \tag{4}$$

For the estimation of the unknown parameters we use an iterative procedure known as *alternating regressions*, which was originally proposed by Wold (1966) and used in the context of bilinear models by de Falguerolles and Francis (1992). The idea is very simple: if we take the row index i in the model equation (3) fixed and consider the parameters b_j and $\lambda_{j\cdot}$ as known for all j, then we see that a regression with intercept of the ith row of the two-way table on the k vectors of loadings yields estimates for a_i and the vector of scores $f_{i\cdot}$. Reversely, if we take j fixed and suppose that a_i and $f_{i\cdot}$ are known for all i, and regress the jth column on the k vectors of scores, then we can update the estimates for b_j and $\lambda_{j\cdot}$. To make things robust, we will of course use a robust regression method. Working with the criterion (4) results in performing L_1 regressions. Unfortunately, L_1-regression is sensitive to leverage points. Therefore we propose a weighted L_1-regression, corresponding to minimizing

$$\sum_{i=1}^{n}\sum_{j=1}^{p}|y_{ij}-\hat{y}_{ij}(\theta)|w_i(\theta)w_j(\theta). \tag{5}$$

The weights will downweight outlying vectors of loadings and scores. Since the true loadings and scores are unobserved, w_i and w_j depend on the unknown parameters, and will be updated at each iteration step in the alternating regression procedure.

In Section 2 of the paper we will discuss in more detail the algorithm we used, while in the last section we illustrate our approach. Moreover, many other simulations and experiments showed that the proposed procedure is highly robust and converges quite fast. A documented S-plus program of the implemented algorithm is freely available at *http://www.statistik.tuwien.ac.at/public/filz/research.html*.

2. Algorithm

We will describe an iterative algorithm for obtaining the model with both additive plus multiplicative fit. Pure additive or multiplicative fits can be obtained in the same vein. In order to identify the additive part of the model we set the restrictions

$$\operatorname*{med}_i(a_i)=\operatorname*{med}_j(b_j)=0 \quad \text{and} \quad \operatorname*{med}_i(f_{il})=\operatorname*{med}_j(\lambda_{jl})=0, \tag{6}$$

for $l=1,\ldots,k$. The importance of the lth term in the factor model is measured by the magnitude of σ_l. We will arrange them in descending order. To identify them we demand that both loadings and scores are normalized, that is

$$\sum_i f_{il}^2=\sum_j \lambda_{jl}^2=1, \tag{7}$$

for $l=1,\ldots,k$. Note that there is in principle no orthogonality restriction between scores f_{il} and loadings λ_{jl}, and that we have complete symmetry in the indices i and j. Therefore we may assume without loss of generality that $n\geq p$.

It is important to take starting values for the estimators which are robust, otherwise the procedure will loose its virtues. We set $a_i^{(0)}=\text{med}_j(y_{ij})$ and $\mu^{(0)}=\text{med}_i\, a_i^{(0)}$ $(i=1,\ldots,n)$ to start up the iteration process. We also need to initialize the score matrix $f_{il}^{(0)}$ $(l=1,\ldots,k)$, which can be done by computing scores based on a robust principal component analysis of the data, and center and scale them afterwards to meet conditions (6) and (7). Croux and Ruiz-Gazen (1996) proposed a fast procedure for a robust PCA.

Suppose now that we are in step t (≥ 1) of the algorithm and that $a_i^{(t-1)}$, $\mu^{(t-1)}$, and $f_{il}^{(t-1)}$ are available. Take column j fix and consider the regression model

$$y_{ij} - \mu^{(t-1)} - a_i^{(t-1)} = b_j + \sum_{l=1}^{k} f_{il}^{(t-1)} \eta_{jl} + \varepsilon_{ij}^{(t)} \quad (i = 1, \ldots, n) \tag{8}$$

with unknown parameters $\eta_{jl} = \sigma_l \lambda_{jl}$ and intercept b_j. For every index i we compute now the corresponding weight w_i, which downweights rows with outlying score vectors. Herefore we compute robust Mahalanobis distances based on the Minimum Volume Ellipsoid (Rousseeuw and van Zomeren, 1990) for the collection of score vectors $f_{i\cdot}^{(t-1)} = (f_{i1}^{(t-1)}, \ldots, f_{ik}^{(t-1)})^\top$, yielding a sequence of robust distances $\text{RD}_1, \ldots, \text{RD}_n$. Every distance RD_i measures how far the k-dimensional vector $f_{i\cdot}^{(t-1)}$ is from the "center" of the data cloud formed by all n score vectors. Afterwards we set

$$w_i = \min(1, \chi^2_{k,0.95}/\text{RD}_i^2) \quad \text{for } i = 1, \ldots, n.$$

This choice for the weights was suggested by Simpson et al (1992). The final regression estimator is then obtained by minimizing

$$\sum_{i=1}^{n} w_i |y_{ij} - \mu^{(t-1)} - a_i^{(t-1)} - (b_j + \sum_{l=1}^{k} f_{il}^{(t-1)} \eta_{jl})|,$$

which can be done by usual L_1-regression. As solution we obtain estimates $\eta_{jl}^{(t)}$ and $b_j^{(t)}$. We repeat this procedure for every column, which implies that we need to perform p times an L_1-fit, and this at every iteration step. Fortunately, this takes almost no time. Note that the weights w_i only need to be computed once every iteration step.

The estimates $\eta_{jl}^{(t)}$ still needs to be decomposed. This can be done by writing

$$\eta_{jl}^{(t)} = \lambda_{jl}^{(t)} \sigma_l^{(t)} + \operatorname*{med}_j \eta_{jl}^{(t)}, \tag{9}$$

with

$$\sigma_l^{(t)} = \|\eta_{\cdot l}^{(t)} - \operatorname*{med}_j \eta_{jl}^{(t)}\| \quad \text{and} \quad \lambda_{jl}^{(t)} = (\eta_{jl}^{(t)} - \operatorname*{med}_j \eta_{jl}^{(t)})/\sigma_l^{(t)}.$$

The last term of (9) enters the additive part

$$a_i^{(t)} = a_i^{(t-1)} + \sum_{l=1}^{k} f_{il}^{(t-1)} \operatorname*{med}_j \eta_{jl}^{(t)}.$$

We end the first part of iteration step t by updating the overall effect $\mu^{(t)} = \mu^{(t-1)} + \text{med}_i\, a_i^{(t)} + \text{med}_j\, b_j^{(t)}$ and centering row and column effects around their median in order to fulfill restrictions (6) and (7).

In the second part of the tth step of this alternating regression algorithm we consider for each row $i = 1, \ldots, n$ the regression model

$$y_{ij} - \mu^{(t)} - b_j^{(t)} = a_i + \sum_{l=1}^{k} \lambda_{jl}^{(t)} \xi_{il} + \varepsilon_{ij}^{(t)} \quad \text{for } j = 1, \ldots, p \tag{10}$$

with unknown parameters $\xi_{il} = f_{il}\sigma_l$ and a_i and where $\mu^{(t)}, b_j^{(t)}$ and $\lambda_{jl}^{(t)}$ are obtained from the first part and are treated as fix. All estimates for step t are now obtained and updated in exactly the same way as above. The weights w_j are now based on the robust distances computed from the loadings vectors. The iterative algorithm will be stopped if there is no essential decrease in the objective function any more. We have no formal proof for the convergence of the algorithm, but according to our experience convergence can be expected after 10 to 20 iteration steps.

A graphical display of the interactions between rows and columns is obtained as an additional output of our program. A constant α for scaling variables and objects can be chosen. Besides that, there is the option of orthogonalizing the finally obtained factors and scores. The program also allows to use other regression estimators, like the *Least Trimmed Squares* estimator. Taking the non robust *Least Squares* estimator corresponds with the classical approach of Gabriel (1978). Our experience however is that the weighted L_1-estimator gives the most satisfying results among the estimators we tried out.

3. Example

As an example we created an artificial data set of size 10×15. We generated the data according to model (3) with randomly generated coefficients. Afterwards, four observations were replaced by severe outliers: cells (1,1), (1,2), (4,4), and (8,8). We fitted an additive and multiplicative model with the described robust procedure and with the usual LS-approach. As benchmark, we also computed an LS-fit on the original, clean data. In Figures 1–3 we plotted the corresponding biplots. The arrows indicate the positions $(\hat{\lambda}_{j1}, \hat{\lambda}_{j2})$ for $j = 1, \ldots p$ of the columns (also called variables here), while the rows are also indicated at positions $(\hat{f}_{i1}, \hat{f}_{i2})$ for $i = 1, \ldots n$. We also plotted three dimensional barplots of the residuals $\hat{\varepsilon}_{ij}$, which are useful in detecting outliers. The idea is that the biplot represents the general interaction structure between rows and columns, while outliers show up in the barplot.

We clearly see that the robust biplot represents the structure of the good observations, without being influenced too much by the outliers. It resembles the usual biplot based on the clean data, showing the robustness of our procedure. The four outliers are clearly indicated by the barplot.

The classical biplot is completely determined by the four outliers. Information about the other 146 cells can hardly be obtained from the biplot. The barplot of the LS-based residuals only retrieves cell (4,4) as an outlier, while some other cells, like (4,1), have a huge residual while not being true outliers. Finally, the barplot of the residuals of the LS-fit based on the clean data does not reveal any outliers, as it should be.

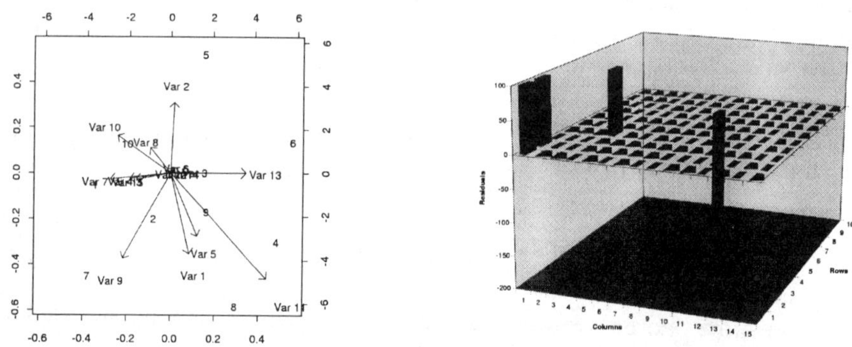

Figure 1: *Biplot and barplot of the residuals for the robust procedure based on the contaminated data*

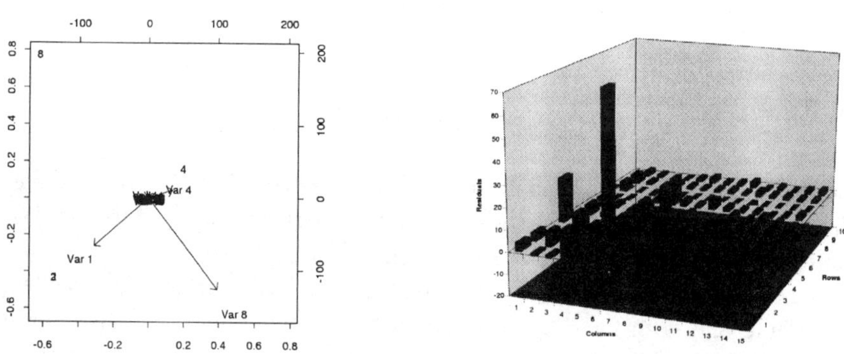

Figure 2: *Biplot and barplot of the residuals for the LS-procedure based on the contaminated data*

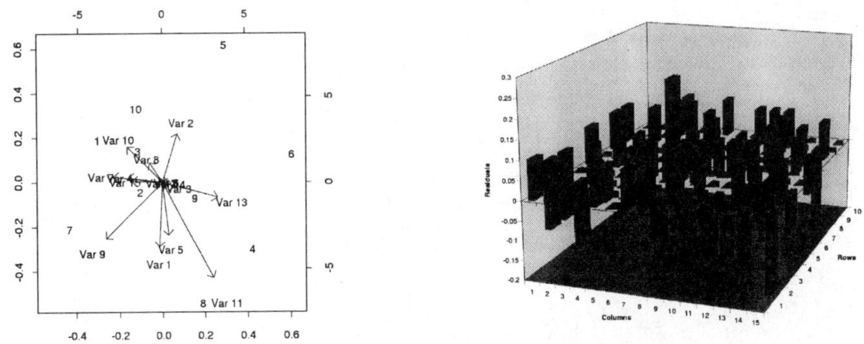

Figure 3: *Biplot and barplot of the residuals for the LS-procedure based on the clean data*

References

Croux, C. & Ruiz-Gazen, A. (1996). A fast algorithm for robust principal components based on projection pursuit, in: *Proceedings in Computational Statistics, COMPSTAT 1996*, Prat, A. (Ed.), Physica-Verlag, Heidelberg, 211-216.

Daigle, G. & Rivest, L.-P. (1992). A robust biplot, *The Canadian Journal of Statistics*, 20, 241-255.

de Falguerolles, A., & Francis, B. (1992). Algorithmic approaches for fitting bilinear models, in: *Computational Statistics, COMPSTAT 1992*, Dodge, Y. & Whittaker, J. (Eds.), Vol. 1, Physica-Verlag, Heidelberg, 77-82.

Gabriel, K.R. (1971). The biplot graphical display of matrices with applications to principal component analysis, *Biometrika*, 58(3), 453-467.

Gabriel, K.R. (1978). Least squares approximation of matrices by additive and multiplicative models, *Journal of the Royal Statistical Society B*, 40(2), 186-196.

Gower, J. & Hand, D. (1996). *Biplots*, Chapman & Hall, New York.

Hoaglin, D., Mosteller, F. & Tukey, J. (1983). *Understanding Robust and Exploratory Data Analysis*, Wiley & Sons, New York.

Hubert, M. (1997). The breakdown value of the L_1 estimator in contingency tables, *Statistics and Probability Letters*, 33, 419-425.

Rousseeuw, P.J. & van Zomeren, B. C. (1990). Unmasking multivariate outliers and leverage points, *Journal of the American Statistical Association*, 85, 633-639.

Simpson, D.G., Ruppert, D. & Carroll, R.J. (1992). On one-step GM estimates and stability of inferences in linear regression, *Journal of the American Statistical Association*, 87, 439-450.

Wold, H. (1966). Nonlinear estimation by iterative least squares procedures, in: *A Festschrift for F. Neyman*, David, F. N. (Ed.), John Wiley, New York, 411-444.

Graphical factor analysis models: specification and model comparison

Paolo Giudici
Dipartimento di Economia Politica
e Metodi Quantitativi, e-mail: giudici@unipv.it
Università di Pavia
Elena Stanghellini
Dipartimento di Scienze Statistiche
Università di Perugia, email: stanghel@stat.unipg.it

Abstract: We introduce a graphical factor analysis model as a graphical Gaussian model with latent variables satisfying a set of conditional independence constraints. After a brief introduction of the factor analysis model, we generalise the class of such models by allowing the concentration matrix of the residuals to have non-zero off-diagonal elements. The study of the associations left unexplained by the latent factors allows a better interpretation of the model. We concentrate on models with one latent variable, for which the identifiability condition is well established. A real data example is then presented to clarify the ideas. Two model selection procedures are presented, one based on the calculations of deviance differences and the other based on the calculation of the posterior probability of the model. Given the analytical intractability of the latter, we propose a Markov Chain Monte Carlo method to approximate both the model probabilities and the inferences on the quantities of interest.

Key words: Factor analysis models, Graphical Gaussian models, Identifiability, Markov Chain Monte Carlo methods, Model selection.

1. Model specification

Results concerning latent variables may be usefully applied in the graphical modelling context. The role and relevance of latent variables depend upon their interpretation. We can distinguish between situations where latent variables have a substantive interpretation as hidden variables and situations where they are merely a computational artefact.

More precisely, the first status typically derives from psychometric factor analysis contexts, where the aim is to measure underlying concepts with a psychological meaning. The second status is less established and essentially deals with the idea that the joint distribution of the observed random variables

can be conveniently simplified by introducing some further random variables which are not observed and typically do not have any 'a priori' substantive meaning. A noticeable example of the latter occurs in exploratory factor analysis, where the aim is to find the smallest number of latent variables such that the observed variables become independent conditionally on them.
We consider the factor analysis model:

$$X = \Lambda \xi + \delta \tag{1}$$

where X is a p×1 vector of observed variables, Λ is a p×q matrix of factor loadings, ξ is a q dimensional vector of latent variables, and δ is a p×1 vector of residuals. We assume δ has the multivariate Gaussian distribution $N(0,\Theta)$ with Θ a positive definite symmetric matrix. We also assume $E(\delta\xi') = 0$. The X and ξ variables are written as deviations from their mean, and $E(\xi\xi')=\Phi$. To avoid a trivial identification problem we set elements of the main diagonal of this matrix equal to one. Under these assumptions it turns out that the covariance matrix of X is:

$$\Sigma=\Lambda\Phi\Lambda' + \Theta. \tag{2}$$

In the classical factor analysis framework, all the associations among the observed variables are explained by the unobserved factors, so that Θ is assumed to be a diagonal matrix. Here we will see the classical factor analysis model as a graphical Gaussian model, with the peculiarity that some random variables are not observed. In this sense, the concentration matrix of residuals Θ will be allowed to have non-zero off-diagonal elements. The resulting model is named a *graphical factor analysis model.*
When latent variables have a substantive meaning, the stucture of associations among the residuals helps in improving the interpretation of the model by giving information about the substantive correlations unexplained by the unobserved variables. On the other hand, when latent variables are computational artefacts, such as in exploratory factor analysis, this procedure can be seen as an intermediate step between fitting a model with one factor and a model with more factors. We also believe that this model may also be relevant for *highly structured stochastic systems* where latent variables may help to elicit conditional independencies.
A relevant problem in specification of a graphical factor analysis model is the *identifiability* of the parameters of the model. If (2) admits a unique solution (or at least a finite number of solutions) in Λ, Φ and Θ, then the model is globally identified. Sufficient and/or necessary conditions concerning identifiability are available in the literature (see Bollen, 1990, for a review). However, they all concern models where Θ is a diagonal matrix. A sufficient condition for identifiability of a single-factor (q=1) model with correlated residuals has been

proposed by Stanghellini (1997). Such a condition can be extended to a generic number of factors, and we are currently examining this research issue.

3. Model comparison

Let $\{G^1, \ldots, G^m\}$ be a collection of graphical models (see e.g. Lauritzen, 1996), describing alternative association structures for $X_V=(X,\xi)$, where ξ are the unobserved random variables. Consider a random sample of n p-variate observations from P_G, $X^{(n)} = x^{(n)}$. By *quantitative* learning we mean that, having fixed G, the evidence is used to estimate the unknown parameter Σ_G. On the other hand, *structural* learning has the objective of establishing which graphs are better supported by the data and the prior information available.
Our objective is to perform structural learning. This will be done in two ways: 1) by means of a deviance-based sequential testing procedure; 2) by comparing the posterior probability of the graphs, corresponding to alternative conditional independence constraints.
When only one latent factor is considered, the graphical factor analysis model introduced in Section 1 specifies that $p(x|\xi) = N(\lambda\ \xi, \Theta_G)$ and $p(\xi) = N(0,1)$, where Θ_G is such that the elements of its inverse, say ϑ^{ij} satisfy the pairwise Markov property:

$$\vartheta^{ij}=0 \Leftrightarrow i \neq\sim j[G]$$

where the symbol $i \neq\sim j[G]$ means that i and j are not adjacent in the graph G.
Concerning the proposed Bayesian model selection procedure, we need to derive the posterior probability of each graph. For each graph G such a probability is defined by Bayes' theorem: $p(G|x^{(n)}) \propto p(x^{(n)}|G)\, p(G)$, where the prior probability $p(G)$ will be taken as uniform. We thus need to calculate the marginal likelihood $p(x^{(n)}|G)$. This requires the specification, for each G, of a prior distribution $p(\Theta|\ G)$ on the $p(p+1)/2$ covariance parameters characterising the graphical factor model corresponding to G. This will be done following the approach in Giudici and Green (1997), taking Θ as Hyper Inverse Wishart with parameters $\alpha=p$ and $\Phi=I$. Finally, as a prior distribution on the loading coefficients, we shall take $\lambda_i \sim N(l, m)$, for $i=1,\ldots,p$ and (l, m) constants to be specified.

3. Computational methods

We performed the ML estimates via the ECM algorithm (Meng and Rubin, 1994). Details of the implementation of the ECM algorithm for the models we consider are given in Stanghellini (1995). Each model is tested against the model with no constraints on Σ via the likelihood ratio (deviance) test.

On the other hand, the calculation of the marginal likelihood $p(x^{(n)}|G)$ cannot be performed analytically, because of the presence of the latent variables, which need to be integrated out. There are other reasons which suggest considering an alternative, simulation-based approach. For instance: a) in realistic situations (such as probabilistic expert systems), the number of random variables considered is very high and, thus, the number of possible graphs forbids an exhaustive model search; b) it may be important to derive quantitative inferences on functions of Σ, such as the partial correlation coefficients.

We propose, but do not detail here, for lack of space, a Markov Chain Monte Carlo method to derive approximately both the posterior model probabilities and quantitative inference on *any* quantities of interest. The design of such a sampler is based on a proposal by Green (1995), known as *reversible jump Markov Chain Monte Carlo method.* We extend recent work of Giudici and Green (1997), by including latent variables in graphical Gaussian models. Briefly, our algorithm contemplates, at each stage of the Markov Chain, a systematic scan through three possible moves: a) adding/dropping an edge from the current graph, yet remaining within the class of decomposable models; b) updating the variance-covariance matrix Θ_G; c) updating the latent parameters ξ and λ. For each of the moves, we simulate a candidate (proposal) new value of either G, Θ or (ξ,λ) and then accept it with probability determined by the Metropolis-Hastings acceptance ratio (see e.g. Green, 1995). The algorithm is very fast, due to simple proposal distributions adopted and the localisation of all the Metropolis-Hastings ratios on the cliques of the current graph.

4. Hospital recoveries

We analyse a set of variables concerning functional ability indices (see McDowell and Newell, 1987, for an introduction to such measures). The sample consists of 70 people recovering in a hospital. The average age of the people is 82.71. The interest is in studying the relationships between different psycho-cognitive attitudes. The measures are from 1 to 7 (1 is total dependence) and are assigned to each person by the physicians carrying out the study, after a follow-up period of 6 months. Only the people who completed the study are included in the sample. The variables concern the people's attitudes towards: self-care (Y_1), comprehension (Y_2), expression (Y_3), interaction with other people (Y_4), solving problems (Y_5) and memory (Y_6).

The covariances and partial correlation coefficients are reported in Table 1. We remark that there is a positive correlation between the age and the measures, which is possibly due to the process by which the sample is selected. To check the validity of the linearity assumption of the Y variables we performed the exploratory battery of tests as suggested by Cox and Wermuth (1994). There is a small non-linearity involving the Y_5 variable, which can be removed by taking the log of such a variable.

There is an unobserved heterogeneity among the people, depending on the varying level of severity of their illness. Therefore, we assume that there is an underlying unobserved variable which takes into account the severity of their illness. The analysis of the association is then performed after conditioning on the latent variable. As explained in Section 1, this coincides with fitting a graphical Gaussian model on the concentration matrix Θ^{ij} of the residuals.
We performed our analyses both in the classical (MLE) and Bayesian framework.

Table 1: *Marginal covariances (lower triangle) and partial correlations (upper triangle).*

Variables	Y_1	Y_2	Y_3	Y_4	Y_5	Y_6
Y_1	3.539	-0.060	0.114	0.237	0.283	-0.046
Y_2	2.088	3.307	0.719	-0.085	-0.094	0.425
Y_3	2.352	3.126	3.463	0.410	0.068	-0.079
Y_4	2.618	2.784	3.083	3.761	0.213	0.106
Y_5	2.283	2.347	2.477	2.662	2.901	0.551
Y_6	2.346	3.017	2.978	2.997	2.867	3.806

We first comment on the results of the former. The graphical factor model which exhibits the best fitting (deviance=5.89, with 6 degrees of freedom) has ϑ^{23}, ϑ^{26} and ϑ^{56} significantly different from zero. This model is the result of a forward selection procedure, starting with a single-factor with Θ diagonal (deviance 52.11, with 9 degrees of freedom, showing that one factor does not explain all the correlations among the observed variables) and adding at each step the most significant edge. The sufficient condition for identifiability of the graph (see Stanghellini, 1997) is satisfied, therefore the selected model is identified.
We observe that the Θ matrix is block-diagonal, showing that the Y_1 and Y_4 variables, measuring the self-care and the interaction with other people, are independent from the other measures, given the latent factor. The other variables remain correlated also after conditioning on the latent variable, i.e., in Θ there are no more marginal independencies. However, these correlations are in large part explained via the common associations of these variables with Y_2 and Y_5 measuring, in order, the comprehension and the capability of solving problems of the person. As already noted, the selected model can be seen as an intermediate step between fitting a single-factor model and other factor analysis models that involve a higher number of latent variables. This analysis points out which associations are left unexplained by the first factor, therefore giving insights about the meaning of a second factor. We also note that the conditional independence graph of the observed variables can be derived by marginalising on the latent variables. This marginalisation can be performed using graphical rules (see Cox and Wermuth, 1996, Ch. 8 for details).
The performed Bayesian structural learning, based on 100,000 MCMC iterations plus 10,000 initial ones which are discarded (burn-in period), shows a good stability of the simulated results. As a general finding, the models selected

are less parsimonious. This result, which indeed occurs also for the marginal graphical model without latent variables, mainly depends on the assumed prior distribution, which particularly favours the independence graph, therefore conflicting with the available evidence.
The best model receives about 54% of the posterior probability and contains 9 conditional associations of the Y variables, given the unobserved one, expressed by the ϑ^{14}, ϑ^{15}, ϑ^{23}, ϑ^{26}, ϑ^{34}, ϑ^{35}, ϑ^{36}, ϑ^{45} and ϑ^{56} parameters significantly different from zero. It is not identified. The second best model, with about 27% of the posterior probability, is more parsimonious, with ϑ^{15}, ϑ^{23}, ϑ^{26}, ϑ^{34} and ϑ^{56} different form zero. It is identified. Both models contain the one selected via the ML procedure. This second model contains two more conditional associations on the Y variables then the model selected via ML. These concern associations between the self-care (Y_1) and solving problems (Y_5) capabilities, and between comprehension (Y_3) and interaction (Y_4) capabilities. Notice that the two models alone receive about 81% of the posterior probability, a result which is remarkable having considered a total of about 26,000 candidate decomposable models.

References

Bollen, K.A. (1989). *Structural Equations with Latent Variables*, Wiley, New York.

Cox, D.R. & Wermuth, N. (1994). Test of linearity, multivariate normality and the adequacy of linear scores, *Applied Statistics*, 347-355.

Cox, D.R. & Wermuth, N. (1996). *Multivariate Dependencies. Models, analysis and interpretation*, Chapman & Hall, London.

Giudici, P. & Green, P.J. (1997). Decomposable graphical Gaussian model determination, *Technical report.*

Green, P.J. (1995). Reversible jump Markov chain Monte Carlo computation and Bayesian model determination, *Biometrika*, 82, 711-732.

Lauritzen, S.L.(1996). *Graphical Models*, Oxford University Press, Oxford.

McDowell, I. & Newell, C. (1987). *Measuring Health. A guide to Rating Scale and Questionnaires*, Oxford University Press, Oxford.

Meng, X.L. & Rubin, D.B. (1993). Maximum likelihood via the ECM algorithm: A general framework, *Biometrika*, 80, 2, 267-278.

Stanghellini, E. (1995). Modelli grafici gaussiani in presenza di fattori latenti. Un'applicazione allo studio delle misure di autosufficienza degli anziani, Ph.D thesis, Università di Firenze.

Stanghellini, E. (1997). Identification of a single-factor model using graphical Gaussian rules, *Biometrika*, 84, 1, 241-244.

Graphical Analysis of Fully Ranked Data

Sang-Tae Han[1], Jonghoo Choi[2]

[1] Dept. of Mathematics, Hoseo University, Asan, 336-795, Korea, e-mail : sthan@dogsuri.hoseo.ac.kr

[2] Dept. of Applied Statistics, Korea University, Chochiwon, 339-700, Korea, e-mail : jchoi@tiger.korea.ac.kr

Abstract: The aim of this study is to extend Nishisato's(1980) quantification method of fully ranked data, using Gabriel's(1971) biplot technique for multivariate data matrix. Specifically, we will propose a row plot, in which the interdistance between two plot points can be interpreted as approximation of the (squared) rank distance between two rankings given by corresponding judges in Spearman's sense. Similarly, we will also propose a column plot for objects with a sensible relationship to the row plot for judges.

Key words: quantification, row plot, column plot, biplot.

1. Introduction

Ranked data are widely used in the area of social sciences, for instance in polls and preference surveys, in which a number of objects(or stimuli) are evaluated and ranked by a panel of judges(or subjects) according to their preference. Ranked data drew attention of many researchers such as Nishisato(1980), Critchlow(1985), Baba(1986) and Diaconis(1988) among others.

Let r_{ij} denote the rank given to the object $j(= 1, ..., p)$ by the judge $i(= 1, ..., n)$, and write $R = \{r_{ij}\}$. Although this matrix notation is conventional, we will use other coding scheme for fully ranked data in this study. The coding scheme is $S = \{s_{ij}\}$ with n rows and p columns, where

$$s_{ij} = r_{ij} - (p+1)/2.$$

The S notation will be useful for later use of matrix decomposition, since redundant row averages, (p+1)/2, of the raw data matrix R are removed in S. In Spearman's sense, the squared rank distance between two rows(or judges) i and i' is defined by

$$d_S^2(i,i') = \sum_{j=1}^{p}(s_{ij} - s_{i'j})^2.$$

2. Quantification of Fully Ranked Data

For lower dimensional reduction technique of multivariate data, we will resort to Lebart, Morineau, and Warwick(1984; Chapter 1). The rows s_i of S can be considered as vectors in $\Re^p$. Let v be a unit vector in $\Re^p$. The magnitude of projection vector of s_i on v is equal to $s_i'v$. Our objective is to maximize the sum of squared inner products. Then, the optimization problem can be formulated as

$$\max_v \sum_{i=1}^{n}(s_i'v)^2, \qquad \text{subject to } v'v = 1.$$

Adopting Lagrangian multiplier method, principal component(PC) reduction can be derived from an eigensystem

$$(S'S)v = \lambda^2 v, \qquad v \in \Re^p.$$

More efficiently, biplot of fully ranked data can be obtained from singular value decomposition(SVD) of the data matrix S. Let

$$S = UDV',$$

where U is the $n \times p$ matrix with orthonormal columns, $V = (v_1, v_2, \cdots, v_p)$ is the $p \times p$ orthogonal matrix, and D is the $p \times p$ diagonal matrix with singular values, $\lambda_1 \geq \cdots \geq \lambda_{p-1} > (\lambda_p = 0)$ as its diagonal elements. Then two-dimensional row plot points are given by the rows of

$$G^*_{(2)} = S(v_1 : v_2) = UDV'(v_1 : v_2) = U_{(2)}D_{(2)},$$

where $U_{(2)}$ is the $n \times 2$ submatrix of U and $D_{(2)}$ is the 2×2 diagonal submatrix of D. Then the interpoint distances between two points in the row plot are approximations of squared Spearman rank distance between corresponding rankings given by judges with overall goodness-of-approximation

$$GOA_{row(2)} = 1 - ||SV - (G^*_{(2)} : O_{n\times(p-2)})||^2/||SV||^2.$$

It turns out that

$$GOA_{row(2)} = \{\lambda_1^2 + \lambda_2^2\}/\{\lambda_1^2 + \cdots + \lambda_p^2\}.$$

For the plot of columns(or objects), we use the same PC axis vector v and a group of hypothetical supplementary rows(or judges). Specifically, to locate the first object for instance, consider $(p-1)!$ rankings

$$(1, 2, 3, \cdots, p), (1, 2, p, \cdots, 3), (1, 3, 2, \cdots, p), \cdots,$$

which give rank 1 to the first object. Then the centroid of these rankings is given by

$$c_1 = (1, (p+2)/2, (p+2)/2, \cdots, (p+2)/2)'.$$

When the size p vector c_1 is projected on the PC axis vector v, it is positioned at $c_1'v$. Similarly , c_j are defined for $j = 2, \cdots, p$ and their projections on v can be carried out individually. Finally, the p objects can be positioned, say in a column plot, at

$$(c_1'v_1, c_1'v_2), (c_2'v_1, c_2'v_2), (c_3'v_1, c_3'v_2), \cdots, (c_p'v_1, c_p'v_2).$$

For the column plot, first note that

$$c_j = (p+2)/2\ (1, \cdots, 1)' - p/2\ (0, \cdots, 1, \cdots, 0)',$$

and that

$$\begin{aligned} c_j'v &= (p+2)/2\ (1, \cdots, 1)v - p/2\ (0, \cdots, 1, \cdots, 0)v \\ &= -p/2\ (0, \cdots, 1, \cdots, 0)v, \end{aligned}$$

since $S\ \mathbf{1} = 0$ for $\mathbf{1} = (1, \cdots, 1)'$, by coding definition, implies that

$$U'(UDV'\mathbf{1}) = 0 \quad \text{or} \quad DV'\mathbf{1} = 0,$$

or $v_j'\mathbf{1} = 0$ for $j = 1, ..., p-1$. Hence the column points for objects in the two dimensional plot are given by the rows of

$$\begin{bmatrix} c_1' \\ \vdots \\ c_j' \\ \vdots \\ c_p' \end{bmatrix} [v_1 : v_2] \;=\; -p/2\ I_p[v_1 : v_2] \;=\; -p/2\ [v_1 : v_2].$$

Therefore, two-dimensional column plot points can be obtained from the rows of

$$H^*_{(2)} = V_{(2)},$$

where $V_{(2)}$ is the $n \times 2$ submatrix of V. Then, the goodness-of-approximation of the two dimensional column plot is given by

$$GOA_{col(2)} = 1 - ||V_{[p]} - (V_{(2)} : O_{p\times(p-3)})||^2/||V_{[p]}||^2 = 2/(p-1),$$

where $V_{[p]}$ is the $p \times (p-1)$ submatrix of V. The biplot, produced by combining row and column plots, can be interpretable as follows. First, note that S can be expressed as

$$S = (UD)V' \cong (U_{(2)}D_{(2)})V'_{(2)} = G^*_{(2)}H^{*\,\prime}_{(2)},$$

For s_{ij}, the (i,j) element of S, we can write

$$s_{ij} \cong g'_{i(2)}h_{j(2)}, \quad i = 1, \cdots, n, \quad j = 1, \cdots, p,$$

where $g'_{i(2)}$ and $h'_{j(2)}$ are i-th and j-th row vectors of $G^*_{(2)}$ and $H^*_{(2)}$. Thus the raw data elements are recovered by the inner products of row and column plot vectors.

3. A Numerical Example

To illustrate the proposed quantification plots, consider a ranked data from Nishisato and Nishisato(1984) in which 31 judges were asked to rank ten government services(and facilities) according to the order of "being satisfactory". The list of services or objects are:

A = public transit,	B = postal service,
C = medical care,	D = sports/recreational facilities,
E = police protection,	F = public libraries,
G = street cleaning,	H = restaurants,
I = theatres,	J = planning and development,

to which are given the rank 1 for "most satisfactory",..., the rank 10 for "least satisfactory".

Row plot by quantification is shown in Figure 1 with the goodness-of-approximation 60%. Also, column plot by quantification are shown in Figure 2 with the goodness-of-approximation 22%. By superimposing the row and column plots, we obtain the biplot of fully ranked data.

In Figure 1, the largest cluster of judges is found in the right-side region along the first axis, of which corresponding position in Figure 2 is assigned to the object B(postal service). It means that this group of judges agree to put B as "least satisfactory". Also, in Figure 1, the second largest cluster of judges is located in the upper-side region along the second axis. Their opposite position in Figure 2 is occupied by the objects I(theatres) and H(restaurants). So, we may interpret that the second largest group of judges evaluate theatres and restaurants very highly.

Figure 1. *Row plot for thirty-one judges by quantification* $(GOA_{row(2)} = 60\%)$

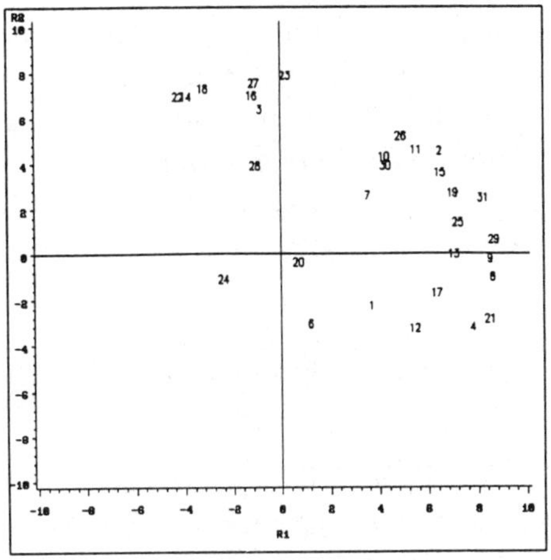

Figure 2. *Column plot for ten government services by quantification* $(GOA_{col(2)} = 22\%)$

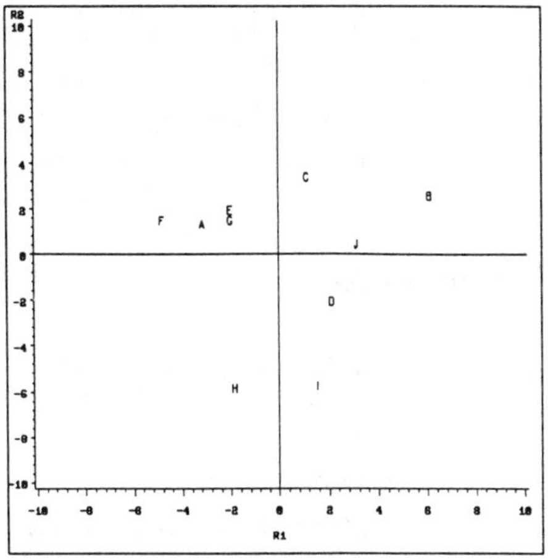

4. Concluding Remarks

Nishisato(1980) considered two types of ranked data, paired comparison type and general ranking type. He considered the typical aregument of dual scaling. But, he did not plot for judges. Also, he did not consider the geometry of the ranked data matrix. Taguri, Hiramatsu, Kittaka and Wakimoto(1976) suggested a graphical representation of rank correlation coefficient which enables to show graphically the correlations between objective variable and the several explanatory variables and exhibits changes of correlations in part. Also, they defined a measure named as the area ratio correlation coefficient to represent the degree of correlation based on the graph which they call the linked vector pattern. Baba et al.(1984) proposed a linked rank graph and a rank graph. The former represents the distribution of ranks, while the latter represents the average ranks and the degree of concordance. Their motivations and ideas behind come from the linked vector pattern. The relations among the three graphs are as follows: the rank graph is a result from the drawing of the linked rank graph, and the concordance chart is obtained by connecting the vectors on the rank graph. These graphs give us good information about the typical rank given to objects. However, they did not consider the graph of judges. Baba(1986) also proposed a graphical method for rank test, which is based on the rank graph. He showed that in case of large sample, the null hypothesis that ranks are given at random to each object(or item) can be tested visually by the rank graph with a critical ellipse. Other graphical method for ranked data include multidimensional scaling, minimal spanning trees and nearest neighbor graphs as discussed by Diaconis(1988). The method we proposed is such a nice graphical tool that subjects(or judges) as well as objects can be plotted to show off mutual relationships. One particular merit in the row plot is that Spearman's distances between rankings given by judges are approximately preserved. Moreover, the methodology can be used for segmentation of judges(or observations). By modifying coding schemes, we may produce biplot in which Kendall's rank distances instead of Spearman's are preserved approximately.

References

Baba, Y. (1986). Graphical analysis of rank data. *Behaviormetrika,* 19, 1-15.

Baba, Y., Wakimoto, K., Endou, K., Kaneko, T, Iida, T. and Nonaka, T.(1-984). Graphical representation of ranks and its application(in Japanese). *Proceedings of the Institute of Statistical Mathematics,* 32, No. 1., 15-23.

Critchlow, D. E. (1985). *Metric Methods for Analyzing Partially Ranked Data.* Lecture Notes in Statistics, No. 34. Springer, New York.

Diaconis, P. (1988). *Group Representations in Probability and Statistics.* Lecture Notes-Monograph Series, Vol. 11. Institute of Mathematical Statistics. CA: Hayward.

Gabriel, K. R. (1971). The biplot graphics of multivariate matrices with applications to principal component analysis. *Biometrika,* 58, 453-467.

Lebart, L., Morineau, A. and Warwick, K. (1984). *Multivariate Descriptive Statistical Analysis: Correspondence Analysis and Related Techniques for Large Matrices.* Wiley, New York.

Nishisato, S. (1980). *Analysis of Categorical Data: Dual Scaling and Its Applications.* Mathematical Expositions, No. 24, University of Toronto Press, Toronto.

Nishisato, S. and Nishisato, I. (1984) *An Introduction to Dual Scaling.* MicroStats, Toronto.

Taguri, M., Hiramatsu, M., Kittaka, T. and Wakimoto, K. (1976). Graphical representation of correlation analysis of order data by linked pattern. *Journal of Japanese Society of Computational Statistics,* 6, No. 2, 17-25.

Improper Solutions in Exploratory Factor Analysis: Causes and Treatments

Yutaka Kano
Faculty of Human Sciences, Osaka University
Suita, Osaka 565, Japan. email: kano@hus.osaka-u.ac.jp

Abstract: There are many causes of occurrence of improper solutions in factor analysis. Identifying potential causes of the improper solutions gives very useful information on suitability of the model considered for a data set.

This paper studies possible causes of improper solutions in exploratory factor analysis, focusing upon (A) sampling fluctuations, (B) model underidentifiable and (C) model unfitted, each having several more detailed items. We then give a checklist to identify the cause of the improper solution obtained and suggest a method of reanalysis of the data set for each cause.

Keywords: Covariance Structure Analysis, Checklist, Underidentifiability, Sample Fluctuations.

1. Factor Analysis Model and Improper Solution

In factor analysis, an observed random p-vector $\boldsymbol{x}$ is assumed to have the following form: $\boldsymbol{x} = \Lambda \boldsymbol{f} + \boldsymbol{u}$, where $\Lambda = (\lambda_{ij})$ is a $p \times k$ matrix of factor loadings, $\boldsymbol{f} = [F_1, \cdots, F_k]'$ is a k-vector of common factors, $\boldsymbol{u} = [U_1, \cdots, U_p]'$ being a p-vector of unique factors. Here k is the number of factors. Assume further that $\text{Var}(\boldsymbol{f}) = I_k$, $\text{Cov}(\boldsymbol{f}, \boldsymbol{u}) = 0$, $\text{Var}(\boldsymbol{u}) = \Psi = \text{diag}(\psi_1, \cdots, \psi_p)$. The covariance matrix Σ $(= (\sigma_{ij}))$ of the observed vector $\boldsymbol{x}$ is representable as $\Sigma = \Lambda\Lambda' + \Psi$. Each diagonal element ψ_i of Ψ is a variance of U_i, so that it should be estimated as a positive value. It is said to be an improper solution or a Heywood case when some elements ψ_i are not positively estimated.

2. Cause, Identification and Treatment

Following van Driel (1978), we distinguish among three types of causes of improper solutions as in Table 1: (A) sampling fluctuations, (B) model underidentifiable, and (C) model unfitted. We shall make brief comments on these causes. Since the parameter space of ψ_i is the finite interval $(0, \sigma_{ii})$ and estimation methods naturally do not require estimates $\hat{\psi}_i$ to be in the interval, $\hat{\psi}_i$ can be outside the parameter space or can be at its boundary at a positive probability, because of sampling variations. A typical example is an improper solution which takes place in a simulation study under a true identified model. Ihara and Okamoto (1985) compared estimation methods such as ML and LS in terms of frequency of improper solutions due

$$
\begin{cases}
\text{A: sampling fluctuation} \\
\text{B: underidentifiability} \begin{cases}
\text{B1: } \Lambda = \left[\begin{array}{cc} & \lambda_{1k} \\ \Lambda_1 & 0 \\ & \vdots \\ & 0 \end{array}\right] & \begin{pmatrix}\text{only \underline{one} nonzero} \\ \text{element in a col-} \\ \text{umn of } \Lambda\end{pmatrix} \\
\text{B2: } \Lambda = \left[\begin{array}{cc} & \lambda_{1k} \\ & \lambda_{2k} \\ \Lambda_1 & 0 \\ & \vdots \\ & 0 \end{array}\right] & \begin{pmatrix}\text{only \underline{two} nonzero} \\ \text{elements in a col-} \\ \text{umn of } \Lambda\end{pmatrix} \\
\text{B3: others}
\end{cases} \\
\text{C: factor model unfitted} \begin{cases}
\text{C1: some true unique variances } \psi_i < 0 \\
\text{C2: inconsistent variables } X_i \text{ included} \\
\text{C3: others (e.g., minor factors)}
\end{cases} \\
\text{D: others (e.g., outliers)}
\end{cases}
$$

Table 1: *Types of causes of improper solutions*

to sampling fluctuations. Anderson and Gerbing (1984), Boomsma (1985) and Gerbing and Anderson (1985) have studied how model characteristics influence on the frequency of occurrence of improper solutions in the context of confirmatory factor analysis. In my experience, however, improper solutions due to sampling fluctuations are not so often met in practice. When sampling fluctuations cause an improper solution, it may be useful to constrain uniqueness estimates $\hat{\psi}_i$ to be nonnegative. Gerbing and Anderson (1987) discussed interpretability of constrained estimates for improper solutions caused by sampling fluctuations in confirmatory factor analysis.

The causes (B) and (C) are important in model inspection. Anderson and Rubin (1956) gave a necessary condition for identification that there be at least three nonzero elements in each column in Λ. The cases (B1) and (B2) in Table 1 violate the necessary condition.

In (B1), the parameters (λ_{1k}, ψ_1) can take any values as far as they meet $\lambda_{1k}^2 + \psi_1 = \sigma_{11} - \sum_{r=1}^{k-1} \lambda_{1r}^2$, so that ψ_1 can be negative. The location of the nonzero loading in the k-th column is arbitrary. The k-th common factor is not a common factor but a unique factor. Thus, it is actually a $(k-1)$-factor model. Accordingly, an improper solution takes place when the number of factors is overestimated. Why is it overestimated? In many cases, a test of goodness of fit suggests it, that is, the test rejects a $(k-1)$-factor model. There would be two potential cases yielding this anomaly ($(k-1)$-factor model is rejected; examination of a k-factor solution suggests a $(k-1)$-factor model):

(B11) Sample size n is large enough to reject reasonably well-fitted models; in other words, because of large samples, the statistical test becomes too sensitive to small deviation from the model, possibly caused by minor factors;

(B12) Distribution assumptions such as normality are violated, and then the distribution of test statistics can not be approximated by a chi-square distribution.

The both cases have been pointed out in the context of covariance structure analysis. For (B11), researchers are advised to use goodness of fit indices such as GFI, CFI and RMSEA to measure the distance of the population from the model (e.g., Jöreskog and Sörbom 1993 section 4.5.2; Bentler 1995 chapter 5). They would then get useful information concerning acceptability of the model. The cause (B11) is also interpreted as an effect of minor factors. In (B12), researchers can take elliptical theory or a type of asymptotically distribution-free (ADF) method. Kano (1990) suggests a noniterative estimation procedure which prevents a unique factor from being reinterpreted as a common factor.

Checking Item	Cause of Improper Solutions			
	A	B1	B2	C1
(1) Does iteration converge?	yes	no	no	yes
(2) Is the solution stable? (not depending largely on estimation methods, starting values, nor optimization algorithms)	yes	unstable in one X_i	unstable in two particular X_i's	yes
(3) Are SE's† of $\hat{\psi}_i$ almost the same in magnitude?	yes	one large SE	one or two large SE('s)	yes
(4) Does the confidence interval of ψ_i, not positively estimated, contain zero?	yes	no	no	no
(5) Are residual elements of $S-(\hat{\Lambda}\hat{\Lambda}'+\hat{\Psi})$ almost the same in magnitude?	yes	yes	no for solution reducing k by one	yes

Table 2: *Checklist for identifying the cause of improper solutions. See text for examining C2. †SE denotes standard error.*

In (B2) an improper solution occurs in the first or second variables only, because the loadings λ_{1k} and λ_{2k} can take any value as long as they satisfy $\lambda_{1k}\lambda_{2k} = \sigma_{12} - \sum_{r=1}^{k-1} \lambda_{1r}\lambda_{2r}$. One can not make exploratory factor analysis when the population factor loading matrix Λ has the form (B2). In the case, the researchers can not help making a constraint to remove the indefiniteness, such as $\lambda_{1k} = \lambda_{2k}$ or $\psi_1 = \psi_2$. It is optional whether to impose $\lambda_{ik} = 0$ $(i = 3, \cdots, p)$. See Kano (1997) for details.

When an improper solution occurs due to (B), underidentifiability, it often happens that iteration does not terminate; the solution depends on starting values, estimation methods (e.g., ML, GLS, LS) or optimization algorithms. In (B1) the location of only one nonzero element, or the variable in which $\psi_i < 0$ is arbitrary and it also depends on starting values etc. As noted above, negative estimates can appear at two particular variables for (B2). These observations will distinguish between (B1) and (B2).

For (C1) or (C2), researchers have to remove all variables inconsistent with the model considered. They can remove the variables with negative unique variances in the case (C1). In (C2) we could examine residuals to identify inconsistent variables. A more sophisticated manner would be to take a likelihood ratio test approach, developed by Kano and Ihara (1994).

The case (D) contains all causes other than (A)-(C). Outliers in samples may cause improper solutions, as pointed out by Bollen (1987). Existence of outliers is classified in the case (D). The other cases in (D) are yet unknown and still need to be studied.

In Table 2, we summarize as a checklist how to identify the cause of improper solutions. Table 3 presents the method of reanalysis for each cause of improper solutions.

Cause	Treatment
A	Obtain a boundary solution with all $\psi_i \geq 0$
B11	Refer to goodness of fit indices such as GFI and CFI
B12	Apply an ADF type of estimation method
B2	Estimate under constraint such as $\lambda_{1k} = \lambda_{2k}$ or $\psi_1 = \psi_2$
C1	Remove the variable X_i with $\psi_i < 0$
C2	Remove inconsistent variables

Table 3: *Treatment after identifying the cause of improper solutions*

3. Example

Maxwell (1961) conducted maximum likelihood factor analysis of each of two samples of 810 normal children, and 148 neurotic children attending a psychiatric clinic, where the first five items of the samples are cognitive tests and the other five are inventories for assessing orectic tendencies (see Table

7 for items). He found that the iterative process for obtaining the MLE does not terminate and the communality of the eighth variable approaches to one for 4-factor model for the normal sample whereas a 3-factor model successfully analyzes the sample of neurotic children. He concluded that the method of maximum likelihood along with goodness-of-fit testing does not always perform well. It later turned out that the 4-factor solution for the normal sample is improper (e.g., Jöreskog 1967).

Here, we shall take the sample of normal children to illustrate our procedure for improper solutions. In the sample, Sato (1987), among others, reported that the improper solution depends on initial estimates for iteration and gave three different improper solutions with $\hat{\psi}_6$, $\hat{\psi}_8$ or $\hat{\psi}_9$ negative, respectively, and that it is difficult to achieve convergence in iteration when uniqueness estimates are not constrained to be nonnegative. Table 4 shows MLE for uniqueness and their standard errors, for each case of $0 \le \psi_i < \infty$ and $-\infty < \psi_i < \infty$. The analysis was made with a covariance structure program Eqs, developed by Bentler (1995).

		goodness-of-fit		ψ_1	ψ_2	ψ_3	ψ_4
		χ^2_{11}-value	P-value				
$0 \le \psi_i < \infty$	MLE	18.447	0.072	384	621	302	638
	SE			037	059	039	037
$-\infty < \psi_i < \infty$	MLE	14.589	0.202	397	629	298	644
	SE			037	056	040	037

		ψ_5	ψ_6	ψ_7	ψ_8	ψ_9	ψ_{10}
$0 \le \psi_i < \infty$	MLE	352	778	287	000	690	599
	SE	096	040	052	003	037	050
$-\infty < \psi_i < \infty$	MLE	324	802	275	−276614	725	609
	SE	108	042	046	156	040	044

Table 4: *Uniqueness estimates* (MLE) *and their standard errors* (SE) *in Maxwell's data* ($n = 810$; $k = 4$). *Values are multiplied by 1000.*

Obviously, the cause of the improper solution is not "A: sampling fluctuations." Maybe we should consider "B: identifiability" as a possible cause. Table 5 shows the list of top five (standardized) residuals in absolute value in 3-factor solution. There is no salient residual in the list, which implies that (B2) is not the cause. To conclude that the cause is (B1), we have to examine (B11) and (B12), but unfortunately, we can not check normality of observations because raw data are not available. We would say that the sample size $n = 810$ is so large that the power of the goodness-of-fit test is raised too much. In fact, other fit indices indicate reasonable fit of the 3-factor model, for instance, GFI=0.981; CFI=0.973. As a conclusion,

X9-X8	X8-X6	X10-X8	X8-X4	X9-X6
0.095	0.084	-0.050	0.046	-0.044

Table 5: *Top five residuals in 3-factor solution*

a probable cause of the improper solution is (B1) and the analysis here suggests a 3-factor model for the sample of normal children, as well as for the sample of neurotic children.

Any model is nothing but an approximation to reality, and deviation of a model from reality always exists. Statistical test can detect the deviation even when it is very small, provided that the sample size gets large. There are two considerations: (i) one employs a slightly misspecified model if one considers the deviation as just an error and so negligible; (ii) one rejects the model and finds a suitable treatment to reduce the deviation.

The treatment above for the improper solution of the normal sample is based on the consideration (i). There is an alternative story based on (ii). The cause of the improper solution is then (C) in this story. The key can be found in the list of the residuals in Table 5.

In Table 5, we can find that the top four residuals are related to the eighth variable. This indicates the possibility that the eighth variable be inconsistent with the model under consideration. The inconsistency could be a cause of the rejection of the 3-factor model. Table 6 shows the goodness-of-fit chi-square test statistics of 10 models, each of which is formed by removing one of 10 variables. The only accepted model is the one in which the eighth variable is removed. As a result, the eighth variable can be regarded as inconsistent with the 3-factor model. See Kano and Ihara (1994) for details.

	Variable Deleted									
	1	2	3	4	5	6	7	8	9	10
LRT	35.32	37.38	48.28	65.37	21.63	55.89	40.73	14.58	45.68	38.47

Table 6: χ^2_{12} *values of 10 models after deletion of one variable.* ($\chi^2_{12}(.05) = 18.55$)

Table 7 shows the MLE in 3-factor model after deletion of the eighth variable.

	Factor 1	Factor 2	Factor 3	Communal.
COGNITIVE TESTS				
X1 Verbal Ability	574	412	323	603
X2 Spatial Ability	095	585	139	371
X3 Reasoning	406	697	225	702
X4 Numerical Ability	325	487	117	356
X5 Verbal Fluency	780	243	092	675
ORECTIC TENDENCIES				
X6 Neuroticism Questionnaire	104	175	396	198
X7 Way to be different	228	143	808	725
X8 Worries and Anxiety	---	---	---	---
X9 Interests	103	180	482	275
X10 Annoyances	000	028	625	391

Table 7: *3-factor solution, rotated by normalized-VARIMAX, of 9 variables after deletion of the 8th variable.* χ^2_{12}*-value*= 14.58 *(*$n = 810$*), P-value=.2712. Estimates are multiplied by 1000.*

We have here suggested two possibilities for treatment of the improper solution in Maxwell's data. Which approach is to be taken, in others words, whether the deviation of the 3-factor model using 10 variables can be considered small enough or not, may depend on researchers and also on interpretability of those two results of the analyses.

4. Remarks

Users may not implement the procedure described above, if they use usual exploratory factor analysis (EFA) programs only. It is absolutely necessary to use programs with which the user can (i) do analysis under no constraint on ψ_i, that is, ψ_i can take negative values; (ii) get standard errors of estimates, particularly, of $\hat{\psi}_i$; (iii) specify starting values; and (iv) get (standardized) residuals, $S - (\hat{\Lambda}\hat{\Lambda}' + \hat{\Psi})$. For this, covariance structure analysis (CSA) programs (e.g., Amos, Eqs, Lisrel) are very useful, although they do not have the option of factor rotation. Researchers are recommended to use both CSA and EFA programs to make exploratory factor analysis in a proper way.

References

Anderson, J. C. and Gerbing, D. W. (1984). The effect of sampling error on convergence, improper solutions, and goodness-of-fit indices for maximum likelihood confirmatory factor analysis. *Psychometrika*, **49**, 155-173.

Anderson, T. W. and Rubin, H. (1956). Statistical inference in factor analysis. *Proceedings of the Third Berkeley Symposium on Mathematical Statistics and Probability,* **5**, 111-150. Berkeley: University of California Press.

Bentler, P. M. (1995). *EQS Structural Equations Program Manual.* Multivariate Software, Inc.: CA.

Bollen, K. A. (1987). Outliers and improper solutions: A confirmatory factor analysis example. *Sociological Meth. Res.*, **15**, 375-384.

Boomsma, A. (1985). Nonconvergence, improper solutions, and starting values in LISREL maximum likelihood estimation. *Psychometrika,* **50**, 229-242.

Gerbing, D. W. and Anderson, J. C. (1985). The effect of sampling error and model characteristics on parameter estimation for maximum likelihood confirmatory factor analysis. *Multi. Behav. Res.*, **20**, 255-271.

Gerbing, D. W. and Anderson, J. C. (1987). Improper solutions in the analysis of covariance structures: Their interpretability and a comparison of alternative respecifications. *Psychometrika,* **52**, 99-111.

Ihara, M and Okamoto, M. (1985). Experimental comparison of least-squares and maximum likelihood methods in factor analysis. *Statist. Probab. Lett.*, **3**. 287-293.

Jöreskog K. G. (1967). Some contributions to maximum likelihood factor analysis. *Psychometrika,* **32**, 443-482.

Jöreskog, K. G. and Sörbom, D. (1993). *LISREL 8: Structural Equation Modeling with the SIMPLIS Command Language* . Scientific Software International, Inc.: Chicago, IL.

Kano, Y. (1990). Noniterative estimation and the choice of the number of factors in exploratory factor analysis. *Psychometrika,* **55**, 277-291.

Kano, Y. (1997). Exploratory factor analysis with a common factor with two indicators. *Behaviormetrika,* **24**, 129-145.

Kano, Y. and Ihara, M. (1994). Identification of inconsistent variates in factor analysis. *Psychometrika,* **59**, 5-20.

Maxwell (1961). Recent trends in factor analysis. *Jour. Roy. Statist. Soc.*, **A124**, 49-59.

Sato, M. (1987). Pragmatic treatment of improper solutions in factor analysis. *Ann. Inst. Statist. Math.*, **39**, 443-455.

Van Driel, O. P. (1978). On various causes of improper solutions in maximum likelihood factor analysis. *Psychometrika,* **43**, 225-243.

PART IV

Symbolic and Textual Data Analysis

- **Symbolic Data Analysis**
- **Textual Data Analysis**

Symbolic Clustering Of Probabilistic Data

Paula BRITO
Faculdade de Economia do Porto
Rua Dr. Roberto Frias
4200 PORTO
PORTUGAL
e-mail : mpbrito@fep.up.pt

Abstract : We start by recalling the basic clustering algorithm for symbolic clustering using the hierarchical and pyramidal models. This algorithm allows to cluster a set of assertion objects and provides a clustering structure where each class is represented by an assertion object, generalising all its members. We then show how to extend this algorithm so as to take into account probability distributions on discrete variables. This extension is made by suitably adapting the generalisation step and the generality degree used by the algorithm.

Key-Words : Symbolic Data Analysis, Hierarchical Clustering, Pyramidal Clustering, Conceptual Clustering, Probabilistic Data

1. The Clustering Models

The proposed methods use as clustering models the hierarchical and the pyramidal structures. Let Ω be the (finite) set of objects being clustered.

A hierarchy on Ω is a family H of non-empty subsets of Ω, such that :

$$\forall a \in \Omega, \{a\} \in H \quad ; \quad \Omega \in H$$
$$\forall h, h' \in H, \quad h \cap h' = \emptyset \ \text{ or } \ h \subseteq h' \ \text{ or } \ h' \subseteq h$$

An hierarchy is hence a set of nested partitions of Ω.
The pyramidal model (Diday (1984, 1986), Bertrand (1986, Bertrand, Diday (1985)) generalises hierarchies by allowing non-disjoint classes at each level.
A pyramid is defined as a family P of non-empty subsets of Ω, such that :

$$\forall a \in \Omega, \{a\} \in P \quad ; \quad \Omega \in P$$
$$\forall p, p' \in P, \quad p \cap p' = \emptyset \ \text{ or } \ p \cap p' \in P$$
there exists an order θ / each element of P is an interval of θ

A pyramid is a family of nested overlappings on Ω, subjected to the condition of a total linear order. Hence, a pyramid provides both a clustering and a seriation on the data.

Pyramids can be considered as being halfway between hierarchies and lattices : the pyramidal model, leading to a system of clusters which is richer than that produced by hierarchical clustering, allows for the identification of concepts that the hierarchical model would not identify ; the existence of a compatible order on the objects leads, however, to a structure which is much simpler than lattices (absence of crossing in the graphical representation) and hence much easier to interpret.

2. Symbolic Hierarchical-Pyramidal Clustering The Basic Algorithm

An algorithm for symbolic hierarchical-pyramidal clustering has been proposed (Brito (1991, 1994)). This algorithm allows to cluster a set of symbolic assertion objects. The criterion that guides the formation of clusters is the duality intention-extension : a cluster is formed if it can be represented by a complete assertion object (conjunctive description) whose extension is the cluster itself. Recall that an assertion is complete if it describes exhaustively its extension, and if it is the less general assertion to fulfil this condition.
The proposed algorithm proceeds in a bottom-up way :
Starting with the one-object clusters $\{a_i\}$, i = 1,...,n at each step, form a cluster p union of $p_1, \ldots, p_j$ and represented by s such that

a) $p_1, \ldots, p_j$ can be merged together :
 - if the structure is a hierarchy : none of them has been aggregated before ;
 - if the structure is a pyramid : none of them has been aggregated twice, and p is an interval of a total order θ on Ω

b) s is complete

c) Generalisation step : s is more general than $s_1, \ldots, s_j \Rightarrow$ choose $s = s_1 \cup \ldots \cup s_j$

d) $ext_\Omega\, s = p$

Recall that if

$s_1 = [y_1 = V_1] \wedge \ldots \wedge [y_p = V_p]$ and $s_2 = [y_1 = W_1] \wedge \ldots \wedge [y_p = W_p]$ then

$$s_1 \cup s_2 = [y_1 = V_1 \cup W_1] \wedge \ldots \wedge [y_p = V_p \cup W_p].$$

The non-unicity of the clusters meeting these conditions lead to the definition of a numerical criterion, which was called "generality degree" and which measures the part of the description space covered by an assertion :
Let $a = \wedge [y_i = V_i]$, $V_i \subseteq O_i$, and admit that O_i is bounded, $1 \leq i \leq p$ (this is not a major restriction, since the set Ω to be clustered is always finite). The generality degree of a is defined as :

$$G(a) = \prod_{i=1}^{p} \frac{c(V_i)}{c(O_i)} = \prod_{i=1}^{p} G(e_i)$$

$$\text{if} \quad a = \wedge e_i$$

where c stands for cardinal if O_i is discrete and interval length if O_i is continuous. Notice that $G(a \cup b)$ is **NOT** a dissimilarity measure between a and b.
The cluster to be formed should meet conditions a)-d), and minimise G(s).

An indexed hierarchy is a couple (H,f), where H is a hierarchy and f a mapping , $f: H \rightarrow R^+$, such that :

i) $f(p) = 0 \Leftrightarrow card(p) = 1$;
ii) $\forall p,q \in H, p \subseteq q \Rightarrow f(p) \leq f(q)$

As in the case of hierarchies, a height is associated to each cluster of a pyramid, defining an *indexed pyramid* :
Let f be a mapping defined on a pyramid P and taking values in R^+. (P,f) is said to be a weakly indexed pyramid if

i) $f(p) = 0 \Leftrightarrow card(p) = 1$;
ii) $\forall p,q \in P, p \subseteq q \Rightarrow f(p) \leq f(q)$
iii) $\forall p,q \in P, p \subseteq q$ and $f(p) = f(q) \Rightarrow$
$\exists p_1, p_2 \in P : p_1 \subseteq p,\ p_2 \subseteq p,\ p_1 \cap p_2 = p$

We index the hierarchy or pyramid by G(s), that is, if s is the assertion object associated to cluster p, than the height of p is G(s). G is strictly increasing, so it meets conditions ii) and iii). Condition i) is not met, since $G(s) > 0 \ \forall$ s.
The algorithm comes to an end when either
- a cluster corresponding to the whole set Ω is formed or
- no cluster meeting conditions a) and d) can be formed ; in this case we obtain an incomplete hierarchy / pyramid.

3. Symbolic Clustering of Probabilistic Data

The aim is to extend the method so as to take into account probability / frequency distributions on discrete variables, that is, so that it allows to cluster assertions which comprehend probabilistic elementary events of the form :

$$[\, y_i = \{ m_1 (p_1), \dots, m_k (p_k) \} \,]$$

where $\{m_1, \ldots, m_k\}$ is the set of categories of variable y_i, $0 \le p_i \le 1$ and $p_1 + \ldots + p_k = 1$.

This type of symbolic objects arises in describing individuals for which variable y_i is a random variable. If the corresponding probability distribution is known, then $p_j = \text{Prob}(y_i = m_j)$, j=1,...,k. Otherwise, these probabilities are estimated by sample frequencies (which implies the repeated observation of variable y_i on each individual). As an example, we can consider the clustering of entities for which repeated measurements on some physical or chemical parameters (variables) have been done.
This extension leads to some adaptation changes in two steps of the algorithm :

- the generalisation step ;
- the computation of the Generality Degree.

Generalisation of two probabilistic elementary events can be done in two ways, each with a particular semantics :

a) <u>Generalisation by the Supremum</u> :
In this case, generalisation of two elementary events is to be done by taking, for each category m_j the supremum of its probabilities :

$$[y_i = \{m_1^1(p_j^1), , m_k^1(p_k^1)\}] \quad \cup \quad [y_i = \{m_1^2(p_1^2), , m_k^2(p_k^2)\}] \quad =$$

$$[y_i = \{m_1(p_1), , m_k(p_k)\}] \quad \text{where } p_j = \max\{p_j^1, p_j^2\}, j=1,\ldots,k.$$

Notice that the generalised elementary event is no longer probabilistic, in the sense that the obtained distribution is no longer a probability (or frequency) distribution : $p_1 + \ldots + p_k \ge 1$.
The extension of such an elementary event e , as concerns variable y_i, is the set of probabilistic assertions a defined by

$$\text{ext } e = \{a \,/\, p_j^a \le p_j, j=1, \ldots, k\}$$

and consists of probabilistic assertions such that for each category of variable y_i, the probability (frequency) of its presence is <u>at most</u> p_j .
The Generality Degree of an elementary event e is then defined as :

$$G_1(e) = \frac{\sum_{j=1}^{k} p_j}{k}$$

b) Generalisation by the Infimum :

In this case, generalisation of two elementary events is to be done by taking, for each category m_j the infimum of its probabilities :

$$[y_i = \{m_1^1(p_1^1),..., m_k^1(p_k^1)\}] \quad \cup \quad [y_i = \{m_1^2(p_1^2),..., m_k^2(p_k^2)\}] \quad =$$

$$[y_i = \{m_1(p_1), \;, m_k(p_k)\}] \quad \text{where } p_j = \min\{p_j^1, p_j^2\}, j=1,...,k.$$

The generalised elementary event is no longer probabilistic : in this case,

$$p_1 + ... + p_k \leq 1.$$

The extension of such an elementary event e , as concerns variable y_i, is the set of probabilistic assertions a defined by

$$\text{ext } e = \{a \,/\, p_j \leq p_j^a, j=1, ..., k\}$$

and consists of probabilistic assertions such that for each category of variable y_i, the probability (frequency) of its presence is at least p_j .

The Generality Degree of e is then defined as :

$$G_2(e) = \frac{\sum_{j=1}^{k}(1 - p_j)}{k}$$

In each case, ext e $\supseteq$ ext e_1 and ext e $\supseteq$ ext e_2, that is, $e = e_1 \cup e_2$ is more general than both e_1 and e_2. Also, $G(e) \geq G(e_1)$ and $G(e) \geq G(e_2)$, that is, the generality degree is increasing with respect to generalisation.

Example :

Let y_i be the variable "colour", and O_i = {blue, green, violet}.

Let e_1 = [colour = {blue (0.4) , green (0.4), violet (0.2) }] and

e_2 = [colour = {blue (0.5) , green (0.5) }]

$G_1(e_1) = G_1(e_2) = 1/3$; $G_2(e_1) = G_1(e_2) = 2/3$

Supremum case :

$e = e_1 \cup e_2$ = [colour = {blue (0.5) , green (0.5), violet (0.2) }] ;

$p_1 + p_2 + p_3 = 1.2 > 1$; $G_1(e) = 1.2 / 3 = 0.4$

Infimum case :

$e = e_1 \cup e_2$ = [colour = {blue (0.4) , green (0.4) }]

$p_1 + p_2 + p_3 = 0.8 < 1$; $G_2(e) = 2.2 / 3$

Conclusion

In this paper we have proposed a symbolic clustering method for probabilistic (frequency) data. The method is an extension of the previously developed symbolic clustering method, using the hierarchical or the pyramidal models as clustering structure. Two different semantics have been considered in generalising probability distributions. The obtained clusters are represented by symbolic objects which allow for an interpretation, in the considered semantics.

References

BERTRAND, P. (1986), *Etude de la Représentation Pyramidale*, Thèse de 3ème cycle, Univ. Paris-IX Dauphine

BERTRAND, P.; DIDAY, E. (1985), A visual representation of the compatibility between an order and a dissimilarity index : The Pyramids, *Computational Statistics Quarterly*, Vol. 2, Issue 1, 1985, pp. 31-42

BRITO, P. (1991), *Analyse de Données Symboliques. Pyramides d'Héritage.* Thèse, Mathématiques de la Décision, Univ. Paris-IX Dauphine.

BRITO, P. (1994), Use of Pyramids in Symbolic Data Analysis, in : *Proc. of the 4th Conference of the International Federation of Classification Societies*, IFCS-93, Diday, E. et al . (Eds.), Springer-Verlag, pp. 378-386

BRITO, P., (1995), Symbolic Objects : Order Structure and Pyramidal Clustering, *Annals of Operations Research*, 55, pp. 277-297.

DIDAY, E. (1984), Une représentation visuelle des classes empiétantes : les pyramides, *Rapport de Recherche INRIA* N° 291, Avril 1984, INRIA, Rocquencourt, Le Chesnay

DIDAY, E. (1986), Orders and overlapping clusters by pyramids, in : *Multidimensional Data Analysis Proc.*, J. De Leeuw et al. (Eds.), DSWO Press, Leiden

DIDAY, E. (1987), The symbolic approach in clustering and related methods of data analysis: the basic choices, in : *Proc. of the First Conf. of the International Federation of Classification Societies,* Technical Univ. of Aachen, Bock, H. (Ed.)

DIDAY, E. ; BRITO, P. ; MFOUMOUNE, E., Learning Probabilistic Models by Conceptual Pyramidal Clustering, in : *Progress in Artificial Intelligence*, Filgueiras, M. & Damas, L. (Eds.), Springer-Verlag, pp.362.

DIDAY, E. ; EMILION, R. (1996), Capacities and Credibilities in Analysis of Probabilistic Objects, in : *Ordinal and Symbolic Data Analysis,* Diday, E., Lechevallier, Y. & Opitz, O. (Eds.), Springer-Verlag , pp.13-30

DIDAY, E. ; EMILION, R. (1997), Treillis de Galois Maximaux et Capacités de Choquet, *Compte-Rendu à l'Académie des Sciences de Paris*, Série I (Analyse Mathématique), t. 325, N° 3.

Statistical proximity functions of Boolean symbolic objects based on histograms

Francisco de A. T. de Carvalho and Renata M. C. R. de Souza
Departamento de Estatística - CCEN / UFPE
Av. Prof. Luiz Freire s/n - Cidade Universitária
CEP: 50740-540 Recife - PE BRASIL
Fax: 55-81-2718422 and E-mail: fatc@npd.ufpe.br

Abstract: In this paper we make a synthesis between the Ichino and Yaguchi (1994) and Moore (1991) metrics to obtain a new logical proximity function between Boolean symbolic objects. Then, we use histograms defined on these objects to obtain a statistical one. For both logical and statistical proximity functions, we study its properties and we present examples to illustrate the usefulness of our approach.

Keywords: Symbolic Objects, Proximity Functions, Histograms.

1. Introduction

In conventional exploratory data analysis each variable takes a single value. Nowadays statistical data bases are able to store more general data spreading from single value to interval or set of values. Such data set are called Boolean symbolic objects. To extend methods of usual data analysis to these objects is the main aim of the Symbolic Data Analysis (Diday (1991)).

Proximity functions play a major role in exploratory data analysis. For example, clustering and ordination may have as input data a matrix of proximities. Proximity functions between Boolean symbolic objects can be found in Gowda and Diday (1991), Ichino and Yaguchi (1994), De Carvalho (1994) and Gowda and Ravi (1995).

Marcotorchino (1991) makes a distinction between logical and statistical proximity functions. Logical proximity functions are based on local informations (i.e., informations given only by the descriptions of the pair of objects considered) while statistical ones are based also on statistical informations concerning the whole set of objects.

In this paper we propose new logical proximity functions between Boolean symbolic objects which are based on symbolic operators (join and conjunction) and on a positive measures (description potential) defined on these objects. Then, we combine histograms, defined on these objects (De Carvalho (1995)), with the logical proximity functions to obtain statistical proximity functions. For both logical and statistical proximity functions, we study its properties and we present examples to illustrate the usefulness of our approach.

2. Symbolic objects

Let Ω be the set of individuals and $\omega \in \Omega$ an individual. A variable is a function y_i: $\Omega \rightarrow O_i$ where O_i is the set of values that y_i may take. A variable may take no values, a single value or several values (a discrete set or an interval) for a symbolic object. Let $\{y_1, \ldots, y_p\}$ be a set of p variables.

A Boolean symbolic object (Diday (1991), denoted as $a = [y_1 \in A_1] \wedge \ldots \wedge [y_p \in A_p]$, where $A_i \subseteq O_i$, $\forall i \in \{1, \ldots, p\}$, expresses the condition, "Variable y_1 takes its values in A_1 *and* ... *and* variable y_p takes its values in A_p". We define a mapping associated to it as a: $\Omega \rightarrow \{1,0\}$ such that $a(\omega) = 1 \Leftrightarrow y_i(\omega) \in A_i$, $\forall i \in \{1, \ldots, p\}$. The extension of a on Ω is defined as $ext_\Omega(a) = \{\omega \in \Omega \,/\, a(\omega) = 1\}$.

Example. Consider the Boolean symbolic object a = [size $\in$ [0, 15]] $\wedge$ [Colour $\in$ {red, white}]. An individual ω is such that $a(\omega) = 1$ if, and only if, its size is between 0 and 15 and its colour is red or white.

Operations.

Let $a = [y_1 \in A_1] \wedge \ldots \wedge [y_p \in A_p]$ and $b = [y_1 \in B_1] \wedge \ldots \wedge [y_p \in B_p]$ two Boolean symbolic objects.

The following operations (Diday (1991), Ichino and Yaguchi (1994)) may be defined:

a) The join is defined as $a \oplus b = [y_1 \in A_1 \oplus B_1] \wedge \ldots \wedge [y_p \in A_p \oplus B_p]$, where

- if y_i is quantitative or ordinal qualitative, $A_i \oplus B_i = [\min(A_{iL}, B_{iL}), \max(A_{iU}, B_{iU})]$, where (A_{iL}, B_{iL}) and (A_{iU}, B_{iU}) are the lower bound and the upper bound of the intervals A_i and B_i, respectively;
- if y_i is nominal qualitative, $A_i \oplus B_i$ is the union between A_i and B_i.

b) The conjunction is defined as $a \wedge b = [y_1 \in A_1 \cap B_1] \wedge \ldots \wedge [y_p \in A_p \cap B_p]$, where $A_i \cap B_i$, $i \in \{1,\ldots,p\}$, is the intersection between the sets A_i and B_i.

Description potential.

Let $a = \wedge_{i=1}^{p} a_i = \wedge_{i=1}^{p} [y_i \in A_i]$, $a_i = [y_i \in A_i]$, a Boolean symbolic object.

The description potential of a (De Carvalho (1994)), denoted $\pi(a)$, is the volume of the Cartesian product $A_1 \times \ldots \times A_p$ formed by the individual descriptions given by a, and it is calculated by the following expression:

$$\pi(a) = \prod_{i=1}^{p} \mu(A_i) \tag{1}$$

where

$$\mu(A_i) = \begin{cases} \text{cardinal}(A_i), \text{ if } y_i \text{ is qualitative or discrete quantitative} \\ \text{range}(A_i), \text{ if } y_i \text{ is quantitative continuous} \end{cases} \tag{2}$$

Example. Let a = [Colour $\in$ {black, white, red}] $\wedge$ [Height $\in$ [50,150]]. So, $\pi(a) = 3\ (150 - 50) = 300$.

3. Proximity functions

Let $a_1,\ldots, a_n$ a set of n Boolean symbolic objects and let $a = \wedge_{i=1}^{p} a_i = \wedge_{i=1}^{p} [y_i \in A_i]$, $a_i = [y_i \in A_i]$ $\forall i \in \{1,\ldots,p\}$, and $\hat{a} = \wedge_{i=1}^{p} \hat{a}_i = \wedge_{i=1}^{p} [y_i \in \hat{A}_i]$, $\hat{a}_i = [y_i \in \hat{A}_i]$ $\forall i \in \{1,\ldots,p\}$, where $a, \hat{a} \in \{ a_1,\ldots, a_n\}$, two Boolean symbolic objects. The dissimilarity between a pair of Boolean symbolic object will be calculated by a comparison between them according to each variable and by the aggregation of these comparisons.

Logical comparison functions

We define the logical dissimilarity between the pair of Boolean symbolic objects $(a, \hat{a})$ for the variable y_i as

$$\Phi_{1k\gamma}(a_i,\hat{a}_i)=\left\{\frac{1}{2}\left[\left(\phi_{1\gamma}(a_i,\hat{a}_i)\right)^k+\left(\phi_{2\gamma}(a_i,\hat{a}_i)\right)^k\right]\right\}^{\frac{1}{k}} \tag{3}$$

where $k \in \{1,2,\ldots\}$, $0 \le \gamma \le 0.5$ and

$$\phi_{1\gamma}(a_i,\hat{a}_i)=(1-2\gamma)\mu(A_i\cap\overline{\hat{A}_i})+\mu(\overline{A_i}\cap\hat{A}_i)+\mu(\overline{A_i}\cap\overline{\hat{A}_i}\cap(A_i\oplus\hat{A}_i)),$$

$$\phi_{2\gamma}(a_i,\hat{a}_i)=\mu(A_i\cap\overline{\hat{A}_i})+(1-2\gamma)\mu(\overline{A_i}\cap\hat{A}_i)+\mu(\overline{A_i}\cap\overline{\hat{A}_i}\cap(A_i\oplus\hat{A}_i)).$$

Two normalised versions of equation (3), where $k \in \{1,2,\ldots\}$ and $0\le\gamma\le0.5$, are:

$$\Phi_{2k\gamma}(a_i,\hat{a}_i)=\left\{\frac{1}{2}\left[\left(\frac{\phi_{1\gamma}(a_i,\hat{a}_i)}{\mu(O_i)}\right)^k+\left(\frac{\phi_{2\gamma}(a_i,\hat{a}_i)}{\mu(O_i)}\right)^k\right]\right\}^{\frac{1}{k}} \tag{4}$$

$$\Phi_{3k\gamma}(a_i,\hat{a}_i)=\left\{\frac{1}{2}\left[\left(\frac{\phi_{1\gamma}(a_i,\hat{a}_i)}{\mu(A_i\oplus\hat{A}_i)}\right)^k+\left(\frac{\phi_{2\gamma}(a_i,\hat{a}_i)}{\mu(A_i\oplus\hat{A}_i)}\right)^k\right]\right\}^{\frac{1}{k}} \tag{5}$$

Proposition 1. The proximity function $\Phi_{jk\gamma}$, $j \in \{1,2,3\}$, is a metric.

Remark. Taking $k = 1$, in equations (3) and (4), we get Ichino and Yaguchi (1994) metrics. Also, if all the variables take interval as values and if we take $\lim_{k\to\infty} \Phi_{1k\gamma}$ we get Moore (1979) metric.

Statistical comparison functions.

Let $\hat{a} = \wedge_{i=1}^{p} \hat{a}_i = \wedge_{i=1}^{p} [y_i \in \hat{A}_i]$, $\hat{a} \in \{ a_1,\ldots, a_n\}$. Let $m \in O_i$ a modality of a discrete variable y_i. The *relative weight* of m is defined as (De Carvalho (1995)):

$$\mathrm{RW(m)} = \left(\sum_{j=1}^{n} \frac{\delta(\mathrm{m})}{\mathrm{cardinal}(\hat{\mathrm{A}}_{\mathrm{i}})} \right) \Big/ \mathrm{n},\ \delta(\mathrm{m}) = \begin{cases} 0, \text{ if m} \notin \hat{\mathrm{A}}_{\mathrm{i}} \\ 1, \text{ if m} \in \hat{\mathrm{A}}_{\mathrm{i}} \end{cases} \tag{6}$$

Let y_i a continuous variable et let $I_1, \ldots, I_{n_i}$ n_i intervals which are a partition of O_i. The *relative weight* of I_α, $\alpha \in \{1, \ldots, n_i\}$, is defined as (De Carvalho (1995)):

$$\mathrm{RW}(\mathrm{I}_\alpha) = \left(\sum_{j=1}^{n} \frac{\mathrm{Range}(\mathrm{I}_\alpha \cap \hat{\mathrm{A}}_{\mathrm{i}})}{\mathrm{Range}(\hat{\mathrm{A}}_{\mathrm{i}})} \right) \Big/ \mathrm{n} \tag{7}$$

Let $\hat{a}_i = [y_i \in \hat{A}_i]$ and let

$$\mathrm{H}(\hat{\mathrm{A}}_{\mathrm{i}}) = \begin{cases} \displaystyle\sum_{\mathrm{m}/\mathrm{m}\in\hat{\mathrm{A}}_{\mathrm{i}}} \frac{1}{\eta(\mathrm{m})}, \text{ if } \mathrm{y}_{\mathrm{i}} \text{ is discrete,} \\ \text{where } \eta(\mathrm{m}) = \begin{cases} 1, \text{ if RW(m)} = 0 \\ \mathrm{RW(m)}, \text{ if RW(m)} \neq 0 \end{cases} \\ \\ \displaystyle\sum_{\mathrm{I}_\alpha/\mathrm{I}_\alpha\subseteq\hat{\mathrm{A}}_{\mathrm{i}}} \frac{\mathrm{Range}(\mathrm{I}_\alpha)}{\eta(\mathrm{I}_\alpha)}, \text{ if } \mathrm{y}_{\mathrm{i}} \text{ is continuous,} \\ \text{where } \eta(\mathrm{I}_\alpha) = \begin{cases} 1, \text{ if } \mathrm{RW}(\mathrm{I}_\alpha) = 0 \\ \mathrm{RW}(\mathrm{I}_\alpha), \text{ if } \mathrm{RW}(\mathrm{I}_\alpha) \neq 0 \end{cases} \end{cases} \tag{8}$$

The function H defined above gives a weight (which is the inverse of the relative weight according to equations (6) and (7)) to each value or interval which is in $\hat{A}_i$. Consequently, a value which is rarely observed (small relative weight) will have a big weight.

An extreme case arrive when a value or an interval has null relative weight. This means that this value or interval is not on the description of any Boolean symbolic object. In that case, the function H does not furnishes a weight to this value or interval.

We define the statistical dissimilarity between the pair of Boolean symbolic objects $(a, \hat{a})$ for the variable y_i as

$$\Phi_{4k\gamma}(\mathrm{a}_{\mathrm{i}}, \hat{\mathrm{a}}_{\mathrm{i}}) = \left\{ \frac{1}{2} \left[\left(\phi_{3\gamma}(\mathrm{a}_{\mathrm{i}}, \hat{\mathrm{a}}_{\mathrm{i}}) \right)^{k} + \left(\phi_{4\gamma}(\mathrm{a}_{\mathrm{i}}, \hat{\mathrm{a}}_{\mathrm{i}}) \right)^{k} \right] \right\}^{\frac{1}{k}} \tag{9}$$

where $k \in \{1,2,\ldots\}$, $0 \leq \gamma \leq 0.5$ and

$$\phi_{3\gamma}(\mathrm{a}_{\mathrm{i}}, \hat{\mathrm{a}}_{\mathrm{i}}) = (1-2\gamma)\mathrm{H}(\mathrm{A}_{\mathrm{i}} \cap \overline{\hat{\mathrm{A}}}_{\mathrm{i}}) + \mathrm{H}(\overline{\mathrm{A}}_{\mathrm{i}} \cap \hat{\mathrm{A}}_{\mathrm{i}}) + \mathrm{H}(\overline{\mathrm{A}}_{\mathrm{i}} \cap \overline{\hat{\mathrm{A}}}_{\mathrm{i}} \cap (\mathrm{A}_{\mathrm{i}} \oplus \hat{\mathrm{A}}_{\mathrm{i}})),$$

$$\phi_{4\gamma}(a_i,\hat{a}_i) = H(A_i \cap \bar{\hat{A}}_i) + (1-2\gamma)H(\bar{A}_i \cap \hat{A}_i) + H(\bar{A}_i \cap \bar{\hat{A}}_i \cap (A_i \oplus \hat{A}_i)).$$

Two normalised versions of equation (9) are:

$$\Phi_{5k\gamma}(a_i,\hat{a}_i) = \left\{ \frac{1}{2}\left[\left(\frac{\phi_{3\gamma}(a_i,\hat{a}_i)}{H(O_i)} \right)^k + \left(\frac{\phi_{4\gamma}(a_i,\hat{a}_i)}{H(O_i)} \right)^k \right] \right\}^{\frac{1}{k}} \quad (10)$$

$$\Phi_{6k\gamma}(a_i,\hat{a}_i) = \left\{ \frac{1}{2}\left[\left(\frac{\phi_{3\gamma}(a_i,\hat{a}_i)}{H(A_i \oplus \hat{A}_i)} \right)^k + \left(\frac{\phi_{4\gamma}(a_i,\hat{a}_i)}{H(A_i \oplus \hat{A}_i)} \right)^k \right] \right\}^{\frac{1}{k}} \quad (11)$$

where $k \in \{1,2,\dots\}$, $0 \le \gamma \le 0.5$.

Aggregation functions

We define the global dissimilarity between the pair of Boolean symbolic objects $(a, \hat{a})$ as

$$d_r(a,\hat{a}) = \frac{1}{p}\left[\sum_{i=1}^{p} \{\Phi_{jk\gamma}(a_i,\hat{a}_i)\}^r \right]^{\frac{1}{r}} \quad (12)$$

where $j \in \{1, 2, 3, 4, 5, 6\}$, $k \in \{1, 2, \dots\}$ and $r \in \{1, 2, \dots\}$

Remark. This global dissimilarity will be logical or statistical according to the comparison function choose in (12).

Proposition 2. The proximity function d_r is a metric if we choose $\Phi_{jk\gamma}$, $j \in \{1,2,3\}$, in (12).

4. Discussion

From the functions $\phi_{1\gamma}$ and $\phi_{2\gamma}$, we may see that the logical dissimilarity functions defined on equations (3), (4) and (5) are based on a measure of the values which expresses the differences between the pair of Boolean symbolic objects for the variable y_i ($\mu(A_i \cap \bar{\hat{A}}_i)$ and $\mu(\bar{A}_i \cap \hat{A}_i)$) and on a measure of the values which appear when y_i is qualitative ordinal or quantitative and when A_i and $\hat{A}_i$ are disjoint ($\mu(\bar{A}_i \cap \bar{\hat{A}}_i \cap (A_i \oplus \hat{A}_i))$).

As the logical dissimilarity functions, the statistical dissimilarity functions defined on equations (9), (10) and (11) are also based on a measure of the values which expresses the differences between the pair of Boolean symbolic objects for the variable y_i ($H(A_i \cap \bar{\hat{A}}_i)$ and $H(\bar{A}_i \cap \hat{A}_i)$) and on a measure of the values which appear when y_i is qualitative ordinal or quantitative and when A_i and $\hat{A}_i$ are disjoint $(H(\overline{A}_i \cap \overline{\hat{A}}_i \cap (A_i \oplus \hat{A}_i)))$ but now weighted by the inverse of the relative weight.

In the case of quantitative variables, to obtain a set of intervals forming a partition of its observation set, we propose to order the set of the lower bound and upper bound of the intervals of each symbolic object. Each interval is formed by two subsequent bounds. The idea is to get intervals which are strictly included on the set of values which is the description of each symbolic object.

In order to corroborate the efficacy of our approaches, several simulation studies with both logical and statistical dissimilarities were made. The effectiveness of our proximity functions were illustrated by the results given by a hierarchical conceptual algorithm based on simple distance calculation (Ichino and Yaguchi (1994)) applied on different data sets (chemical, biological and others). All the simulations made seems corroborate the approaches.

In these approaches, a great logical dissimilarity is get when the number of differences and the separation between the objects is important as waited, whereas a great statistical dissimilarity is get also when these differences and separation are rarely observed on the symbolic objects of the Knowledge base.

References

De Carvalho, F. A. T (1994). Proximity coefficients between Boolean symbolic objects. In: *New Approaches in Classification and Data Analysis*, Diday, E. et al (eds.), 387 - 394, Springer - Verlag, Berlin - Heidelberg.

De Carvalho, F. A. T (1995). Histograms in symbolic data analysis. *Annals of Operations Research*, 55, 299-322.

Diday, E. (1991). Des objets de l'analyse de données à ceux de l'analyse de connaissances. In: *Induction symbolique et numérique à partir de données*, Diday, E. et Kodratoff, Y. (eds.), 9 - 75, Cepadues Editions, Toulouse.

Marcotorchino, F. (1991). La classification automatique aujourdhui: bref aperçu historique, applicatif et calculatoire. *Publication Scientifiques et Techniques d'IBM France* 2, 35-94.

Gowda, K. C. and Diday, E. (1991). Symbolic clustering using a new dissimilarity measure. *Pattern Recognition* 24 (6), 567 - 578.

Gowda, K. C. and Ravi, T.V. (1995). Agglomerative clustering of symbolic object using the concepts of both similarity and dissimilarity. *Pattern Recognition Letters* 16 , 647 - 652.

Ichino, M. and Yaguchi, H. (1994). General Minkowsky metrics for mixed features type data analysis. *IEEE Transaction on System, Man and Cybernetics* 24, 698 - 708.

Moore, R. E. (1979). *Interval Analysis*. Prentice Hall Inc., Englewood Cliffs, N. J.

Vertices Principal Components Analysis With an Improved Factorial Representation

A. Chouakria , E. Diday and P. Cazes
Université Paris IX Dauphine, LISE-CEREMADE
Place du Maréchal de Lattre de Tassigny F-75016, Paris France.
Ahlame.Chouakria@inria.fr, Chouakri@ceremade.dauphine.fr

Abstract: We propose, in the framework of the Symbolic Data Analysis (Diday 89), a new method which takes as input an array of objects described by interval data (Cazes & al. 97). In the factorial space, objects are described by an interval type principal components. It is therefore, represented in the factorial plane by a rectangle, a segment or a point . The interpretation parameters are generalised accordingly. An additional iterative subroutine is proposed to give a rectangular representation which takes into account the relative contributions of the objects. An application on ``face's recognition'' is provided to illustrate the effectiveness of the proposed whole method.

Keywords: Principal component analysis, interval data, Symbolic Data Analysis.

1. Introduction

Where do the interval data come from ? An interval can be the value taken by an attribute describing a database's object, or the value taken by a slide describing an expert system's frame, or merely the value taken by a feature describing a class obtained by an automatic classification algorithm. An interval can also be the result of an imprecise measure, a confidence interval, or the summary of a set of observations, etc. Semantically, an interval expresses imprecision or variability knowledge.

Intervals expressing an imprecise knowledge, are often of little span and denoted $[x - \delta, x + \delta]$ where, x is the observed value and δ is the imprecision of the measuring instrument. Intervals expressing variability knowledge, generally summarise a set of values. For instance, the description of the number of petals of the object plant, which has many flowers, is the interval which encloses the numbers of petals of all the flowers of the plant.

We wish to extend principal component analysis to interval data. We suppose that all the information at our disposal is the bounds of the intervals with no other consideration towards supposed probability distributions. If so many data are available, other data analysis methods would be more appropriate. However, in any case the synthetic power of interval representation should be worth considering. Our Vertices Method applies the dimensionality reduction of Principal Component Analysis while coping with the specific semantics expressed by interval data.

2. Interval Data

Let $H_1,\dots,H_m$ be m objects described by q features $X_1,\dots,X_q$ of interval type.

$$X_H = \begin{pmatrix} H_1 \\ \dots \\ H_m \end{pmatrix} = \begin{pmatrix} [\underline{x_{11}},\overline{x_{11}}] & \cdots & [\underline{x_{1q}},\overline{x_{1q}}] \\ \vdots & \ddots & \vdots \\ [\underline{x_{m1}},\overline{x_{m1}}] & \cdots & [\underline{x_{mq}},\overline{x_{mq}}] \end{pmatrix} \quad \textbf{(1)}$$

where, $[\underline{x_{ij}},\overline{x_{ij}}]$ is the interval taken by the feature X_j for the object H_i and $\underline{x_{ij}}$, $\overline{x_{ij}}$ are, respectively, the lower and the upper bounds of feature X_j for object H_i. An interval $[\underline{x_{ij}},\overline{x_{ij}}]$ is called trivial if it is reduced to a point : $\underline{x_{ij}} = \overline{x_{ij}}$. Let q_i be the number of no trivial intervals describing the object H_i. Such object can be represented in the descriptive space, defined by the q features X_j, by an hyper-parallelogram-rectangle (called simply hyper-cube in the sequel) of $n_i = 2^{q_i}$ vertices. The lengths of the sides of the hyper-cube are given by the span of the intervals of the corresponding descriptive features. We use in the following, crudely the term hyper-cube for referring a point ($q_i = 0$), a segment ($q_i = 1$), a rectangle ($q_i = 2$), etc. We propose to describe each object H_i defined in **(1)** by a matrix $X_{H_i}(x_{S^i_k j})$ of n_i rows and q columns, where $x_{S^i_k j}$ is the value taken by the feature X_j for the k-th vertex of the hyper-cube H_i. The matrix X_{H_i} gives the description of $\{S^i_1,\dots,S^i_{n_i}\}$ the set of the vertices of the hyper-cube H_i by the q features X_j.

3.Objects and Vertices weighting

Let p_i be the weight of the object H_i. The main idea of the Vertices Method is to distribute all the mass of the object over it's vertices. Let x^o_{ij} be the reference value that best represents the interval $[\underline{x_{ij}},\overline{x_{ij}}]$. If the distribution inside the interval is known x^o_{ij} is the mean, the mode... of the distribution. If no information is available, x_{ij}^{o} is taken as the centre of the interval. We determine the probabilistic coefficients $\underline{p_{ij}}$, $\overline{p_{ij}}$ of , respectively, the bounds $\underline{x_{ij}}$ and $\overline{x_{ij}}$ so that x_{ij}^{o} becomes the gravity centre of the bounds $\underline{x_{ij}}$ and $\overline{x_{ij}}$: $\underline{p_{ij}} + \overline{p_{ij}} = 1$ and $\underline{p_{ij}}\ \underline{x_{ij}} + \overline{p_{ij}}\ \overline{x_{ij}} = x_{ij}^{o}$. We define the weight $p_{S_k^i}$ of the vertex S_k^i as the product of the probabilistic coefficients of the associated bounds : $p_{S_K^i} = \prod_{j=1}^{q_i} p\left(x_{S_{Kj}^i}\right) p_i$; with with, $p(x_{S_K^i}) = \underline{p_{ij}}$ if $x_{S_{kj}^i} = \underline{x_{ij}}$ and $p(x_{S_K^i}) = \overline{p_{ij}}$ if $x_{S_{kj}^i} = \overline{x_{ij}}$. We can consider other operators than the product.

4. Principal components scores

Let X ($x_{S_{kj}^i}$) be the data matrix of (n=$\sum_{i=1}^{m} n_i$) rows and q columns, to be diagonalised. We note F($pc_{S_{kj}^i}$) the matrix of (n= $\sum_{i=1}^{m} n_i$) rows and p columns giving the coordinates of the vertices of the hyper-cubes, on the p first principal components. The coordinate of the hyper-cube H_i on the j-th principal component of interval type is : $\underline{pc_{ij}} = \min_{k=1}^{n_i} pc_{S_{kj}^i}$ and $\overline{pc_{ij}} = \max_{k=1}^{n_i} pc_{S_{kj}^i}$.

Let Vs be the variance-covariance matrix of general term $v_{jj}^{s} = \sum_{i=1}^{m} \sum_{k=1}^{ni} p_{S_k^i} (x_{S_{kj}^i})^2$, it can be defined as the sum of two matrices Vc and E. Vc of general term $v_{jj}^{c} = \sum_{i=1}^{m} p_i (x_{ij}^{o})^2$ is the variance-covariance matrix of numerical features obtained by substituting a single central value to each interval. E of general term $e_{jj} = \sum_{i=1}^{m} pi\, \underline{p_{ij}}\, \overline{p_{ij}} (\overline{x_{ij}} - \underline{x_{ij}})^2$ is a diagonal matrix which expresses the variability due to the interval data : $v_{jj}^{s} = v_{jj}^{c} + e_{jj}$

5. Generalised interpretation parameters

The parameters of interpretation are easily extended to the objects H_i. To measure the relative contribution of the object H_i. according to the j-th

factorial axis, let us introduce : $Cor(H_i, u_j) = \frac{p_{S_k^i}}{p_i} \sum_{k=1}^{n_i} \frac{pc_{S_{kj}^i}^2}{d^2(S_k^i, G)}$

which indicates the mean of the square cosines of the angles made by the n_i vertices and the j-th factorial axis. We denote the relative contribution of S_k^i according to the j-th axis $Cor(S_k^i,u_j)=\sum_{k=1}^{n_i}\frac{pc_{S_k^i j}^2}{d^2(S_k^i,G)}$. Similarly, we measure the contribution of H_i to the inertia λ_j of the j-th factorial axes: $Ctr(H_i,u_j)=\frac{\sum_{k=1}^{n_i} p_{S_k^i}\, pc_{S_k^i j}^2}{\lambda_j}$ and to the total inertia I_T of the n vertices of the m objects : $Inr(H_i)=\frac{\sum_{k=1}^{n_i} p_{S_k^i} d^2(S_k,G)}{I_T}$

6. More precise factorial representation

The rectangle ($[\underline{pc_{ij}},\overline{pc_{ij}}],[\underline{pc_{il}},\overline{pc_{il}}]$) representing the object H_i, in the factorial plan (P_{c_j},P_{c_l}) , is an over-cover of the hyper-cube projection. It is built from all the vertices S_k^i. without regards to the relative contribution of the vertices on the factorial plan (P_{c_j},P_{c_l}). We propose an iterative subroutine which gives for each pair of principal components (P_{c_j},P_{c_l}), and for each level α ($0\le\alpha\le 1$), the representation of the object H_i, by a rectangle (point or segment) built from the vertices S_k^i whose relative contributions $Cor(S_k^i)=Cor(S_k^i,u_j)+Cor(S_k^i,u_l)$ according to the j-th and the l-th factorial axes, are greater or equal to α. In the present work the level α is a priori fixed by an expert. Our current studies aim to define the level α in terms of the quality representation of the objects. The proposed subroutine, allows to estimate the accuracy of the overlap, and of the closeness of the objects, in the factorial plane. Let $[\underline{pc_{ij}^{\alpha}},\overline{pc_{ij}^{\alpha}}]$ be the coordinate of the object H_i, on the j-th axis of the factorial plane $(P_{c_j},P_{c_l})^{\alpha}$ of the level α :

$$\underline{pc_{ij}^{\alpha}}=\min_{k=1}^{n_i}\left\{pc_{S_k^i j}\ /\ Cor(S_k^i)\ge\alpha\right\},\ \overline{pc_{ij}^{\alpha}}=\max_{k=1}^{n_i}\left\{pc_{S_k^i j}\ /\ Cor(S_k^i)\ge\alpha\right\}$$

7. Face recognition application

Data description : The face recognition data (Leroy & al. 96) describes 27 images of 9 human faces : {Fra (o), Hus (Δ), Inc(+), Isa ($\times$), Jpl($\Diamond$), Kha,(∇), Lot($\otimes$), Phi($*$), Rom (<+>) }. Three images per person are taken. The 27 images are described by 6 features X_j of interval type. Each feature X_j measures a distance (by a number of pixel) between two particular points of the face (figure 1). The input data can be summarised in a table of 27 rows and 6 columns. The interval feature expresses the imprecision measure due to the rotation, the deviation, the removal, etc. of the face while taking the image.

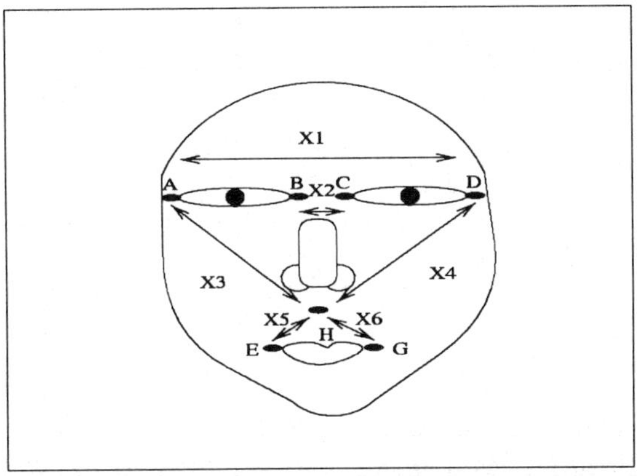

Figure 1: Face measures

Results : The Vertices Method applied to the previous described data draws off three distinct classes of similar human faces : {Hus, Jpl, Phi}, {Fra, Inc} and {Kha, Lot, Isa, Rom}. Each image is represented, in the factorial plane, of the two first principal components by a rectangle, a segment or a point. The representation of the 27 human faces images, in the factorial plane of the two first principal components, of the level α = 0.2 is given in the figures 2 (a). A more precise representation of the level α = 0.6 is given in the figure 2 (b). We can say that objects which are represented by horizontal segments are characterised by high variation on the most correlated features with PC1, whereas the objects represented by vertical segments are characterised by high variation on the most correlated features with PC2.

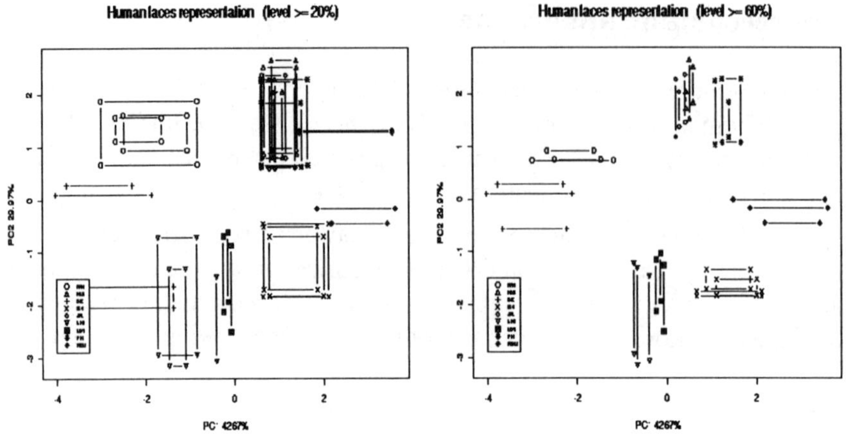

Figure 2: (a) (b)

8. References

Cazes, P. & Chouakria, A. & Diday, E, Schecktman, Y. (1997). Extension de l'analyse en composantes principales à des données de type intervalle. in : *Rev. Statistique. Appliquée,* XLV, 5-24.

Diday, E. (1989). Data Analysis, Learning Symbolic and Numeric knowledge, *Nova Science*, Antibes, France.

Leroy, B. & Chouakria, A. & Herlin, I. & Diday, E. (1996). Approche géometrique et classification pour la reconnaissance de visage. *RFIA'96, Reconnaissance des Formes et Intelligence Artificielle,* 548-557.

On the Complexity of Computation with Symbolic Objects using Domain Knowledge

Marc Csernel
INRIA-Rocquencourt, Domaine De Voluceau BP 105,
78153 Le Chesnay Cedex, France. E-mail: Marc.Csernel@inria.fr

Abstract: Instead of representing data as a point within the description space, Symbolic Objects represent them as hyper-rectangles, to take into account some variability within the description. They also enable to add some domain knowledge represented by rules which reduce the description space. Unfortunately, this supplementary knowledge induces most of the time a combinatorial growing of the possible calculus time. In a preceding paper we presented a method leading to a decomposition of symbolic objects into a Normal Symbolic Form (NSF) which allows an easier calculation. In this paper we will detail the complexity of computation, comparing the usual method and the one induced by the NSF.

Key words: Symbolic Objects, rules, complexity

1. Introduction

Constrained Boolean Symbolic Objects defined by Diday (1991) are better adapted than usual objects of data analysis to describe classes of individuals such as populations or species, being able to take into account variability. They are expressed by a logical conjunction of elements called elementary events, and each of this elementary event represents a set of values associated with a variable. Each Boolean symbolic object describes a volume which is a subset of the Cartesian product of the description variables.

Some domain knowledge can be added to a set of symbolic objects by different kinds of rules which express dependencies between the variables.

These rules reduce the description space, they interfere greatly on the different computation between symbolic objects.

We shall focus on the computation of a useful measure function: the description potential (De Carvalho 1997) of each object. We define the description potential as the part of the volume described by a symbolic object which is coherent, i.e. where all the values are satisfying all the dependencies rules.

Until now the methods used to compute description potential were taking rules into account by computation of the incoherent part of each object. This computation is combinatorial with the number of rule.
In a preceding paper (Csernel & De Carvalho 1997) we presented a method leading to a decomposition of symbolic objects into a Normal Symbolic Form which lead to an easier computation.
We will see that most of the time we get a polynomial time of computation.

2. Boolean Symbolic Object (S.O.)

Let Ω be a set of elementary objects generally called "individuals", described by p variables y_i where $i \in \{1,...p\}$. Let O_i be a set of observations where the variable y_i takes its values. The set $O = O_1 x O_2 ... x Op$ is then called the description space.
An elementary event e_i ,denoted by the symbolic expression $ei = yi \in Vi$ where $i \in \{1,..p\}$, $Vi \subseteq Oi$, expresses that "the variable y_i takes its values in Vi".
A Boolean symbolic object a, is a conjunction of elementary events of the form: $a = \wedge \{e_i = [y_i \in Vi] \}$ where the different $Vi \subseteq Oi$ represent the intention of a set of individuals $C \subseteq \Omega$. It can be interpreted as follows: the values of y_i for any individuals $c \in C$ are in Vi.
Example : a = [colour ∈ *{blue, red}*] ∧ [size ∈ *{small, medium}*]
means that the colour of *a* is red or blue and the size small or medium.
SO can be associated with a domain knowledge which constraints the description given by the objects. Such constraints are expressed as dependencies between the variables. We take into account two kinds of dependencies expressed by rules.
- The first one, called hierarchical dependencies is of the form:
  if A $\in \{a_1,.. a_j,..a_n\}$ *then B has No Sense*, as in :
  if wings ∈ {Absent} then wings_color = No_Sense (r1)
- The other, called logical dependencies, is of the form:
  if A $\in \{a_1,.. a_j,..a_n\}$ *then B* $\in \{b_1,.. b_k..,b_m\}$ as in :
  if wings_color ∈ {red} then Thorax_color ∈ {blue} (r2)

The term No_Sense means that the variable does not exist, hence it's value is not applicable. Both of these constraint reduce the description space, but the first one reduces the number of dimension, the second not. We will call premise variable a variable which is present within a premise.

3. The Description Potential (D.P.)

The description potential is the measure of the coherent part of the hypervolume described by a symbolic object. By coherent we means not in contradiction with

the rules. The D.P. is a good measure to compare symbolic object, mostly used (but not only) in distance computation. If the computation of the D.P. is straight when no rules occurs, it become combinatorial when rules are present, because one have to drop out of the D.P. the volume represented by each rule, but to put back the volume represented of the intersection of the rules two by two, the to drop out the volume of the intersection of the rules three by three and so on.
So the computation is represented by the following formulae due to De Carvalho, and related to the Poincarre Theorem, where $\pi(a)$ represent the D.P. of an object, *V(a)* the hypervolume described by the object *a* and r_i the different rules which constrained the objects.

$$\pi(a) = V(a) - \pi(a \wedge (\neg(r_1 \wedge \ldots \wedge r_t))) \text{ and}$$
$$\pi(a \wedge (\neg(r_1 \wedge \ldots \wedge r_t))) = \Sigma_{j=1}^{t} \pi(a \neg r_j) - \Sigma_{j<k} \pi((a \neg r_j) \wedge r_k) + \ldots \quad (1)$$
$$+ (-1)^{T-1} \pi(\ldots(a \neg r_1) \wedge r_2) \ldots) \wedge r_t)$$

The complexity of the calculation of the description potential of a constrained Boolean symbolic object is exponential on the number of rules and linear on the number of variables of each connected graph of dependencies.

4. The Normal Symbolic Form (NSF)

To avoid this huge computation we were induced (Csernel & de Carvalho 1997) to define a normal symbolic form, which consist roughly in a cutting of the description space in different subspaces, each of them containing a premise variable and all the conclusion variables linked with by rules.
Then, each of the different subspaces will be cut into different slices where all values of the premise variable will lead to the same conclusion. Only the coherent part of each object will be represented. We say that a knowledge base is under Normal Symbolic Form (NSF) if the following conditions are satisfied.
- **First NSF condition**: If no dependency occurs between the variables belonging to the same array, or if a dependency occurs between the first variable V1 and all the others. Then V1 must be a premise variable, and all the other variable some conclusion variables linked with V1 by a rule.
- **Second NSF condition**: If all the values expressed by the premise variable of one object lead to the same conclusion (or absence of conclusion)
The NSF is applicable only if the dependencies form a tree or a set of trees. Most of the time a symbolic object has to be decomposed to follow the NSF, as we can see in the following example,

Table 1 original table

	wings	wings_color	Thorax_color	Thorax_size
a_1	{absent,present}	{red, blue}	{blue, yellow}	{big, small}
a_2	{absent,present}	{red, green}	{blue, red}	{small}

The previous array represents two Boolean symbolic objects called *a1* and *a2*, the dependencies rules r1,r2 are associated with the definition.

if wings = Absent then wings_color = No_Sense. (r1)

it if wings_color = red then Thorax_color = blue. (r2)

This rules induces the following dependency tree.

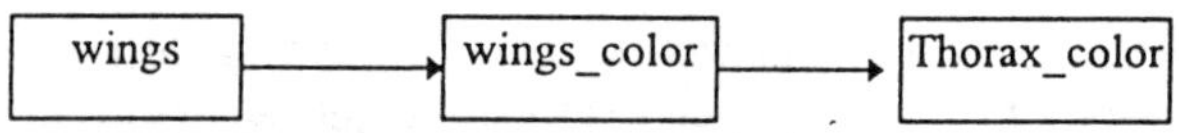

The description of the objects $a_1$and a_2 representing two different (imaginary) insect species is not NSF.

The description has to be transformed in the sequence of the three following tables to be NSF. The transformation process follows the dependency tree induced by the rules. Each premise variable induces a new secondary table where the premise variable is the first variable and the other variables are the different conclusion variables linked with.

In these tables the upper left corner contains the table name, a new kind of column appears where the values are integers referring to a line in another table with the same name as the column. The column D.P. will be used in a following paragraph.

The main table refers to the initial table. Table 1 and 2 refers to secondary tables, In each secondary table, a double line separates the lines where the first variable verifies the premise, from the lines where the first variable does not verify it.

main table

	wings...	Thorax_size	D.P.
a1	{ 1, 3}	{big, small}	10
a2	{2,4}	{small}	5

table 1

wings ...	wings	colour	D.P.
1	absent	{4}	2
2	absent	{5}	2
3	present	{1, 2 }	3
4	present	{1, 3 }	3

table 2

colour	wings_color	Thorax_color	D.P.
1	{ red }	{blue }	1
2	{ blue }	{ blue, yellow }	2
3	{ green }	{blue, red }	2
4	NS	{blue, yellow }	2
5	NS	{ blue, red }	2

We have now three tables instead of a single one, but only the valid parts of the objects are represented: now, the tables include the rules.

So under NSF the D.P. computation is nearly immediate. Its a recursive process,

one has to compute first the D.P of the object included in the table corresponding to the leafs of the dependence tree, then compute the D.P. of the tables corresponding to an upper level in the tree using the results previously computed.
So we first compute the D.P. of the elements of the *table2*, then we compute the D.P. *table1* by summing up (for the column color) the D.P. computed in the lines of the *table2* and so on. **For example :** Line 1 of table one refers to line 1 and 2 of table 2. The D.P. is the sum of the potential described by these two lines : 1 + 2 = 3. In the same way the potential of *a1* is obtained by multiplying the sum of D.P due to the line 1 and 3 of *table 1* by the D.P due to the variable Thorax -size (2) giving 10.

5. Complexity using NSF

When one is using the NSF, the estimation of the complexity has to be divided in two parts:
- the first one concerns the NSF transformation, and has to be computed only once.
-the second one concerns the computation itself.

Estimation of the complexity of the NSF transformation is depending on the quality of the rules provided, and we can not detail it precisely because of the lack of place, but it can be appreciated by the factorisation factor F.

$$F = \frac{\text{Average size of secondary table}}{\text{Nb of initial Objects}}$$

The main complexity of the NSF transformation is due to a factorisation process which limits the growing of the tables. This process is repeated for each *(T)* secondary table, and its complexity is the same that for a sort, N^2 roughly speaking. So the complexity of the NSF transformation can be estimated to $TV(N*F)^2$ where T is the number of premise variables, N the number of objects, V the number of variables and F the Factorisation factor. For each real examples we have treated, we always had $F < 1$ then the complexity is polynomial.

But it's not always the case, and with special kinds of rules, F can be greater than one. In this case we will not estimate F but S the size of the greatest table which can occur. In the worst case the size of each secondary table is the double of the size of its main table. As the NSF transformation is induced by dependencies tree, S will be influenced by the shape of the tree.

- If the tree is well balanced we will have $S = N*2^{log\ (T)} = N*T$, T being the number of premise variables, because $log(T)$ is the average deep of a well balanced tree, and the 2 factor is due to the fact that in the worst case, each secondary table double the number of lines of its main table. The transformation is still polynomial.

- If the tree is **NOT** well balanced then the deep of the tree is *T*, the size *S* of the greatest table will be 2^T, this very worst case can happen only if a conjunction of rare circumstances occurs. The complexity is still lower than is the usual case because it depends on the number of premise variables *T* which is smaller than the number of rules.

Considering the data in NSF, then the complexity of the computation itself is linear following the number of variables and the number of rules, and polynomial $(N*F)^2$ following the number of objects: $(F*N)^2*V*T$.

6. Conclusion

The decomposition of symbolic objects following the Normal Symbolic Form induce an easy way to take dependencies rules into account to deal with symbolic objects. An our first application trial we obtain a reduction of about 90% of our computational time, including the NSF transformation process. On real data, we never observed a size growing of the data, so we always obtained a polynomial computation time. We need to carry our work on simulated data to determine more precisely the performances of NFS and to see more precisely the influence of the rule form on these performances.

References

Csernel M. et De Carvalho, F.A.T. (1997): Towards a Normal Symbolic Form. In: Data Science, Classification and Related Methods, Hayshi, C. et al (eds.), Springer-Verlag, Tokyo, Japan. (accepted to be published)

De Carvalho, F.A.T. (1997):Extension based proximities between constrained Boolean symbolic objects. In: *Data Science, Classification and Related Methods*, Hayshi, C.et al (eds.), Springer-Verlag, Tokyo, Japan. (accepted to be published)

Diday, E. (1991): Des objets de l'analyse de données à ceux de l'analyse de connaissances. In : *Induction Symbolique et Numérique à partir de Données,* Diday, E. et Kodratoff (eds.), 9-75, Cepadue Editions, Toulouse.

Ichino M, Yaguchi H(1994), Generalized Minkowski Metrics for Mixed Feature-Type Data Analysis. *IEEE Transaction on Systems, Man and Cybernetics* **24**,4,698-708

Symbolic Data Analysis: a Mathematical Framework and Tool for Data Mining

Edwin Diday
University Paris 9 Dauphine, Ceremade. Pl. Du Mle de L. de Tassigny. 75016 Paris, FRANCE

Abstract: Knowledge extraction from large data bases is called "Data Mining". In Data Bases the descriptions of the units are more complex than the standard ones due to the fact that they can contain internal variation and be structured. Moreover, symbolic data happen from many sources in order to summarise huge sets of data. They need more complex data tables called "symbolic data tables" because a cell of such data table does not necessarily contain as usual, a single quantitative or categorical values. For instance, a cell can contain several values linked by a taxonomy . The need to extend standard data analysis methods (exploratory, clustering, factorial analysis, discrimination,...) to symbolic data table is increasing in order to get more accurate information and summarise extensive data sets contained in Data Bases. We define "Symbolic Data Analysis" (SDA) the extension of standard Data Analysis to such tables. "Symbolic objects" are defined, in order to describe in an explanatory way classes of such units. They constitute an explanatory output of a SDA and they can be used as queries of the Data Base. A symbolic object is "complete" if its "extent" covers exactly the class that it describes. The set of complete symbolic objects constitutes a Galois lattice. The SDA tools developed in the European Community project "SODAS" are finally mentioned.

Key words: Symbolic Data Analysis, Conceptual Analysis, Complex data, Data Mining, Partitioning, Hierarchical and Pyramidal clustering, Lattices.

1. Introduction

In a census of a country, each individual of each region is described by a set of numerical or categorical variables given in several relations of a Data Base. In order to study the regions, we can describe each of them summarising the values taken by the individuals which leave in, by means, intervals, subsets of categorical values, histograms, probability distributions, depending on the concerned variable. In such a way we obtain a "symbolic data table" where each row defines the "description" of a region and each column is associated to an original variable. An

extension of standard Data Analysis to such data table is one of the aim of what we have called "Symbolic Data Analysis".
Another important aim is to obtain (or "mine") explanatory results by extracted, which we have called "symbolic objects". The description of a region is called "intent", the set of individuals which satisfy this intent is called "extent". A "symbolic object" associated to a region is defined by its intent and by a way of finding its extent. Its syntax has an explanatory power. For instance, the symbolic object defined by the following expression:
$a(w) = [age(w) \in [30, 35]\ ] \wedge [Number\ of\ children(w) \leq 2]$. It gives at the same time the intent of a class of individuals by the description d = ([30, 35], 2} and allows to calculate its extent. It means that an individual "w" satisfies this intent if his age is between 30 and 35 years old and he has less than 2 children. If we have a set of such classes of individuals, their associated descriptions define a higher level symbolic data table. Using this new data table we can obtain classes (of classes) which can be associated to higher level symbolic objects. For instance, the following symbolic object $b(w) = [age(w) \subseteq [25, 40]\ ] \wedge [Number\ of\ children(w) \leq 3]$ contains in its extent the preceding class of individuals considered as a higher level unit. In this paper, we show how Data Bases constitutes a natural source of symbolic data tables. Thus symbolic objects are then introduced in order to describe classes of units described by rows of such data tables. We give a formal definition of symbolic objects and some properties of them. Finally, some tools developed in the "Sodas" European Project are presented.

2.The input of a symbolic data analysis

"Symbolic data tables" constitute the main input of a Symbolic Data Analysis . They are defined in the following way. Columns of the input data table are « variables » which are used in order to describe a set of units called "individuals". Rows are called « symbolic descriptions » of these individuals because they are not as usual, only vectors of single quantitative or categorical values. Each cell of this « symbolic data table » contain data of different types:
(a) Single quantitative value : for instance, if « height » is a variable and w is an individual : height(w) = 3.5. (b) Single categorical value: for instance, Town(w) = London.
(c) Multivalued: for instance, in the quantitative case height(w) = {3.5, 2.1, 5} means that the height of w can be either 3.5 or 2.1 or 5. Notice that (a) and (b) are special cases of (c).
(d) Interval: for instance height(w) = [3, 5], which means that the height of w varies in the interval [3, 5].
(e) Multivalued with weights: for instance a histogram or a membership function (notice that (a) and (b) and (c) are special cases of (e) when the weights are equal to 1).
Variables can be:

(g) Taxonomic: for instance, « the colour is considered to be "light" if it is "yellow", "white" or "pink" .

(h) Hierarchically dependent : for instance, we can describe the kind of computer of a company only if it has a computer, hence the variable "does the company has computers? " and the variable " kind of computer" are hierarchically linked.

(i) With logical dependencies, for instance: « if age(w) is less than 2 months then height(w) is less than 10 ».

3. Source of Symbolic Data:

Symbolic data happen from many sources in order to summarise huge sets of data. They result from expert knowledge (scenario of traffic accidents, type of unemployment, species of insects ...), from the probability distribution , the percentiles or the range of any random variable associated to each cell of a stochastic data table, from Data Analysis (factorial analysis, clustering, neural networks, ...) of standard data table in order to summarise and explain the results, from time series (in describing intervals of time), from confidential data (in order to hide the initial data by less accuracy), etc. They result also, from Relational Data Bases in summarising the answer to a query, or in order to study a set of units which description need the merging of several relations as it is shown in the following example.

Example: We have two relations of a Relational Data Base defined as follows. The first one called "delivery" is given in table 1. It describes five supplying by the name of the supplier, its
company and the town from where the supplying is coming.

Delivery	Supplier	Company	Town
Liv1	F1	CNET	Paris
Liv2	F2	MATRA	Toulouse
Liv3	F3	EDF	Clamart
Liv4	F1	CNET	Lannion
Liv5	F3	EDF	Clamart

Table 1: *Relation "Delivery"*

The supplying are described by the relation "Supplying" defined in the following table 2.

Supplying	Supplier	Town
FT1	F1	Paris
FT2	F2	Toulouse
FT3	F1	Lannion
FT4	F3	Clamart
FT5	F3	Clamart

Table 2: *Relation "Supplying"*

From these two relations we can deduce the following data table 3, which describes each supplier by his company, his supplying and his towns:

Supplier	Company	Supplying	Town
F1	CNET	FT1, FT3	½ Paris, ½ Lannion
F2	MATRA	FT2	Toulouse
F3	EDF	FT4, FT5	Clamart

Table 3: *Relation "Supplier" obtained by merging the relations "Delivery" and "Supplying".*

Hence, we can see that in order to study a set of suppliers described by the variables associated to the two relations we are naturally conducted to take account of the four following conditions which characterise symbolic data:
i) Multivalued: this happen with the variables "Supplying" and "Town" which can have several values in the table 3.
ii) Multivalued with weights: this is the case for the towns of the supplier F1. The weights ½ means that the town of the supplier F1 is Paris or Lannion with a frequency equal to ½.
iii) Rules: some rules have to be given as input in addition to the data table 3. For instance, "if the town is Paris and the supplier is CNET, then the supplying is FT1.
iv) Taxonomy: by using regions we can replace for instance {Paris, Clamart} by " Parisian Region ".

4. Main output of Symbolic Data Analysis algorithms:

Let Ω be a set of individuals, D a set of descriptions, « y » a mapping defined from Ω into D which associates to each $w \in \Omega$ a description $d \in D$ from a given symbolic data table. We denote by R, a relation defined on D. It is defined by a subset of D×D. If $(x,y) \in DxG$ we say that x and y are connected by R and this is denoted by xRy . The characteristic mapping of R is h_R: $D \times D \rightarrow \{0,1\}$ such that $h_R(x,y) = 1$ iff $(x,y) \in D \times D$. We generalise the mapping h_R by the mapping H_R: $DxD \rightarrow L$ and we denote $[d' \, R \, d] = H_R(d,d')$ the result of the "comparison" of d and d' by H_R. We can have L ={true, false}, in this case [d' R d] = true means that there is a connection between d and d'. We can also have L= [0, 1] if d is more or less connected to d'. In this case, [d' R d] can be interpreted as the "true value" of xRy or " the degree to which d is in relation R with d' (see in Bandemer and Nather (1992), the section 5.2 on fuzzy relations.).

For instance, $R \in \{=, \equiv, \leq, \subseteq\}$ or is an implication, a kind of matching, etc. R can also use a set of such operators.

We say that a description $d \in D$ is "coherent" if it is compatible with the conditions (h) and (i) (defined in section 2). For instance, a description which expresses the fact that "sex" is "male" and "number of deliveries" is 1, is not coherent. These descriptions constitute a set of objects to which any symbolic data analysis algorithm can be applied. That is why, we use the word « object » to denote any coherent description . A coherent description of an individual, is called « individual object ». It can be also the description of a class of individuals, in this case it is an "intensional object". For instance, the coherent description of a scenario of accidents, of a class of failures, etc. is an intensional object. A « symbolic object » is defined both by an object and a way of comparing it to individual objects. More formally, its definition is:

Definition of a symbolic object

A symbolic object is a triple s = (a, R, d) where R is a relation between descriptions, d is a description and «a» is a mapping defined from Ω in L depending on R and d.

Symbolic Data Analysis concerns usually classes of symbolic objects where R is fixed, "d" varies among a finite set of coherent descriptions and a(w) = H(y(w),d) where H is a mapping $DxD \rightarrow L$. For instance, a(w) = [h(y(w)) R d] where h can be for instance, a filter (see an example thereunder). There are two kinds of symbolic objects:

- « Boolean symbolic objects » if [y(w) R d] $\in$ L = {true, false}. In this case, if $y(w) = (y_1,...,y_p)$, the y_i are of type (a) to (d), defined in section 1.

Example:

y = (colour, height), d = ({red, blue, yellow}, [10,15]), colour(u) = {red, yellow}, height(u) = {21}, R = ($\subseteq$, $\subseteq$), a(u) = [colour(u) $\subseteq$ {red, blue, yellow}]$\vee$ [height(u) $\subseteq$ [10,15]]= true $\vee$ false = true. If h(y(w)) = (colour(w), $\emptyset$), h is a filter as a(w) = [h(y(w)R d] = [colour(w) $\subseteq$ {red, blue, yellow}]$\vee$ [$\emptyset$ $\subseteq$ [10,15]]= [colour(w) $\subseteq$ {red, blue, yellow}] and we get also a(u) = true.

- « modal symbolic objects » if [y(w) R d] $\in$ L = [0,1]. In this case, the y(w) are of type (e). Thereunder, we give an example of such symbolic object.

Syntax of symbolic objects in the case of "assertions":

If the initial data table contains p variables we denote $y(w) = (y_1(w),..., y_p(w))$, $D = (D_1,...,D_p)$, $d \in D$: $d = (d_1,..., d_p)$ and $R' = (R_1,...,R_p)$ where R_i is a relation defined on D_i. We call « assertion » a special case of a symbolic object defined by s = (a,R,d) where R is defined by $[d' R d] = \wedge_{i=1,p} [y_i(w) \ R_i \ d_i]$ and "a" by: a(w) = [y(w) R d]. Notice also that we have s considering the expression $a(w) = \wedge_{i=1,p} [y_i(w) \ R_i \ d_i]$. Hence, we can say that this explanatory expression defines a symbolic object.

For example, a Boolean assertion where SPC means « socio-professional-category » is:
$a(w) = [age(w) \subseteq \{12, 20, 28\}] \wedge [SPC(w) \subseteq \{employee, worker\}]$. If the individual u is described in the original symbolic data table by $age(u)=\{12, 20\}$ and $SPC(u) = \{employee\}$ then: $a(u) = [\{12, 20\} \subseteq \{12, 20, 28\}] \wedge [\{employee\} \subseteq \{employee, worker\}] = true$
If the variables are multivalued and weighted, an example is given by:
$a(w) = [age(w)\ R_1\ \{(0.2)12, (0.8)\ [20, 28]\}] \wedge [SPC(w)\ R_2\ \{(0.4)employee, (0.6)worker\}]$ where for instance the "matching" of two probability distributions is defined for two discrete probability distributions r and q of k values by: $r\ R_i\ q = \sum_{j=1,k} r_j q_j e^{(r_j - \min(r_j, q_j))}$.
Extent of a symbolic object s: in the Boolean case, the extent of a symbolic object is denoted Ext(s) and defined by the extent of a, which is: $Extent(a) = \{w \in \Omega / a(w) = true\}$. . In the modal case, given a threshold α, it is defined by $Ext_\alpha(s) = Extent_\alpha(a) = \{w \in \Omega / a(w) \geq \alpha\}$.
Order between symbolic objects: if r is a given order on D, then the induced order on the set of symbolic objects denoted by r_s is defined by : $s_1\ r_s\ s_2$ iff $d_1\ r\ d_2$.
If R is such that [d R d']= true implies d r d', then $Ext(s1) \subseteq Ext(s2)$ if s1 r_s s2 . If R is such that [d R d']= true implies d' r d then $Ext(s2) \subseteq Ext(s1)$ if s1 r_s s2.
Tools for symbolic objects: tools between symbolic objects (Diday (1995)) can be needed such as similarities (F. de Carvalho (1998)), matching, merging by generalisation where a t-norm or a t-conorm (Schweizer, Sklar (1983) and Diday, Emilion (1995)) denoted T can be used, splitting by specialisation (Ciampi et al. (1995)). Under some assumption on the choice of R and T it can be shown that the underlying structure of a set of symbolic objects is a Galois lattice (Brito(1994), Polaillon, Diday (1997)), where the vertices are closed sets defined by « complete symbolic objects ». More precisely, the associated Galois correspondence is defined by two mappings F and G:
-F: from $P(\Omega)$ (the power set of Ω) into S (the set of symbolic objects) such that $F(C) = s$ where $s = (a, R, d)$ is defined by $d = T_{c\in C}\ y(c)$ and so $a(w) = [y(w)\ R\ T_{c\in C}\ y(c)]$, for a given R. For example, if $T_{c\in C}\ y(c) = \cup_{c\in C}\ y(c)$, $R \equiv$ « $\subseteq$ », $y(u) = \{pink, blue\}$, $C = \{c, c'\}$, $y(c) = \{pink, red\}$, $y(c') = \{blue, red\}$, then
$a(u) = [y(w)\ R\ T_{c\in C}\ y(c)] = [\{pink, blue\} \subseteq \{pink, red\} \cup \{blue, red\}\}) = \{pink, red, blue\}] = true$ and $u \in Ext(s)$.
-G: from S in $P(\Omega)$ such that: $G(s) = Ext(s)$.
A « complete symbolic object » s is such that $F(G(s)) = s$. Such objects can be selected from the Galois lattice but also, from a partitioning, a hierarchical or a pyramidal clustering, from the most influential individuals to a factorial axis, from a decision tree, etc.
We can consider four kinds of advantages given by symbolic objects. First, they give a summary of the original symbolic data table in an explanatory way, (i.e. close to the initial language of the user) by expressing descriptions based on properties concerning the initial variables. Second, they can be easily transformed in term of query of a Data base. Third, by being independent of the initial data

table they are able to identify any matching individual described in any data table. Fourth, they are able to give a new symbolic data table of higher level on which a symbolic data analysis of second level can be applied.

5. What standard statistical methods are not able to do that symbolic data analysis can do?

-Starting as input with a symbolic data table.
-Using generalisation processes during the algorithms-Giving explanation of the results in a language close from the one of the user by symbolic objects as output.
- Giving graphical description taking account on the internal variation of the symbolic objects.
In the European Community Project SODAS (« Symbolic Official Data Analysis System») the following methods are developed by its members:
- Principal Component, Correspondence Analysis and Discriminate Factorial Analysis of a symbolic data table,
- extension of elementary descriptive statistic (central object, histograms, dispersion, co-dispersion, etc. from a symbolic data table) to symbolic data.
- mining symbolic objects from the answers to queries of a relational data base ,
- partitioning, hierarchical or pyramidal clustering of a set of individuals described by a symbolic data table such that each class be associated to à complete symbolic object.
- dissimilarities between boolean or probabilistic symbolic objects,
- extension of decision trees on probabilistic symbolic objects, extension of a Parzen discrimination method to classes of symbolic objects,
- generalisation by a disjunction of symbolic objects of a class of individuals described in a standard way.
- inter-active and ergonomic graphical representation of symbolic objects.

Conclusion

Our general aim can be stated in the following way: mining symbolic objects in order to summarise huge data sets and then, analyse them by Symbolic Data Analysis. As the underlying lattice of symbolic objects often becomes too large, other methods which provide also symbolic objects have to be used. The need to extend standard data analysis methods (exploratory, clustering, factorial analysis, discrimination,...) to symbolic data tables is increasing due to the expansion of information technology. This need, has led to an European Community project called SODAS (Hebrail (1996)) for a « Symbolic Official Data Analysis System» in which 15 institutions of 9 European countries are concerned. Three Official Statistical Institutions are involved in this project: EUSTAT (Spane), INE (Portugal) and ONS (England). An example of application proposed on their

Census data consists in finding clusters of unemployment people and their associated mined symbolic objects in a country, calculating its extent in the census of another country and describing this extent by new symbolic objects in order to compare the behaviour of the two countries.

References

Brito P. (1994) "Order structure of symbolic assertion objects". IEEE TR. on Knowledge and Data Engineering Vol.6, n° 5, October.

Bandemer H., Nather W. (1992) "Fuzzy Data Analysis". Kluwer Academic Publisher.

De Carvalho F.A.T. (1998) "New metrics for constrained boolean symbolic objects" Proc. KESDA'98, Eurostat. Luxembourg.

Ciampi A., Diday E., Lebbe J., Périnel E., Vigne (1995) R. " Recursive partition with probabilistically imprecise data". OSDA'95. Editors: Diday, Lechevallier, Opitz Springer Verlag (1996).

Diday E., Emilion R. (1995) "Lattices and Capacities in Analysis of Probabilist Objects". Proceed. of OSDA'95 (Ordinal and Symbolic Data Analysis). Springer Verlag Editor (1996).

Diday E. (1995) " Probabilist, possibilist and belief objects for knowledge analysis " .Annals of Operations Research . 55, 227-276.

Hebrail G. (1996) " SODAS (Symbolic Official Data Analysis System) ". Proceedings of IFCS'96, Kobe , Japan. Springer Verlag.

Pollaillon G., Diday, E. (1997) " Galois lattices of symbolic objects " Rapport du Ceremade University Paris9- Dauphine (February).

Schweizer B. , Sklar A. (1983) " Probabilist metric spaces ". Elsever North-Holland, New-York.

Symbolic Kernel Discriminant Analysis

Sandrine Lissoir, Jean-Paul Rasson
Département de Mathématique
Facultés Universitaires Notre-Dame de la Paix
Rempart de la Vierge, 8, B5000 Namur, Belgique
sandrine.lissoir@fundp.ac.be, jpr@math.fundp.ac.be

Abstract : Current technological progress in Hardware, Data Bases and Object Oriented languages implies the manipulation, stock and representation of objects with more and more complex data. The notion of Symbolic Objects is introduced on the base of Diday's work and the necessity to be adapted to this notion appears for most recent classification methods. The aim of this paper is the adaptation of the classical Bayesian discrimination rule to the Symbolic Objects problematic. This will be performed by the a priori probabilities' estimation and by a kernel density estimation.

Keywords : Symbolic objects, training sets, a priori and a posteriori probabilities, window's bandwidth, kernel density estimation, bayesian discrimination rule, EM-like algorithm.

1. Introduction

Kernel density estimation is a tool which allows the statistician to construct a density on any sample of data without any a priori probabilistic hypothesis. Recent references are numerous (e.g. books by Hand [6], Silverman [9], Devroye [2], ...) These methods compute a weighted sum of kernels centered on each data point. Examples of kernels density estimations are to be found essentially for quantitative (discrete or continuous) and qualitative data. For mixed data of both previous types, the method of "product kernels" is suggested (Hand [6]).

The kernel discrimination analysis tries to give a solution to the following problem : "given g samples belonging to g classes, and using them as training sets for these classes, how can we affect one new data point (or several) to one of these classes?" The classical Bayesian discrimination rule which is "affect the new data x to that component $p_k f_k$ of the decomposed global density $f(x)=\sum_{k=1}^{g} p_k f_k(x)$ for which $p_k f_k$ is maximum", needs as most of classification methods to be adapted to the Symbolic Objects problematic (Diday [3][4][5]), in which we can have information of different types: quantitative, qualitative, probabilistic, of interval type, ... within a conjunction of theses.

2. The decision rule

Let V be the observation space (the symbolic objects space). The identification of the group from which an observation $x \in V$ is issued implies the definition of an application d, called decision rule: $d: V \rightarrow \{1, ..., g\} : x \rightarrow d(x)$.
This rule depends of the training points $x_{11}, ..., x_{1n1}, ..., x_{g1}, ..., x_{gng}$ and is defined by a minimization of the error rate. The rule d determines a partition $(P_1, ..., P_g)$ of V: $P_k = \{ x \mid d(x)=k \},\ 1 \le k \le g$.
The rule can be defined using g functions $h_k(x)$, $k=1, ..., g$. The function $h_k(x)$ associated with the k-th population gives an idea of the "similarity" between x and the individuals $x_{k1}, ..., x_{knk}$ of the k-th training sample. We have:

$$P_k = \{ x \mid h_k(x) \ge h_i(x)\ \forall i = 1, ...g \},\ 1 \le k \le g.$$

3. The risk vector

During the affectation of observations to the different classes, some errors can occur. Let p_{ki}, $1 \le k \le g$, $1 \le i \le g$, be the probability to affect an individual of group k to group i. Given that some errors can be less important than others, a cost c_{ki}, $1 \le k \le g$, $1 \le i \le g$, can be associated with the affectation to group i of an observation from group k. Note that $\forall k,\ c_{kk} = 0$, no cost being attributed to a correct affectation.
For each population k, $1 \le k \le g$, we define the risk $R_k(d)$ as the expected value of the cost attributed to a wrong affectation of individuals of the group:

$$R_k(d) = \sum_{i=1}^{g} c_{ki}\, p_{ki}\ ,\ 1 \le k \le g.$$

Note that the choice of the best rule is a critical problem.

4. The bayesian rule

Let's suppose that the a priori probabilities p_k , $1 \le k \le g$, of belonging to the population k, are known. The bayesian risk is:

$$R(d) = \sum_{k=1}^{g} p_k\, R_k(d).$$

Assuming that the data from population k are distributed following a density function $f_k(.)$, $1 \le k \le g$, we have:

$$R_k(d) = \sum_{i=1}^{g} c_{ki}\, p_{ki} = \sum_{i=1}^{g} c_{ki} \int_{P_i} f_k(x)\, dx\ ,\ 1 \le k \le g,$$

and then:

$$R(d) = \sum_{k=1}^{g} p_k \sum_{i=1}^{g} c_{ki} \int_{P_i} f_k(x)\,dx = \sum_{i=1}^{g} \int_{P_i} (\sum_{k=1}^{g} p_k\, c_{ki}\, f_k(x))\,dx.$$

$R(d)$ is minimal for the partition $P = (P_1, ..., P_g)$ defined by:

$$P_i = \{ x \mid \sum_{k=1}^{g} p_k\, c_{ki}\, f_k(x) \leq \sum_{k=1}^{g} p_k\, c_{kj}\, f_k(x) \quad \forall j=1, ..., g \}.$$

The costs associated to the classification errors being generally supposed equal: $c_{ki} = c \ \forall k, i = 1, ..., g, k \neq i$ and $c_{kk} = 0 \ \forall k=1, ..., g$, we have:

$$\sum_{k=1}^{g} p_k\, c_{ki}\, f_k(x) = c \sum_{k=1}^{g} p_k\, f_k(x) - c\, p_i\, f_i(x)\,,$$

and finally, leaving the constant terms out:

$$P_i = \{ x \mid -p_i\, f_i(x) \leq -p_j\, f_j(x) \quad \forall j \} = \{ x \mid p_i\, f_i(x) \geq p_j\, f_j(x) \quad \forall j \}.$$

The discriminant functions defining the Bayesian rule are then $h_k(x) = p_k\, f_k(x)$. The Bayesian rule can be interpreted by the affectation of an individual x to the population k maximizing the a posteriori probability $q_k(x) = \dfrac{p_k f_k(x)}{\sum_{i=1}^{g} p_i f_i(x)}$.

If there are more than one group maximizing $p_k\, f_k(x)$ ($\exists\, i{\neq}k : p_i\, f_i(x) = p_k\, f_k(x) \geq p_j\ f_j(x)\ \forall j$), some choices are possible. The individual x can be randomly affected to one of these groups, can be affected to the group j ($j{=}min(i,k)$), can be let unclassified...

Note that the maximum likelihood rule corresponds to the Bayesian one in the case of equal a priori probabilities and equal error costs.

5. The density estimates

The classification problem is then solved when the densities $f_k(x)$, $1 \leq k \leq g$, are known. Actually, these densities are unknown but a sample of each population is given. These samples will be used to estimate the g densities functions.

In discriminant analysis of classical data, hypotheses of multivariate normality can be made concerning the densities. The sample data can be supposed to be issued from g normal multivariate populations of means μ_k, $1 \leq k \leq g$, and of equal covariance matrices ($S_k = S \quad \forall k=1, ..., g$). The training sets are used to estimate the parameters:

$$\hat{\mu}_k = \bar{x}_k = \sum_{i=1}^{n_k} x_{ki} \quad \forall k=1, ..., g$$

$$S = \frac{\sum_{k=1}^{g} \sum_{i=1}^{n_k} (x_{ki} - \bar{x}_k)(x_{ki} - \bar{x}_k)^T}{\sum_{k=1}^{g} n_k - g}.$$

These estimators are plugged in the densities expressions and applying the maximum likelihood rule, we obtain the Fisher's linear discrimination rule.
Note that $\forall\ k \neq i \in \{1, \ldots, g\}$, we have an equivalence between

$$f_k(x) \geq f_i(x) \text{ and } \{x - \frac{1}{2}(\bar{x}_k - \bar{x}_i)\}^T S^{-1}(\bar{x}_k - \bar{x}_i)\} \geq 0,$$

which is the equation of an hyperplane. That is why we talk about a linear discrimination rule.
In the quadratic approach, the covariance matrices are not supposed equal and the parameters are estimated by:

$$\hat{\mu}_k = \bar{x}_k = \sum_{i=1}^{n_k} x_{ki} \quad \forall k=1, \ldots, g$$

$$S_k = \frac{\sum_{i=1}^{n_k} (x_{ki} - \bar{x}_k)(x_{ki} - \bar{x}_k)^T}{n_k - 1} \quad \forall k=1, \ldots, g.$$

After replacement by the parameters in the densities expressions, the equations:

$$f_k(x) \geq f_i(x), \ \forall\ k \neq i \in \{1, \ldots, g\}$$

lead to quadratic forms.
If no hypotheses are made about the densities, non parametric methods must be used to obtain density estimates. The kernel method is one of them.

6. The uniform kernel method

The use of the uniform kernel density estimate produces for d-dimensional quantitative data :

$$\hat{f}_k(x) = \frac{1}{n_k (2h_k)^d} \sum_{i=1}^{n_k} K\left(\frac{x - x_{ki}}{h_k}\right)$$

where . $x_{k1}, \ldots, x_{kn_k}$ are the n_k training set points of the kth population

. h_k is the window bandwidth

. $K(.)$ is the uniform kernel : $K(y) = \begin{cases} 1 & \text{if } \|y\|_\infty < 1 \\ 0 & \text{else} \end{cases}$

Roughly speaking, $\sum_{i=1}^{n_k} K\left(\frac{x - x_{ki}}{h_k}\right)$ counts the number of points of the training set of population k, which are at a distance less than h_k from y (ie in a hypercube of size $2\,h_k$).

The choice of the h_k is crucial. According to Rasson and Granville [8], that choice has to be made by minimizing the estimated error rate (e.g. by leaving-one-out estimation of it).
As stated in the literature, the choice of the kernel does not seem to be critical. For example, other kernels are proposed by Silverman [9] to solve problems implying mixed classical data including binary and continuous components:

$$K(y) = \lambda^{k_1 - d_1(x,y)} (1-\lambda)^{d_1(x,y)} h^{-k_2} \Phi\left(h^{-1} d_2(x,y)(2\pi)^{\frac{1-k_2}{2}}\right)$$

where . d_1 is the distance between the binary components :
$(d_1(x,y) = (x-y)^T(x-y))$
. d_2 is the euclidian distance between the continuous components
. Φ is the normal density fonction
. λ and h are smoothing parameters
. k_1 and k_2 are the numbers of components of binary and continuous types respectively.

Among all possible kernels, the uniform one has been chosen for its computational and interpretation easiness.

7. Using kernel density measures for symbolic data

The building of such a uniform kernel can be generalized to symbolic objects. For this purpose, the important thing is to have a dissimilarity measure between symbolic objects. If no such dissimilarity is available, it is also possible to work with N dissimilarity measures (one measure for each variable), or at least with K dissimilarity measures (one for each kind of data involved in the symbolic objects).
As an example, consider that the data are symbolic objects including 3 qualitative multi-valuated variables, 2 quantitative interval variables and 2 modal variables. The density estimation can be generalized either using 1 dissimilarity measure between objects, either using 7 dissimilarity measures (one for each variable), or using 3 dissimilarity measures (one for the qualitative multi-valuated type, one for the quantitative interval type and one for the modal type). The dissimilarity choices are numerous (Ichino[7], De Carvalho[1]).
In the first case, the density estimation is simply performed by counting the points of the training set of each population being in the hypercube. In the second case, a kernel is built for each variable using the appropriate dissimilarity measure. Then the global kernel for the symbolic object is the product of the N kernels. The density estimation is then performed by counting the points of each training set being in all "intervals". In the last case, a kernel is built for each type. Then the global kernel for the symbolic object is the product of the kernels built on each type of data. In that last case, the density estimation

is performed by counting the points of each training set being in all "hypercubes".
Note that the density estimators used here for symbolic data do not refer to classical probability densities, as no measure is available on the symbolic objects space. They are simply indicators of the concentration in a given neighbourhood.

8. A priori estimates

Estimates of the a priori probabilities p_k are also needed. We choose $\hat{p} = (\hat{p}_1, \dots, \hat{p}_g)$ maximizing the likelihood function :

$$L(p) = \prod_{y} \sum_{k=1}^{g} p_k \hat{f}_k(y)$$

where the product is taken over all points to be classified.
The EM-like algorithm computes iteratively the optimum. Beginning with the uniform proportions $\hat{p}_k(0) = \frac{1}{g}$, the EM-like algorithm uses the recursion

$$\hat{p}_k(t+1) = \frac{1}{n} \sum_{y} \frac{\hat{p}_k(t)\hat{f}_k(y)}{\sum_{j=1}^{g} \hat{p}_j(t)\hat{f}_j(y)}$$

where . n is the number of points to be classified
. $\sum_{y}$ is taken over all these points.

Here, the density estimates are kept constant.

9. Output data

Using the given training sets, the algorithm computes, for each input symbolic object x, the estimates of $\hat{f}_k(x)$, $k = 1, \dots, g$. Thanks to the a priori probabilities estimates, we obtain, for each x, the following values:

$$\hat{p}_k \hat{f}_k(x), \quad k = 1, \dots, g.$$

By the normalization of these values we finally have the set of a posteriori probabilities of belonging to each class :

$$\frac{\hat{p}_k \hat{f}_k(x)}{\sum_{i=1}^{g} \hat{p}_i \hat{f}_i(x)}, \quad k = 1, \dots, g.$$

10. Conclusion

Currently, it is of primary importance, for any field of work, to be easily adaptable to the other fields progress. Classical distance measures do not apply to data of symbolic type. New distance notions are elaborated, which make possible the adaptation of most Data Analysis methods. So with a well chosen distance between objects, Symbolic Kernel Discriminant Analysis can be an efficient tool in the field of supervised classification of complex data.

11. References

[1] De Carvalho, F. : Un indice de dissimilarité entre objets symboliques booléens basé sur l'extension. *Internal report.* Universidade Rural de Pernambuco.

[2] Devroye, L. (1985) : *Non parametric Density Estimation : The L_1 view.* Wiley, New York.

[3] Diday, E. (1989) : Data Analysis, Learning Symbolic and Numeric knowledge. In *Nova Science*, Antibes, France.

[4] Diday, E. (1988) : The symbolic approach in clustering. In *Classification and related methods of Data Analysis*, Bock, H.H., Amsterdam : Elsevier Science.

[5] Diday, E. (1989) : Knowledge representation and symbolic data analysis. *Proceedings of the Second International Workshop on Data, Expert knowledge and Decision, Hamburg.*

[6] Hand, D.J. (1982) : *Kernel Discriminant Analysis*, Wiley, Chichester.

[7] Ichino, M. (1994) : Generalized Minkowski metrics for mixed feature type data analysis. *Trans. Systems, Man and Cybernetics*, Vol. 24, n° 4.

[8] Rasson, J.P., Granville, V. (1996) : Geometrical tools in classification. *Computational Statistics and Data Analysis.* Vol. 23, 105-123.

[9] Silverman, B.W. (1986) : *Density estimation for Statistics and Data Analysis.* Chapman and Hall, London.

Decision Trees and Uncertain Inference

Emmanuel Périnel
ENSAR - INSFA, 65 rue de Saint-Brieuc, 35042 Rennes cedex
INRIA Rocquencourt, Domaine de Voluceau, 78153 Le Chesnay cedex

Summary: The paper presents a general overview of recent methodology concerned with the problem of growing a decision tree in the presence of uncertainty or imprecision. One can identify two main approaches in the literature. The first aims at "softening" the hard recursive partition process through the use of fuzzy splits. The second proposes to handle the uncertainty by using more complex representation languages. Both approaches, which can be related in a probabilistic framework, produce in a general way more flexible and robust classification rules.

Key words: decision or classification tree, imprecision and uncertainty, soft split, imprecise description, symbolic data analysis.

1. A sensitive methodology

Binary segmentation (or decision tree, recursive partition, tree growing,...) aims at building a tree-structured classification rule for a given partitioned set of objects described by predictors. In spite of many successful applications, both in the Artificial Intelligence community (Quinlan 93, Kononenko *et al.* 84) and the statistical community (Breiman *et al.* 84, Ciampi 92), several authors have emphasized the particular sensitivity of these methods in the case of an imprecise data set. For example, (Quinlan 86) highlighted the general degrading of the performances of a decision tree from a predictive or a descriptive point of view caused by introducing various levels of noise in the data. One can see a double origin for this problem: on the one hand, the "hard" recursive partition process; on the other hand, the difficulty of handling any form of imprecision (measurement errors, subjective judgement of an expert,...) during the coding phase. These remarks have led to two main approaches that are now presented.

2. Fuzzy or soft splits

The objective of the soft splits is to introduce more flexibility during the recursive partition process. This kind of split is particularly suitable for two reasons. First, the assignment of the objects does not change in an important way when the observations, because of an imprecise context, are slightly modified. Secondly, it permits one to give more importance to the observations satisfying in a better way the properties associated with the split. The general principle of the soft split is represented in figure 1.

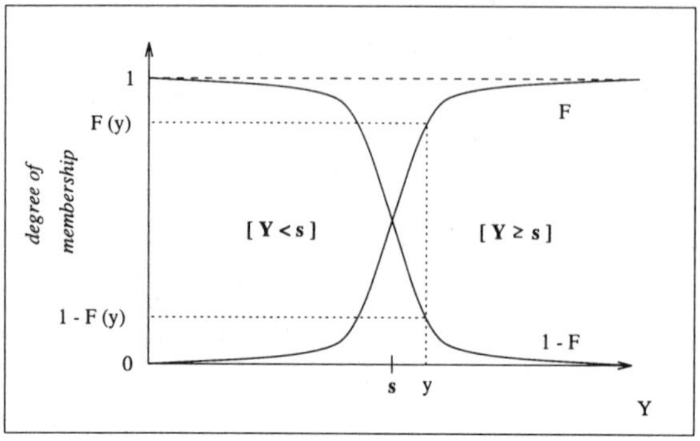

Figure 1: *Principle of a soft split.*

Here, the function F may be defined for example to be a logistic function:

$$F(y) = \frac{\exp(ay+b)}{1+\exp(ay+b)} \tag{1}$$

where parameters a and b (representing respectively the cutting point and the softness of the split) are either estimated automatically by the method (for example: (Quinlan 90) or (Ciampi *et al.* 96)) or given by an expert (for example (Weber 92)). These approaches have been widely used within fuzzy decision tree algorithms (see (Yuan *et al.* 95) for more details). There, in most of cases, fuzzy splits are represented by triangular or trapezoidal membership functions because their easy building and interpretation by the user.

3. Explicit imprecise observations

In applications, we often face imprecise observations that are, afterwards, reduced to single values in order to avoid having complex codings. In symbolic data analysis (Diday *et al.* 97), it is proposed to extend the standard "deterministic" description of an object (that is, a vector of values) to a set of random variables, each of them being associated with more complex descriptions such as disjunction of values or probability functions. More precisely, if the description space is based on P predictors $Y_1, Y_2, ..., Y_P$, the studied population will be composed of objects i whose description is given by

$$(Y_{i1} \sim f_{i1}, Y_{i2} \sim f_{i2}, ..., Y_{iP} \sim f_{iP}) \tag{2}$$

where f_{ij} represents the probability function describing the imprecision of the object i for the predictor j. For example, if one does not know precisely the value of a descriptor Y, it is possible to take the imprecision into account by defining Y to be a random variable distributed according to a uniform probability function f on $[a, b]$ (figure 2).

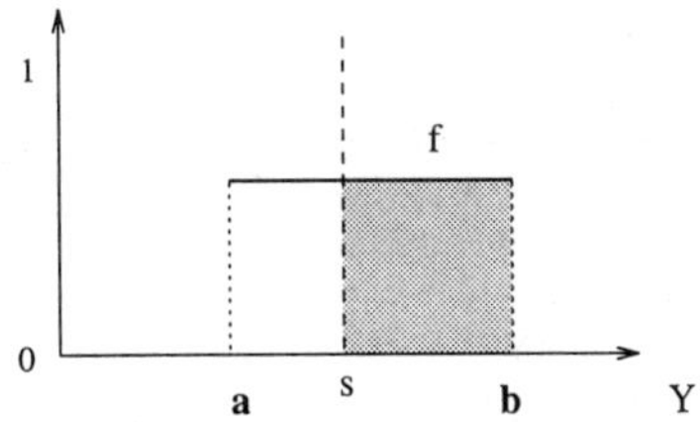

Figure 2: *Assignment of a probabilistically imprecise description.*

In this case (see (Ciampi *et al.* 96)), the observation described by f would be assigned to the region $[Y \geq s]$ with probability

$$p = \int_{[Y \geq s]} f(y).dy = \frac{b-s}{b-a} \tag{3}$$

In fuzzy logic, explicitly fuzzy initial data have been much less studied than classical data (usually continuous) which are "fuzzified" through the use of soft splits. Nevertheless, and unlike the probabilistic approach, a non negligible attention has been paid to the problem of taking into account a fuzzy membership of the objects to the classes of the prior partition (a problem which concerns many real applications).

4. A connection between the two approaches

Both previous approaches lead to the same notion of a soft or flexible assignment: it consists in substituting the usual boolean degree of membership to the left or right node by a number varying in the interval $[0, 1]$.
Moreover, in spite of their apparently distinct motivations and principles, these can be clearly related within the following framework. We will here limit ourselves to a probabilistic approach. Let us first consider the problem of the assignment of N observations $y_1, ..., y_N$ through the binary split $[Y < s]/[Y \geq s]$ (figure 3). I_s is representing the usual "hard" split while F_s denotes a soft split (with $F_s(s) = 0.5$). On the basis of $y_1, ..., y_N$, we build a set of imprecise observations $f_{y_1}, ..., f_{y_N}$ centered on y_i and with $f' = F$.
We have then the following result: *applying the soft split F_s on the precise observations $y_1, ..., y_P$ is strictly equivalent to use the usual split I_s on the set of imprecise observations f_{y_i}.* This can be formally written through the following equalities:

$$P[Y_i > s] = \int_s^{+\infty} f_{y_i}(y).dy = F_s(y_i)$$

In other words, using for instance a soft split associated to a large "degree of softness" is equivalent to consider that we have a classical split applied to observations which are all blurred by the same very large imprecision (that is, the probability functions f_{y_i} associated to large standard errors). As a consequence, if a large variability of the slope of the split does not change

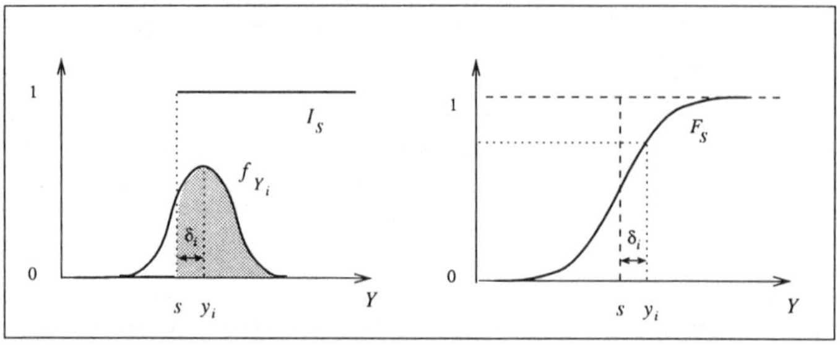

Figure 3: *Connection between a soft split and imprecise observations.*

in an important way the quality of the split, we can deduce that the split is robust in the sense that it is not affected by very imprecise observations.
Of course, if we consider a set of observations which could be affected by *different* levels of noise or imprecision, it is not possible anymore to draw a parallel between these two approaches.

As shown by several authors, advantages of these two approaches may be firstly, to permit the selection of more robust predictors during the growing of the tree; secondly, to improve the classification accuracy in preventing too many arbitrary assignments, such as can happen when a high level of uncertainty is present.

Finally, note that, in a probabilistic context (for both previous approaches), the modeling of a decision rule associated with a classification tree is now expressed as follows:

$$P(c\,|y) = \sum_{t=1}^{T} P_t(y)\,.\,P(c\,|t) \tag{4}$$

where $P_t(y)$, which denotes the probability for an observation y to belong to the node t.

5. Semantical aspects

In the previous section, which was only concerned with a probabilistic context, we were not interested immediatly in giving a sense to the fuzziness of the split but first of all, in statistical considerations: looking for more robust splits, which are suitable for noisy contexts, and well-adapted to a given data configuration. As remarked by (Yuan *et al.* 95), decision trees developped within the fuzzy logic theory are above all interested in another kind of uncertainty called *cognitive uncertainty.* This uncertainty *"deals with phenomena arising from human thinking, reasoning, cognition and perception processes".* This kind of uncertainty has been generally employed to model and to give a clear interpretation to the fuzzy splits. Such splits are especially employed in

order to represent the notion of overlapping between the linguistic terms used by an expert.
In comparison, the assignment process in a probabilistic context (resulting either from a soft split or an imprecise observation) relies on a clearly distinct idea: even if it leads also to the calculus of membership degrees of an object to the left or right nodes, it is assumed that the object actually belongs to a single node; but because of the uncertainty affecting the observation (or the split), this membership cannot be computed in a deterministic way.

About validation. Even if bridges between the different axiomatics of uncertainty have been proposed (see for instance (Drakopoulos 95)), the question of estimating the predictive performances of a decision tree as not been as much studied as in the statistical domain (validation is often submitted to the expert). Many works interested in that aspect rely on the use of fuzzy probabilities which permit to employ the classical results of the bayesian decision theory.

5. Discussion: symbolic data analysis and decision trees

Symbolic data analysis offers a quite suitable framework in order to study the problem of recursive partition (or decision tree induction, with respect to the terminology used in supervised learning) in the case of more complexe data. The main reasons why the formalism of the symbolic objects is particularly convenient may be summarized as follows:

(i) Any decision tree structure may be described by an "organized" list of symbolic objects. Indeed, each terminal node built by the recursive partition may be formally described by a symbolic object, called assertion, that is, a knowledge about the initial data set. The sub population associated to this node corresponds to the *extension* of the symbolic object; while its description (the conjunction of successive properties that lead to the node) forms its *intension*; for example, the description of a leaf t is usually written as

$$s_t \quad = \quad \bigwedge_{j \in J_t} [Y_j = V_t^j] \qquad (5)$$

where J_t is the subset of the predictors Y_j that characterise the leaf t, and V_t^j is the description attached to Y_j.

(ii) its logical representation permits to represent, in a natural way, the different sub populations obtained through an iterative process (note that the various terminal nodes are not necessarily characterised by the same predictors);

(iii) Each assertion s_t characterising a sub population is a conjunction of properties (coming from the binary questions) which are, in general, based on *multivaluate* descriptions V_t^j; this kind of description is well adapted to represent the natural variability inside a population of objects.

Because the language of symbolic objects permits also to represent the initial data set (including for instance imprecise descriptions with the help of probabilistic descriptions), the tree growing procedure may be presented in the general following way: it consists in searching iteratively for an organized set of symbolic objects (of kind (5)) which, at each step, summarizes "as well as possible" the first list of symbolic objects (the initial data). To do that, one should of course define a suitable measure of affinity between two sets of symbolic objects, depending principally on the kind of descriptions that are studied. For example, in (Ciampi *et al.* 96), (Périnel 96), because the aim was to summarize a set of probabilistically imprecise descriptions, the affinity measure was built as an extension of the classical statistical information measure.
The main advantage of presenting the recursive partition problem as above (that is, more from a knowledge extraction point of view), is that it permits to take into account of more complex descriptions: on one hand, in order to describe the sub populations built by the partitioning (here, these can be obtained on the basis of probabilistic soft splits); on the other hand, in order to take into account complex data (described as in (2)) expressing for instance the notions of noise, doubt or imprecision.

References

Breiman, L., Friedman, J.H., Olshen, R.A. and Stone, C.J. (1984). Classification and regression trees. Belmont: Wadsworth.

Ciampi, A. (1992). Constructing prediction trees from data: the RECPAM approach. *Proceedings from the Prague 1991 University Summer School on computational aspects of model choice*, 105–152, Physica Verlag, Heidelberg.

Ciampi, A., Diday, E., Lebbe, J., Périnel, E. and Vignes, R. (1996). Recursive partition with probabilistically imprecise data. In: *Ordinal and Symbolic Data Analysis*, 201-212, Diday E. *et al.* editors, *Springer-Verlag.*

Diday, E. and Émilion, R. (1997). Treillis de Galois maximaux et Capacités de Choquet. *C.R. Acad. Sci. Paris, Analyse mathématique*, t. 324, série 1.

Drakopoulos, J.A. (1995) Probabilities, possibilities and fuzzy sets, *Fuzzy set and systems*, vol. 75, 1, pp. 1–17, october.

Kononenko, I., Bratko, I. and Roskar, E. (1984). Experiments in automatic learning of medical diagnostic rules. *Technical Report, Jozef Stefan Insitute*, Ljubjana, Jugoslavia.

Périnel, E. (1996). Méthodes de segmentation et analyse des données symboliques. Le cas de données probabiliste imprécises. *Thèse de l'Université Paris* IX-*Dauphine.*

Quinlan, J.R. (1986). The effect of noise on concept learning. In Michalski, Carbonel & Mitchell (Eds.), *Machine Learning: an artificial intelligence approach.* San Mateo, CA: Morgan Kaufman.

Quinlan, J.R. (1990). Probabilistic decision trees. In: *Machine Learning* III, Kodratoff Y., Michalski R. (eds.), 140-152.

Quinlan, J.R. (1993). *C4.5 Programs for Machine Learning.* Morgan Kaufman, San Mateo, California.

Weber, R. (1992). Fuzzy-ID3: a class of methods for automatic knowledge acquisition. *Proc. 2nd Internat. Conf. on fuzzy logic and neural networks*, pp. 265–268, 17–22 July, Iizuka, Japan.

Yuan, Y. and Shaw, M.J. (1995). Induction of fuzzy decision trees. *Fuzzy Sets and Systems*, 69, 125–139.

Interpretation and Reduction of Galois Lattices of Complex Data

G.Polaillon
Lise-Ceremade, Université Parix IX Dauphine
Place du Mal de Lattre de Tassigny, 75775 Paris Cedex 16, France
polail@ceremade.dauphine.fr

Abstract: Galois lattices are a kind of conceptual classification. Galois lattices theory and construction have been extended from binary tables to multivariate, interval and histogram data with missing values via the Symbolic object formalism. The Galois lattice of all "complete symbolic objects" is a possible output, but isn't often legible. First we propose the reduction of the lattice by a pyramid or a hierarchy using numerical and symbolic criteria. Then we propose to transform the lattice into an identification tree. The last proposed solution is the extent of Duquenne's minimal set of rules.

Keywords : Galois Lattice, Symbolic Objects, Reduction and Interpretation

This article is organised as follows : first, we introduce Symbolic Data Analysis; Then, we remind the principle of Galois lattices of complex data; Finally, we present some possible symbolic Galois lattice reductions.

1. Complex data table formalism

Standard data analysis methods extent to complex data is called Symbolic Data Analysis [Diday 91]. Symbolic objects help to represent and analyze real data, and allow us to treat directly complex tables such that any case could contain a disjunction of values, an interval of values, an histogram of frequencies or are empty.

1.1 Individuals

The description of an individual ω is represented by $\omega_s = (y_1(\omega), ..., y_p(\omega))$ where each variable y_i, i=1,...,p, is an application from Ω to $P(O_i)$, where Ω is the set of individuals, O_i the description space for each variable y_i, and $P(O_i)$ the power set of O_i. The variable $y_i(\omega)$ may be quantitative (discrete or continuous) or qualitative (nominal) and may be represented as a set of values, an interval of values, or an histogram of values.

1.2 Classes

Symbolic objects help to formalize classes.
Let $\mathcal{C}$ be the set of all possibles descriptions : $\mathcal{C} = P(O_1) \times ... \times P(O_p)$. For i=1...p, we define an application on descriptions :

$$\begin{aligned} d_i : \ & \mathcal{C} \rightarrow P(O_i) \\ & c \mapsto d_i(c) \end{aligned}$$

where d_i(c) may be quantitative (discrete or continuous) or qualitative (nominal) and represents the description as a set of values, an interval or an histogram of values.

Let R be a relation between three sets : the object set, the description set and the variable set. This relation is defined by the operators : R $\in$ $\{=,\subseteq,\supseteq,\in,\ni,\rightarrow,...\}$. The expression y_i(w)Rd_i(c) is a predicate which concerns the variable y_i, the individual w, the description c, and R is one of the previous operator. This predicate is true if the relation is true.

Definition 1 *A symbolic object is a triplet (a,R,d) where R is a relation, d is a description, and a is an application linking the individual, the relation and the description.*

Different classes of symbolic objects exist and we remind here the basic definition of an assertion object.

Definition 2 *Let R et d be fixed; an assertion object a is a symbolic object completely defined by the following application :*

$$\begin{aligned} a : \ & \Omega \rightarrow \{true, false\} \\ & \omega \mapsto a(w) = [y_1(\omega)Rd_1] \wedge ... \wedge [y_p(\omega)Rd_p] \end{aligned}$$

such that $a(\omega) = true$ *if and only if* $\forall i = 1, ..., p,\ y_i(\omega)Rd_i$.

Definition 3 *The extent of a on* Ω *is defined as :*

$$ext_\Omega(a) = \{\omega \in \Omega / a(\omega) = true\} = a^{-1}(true)$$

2. Galois lattice of complex data

Galois lattices help in many domains to represent, interpret and identify data [Wille 90, Duquenne & Guigues 86, Godin & Missaoui 94]. Classically, complex data are recoded before treatment.
Galois lattices construction has been extended from binary case to more complex data by incremental and non-incremental algorithms [Polaillon & Diday 96]. We obtain Galois lattices where each node is a "complete assertion object" [Diday 91], which generalizes the intent of a binary concept in

	Var 1	Var 2
Ind.1	[0.75,0.8]	[1,2]
Ind.2	[0.6,0.8]	[1,1]
Ind.3	[0.5,0.7]	[2,2]
Ind.4	[0.72,0.73]	[1,2]

Table 1: Exemple of data

the case of more complex data. Two kinds of Galois lattices can be obtained depending on our aim : if we are interested by classes described by properties satisfied by all individuals belonging to the class (intersection lattice) or satisfied by at least one of them (union lattice).
Galois lattices have the following advantages : we obtain all possible "complete assertion objects", each node is automatically describe by one of them and his extent, they are ordered by a generalization/specialization relation.

We consider the table 1 with interval data as an illustrative example, but algorithms work with other kind of data (multivalued, histograms) :

Figure 1 gives the intersection lattice : nodes are described by their extent and new objects (found by considering inheritance).

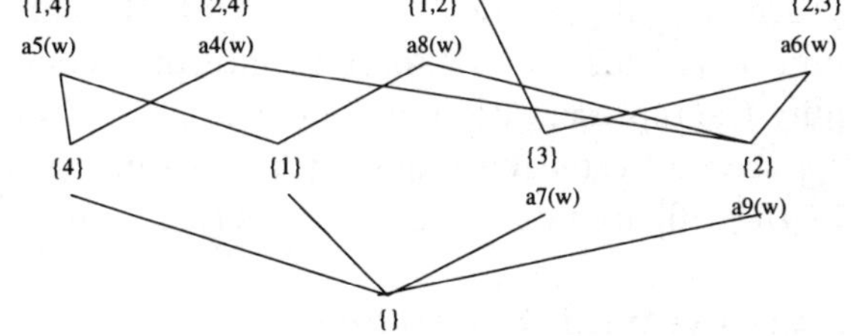

Figure 1: Intersection lattice

Corresponding objects, without inherited properties, are :
a1=true
a2=[$y2$ ⊇[2,2]]
a3=[$y2$ ⊇[1,1]]
a4=[$y1$ ⊇[0.72,0.73]]

a5=[$y2$ ⊇[1,2]]
a6=[$y1$ ⊇[0.6,0.7]]
a7=[$y1$ ⊇[0.5,0.6]]
a8=[$y1$ ⊇[0.75,0.8]]
a9=[$y1$ ⊇[0.7,0.72]] ∧ [$y1$ ⊇[0.73,0.75]]

The complete object corresponding to each node is calculated by the conjunction of the new object of the node and objects inherited from predecessors. For example, complete object for class {4} is a4 ∧ a5∧ a3∧ a2.

Figure 2 gives the union lattice : nodes are described by their extent and their complete objects.

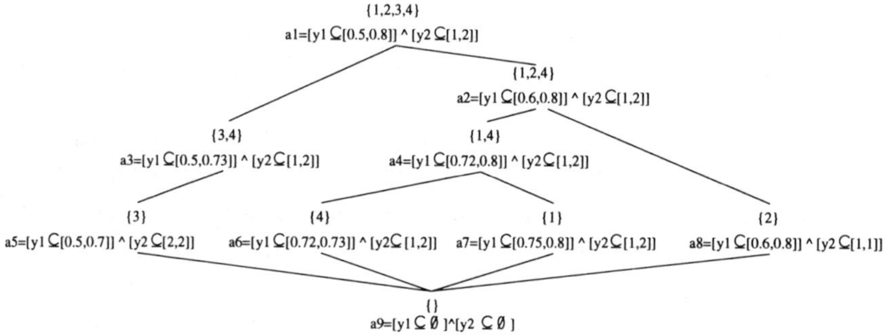

Figure 2: Union lattice

3. Galois lattice interpretation

Data analysis with Galois lattices leads us to two problems : the size of Galois lattices grows up with the number of individuals and attibutes, and, there is no automatic aid for a non expert to interpret Galois Lattices. So we have to simplify the representation of Galois lattices. [Godin & Missaoui 94, Guénoche 93] are interested by reducing lattices into binary trees, and [Duquenne & Guigues 86] are interested by extracting a minimal set of rules.

3.1 Reduction via pyramid, hierarchies

Galois lattices have been constructed by mathematical theorems, but the sense of data is not taken into account.

We compare two kinds of algorithms, which are applied directly on the two lattices (intersection and union lattices) :

- This algorithm is based on a similarity successor/predecessor (this algorithm was first proposed by [Bournaud 96]):
  To obtain a hierarchy from a Galois lattice, for each node (starting

with specific complete objects), we choose the "nearest" predecessor according to a distance and we suppress links with other predecessors. At the end, we suppress nodes which are not linked at least with one individual.

- This algorithm is an extent of the hierarchical classification (first proposed by [Guénoche 93]) and the pyramidal classification :
  To obtain a hierarchy : We start with specific complete objects, while objects are active, we look for the most similar pair of objects, which are desactived, we find their common parent, which is actived, and links are updated.
  To obtain a pyramid : The algorithm is quite the same. Differences are : objects can be associated two times to other objects and order contraints must be respected (see the initial algorithm of [Bertrand & Diday 90]).

Many distances between symbolic objects can be found in the litterature, but we have specificities that are never taken into account. For example, we can not choose the same distance in the union and intersection case, missing values have particular interpretation, the "nearest" predecessor is the one which fits at best, we have to take into account the order induced by the lattice... We define a new distance, which includes numerical and symbolic criteria. The numerical criterion is based on [Gowda & Diday 94] dissimilarity by changing the position coefficient (we take into account the upper bound, otherwise similar situations can lead to different results) and including the case of missing values. The symbolic criterion takes into account the lattice order by the use of the distance proposed by [Leclerc 94].

We obtain a hierarchy or a pyramid without choosing an agregation criterion.

With the previous example, from the intersection lattice, we obtain figure 3.

3.2 Reduction via identification tree

It is a descending approach which is different from the divisive approach as we work on the Galois lattice. Many discriminant criteria can be found in the litterature [Guénoche 93, Ziani96]. The advantages of our method are that we work with complete objects (we have all robust descriptions) and we already have all possible variables splits (we don't need other criteria for splitting each variable domain and choose the best one).

We compare two algorithms :

- this algorithm is close to the divise approach :

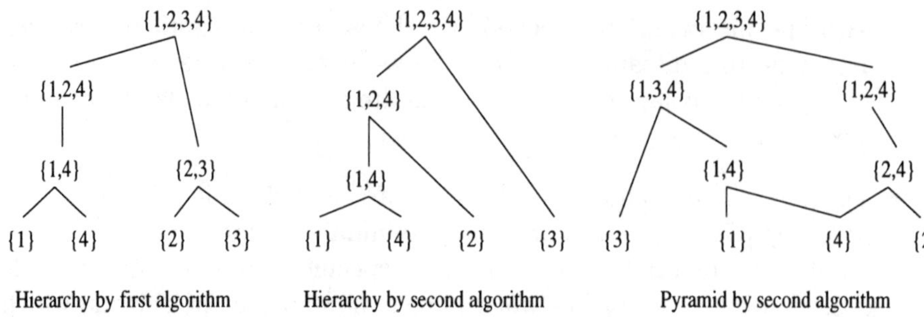

Figure 3: Trees obtained with ascendant approach

Starting at the top of the lattice, given an optimal criterion, for each successor node, we choose those which satisfy the criterion, and we continue until bottom nodes are reached.

- this algorithm is quite intuitive and is inspired by neural networks: Starting at the top of the lattice, we choose successor nodes where individuals have greatest probabilities to go, and we continue until bottom nodes are reached. This method requires a bootstrap validation.

We obtain directly an identification tree which works on all variables at the same time, ignoring a priori classes and where nodes are complete assertion objects.

The previous example is too small to apply the first algorithm, but with the second algorithm, we obtain the figure 4.

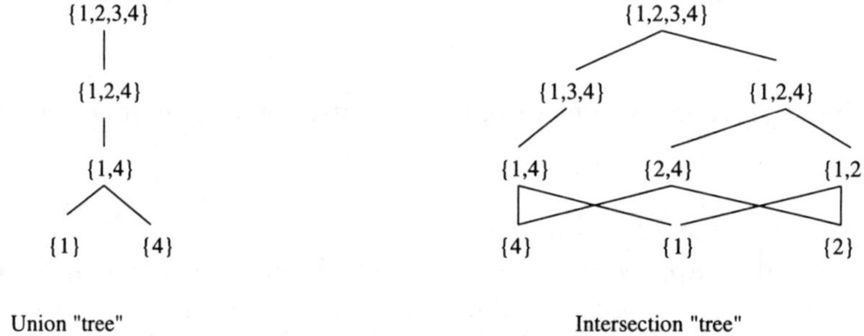

Figure 4: Trees obtained with the probabilistic descending approach

3.3 Duquenne's rules

The reduction and the interpretation of the lattice by a minimal set of rules

between attributes in the binary case have been proposed by [Duquenne & Guigues 86].
In the case of multivaluate data, interval data, histogram data, the extent of the rules extraction is evident, as we work on inclusion of extents.

The property used in the demonstration is immediatly extended : the implication $A \rightarrow$ B is true if and only if A' $\subseteq$ B', where A, B denotes a set attributes, and A', B' denotes objects having A, B. In the case of complex data, A, B denotes a set attributes values and A', B' are extent of objects composed by descriptions A, B, with relation R$\equiv\supseteq$. Immediatly, we have the same property. We can use Duquenne's inference rules (B,C,D denotes attributes) in order to find the minimal set of rules :

- B $\rightarrow$ D is a consequence of B $\rightarrow$ C and C $\rightarrow$ D
- B$\cup$ X $\rightarrow$ D$\cup$ X is a consequence of B $\rightarrow$ D
- B $\rightarrow$ D is true when B $\supseteq$ D

From a complex table, we obtain directly rules between complex data.

By treating the complete intersection lattice of figure 1, we obtain the following rules :
a8 $\rightarrow$ a3
a4 $\rightarrow$ a3
a9 $\rightarrow$ a4, a6
a3, a6 $\rightarrow$ a4, a8
a4, a8 $\rightarrow$ a9
a7 $\rightarrow$ a2, a6
a2, a6 $\rightarrow$ a7
a5 $\rightarrow$ a2, a3
a2, a3 $\rightarrow$ a5

4. Conclusion

We treat directly complex data tables. Galois lattices give us all "complete assertion objects", which represent a robust mathematical space to work on. By taking in account different numerical criteria (dissimilarities on complex objects, discriminant distances...) and lattices properties, it is possible to extract a hierarchy, a pyramid or an identification tree from Galois lattices. With classical criteria, we find the same results found in the litterature, but the difference is that we are sure to obtain robust complete objects. By taking into account properties of the lattice, it is possible to summarize the Galois lattice as a minimal set of rules between complex attributes directly. The perspective of this work is to validate all approachs, may be by comparing them to each other.

References

Bertrand, P., Diday, E. (1990): Une généralisation des arbres hiérarchiques : les représentations pyramidales. *Rev. Statistique Appliquée*, XXXVIII(3), 53-78.

Bournaud, I. (1996): *Regroupement conceptuel pour l'organisation des connaissances.* Thèse Paris VI.

Diday, E. (1991): Des objets de l'analyse des données à ceux de l'analyse des connaissances. In: *Induction Symbolique et Numérique à Partir des Données*, Diday, E. & Kodratoff, Y. (eds), Cepadues Editions, Toulouse, 9-75.

Duquenne, V. & Guigues, J.L. (1986): Familles minimales d'implication informatives résultant d'un tableau de données binaires. *Math. Sci. Humaines*, 95, 5-18.

Godin, R. & Missaoui, R. (1994): An incremental concept formation approach for learning from databases. *Theoretical Computer Science*, 133, 387-419.

Gowda, K.G., Diday, E. (1991): Symbolic clustering algorithms using similarity and dissimilarity measures. *Pattern Recognition*, 24,6, 567-578.

Guénoche, A. (1993): Hiérarchies conceptuelles de données binaires. *Math. Inf. Sci. Hum.*, 121, 23-34.

Leclerc, B. (1994): Medians for weight metrics in the covering graphs of semilattices. *Discrete Applied Mathematics 49*, 281-297.

Polaillon, G. & Diday, E. (1996): Galois lattices of complete assertion objects in symbolic data analysis, *Cahiers du Ceremade (URA CNRS 749) N° 9631*, Université Paris IX Dauphine.

Polaillon, G. & Diday, E. (1998): Reduction of symbolic galois lattices via hierarchies and pyramid, *Proceedings of KESDA'98*, Luxembourg.

Wille, R. (1990): Concept lattices and conceptual knowledge systems. *Preprint 1340*, Technische Hochschule Darmstadt.

Ziani, D. (1996): *Sélection de variables sur un ensemble d'objets symboliques : traitement des dépendances entre variables.* Thèse Paris IX Dauphine.

Proximity function for clustering of software objects

Petko Valtchev
INRIA Rhône-Alpes
655 av. de l'Europe, 38330 Montbonnot Saint-Martin, France
e-mail: Petko.Valtchev@inrialpes.fr

Abstract: The analysis of object-oriented datasets is a challenging task since significant differences exist between the usual data formats and software objects. An approach towards the definition of an object proximity measure is presented here which fits the basic principles of the object paradigm.

Keywords: Proximity computation, Automatic clustering

1. Introduction

The ever growing demand of analysis for complex real-world datasets (Diday 87) has put the emphasis on issues related to the integration between data sources (e.g. databases or knowledge bases) and analysis tools (Piatetsky-Shapiro *et al* 91). Our own study focuses on the application of automatic clustering within object knowledge bases (KB). An object base constitutes a model of a real-world domain where domain relevant knowledge is expressed through objects organized into class taxonomies. Clustering may be helpful for the task of class building when only an unstructured set of objects is available. It requires a consistent proximity function for objects, i.e. one which processes objects within their original context, the KB, and with respect to all the relevant information contained in the KB.

In this paper we discuss the features of the object formalism which make two objects similar and provide effective means for measuring the induced proximities. Moreover we address the problem of combining the contribution of the different object aspects into a global object proximity measure. The described function is complete, universal and flexible, and therefore enables sensible application of classical clustering techniques on objects.

2. Object based knowledge representation

Object formalisms concentrate the knowledge about a particular element of the modeled domain into a single expression, a *structured object*. Moreover, they distinguish between individual knowledge, expressed as objects and generic knowledge, modeled through object classes. In the following, to illustrate the object paradigm we provide an example of a KB modeling a real-estate domain. The KB the description language of TROPES (Sherpa project 95), an object representation system developed by our research team.

2.1. Objects

A *structured object* may be seen as a list of valued *fields* (attributes) where each field embodies a particular aspect of the modeled element. For example, an object representing a flat will have fields like rent, standing, rooms, owner, etc. In TROPES, objects are divided into disjoint families, called *concepts* (e.g. `flat`, `human`, `room`). A concept defines the structure, i.e. the set of fields, of the belonging objects, its *instances* : the instances identifiable as `flat#5` and `flat#18` both have a `rent`, an `owner`, etc. Fields model features of three kinds: *properties* are element own features (e.g. rent, standing), *components* introduce element parts which are objects themselves (e.g. flat and its rooms), *links* model other relations between elements different from composition (e.g. ownership in case of the `owner` field). TROPES uses a *strong typing*, i.e. each field is assigned a type, i.e. set of admissible values and operations. Property fields are typed by simple data types : here `rent` and `age` are both of integer type. Relational fields are typed by concepts : in the real-estate base, `rooms` has the `room` concept as a type and `owner` `human`. Finally, multi-valued fields of two kinds: *sets* and *lists* are admitted. Thus, `rooms` is a set-valued attribute, for example it relates `flat#5` to the respective set of `room` instances (noted `flat#5.rooms = {room#24, room#31, room#6}`).

Figure 1: *Example of two sets of related objects corresponding to* `flat#5` *and* `flat#18`*: flat owners,* `human#36` *and* `human#62`*, and room sets are included. Relational fields are drawn as labeled arrows pointing at the target object whereas properties are directly cited.*

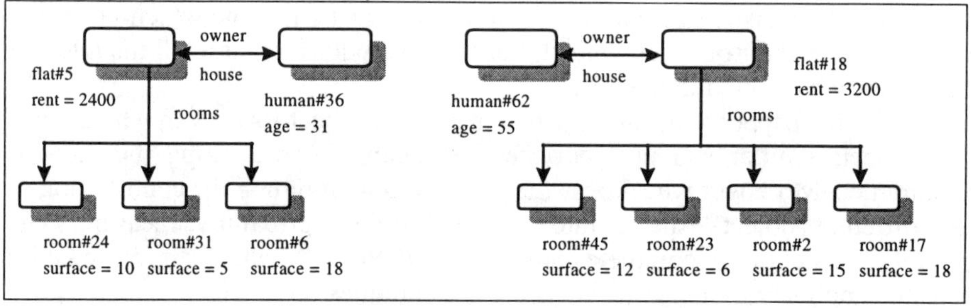

2.2. Classes

Classes describe sets of objects (in TROPES sub-sets of concept instances) intentionally, i.e. by providing information about object fields: types and value constraints. Field constraints on properties basically define sub-domains of the field domain fixed by the concept. Imagine that `kitchen` restricts the `surface` field, originally fixed in `room` to a positive integer by : `kitchen.surface = [9 15]` (to read surface of a kitchen is between 9 and 15 m^2). Relational fields are constrained by providing a class of the type concept.

The set of all constraints constitute a necessary condition for membership in a class: instance field values must satisfy all those restrictions. Thus, `room#24` and `room#45` being members of `kitchen` within the `room` concept (see Fig.2) both satisfy the constraint on `surface`. Classes of a concept are organized into a taxonomy by *specialization*, i.e. `a-kind-of` relation distinguishing a super-class and a sub-class. Specialization implies both set inclusion of class extensions and growing strength of value restrictions. An object is attached to a single class (`is-a` relation or *instanciation*), but is member of all its super-classes. This means `room#24` and `room#45` are instances of the `service` class as well. In sum, a KB should be seen as a set of inter-related objects described by fields and belonging to various classes. Moreover, classes express taxonomic knowledge about their instances which is complementary to object descriptions.

Figure 2: *Example of object instanciation: the taxonomy of the* `room` *concept is drawn with the attachment class for each room.*

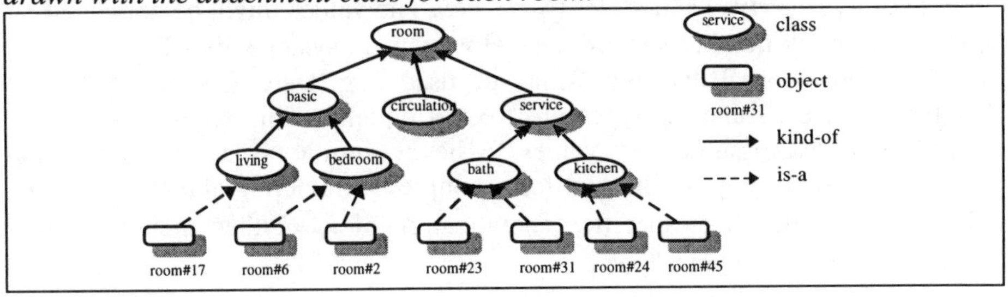

2.3. Analysis issues

Classes group objects of identical structure and with field values laying within restricted domains, that is class instances are *similar* in a way. Therefore, given an unstructured set of objects, when homogeneous object clusters appear within it one may expect those clusters to represent classes. The main topic is thus to formally state a criterion for object similarity with respect to object paradigm features.

Objects are thoroughly characterized by the set of their field values, simple values or other objects, and by the classes they belong to. With respect to what has been said previously, measuring a proximity between a couple of objects requires effective evaluation means for elementary proximities on properties, components, links and classes.

Besides, whereas properties are local to an object, the relational fields introduce dependency between objects (e.g. between a `flat` and its `owner`). Actually, the set of all objects related to a given object by chains of relational attributes may be seen as a network (see Fig.1 for an example). Both nodes and arcs in the network are labeled : nodes by the concept and, possibly, by an attachment class, whereas arcs by the field name. The network represents the smallest context in which an object appear. Therefore, a sensible proximity measure for objects should combine the contribution of each element of the respective networks. Please notice that what is needed is a proximity for nodes rather than for entire networks. In the next section we present an object proximity which computes contributions of related objects with respect to the length of the chain linking them to the compared objects.

3. Object proximity measure

Let C be a concept, e.g. flat, with n fields and let $dom(a_i)$ denote the domain and $type(a_i)$ the type of a field a_i. For a couple of concept instances o and o', let $o.a_i=v$ and $o'.a_i=v'$ denote the respective field values and δ_i^f the field-specific proximity function. Notice that all real-valued functions below are given in a dissimilarity form and normalized.

3.1. Elementary proximities

In case of a *property,* v and v', let $type(a_i)=T$. The values might be compared *ad hoc* to T (e.g. by number subtraction). However, in systems like TROPES, non-standard and external data types may be used (e.g. Date, Chemical element, Nucleic acid, etc.) and all types are treated equally. This requires a generic comparison : we assume two values to be as similar as a type expression generalizing both (e.g. an interval for a couple of numeric values) is specific. Thus, let τ be the most specific generalization of v, v', then their proximity is the relative size, noted $\| \, \|$, of τ[1] :

$$\delta_T^{pr}(v,v') = \frac{\|\tau\|}{\|dom(a)\|}.$$

On Fig.1, the difference between the rents of both flats is $\delta_{rent}^{pr}(2400,3200) = 0{,}17$ with $\|dom(\texttt{rent})\| = 4800$ and $\tau = [2400,3200]$.
For a relational field a_i, v and v' are instances of the concept $C'=type(a)$. The more similar those objects, the greater the proximity of o and o'. Both for link and component fields, the proximity of the referred objects, within their own concept (see next subsection) is substituted to δ_i^f : $\delta_{C'}^f(v,v') = d_{C'}^o(v,v')$.
Still another aspect of object comparison is the processing of multi-valued attributes a_i, (e.g. the `rooms` attribute). In this case v and v' are collections of elements for which a proximity δ_i^f is assumed available. As collections may have variable length one needs a matching between their members. The matching should, in addition, optimize the sum of the pair-wise proximities δ_i^f of matched collection members. The exact procedure of finding an optimal matching $M_{opt}(v,v')$ for *sets* and *lists* may be found in (Valtchev *et al* 97) and is omitted here for space reasons. The collection proximity is the ratio of total

[1] In (Valtchev *et al* 96) a slightly different definition based on graph distance is given.

proximity of the matching (possibly corrected by the unmatched elements) and the maximal collection size :

$$\delta^m(v,v',\delta_i^f,M_{opt}(v,v')) = \frac{\sum_{(e,e')\in M_{opt}(v,v')} \delta_i^f(e,e') + |\,l-l'\,|}{\max(l,l')}$$

where l and l' are the sizes of v and v' respectively. For example, an optimal matching for `flat#5.rooms` and `flat#18.rooms` with δ_i^f equal to δ_{room}^{cl} is M_{opt} = ((`room#24`, `room#45`),(`room#31`, `room#23`),(`room#6`, `room#2`)) with δ^m of 0.5 (`room#17` is unmatched).

The proximity of a couple of objects o and o' with respect to their attachment classes c and c' may be evaluated according to the same principle of least generalization. Thus, c and c' are as similar, as their common super-class $\bar{c}$ within the taxonomy H of **C'** is specific. The effective measure $\delta_C^{cl}(o,o')$, described in (Valtchev *et al* 96), is based on the graph distance, its value is proportional to the length of the shortest path linking o and o' in the taxonomy H, or, what is equal, the total number of intermediate classes between each of the objects and $\bar{c}$. For example, $\overline{\delta_C^{cl}}(\text{room\#6}, \text{room\#2}) = 0.33$ since both objects share the attachment class `bedroom`, whereas $\overline{\delta_C^{cl}}(\text{room\#17}, \text{room\#2}) = 0.66$ with respect to the common super-class `basic` (see Fig.2).

3.2. Object-level function

Suppose the field functions `cons` and `nat` return the constructor (`one` for single-valued fields, `list` or `set`) and the nature (`property`, `component` or `link`) of an object field. Notice that all real-valued functions below are normalized.

Given a concept C the *object dissimilarity function* $d_C^o: C \times C \rightarrow \Re$ is an additive function of the class dissimilarity δ_C^{cl} and the elementary proximities on each field δ_i^f . For a couple of objects o and o',

$$d_C^o(o,o') = \lambda_0 * \delta_C^{cl}(o,o') + \sum_{i=1}^{n} \lambda_i * \delta_i^f(o.a_i, o'.a_i) \tag{1}$$

where λ_i are the user-specified field weights ($\sum_{i=0}^{n} \lambda_i = 1$). Please notice that if no taxonomy is available on C then δ_C^{cl} and λ_0 are set to *0*. Next, δ_i^f on a field a_i depends on its constructor:

$$\delta_i^f(v,v') = \begin{cases} \delta^t(v,v') & \text{cons}(a_i) = \text{one} \\ \delta^m(v,v',\delta^t,M_{opt}(v,v')) & \text{cons}(a_i) \in \{\text{set}, \text{list}\} \end{cases}$$

Here δ^m is a collections dissimilarity and δ^t is a generic function further instanciated with respect to the field nature :

$$\delta^t(v,v') = \begin{cases} \delta_T^{pr}(v,v') & \text{nat}(a_i) = \text{property, type}(a_i) = T \\ d_{C'}^o(v,v') & nat(a_i) \in \{\text{link, component}\}, \text{type}(a_i) = C' \end{cases}$$

Here δ_T^{pr} is the elementary proximity function on whereas $d_{C'}^o$ is the (also normalized) object proximity on concept C' as defined by (1).

3.3. Computation issues

Given a couple of objects o and o', d_C^o may be only computed if all values of δ_i^f are known in (1). Property and for component proximities are always possible to calculate prior to the object proximity computation. In contrast, links may introduce circuits in the networks thus making $d_C^o(o,o')$ recursively depend on itself. For example, if for a link a_i there is a field a'_j in *C'=type*(a_i) describing the inverse relation, i.e. for all e in C', $o.a_i = e$ iff $e.a'_j = o$. The couple a_i, a'_j establishes a cyclic dependency between d_C^o and $d_{C'}^o$. In the real-estate KB, the fields `owner` and `house` constitute such a couple between `house` and `human`. Cycles may occur where more than two objects are involved. To find the effective values of the pair-wise proximities, the formula in (1) may be seen as an equation where those proximities are variables (Bisson 95). Thus, picking up all the formulas for object couples which take part in circuits within the networks of o and o' we obtain a system of equations. For each equation, the coefficients are field weights and the variables object proximities for couples of objects which are related to o and o' by the *same* chain of links. The right-hand side member is the local part of the respective proximity, which is made up of the known values : property, component and class proximities (if any). The obtained system is square (m variables and m equations) and linear provided no multi-valued links exist in a circuit. In this case, the system matrix is *diagonal dominant* so the system has a unique which could be found by an iterative method. In case of multi-valued links, all the equations remain linear, but a non-determinism is introduced due to the matching in the computation of δ^m. This amounts to solving several linear equation systems in parallel, one per matching, which would be too expensive. Instead, a heuristics may be used to make the iterative method fit the new situation : a matching could be carried out after each iteration (for details the reader is invited to look in (Bisson 95)).

4. Example

Suppose one has to compute the proximity of the flats on Fig.1. As indicated, there is a circuit between `flat#5` and `human#36` on the one hand and `flat#18` and `human#62` on the other hand. First, we assign equal weights to the field functions in all three concepts. Thus, `age` and `house` are weighted 0.5 in `human` (no taxonomy is given), `rent`, `owner` and `rooms` score to 0.33 in `flat`, whereas `surface` is weighted 0.5 in `room` due to the class dissimilarity δ_C^{cl} which weight is 0.5 too. Next, let the ranges for the fields `age`, `surface` and `rent` be 80, 20 and 4800 respectively. The object proximity d^o_{room} may be computed independently from the linear system. For example, $d^o_{room}(\text{room\#24}, \text{room\#45}) = 0.5 * (0.1 + 0.33) = 0.22$.

Table 1 : *Values of* d^o_{room}

d^o_{room}	`room#45`	`room#23`	`room#2`	`room#17`
`room#24`	0.22	0.43	0.63	0.7
`room#31`	0.5	0.19	0.75	0.83
`room#6`	0.65	0.8	0.24	0.5

Using the above values for d^o_{room} we obtain the same optimal matching between room collections M_{opt} as in the previous section. This time however the collection proximity δ^m is 0.45.

Finally, the proximity of owners according to their age is $\delta^f_{age}(31{,}55) = \delta^{pr}_{\text{int}}(31{,}55) = 0{,}3$ and the proximity of flat rents is $\delta^f_{rent}(2400{,}3200) = \delta^{pr}_{\text{int}}(2100{,}3200) = 0{,}17$. The linear system can now be composed with two variables : x_1 standing for d^o_{flat}(`flat#5, flat#18`) and x_2 for d^o_{human}(`human#36, human#62`).

$$\begin{cases} x_1 = 0.33 * (x_2 + 0.45 + 0.17) \\ x_2 = 0.5 * (x_1 + 0.3) \end{cases}$$

After a resolution, we obtain $x_1 = 0.304$ and $x_2 = 0.308$.

5. Conclusion

In the paper, the object formalism specific features have been discussed with regard to their contribution to object proximity. The results of the analysis have been used in the definition of an object proximity model. The proposed measure presents the following advantages. First, it is complete with respect to object description: potentially all the available knowledge about object interpretations may be taken into account. Next, the function respects the fundamental principles of the formalism like hierarchical organisation, data abstraction, object composition, etc. In the same time, the dissimilarity is tunable, i.e. parameters can be fed-up to adapt the function to a particular domain and situation. The behavior of the function on datasets of complex structure and greater size is still to be tested.

References

Bisson, B. (1995). Why and how to define a similarity measure for object-based representation systems, in: *Towards Very Large Knowledge Bases*, N.J.I. Mars (Ed.), IOS Press, Amsterdam, 236-246.

Diday, E (1987). The symbolic approach in clustering and related methods of data analysis: the basic choices, in: *Proceedings of the First IFCS Conference*, Bock, H. (Ed.).

Piatetsky-Shapiro, G & Frawley, W. (1991). Knowledge discovery in databases, AAAI Press.

Sherpa project (1995). Tropes 1.0 reference manual, INRIA Rhône-Alpes, Grenoble (FR).

Valtchev, P. & Euzenat, J. (1996). Classification of concepts through products of concepts and abstract data types. in: *Ordinal and symbolic data analysis*, Lechevallier, Y. & Diday, E. & Dr Opitz, O. (Eds.), Springer, 3-12.

Valtchev, P. & Euzenat, J. (1997). Dissimilarity measure for collections of objects and values. in: *Proceedings of the 2nd Symposium on Intelligent Data Analysis*, LNCS, vol 1280, Cohen, P. & Liu, X. & Berthold, M,. (Eds.), Springer, 259-272.

Textual data analysis for open-questions in repeated surveys

Simona Balbi
Dip.to di Matematica e Statistica, Univ. di Napoli "Federico II"
Via Cintia, Complesso Monte Sant'Angelo, 80126 Napoli
e-mail:sb@unina.it

Abstract: The paper addresses the problem of analysing answers to an open question observed in different waves of a repeated survey. Multiway data analysis techniques can offer interesting tools for doing that. Unfortunately, dealing with textual data some peculiar problems arise. The aim of this paper is to propose the use of non-symmetrical data analysis techniques in order to follow lexical behaviours with respect to a set of explanatory numerical variables, defining groups through time. Furthermore, attention is paid to the definition of a conjoint vocabulary.

Keywords: Principal matrices, Graphical Displays, Textual forms.

1. Introduction

One of the most common textual data analysis techniques consists in an ordinary Correspondence Analysis performed on a lexical table (Lebart, Salem, 1994), built by cross-classifying a lexical variable with a categorical (or categorised) variable. Dealing with a questionnaire, generally the answers to an open question are cross-classified with the answers to a close-ended question. Balbi (1995) proposes the use of non-symmetrical correspondence analysis (Lauro, D'Ambra, 1984) in order to emphasise the different roles played by the two variables/questions. As a consequence of the chosen method, distances among occurrences on a factorial plane are measured in an usual Euclidean metric, which takes into account the marginal frequencies. Thus the identified structure is strongly influenced by common words.
The aim of this paper is to extend the non-symmetrical approach to the case we deal with open-ended questions in repeated surveys. The idea is to distinguish groups of individuals (pseudo-panel) identified by time-invariant categorical variables and to follow their behaviour in time, by displaying trajectories on a principal plane obtained by means of a Principal Matrices Analysis onto a Design Matrix (Balbi, Lauro, Scepi, 1994). One of the most interesting problems in such a kind of approach consists in a proper definition of the *vocabulary*. It is essential to understand how words change their characterisations through time. Textual forms have to be deeply investigated.

Tools useful in building one-time vocabulary, such as regions defined by bootstrap techniques (Balbi, 1995), can be applied in order to compare the same form observed in different occasions.

After introducing the peculiarity of the data structure (Sec. 2), the principal methodological tools used are presented in order to suggest a strategy for analysing the answers to open questions in repeated surveys (Sec. 3). The problem of defining a proper vocabulary is faced in Section 4. The paper ends with an application in a slight different context: the comparison of different advertisement campaigns on Italian press related to the life cycle of a product, the *cellular telephone*, from elitist consumers to a mass market (Sec. 5).

2. The Data Structure

In many countries there is a severe lack of panel surveys, while a notable amount of information is available through repeated independent cross-sections. Thus, the same questionnaire, usually consisting in Q_1 close questions and Q_2 ($\ll Q_1$) open questions, is submitted to T different samples, not necessarily of the same size, in T different occasions. In the following we consider the case in which $Q_2 = 1$, and the answers to this question have to be analysed by means of a textual data analysis technique, instead of recoding them *a posteriori*.

Thus the data structure at hand consists in $\boldsymbol{X}_1, \ldots, \boldsymbol{X}_t, \ldots, \boldsymbol{X}_T$ $(n_t, w_t + m)$ matrices (with $t = 1, \ldots, T$). The total number of individuals is $n = \sum_{t=1}^{T} n_t$. Each $\boldsymbol{X}_t$ is partitioned into two submatrices: the lexical matrix $\boldsymbol{V}_t$ (n_t, w_t), individuals by the w_t forms used in answering the open question and an indicator (n_t, m) matrix $\boldsymbol{Y}_t$, containing the answers of the same individuals to a subset q_1 of the Q_1 close-ended questions in disjunctive coding, with $m = \sum_{j=1}^{q_1} m_j$, being m_j the number of categories of the j-th close question considered. For example, we can consider two demographic variables: *sex* and *class of age* in three categories: *young*, *adult*, *aged*, being $\boldsymbol{Y}_t$ a $(n_t, 6)$ matrix.

As the interest of the analysis usually consists in understanding the relation between vocabulary and sociodemographic characteristics of the individuals, we fix the latter as a structure common to all the occasions. If those variables are the stratification factors in the sampling plan, each wave of the survey can be seen as a replication of the experiment under the same conditions. This approach was proposed by Balbi, Lauro, Scepi (1994), dealing only with close questions, but hypothesising a dependence structure in the questionnaire.

Thus we transform each $\boldsymbol{X}_t$ in an $\boldsymbol{A}_t$ (w_t, m) matrix, by aggregating individuals according with their own condition, defined by a combination of the socio-demographic variables. The so obtained three-way structure consists in T (w_t, m) $\boldsymbol{A}_t$ matrices, suitable to be analysed by an *interstructure - compromise - intrastructure* method, such as STATIS (Escoufier, 1987) or Principal Matrices Analysis (Rizzi, 1989).

Such a kind of methods allows to build a reference structure where to compare the behaviour of one word through time, or to study how the so defined groups of individuals change their vocabulary from one occasion to another.

3. The Analysis

As in one-time textual data analysis, each $\boldsymbol{A}_t$ can be seen as a contingency table, which presents in its general cell a_{ij} the number of times the *i*-th word appears in the answers given by the *j*-th group of individuals. In STATIS-like method (see, *inter alia*, Glaçon, 1981), when we deal with contingency tables, they are transformed in the corresponding Burt matrices, in order to compute, for each couple of matrices, the Rv coefficients (Robert, Escoufier, 1976), general elements of the *interstructure* matrix $\boldsymbol{S}$.
After the spectral decomposition of $\boldsymbol{S}$, which enables to represent each $\boldsymbol{A}_t$ as a point on a factorial plane, the *compromise* matrices $\boldsymbol{C}_\alpha$ are built: $\boldsymbol{C}_\alpha = \sum_{t=1}^{T} u_{\alpha t} \boldsymbol{A}_t$, where $u_{\alpha t}$ is the *t*-th element of the first eigenvector corresponding to the α-th eigenvalue of $\boldsymbol{S}$, in decreasing order.
Following Balbi (1995), a non symmetrical correspondence analysis (NSCA, in the following) is performed on c (α=1, ..., c) $\boldsymbol{C}_\alpha$. The choice of a proper c (i.e. the number of principal matrices considered) is determined, like the number of components in principal component analysis, by using classical criteria, such as the scree-test, or looking at the percentage of $\boldsymbol{S}$ trace, relative to its first eigenvalues. As in the present, when the variable defining the third way is time, the decomposition of the total variability in principal component of decreasing importance can be read, using time series language (D'Alessio, 1989). Thus the first matrix explains the main structure of the analysed phenomenon, while the second one a cyclic component, then a seasonal one, and so on.
In the last step, the so called *intrastructure* step, original elements (words and categories of the sociodemographic variables), observed in the T different occasions, are represented on the NSCA first factorial plane of each $\boldsymbol{C}_\alpha$, being projected as supplementary points (Lebart, Morineau, Piron, 1995), and trajectories for homologue elements, observed in the T times, can be drawn.
The previous analysis has to be performed on matrices having the same structure. Grouping individuals according to some time-invariant characteristics overcomes the problem for columns. Furthermore, a second point has to be considered for rows: how to deal with different vocabularies, in order to build a common *corpus* by texts belonging to the different waves of the repeated survey. Till now we have spoken of *words*, now we have to choose the units for the textual data analysis. Thus we have to put $w_t = w$, $\forall t \in \{1, \ldots, T\}$, i.e. we have to build and *consolidate* the vocabulary we want to analyse.
Differently from one-time textual data analysis, in which this step is not always necessary and it is preferable not to carry it out too early, dealing with multiway

textual structure we need to perform a consolidation procedure in the preliminary step of the analysis.

4. Building a Common Vocabulary

In textual data analysis, the first question to be solved consists in choosing the statistical units. Vocabularies are generally built on the basis of grammatical rules (e.g. nouns are considered in their singular form) which can differ in different languages. Such a kind of standardisation of forms means a heavy recoding of the data, and can destroy some interesting information. Lebart (1981) suggests the use of *graphical forms* (i.e. series of non-delimiting characters, *g. f.*'s, in the following): they are easier to be stored and not influenced by recoding (Lebart, Salem, Berry, 1991). Bolasco (1994) proposes the use of *textual forms*, defined as a mixture of pure *g. f.* 's, lexemes derived from *g. f.* 's, segments (i.e. consecutive *g. f.* 's), in order to both eliminate any ambiguities and preserve the semantic variability in the analysed text, as well as merge all the elements which are semantically invariant. In other words, the contextual aspects have to be taken into account.
Dealing with data observed in different times, vocabulary has to be properly consolidated by declaring the equivalencies among *g. f.*'s, which correspond to one word in the further analysis. This pre-treatment requires extreme attention, because equivalencies valid in one period can be misleading in a different moment. On the contrary, willing to use the data as a whole, some equivalencies can be imposed in order to save the comparability among themes, observed in different occasions.
All this means that a vocabulary common to all the different waves of the survey has to be built. As pointed out in a previous paper (Balbi, Esposito, 1997), sometimes this may seem to be artificial and may happen that interesting forms, not present in all the occasions, have to disappear. The preservation of the textual forms in order to compare the behaviours in the different occasions of the groups/words on the same factorial plane seems to justify this choice.
In doing that, tools proposed in one-time analysis (e.g., the so called *iso-frequency* criterion proposed by Bolasco, 1993, or the *bootstrap convex hulls* co-ordinates proposed in Balbi, 1995) seem to be very useful.
The *iso-frequency* criterion is interesting for nouns (and adjective in languages in which, e.g., the masculine form differs from the feminine one) in order to decide if we have to distinguish singular and plural forms: if the two forms appear almost an equal number of times, they can be legitimately merged.
The use of *bootstrap convex hulls* can be extremely useful in all the cases we want to evaluate if different graphical forms of the same morpheme have actually a different representation on factorial planes and analogously for synonyms. In Fig. an example dealing with verbs: analysing the publicity for computers in 1995 on Italian political magazines. The problem of considering

the only infinitive form for the verb *avere* (to have) was solved by observing the different use of the first person plural form by the leader IBM and the second person by the other producers.

Figure 1: *Bootstrap convex hulls for factorial coordinates: computers advertisements on Italian press in 1995*

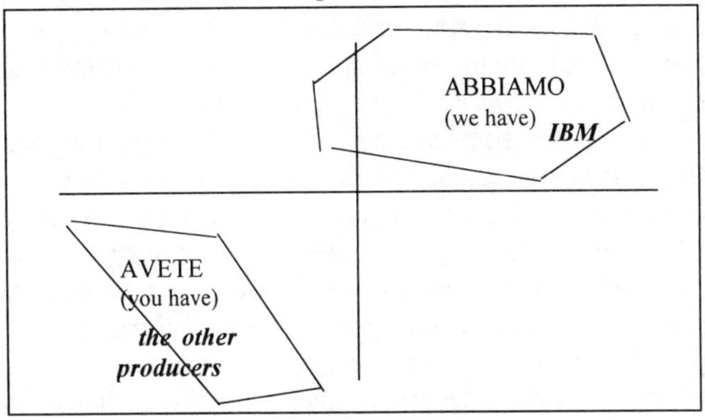

Dealing with *T* matrices, convex hulls can be obtained by the analysis of compromise. The projections on its first factorial plane of words belonging to different times, together with the form and the extension of the convex hulls, can give useful suggestions in building *textual forms* through time.

5. The Cellular Telephone on Italian Press: From Élite to Mass

In the Eighties, in Italy, the *cellular telephone* was the symbol of success, a tool for busy, rich and powerful men. The begin of Nineties and the end of *yuppism* meant that the cellular telephone market had to deeply change. The old image of the product had to be destroyed: cellular telephone should become an easy tool for young people, and for women, to be used for chatting and not for business. Thus, it is interesting to understand how the advertisement campaigns appeared on Italian press have participated to the process of transformation from an élite market to a mass one. Additionally, it is worth to notice the influence of the new Italian economic policy, characterised by the privatisation of some services, such as telephone, and the end of the monopoly of State. In 1995, the market of cellular telephones connections became a duopoly, private (OMNITEL) against public (TIM, before SIP), both using the same physical net. Furthermore, technological developments are quickly changing the characteristics of the product. All these aspects make the study of the evolution of the language in this market especially interesting.

In order to understand the evolution of the language with respect to the new target, we have considered the advertisements by taking into account the kind of press where they appeared (*popular* or *specialised*, i. e. political or technical magazines) and the style of presentation, i.e. informative (*technical*) or imaginative (*descriptive*). The dependent variables are represented by words. The three occasions are represented by the three years (1993, 1996, 1997). The explanatory categories are the time-invariant features of messages.
Recall that words should be the same in the occasions, thus the first problem in the analysis consists in building the three matrices $\boldsymbol{A}_{93}$, $\boldsymbol{A}_{96}$ and $\boldsymbol{A}_{97}$. A preliminary recoding has been adopted, by mainly using lexemes, instead of graphical forms. In some cases, the root has been taken into account (e. g. *innovazione* and *innovativo/a/e/i*, in order to take into account the theme of technological innovation). In other cases, when a *graphical form* is very frequent (e. g. *telefono* or *telefono-cellulare*), we have referred to the graphical form. Additionally, topics like services for calling abroad have been merged under the label *Europa*, or the reference to the name of each company under the label *marca*. Words with a length smaller than three characters have been ignored. When a *g.f.* was in two of the three matrices, with high frequencies, it was introduced in the third one, by a row of 0's.
The analysis of *interstructure* (Fig. 2) shows a strong opposition between the campaigns in 1993 on the left and 1996/1997 on the right, suggesting the use of one *compromise* (c=1).

Figure 2: *Interstructure first factorial plane*

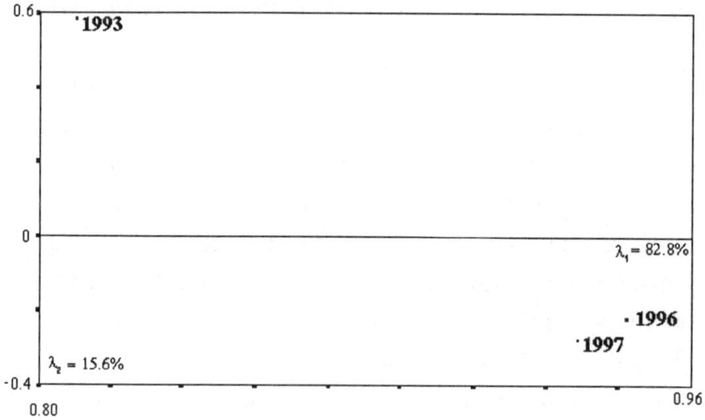

The NSCA first factorial plane (explaining the 80% of the dependence of the vocabulary on the explanatory categories) shows a first axis (46%) characterised by the language used on the popular press, while a second one (34%) on the specialised magazines (Fig. 3). It is interesting to underline (Fig. 4) the opposition of *telefonino* on the right (i.e. more technical advertisements) and *telefono* on the left. The advertisements using rhetoric and/or iconic tools ("descriptive") use ordinary references: *conversazione* (talk), *fare* (to do), *più*

(more), but they are also characterised by the more technical words, i. e. *GSM* and *stand-by*. On the contrary, the more informative messages pay more attention to price (*lire*, *-mila*). For specialised magazines too, the conditions of payment appear on technical advertisements, while *telefono cellulare*, and *telefonino*, near to *semplice* are on the "more descriptive" side.

Figure 3: *Compromise first factorial plane: the explanatory categories*

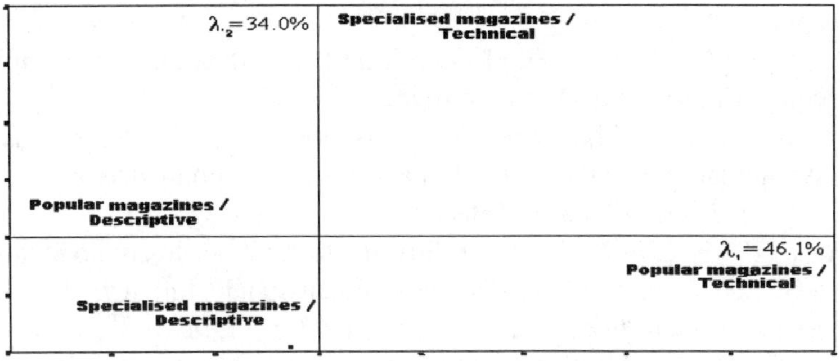

Figure 4: *Compromise first factorial plane: words*

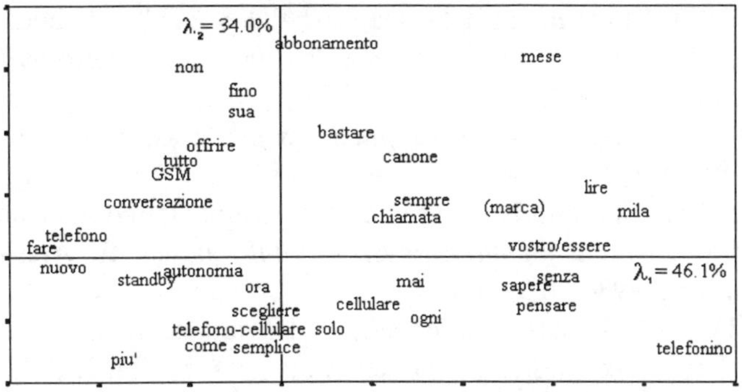

The last step, the *intrastructure*, allows to compare the changing of the coordinates of some words through time. Fig. 5 shows some examples.

Figure 5: *Trajectories for words: e.g.* ***"marca"*** *and* ***"telefonino"***

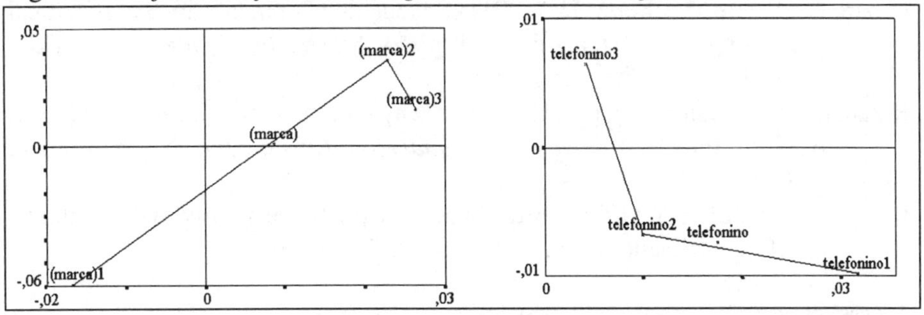

References

Balbi S. (1995). Non symmetrical correspondence analysis of textual data and confidence regions for graphical forms. In: Bolasco S., Lebart L., Salem A. (eds.). *JADT 1995*, Roma: CISU, 2, 5-12.

Balbi S., Esposito V. (1997). Comparing advertising campaigns by means of textual data analysis with external information. In: *JADT'98*. Nice (to appear).

Balbi S., Lauro N. C., Scepi G. (1994). A multiway data analysis technique for comparing surveys. *Methodologica*, 3, 79-90.

Bolasco S. (1993). Choix de lemmatisation en vue de reconstruction syntagmatiques du texte par l'analyse de correspondance. In: Anastex S. J. (ed.), *JADT'93*. Paris: Telecom.

Bolasco S. (1994). L'individuazione di forme testuali per lo studio statistico dei testi con tecniche di analisi multidimensionale. In: *Atti della XXXVII riunione scientifica della S.I.S.*. Roma: CISU, 2, 95-103.

D'Alessio G. (1989). Multistep principal component analysis in the study of panel data. In: R. Coppi & S. Bolasco (eds.) *Multiway data analysis*. Amsterdam: North-Holland, 375-381.

Escoufier Y. (1987). Three-mode data analysis: the STATIS method. In: B. Fichet & N.C. Lauro (eds.) *Methods for Multidimensional Data Analysis*, ECAS, 259-272.

Glaçon F. (1981). *Analyse conjointe de plusieres matrices*, Thèse du 3-ème cycle, Grenoble.

Lauro N.C., D'Ambra L. (1984). L'analyse non symétrique des correspondances. In E. Diday *et al.* (eds.), *Data Analysis and Informatics*, III, Amsterdam: North-Holland, 1984, 433 - 446.

Lebart L. (1981). Vers l'analyse automatique des textes: le traitement des réponses libres aux questions ouvertes d'une enquête. In: J. P. Benzécri & collaborateurs, *Pratique de l'Analyse des Données. Linguistique & Lexicologie*. Paris: DUNOD, 414-419.

Lebart L., Morineau A., Piron M. (1985). *Statistique exploratoire multidimensionnelle*. Paris: DUNOD.

Lebart L., Salem A. (1994). *Statistique textuelle*. Paris: DUNOD.

Lebart L., Salem A., Berry L. (1991). Recent developments in the statistical processing of textual data. *Applied Stochastic Models and Data Analysis*, 7, 47-62.

Rizzi A. (1989). On the synthesis of three-way data matrices. In: R. Coppi & S. Bolasco (eds.) *Multiway data analysis*. Amsterdam: North-Holland, 143-154.

Robert P., Escoufier Y (1976). A unifying tool for linear multivariate methods: the Rv coefficient. *Applied Statistics*. 25, 257-265.

Acknowledgement: *The present paper is supported by a CNR grant .*

Three-way textual data analysis

Mónica Bécue-Bertaut
Universitat Politècnica de Catalunya.
Pau Gargallo, 5 - 08028 Barcelona. Spain. e-mail : monica@eio.upc.es

Abstract : In this work, we present methods, stemming from particular three-way data analysis methods, to tackle two specific problems arising when analysing open-ended answers in surveys. The first problem consists in the comparison of the answers given to a same question in different surveys ; the second one is the study of the set of categories or individuals as described by their answers to several open questions. Specific software routines have been devised.

Keywords : textual data, open-ended questions, three-way data analysis.

1. The basic problem

Multidimensional descriptive statistical methods such as principal axes (principal component analysis and correspondence analysis) and cluster analysis methods constitute a tool for analyzing responses to one single question in surveys (see Lebart and Salem, 1997).

The answers to closed-end questions lead to a classical (n,p) matrix **X**. To analyse open questions, the starting point is to consider the answers to each open question as represented by a (n,m) matrix **Y** whose i-th row contains the vector whose m components are the frequencies of the j words in the i-th individual answer. **Y** is called *lexical table*. When individuals are grouped according to a socioeconomic characteristic, the rows are the individual categories, and **Y** is called *aggregated lexical table*. Usually, only the most frequent words are kept.

When several open questions are present, simultaneous analysis of the responses requires specific tools allowing for systematic comparisons. A similar problem arises from the repetition of the same question in different surveys. We propose here some methods stemming from three-way data analysis.

Different aspects can be studied : the relationships between the structures induced by the questions ; the links between words used to answer different questions ; the detection of individuals or individual categories with similar responses to the whole set of questions or only to some of them ; the evolution of the responses relating to a same question, etc..

From now on, we will focus on two main problems : 1) the comparison of the answers given by *different samples* to the same question ; 2) the study of the set of categories (or individuals) as described by their answers *to several open questions*.

2. Comparative analysis of the answers given by different surveys

2.1 Example

As an example, we will use a corpus of responses to a permanent survey, carried out twice a year using independent samples (Lebart, 1987). This survey was conducted by means of both closed-end questionnaire and open questions. We will process the answers to the question "What are the reasons that might cause a couple or a woman to hesitate having children?" as given by two annual surveys (1986 and 1988) of 2000 individuals. As grouping criteria, we use the age of the respondent, split into five levels (less than 25, 25 to 34, 35 to 49, 50 to 64, 65 years and over).

2.2 Methodology

2.2.1 Data tables

In the case of a same question asked at different periods, in different regions or, more generally, to different subpopulations, the set of words or "vocabulary" used by the corresponding samples can be considered similar. So we have to deal with a sequence of two-way tables $\mathbf{Y}_t$, defined by the same couple of variables (lexical variable and socioeconomic variable) and indexed by survey number t, t=1,...T (figure 1).

Figure 1 : *Sequence of tables* $\boldsymbol{Y}_t$

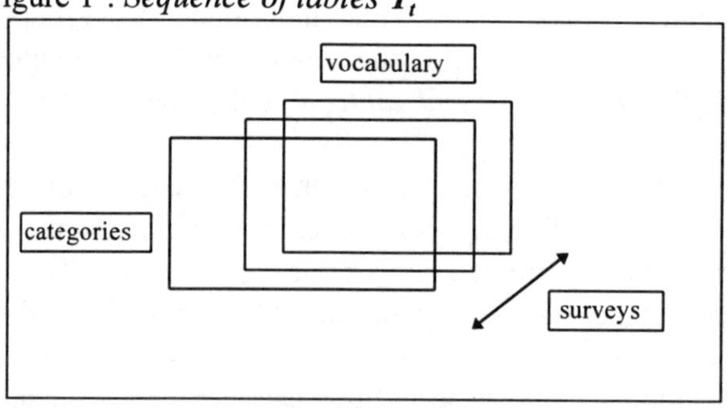

At that stage we find ourselves within the scope of three-way data methods. For a synthesis of these methods see, for example, Carlier *et al.* (1988) and Kiers (1989). Escofier (1987), Escofier and Pagès (1993) and Abdessemed and Escofier (1996) present tools to compare data tables. With regard to the evolution of contingency tables, Carlier and Ewing (1992), Carlier and Kroonenberg (1996) and van der Heijden (1987) can be consulted.

2.2.2. Corpus segmentation and tables building

A first information is obtained when identifying and counting up the words being used. Their frequency in each survey are compared, and the significant differences recorded. In the example, very few words offer a significant frequency difference between samples. We can note that the frequency of the words corresponding to refusals (expressed using the expressions *"I don't know"*, *"I don't get you"*, or *"no idea"*) are falling from the first survey to the second.

In a second step, the answers are grouped according to the values of a categorical variable, and the aggregated lexical tables corresponding to each sample are built up, keeping only the most frequent words. The main goal is to highlight the common structure of these tables, and also to study the deviations of each table from this common structure.

To achieve this goal, the tables $\mathbf{Y}_t$ are juxtaposed into a two-way table, $\mathbf{Y}_J$. There are two options, depending on which dimension, vocabulary or categorical variable, is chosen as a common dimension. The choice depends on the initial problem and also on the textual data nature. In general, we want to compare the *structures* of both categories sets, as induced by the words selected among a common vocabulary. If the most of the words are used with frequencies similar in the different surveys, it is legitimate to assume that a common vocabulary really exists, as it is the case in our example.

It implies that the vocabulary constitutes the common dimension, and the categories are repeated as many times as the samples are. The general term of the juxtaposed table is denoted f_{ijt} ; f_{ijt} denotes the relative frequency of word j, in category i and survey t.

Another table is also built up : the *sum-table* $\mathbf{Y}_S$. Such table is obtained by adding up the frequencies of the words corresponding to a same category, independently of the sample.

2.2.3 Analysis of the tables

First step : each table $\mathbf{Y}_t$, and also the *sum-table* $\mathbf{Y}_S$, is visualized through correspondence analysis. It allows to pin-point the principal features of the data, and to detect a possible common structure in the different tables $\mathbf{Y}_t$.

Sum-table $\mathbf{Y}_S$ analysis is the analysis of the mean profiles of the categories as computed upon all samples. Projecting the columns-categories of all tables $\mathbf{Y}_t$ as supplementary columns upon the principal axes resulting from the analysis of the sum-table provides us with a visualization, on the same mapping, of the mean structure of the categories, and also of categories structures corresponding to each survey.

In the example, using age as grouping criterion, the three categories structures (corresponding to the sum-table and to each survey), are similar ; these age level categories have ordered coordinates on the principal axis, and the second axis opposes extreme categories (less than 25, 25 to 34 and 65 years and over) to intermediate categories (35 to 49 and 50 to 64 years). We can state that there is a quite regular evolution of the vocabulary with respect to age.
Second step : the second step consists in analysing the juxtaposed table $\mathbf{Y_J}$ through correspondence analysis.

Figure 2 : *Correspondence analysis of juxtaposed table. Principal plane.*

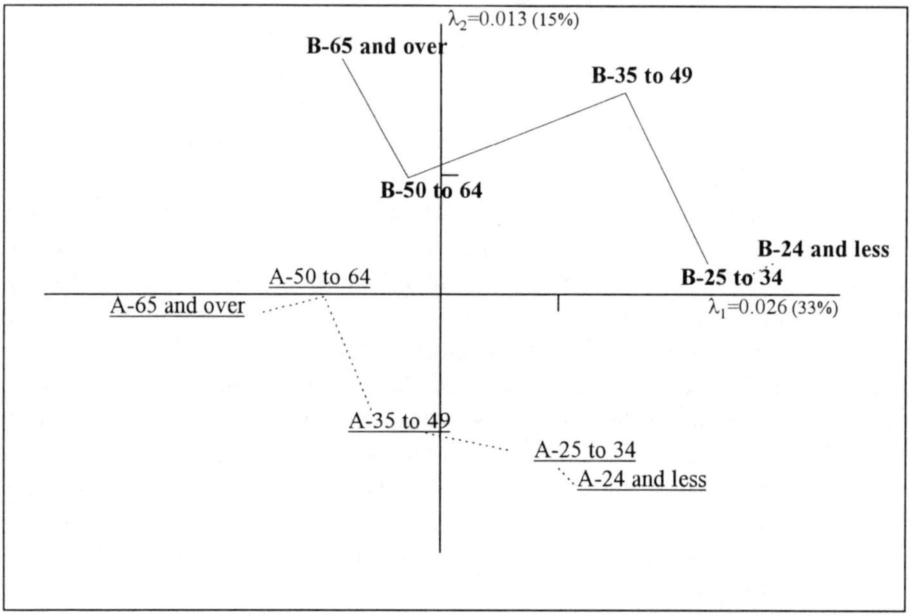

Figure 2 shows the principal plane obtained in this analysis. The initial letters A and B identify categories as respectively belonging to first or second survey. The results depend on the relative importance of the variance of the categories within each survey (within-variance) and on the variance of the surveys centroids (between-surveys variance) (table1).

Table 1 : *Within-surveys and between-surveys variance*

Variance	space	Axis 1	Axis 2
space	0.0856	0.0285	0.0131
within-surveys	0.0689	0.0205 (72%)	0.0054 (41.5%)
between-surveys	0.0167	0.0080 (28%)	0.0077 (58.5%)

Another important interpretation aid consists in the correlations between the principal components of each couple of tables $(\mathbf{Y_J},\mathbf{Y_t})$, t=1,...,T to detect common and specific axes. In the example, the first principal component of the juxtaposed analysis is highly correlated with the respective principal component of each survey analysis.
The first axis is, above all, an axis describing the variance of the categories within each survey, it is an "age axis". It opposes the words used by young categories to those used by old ones.
The second axis depends on both variances, although slightly more on between surveys variances. The following axes depend, above all, on within survey variance. We can note that both categories structures indicate a quite parallel evolution of the vocabulary with age in both surveys.
Third step : another analysis offers helps to visualize the differences existing in the correspondence between socio-economic variable and vocabulary according to the survey. The goal of this analysis is to provide us with a visualization of the internal structure of each survey, but representing all of these structures in a same compromise reference space (figure 3).
The basic idea is the following : denoting $\mathbf{N_t}$, the subcloud of categories-profiles corresponding to the same survey t, each subcloud $\mathbf{N_t}$ is traslated so that its centroid coincides with the global centroid. The whole cloud - resulting of all these subclouds- is analysed using the same chi-square metric than in step 2. It is equivalent to analyse the juxtaposed table by correspondence analysis, but relatively to a particular model (and not relatively to the global independence in accordance to juxtaposed table margins) (Escofier, Pagès, 1993). This model corresponds to the independence between categories and vocabulary within each table $\mathbf{Y_t}$. It can be shown that it is equivalent to process by classical correspondence analysis the table $\mathbf{Y_C}$ whose general entry is :

$$f_{ijt} - \left(\frac{f_{i.t}}{f_{..t}}\right) \cdot f_{.jt} + f_{.j.} \cdot f_{i.t} \tag{1}$$

Upon the principal axis corresponding to this last analysis, it is useful to represent the categories using the projection of the initial categories-rows (i,t) as supplementary elements (Bry, 1996) ; it is the way of each category being located in the centroid (thanks to a constant) of the words, giving them a relative weight corresponding to their importance in the responses of this category ; it allows one to find again a usual result when two-way correspondence analysis is used, but lost in this analysis in the case of the categories-rows. This analysis offers an optimal visualization of both internal structure, in the sense that, in the projection, the maximum internal variance is kept, taking into account both surveys.

Figure 3 shows the first principal plane resulting from this method applied to the example ; with regard to categories, they are represented by the initial rows as supplementary rows, as justified just before.
To interpret the results, the relative importance of within-surveys and between-surveys variances are computed, and the axes common to different surveys are detected, as well as those which are specific to one of them.

Figure 3 : *Principal plane of the analysis of the table* $\mathbf{Y}_C$ *(centring the categories-profiles within each survey)*

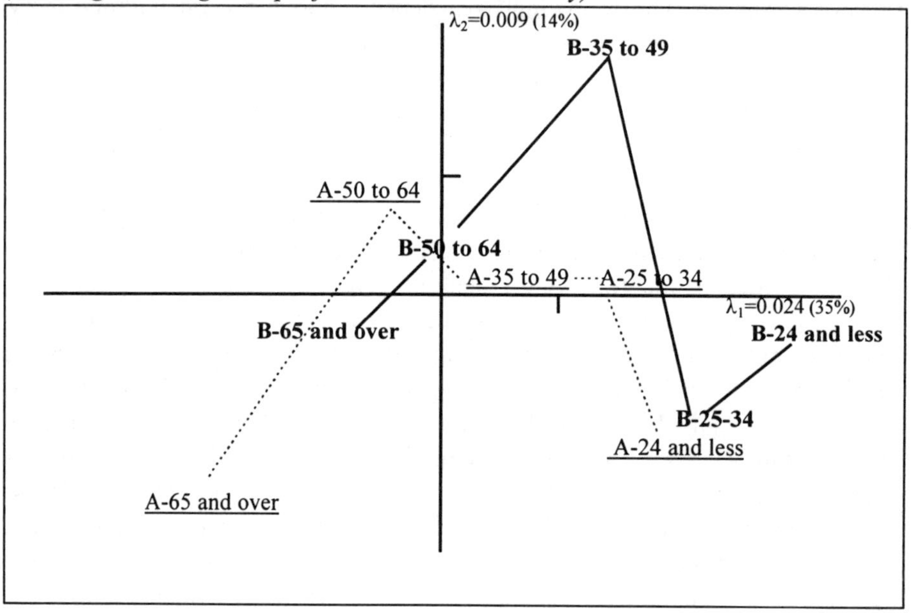

3. Description of individuals or individual categories using the answers to various open questions

When the same individuals answer several open questions, it is interesting to simultaneously study theirs responses. In this case, the individuals are described by several sets of frequency variables.

3.1 Data tables

There are as many "vocabularies" as questions. Each category/individual is described by the frequencies associated with the words within each vocabulary. A table $\mathbf{Y}_q$ is associated to each open question. These tables $\mathbf{Y}_q$ are juxtaposed, so that the rows represent the categories/ individuals and the columns are the words corresponding to the various questions q, q =1,...Q.

3.2 Methodology

In a first step, each table $\mathbf{Y}_q$ is analysed by correspondence analysis (in this case, *sum-table* is meaningless). In a second step, the correspondence analysis of juxtaposed tables $\mathbf{Y}_q$ (figure 4), but centring the words-profiles within each survey, gives us a tool to compare them, using a common reference space. As it has been shown in paragraph 2, it allows one to detect common and specific axes in the different tables $\mathbf{Y}_q$.

Figure 4 : *Juxtaposition of tables* Y_q

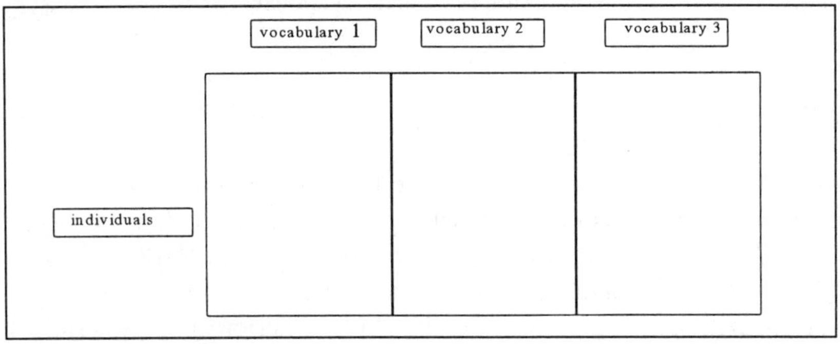

Each category/individual appears once, located at the centroid of the words used in the different answers. Although word sets are different for each question, some words - at least usual function words- will be repeated. In this case, they will appear as many times as they are questions. It provides us with a visualization of the interrelations between words, stemming which words are frequently present in the different answers of a same category/ individual. Comparing the different positions of a same word in the different questions will bring information about their role as elements of the discourse.

To compare the structures of the categories according to the different questions, it is then possible to use the initial tables $\mathbf{Y}_q$ as supplementary rows. It provides us with a representation which superimposes the description of the categories/ individuals as given by each question, and the overall description.

4. Weighting the different tables

A problem arises when a table plays a predominant role and, by itself, determines the first principal axis. When it occurs, it will be necessary to adopt a multiple factor analysis point of view and to reweight the tables $\mathbf{Y}_t$ (or $\mathbf{Y}_q$) to balance their influence on the construction of first axes. But we deal with two-way tables with different row margins and a compromise weighting must be chosen for the rows.

5. Other issues

It is possible to tag the corpus units using a morpho-syntactic analyzer, and to count the frequencies of the syntactic categories in the answers (Salem, 1995). The individuals are then described by the words on the one hand, and by the syntactic categories they use, on the other. A systematic comparison of the structures induced by both sets will provide information about discourse features. The tools here presented offer the possibility to study and to compare how the individuals are described by the content of their responses and, simultaneously, by the syntactic features of these responses.

References

Abdessemed, L. Escofier, B. (1996). Analyse factorielle multiple de tableaux de fréquences ; comparaison avec l'analyse canonique des correspondances, *Journal de la Société de Statistique de Paris*, tome 137, n°2, 3-18.

Bry, X. (1996). *Analyses factorielles multiples*, Economica, Paris.

Carlier, A., Lavit, C., Pagès, M., Pernin, M.O., Turlot, J.-C. (1988). A comparative review of methods which handles a set of indexed data tables, in : *Multiway Data Analysis*, Coppi R., Bolasco S. (Eds), North Holland, Amsterdam, 85-102.

Carlier, A., Ewing, G. (1992). Longitudinal Analysis of Contingency Tables with Application to Demographic Data, *Computational Statistics*, 7, 329-353.

Carlier, A., Kroonenberg, P. M. (1996). Descompositions and biplots in three-way correspondence analysis, *Psychometrika*, 61, 2, 355-373.

Escofier, B (1987). Analysis and Comparison of Different Tables, *Recent Developments in Clustering and Data Analysis*, Débats du Séminaire scientifique franco-japonais, 307-322.

Escofier, B., Pagès, J. (1993). *Analyses factorielles simples et multiples ; objectifs, méthodes et interprétation,* Dunod, Paris.

Kiers, H. A. L. (1989). *Three-way methods for the Analysis of Quantitative and Qualitative Two-way Data*, DSWO Press, Leiden.

Lebart, L. (1987). Conditions de vie et aspirations des français, évolution et structure des opinions de 1978 à 1986, *Futuribles*, septembre 1987, 25-56.

Lebart, L., Salem, A., Berry E. (1997). *Exploring textual data*, Kluwer, Dorfrecht.

Salem, A. (1995). Les unités lexicométriques in : *Analisi Statistica dei Dati Testuali,* Bolasco et al. (Eds.), CISU, Roma, 19-26.

van der Heijden, P. G. M. (1987). *Correspondence Analysis of Longitudinal Categorical Data*, DWSO Press, Leiden.

Classification Problems in Text Analysis and Information Retrieval

Ludovic Lebart
Centre National de la Recherche Scientifique,
Ecole Nationale Supérieure des Télécommunications,
46, rue Barrault, 75013, Paris, France, email: lebart@eco.enst.fr

Abstract: The specific complexity of textual data sets (free answers in surveys, documentary data bases, etc.) is emphasized. Recent trends of research show that classification techniques (discrimination and unsupervised clustering as well) are widely used and have great potential in both Information Retrieval and Text Mining.

Keywords: Classification; clustering; textual corpora; text mining; information retrieval.

1. The scope of multivariate analysis of texts

The amount of information available only in unstructured textual form is rapidly increasing. Classification methods play a major role in the computerized exploration of such corpora. They can contribute to process textual data sets in the three following main domains:

1.1 Producing visualizations and/or groupings of elements (free responses in marketing and socioeconomic surveys, discourses, scientific abstracts, patents, broadcast news, financial and economic reports, literary texts, etc.); looking for associations and patterns (exploratory context of Text Mining, whose ultimate aim is to extract knowledge from large bodies of unstructured textual data).

1.2 Devising decision aids for attributing a text to an author or a period; choosing a document within a database; coding information expressed in natural language.

1.3 Helping to achieve more technical or upstream contributions, such as lexical disambiguation, parsing, selection of statistical units, description of semantic graphs, speech and optical character recognition.

Note that cluster analysis has been involved in these applications ever since the beginning of these investigations (see: Jardine and van Rijsbergen, 1971; Willet, 1988).

2. Complexity and specific features of textual data

2.1 The concepts of variables and observations are more complex than those usually dealt with in most statistical applications. *Variables*, instead of being defined *a priori*, are derived from the text. Examples of (categorical) variables are the following text units: words, lemmas, segments (sequences of words appearing with a certain frequency). In the following, we use the term *word* to designate the textual unit under consideration. In lexicometry terminology, *word* is synonym of *type*, as opposed to *token*, which designates a particular occurrence of a type.

Statistical units (or: observations, subjects, individuals, examples) are generally *documents* (described by their titles or abstracts) in documentary databases, *respondents* (described by their responses to open questions) in surveys, or segments of texts (sentences, context units, paragraphs) in literary applications. Besides the respondents, a second level of statistical units is constituted by the occurrences of words. Some statistical tests may involve counts of occurrences, whereas others deal with counts of documents or respondents. Such duality is often a source of difficulty and misunderstanding.

2.2 Two additional characteristics increase the complexity of the basic data tables: These tables are *large* (thousands of documents, thousands of words), often *sparse* (a document contains a relatively small number of words).

2.3 But the main feature of textual data sets is certainly the enormous amount of available *meta-data.* Every word is allocated several rows in a dictionary. To identify without ambiguity the lemma associated with a word may often require the help of a sophisticated syntactic analyzer. Rules of grammar, semantic networks, obviously constitute basic meta-information.

2.4 Eventually, let us remind that we are dealing with *sequences* of occurrences (or: strings) of items, whose order could be of importance, another non standard situation in multidimensional data analysis. Data analysts are accustomed to dealing with rectangular arrays of nominal, ordinal, or numerical variables. In textual data analysis, the basic data cannot be reduced to such an array. Let us consider the case of the responses of n individuals to an open question.

If $\{ s_1, s_2, \dots , s_v \}$ designates a set of v different elements (the vocabulary, i. e. the set of v *different* words, in the present case), an individual i ($i \leq n$) will be characterized by an *ordered sequence*, with *variable length* $\gamma(i)$: $\{ s_{r(i,1)}, s_{r(i,2)}, .. , s_{r(i,\gamma(i))} \}$, where $1 \leq r(i,k) \leq v$, and $1 \leq k \leq \gamma(i)$.

Note that a word can appear several times in a sequence. r(i,k) is thus the index of the kth word in the response of individual i. The first task of any classification method is then to compute similarities between such *sequences (with variable length) of ordered items with repetition* (see below section 4.1).

3. Clustering of observations and words

3.1 Observations (responses, documents)

The starting point is to consider each observation as described by its lexical profile, i.e. by a vector that contains the frequency of all the selected units in the text (these units could be words, at the outset). In many cases, a textual data set can lead to building a (n,v) contingency table **X** whose general entry (i,j) is the number of occurrences of word j in the text (observation) i. **X** can be easily derived from the non-rectangular array whose general entry is r(i,k) (as defined in section 2.4), but the converse is not true: the information relative to the order of the words within a response is lost in **X**. In most applications, the array r(i,k) is actually much more compact than **X**; thus, a response i containing $\gamma(i)=20$ occurrences out of a vocabulary of 2000 words corresponds to a row r(i,k) of length 20, and to a row x(i,j) of length 2000. A clustering algorithm involving computations of distances directly from the data table, such as the k-means method, can easily be reformulated using r(i,k) instead of x(i,j) in order to take advantage of the sparsity of **X**.

Note that from a given corpus, many different contingency tables can be built, according to various thresholds of frequencies for the words.

However, an usual classification algorithm applied to the rows of **X** could lead to poor or misleading results. As mentioned above, the matrix **X** could be very sparse, many pairs of rows could have no element at all in common (the computational advantage of sparsity is then pointless). Moreover, available meta-data need to be taken into account (syntactic relationships, semantic networks, external corpus and lexicons, etc.) as well as the order of occurrences within each response or text. Section 4 below will suggest some possible improvements.

3.2 Words (columns of matrix **X**)

Classification of words is rarely the final outcome of a text analysis. It is however an important intermediate step, allowing for the definition of new statistical units, thence improving the similarities between observations (see also: section 4).

At the lower levels of a hierarchy, one can find textual co-occurrences within sentences, as well as fixed length text segments and paragraphs. The search for preferential associations is an important factor in applications involving natural language processing (see, e.g.: Lewis and Croft, 1990). It can help to solve some disambiguation problems useful, for example, in the recognition phase following optical scanning of characters.

Note that non-symmetrical measures of local associations between words (e.g.: mutual information index *I(x,y)* resulting from the Information Theory of Shannon, as proposed by Church and Hanks (1990)) entail difficulties with classical clustering algorithms.

The main patterns observed in the first principal subspaces spanned by the first principal axes of a Singular Values Decomposition (or of a correspondence analysis) of matrix **X** are generally marked out by the higher level clusters produced by hierarchical clustering.

4. Enhancing the similarities

4.1 Taking into account the order of items

The use of additional units, such as *repeated segments* (Salem, 1984), can partially enrich the data arrays with information about order of items within texts. The repeated segment approach deals with the blind and automated detection of repeated sequences of words within a given corpus, whether or not these sequences constitute frozen phrases or expressions. The principles of a fast algorithm able to uncover such segments are given in Lebart and Salem (1997).

Direct measures of distances (such as the Levenshtein distance) have been specifically devised for strings (see, e.g.: Coggins, 1983). In this context, clustering has proved to be a crucial step when trying to extract generative grammars from a corpus of strings.

4.2 Using syntactic information

Additional variables obtained from a morpho-syntactic analyzer can be instrumental in making more meaningful the distances between lexical profiles of observations (responses or texts). The main idea is to tag the words depending on their category (nouns, verbs, preposition, etc.), and to complement the p-vector associated to each response or text with these new components of a different nature (see, e.g.: Biber, 1995; Habert and Salem, 1995, Habert et al., 1997).

4.3 Semantic relationships

The semantic information defined over the pairs of statistical units (words, lemmas) is summarized by a graph that can lead to a specific metric structure.

a)The semantic graph can be constructed from an external source of information (e.g.: a dictionary of synonyms, a thesaurus). In such a case, a preliminary lemmatization of the text must be performed. A practical way of taking into account the semantic neighbours consists in complementing the words of a given response with their semantic neighbours (provided with inferior weights). This leads to the transformation: $\mathbf{Y} = \mathbf{X}(\mathbf{I} + \alpha\mathbf{M})$, where **I** is the unit matrix, **M** the binary matrix associated with the semantic graph defined previously, and α a diffusion weight calibrating the importance given to semantic neighbourhoods. Due to this *semantic contamination*, **Y** is less sparse than **X**. The induced metric defined by the matrix: $\mathbf{Q} = (\mathbf{I}+\alpha\mathbf{M})^2$ leads to a new similarity index that can be used for the classification of the subjects.

b) The semantic graph can also be built up according to the associations observed in an external reference corpus, or within the corpus itself (see: Becue and Lebart, 1996). Descriptions of semantic relationships between words through self-organizing maps have been suggested by Ritter and Kohonen (1989). A hierarchical classification of words (characterized by their associates in a thesaurus), complemented with a principal axes visualization of the main nodes, produces also satisfactory descriptions of such huge graphs.

c) Another way of deriving a matrix **M** from the data themselves is to perform a hierarchical classification of words (described by their neighbours in a reference corpus) and to cut the dendrogram at a low level of the index. It can either provides a graph associated with a partition, or a more general weighted graph if the nested set of partitions (corresponding to the values of the index less than a fixed threshold) is taken into account.

5. Discriminant analysis and information retrieval (IR)

In this context, there are several outside sources of information that can be called upon to resolve classification problems: syntactic analyzers, preliminary steps toward gaining an *understanding* of the search, dictionaries or semantic networks to lemmatize and eliminate ambiguities within the search, and possibly artificial corpora that resort to experts. The major discriminant analysis (DA) methods that are suited to large matrices of qualitative data are: *DA under conditional independence*, *DA by direct density estimation, DA by the method of nearest neighbours, DA on principal coordinates, neural networks.*

5.1 Regularized discriminant analysis and latent semantic indexing

Regularization technique strive to make discriminant analysis possible in cases that statisticians deem to be "poorly posed" (hardly more individuals than variables) or "ill posed" (fewer individuals than variables). Correspondence analysis makes it possible to replace qualitative variables (presence or frequency of a word) with numeric variables (the values of principal coordinates), and thus to apply classical DA (linear or quadratic). It thus serves as a "bridge" between textual data (that are qualitative, and often sparse) and the usual methods of DA. But most important, a filtering of information is accomplished by dropping the last principal coordinates. This process strengthens the predictive power of the procedure as a whole (Wold, 1976). These properties are applied in Information Retrieval. For instance Deerwester et al. (1990) suggest, under the name of Latent Semantic Indexing (LSI) using preliminary filtering through singular value decomposition, which is the basis of both correspondence analysis and principal components analysis. For a recent review of several filtering methods, including LSI, see: Yang (1995).

5.2 Textual units and discriminant analysis

The nature and the quality of discriminant analysis can be caused to vary depending upon the choice of the basic variables and the combination of methods:

a) Words chosen can be enhanced with counts of segments.
b) Words, and/or segments› can be selected with the help of a previously established frequency threshold.
c) The text can be lemmatized (with or without elimination of function words) and enriched with syntactic categories.
d) Only the words (segments, lemmas, etc.) that characterize the groups to be discriminated can be selected beforehand.

On the basis of the working vocabulary thus created it is possible to:

e) proceed to a preliminary singular value decomposition (or to a correspondence analysis) of the table **X** (words × observations), and keep only the first axes (filtering and regularizing through SVD).
f) proceed to a preliminary cluster analysis, and work on aggregates of units (Hearst and Pedersen, 1996). In Lebart (1992), clusters are used to take into account variations of density within each category. The approaches of Salton and Mc Gill (1983), Iwayama and Tokunaga (1995), in the framework of discriminant analysis (named in this context "Text Categorization") consist of a combination of clustering and discrimination: in a preliminary phase, clusters are built to mark out the vector space containing the observations, and to limit the number of comparisons of distances during the categorization or assignment step.

All of these alternatives imply a strategy to be developed by the user. Different strategies do exist in the framework of learning theory for using combination of methods (such as *stacking, bagging, boosting*; for a review in the specific context of IR, see: Hull *and al.*, 1996).

References

Bartell B.T., Cottrell G.W., Belew R.K. (1992). Latent semantic indexing is an optimal special case of multidimensional scaling, Proceedings of the 15th Int. *ACM-SIGIR Conf. on Res. and Dev. in Information Retrieval*, Belkin N and al. Ed., 161-167, ACM Press, New York.

Becue M., Lebart L. (1996). Clustering of texts using semantic graphs. Application to open-ended questions in surveys, *Proceedings of the IFCS 96 Symposium*, Kobe, Springer Verlag, Tokyo (in press).

Benzecri J.-P.(1977). Analyse discriminante et analyse factorielle, *Les Cahiers de l'Analyse des Donnees*, II, 4, 369-406.

Biber D. (1995). *Dimensions of register variation.* Cambridge Univ. Press, Cambridge.

Church K.W., Hanks P. (1990). Word association norms, mutual information and lexicography, *Computational Linguistics,* 16, 22-29.

Coggins J. M. (1983). Dissimilarity measures for clustering strings. in: *Time warps, string edit, and macromolecules: The theory and practice of sequence comparison,* Sankoff D. and Kruskal J. B. (eds), Addison Wesley, 311-321.

Deerwester S., Dumais S.T., Furnas G.W., Landauer T.K., Harshman R. (1990). Indexing by latent semantic analysis, *J. of the Amer. Soc. for Information Science,* 41 (6), 391-407.

Fuhr N., Pfeifer U. (1991). Combining model-oriented and description-oriented approaches for probabilistics indexing, *Proceedings of the 14th Int. ACM Conf. on Res. and Dev. in Information Retrieval*, Bookstein A. and al., Ed, ACM Press, New York, 46-56.

Furnas G. W., Deerwester S., Dumais S.T., Landauer T.K., Harshman R. A., Streeter L.A., Lochbaum K.E. (1988). Information retrieval using a singular value decomposition model of latent semantic structure, *Proceedings of the 14th Int. ACM Conf. on Res. and Dev. in Information Retrieval*, 465-480.

Habert B., Salem A. (1995). L'utilisation de catégorisations multiples pour l'analyse quantitative de données textuelles, *Traitement Aut. des Langues*, 36, 1-2, 249-275.

Habert B., Nazarenko A., Salem A. (1997). *Les linguistiques de corpus.* Armand colin, Paris.

Hearst M. A., Pedersen J. O. (1996). Reexamining the cluster hypothesis. Scatter-Gather on retrieval results. *Proceedings of the 19th Annual International ACM SIGIR Conf. on Research and Development in Inf. Retrieval.* Zurich, 76-84.

Hull D. A.., Pedersen J. O., Schutze H. (1996). Method combination for document filtering, in: *ACM / SIGIR'96* (Frei H. P., Harman D., Schauble P., Wilkinson R., eds), Zurich, Switzerland, 279-287.

Iwayama, M., Tokunaga, T. (1995). Cluster-based text categorization: a comparison of category search strategies. in: *ACM / SIGIR'95*, (Fox E.A., Ingwersen P., Fidel R., eds), Seattle, WA, USA, 273-280.

Jardine N., van Rijsbergen C. J. (1971). The use of hierarchical clustering in information retrieval. *Information Storage and Retrieval.* 7, 217-240.

Lebart L. (1982). Exploratory analysis of large sparse matrices, with application to textual data, *COMPSTAT,* Physica Verlag, 67-76.

Lebart L. (1992). Discrimination through the regularized nearest cluster method. *COMPSTAT; Proceedings of the 10th Symposium on Computational Statistics*, Physica Verlag, Vienna, 103-118

Lebart L. (1995). Assessing and Comparing Patterns in Multivariate Analysis. in: *Data Science and its Application.* Escoufier and Hayashi (Eds), Academic Press, Tokyo, 193-204.

Lebart L., Salem A., Berry E. (1991). Recent development in the statistical processing of textual data, *Applied Stoch. Model and Data Analysis*, 7, 47-62.

Lebart L., Salem A., Berry E. (1998). *Exploring Textual Data.* Kluwer, Dordrecht.

Lewis D.D., Croft W.B. (1990). Term clustering of syntactic phrases, *Proceedings of the 13th Int. ACM Conf. on Res. and Dev. in Information Retrieval*, Vidick J.L., (ed), A.C.M.Press, New York, 385-395.

McLachlan G.J. (1992). *Discriminant Analysis and Statistical Pattern Recognition,* Wiley, New-York.

Rajman M., Rungsawang A. (1995). Textual information retrieval based on the concept of distributional semantics. In: *JADT-1995* (*3rd Int. Conf. on Statist. Analysis of Textual Data*), Bolasco S. et al. (eds). CISU, Roma, 237-244.

Ritter H., Kohonen T. (1989). Self Organizing Semantic Maps. *Biol. Cybern.* 61, 241-254.

Salem A. (1984). La typologie des segments répétés dans un corpus, fondée sur l'analyse d'un tableau croisant mots et textes, *Cahiers de l'Analyse des Données,* 489-500.

Salton G. (1988). *Automatic Text Processing : the Transformation, Analysis and Retrieval of Information by Computer,* Addison-Wesley, New York.

Salton G., Mc Gill M.J. (1983). *Introduction to Modern Information Retrieval,* International Student Edition.

Willet P. (1988). Recent trends in hierarchical document clustering: A critical review. *Information Processing and Management,* 24 (5), 577-597.

Wold S. (1976). Pattern recognition by means of disjoint principal component models, *Pattern Recognition*, 8, 127-139.

Yang Y. (1995). Noise reduction in a statistical approach to text categorization. in: *ACM / SIGIR'95*, (Fox E.A., Ingwersen P., Fidel R., eds), Seattle, WA, USA, 256-263.

Text Mining - Knowledge extraction from unstructured textual data

Martin Rajman, Romaric Besançon
Artificial Intelligence Laboratory
Computer Science Dpt
Swiss Federal Institute of Technology

Abstract: In the general context of Knowledge Discovery, specific techniques, called Text Mining techniques, are necessary to extract information from unstructured textual data. The extracted information can then be used for the classification of the content of large textual bases. In this paper, we present two examples of information that can be automatically extracted from text collections: probabilistic associations of key-words and prototypical document instances. The Natural Language Processing (NLP) tools necessary for such extractions are also presented.

Key-words: text mining, knowledge discovery, natural language processing

1. Introduction

The general purpose of Knowledge Discovery is to "extract implicit, previously unknown, and potentially useful information from data" (Frawley, 1991). Due to the continuous growth of the volume of electronic data currently available, automated knowledge extraction techniques become always more necessary to valorize the huge amounts of data stored in the information systems. In addition, as the usual Data Mining techniques are essentially designed to operate on structured databases, specific techniques, called Text Mining techniques, have to be developed to process the important part of the available information that can be found in unstructured textual form.

Text Mining (TM) therefore corresponds to the extension of the more traditional Data Mining approach to unstructured textual data and is concerned with various tasks such as extraction of information implicitly contained in collections of documents or similarity-based structuring and visualisation of large sets of texts.

In sections 2 and 3, we present two examples of information extraction using different TM techniques: automated key-word association extraction, and prototypical document mining. In section 4, a discussion on the relation between these TM techniques and other Textual Data Analysis techniques is presented.

2. Mining for associations

In the case of an indexed document collection, the indexing structures (usually key-word sets) can be used as a basis for information extraction.
In such a framework, one possible goal is to extract significant key-word *associations*.

Let's consider a set of key-words $A = \{w_1, w_2, ..., w_m\}$ and a collection of indexed documents $T = \{t_1, t_2, ..., t_n\}$ (*i.e.* each t_i is associated with a subset of A denoted $t_i(A)$).
Let $W \subseteq A$ be a set of key-words, the set of all documents t in T such that $W \subseteq t(A)$ will be called the **covering set** for W and denoted $[W]$.
Any pair (W, w), where $W \subseteq A$ is a set of key-words and $w \in A \backslash W$, will be called an **association rule** (or simply an **association**), and denoted $(W \Rightarrow w)$.

Given an association rule $R\ :\ (W \Rightarrow w)$,

- $S(R,T) = |[W \cup \{w\}]|$ is called the **support** of R with respect to the collection T ($|X|$ denotes the size of the set X)

- $C(R,T) = \frac{|[W \cup \{w\}]|}{|[W]|}$ is called the **confidence** of R with respect to the collection T.

  Notice that $C(R,T)$ is an approximation (maximum likelihood estimate) of the conditional probability for a text of being indexed by the key-word w if it is already indexed by the key-word set W.

An association rule R generated from a collection of texts T is said to satisfy support and confidence constraints σ and γ if

$$S(R,T) \geq \sigma \text{ and } C(R,T) \geq \gamma$$

To simplify notations, $[W \cup \{w\}]$ will be often written $[Ww]$ and a rule $R\ :\ (W \Rightarrow w)$ satisfying given support and confidence constraints will be simply written as:

$$W \Rightarrow w \qquad S(R,T)/C(R,T)$$

Informally, for an association rule $(W \Rightarrow w)$, such σ/γ constraints can be interpreted as: there exists a significant number of documents (at least σ), for which being related to the topic characterised by the key-word set W implies (with a conditional probability estimated by γ) to be also related to the topic characterised by the key-word w.

As far as the actual association extraction is concerned, the common procedures are usually two-steps algorithms:

- generation of all the key-word sets with support at least equal to σ (*i.e.* all the key-word sets W such that $\|[W]\| \geq \sigma$). The generated key-word sets are called the **frequent sets** (or **σ-covers**);
- generation of all the association rules that can be derived from the produced frequent sets and that satisfy the confidence constraint γ.

The frequent sets are obtained by incremental algorithms that explore the possible key-word subsets, starting from the frequent singletons (*i.e.* the $\{w\}$ such as $\|[\{w\}]\| \geq \sigma\}$) and iteratively adding only those key-words that produce new frequent sets. This step is the most computationally expensive (exponential in the worst case) in the extraction procedure.
The associations derived from a frequent set W are then obtained by generating all the implications of the form $W\backslash\{w\} \Rightarrow w, (w \in W)$, and keeping only the ones satisfying the confidence constraint γ.
Some additional treatment (structural or statistical pruning, redundancy elimination) is usually added to the extraction procedure in order to reduce the number of generated associations.

Association-based Text Mining techniques have been explored by R. Feldman (Feldman, 1996) with the KDT (Knowledge Discovery in Texts) tool on the Reuter corpus. This corpus is a newswire collection containing about 22.000 articles manually indexed with 135 categories form the Economics domain by Reuters Ltd. and Carnegie Group Inc. in 1987.
For such a corpus, association extraction from the key-word sets allows to satisfy information needs expressed by queries. The result of the extraction process is a list of associations, ordered by support and confidence. An example of queries and associated results is given in Table 1.

Table 1: *Examples of associations*

query	:	'*find all associations between a set of countries including Iran and any person*'
result	:	[Iran, Nicaragua, Usa] $\Rightarrow$ Reagan 6/1.000
		...
query	:	'*find all associations between a set of topics including Gold and any country*'
result	:	[gold, copper] $\Rightarrow$ Canada 5/0.556
		[gold, silver] $\Rightarrow$ USA 18/0.692
		...

3. Mining for prototypical documents

Another direction of research for automated information extraction is to apply knowledge discovery techniques to the complete textual content of the documents (in a so-called "full text" approach as opposed to approaches only considering indexing key-words). However, our experiments on the Reuter corpus (Rajman, 1997) have shown that the extraction process does not produce any exploitable results when the standard association extraction techniques are directly applied on the words contained in the documents instead of operating on the already abstract concepts represented by the key-words. Among the extracted associations, some only indicate the presence of domain-dependent compounds (*{wall}* $\Rightarrow$ *street*, *{government prime}* $\Rightarrow$ *minister*, *{treasury secretary james}* $\Rightarrow$ *baker*), while others are simply uninterpretable (*{dollars shares exchange total commission stake}* $\Rightarrow$ *securities*, *{million april management lead issues underwriting denominations}* $\Rightarrow$ *selling*).

A different approach is therefore necessary when full text is considered: prototypical document extraction. A *prototypical document* is informally defined as a document corresponding to an information that occurs in a repetitive fashion in the document collection, *i.e.* a document representing a class of similar documents in the textual base.
The extraction techniques operating in this framework still use the notion of frequent sets, but additional Natural Language (NL) techniques are used to preprocess the data, and identify more significant linguistic entities (*terms*) for the frequent set extraction process.
More precisely, the NL preprocessing, realized in collaboration with R. Feldman's team at Bar Ilan University, was decomposed into two steps: *Part-of-Speech tagging* and *term extraction.*

Part-of-Speech tagging
This process automatically identifies the morpho-syntactic categories (noun, verb, adjective, ...) of words in the documents. Such a tagging allows to filter non-significant words on the basis of their morpho-syntactic category. In our experiments, we used a rule-based tagger designed by E. Brill (Brill, 1992), and restricted the extraction process to operate only on nouns, adjectives and verbs.

Term extraction
This process aims at the identification of the domain-dependent compounds. It allows the mining process to focus on more meaningful co-occurrences, and can be decomposed into:

- term candidates identification (on the basis of structural linguistic information; in our case, morpho-syntactic patterns such as *Noun Noun*, *Noun of Noun*,...);

- term candidates filtering (based on statistical relevance scoring (Daille, 1994)).

For example, the already mentioned sequence of words (found by association extraction on full text) *(Treasury, Secretary, James, Baker)* was tagged as *Treasury/Noun Secretary/Noun James/Noun Baker/Noun* and subsequently identified as one single term *Treasury_ Secretary_ James_ Baker/Noun.*

Mining process

On the basis of the terms resulting from the NL preprocessing step, an algorithm similar to the one described in the previous section was used to extract frequent term sets from the document collection.

The extracted frequent sets were then submitted to several additional treatments in order to determine the prototypical documents:

- To reduce information redundancy, a clustering of the frequent term sets was performed, on the basis of a similarity measure derived from the number of common terms in the sets. The resulting clusters were represented by the union of their constitutive term sets.
- To limit the possible meaning shifts due to variations in the word ordering, the clusters were further split into sets of *term sequences* associated with paragraphs boundaries in the original documents.

An example of the treatments on the Reuter corpus is given in Figure 1.

Figure 1: *The frequent sets processing*

Some frequent term sets (with their frequency) extracted from the Reuter Corpus for a support of 80:

{due available management priced issuing denominations payment_date} 87
{due management issuing denominations luxembourg payment_date} 81
{due management priced issuing combined paying underwriting} 80
{due management selling priced issuing listed} 81
{due priced issuing combined denominations payment_date} 80
{management issuing combined underwriting payment_date} 80
(...)

Resulting cluster:

{due available management priced issuing combined denominations listed underwriting luxembourg payment_date paying} 45

Most frequent sequential decomposition:

(issuing due paying priced) (available denominations listed luxembourg) (payment_date) (management underwriting combined) 41

Prototypical documents are then all the documents (or document parts) that instantiate any of the extracted sequential decompositions of the frequent term set. An example of a prototypical document instantiating the above mentioned decomposition is shown in Figure 2.

Figure 2: *A prototypical document*

<DOC2088>
Nissan_Motor_Co_Ltd "NSAN.T" is **issuing** a 35_billion_yen eurobond **due** March_25 1992 **paying** 5-1/8_percent and **priced** at 103-3/8, Nikko_Securities_Co (Europe) Ltd said.
The non-callable_issue is **available** in **denominations** of one_million Yen and will be **listed** in **Luxembourg**.
The **payment_date** is March_25.
The selling_concession is 1-1/4_percent while **management** and **underwriting combined** pays 5/8_percent.
Nikko said it was still completing the syndicate.

By definition, these prototypical documents are representative of classes of repetitive document structures in the collection of texts. Their main advantage is to provide a usable interpretation scheme for the information extracted from the document collection in the form of frequent term sets, and, as such, they constitute good candidates for a partial synthesis of the information content hidden in a textual base.

4. Related Work

Several other domains concerned with Textual Data Processing (such as Textual Data Analysis or Content Analysis) can provide interesting insights on the techniques presented in this paper.
The problem of frequent set extraction could be for instance partially related to the identification of co-occurrent words (Lafon, 1981), repeated segments (Salem, 1987), or quasi-segments (Becue, 1993), often considered in the domain of Textual Data Analysis. The main difference here is that the Text Mining techniques rely on the use of frequencies of *sets of words* instead of considering co-frequencies of *pairs*.
As far as more sophisticated information extraction is concerned, methods used in Textual Data Analysis (Lebart, 1998) usually rely on a cluster analysis based on the the chi-square distance between the lexical profiles. For each of the resulting clusters of documents, *characteristic words* (Lafon, 1980) (i.e. words with a frequency in the cluster significantly higher

than the one expected according to a predefined probabilistic model) are then extracted. Each of the cluster is then represented by a *characteristic document* which is the document in the cluster that contains the most characteristic words.
The differences between such approaches and prototypical document extraction as described in this paper are essentially of two kinds: (1) prototypical document extraction integrates a more substantial amount of explicit linguistic knowledge, in particular in the preprocessing phase, where morpho-syntactic patterns are used for the extraction of indexing terms; (2) the aims underlying the two methods are in fact quite different: documents characteristic for a cluster identify the information content that is the more discriminant for the cluster relatively to the rest of the document collection. On the opposite, prototypical documents tend to identify repetitive patterns of texts particularly frequent in the document collection, and that will serve to structure its informational content.
The two approaches therefore appear to be rather complementary in the sense that prototypical documents could be thought as kinds of linguistic frames in which the informational content (as identified by the characteristic documents) could be preferentially expressed.

In addition, in order to allow better representativity, a more generic representation could be achieved by using *name entity tagging*, a semantic tagging that allows to identify and generalise certain elements of a sentence. Such a tagging could lead to representations where the variable parts of the prototypical documents would be replaced by concepts.
For instance, on the basis of the results of name entity tagging applied to the document given in Figure 2 (these results were produced by the Alembic tool and provided by Christopher Clifton, from the MITRE NLP group (MITRE, 1997)), the associated document class could be represented by the generic prototypical document presented in Figure 3.

Figure 3: *A generic prototypical document*

<ORGANIZATION> is **issuing** a *<NUMBER>* yen eurobond **due** *<DATE>* **paying** *<NUMBER>* percent and **priced** at *<NUMBER>* , *<ORGANIZATION>* (*<LOCATION>*) said.
The non-callable issue is **available** in **denominations** of *<NUMBER>* and will be **listed** in *<LOCATION>*.
The **payment date** is *<DATE>*.
The selling concession is *<NUMBER>* percent while **management** and **underwriting combined** pays *<NUMBER>* percent.
<ORGANIZATION> said it was still completing the syndicate.

5. Conclusion

We have presented two examples of Text Mining tasks for the extraction of information from collections of textual data: an association extraction method operating on indexed documents, and a prototypical document extraction algorithm that can be applied on plain documents (full text approach). For both tasks, preliminary results have been obtained. Further research will be carried out to explore the use of prototypical documents for the automated synthesis of the information content of document classes in large collections of textual data.

References

Becue M., Peiro R. (1993). Les quasi-segments pour une classification automatique des réponses ouvertes, in *Actes des secondes journées internationales d'analyse des données textuelles*, (Montpellier), ENST, Paris, 310-325.

Brill E. (1992). A simple Rule-Based Part-of-Speech Tagger, in *Proc. of the 3rd Conf. on Applied Natural Language Processing.*

Daille B.(1994). Study and Implementation of Combined Techniques for Automatic Extraction of Terminology, in *Proc. of the 32nd Annual Meeting of the Association for Computational Linguistics.*

Feldman R., Dagan I. and Kloegsen W. (1996). Efficient Algorithm for Mining and Manipulating Associations in Texts, in *Proc. of the* 13th *European Meeting on Cybernetics and Research.*

Frawley W.J., Piatetsky-Shapiro G., and Matheus C.J. (1991). Knowlegde Discovery in Databases : An Overview, in *Knowledge Discovery in Databases*, MIT Press, pages 1–27.

Lafon P. (1980). Sur la variabilité de la fréquence des formes dans un corpus, *Mots*, 1, 127–165.

Lafon P. (1981). *Dépouillements et statistiques en lexicométrie*, Slatkine-Champion, 1984, Paris.

Lebart L., Salem A., Berry L. (1998). *Exploring Textual Data*, Kluwer Academic Publishers, Dordrecht.

MITRE NLP Group (1997). Alembic Language Processing for Intelligence Applications. At URL :
http://www.mitre.org/resources/centers/advanced_info/g04h/nl-index.html

Rajman M. and Besançon R. (1997). A Lattice Based Algorithm for Text Mining. Technical Report TR-LIA-LN1/97, Swiss Federal Institute of Technology.

Salem A. (1987). *Pratique des segments répétés, Essai de statistique textuelle*, Klincksieck, Paris.

PART V

Multivariate and Multidimensional Data Analysis

- Proximity Analysis and Multidimensional Scaling
- Multivariate Data Analysis
- Non Linear Data Analysis
- Multiway Data Analysis

A general formulation of multidimensional unfolding models involving the latitude of acceptance

Guanzhong Luo

South China Normal University, Guangzhou, China
and
Murdoch University, Western Australia

Abstract: The latitude of acceptance (Sherif & Hovland, 1961; Coombs, 1964) is considered as the cornerstone of social judgement theory and attitude measurement (Eagly & Chaiken, 1993). Luo (1998a) introduced a general form of the unidimensional unfolding models in which the latitude of acceptance is explicitly parameterised. This paper generalises this form further into a multidimensional model. It is shown that the structure of multidimensional unfolding models is defined by the operation function and the magnitude of the latitude of acceptance. It is also demonstrated that some well referenced multidimensional scaling models are special cases of the general form introduced in this paper. According to this general form, some new multidimensional unfolding models can be readily specified. This general form provides a systematic understanding of the multidimensional unfolding models, and for applications, it reveals the possibility of developing a general algorithm for the estimation of the parameters of multidimensional unfolding models.

Key words: Unfolding model, Latitude of acceptance, Multidimensional Scaling.

1. Introduction

Coombs (1964) classified the procedures of data collection in attitude measurement into 4 types: *Preference choice, Single stimulus, Stimulus comparison* and *Similarities.* The single stimulus method, which is the simplest, requires respondents to respond positively or negatively to a stimulus (item). Data collected in this way is dichotomous: *Agree* - 1 and *Disagree*: 0. Various probabilistic unfolding models have been developed to analyse this type of data. These models can be classified into two classes: unidimensional and multidimensional. Unidimensional unfolding models parameterise item locations and person locations into a unidimensional real continuum (e. g., DeSarbo & Hoffman, 1986; Andrich, 1988; Hoijtink, 1990; Andrich & Luo, 1993; Andrich,

1996) and multidimensional unfolding models parameterise item locations and person locations into a multidimensional real space (e. g. Rao, 1986; DeSarbo & Rao, 1986). Luo (1998a) synthesised unidimensional probabilistic unfolding models by introducing a general formulation for this class of unfolding models. This paper generalises this general formulation into multidimensional situations.

2. The general form of multidimensional unfolding models

Luo (1998a) introduced a general form of the unidimensional unfolding models as

$$P\{x_{ni} = 1\} = \frac{\Psi(\rho)}{\Psi(\rho) + \Psi(\beta_n - \delta_i)}, \tag{1}$$

where Ψ, which is termed the operational function, has the following properties:

(P1). *Always positive:* $\Psi(t) \geq 0$ for any real t;

(P2). *Monotonic in the positive domain*: $\Psi(t_1) \geq \Psi(t_2)$ for any $t_1 > t_2 > 0$;

and

(P3). *Symmetric about the origin*: $\Psi(-t) = \Psi(t)$ for any real t.

The three parameters of the model of Equation (1) are:

δ_i : the location parameter of item i, $i = 1, \ldots, I$;

β_n : the location parameter of person n, $n = 1, \ldots, N$; and

ρ : the latitude of acceptance parameter, which can be subscripted as an item parameter ρ_i or a person parameter ρ_n .

Graphically, $\delta_i \pm \rho$ are the interception points of the curve $P\{x_{ni} = 1\}$ and $P\{x_{ni} = 0\}$. The significance of Equation (1) is that it parameterises the latitude of acceptance, a well known but little researched concept (Sherif & Hovland, 1961; Coombs, 1964).

For the multidimensional cases, which are studied in *the theory of Multidimensional Scaling* (Borg & Groenen, 1997), the item and person location parameters have K coordinates,

$$\delta_i = (\delta_{i1}, \delta_{i2}, \ldots, \delta_{iK});$$
$$\beta_n = (\beta_{n1}, \beta_{n2}, \ldots, \beta_{nK}).$$

The difference $\beta_n - \delta_i$ in Equation (1) can be replaced by their distance between β_n and δ_i in a K-dimension space

$$\| \beta_n - \delta_i \| = \sqrt{\sum_{k=1}^{K} w_k (\beta_{nk} - \delta_{ik})^2} . \tag{2}$$

Then the general form for multidimensional unfolding models is

$$P\{x_{ni} = 1\} = \frac{\Psi(\rho)}{\Psi(\rho) + \Psi(\| \beta_n - \delta_i \|)} . \tag{3}$$

3. Special cases of the general form

1. The DeSarbo-Hoffman model
DeSarbo and Hoffman (1986) introduced the Spatial Choice Model

$$P\{x_{ni} = 1\} = \frac{1}{1 + \exp(\sum_{k=1}^{K} w_{nk} (\beta_{ni} - \delta_{ik})^2 + c_n)} ; \tag{4}$$

where c_n was termed the *threshold* and w_{nk} was called the *salience or importance* of person n for dimension k. Equation (4) is equivalent to the following form

$$P\{x_{ni} = 1\} = \frac{\exp((\sqrt{-c_n})^2)}{\exp((\sqrt{-c_n})^2) + \exp(\left(\sqrt{\sum_{k=1}^{K} w_{nk} (\beta_{nk} - \delta_{ik})^2}\right)^2)} . \tag{5}$$

Comparing equation(5) with equation (3) leads to

$$\rho = \sqrt{-c_n} ;$$
$$\Psi(t) = \exp(t^2) .$$

2. The DeSarbo-Rao model
Let Φ be the normal distribution $N(0, \sigma^2)$, DeSarbo and Rao (1986) defined a multidimensional unfolding modes as

$$P\{\mathrm{x}_{ni} = 1 \mid \beta_n, \delta_i, c_n\} = \Phi(c_n - \sum_{k=1}^{K}(\beta_{nk} - \delta_{ik})^2). \tag{6}$$

Obviously, when $\beta_n = \delta_i$, Equation (6) has the maximum value of $\Phi(c_n)$; when $\sqrt{\sum_{k=1}^{K}(\beta_{nk} - \delta_{ik})^2} = \sqrt{c_n}$, we have

$$P\{\mathrm{x}_{ni} = 1 \mid \beta_n, \delta_i, c_n\} = \Phi(0) = 0.5 = P\{\mathrm{x}_{ni} = 0 \mid \beta_n, \delta_i, c_n\}. \tag{7}$$

That is, $\sqrt{c_n}$ is the latitude of acceptance parameter in the form of equation (3). To decide the expression of the corresponding operational function Ψ, equating equation (3) ($w_k = 1$) to equation (6) leads to

$$\frac{\Psi(\sqrt{c_n})}{\Psi(\sqrt{c_n}) + \Psi(\sqrt{\sum_{k=1}^{K}(\beta_{nk} - \delta_{ik})^2})} = \Phi(c_n - \sum_{k=1}^{K}(\beta_{nk} - \delta_{ik})^2), \tag{8}$$

and

$$\frac{\Psi(\sqrt{\sum_{k=1}^{K}(\beta_{nk} - \delta_{ik})^2})}{\Psi(\sqrt{c_n}) + \Psi(\sqrt{\sum_{k=1}^{K}(\beta_{nk} - \delta_{ik})^2})} = 1 - \Phi(c_n - \sum_{k=1}^{K}(\beta_{nk} - \delta_{ik})^2). \tag{9}$$

Therefore,

$$\Psi(\sqrt{\sum_{k=1}^{K}(\beta_{nk} - \delta_{ik})^2}) = \Psi(\sqrt{c_n})\frac{1 - \Phi(c_n - \sum_{k=1}^{K}(\beta_{nk} - \delta_{ik})^2)}{\Phi(c_n - \sum_{k=1}^{K}(\beta_{nk} - \delta_{ik})^2)}. \tag{10}$$

Let

$$\Psi(t) = \frac{1 - \Phi(c_n - t^2)}{\Phi(c_n - t^2)}, \tag{11}$$

then it is easy to demonstrate that the properties (P1), (P2) and (P3) are satisfied. The minimum point of Ψ is at $t = 0$:

$$\Psi(0) = \frac{1 - \Phi(c_n)}{\Phi(c_n)}. \tag{12}$$

Because $\Psi(\sqrt{c_n}) = 1$, the model of equation (6) can be re-expressed as

$$P\{x_{ni} = 1 \mid \beta_n, \delta_i, c_n, \delta^2\} = \frac{1}{1 + \Psi(\sqrt{\sum_{k=1}^{K}(\beta_{nk} - \delta_{ik})^2})}. \quad (13)$$

4. Some new multidimensional unfolding models generated according to the general form

1. MDHCM: Multidimensional version of the HCM (Andrich & Luo, 1993)

$$P\{x_{ni} = 1\} = \frac{\cosh(\rho)}{\cosh(\rho) + \cosh(\sqrt{\sum_{k=1}^{K}(\beta_{nk} - \delta_{ik})^2})}. \quad (14)$$

2. MDPARELLA: Multidimensional version of the PARELLA (Hoijtink, 1990).

$$P\{x_{ni} = 1\} = \frac{\rho^{2\gamma}}{\rho^{2\gamma} + (\sqrt{\sum_{k=1}^{K}(\beta_{nk} - \delta_{ik})^2})^{2\gamma}}, \quad (15)$$

where $\gamma > 0$ is a structural parameter specifying the power of the operational function $\Psi(t) = t^{2\gamma}$.

5. Conclusions

The general form of multidimensional unfolding models of this paper provides a systematic understanding of the multidimensional unfolding models. According to the general formulation of multidimensional unfolding models, $P\{x_{ni} = 1\}$ is single-peaked function of person-item distance in the multidimensional space. Therefore, all the results from unidimensional unfolding modelling, which is based on the single-peakedness of its response function, are potentially applicable to the multidimensional form. For purpose of application, it reveals the possibility of developing a general algorithm for the estimation of the parameters of multidimensional unfolding models. This general formulation can also be extended to polytomous situations using the Rating approach or Sequential process (Luo, 1998b).

References

Andrich, D. (1988). The application of an unfolding model of the PIRT type to the measurement of attitude. *Applied Psychological Measurement, 12,* 33-51.

Andrich, D. (1996). Hyperbolic Cosine Latent Trait Models for Unfolding Direct-Responses and Pairwise Preferences. *Applied Psychological Measurement, 20,* 269-290.

Andrich, D. & Luo, G. (1993). A hyperbolic cosine latent trait model for unfolding dichotomous single-stimulus responses. *Applied Psychological Measurement, 17,* 253-276.

Borg, I. & Groenen, P.(1997). *Mordern multidimensional scaling: Theory and application.* New York: Springer.

Coombs, C. H. (1964). *A theory of data*. New York: Wiley.

DeSarbo, W. S. and Hoffman, D. L. (1986). Simple and weighted unfolding threshold models for the spatial representation of binary choice data. *Applied Psychological Measurement, 10,* 247-264.

DeSarbo, W. S. & Rao, V. R. (1986). A new constrained unfolding methodology for product positioning. *Marketing Science, 5*, 1-19.

Eagly, A.H. & Chaiken, S. (1993). *The Psychology of Attitudes.* Fort Worth: Harcourt Brace Jovanovich.

Hoijtink, H. (1990). *PARELLA: Measurement of latent traits by proximity items.* University of Groningen.

Luo, G. (1998a). A general formulation for unidimensional unfolding and pairwise preference models: Making explicit the latitude of acceptance. In press, *Journal of Mathematical Psychology.*

Luo, G. (1998b). Formulations for polytomous probabilistic unfolding models. *Research Report 14,* Social Measurement Laboratory, Murdoch University.

Sherif, M & Hovland, C. (1961). *Social judgment: Assimilation and contrast effects in communication and attitude change.* New Haven, CT: Yale University Press.

The Data Theory Scaling System

Jacqueline J. Meulman[1], Lawrence J. Hubert[2], Willem J. Heiser[3]
[1]Department of Education, Leiden University, email: meulman@rulfsw.fsw.leidenuniv.nl;
[2]Department of Psychology, University of Illinois at Champaign, email: lhubert@s.psych.uiuc.edu;
[3]Department of Psychology, Leiden University, email: heiser@rulfsw.fsw.leidenuniv.nl

Abstract: Optimal scaling is reviewed in the framework of a comprehensive multidimensional data analysis strategy, called the Data Theory Scaling System (DTSS). The optimal scaling methodology has a lot of potential for the analysis of multivariate data that are qualitative, non-normally distributed, or incomplete. Also, the system can easily deal with nonlinear relationships between the variables to be analyzed. A crucial aspect of DTSS is the graphical display of the results of the analysis. The most recent developments of DTSS include the integration of optimal scaling with combinatorial data analysis, e.g., ordering tasks in one dimension, and the fitting of graphs to multivariate data.

Keywords: multivariate analysis, categorical data, optimal scaling, unidimensional scaling, multidimensional scaling, combinatorial optimization, ordering, consensus, networks, graphs.

1. Introduction

Classical statistical methods need to be adapted in various ways to suit the particular characteristics of data obtained in the social and behavioral sciences. The latter are often data that are non-numerical, with measurements recorded on scales that have an uncertain unit of measurement. Data would typically consist of qualitative or categorical variables that describe the units (objects, subjects, individuals) in a limited number of categories. The zero point of these scales is uncertain, the relationships among the different categories is often unknown, and although frequently it can be assumed that the categories are ordered, their mutual distances would still be unknown. The uncertainty in the unit of measurement is not just a matter of measurement error, because its variability may have a systematic component. An important development in multidimensional data analysis has been the optimal assignment of quantitative values to such qualitative scales (optimal scaling), where we distinguish in addition to the traditional interval level • the ordinal scaling level, taking only rank-orders (among categories) into account, and using least squares monotonic regression or monotonic regression splines, and • the nominal scaling level, taking only categorical information into account, and using rank=1 regression (least squares or regression splines) or the centroids approach (rank=p regression, where p denotes the chosen dimensionality in the optimization process). A variable is represented by a set of category points; rank=p regression locates a category point in the center of gravity of the associated objects; rank=1 regression fits category points on a straight line through the origin.

2. Optimal Scaling: A General Data Analysis Strategy

A review focussed on the optimal scaling paradigm needs to address its history. The idea of optimal scaling originates with different sources. Looking at the history of optimal scaling with the rank=p approach, famous early contributions are by Horst (1935), Fisher (1940), Guttman (1941), Burt (1950), and Hayashi (1952). This approach to categorical data analysis leads eventually to (multiple) correspondence analysis (Greenacre, 1984), a literal translation of Benzécri's "analyse des correspondances (multiple)". The class of techniques is also known as dual scaling (Nishisato, 1980; 1994), and homogeneity analysis (Gifi, 1981, 1990). A recent discussion of homogeneity analysis from the psychometric point of view can be found in Heiser and Meulman (1994). A second, and perhaps the major impetus to optimal scaling found its origin in the area of nonmetric multidimensional scaling, as pioneered by Shepard (1962) and Kruskal (1964). In nonmetric multidimensional scaling, optimal scaling of the proximities is typically performed by a monotonic (rank=1) regression. After its remarkable accomplishments in MDS, optimal scaling was soon to be incorporated in techniques for multivariate analysis, notably by Kruskal (1965), Shepard (1966), and Roskam (1968). In the 1970s and 1980s psychometric contributions to the area became numerous; attempts at systematization resulted in the ALSOS system by Young, De Leeuw and Takane (1976, 1978), Young (1981), and the system by the Leiden "Albert Gifi" group. The Albert Gifi (1990) book "Nonlinear Multivariate Analysis" aimed to provide a comprehensive system, combining optimal scaling with multivariate analysis, including statistical developments from the 1970s and 1980s. Since the middle 1980s, the principles of optimal scaling have gradually appeared in the mainstream statistical literature (Breiman and Friedman, 1985; Gilula and Haberman, 1988; Ramsay, 1988; Buja, 1990; Hastie, Tibshirani, and Buja, 1994). The Gifi system is discussed among the traditional statistical techniques in Krzanowski and Marriott (1994). In the 1990s, optimal scaling methods have been extended into a more general framework, called the "Data Theory Scaling System"; its mission is to meet typical concerns in the social and behavioral sciences, both from a substantive perspective as well as the technical point of view, dealing a.o. with • discrete multivariate data • ordinal data • incomplete data • nonlinear relationships between pairs of variables • non-normal distributions • ordering/scaling of response patterns • social network data and other proximity relations. DTSS focusses on the multivariate analysis of qualitative data, including:

• dimension reduction by linear mapping (see, e.g., Heiser and Meulman, 1995); the Gifi system is confined to this aspect of DTSS;

• distance approximation in multivariate data analysis (Meulman, 1992);
• clustering of objects (Heiser, 1993; Heiser and Groenen, 1997), or
• clustering of variables (Meulman and Verboon, 1993; Meulman, 1997);
• graphical display of objects and variables in linear biplots as in Tucker (1960), and Gabriel (1971), optimized for least squares multidimensional scaling of multivariate data (Meulman, 1997), and
•nonlinear biplots as in Gower and Harding (1991), generalized for least squares MDS (Meulman and Heiser, 1993; Groenen and Meulman, 1997);
• the implementation of procedures in SPSS Categories (the 8.0 version includes categorical regression and correspondence analysis with restrictions).
Hence, DTSS can be viewed as a successor to the Gifi system, being much more general. In the sequel of this paper we will discuss the two most recent extensions of DTSS:
• combinatorial data analysis, and
• fitting of graphs or networks.

3. Finding An Optimal Ordering of Objects

Allowing for optimal scaling of the variables in multivariate data was recently combined with the objective of finding an optimal ordering of the objects (Meulman, Hubert and Arabie, 1997). The latter is fundamentally a combinatorial problem and cannot be solved by ordinary optimization techniques used in multidimensional scaling approaches that are based on some form of (sub)gradient method. We are considering the analysis of an $N \times M$ multivariate data matrix, where N denotes the number of objects in the rows of the data matrix and M denotes the number of variables in the columns. In the study of dependence among variables, one usually resorts to some form of principal components analysis. The results of a principal component analysis have been known for quite some time to display characteristics of the objects as well (Gower, 1966), and this is in the form of distances that approximate proximities that would be observed if the Euclidean distances among the rows of the data matrix were inspected. In Meulman (1992), a distance-based PCA was defined through the minimization of a least squares loss function, defined on the distances, and using majorization. In addition, optimal scaling of the variables was included in the procedure. This approach to optimal scaling is basically different from nonmetric MDS, because the latter involves optimal scaling of the proximities instead of the variables (whereof the proximities are derived). If we wish to find a single ordering of the objects, the objective specializes to a one-dimensional scaling task. The associated least squares loss function can be written as

$$\text{STRESS}(\mathbf{Q}; \mathbf{x}; c) = ||D(\mathbf{Q}) + c - D(\mathbf{x})||^2. \tag{1}$$

Here an element of $D(\cdot)$ is a Euclidean distance between two objects; $\mathbf{Q}$ denotes the multivariate data matrix, where the columns are separately scaled according to the chosen level and method. The optimal scaling of the variables is attained by the use of constrained majorization (as in Meulman, 1992). The vector $\mathbf{x}$ gives a one-dimensional scale, with distances as close as possible to the proximities in $D(\mathbf{Q})$; c is a constant. Given a fixed permutation $\varphi(\cdot)$, finding c and $\mathbf{x}$ can be done as in Hubert, Arabie and Meulman (1997), i.e., by an iterative projection approach, adopted from Dykstra (1983). Finding the optimal ordering of the objects in $\mathbf{x}$ is a combinatorial problem. Since gradient or majorization methods have severe local minima problems when dealing with this task, an alternative strategy was used, shown to be successful in Hubert and Arabie (1994), that involves a number of local operations. These are: (a) all pairwise interchanges of objects; (b) all insertions of k consecutive objects between any two existing objects or at the beginning and end of the permutation; (c) all complete reversals of k consecutive objects. These operations are cycled through repeatedly until no change occurs. The combination of optimal scaling of variables with optimal ordering of objects can give a very parsimonous representation of a (large) number of qualitative variables. A very suitable application would be the situation where different persons have judged different options according to a number of different criteria. In the analysis, persons would function as the variables, and the result would be an optimal consensus ordering of the options. Since distance-based MVA can now be applied to problems that are basically combinatorial in nature, we can also consider other representations, by analogy with Hubert, Arabie and Meulman (1997). These tasks include circular unidimensional structures, additive trees, and ultrametrics. These applications are forthcoming.

4. Fitting Graphs to Multivariate Data

Another, and not completely unrelated, recent development in DTSS is the incorporation of spatial representations other than Euclidean, specifically through graphs (networks). In Heiser and Meulman (1997), a prevalent type of categorical multivariate data was considered, consisting of binary variables. An example of the latter in the domain of psychometrics is the collection of scores of examinees on binary items. The aim of the analysis would be to order the N examinees on a latent one-dimensional continuum (scale), to be constructed from the examinees' responses on the M binary items. This objective can more generally be formulated as to form a partially ordered structure among the response profiles. A second example, from numerical taxonomy, consists of presence/absence scores of M attributes in N individuals, where the major task is to construct hierarchical trees. For binary data, the information in the $N \times M$ multivariate data matrix, can be efficiently transformed into a (reduced) profile frequency matrix, where a weight is attached to each profile corresponding to the occurrence of the row profile in the $N \times M$ data matrix. In traditional statistics, binary data are very often presented in the form of a multiway contingency table; next, models are fitted to the cell counts, with respect to the marginals of the table. Currently, the most popular method to analyze this type of categorical data is loglinear analysis. An alternative analysis could be a multiple correspondence analysis of the profile frequency matrix. Geometrically, the row profiles can be regarded as the 2^M corners of a hypercube, where each corner of the hypercube is associated with a mass corresponding to the occurrence of a profile. Multiple correspondence analysis then aims at a simplified representation of the set of profiles. In contrast to the prevailing belief that loglinear analysis is radically different from multiple correspondence analysis because the latter would ignore higher-order interactions, Meulman and Heiser (1997) have shown that distances in the graphical display in multiple correspondence analysis are inverse functions of the odd-ratios that express higher-order interactions. The scaling approach presented in Heiser and Meulman (1997) is a very different spatial alternative to loglinear analysis. In contrast to the graphical models, as reviewed in Whittaker (1990), focussing on the representation of *variables*, the objective is to develop a graph model for the response patterns, i.e., for a *group of objects*, having the particular profile. The cell frequencies in the multiway contingency table are associated with distances between vertices in a graph. The Heiser and Meulman (1997) method is based on the equivalence between the Hamming distance among the corner points of the hypercube (which is a distance of the city-block type computed from binary coordinates) and the shortest path-length distance in a specific type of graph. In the Heiser and Meulman (1997) paper, a weighted least squares method was developed to fit the Hamming distances in the graph to the Hamming distances derived from the data, with weights composed from the pairwise product of the associated profile frequencies. Thus, the following least squares loss function was minimized:

$$\sigma(\mathbf{B}) = \sum_{i<j}^{2^M} p_i p_j (d_{ij}(\mathbf{A}) - (d_{ij}(\mathbf{B}))^2, \tag{2}$$

where p_i and p_j are the profile frequencies of profiles $\mathbf{a}_i$ and $\mathbf{a}_j$, respectively, the $d_{ij}(\mathbf{A})$ are the distances between the profiles, and the $d_{ij}(\mathbf{B})$ are the distances in the graph. An important feature of the Hamming distance is the additivity of distances along the shortest path in the graph (Heiser, 1997). Additivity of distances induces the feature of "betweenness": the distance between two points $\mathbf{b}_i$ and $\mathbf{b}_j$ is equal to the sum of the distances between $\mathbf{b}_i$ and $\mathbf{b}_j$ to an intermediate point $\mathbf{b}_k$. Apart from selecting a best fitting graph, the Heiser and Meulman (1997) method was extended with a clustering feature (see Heiser and Meulman, 1998). When many corners of the hypercube represent only a few objects (profiles with low frequencies), it is useful to allow profile points to be clustered in the representation. Therefore, the observed profiles have to be allocated to a limited number (K) of vertices in the graph. It can be shown that the resulting multidimensional scaling problem that forms the core of the method, is now only of order K, rather than N, as in ordinary multidimensional scaling of multivariate data, or 2^M, the number of vertices in the hypercube, which is equal to the maximum number of rows in the profile frequency matrix (as in Groenen, Commandeur, and Meulman, 1997).

5. Discussion

The integration of the optimal scaling paradigm in state-of-the-art multidimensional data analysis offers a wealth of exciting new methods to explore qualitative multivariate data that goes beyond the "nonlinear multivariate analysis" techniques as advocated in Gifi (1990). Specifically, the matching of optimal scaling with combinatorial data analysis, and the fitting of graphs or networks opens new avenues for modeling systematic relationships among the individuals or objects in data from the social and behavioral sciences.

References

Breiman, L., & Friedman, J.H. (1985). Estimating optimal transformations for multiple regression and correlation. *Journal of the American Statistical Association, 80,* 580-598.

Buja, A. (1990). Remarks on functional canonical variates, alternating least squares methods and ACE. *Annals of Statistics, 18,* 1032-1069.

Burt, C. (1950). The factorial analysis of qualitative data. *British Journal of Psychology, 3,* 166-185.

Dykstra, R.L. (1983). An algorithm for restricted least squares regression. *Journal of the American Statistical Association, 78,* 837-842.

Fisher, R.A. (1940). The precision of discriminant functions. *Annals of Eugenics, 10,* 422-429.

Gabriel, K.R. (1971). The biplot graphic display of matrices with application to principal components analysis. *Biometrika, 58,* 453-467.

Gifi, A. (1990). *Nonlinear multivariate analysis.* Chichester: John Wiley and Sons.

Gilula, Z., & Haberman, S.J. (1988). The analysis of multivariate contingency tables by restricted canonical and restricted association models. *Journal of the American Statistical Association, 83,* 760-771.

Gower, J.C. (1966). Some distance properties of latent roots and vector methods used in multivariate analysis. *Biometrika, 53,* 325-338.

Gower, J.C., & Harding, S.A. (1988). Nonlinear biplots. *Biometrika, 75,* 445-455.

Greenacre, M.J. (1984). *Theory and applications of correspondence analysis.* London: Academic Press.

Groenen, P.J.F., Commandeur, J.J.F., & Meulman J.J. (1997). Distance analysis of large data sets of categorical variables using object weights. *Submitted for publication.*

Groenen, P.J.F. & Meulman, J.J. (1997). *Fitting objects and variables simultaneously in smoothed nonlinear biplots by least squares multidimensional scaling.* Paper presented at the 51st I.S.I. Session, Istanbul.

Guttman, L. (1941). The quantification of a class of attributes: a theory and method of scale construction. In P. Horst et al.(Eds.), *The prediction of personal adjustment* (pp.319-348). New York: Social Science Research Council.

Hayashi, C. (1952). On the prediction of phenomena from qualitative data and the quantification of qualitative data from the mathematicostatistical point of view. *Annals of the Institute of Statistical Mathematics, 2,* 93-96.

Hastie, T., Tibshirani, R. & Buja, A. (1994). Flexible discriminant analysis. *Journal of the American Statistical Association, 89,* 1255-1270.

Heiser, W.J. (1993). Clustering in low-dimensional space. In: O. Opitz, B. Lausen, & R. Klar (Eds.), *Information and classification: Concepts, methods and applications* (pp.145-155). Berlin: Springer-Verlag.

Heiser, W.J. (1997). Fitting graphs and trees. In: C. Hayashi, N. Ohsumi & Y. Baba (Eds.), *Data Science, Classification, and related Methods.* Tokyo: Springer Verlag, (in press).

Heiser, W.J., & Groenen, P.J.F. (1997). Cluster differences scaling with a within-clusters loss component and a fuzzy successive approximation strategy to avoid local minima. *Psychometrika, 62*, 63-83.

Heiser, W.J., & Meulman, J.J., (1994). Homogeneity analysis: exploring the distribution of variables and their nonlinear relationships. In: M. Greenacre, & J. Blasius (Eds.), *Correspondence Analysis in the Social Sciences: Recent Developments and Applications.* New York: Academic Press, 179-209.

Heiser, W.J. & Meulman, J.J., (1995). Nonlinear methods for the analysis of homogeneity and heterogeneity. In: W.J. Krzanowski (Ed.), *Recent Advances in Descriptive Multivariate Analysis.* Oxford: Oxford University Press, 51-89.

Heiser, W.J., & Meulman, J.J. (1997). Representation of binary multivariate data by graph models using the Hamming distance. *Computing Science and Statistics, 29*, 517-525.

Heiser, W.J., & Meulman, J.J. (1998). A clustering method for a distribution of points on a lattice. In: A. Rizzi and M. Vichi (Eds.), *Book of Short Papers of IFCS98, Rome*, (in press).

Horst, P. (1935). Measuring complex attitudes. *Journal of Social Psychology, 6,* 369-374.

Hubert, L.J., & Arabie, P. (1994). The analysis of proximity matrices through sums of matrices having (anti-)Robinson forms. *British Journal of Mathematical and Statistical Psychology, 47*, 1-40.

Hubert, L.J., Arabie, P., & Meulman, J.J. (1997). Linear and circular unidimensional scaling for symmetric proximity matrices. *British Journal of Mathematical and Statistical Psychology, 50,* 253-284.

Kruskal, J.B. (1964). Multidimensional scaling by optimizing goodness of fit to a nonmetric hypothesis. *Psychometrika, 29,* 1-28.

Kruskal, J.B. (1965). Analysis of factorial experiments by estimating monotone transformations of the data. *Journal of the Royal Statistical Society Series B, 27,* 251-263.

Krzanowski W.J. & Marriott, F.H.C. (1994). *Multivariate Analysis, Part I, Distributions, Ordination and Inference,* Edward Arnold, London, 1994.

Meulman, J.J. (1992). The integration of multidimensional scaling and multivariate analysis with optimal transformations of the variables. *Psychometrika, 57,* 539-565.

Meulman, J.J. (1996). Fitting a distance model to homogeneous subsets of variables: Points of view analysis of categorical data. *Journal of Classification, 13,* 249-266.

Meulman, J.J., (1997). A distance-based biplot for multidimensional scaling of multivariate data. In: C. Hayashi, N. Ohsumi & Y. Baba (Eds.), *Data Science, Classification, and related Methods.* Tokyo: Springer Verlag, (in press).

Meulman, J.J., & Heiser, W.J. (1993). Nonlinear biplots for nonlinear mappings. In: O. Opitz, B. Lausen, & R. Klar (Eds), *Information and Classification: Concepts, Methods and Applications* (pp.201-213). Berlin: Springer Verlag.

Meulman, J.J. & Heiser, W.J., (1997). Graphical display of interaction in multiway contingency tables by use of homogeneity analysis. In: M. Greenacre, & J. Blasius (Eds.), *Visual Display of Categorical Data.* New York: Academic Press, (in press).

Meulman, J.J., Hubert, L.J., & Arabie, P. (1997). Distance Analysis of Multivariate Ordinal and Nominal Data: Application to One-dimensional Scaling. Paper Presented at the Annual Meeting of the Psychometric Society, Gatlinburg TN, USA.

Meulman, J.J., & Verboon, P. (1993). Points of view analysis revisited: fitting multidimensional structures to optimal distance components with cluster restrictions on the variables. *Psychometrika, 58,* 7-35.

Nishisato, S. (1980). *Analysis of categorical data: Dual scaling and its applications.* Toronto: University of Toronto Press.

Nishisato, S. (1994). *Elements of Dual Scaling: An Introduction to Practical Data Analysis.* Hillsdale, NJ: Lawrence Erlbaum.

Ramsay, J.O. (1988). Monotone regression splines in action. *Statistical Science, 3,* 425-441.

Roskam, E.E.C.I. (1968). *Metric analysis of ordinal data in psychology.* Voorschoten: VAM.

Shepard, R.N. (1962). The analysis of proximities: Multidimensional scaling with an unknown distance function I. *Psychometrika, 27,* 125-140. II. *Psychometrika, 27,* 219-246.

Shepard, R.N. (1966). Metric structures in ordinal data. *Journal of Mathematical Psychology, 3,* 287-315.

Tucker, L.R (1960). Intra-individual and inter-individual multidimensionality. In: H. Gulliksen & S. Messick (Eds.), *Psychological Scaling: Theory and Applications.* New York: Wiley.

Whittaker, J. (1990). *Graphical Models in Applied Multivariate Statistics.* New York: Wiley.

Young, F.W. (1981). Quantitative analysis of qualitative data. *Psychometrika, 46,* 357-387.

Young, F.W., De Leeuw, J., & Takane, Y. (1976). Regression with qualitative and quantitative variables: An alternating least squares method with optimal scaling features. *Psychometrika, 41,* 505-528.

Young, F.W., Takane, Y., & De Leeuw, J. (1978). The principal components of mixed measurement level multivariate data: An alternating least squares method with optimal scaling features. *Psychometrika, 43,* 279-281.

A Note on Identification and Similarity Models

Hisao Miyano
Department of Cognitive and Information Sciences,
Chiba University, Inage-ku, Chiba, Japan.
e-mail: miyano@cogsci.L.chiba-u.ac.jp

Abstract: New identification and similarity models are derived from a unified viewpoint based on maximum information entropy principle. Relationships among new models and some existing models, especially Ashby's general recognition theory (GRT) model, are also clarified.

Keywords: Information Entropy, Choice Model, Multidimensional Scaling.

1. Introduction

Models for stimulus identification data have been studied and developed as one of the most important and attractive problems in psychology, especially in the fields of mathematical psychology and psychometrics. This is perhaps because stimulus identification is closely related to other basic psychological concepts such as similarity, categorization, preference, and signal detection.

There are at least two types of identification models incorporating multidimensional representation; the MDS choice model (Nosofsky, 1991; Takane & Shibayama, 1992) and the GRT (general recognition theory) identification model (Ashby & Perrin, 1988). I, recently, developed an identification model called the MIP (maximum information principle) identification model (Miyano, 1997), and discussed two issues on these two models. One was on the assumption of choice rule employed in the MDS choice model. The other was on the relationship between the MDS choice model and the GRT identification model. As a result, it was shown that the MIP model could derive the MDS choice model without assuming the Luce's choice rule, and clarified a simple relation between the MDS choice and GRT identification models.

As noted above, the concept of identification is closely related to similarity, since it is often assumed that the greater the similarity between a pair of stimuli, the more likely one will be confused with the other (e.g., Luce, 1963; Tversky & Gati, 1982). Although this does not mean that similarity (confusability) is the only factor affecting confusability (similarity), it seems reasonable to derive similarity model from identification model , as shown by

Ashby & Perrin (1988).

In this paper, we first give a brief summary of the MIP identification model, and then discuss on a problem how it can be used to explain the perceived similarity.

2. Identification Model

Consider the case of a complete stimulus identification task, where subjects are asked to identify given stimuli. Let $S_j, j = 1,2,\ldots,m$ be a stimulus, R_j be a unique response corresponding to S_j, and $P(R_i \mid S_j)$ be the conditional probability with which stimulus S_j is identified as stimulus S_i. According to the MDS choice model, $P(R_i \mid S_j)$ is given by

$$P(R_i \mid S_j) = \frac{w_i \exp(-d_{ji}^a)}{\sum_{k=1}^m w_k \exp(-d_{jk}^a)}, \tag{1}$$

where d_{ji} is the distance between S_j and S_i in some multidimensional psychological space, w_i is the bias associated with R_i, and a is a positive parameter. The MDS choice model is a multidimensional representation of the so-called similarity choice model based on the Luce's choice rule.

In the GRT identification model, $P(R_i \mid S_j)$ is given by

$$P(R_i \mid S_j) = \int_{r_i} f_j(\mathbf{x}) d\mathbf{x}, \tag{2}$$

where f_j is the probability density function of S_i being represented as $\mathbf{x}$ in r-dimensional space $U = R^r$, and r_i is the spatial region in U associated with response R_j. The GRT model is a multidimensional extension of Thurstone model, and outperforms the MDS choice model, while the latter shows considerably good performance for some identification data (Nosofsky, 1991).

As shown in this section, the MIP identification model gives a simple explanation of the relation between the MDS choice and GRT models. The MIP model postulates that stimulus identification process is a two stage process; the first stage is the transduction process which transduces external information into mental representation, and the second is the decision process which determines the response to the mental representation.

The transduction process is modeled under the following two assumptions.

(A-1). The percepts of any single presentation of stimulus can be represented as a point $\mathbf{x}$ in r-dimensional space $U = R^r$.

(A-2). The probability of stimulus S_j being represented as $\mathbf{x}$ in U is determined by a penalty function $\theta_j(\mathbf{x} - \mathbf{x}_j)$ and the maximum information entropy principle subject to the constraint,

$$\int_U p(\mathbf{x} \mid S_j)\theta(\mathbf{x} - \mathbf{x}_j)d\mathbf{x} = k_j, \tag{3}$$

where k_j is a constant inherent to stimulus S_j. Penalty function $\theta_j(\mathbf{x})$ is a non-negative convex function which satisfies (1) $\theta_j(\mathbf{x}) = 0$ iff $\mathbf{x} = \mathbf{0}$, and (2) $\theta_j(\mathbf{x}) = \theta_j(-\mathbf{x})$.

Based on these assumptions, we can derive the probability of S_j being represented as $\mathbf{x}$ in U. Let $p(\mathbf{x} \mid S_j)$ be the probability density of $\mathbf{x}$ when S_j is presented. Then, our problem is to find $p(\mathbf{x} \mid S_j)$ which maximizes information entropy ψ,

$$\psi = -\int_U p(\mathbf{x} \mid S_j)\log p(\mathbf{x} \mid S_j)d\mathbf{x}, \tag{4}$$

subject to the constraint (3).

This problem can be solved easily by using a Lagrange method; that is, we obtain

$$p(\mathbf{x} \mid S_j) = \frac{\exp(-\beta_j\theta_j(\mathbf{x} - \mathbf{x}_j))}{\int_U \exp(-\beta_j\theta_j(\mathbf{x} - \mathbf{x}_j))d\mathbf{x}}, \tag{5}$$

where β_j is a Lagrange parameter determined by (3).

Likewise, the decision process can be modeled by using the maximum information entropy principle. That is, the response probability $P(R_i \mid \mathbf{x})$ can be derived by maximizing the entropy,

$$\psi_r = -\sum_{i=1}^{m} P(R_i \mid \mathbf{x}) \log P(R_i \mid \mathbf{x}), \tag{6}$$

under the constraint

$$\sum_{i=1}^{m} P(R_i \mid \mathbf{x})\{\beta_i\theta_i(\mathbf{x} - \mathbf{x}_i) - \lambda \log b_i\} = \text{constant}, \tag{7}$$

where b_i is the response bias for R_i and λ is a weighting parameter to adjust the effects of the biases; that is, if we assume β is a Lagrange parameter corresponding to (7), then we obtain

$$P(R_i \mid \mathbf{x}) = \frac{b_i^{\beta\lambda}\exp(-\beta\tilde{\theta}_i(\mathbf{x} - \mathbf{x}_i))}{\sum_{k=1}^{m} b_k^{\beta\lambda}\exp(-\beta\tilde{\theta}_k(\mathbf{x} - \mathbf{x}_k))}, \tag{8}$$

where $\tilde{\theta}_i(\mathbf{x} - \mathbf{x}_i) = \beta_i\theta_i(\mathbf{x} - \mathbf{x}_i)$.

The constraint (7), successfully introduced by Buhmann *et al.* (1993) in their clustering algorithm, controls the randomness of the response to $\mathbf{x}$. From (5) and (8), the probability of R_i when S_j is presented is given by

$$P(R_i \mid S_j) = \frac{1}{c_j}\int_U \exp(-\beta_j\theta_j(\mathbf{x} - \mathbf{x}_j))\frac{b_i^{\beta\lambda}\exp(-\beta\tilde{\theta}_i(\mathbf{x} - \mathbf{x}_i))}{\sum_{k=1}^{m} b_k^{\beta\lambda}\exp(-\beta\tilde{\theta}_k(\mathbf{x} - \mathbf{x}_k))}d\mathbf{x}, \tag{9}$$

where c_j is a normalization term defined by

$$c_j = \int_U \exp(-\beta_j\theta_j(\mathbf{x} - \mathbf{x}_j))d\mathbf{x}. \tag{10}$$

It is easy to see that at a very large β_j, $p(\mathbf{x} \mid S_j) = \delta(\mathbf{x} - \mathbf{x}_j)$, and at a very large β, the response R_i becomes deterministic; $P(R_i \mid \mathbf{x}) = 1$ if for any $k \neq i$, $\tilde{\theta}_i(\mathbf{x} - \mathbf{x}_i) - \lambda \log b_i < \tilde{\theta}_k(\mathbf{x} - \mathbf{x}_k) - \lambda \log b_k$, and $P(R_i \mid \mathbf{x}) = 0$, otherwise. The latter property on $P(R_i \mid \mathbf{x})$ gives the same decision bound as that of the GRT choice model defined by the likelihood equation $p(\mathbf{x} \mid S_j)/p(\mathbf{x} \mid S_k) = 1$. In fact, if we assume that $\beta_1 = \beta_2 = \mathrm{K} = \beta_0$ and $b_j = c_j^{1/\lambda\beta_0}$, then the decision bound defined by $\tilde{\theta}_i(\mathbf{x} - \mathbf{x}_i) - \lambda \log b_i < \tilde{\theta}_k(\mathbf{x} - \mathbf{x}_k) - \lambda \log b_k$ coincides with that of the GRT choice model.

Hence, we can assert the following properties.

(1) If β is very large, then the MIP identification model is equivalent to the GRT identification model.

(2) If β_j is very large, then the MIP identification model is equivalent to the MDS identification model.

(3) Shepard-Luce choice rule and exponential similarity function are not indispensable assumptions to the MDS choice model.

3. Similarity Model

As assumed by Ashby *et.al.* (1988), we employ an assumption that perceived similarity is proportional to the probability of a confusion. This implies that the perceived similarity $s(S_a, S_b)$ of S_a to S_b is defined by

$$s(S_a, S_b) = \frac{1}{c_a} \int_U \exp(-\beta_a \theta_a(\mathbf{x} - \mathbf{x}_a)) u(\mathbf{x}; \mathbf{x}_a, \mathbf{x}_b, \beta) d\mathbf{x}, \tag{11}$$

where

$$u(\mathbf{x}; \mathbf{x}_a, \mathbf{x}_b, \beta) = \frac{b_b^{\beta\lambda} \exp(-\beta\theta_b(\mathbf{x} - \mathbf{x}_b))}{b_a^{\beta\lambda} \exp(-\beta\theta_a(\mathbf{x} - \mathbf{x}_a)) + b_b^{\beta\lambda} \exp(-\beta\theta_b(\mathbf{x} - \mathbf{x}_b))}. \tag{12}$$

This model corresponds to the similarity model proposed by Ashy *et.al.*, an extended version of the GRT similarity model (see Appendix A of Ashby & Perrin, 1988). Suppose, for convenience, that the S_b perceptual mean x_b is greater than the S_a mean x_a, then their model for the unidimensional case is given by

$$s(S_a, S_b) = \int_{-\infty}^{\infty} g(x - x_c) f_a(x) dx, \tag{13}$$

where $f_a(x)$ (f_b) is the probability distribution of the perceptual effects elicited by S_a (S_b), x_c is a point at which the likelihood f_a / f_b is 1.0, and g is a smooth continuous function such as a normal cumulative distribution with mean x_c and variance σ^2. Note that if the S_b mean is smaller than the S_a mean, then g must be replaced in (13) by $1 - g$.

For comparison, let us consider the unidimensional case of (11), and assume that the penalty functions are quadratic; that is, $\theta_a = (x - x_a)^2$, and $\theta_b = (x - x_b)^2$. Then, $u(x;x_a,x_b,\beta)$ is given by

$$u(x;x_a,x_b,\beta) = \frac{1}{1 + k_{ab}\exp(-\beta d_{ab}(x - x_0))}, \quad (14)$$

where $k_{ab} = (b_a / b_b)^{\beta\lambda}$, $d_{ab} = 2(x_a - x_b)$, and $x_0 = (x_a + x_b)/2$. From this result, we may expect that the MIP similarity model has the same properties as the extended GRT similarity model, since the shape of logistic function is very similar to the normal distribution.

Hence, we can assert that our similarity model is a natural realization of the above GRT similarity model, and β acts like the unit of measurement which will become small when subjects judge the similarities among stimuli never being confused.

4. Conclusions

In this paper, we have shown that the MDS choice model can be derived without assuming the Luce's choice rule and exponential similarity function, and that there exists a close relationship between the MDS choice model and the GRT identification model. It also has been demonstrated that our MIP identification model gives a natural extension of the GRT similarity model.

While the general recognition theory is well known as a very powerful and general theory of recognition, we believe that our approach based on the maximum information entropy at least will contribute to the further development of that theory.

References

Ashby, F. G., & Perrin, N. A. (1988). Toward a unified theory of similarity and recognition, *Psychological Review*, 95, 124-150.

Buhmann, J., & Kuhnel, H. (1993). Complexity optimized data clustering by competitive neural networks, *Neural Computation*, 5, 75-88.

Haken, H. (1988). *Information and Self-Organization*, Springer, New York.

Luce, R. D. (1963). Detection and recognition, in: *Handbook of Mathematical Psychology*, R. D. Luce, R. R. Bush, & E. Galanter (Eds.), Wiley.

Miyano, H. (1997). Identification model based on maximum information principle, submitted.*

Nosofsky, R. M. (1991). Stimulus, bias, asymmetric similarity, and classification, *Cognitive Psychology*, 23, 94-140.

Takane, Y., & Shibayama, T. (1992). Structures in stimulus identification data, in: *Multidimensional Models of Perception and Cognition*, F. G. Ashby (Ed.), Lawrence Erlbaum.

Tversky, A., & Gati, I. (1982). Similarity, separability, and the triangle inequality, *Psychological Review*, 89, 123-154.

Generalized Distance Measures for Asymmetric Multivariate Distributions

Marco Riani and Sergio Zani
Istituto di Statistica, Facoltà di Economia, Univ. of Parma, Italy

Abstract:[1] In this paper we suggest a non parametric generalization of the Mahalanobis distance which enables to take into account the differing spread of the data in the different directions. The output is an easy to handle metric which can be conveniently used both in an exploratory stage of the analysis for the detection of multivariate outliers and successively as a tool for non parametric discriminant analysis, multidimensional scaling and cluster analysis. In addition, the use of this metric can provide information about multivariate transformations and multiple outliers.

Keywords: Mahalanobis distance, equidistance contours, convex hull peeling, B-spline smoothing.

1 Introduction

The distance concept plays an important role in many topics of multivariate analysis. One of the most widely used distance measures is the one of Mahalanobis (d_M) (e.g. Dasgupta, 1993). This distance is appropriate for use in sample spaces where there exist differential variances and correlations between variables. In this metric, contours of equi-distance in the p-dimensional space are p-dimensional hyperellipsoids. Consequently, when we use d_M we implicitly assume that the spread of the data in the different directions is symmetric. Therefore, in presence of highly asymmetric data the use of this distance seems questionable. More generally, in a preliminary stage of the analysis it seems preferable to use a metric which does not assume an underlying distribution. Finally, even if the ellipticity hypothesis is satisfied, in order to construct d_M we still have to face the problem of estimating the means of the variables and the covariance matrix. The breakdown point of d_M is 0 and the presence of multiple outliers can cause masking and swamping problems (e.g. Barnett and Lewis, 1994).

The purpose of this paper is to suggest a simple generalization of d_M which: (a) is non parametric, (b) is robust to the presence of atypical observations, (c) keeps into account the differing spread of the data in the different directions, (d) reduces to the usual d_M when the ellipticity hypothesis is satisfied and the data are not contaminated.

[1]The program for computing the non parametric distance suggested in this paper is available upon request. We can be contacted at `statdue@ipruniv.cce.unipr.it` or `zani@ipruniv.cce.unipr.it`. The authors wish to thank Aldo Corbellini for programming help.

2 Description of the method

First let us consider the case in which the number of variables (p) is equal to 2. Our approach starts defining a non parametric bivariate central region and a robust centroid. As pointed out by Riani *et al.* (1998), a natural and completely non parametric way for defining a central region in $\mathbb{R}^2$ is through the use of the so called convex hull peeling. Successive convex hulls are peeled until the first one is obtained which includes not more than 50% of the data (and so asymptotically half of the data). This 50%-hull is smoothed using a B-spline, constructed from cubic polynomial pieces, which uses the vertices of the 50%-hull to provide information about the location of the knots. (From now on this spline will be called 50%-spline). Zani *et al.* (1998) discuss several choices of a robust centroid. In this work we use the intersection of the two least squares lines built with the observations which lie inside or at the boundary of the 50%-spline. As an illustration of the suggested approach let us consider the data referred to 160 European regions reported in Figure 1. On the x-axis we have the Index Numbers of GNP per inhabitant,

Figure 1: Bivariate boxplot of Unemployment rate in % versus Index Numbers of GNP per inhabitant for 160 European regions (50% and 99% contours).

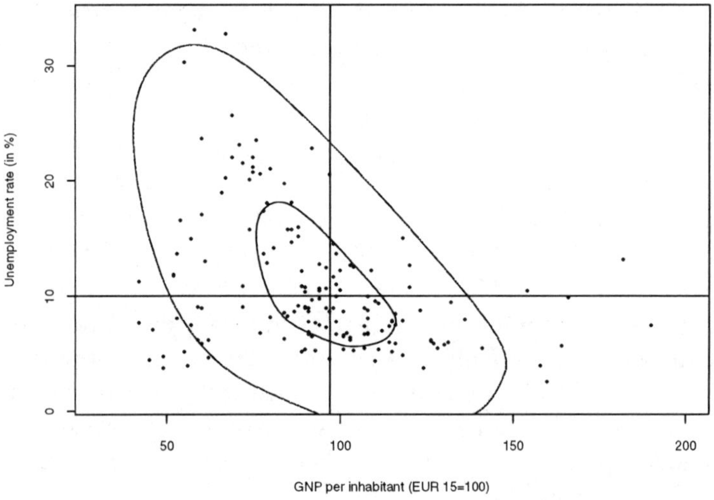

PPS (EUR 15=100). On the y-axis we have the unemployment rate in % (Source: Eurostat, REGIO, 1996). Figure 1 also reports the 50%-spline. As emerges clearly from the plot, the spread of the data in the differing directions is different, therefore the traditional approach based on d_M does not seem to be appropriate. Two straight lines in correspondence of the bivariate centroid have also been drawn. Zani *et al.*, (1998) show that in order to obtain an outer contour which under the assumption of bivariate normality leaves outside a percentage of observations close to 1%, we must multiply the distance from the

50%-spline to the robust center by 1.68. Using this coefficient we obtained the outermost contour reported in Figure 1. The units which lie outside the bivariate contour can be interpreted as atypical. The 50%-spline and the outermost contour can be interpreted as non parametric equidistance contours from the center. In this way we can take into account non parametrically the differing spread of the data. This leads us to define a new metric. Let us consider separately the distance of one point from the centroid and the distance between two generic points.

Distance from the centroid in $\mathbb{R}^2$

In our metric the observations which lie on the 50%-spline have the same distance from the centroid. In order to define a measure unit, without loss of generality we can set equal to 1 the distance from the center of a point which lies on the 50%-spline. The distance of every other observation can be based on the former unit of measure. For example, in Figure 2 let us consider the

Figure 2: Example of computation of the distance from the robust centroid for 4 of the European regions reported in Figure 1 (Standardized variables).

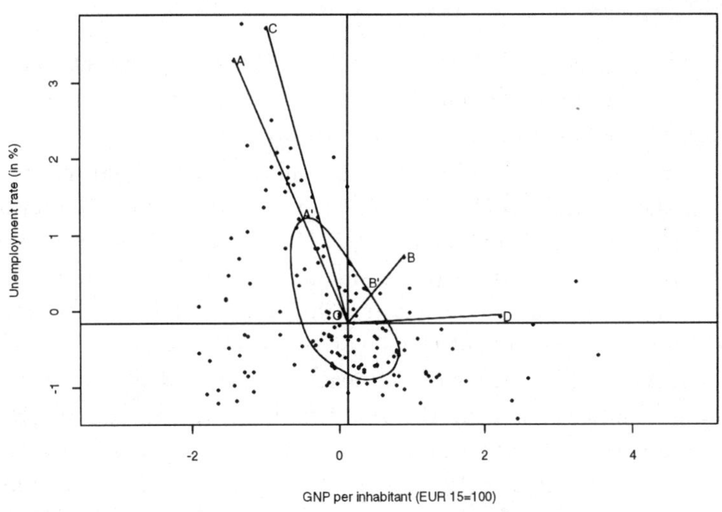

straight line $\overline{OA}$ passing through the robust centroid and point A. Let point A' be the intersection of segment $\overline{OA}$ with the 50%-spline. The distance $\overline{OA}$ in this case is 2.55 times the distance $\overline{OA'}$. Therefore in our metric point A lies at a distance 2.55 from the origin. In general: the distance from the centroid of every point K depends on the ratio between $\overline{OK}$ and $\overline{OH}$ where H is the intersection of the straight line passing through $\overline{OK}$ with the 50%-spline. In Figure 2 units A e B respectively correspond to the regions Extremadura (E) and Uusimaa (FIN). Note that using the standardized Euclidean distance point B would be much closer to the center than point A. Using our metric the ratio $\overline{OB}/\overline{OB'}$ is 2.50. This implies that if we take into account the differing spread of the data in the different directions these two regions have

Figure 3: Plot of the points of Figure 2 in the transformed space using the suggested metric (The two superimposed circumferences define equidistant contours from the robust centroid).

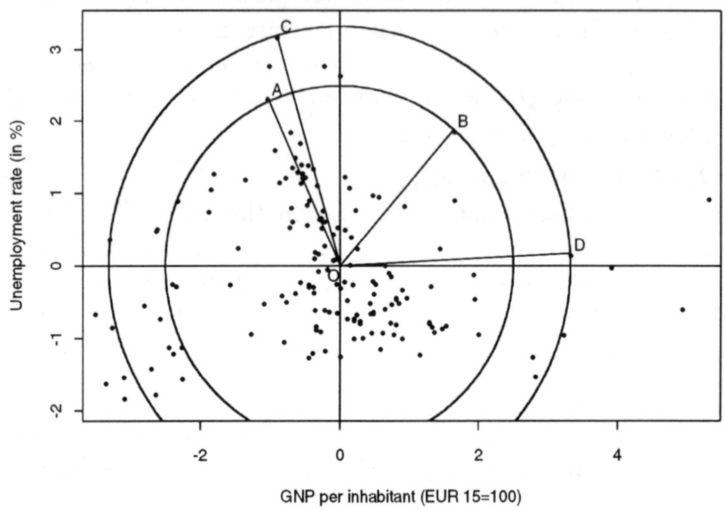

approximately the same distance. Following similar arguments we can claim that point C (Ceuta y Melilla, E) has a distance from the centroid (3.32) which is exactly equal to that of point D (Bremen, D).

In Table 1 we compare for units A,B, C and D our non parametric distance (d_{RZ}) from the centroid with the Euclidean one (d) (using standardized variables) and the one of Mahalanobis (d_M). This table shows that for units A and C the values of d_{RZ} and d_M are smaller than those produced by the Euclidean distance. The opposite happens for units B and D. In fact d_{RZ} and d_M correctly keep into account the negative correlation ($r = -0.359$) between the two variables. The values of our metric compared to d_M are smaller for units A and C and greater for the two remaining units. Our metric is based on non parametric contours and enables to reduce (increase) the distance for the units located in the directions where the spread of the data is high (low). The assumption of d_M of symmetric spread usually is not met in practise. For example, in the data reported in Figure 1 it is clear that the spread of the data in the South-West direction is much more evident than that in the North-East direction. The ellipses (here not reported for lack of space) which represent the equidistance contours in the Mahalanobis metric, treat in the same way the units located South-West with those located North-East. This explains why for example, the value of d_{RZ} for unit D is much bigger than that reported by d_M.

Remark 1: If the underlying hypothesis of elliptic distribution is true the 50%-spline tends to become an ellipse (Atkinson and Riani, 1997). In addition, convex hull peeling is invariant under linear transformations of the data.

Table 1: Squared distances from the centroid of four regions using different metrics.

	d^2_{RZ}	d^2_M	d^2
A: Extremadura	6.528	11.070	13.041
B: Uusimaa	6.275	2.064	1.331
C: Ceuta y Melilla	11.050	14.139	14.961
D: Bremen	11.038	5.573	4.935

Therefore the suggested metric simply reduces to the d_M when the ellipticity hypothesis is satisfied and the data are not contaminated.

Remark 2: In order to calculate the former distances, in our metric we simply need an estimate of the centroid but we do not have to estimate either dispersion or correlation parameters.

Remark 3: It is possible to use a different spline contour (for example a 60%-spline or 75%-spline). With this choice the shape of the inner region may adapt more to the spread of the units lying far from the centroid. However this results in a decrease of robustness.

Figure 3 shows the 160 European regions of Figure 1 after eliminating non parametrically the differing spred of the data in the various directions. In this new space both unit C and D lie on the same circumference centered on the robust centroid whose radius is 3.32. Similarly, units A and B lie approximately on the circumference whose radius is 2.5.

Distance between two points in $\mathbb{R}^2$

In our metric the splines which define the equidistant contours are transformed into circles. This implies, for example, that the units which have a distance equal to 3.32 from the robust center after the transformation lie on a circle with radius 3.32. In the transformed space (Figure 3) Euclidean distances referring to different directions are directly comparable because we have removed (non parametrically) the different spread of the data in the various directions. Therefore our metric is equal to the Euclidean distance between two points in the transformed space.

Extensions to p-dimensional data

In presence of p-dimensional data the situation becomes more complicated because we cannot rely on the graphical representation anymore. With the purpose of defining a distance measure which, with well behaved data, reduces to the d_M we have to consider pairs of variables such that the sum of the marginal bivariate d_M is equal to the overall d_M using the whole set of p-variables. It is possible to show that the p-variate d_M can be expressed as the sum of $p/2$ bivariate d_M if p is even and as the sum of $[p/2]$ bivariate d_M plus a univariate d_M if p is odd. For example if $p = 3$, (considering for simplicity the variables in standardized form) using the matrix inversion lemma it is possible to prove that the squared Mahalanobis distance of the trivariate vector $\mathbf{x} = (x, y, z)'$ from $\mathbf{0}$ can be decomposed as:

$$\mathbf{x}'\mathbf{\Sigma}^{-1}\mathbf{x} = \frac{x^2 + y^2 - 2\rho_{xy}xy}{1-\rho_{xy}^2} + \frac{\left[z\sqrt{1-\rho_{xy}^2} - x\rho_{xz.y}\sqrt{1-\rho_{yz}^2} - y\rho_{yz.x}\sqrt{1-\rho_{xz}^2}\right]^2}{|\mathbf{\Sigma}|} \tag{1}$$

where $\mathbf{\Sigma}$ is the correlation matrix and $\rho_{jl.k}$ denotes the partial correlation coefficient between variables j and l given variable k.

The first term on the right hand side of equation (1) is nothing but the squared marginal d_M between the first two variables. The second term can be shown to be the squared d_M from 0 of the univariate variable $z|(x, y)$:

$$z|(x,y) = z - \frac{[(\rho_{xz} - \rho_{xy}\rho_{yz})x + (\rho_{yz} - \rho_{xy}\rho_{xz})y]}{1-\rho_{xy}^2} \tag{2}$$

In the case of a univariate random variable the central region is given by the interquartile range and the robust center is the median. Similarly to the bivariate case, we can set equal to 1 the distance between the first (third) quartile and the median and we can use this length for defining the other distances. With 3 variables, therefore, we initially have to compute our distance using the first two variables, then we have to consider the transformed variable $z|(x, y)$. If the ellipticity hypothesis is satisfied and the data are not contaminated the simple sum of these two distances reduces to the global d_M. With elliptic distributions it is immaterial the order in which the variables are considered. Finally, in presence of asymmetric data we end up with a metric which adapts non parametrically to the various spread of the data in the different directions.

References

Atkinson, A. C., and Riani, M. (1997), Bivariate Boxplots, Multiple Outliers, Multivariate Transformations and Discriminant Analysis: the 1997 Hunter Lecture, *Environmetrics*, 8, 583-602.

Barnett, V. and Lewis, T. (1994), *Outliers in Statistical Data*, Wiley, N. Y.

Dasgupta, S. (1993), "The Evolution of the D^2-Statistic of Mahalanobis", *Sankhyā*, special volume dedicated to the memory of P.C. Mahalanobis, 442-459.

Riani, M., Zani, S. and Corbellini, A. (1998), Robust Bivariate Boxplots and Visualization of Multivariate Data, in: I. Balderjahn, R. Mathar and M. Schader (Eds.), *Classification, Data Analysis and Data Highways*, Springer Verlag, Berlin, 93-100.

Zani, S., Riani, M. and Corbellini, A. (1998), Robust Bivariate Boxplots and Multiple Outlier Detection, forthcoming in *Computational Statistics and Data Analysis.*

Obtaining Disparities as Slopes of Greatest Minorant Convex

Teresa Rivas Moya
Facultad de Psicología, Campus de Teatinos s.n., 29071 Málaga (España)
Universidad de Málaga
e-mail:moya@uma.es

Abstract: This paper obtains the disparities using concepts of isotonic regression applied to the Euclidean distance and to nonmetric multidimensional scaling model. The disparities are shown given to be by the slopes of the Greatest Minorant Convex (GMC), and are calculated by example. This justifies their inclusion in the goodness of fit (stress), which measures the discrepancies between the initial and fitted configuration, and it is used for procedures (ALSCAL and MDS) which fit the model.

Keywords : Nonmetric multidimensional scaling, Isotonic regression, Disparities, Goodness of fit, Stress.

1. Disparities as slopes of segments of Greatest Minorant Convex

Given a set I of n objects and Δ a dissimilarity matrix, δ being defined on Ω. The order over Δ induces an order over Ω, and in turn over I, so that the pair "(i,j) is *less dissimilar* than (i',j')", and is shown as $(i,j) \leq (i',j')$ if, and only if,

$$\delta(i,j) \leq \delta(i',j') \ \forall \ (i,j) \in \Omega \tag{1}$$

Given the order "$\leq$" over Ω, an initial configuration is obtained. From it the nonmetric euclidean MDS model obtains the coordinates of n points $X_1, X_2, \ldots, X_n$ in a euclidean space of p reduced dimension, $p<n$, so that the distances between each pair of points of configuration, d_{ij}, are approximated as much as possible to the dissimilarity between the pairs of objects, δ_{ij}. That is, if the distances between points of configuration are a monotone transformation of dissimilarities between objects, then

$$if \ \delta(i,j) \leq \delta(i',j') \Rightarrow d_{ij} \leq d_{i'j'} \tag{2}$$

This function which changes dissimilarities into distances is monotone, non fixed before fitting the model. That is $d_{ij} = f(\delta_{ij})$ for all $(i,j) \in \Omega$, f being a priori an unknown monotone function. In practice, dissimilarities given on an ordinal scale cannot be perfectly transformed into distances by a monotone transformation, so that (2) is not usually satisfied. A measure of degree in which this relation is not reached will be defined from disparity function, $\hat{d}$, which must verify

$$if\ \delta(i,j) \leq \delta(i',j') \text{ then } \hat{d}_{ij} \leq \hat{d}_{i'j'}\ \forall (i,j) \in \Omega \qquad (3)$$

and differs as little as possible from the distances between points of configuration by method of Least Squares.
This paper shows that disparity function is given by the isotonic regression, $\hat{d}$, of a distance function d. The isotonic regression of distance function is defined from basic concepts of isotonic regression: Isotonic function on X, Class of Isotonic functions, Dual Cone, Cumulative Sum Diagram (CSD), Greatest Minorant Convex (GMC), etc. which were formulated and adapted to the problem of nonmetric scaling in Rivas (1989) from the work of Barlow, Bartholomew, Bremmer, Brunk (1972).
From these concepts, the two following theorems prove that a necessary and sufficient condition for $\hat{d}$ to be an isotonic regression of distance d is that $\hat{d}$ has to be the function which associates each $x \in X$ with the slope of segment whose endpoint is this point of GMC.

Theorem 1. Necessary Condition

If $\hat{d}$ is isotonic regression of d, then it is the slope of GMC. That is, it is the function which minimizes the length of all chords joining the origin to endpoint of CSD, and lying below the CSD.
The length of a segment of chord joining two consecutive points (W_{ij}, F_{ij}) and (W_{ij-1}, F_{ij-1}) in the euclidean space is given as

$$l_{ij} = \sqrt{(F_{ij} - F_{ij-1})^2 + (W_{ij} - W_{ij-1})^2} = \sqrt{(f_{ij}^2 + 1)w_{ij}^2}\ \forall (\mathrm{i},\mathrm{j}) \in \Omega \qquad (4)$$

The following problem must be solved

$$min\left\{ \sum_{(\mathrm{i},\mathrm{j})\in\Omega} (f_{ij}^2 + 1)^{1/2} w_{ij} \right\} \qquad (5)$$

in the class of functions satisfying: a) $F_{ij} \leq D_{ij}$ for all $(i,j) \in \Omega$, that is, the chord starts from P_{00} and passes through or below each point P_{ij} ; and b) $F_{in} = D_{n-1,n}$ the chord passes through $P_{n-1,n}$ of *CSD*
This is changed into an equivalent problem associated with the convex cone of a new set of functions $f' = d - f$ on X. Then, (4) must be minimized in the class of functions $\{f: X \rightarrow R;\ d - f \in \pi_0^*\}$, π_0^* being the dual cone of convex cone π_0 and the conditions a) and b) are substituted by a') and b'),

a') $\sum_{q=i}^{n-1} \sum_{h=j}^{n} f'_{qh}\, w_{qh} \leq 0 \quad (i,j) \in \Omega$

b') $\sum_{q=i}^{n-1} \sum_{h=j}^{n} f'_{qh}\, w_{qh} = 0$

This problem of minimization is solved for any convex function in the proposition given in Barlow et al. (1972, p.50). Minimization of (5) is obtained as a direct consequence of this proposition : d and w are given real functions on X, being $w_{ij} > 0$ for $(i,j) \in \Omega$. Let ϕ be a convex function and ϕ' a determination of its derivative, and π_0 being the class of isotonic functions with respect to the order induced by δ over Ω. Then $\hat{d}$ minimizes the length of chords lying below CSD, $\sum_{(i,j)\in\Omega} \left(f_{ij}^2 + 1\right)^{1/2} w_{ij}$, in the class of functions f so that $d - f \in \pi_0^*$.

Proof : $\sum_{(i,j)\in\Omega} \left(f_{ij}^2 + 1\right)^{1/2} w_{ij} = \sum_{(i,j)\in\Omega} \phi\left(f_{ij}\right) w_{ij}$ must be minimized.

Given that π_0 is the class of functions f such that $d - f \in \pi_0^*$, it must be shown that $\hat{d}$ satisfies

$$\sum_{(i,j)\in\Omega} \left(\hat{d}_{ij} - d_{ij}\right)\left\{\phi'\left(\hat{d}_{ij}\right)\right\} w_{ij} = 0 \qquad (6) \qquad\qquad \sum_{(i,j)\in\Omega} \left(f_{ij} - d_{ij}\right)\left\{\phi'\left(\hat{d}_{ij}\right)\right\} w_{ij} \geq 0 \qquad (7)$$

Given that any real function β verifies $\sum_{(i,j)\in\Omega} \left(d_{ij} - \hat{d}_{ij}\right) \beta\left(\hat{d}_{ij}\right) w_{ij} = 0$

(Barlow et al. 1972, p. 35), in particular substituting $\phi' = \beta$, then (6) is satisfied.

In addition (7) is satisfied. As $\hat{d}$ is isotonic and ϕ' is non decreasing, then $\phi'(\hat{d})$ belongs to a class of functions π_0. Given f, $d - f \in \pi_0^*$ and by definition of dual cone the following inequality is given

$$\sum_{(i,j)\in\Omega}(d-f)(x_i x_j)\phi'(\hat{d}(x_i x_j))w(x_i x_j) \leq 0$$

Therefore, $$\sum_{(i,j)\in\Omega}(f-d)(x_i x_j)\phi'(\hat{d}(x_i x_j))w(x_i x_j) \geq 0 \qquad (7)$$

Theorem 2. Sufficient condition

A solution representing the objects of I is given as (X,d). If $\hat{d}$ is the slope of GMC, then $\hat{d}$ is an isotonic regression of d. That is, any isotonic function on X gives

$$\sum_{(i,j)\in\Omega}\left(d_{ij}-f_{ij}\right)^2 w_{ij} \geq \sum_{(i,j)\in\Omega}\left(d_{ij}-\hat{d}_{ij}\right)^2 w_{ij}.$$

In addition, there is only one isotonic regression.

Proof : Adding and subtracting $\hat{d}_{ij}$ from $d_{ij} - f_{ij}$

$$\sum_{(i,j)\in\Omega}\left(d_{ij}-f_{ij}\right)^2 w_{ij} \geq \sum_{(i,j)\in\Omega}\left(d_{ij}-\hat{d}_{ij}\right)^2 w_{ij} + \sum_{(i,j)\in\Omega}\left(\hat{d}_{ij}-f_{ij}\right)^2 w_{ij}$$

The difference between the two terms is

$$2\sum_{(i,j)\in\Omega}\left(d_{ij}-\hat{d}_{ij}\right)\left(\hat{d}_{ij}-f_{ij}\right)w_{ij} \qquad (8)$$

By the partial summation of Abel´s formula (Apostol, 1979) and the properties of GMC, this term is greater or equal to 0 for any isotonic function on X (proved by Rivas (1989, pp. 174-175)). Therefore

$$2\sum_{(i,j)\in\Omega}\left(d_{ij}-\hat{d}_{ij}\right)\left(\hat{d}_{ij}-f_{ij}\right)w_{ij} = \sum_{i=1}^{n-2}\sum_{j=2}^{n-1}\left[\left(f_{ij}-f_{ij-1}\right)-\left(\hat{d}_{ij}-\hat{d}_{ij-1}\right)\right]\left(D_{ij-1}-\hat{D}_{ij-1}\right)$$
$$+\sum_{i=1}^{n-2}\left[\left(f_{i+1,i+2}-f_{in}\right)-\left(\hat{d}_{i+1,i+2}-d_{in}\right)\right]\left(D_{in}-\hat{D}_{in}\right)+\left(\hat{d}_{n-1n}-f_{n-1n}\right)\left(D_{n-1n}-\hat{D}_{n-1n}\right)$$

Considering $D_{00} = \hat{D}_{00} = f_{00} = \hat{d}_{00} = 0$, the last term of the above equality is 0 by property (1) of GMC, in that the last points of CSD and GMC are coincident. Thus $D_{n-1n} = \hat{D}_{n-1n}$ and, applying property (2), $\hat{d}_{ij} - \hat{d}_{ij-1} = 0$ and $\hat{d}_{i+1,i+2} - d_{in} = 0$, then

$$2 \sum_{(i,j)\in\Omega} \left(d_{ij} - \hat{d}_{ij}\right)\left(\hat{d}_{ij} - f_{ij}\right) w_{ij} = \sum_{i=1}^{n-2} \sum_{j=2}^{n-1} \left(f_{ij} - f_{ij-1}\right)\left(D_{ij-1} - \hat{D}_{ij-1}\right) + \sum_{i=1}^{n-2} \left(f_{i+1,i+2} - f_{in}\right)\left(D_{in} - \hat{D}_{in}\right)$$

Again applying property (2) $\hat{D}_{ij-1} < D_{ij-1}$ and $\hat{D}_{in} < D_{in}$, and f being isotonic, in that $f_{ij} - f_{ij-1} \geq 0$ and $f_{i+1i+2} - f_{in} \geq 0$, then (8) is greater or equal to 0.
Suppose that another isotonic regression d' on X exists. For any isotonic function f must be

$$\sum_{(i,j)\in\Omega} \left(d_{ij} - f_{ij}\right)^2 w_{ij} \geq \sum_{(i,j)\in\Omega} \left(d_{ij} - d'_{ij}\right)^2 w_{ij}$$

By the previous theorem, given the isotonic function $\hat{d}$,

$$\sum_{(i,j)\in\Omega} \left(d_{ij} - d'_{ij}\right)^2 w_{ij} \geq \sum_{(i,j)\in\Omega} \left(d_{ij} - \hat{d}_{ij}\right)^2 w_{ij} + \sum_{(i,j)\in\Omega} \left(\hat{d}_{ij} - d'_{ij}\right)^2 w_{ij}$$ must be verified.

It is right if

$$\sum_{(i,j)\in\Omega} \left(d_{ij} - \hat{d}_{ij}\right)\left(\hat{d}_{ij} - d'_{ij}\right) w_{ij} \geq 0 \Leftrightarrow \left(d_{ij} - \hat{d}_{ij}\right) \geq 0 \text{ and } \left(\hat{d}_{ij} - d'_{ij}\right) \geq 0$$

but $\hat{d}_{ij} > d'_{ij}$ contradicts $\hat{d}$ being an isotonic regression of d, then $\hat{d}_{ij} = d'_{ij}$.

Based on these results, the known goodness of fit measure of nonmetric MDS model is defined.
The disparities (images of disparity function) have been obtained by application of isotonic regression to distances between points of X configuration that represent the objects in a space of p dimension. From distance function d on X and the order '$\leq$' over Ω, the isotonic regression of d has given $\hat{d}: X \rightarrow R^+$ whose images satisfy

$$\hat{d}(x_i x_j) \leq \hat{d}(x_i, x_{j+1}) i: 1,\ldots,n-2;\ j: 2,\ldots,n-1; \quad \hat{d}(x_i x_n) \leq \hat{d}(x_{i+1} x_{i+2}) i: 1,\ldots,n-2$$

These values preserve a perfect monotone relation with dissimilarities and, simultaneously, the differences between disparities and distances are minimized.

If all the distances between points are different, that is, weight function w is identical to 1, a global deviation of isotony is given as

$$S(X,p,\Delta)=\sum_{(i,j)\in\Omega}\left(\hat{d}(x_i,x_j)-d(x_i,x_j)\right)^2$$

This measure is known in nonmetric scaling as *raw Stress* (Coxon, 1982; pp. 55-58), and shows the degree to which the fitted configuration represents the dissimilarities. But it is not invariant to changes of scale of configuration. To avoid this inconvenience, raw Stress is divided by a *scaling factor*. The most usual scaling factors give the measures of goodness of fit known as *Stress* S_1 and S_2.

2. Obtaining disparities by an example

Column 4 gives the fitted distances d_{ij} obtained from symmetric dissimilarites between pairs of 5 objects, (table 1, Column 1), (Coxon;1982, p.62). The order over pairs of objects (i,j) is induced by dissimilarities δ_{ij} (Column 2) . In Column 3, weight w_{ij} associated with each pair of objects is 1 (because each pair has a different dissimilarity). Applying the above concepts, $P_{ij}=(W_{ij},D_{ij})$ for all $(i,j)\in\Omega$ of CSD are obtained (Column 7).

Table 1: *Points of CSD and GMC and slopes of segments of GMC*

(i,j)	δ_{ij}	w_{ij}	d_{ij}	W_{ij}	D_{ij}	P_{ij}	$\hat{d}_{ij}$	$\hat{W}_{ij}$	$\hat{D}_{ij}$	$\hat{P}_{ij}$
(5,1)	1	1	3	1	3	(1,3)	3	1	3	(1,3)
(5,1)	2	1	6(*)	2	9	(2,9)	4.5(*)	2	7.5	(2,7.5)
(2,1)	3	1	3(*)	3	12	(3,12)	4.5(*)	3	12	(3,12)
(3,2)	4	1	5	4	17	(4,17)	5	4	17	(4,17)
(4,2)	5	1	8	5	25	(5,25)	8	5	25	(5,25)
(3,1)	6	1	10	6	35	(6,35)	10	6	35	(6,35)
(4,3)	7	1	13(**)	7	48	(7,48)	11(**)	7	46	(7,46)
(5,2)	8	1	11(**)	8	59	(8,59)	11(**)	8	57	(8,57)
(5,3)	9	1	9(**)	9	68	(9,68)	11(**)	9	68	(9,68)
(4,1)	10	1	15	10	83	(10,83)	15	10	83	(10,83)

Pairs of objects (5,1), (2,1), (4,3), (5,2), (5,3) given in Column 4 marked by (*) in Table 1, do not satisfy the monotone relation (see Figure 1). Thus, distances must be fitted by isotonic regression.

The remaining points of CSD satisfy $\hat{W}_{ij} = W_{ij}$ $\hat{D}_{ij} = D_{ij}$, and thus $\hat{P}_{ij} = P_{ij}$. The disparities and new coordinates of points on GMC $\hat{P}_{ij} = \left(\hat{W}_{ij}, \hat{D}_{ij}\right)$ are given in Column 11.

The CSD is given in Figure 1, plotting $P_{ij} = \left(W_{ij}, D_{ij}\right)$, for all $(i, j) \in \Omega$. It can be appreciated that the GMC corrects the lack of monotony on the points which do not satisfy it when they are placed on the CSD.

Figure 1. *Plot of CSD and GMC*

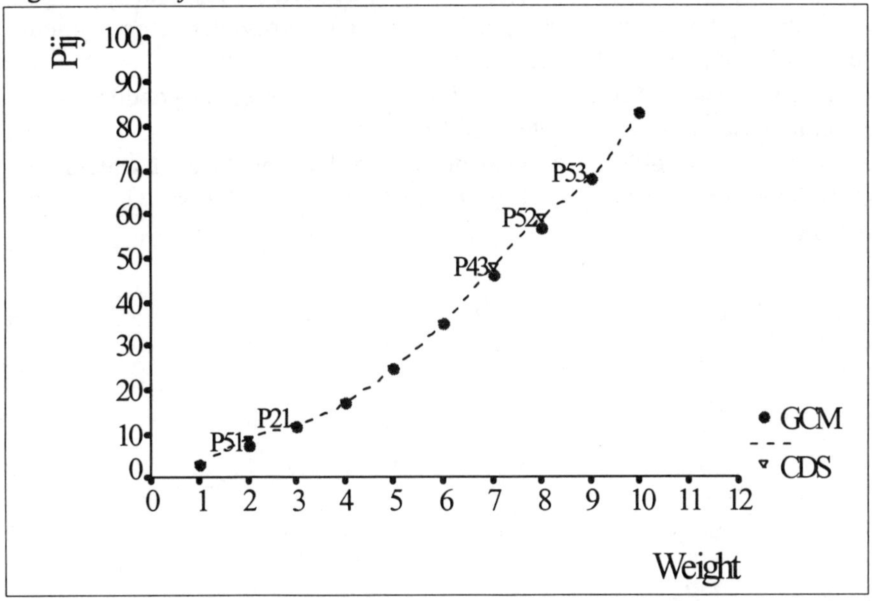

The disparities associated with distances which do not satisfy the monotone relation have been obtained. The slope of segments associated with distances 6, 3, 9,13 and 11 lead to the squared minima discrepancies $\left(d_{ij} - \hat{d}_{ij}\right)^2$ defining the goodness of fit. These measures being $S_1 = 0{,}122$ and $S_2 = 0{,}28$.

3. Conclusions

Contrary to other Statistical models where the theoretical model is developed before the algorithm to fit it, in Multidimensional Scaling the algorithms have arisen and then the papers studying their characteristics and properties. This paper gives statistical support to procedures based on isotonic regression that fit this model. It should serve as a basis for further study when a dissimilarity matrix contains equal or asymmetric data. It can also be extended to other models such as nonmetric individual differences scaling. The usefulness of disparity function to preserve the order of data, given on an ordinal scale, by a monotone function should be noted.

References

Apostol, T. M. (1979). *Cálculo con funciones de una variable.* Barna. Reverte

Barlow, R.E., Bartholomew, D.J., Bremmer, J.M. & Brunk,H.D. (1972). *Statistical Inference under order restrictions.* London: Wiley

Coxon, A.P.M.(1982). *The user's guide to multidimensional scaling.* London: Heinemann Educational Books

Kruskal, J.B. (1964a,). Mutidimensional scaling by optimizing goodness of fit to a nonmetric hipoteses. *Psychometrika*, 29, 1-27.

Rivas, T. (1989). Medida de bondad del ajuste del modelo de Escalamiento Multidimensional Euclídeo. Unpublished Ph. D. Thesis. Universidad Málaga.

Non-parametric regression models for the conjoint analysis of qualitative and quantitative data

Simone Borra - Agostino Di Ciaccio
Facoltà di Economia - Università di Urbino

Abstract: We considered some recent non-parametric regression models to jointly analyze qualitative and quantitative data. A comparison of the different methods was done from a theoretical points of view, using a unified approach. The capabilities of the methods to model a strongly non-linear simulated data-set was presented.

Key words: non-parametric regression, non-linear methods, analysis of mixed data, neural networks, projection pursuit regression, MARS, optimal scaling.

1. Introduction

The aim of this paper is to analyze some recent non-parametric non-linear methods devoted to regression analysis. We are particularly interested in analyzing the capabilities of these methods with respect to the conjoint analysis of qualitative and quantitative data from empirical and theoretical points of view. The methods considered can be used to analyze a large set of variables even with a moderate number of statistical units. No hypothesis will be made on the distribution of the variables and the analysis will be essentially explorative.

With a large number of variables we have the problem known as the "curse of dimensionality", especially if we want to analyze interactions among variables. To overcome this difficulty, all the methods use a set of low dimensional functions.

We will not consider local-methods such as Kernel methods, due to the difficulties in applying these methods to high dimensional data with a not very large number of units. All the methods analyzed, as will be shown later, can be considered to be a special case of the general expression:

$$Y = \alpha_0 + \sum_{m=1}^{M} \alpha_m \, \mathrm{B}_m(\mathbf{X}, \vartheta_m) + \varepsilon \tag{1}$$

where **X** is a vector of P explicative variables.

The methods differ with respect to the definition of the set $\{B_m\}$ of basis functions, this allows a unified view of the different approaches.

2. Transforming variables inside additive models

The first approach we consider uses the additive model:

$$Y = \alpha_0 + \sum_{i=1}^{P} \alpha_i f_i(X_i) + \varepsilon \tag{2}$$

where α_i (i=0,1...P) are parameters to estimate, P is the number of predictors X_i, f_i are non-specified transformation functions. The presence of the parameters α_i is due to normalization constraints on f_i. The expression (2) can be obtained from (1) setting M=P and using univariate transformation functions.
In the Optimal Scaling approach (Gifi 1990) with quantitative variables, the f_i can be spline transformations, meanwhile for qualitative variables we put $f_i(x_i)=\sum_j \delta_{ij}\gamma_j$ where δ_{ij} are the dummy variables associated to the categories, and γ_j are parameters to be estimated.
The analysis of ordinal qualitative data can be easily included in (2) using the dummy variables of order that allow the assignment of an increasing coefficient to the categories of the ordinal variable (see Di Ciaccio 1988).
Hastie & Tibshirani (1990) proposed a similar approach, but in a probabilistic framework, known as Generalized Additive Model (GAM). Setting $\eta = \mathrm{E}(Y / X_1 X_2 \dots X_P)$, GAM can be written as:

$$l(\eta) = \alpha_0 + \sum_{i=1}^{P} \alpha_i f_i(X_i) \tag{3}$$

where l is a given link function and the parameters α_i can be absorbed by the functions f_i if these are not constrained. These functions are estimated by maximum likelihood using a Backfitting algorithm, while the Optimal Scaling approach uses an alternate least squares algorithm.
The additive model approaches can include interactions among variables only by introducing explicitly interactive terms like $f_{1,2}(X_1, X_2)$, usually in the simple form $f_{1,2}(X_1 \cdot X_2)$. This can be done in a *hierarchical* approach, that allows the distinction between the contribution of main effects and interactions. Modelling interactions with these approaches is usually quite difficult.

3. Projection methods

Some techniques, rather than considering transformations of the original data, use transformations of optimal projections of data on low dimension subspaces (usually ≤ 2). The only best-known method is the Projection Pursuit Regression model (PPR) (Friedman & Stuetzle 1981), which, given one dependent variable Y, can be formulated as:

$$Y = \alpha + \sum_{m=1}^{M} \alpha_m f_m \left(\gamma_{0m} \sum_{i=1}^{P} \gamma_{im} X_i \right) + \varepsilon \tag{4}$$

where f_m are normalized one-dimensional smooth functions of the one dimensional projection $\gamma_m^T \mathbf{X}$ and α=E(Y | **X=x**). Obviously, when the number of explanatory variables P is high, the PPR model is more parsimonious compared to the additive model, which is additive in its predictor effects. Furthermore the PPR can deal with interactions of predictors but the interpretation is usually difficult. A smooth multiple additive regression technique, SMART, was proposed as a generalization of (4) for K dependent variables (Friedman 1985). Analysis of dichotomous or polychotomous explanatory variables could be included in (4) by means of dummy variables. The analysis of ordinal qualitative data is not considered in literature but it can be easily included in PPR using the dummy variables of order. Using this approach we have to impose a limit on the search for the optimal directions γ into a region where some elements γ_{im} are non-negative.

4. Neural Network

A well-known form of neural network is the feed forward neural network (NN) where nodes are organized in different layers, one for input nodes (the explanatory variables), one or more intermediate layers for hidden nodes and one layer for output nodes (the dependent variables). Furthermore, the only permissible connections are between nodes in consecutive layers and directed upwards. The simplest architecture of a feed forward neural network has only one single hidden layer and one output node :

$$Y = g\left(\alpha_0 + \sum_{m=1}^{M} \alpha_m f_m\left(\gamma_{0m} + \sum_{i=1}^{P} \gamma_{im} X_i\right)\right) + \varepsilon \qquad (5)$$

where X_i is i-th explanatory variable, Y is the dependent variable, γ_{im} denotes a weight for the connection between the i-th explanatory variable and the m-th hidden node, and α_m a weight between m-th hidden node and the output node.

Expression (5) is very general; in fact introducing specific constrains we could obtain different non-parametric models (see Cheng, Titterington 1994) like PPR (when g is the identity function and f_m are not specified) or GAM (M=P, γ_{0m}=0, γ_{im} is the Kronecker δ_{im} and g is the identity function). However, while in PPR and GAM models we consider M different functions f_m determined from the data, in neural networks the f_m's are M same functions fixed by the user (generally chosen to be monotone, f.e. sigmoidal functions or hyperbolic tangent) and g is usually a linear or sigmoidal or threshold function.

Another difference of NN with projection method and other methods based on recursive partition is that for NN all of the parameters are estimated simultaneously by optimizing some measure of fit to the training data. The goal of fitting depends on the right choice of several elements: architecture (number of nodes and layers), learning (rate of learning and momentum coefficient), algorithm (generalized delta rule, conjugate gradients, quasi-Newton algorithms etc.).

If we analyze data sets including qualitative variables it is useful to distinguish between them with respect to nominal scale or ordinal scale. In the first case, one dichotomous input could correspond to each class; in the second case we might consider the dummy variables of order as described in the previous paragraph. In this case we should impose non-negativity constraints on the weights corresponding to the input dummy variables. The connection between NN and PPR is not only formal because under mild assumptions they have same property of

convergence and approximation to continuous functions on compact subsets of R^p (Cybenko, 1989; Jones, 1992). However when the number of input variables is high and the variables are highly correlated, PPR turns out to be more parsimonious then NN in terms of parameters (Ripley, 1994a).
Feed forward NN are also widely used in discrimination analysis (Ripley 1996, Schumacher et al. 1996). In this case the output variables are dummy indicators of classes and a logistic function could be chosen as g to have values of output nodes in [0,1]. The number of weights in NN is very relevant so often we obtain weakly identifiable solutions.

5. Methods based on recursive partition regression

Recursive Partition Regression (RPR) is a powerful method (Breiman, Friedman, Olshen, Stone 1984). This approach is very useful when there is a significant interaction structure in the predictors. It is based on a binary decision tree where knots split the data into two groups which are mostly homogeneous with respect to the response. The corresponding CART algorithm arranges a very large tree and by cross-validation then prunes it back to a reasonable size. We can write the RPR as:

$$Y = \sum_{m=1}^{M} \alpha_m \mathrm{I}(\mathbf{X} \in \mathrm{R}_m) + \varepsilon \tag{6}$$

where I(.) is the indicator function, R_m are disjoint regions of the covariate space and the α_m coefficients are estimated by the means of Y in R_m. The fundamental limit of recursive partitioning models is the lack of continuity: the model produced is piecewise constant and sharply discontinuous at sub-region boundaries.
Multivariate Adaptive Regression Splines (MARS) (Friedman 1991) can be seen as a generalization of the RPR. To span the space of univariate q-order splines we can use the basis (one-sided truncated power basis functions) 1, $\{X^j\}_1^q$, $\{(X - t_k)_+^q\}_1^K$, where t_k are the knot locations, K indicates the number of knots and the subscript + indicates a value of zero for negative values of the argument. In particular MARS uses a truncated two-sided power basis with q=1 to construct the multivariate spline basis functions

$$\mathrm{B}_m(\mathbf{X}) = \prod_{k=1}^{K_m} \left[s_{km} \left(X_{v(k,m)} - t_{km} \right) \right]_+ \tag{7}$$

where $v(k,m)$ label points to the explanatory variables used in the k-th factor of the m-th product; s_{km} takes on a value of +1 or -1 and indicates the (right/left) sense of the associated function. With this definition of basis functions, the MARS model can be written as in (1). Because basis elements are products of a finite number of univariate linear splines, this approach produces a piecewise linear function which does not possess continuous derivatives. In the final step, MARS uses the previous model to derive an analogous piecewise-cubic basis with continuous first derivatives. The algorithm is a stepwise procedure with an automatic selection of basis functions. Neural Network, in which the back-propagation optimize simultaneously over all basis elements, takes the opposite approach.

RPR is invariant with respect to monotone transformations of explanatory variables. This relevant property simplifies the analysis of ordinal qualitative variables, which can be inserted into the analysis by simply assigning a progressive value to the categories. We can adopt the same approach with MARS.
Rearranging the terms of the model we can write MARS as the more interpretable expression:

$$Y = \alpha_0 + \sum_{K_m=1} f_i(X_i) + \sum_{K_m=2} f_{ij}(X_i, X_j) + \sum_{K_m=3} f_{ijk}(X_i, X_j, X_k) + \ldots\ldots\ldots\ldots + \varepsilon \qquad (8)$$

MARS was successively extended to analyze qualitative nominal data, considering its relation with Recursive Partitioning Regression (RPR) (Friedman 1993).
Let A_l a subset of modalities of a nominal variable X_j, the basis functions of X_j will assume the form $I(X_j \in A_l)$. Then the MARS model with ordinal and nominal variables becomes (with q=1):

$$Y = \alpha_0 + \sum_{m=1}^{M} \alpha_m \prod_{l=1}^{C_m} I(X_{v(l,m)} \in A_{lm}) \prod_{k=1}^{K_m} \left[s_{km} \left(X_{v(k,m)} - t_{km} \right) \right]_+ + \varepsilon \qquad (9)$$

6. Comparison on noisy artificial data

In this section, an application of the methods to an artificial data-set is presented. In this way we can easily evaluate the fit to the underlying generation model. We use 4 variables: 2 quantitative variables X and Y, 1 categorical variable V with 6 modalities {a,b,c,d,e,f} and one variable Z defined as

$$\begin{aligned} Z &= X + \sin(2\pi XY) && \text{if } V = a \text{ or } V = c \\ Z &= X + \log(Y) + \cos(\pi XY)^2 && \text{if } V = b \text{ or } V = d \\ Z &= X + \log(Y) - 4XY && \text{if } V = e \text{ or } V = f \end{aligned} \qquad (10)$$

The variables X, Y and V were independently generated from uniform distributions; X and Y assume values in the range (0,1).
Note that, given the category of V, we can represent Z=f(X,Y/V) in a 3-dimensional space. The first graph of figure 1, the first graph of figure 2 and figure 3, give the 3 different graphical representations of the function Z=f(X,Y/V). Then we extracted a sample of 100 observations and added a Normal error giving a rate signal/noise = 3/1.
The artificial data obtained shows a strong interaction among all the explanatory variables, although we have a small problem with only 3 explicative (mixed) variables which can be examined graphically conditioned on the category of V. On the other hand we have few perturbed observations (15-18) for each category of V and it is very difficult to discover the true underlying functions by means of a simple graphical analysis. The function f(X,Y,V) considered in the artificial data is quite "smooth". This is a requirement in a non-parametric approach: the only way a non-parametric method has to distinguish between signal and noise is to suppose smoothness of the signal. Function f(X,Y,V) should varies smoothly while

Figure 1

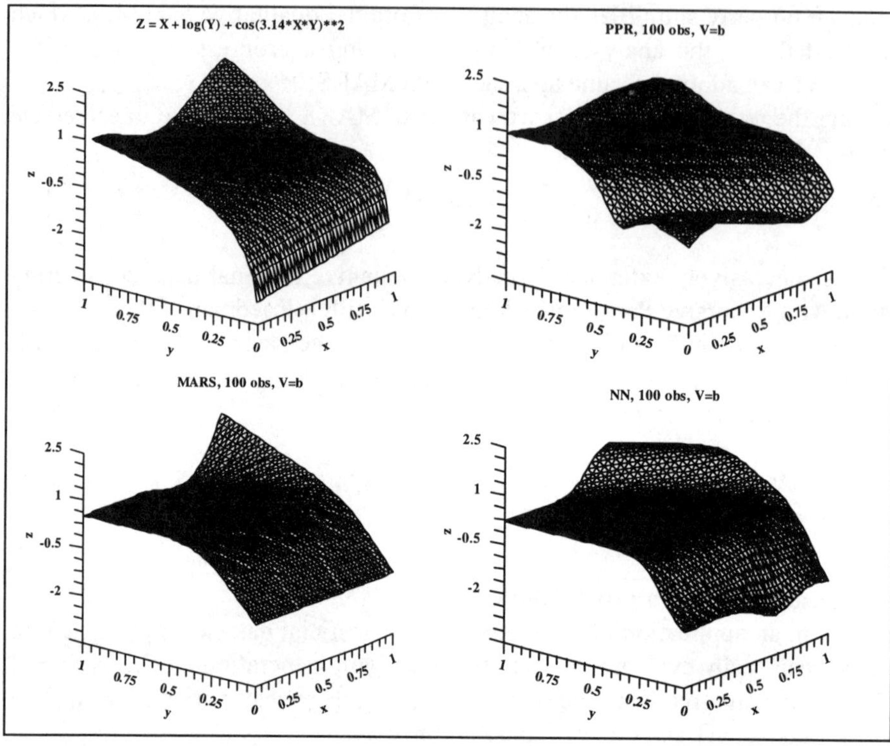

Figure 2

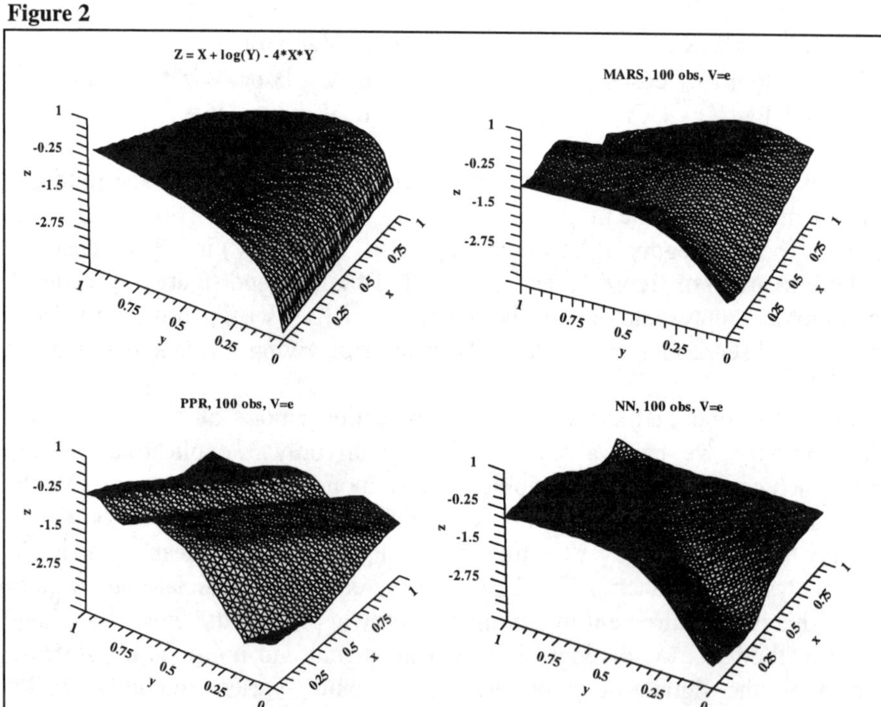

Figure 3

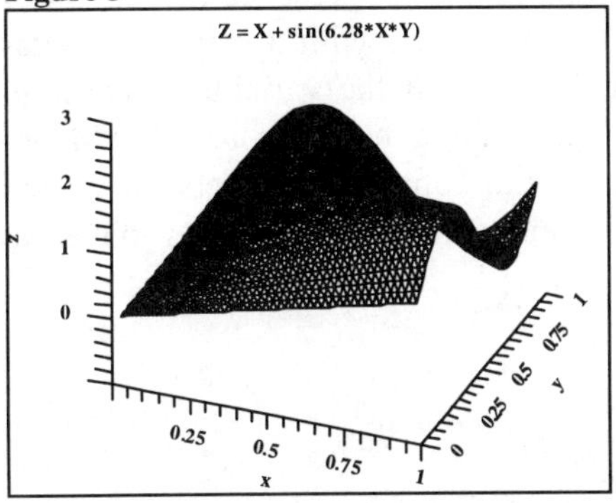

Figure 4

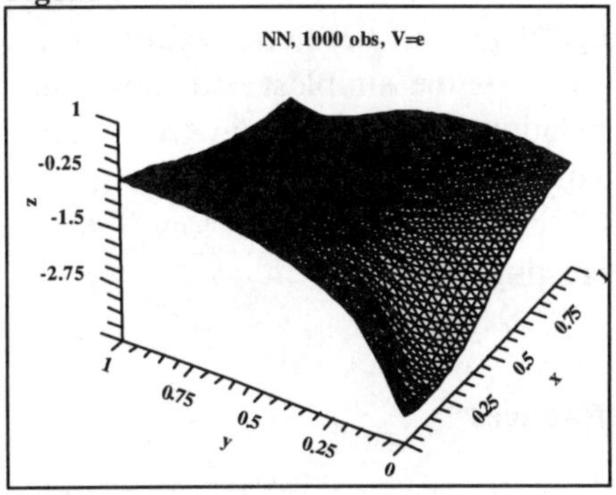

noise should not. A non-parametric method is effective if it can well approximate the smooth true function filtering the rough noise.

We try applying Projection Pursuit, Neural Networks and MARS, which are able to model interactions among variables, to this data. The application of the feedforward NN required to search an appropriate architecture to connect the 8 input nodes (two input nodes for X, Y and six dichotomous inputs for the categories of V) to the output node (perturbed Z). We found an optimal solution with a fully connected network with 3 hidden layers having 5, 4, 3 nodes respectively, i.e. 88 parameters. The optimal solution for PPR was obtained for M=3, while the optimal solution of MARS includes 6 basis functions and the interactions (X,V), (Y,V), (X,Y,V). In Figure 1 and 2 we show, with perspective plots, the true f(X,Y/V=b) and f(X,Y/V=e) with the corresponding estimated functions for each method. Examining these graphs we can note a better performance of MARS on our data-set. The other two methods are equivalent and less effective than MARS, although the simpler and more parsimonious PPR model should be preferred to NN.

We considered also a larger data-set with 1000 observations to evaluate the performance of the methods on larger sample. Applying NN to this data, we used a more complex feed forward NN with 10 nodes for the first hidden layer, 8 nodes for the second and 4 for the third (219 parameters). With this larger sample, NN increases the capability to identify the true underlying function. The fit of Z=f(X,Y/V=e) represented in Figure 4 show a clear improvement while the increased sample size is not so useful with the other two methods.

7. Conclusion

Every non-parametric method we have analyzed has some weakness and some capabilities which we observe theoretically and which we find again by empirical means, analyzing simulation data. NN represents an interesting and general class of non-parametric methods, but it is difficult to interpret, requires much training data and some experience to avoid local minima. MARS is a very powerful method, particularly suitable to analyzing interactions between variables. However, for high complex non-additive data with very large samples it can be outperformed by NN.

This method can also be more advantageous than MARS in presence of outliers or a high degree of multicollinearity in the predictors, always with a very large data-set. PPR is the simplest and most interpretable model if the variables have high correlations and we obtain M<3. The Optimal Scaling methods and GAM are usually outperformed by the previous methods with complex data-sets but in the case that variables have low interaction levels these methods can have an interesting interpretability.

References

Breiman L., Friedman J.H., Olshen R.A., Stone C.J. (1984). *Classification and regression trees*, Blemont, CA:Wadsworth.

Cheng B., Titterington D. M. (1994). Neural Networks : a review from a statistical perspective (with discussion), *Statistical Science*, 9, 2-54.

Cybenko G., (1989). Approximation by superpositions of a sigmoidal function, *Math. Control Syst. Sign.*, 2, 303-314.

Di Ciaccio A. (1988). Some considerations on the quantification of categorical data, Research-Report, Department of Data Theory, University of Leiden.

Friedman J.H. (1993). Estimating functions of mixed ordinal and categorical variables using adaptive splines, in: *New Directions in Statistical Data Analysis and Robustness*, Morgenthaler S. et al. (Eds.), Birkäuser Verlag, Berlin.

Friedman J.H. (1991). Multivariate adaptive regression splines (with discussion), *The Annals of Statistics*, Vol.19, 1, 1-141.

Friedman J.H. (1985). Classification and multiple regression through projection pursuit, Tech. Rep. 12, Lab. Comp. Stat., Dep. of Statistics, Stanford University.

Friedman J. H., Stuetzle W. (1981). Projection Pursuit Regression, *J.A.S.A.*, Vol. 76, N.376.

Gifi A. (1990). *Nonlinear multivariate analysis*, Wiley , Chicester.

Hastie T., Tibshirani R. (1990). *Generalized Additive Models*, Chapman and Hall, London.

Jones L.K. (1992). A simple lemma on greedy approximation in hilbert space and convergence rates for projection pursuit regression and neural network training, *The Annals of Statistics*, Vol.20, N.1, 608-613.

Ripley B.D. (1996). *Pattern recognition and neural networks*, Cambridge University Press.

Ripley B.D. (1994a). comment on: Neural Networks: a review from a statistical perspective (with discussion), Cheng B., Titterington D. M., *Statistical Science*, 9, 2-54.

Ripley B.D. (1994b). Neural Networks and Related Methods for Classification (with discussion), *J. R. Statist. Soc.*, B, Vol.56, N. 3, 409-456.

Schumacher M., Roßner R., Vach W. (1996). Neural Networks and logistic regression: Part I & II, *Computational Statistics & Data Analysis*, 21, 661-701.

Local multivariate analysis

Pierre-André Cornillon[†], Robert Sabatier[‡]
† Unité de biométrie, INRA - ENSA.M - UM II, Place Paul Viala, 34060 Montpellier, France, e-mail: pac@helios.ensam.inra.fr
‡ Laboratoire de Physique Moléculaire et Structurale, 15 av. Ch. Flahaut 34060 Montpellier, France, e-mail: sabatier@pharma.univ-montp1.fr.

Abstract: We present two methods to explain linearly a set variables Y by another set X through a neighbourhood relationship matrix, with non-independent observations. We establish a link between these two methods and the regression framework with autocorrelated error.

Keywords: Local analysis; Multivariate analysis; Partial Least Square; Principal Component Analysis with respect to Instrumental Variables.

1 Introduction

Explaining one or more dependent response variables $(Y_1, Y_2 \cdots Y_q)$ by severals variables $X_1, X_2 \cdots X_p$ is common in such various fields as biology, behaviourial or social sciences. Numerous articles or books are devoted to both theoretical (e.g. Rao and Toutenburg, 1995) and applied (e.g. Draper and Smith, 1981 or Haining, 1990) aspects of this vast subject. For example Multivariate linear models is an attempt to solve this sort of problems using linear combination of the dependent variables to explain Y. From this linear point of view it is possible to derive numerous methods like Partial Least Squares (PLS) regression (e.g. Wold, 1985) or Principal Component Analysis with respect to Instrumental Variables (PCAIV). An extensive bibliography of this latter one can be found in van der Burg and de Leeuw (1990) including the historical paper of Rao (1964). We focus our interest on non-independent case of these two methods. Although classical framework enables us to tackle non-independence, through non diagonal variance matrices of error, we introduce a different approach of modeling non-independence between observations. In the following we shall write n for the number of theses observations.

2 Euclidean framework

Following Lebart (1969) we introduce a matrix M, such that the element M_{ij} reflect the neighbouring relationship between the observation i and the observation j. These coe̎fficients are chosen by the user to belong to $[0, 1]$. For instance if data are measured at different depth (in a lake for instance) we can use a linear boolean neighbourhood matrix constructed as shown below :

Figure 1: *Linear neighbourhood*

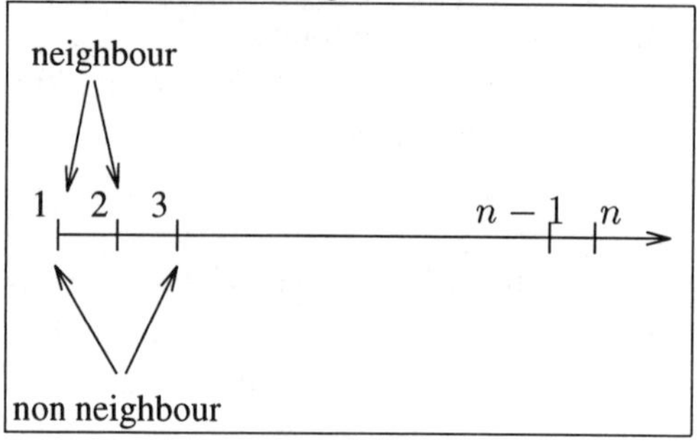

Other more complex structures of this matrix M can be found in Cliff and Ord (1981), for instance the coefficients M_{ij} could be function of the distance between observations.
We adopt in the present paper the geometrical approach using metrics to compute distances between objects and variables, see for instance Escoufier (1987).
Let Y be the $n \times q$ data matrix, Q be a $q \times q$ symmetric definite positive matrix defining a scalar product on the space of dependent variables, and $D = diag(p_1, p_2, \cdots, p_n)$ a n-dimensional diagonal matrix with positive element summing to 1. Recall that the empirical total variance of the kth dependent variable can be written as:

$$V(Y_{.k}) = \frac{1}{2}\sum_{i,j} p_i p_j \left(Y_{ik} - Y_{jk}\right)^2.$$

If we are just interested in the local structure of variations, we can introduce the neighbourhood structure defining the local variance (Lebart, 1969, Banet and Lebart, 1984) which is closely related to the Geary index (Geary, 1954):

$$V_L(Y_{.k}) = \frac{1}{2}\sum_{i,j} p_i p_j M_{ij} \left(Y_{ik} - Y_{jk}\right)^2.$$

From this we can also write the local covariance as:

$$cov_L(Y_{.k}, Y_{.l}) = \frac{1}{2}\sum_{i,j} p_i p_j M_{ij} \left(Y_{ik} - Y_{jk}\right)\left(Y_{il} - Y_{jl}\right).$$

Within this framework we shall redefine PLS regression and PCAIV on a local point of view in the following sections. Finally we shall establish a correspondence between this approach and a regression one using autocorrelated errors.

3 Local PLS regression

Partial least squares regression is a method of analysis useful for designs with more explanatory variables than samples. It can easily deal with multicollinearity in the matrix X. The aim is to predict the response by a model based on linear transformations of the explanatory variables, and to construct a regression model of the following type:

$$\widehat{Y}_{.j}^{(m)} = \beta_0^{(m)} + \beta_1^{(m)} Z_1 + ... + \beta_m^{(m)} Z_m,$$

where m is the dimension of the model,

$\widehat{Y}_{.j}$ is the prediction of the jth variable of Y

The latent variables $Z_1 \cdots Z_m$ are linear combinations of the explanatory variables $X_1 \cdots X_p$. They are chosen such that they are uncorrelated to each other. Because we are only interested in local variations, we search latent variables locally uncorrelated to each other. For completeness we can also define another scalar product in the vector space $\mathbb{R}^p$ of observations of the independent variables by a definite positive matrix N. We can write the local PLS regression procedure as follows:

Table 1: *Local PLS model of dimension* m

$Y^{(1)} \Leftarrow Y$ and $X^{(1)} \Leftarrow X$
for $k = 1$ **to** m
$Z_k \Leftarrow Xu^{(k)}$, with
$Y^{(k)\prime} D(D^* - MD) X^{(k)} N X^{(k)\prime} D(D^* - MD) Y^{(k)} Q u^{(k)} = \lambda u^{(k)}$,
where $D^* = diag(p_i^*)$ and $p_i^* = \sum_{j=1}^n M_{ij} p_j$,
and λ is the first (biggest) eigen value of the operator.
$X^{(k+1)} \Leftarrow (\mathrm{Id}_n - Z_k(Z_k{}' D(D^* - MD) Z_k)^{-1} Z_k{}') X$
$Y^{(k+1)} \Leftarrow (\mathrm{Id}_n - Z_k(Z_k{}' D(D^* - MD) Z_k)^{-1} Z_k{}') Y$
end for

It can be easily seen that local PLS regression is just a special case of PLS regression with a new scalar product in the space $\mathbb{R}^n$ of variables. The matrix defining this scalar product is written as $D(D^* - MD)$ and take into account the neighbourhood relationship between the observations.
The dimension m of the model can be checked out by using a cross validation procedure. One observation (one line, for instance the ith line) is taken out from X and Y. The model at stage k is then fitted to this reduced data set. The squared difference between the predicted value according to model dimension k, $\widehat{Y}_{(-i)j}^{(k)}$ and the actual value of jth variable of Y is added to the jth $PRESS(k)$. The total sum of squares of predictions minus observations for every deleted line is a measure of predictive power:

$$PRESS_{(k)j} = \frac{1}{n} \sum_{i=1}^{n} \sum_{j=1}^{p} (Y_{ij} - \widehat{Y}_{(-i)j}^{(k)})^2.$$

The size m of the model for j^{th} variable is then chosen to achieve the minimum of $PRESS_{(k)j}$

We can apply this method to the Californian data set (Upton and Fingleton, 1985, p.273). These data relate to islands and to mainland localities along the Californian coast. The variables measured were: Y the number of species, X_1 the area in squares miles, X_2 the maximum elevation in feet and X_3 the latitude (degrees North). An ordinary PLS of size 1 re-expressed as an ordinary regression leads to the following model:

$$\widehat{Y^{(1)}} = \alpha_1 X_1 + \alpha_2 X_2 + \alpha_3 X_3.$$

The parameter estimates are:

Table 2: *parameter estimates of PLS model fitted to the Californian data*

$\widehat{\alpha_1}$	$\widehat{\alpha_2}$	$\widehat{\alpha_3}$
0.379	0.495	0.374

We can use the 'two-part distance-based' weight matrix M defined in Upton and Fingleton (p.291) as the squared inverse of the distance between two location up to a cut-off distance (taken to be 187.5 miles). Beyond this limit weights are taken to be zero. These weights are then normalized to have the sum of each line equal to 1. A Geary test can be performed on the residuals of PLS regression. It shows that they are not independent (p-value < 0.01) and a local PLS regression could be used to modelize this fact. The fitted model becomes:

Table 3: *parameter estimates of local PLS model fitted to the Californian data*

$\widehat{\alpha_1}$	$\widehat{\alpha_2}$	$\widehat{\alpha_3}$
0.434	0.484	0.065

We can see that the influence of latitude is far less important. The residuals take into account the location dependency and the fitted model no longer considers the latitude as a good explanatory variable. A look on the PRESS' variations confirms our selection of an unidimensional model for this example.

4 Local PCAIV

Local PCAIV is another way to explain the local variation of the dependent variables $Y_1 \cdots Y_q$ by a linear transformation of the explanatory variables $X_1 \cdots X_p$. We have to recall first the expression of local inertia I_L as a natural extension of the local variance (1):

$$I_L = \sum_{i,j} p_i p_j M_{ij} \|Y_{i.} - Y_{j.}\|_Q^2$$

where $Y_{i.} \in \mathbb{R}^q$ is the ith row of the matrix Y.

We can define local PCAIV as finding the linear transformation $\tilde{H}$ which realizes the minimum of the local inertia of discrepancy between Y and its explanation by a linear transformation:

$$\tilde{H} = \arg\min_H \sum_{i,j} p_i p_j M_{ij} \, \|(Y_{i.} - HX_{i.}) - (Y_{j.} - HX_{j.})\|_Q^2 \, .$$

From this definition it can be found that:

$$\begin{aligned} \tilde{H} &= Y'DEX(X'DEX)^{-1}, \\ &\quad \text{where } E = (D^* - MD) \\ &\quad \text{and } D^* = diag(p_i^*) \text{ and } p_i^* = \sum_{j=1}^{n} M_{ij} p_j . \end{aligned}$$

This is a particular case of PCAIV with a new scalar product in the space $\mathbb{R}^n$ of variables, defined by DE as it is in the local PLS regression. Although this method is a linear explanation of Y by X for the local variations, the obtained model is obviously different from local PLS regression. Local PCAIV can be defined as searching an optimal metric in the space of observations of the latent variables $\mathbb{R}^p$. We can write this metric N as

$$N = (X'DEX)^{-1}X'DEYQY'DEX(X'DEX)^{-1},$$

and of course an *a priori* choice of the metric in $\mathbb{R}^p$, like in local PLS regression is not possible.
Because of the projection on the space spanned by the column of X, multi-collinearity within X must be dealt by the use of a Moore Penrose inverse instead of the normal inverse. In this case, a better predictive power is achieved by PLS regression. We will illustrate this method using a data set from ecology.

5 Local multivariate analysis and autoregressive errors

To take into account the non-independence of the observations in a regression framework, it is sometimes useful to assume that the error term is autocorrelated. Several classical models of autocorrelation exist and for a complete description of them we refer to Cliff and Ord (1981). If a conditional autoregressive model (CAR) is chosen, then the variance of error can be written as:

$$\begin{aligned} Var(\varepsilon) &= \sigma^2(\mathrm{Id}_n - \rho W)^{-1} \\ &\quad \text{where } (\sigma, \rho) \text{ are parameters of the CAR errors model,} \\ &\quad W \text{ are the known coefficients of autoregression.} \end{aligned}$$

The least square estimator is then the projection of Y on the vector space spanned by the column of X with respect to the metric $Var(\varepsilon)^{-1}$. It can be found easily that, if the weights of the observations $\{p_i\}$ are chosen equal to $\frac{1}{\sigma}$ and $M =$

ρW, then using local multivariate methods or an CAR error term in a regression framework are equivalent in the sense that they use both the same metric. This particular metric can also be found in local Principal Component Analysis or in a local Correspondence Analysis (Thioulouse, Chessel and Champely, 1995). All these analysis build a local multivariate data analysis framework, following the paper of Lebart (1969).

References

Aluja Banet, T. and Lebart, L. (1984). Local and partial principal component analysis and correspondence analysis. In *COMPSTAT*, pages 113–118.

Cliff, A. D. and Ord, J. K. (1981). *Spatial processes : models and applications.* Pion, London.

Draper, N. and Smith, H. (1981). *Applied regression analysis.* John Wiley and Sons, New-York, 2 edition.

Escoufier, Y. (1987). Principal components analysis with respect to instrumental variables. In *European courses in advanced statistics*, pages 27–28. University of Napoli.

Geary, R. (1954). The contiguity ratio and statistical mapping. *The incorporated statistician*, 5:115–145.

Haining, R. (1990). *Spatial data analysis in the social and environmental sciences.* Cambridge university press.

Lebart, L. (1969). Analyse statistique de la contiguïté. *Publication de l'institut de statistiques de l'université de Paris*, 28:81–112.

Rao, C. R. (1964). The use and the interpretation of principal component analysis in applied research. *Sankhyā*, 26:329–358.

Thioulouse, J., Chessel, D., and Champely, S. (1995). Multivariate analysis of spatial patterns : a unified approach to local and global structures. *Environmental and ecological statistics*, 2:1–14.

Upton, G. J. G. and Fingleton, B. (1985). *Spatial analysis by example*, volume 1. John Wiley and Sons, New-York, 2 edition.

van der Burg, J. and de Leeuw, J. (1990). Nonlinear redundancy analysis. *British journal of mathematical and statistical Psychology*, 43:217–230.

Wold, H. (1985). Partial least squares. In Kotz, S. and Johnson, N. L., editors, *Encyclopedia of statistical sciences*, volume 6, pages 581–591. John Wiley and Sons.

Non-parametric Regression and Non-linear Data Analysis: an overview

D'Ambra L.*, Lombardo R.**, Tessitore G. *
* Department of Mathematics and Statistics- University of Naples -Italy-
e-mail: dambra@unina.it, gt@dmsna.dms.unina.it
** Institute of Management and Quantitative Research- The S.U.N.-Italy-
e-mail: lombardo@dmsna.dms.unina.it

Abstract: This paper presents an overview of the main non-parametric approaches in regression and multidimensional data analysis. Most of the techniques will be introduced in compact form just to give the essential idea which are relevant to non-parametric methods.

Key words Non-parametric Regression, Non linear Principal Component Analysis, Optimal Spline transformation.

1. Introduction

In literature (Stone 1985) the basic aspects of a statistical model are: flexibility, dimensionality and interpretability. The flexibility is the model capacity to give unbiased estimates which is extremely helpful in a preliminary and exploratory data analysis. The dimensionality is strictly related with the variance estimate, in particular, "curse of dimensionality" means that exponentially increasing data points are requested to populate spaces of augmenting dimensions. The interpretability represents the capacity of the model to point out data structure. The parametric models often do not meet all.

The term nonparametric refers to the flexible functional form of the regression curve Härdle (1989). In non-parametric regression models, the experimenter looks at flexibility properties of the regression function in fitting the data. In contrast under a parametric model only one possible family of curves is chosen and the data information is restricted to the assumed parametric form. It should be noted that, even though parametric and non-parametric models represent distinct different approaches to regression analysis, non-parametric techniques can be used to asses the validity of parametric ones. In low dimensional settings, global parametric modeling has been successfully generalised using paradigms-piecewise and local parametric fitting and roughness penalty methods (the most popular are based on splines). In higher dimensional settings, a simultaneous estimation of a large number of parameters is not conceivable fixing them all externally to the procedure and

only non-parametric regression techniques are able to give flexible function estimates providing a versatile pre-screening method for outliers.
In multidimensional data analysis (*MDA*), the non-parametric procedures have found adhesion with the introduction of non linear versions of the more familiar linear techniques. So firstly the *Gifi System* (Gifi 1990) is introduced on the idea to optimise an objective loss function by means of Alternating Least Squares (*ALS*) algorithms. The least squares estimates gives the parameters of one subset while assuming constants the others. In this complex data analysis system the authors pay attention to the optimal estimates of the model parameters. Subsequently in the literature more attention has been devoted to the optimal estimates of non linear transformation parameters of the original variables (*non linear Principal Component Analysis,* van Rijckevorsel 1987, van Rijckevorsel *et al.* 1988, van Rijckevorsel *et al.,*1993 and Tessitore *et al.*, 1996) and of the principal components (*Adaptive Principal Surfaces,* LeBlanc *et al.*, 1994, *Principal Surface Constrained Analysis*, Lombardo *et al.* 1997a,b,c).

2. Non-parametric Regression Models

In this section we provide a brief description of the most popular approaches to non parametric regression problem (such as kernel, smoothing spline and least-squares spline estimators) and an overview of several promising new nonparametric estimation techniques for analysis of high dimensional data.
A non parametric linear estimator of a simple regression curve will have the form $E(y \mid x) = \sum_{i=1}^{n} K(x; x_i; \lambda) y_i$ where $K(\cdot, x_i; \lambda)$ is a set of weight functions which depend on λ. A simpler method of choosing the weights is to use a symmetric unimodal probability density function *K* defined on [-1;1] for some real number $\lambda > 0$ (λ, called the bandwidth, governs the amount of averaging and controls the peakedness of the function) and estimate E(y|x) by kernel estimators (Dirichlet kernel form of classical Fourier series estimators, Eubank 1988): $E(y|x) = \sum_{i=1}^{n} K[\lambda^{-1}(x - x_i)] y_i \Big/ \sum_{i=1}^{n} K[\lambda^{-1}(x - x_i)]$. Unfortunately the lack of data at the boundaries causes a boundary bias problem for kernel estimators. Differently spline functions (regression or smoothing spline) are suitable for their property of flexibility. The spline function is a linear combination of a suitable set of piecewise function (usually Basis splines, Eubank 1988), defined on each interval; a non-negative combination of integrals of B.splines defines monotone splines denoted as I-splines. A spline knot sequence (number and location) partitions the domain of a continuous variable into a limited number of adjoining intervals. The dimension of a spline function depends on the degree and the number of interior knots. Concerning smoothing spline, the regression function estimator, which belongs to the Sobolev space $W_2^m[0,1]$, is: $\mathbf{arg\,min}\, 1/n \sum_{i=1}^{n} (y_i - f(x_i))^2 + \lambda \int_0^1 f^{(m)}(x)^2 dx \quad \lambda > 0$, where x_i are in [0,1],

the solution is unique and belongs to the space of univariate natural spline. Knots are located at distinct data points and the parameter λ, selected through visual inspection of the data, controls the trade-off between the fit to the data and the smoothness of the estimator. By the way they are used, smoothing splines mainly differ from regression splines. With respect to regression or least-squares spline, the tuning parameter λ (knot sequence) in the univariate setting is estimated using least-squares.

Let $E(y\,|\,\mathbf{x}) = \sum_{j=1}^{p} \beta_j x_{ij}$, $\mathbf{x} \in R^p$; $i = 1,...,n$ be the standard linear regression model. The estimation of this model in a non parametric setting could be accomplished by using Laplacian smoothing spline or multivariate kernel estimators, but problems arise when p is large increasing variance for increasing dimensionality (curse of dimensionality) and the lack of interpretability of multivariate estimators. An alternative to multivariate smoothers is to consider the following popular generalisations:

i) The y mean depends on explicative variables in additive way (Breiman 1993). This interpretation allows the substitution of x_j with p simple functions of x_j estimated using the univariate estimators (circumventing the curse of dimensionality).

ii) The y mean depends on one linear combination of predictors. The mean changes along one direction in the space R^p and it is constant on the orthogonal directions.

In the ***i***) interpretation are relevant *Generalised Additive Models* (*GAM*, Hastie *et al.* 1986):

$$g(\mu(\mathbf{x})) = \alpha + \sum_{j=1}^{p} f_j(x_{ij}) \qquad \text{with } E[f_j(x)] = 0$$

where $\mu(\mathbf{x}) = E(y\,|\,\mathbf{x})$ and the function $g(\cdot)$, assumed note, is called the link function. Estimation of f_j in *GAM* be based on the *backfitting algorithm* of Friedman *et al.* (1981), which is a Gauss-Seidel type procedure and implements at each step an univariate smoother. It forms the "inner loop" of the *Alternating Conditional Expectation Model* (*ACE*, Breiman *et al.* 1985) algorithm. *ACE* is a non-parametric generalization of the additive model that fits an additive model as part of an alternating estimation procedure.. In *ACE* the response variable y and the predictor variables $x_1,...,x_p$ are replaced by $\theta(y)$ and $\phi_1(x_1)....\phi_p(x_p)$.

The model is $E[\theta(y)] = \sum_{j=1}^{p} \phi_j(x_{ij})$. *ACE* provides a versatile method for estimating the maximal correlation between two variables because the functions θ and $(\phi_j)_{1\le j\le p}$ are estimated by maximising the square correlation coefficient. The disadvantage of this method (Ramsay 1988) is that when explicative variables are correlated the algorithm results very sensitive to their order, further it seems more suitable for a correlation than regression analysis. This method has been proposed to overcome the limit of the *GAM* consisting in inability to share the additive from the interactive effects of the predictors. If the smoother refers to global univariate least squares fits, then backfitting converges to the multivariate least squares solution. Furthermore we remember the algorithm *alternating variance estimation* (*AVAS*, Tibshirani 1988) which improves *ACE* using an asymptotic stability variance criteria. Another approach suggested by Stone (1985) for estimating *GAM* is to use least-squares splines.

In the ***ii***) interpretation, an interesting generalisation is the Projection Pursuit Regression (*PPR*, Friedman *et al.* 1981, Buja *et al.* 1989, Huber 1985) expressed in the model:

$$E(y \mid \mathbf{x}) = \sum_{k=1}^{K} f_k\left(\alpha_k^t \,\mathrm{x}\right).$$

where **x** is a *p*-dimensional variable, the α_k are direction vectors onto which the data are projected, the number *K* of projections is a parameter of the procedure and each f_k is an arbitrary one-dimensional function. The *GAM* is a special case of *PPR* in which there are exactly *p* directions fixed at the coordinate directions (it is less general than *PPR* but it is more easily interpretable). Friedman *et al.* (1983), and Ramsay (1988) propose splines to approximate the functions in *PPR*. *Smooth Multiple Additive Regression Techniques* (*SMART,*1984) represents an extension of *PPR* to the case of more response variables. The outlined above, non-parametric function estimation based on univariate functions is an important forward step and (especially for additive modeling) has met practical success solving the curse of dimensionality problem. Regrettably this approach has some limitations as a general method for estimating a large number of functions. So the further step has been the development of strategies for the high dimensional approximation based on adaptive computation. Adaptive algorithms for function approximation have been developed using both two paradigms, recursive partitioning (Breiman et al. 1984) and projection pursuit selection (Friedman 1981, Friedman 1983). A flexible regression modeling of high dimensional data was proposed by Friedman (*Multivariate Adaptive Regression Splines*, *MARS* 1991). The model takes the form of an expansion in product spline basis functions, where the parameters associated to the basis are automatically determined by a recursive partitioning approach. The result is satisfactory with respect to modeling both additive and interactive relationships.

3. Non-linear Multidimensional Data Analysis

Principal Component Analysis (*PCA*) can be generalised in different directions to yield *Non Linear Principal Component Analysis* (*NL-PCA*, van Rijckevorsel 1987, van Rijckevorsel *et al.* 1988).). The incorporation of non linear transformations in *PCA* has quite a long history and can be found among others in Gifi (1990). In Gifi system *Homogeneity Analysis* (*HO*), also known as *Multiple Correspondence Analysis*, is a key method.

This analysis is performed by the minimization of a specific loss function. On the basis of the restriction imposed on the *HO* loss, all the kinds of analysis may be performed and accomplished by algorithm of the the *Alternating Least*

Squares (*ALS*) type. In each *ALS* iteration two steps are alternated, in the first step the optimum basis (corresponding with scores for the individuals or objects) are computed for given values of the transformations (categories quantifications or transformation coefficients), in the second step new values for the optimum transformations are calculated for the given basis, previously computed. Alternating these steps obviously produces a decreasing loss function values sequence. Being the transformation not a priori determined, the procedure itself can be considered non parametric. We briefly introduce *NL-PCA*. Let **X** be the object scores (n,m) matrix, $\mathbf{Y}_j$ and $\mathbf{G}_j$ respectively the (k_j,m) weighting matrix and the (n,k_j) coding matrix $(j=1,...,p)$, with p variables, n objects, m components, and k_j basis functions. The loss function in *NL-PCA* (Gifi 1990, van Rijckevorsel 1987) is:

$$LOF(\mathbf{X};\mathbf{Y}) = p^{-1} tr \sum_{j=1}^{p} \left(\mathbf{X} - \mathbf{G}_j \mathbf{Y}_j\right)^T \left(\mathbf{X} - \mathbf{G}_j \mathbf{Y}_j\right).$$

This equation measures the departure of the transformed variables $\mathbf{G}_j\mathbf{Y}_j$ $(j=1,...,p)$ from an hypothetical **X**, synthesis and linear combination of the original variables, often called the "components". It is minimised with respect to the **X** and $\mathbf{Y}_j$ matrices by the use of an *ALS* algorithm. In reality the importance of non-parametric *MDA* has been stated when more attention has been paid for non linear transformation parameters. In particular this has concerned the parameters of the transformation splines (the degree and the knot sequence). The optimal knot location can be regarded as a choice of the model (setting externally parameters, for example a uniform knot placement) or as a free parameter to be optimised (it can influence the fit of a subsequent model). Splines have been used to transform the original variables in *NL-PCA* (*Fuzzy Coding*, van Rijckevorsel *et al.* 1988, van Rijckevorsel 1987, and for the use of I-splines: Besse *et al.* 1995, Ferraty 1997), in the *PCA on Instrumental Variables* (Durand 1993), in *NL Non Symmetric Correspondence Analysis* (Verde 1992), in *Discriminant Analysis* (Hastie *et al.* 1994), in the distance approach to *MDA Principal Distance Analysis* (whose objective falls within the *Multidimensional Scaling* domain (Meulman 1986,1992) and the components in *Adaptive Principal Surface* (LeBlanc *et al.*, 1994), in *Constrained PCA* (Lombardo *et al.* 1997a,b,c). In order to get an optimal transformation of the original variables both coherent with the purpose of the analysis and the data structure, van Rijckevorsel *et al.* (1993) propose an adaptive procedure for the non-parametric detection of the optimal knot sequence, using the recursive partitioning approach. Furthermore to find a good compromise between the optimality of the function approximation of the original variables and the over-fitting problems both tied to increasing number of parameters, Tessitore *et al.* (1996) propose an algorithm to detect the optimal number of knots using the *Generalised Cross Validation* criterion (Craven *et al.* 1979). Lauro *et al.*(1996) extend the non-parametric optimal knot location detection to the multivariate coding through multivariate B-splines. In the context of non linear *PCA* by means splines smoothers: smoothing splines, hybrid splines, monotone smoothing splines and monotone hybrid splines, Ferraty (1997) proposes smooth optimal transformations of the original variables, which maximises the explained variance of the transformed data. A bootstrap procedure adapted to the model leads to an optimal choice of the smoothing parameter. Durand (1993) proposes *Additive Spline PCA on Instrumental Variables* (*ASPCAIV*) which combines features of multi-response additive spline regression analysis and *PCA* via a spline transformation of the predictor matrix. Dimension reduction occurs after a linear smoothing procedure has been applied. The associated smoother matrix is the projection matrix which is to be compared with that of the least squares splines. Although projections are processed on spaces of smaller dimensions, *ASPCAIV* is however sensitive to scarcity of data and multicollinearity. Hastie *et al.* (1994) propose non-parametric versions of *Discriminant Analysis* (*DA*). Linear *DA* is a tool for multigroup classification, it can be seen as equivalent to multi-response linear regression using optimal scorings to represent the groups. By replacing linear regression by any non-parametric regression method different non-parametric versions are obtained. In this way any multi-response regression technique (such as *MARS*) can be post-processed to improve its classification performance. Furthermore very interesting it has been the extension of non-parametric adaptive procedures to the principal components. The main aim of these recent techniques is the optimal transformation of the principal components in order to represent objects taking into account the variables interactions by computing multivariate splines. While the transformed original variables are still considered univariate, the non linear components are now considered both as single and interactive variables. Le Blanc *et al.* (1994) develop a non linear generalization of *PCA* for dimensionality reduction. A *Principal Surface* (*PS*) of the data is constructed adaptively using *MARS* procedure. *PS* are multivariate splines obtained as tensor products of univariate transformation splines of the single components by means of the following formula: $\mathbf{f}(\mathbf{c}) = \beta_0 + \sum_{m=1}^{M} \beta_m \prod_{k=1}^{K_m} \mathbf{b}_{(k,m)}\left(\mathbf{c}_{j(k,m)}\right)$ where β are the coefficients for the construction of the spline transformation; M is the number of basis functions of the multivariate spline; K_m is the basis number of the univariate spline to construct each tensor; $\mathbf{b}$ is the basis function of the generic component $\mathbf{c}_{j(k,m)}$, on the partition of the multivariate domain, used in the kth term of the mth product. The optimal *PS* is constructed adaptively using the *MARS* procedure. In order to detect the optimal number of components used to construct the *PS* (space dimension problem) the *GCV* criterion is considered. In the frame of constrained *PCA* Lombardo *et al.* (1997b) baptise the analysis *Principal Surfaces Constrained Analysis* (*PSCA*) in the way it relaxes linearity assumption of principal components, permitting to represent data on *PS* optimally computed by an adaptive procedure. They look for the optimal *PS* representation optimising the knot location of the multivariate transformation splines through a modified version of the recursive partitioning procedure (Friedman 1991). As the non linearity invests the components and not anymore the original variables the knot detector optimises the following criterion:

$E\left[\sum_{j=1}^{p}\left(\mathbf{x}_j - f(\mathbf{c}_j)\right)^2\right]$. The vector $\mathbf{x}_j$ (the jth column of **X**) and each $f(\mathbf{c}_j)$ is a non linear parameterised function mapping R^p into R. The resulting "multidimensional" representation allows to investigate in depth the similarities among objects due to a particular variable (classification variable). Furthermore in order to avoid over-fitting problems related to the number of transformation parameters, Lombardo *et al.* (1997c) propose an algorithm for the optimal number of knots detection. At the end, to complete this overview, it is worthwhile to remind *Additive Spline Partial Least Squares* (*ASPLS,* Durand & Sabatier 1997). The *ASPLS* regression consists in carrying out a *PCA* of predictors matrix, using spline transformation for principal components which play the role of explanatory variables, allowing an additive approximation to the response variables. In conclusion, this paper provides a brief summary of the most selected topics concerning non-parametric procedures in regression and multidimensional data analysis, in order to sketch that theory part opening new development and perspectives.

References

Craven P. & Wahba G. (1979). Smoothing Noisy Data with Spline functions.Estimating the Correct Degree of Smoothing by the Method of Generalised Cross-Validation. *Numerische Mathematik*, 31.

Breiman L. (1993) Fitting additive models to regression data. Diagnostics and alternative views. *Computational Statistics & Data Analysis* 15.

Buja A., Hastie T.J., Tibshirani R.J. (1989) Linear smoothers and additive models. *Annals of Statistics* 17.

Besse P. & Ferraty F. (1995) A Fixed Effect Curvilinear Model. *Computational Statistics*, 10.

Breiman L. & Friedman J.H. (1985) Estimating optimal transforming for multiple regression and correlation. J.A.S.A. vol.80.

Breiman L., Friedman J.H., Olshen R., Stone C. (1984) Classification and Regression Trees. *Wadworth, Belmont. Calif.*

Durand J.F. (1993) Generalized Principal Component Analysis with respect to Instrumental variables via univariate spline transformations. *C.S & D.A.*, 16.

Durand J.F.& Sabatier R.(1997)Additive Splines for PLS regression. *J.A.S.A.*

Eubank R.L.(1988)*Spline Smoothing and Non-parametric Regression.* Dekker.

Ferraty (1997) Estimations de transformations optimales en ACP curvilièaire.Re.*Statistique Appliquèe*,XLV(1).

Friedman J.H., Stüetzle W. (1981) Projection Pursuit Regression .*J.A.S.A.*,vol.76.

Friedman J. H. (1984) SMART users'guide.*Tech Rep. LCM001, Dept. Of Statistics, Stanford University.*

Friedman J.,Grosse E., Stüetzle W. (1983) Multidimentional additive Spline approximation. *SIAM,J.Sci.Stat.Comput* 4.

Friedman J.H.(1991) Multivariate Adaptive Regression Splines.*The Annals of Statistics,*19.

Gifi A. (1990) *Non Linear multivariate analysis.* Chichester: Wiley.

Härdle W. (1989) *Applied non-parametric regression. Cambridge Univ. Press.*

Hastie T. & Tibshirani R.. (1986). Generalized Additive Models. *Statistical Science,* n.1.

Hastie T., Tibshirani R., Buja A. (1994). Flexible Discrimination Analysis by Optimal Scoring *J.A.S.A.* vol.89.

Huber P.J. (1985). Projection pursuit. *The Annals of Statistics,* vol.13.

LeBlanc M. & Tibshirani R.(1994)Adaptive Principal Surfaces. *J.A.S.A.*,vol.89.

Lauro N.C.,Tessitore G.,Verde R. (1996) Multivariate B-spline coding. *IFCS.*

Lombardo R. & Tessitore G. (1997a) Adaptive Principal Surface in Constrained Principal Component Analysis. *Contributed papers IFCS classification and data analysis group.*

Lombardo R. & Tessitore G. (1997b) *PS* Constrained Analysis. (*submitted*)

Lombardo R., D'Ambra L., Tessitore G. (1997c) An algorithm for detecting the number o knots in Constrained Principal Component Analysis. *Contributed papers* vol. ISI 1997.

Meulman J.J. (1986) A distance approach to non linear multivariate analysis. Leiden:DSWO Press

Meulman J.J. (1992) The integretation of Multidimensional Scaling and Multivariate Analysis with Optimal Transformations. *Psychometrika, vol.*57.

Ramsay J.O.(1988) Monotone regression spline in action. *Statistical Science,*3.

Rijckevorsel J. L.A. van (1987). *The application of horseshoes and fuzzy coding in multiple correspondence analysis.* Leiden: DSWO Press.

Rijckevorsel J. L.A.van & de Leeuw J (1988) *Component and Correspondence Analysis*. Chicester: Wiley.

Rijckevorsel J. L.A. van & Tessitore G. (1993) An algorithm for Multivariate Adaptive Component and Correspondence Analysis (MACCA). *49th ISI session, Contributed Papers 2.*

Stone C. (1985) Additive Regression and others non-parametric models. The *Annals of Statistics* N.13.

Tessitore G. & Balbi S. (1996) An algorithm for detecting the number of knots in Non linear Principal Component Analysis. *Compstat*

Thisted A.R. (1988) Elements of Statistical Computing: Numerical computation, *Chapman & Hall.*

Verde R. (1992) Non symmetrical Correspondence Analysis: a non linear approach. Statistica Applicata, vol.4, n.4.

Acknowledgements *This paper was supported by project CNR (1997) «Nuove Metodologie per l'Analisi dei Dati a tre indici», responsible prof. Luigi D'Ambra. The authors thank the referees for the helpful comments.*

Multivariate L_1 Mean

Yadolah Dodge and Valentin Rousson

Statistics Group, University of Neuchâtel, Switzerland

Summary: The center $\boldsymbol{\mu}$ of a multivariate data set $\{\mathbf{x}_1, ..., \mathbf{x}_n\}$ can be defined as minimizer of the norm of a vector of distances. In this paper, we consider a vector of L_1-distances $\mathbf{y}_1' = (||\mathbf{x}_1 - \boldsymbol{\mu}||_1, ..., ||\mathbf{x}_n - \boldsymbol{\mu}||_1)$ and a vector of L_2-distances $\mathbf{y}_2' = (||\mathbf{x}_1 - \boldsymbol{\mu}||_2, ..., ||\mathbf{x}_n - \boldsymbol{\mu}||_2)$. We define the L_1-median and the L_1-mean as the minimizers of respectively the L_1- and the L_2-norm of $\mathbf{y}_1$; and then the L_2-median and the L_2-mean as the minimizers of respectively the L_1- and the L_2-norm of $\mathbf{y}_2$. In doing so, we obtain four alternatives to describe the center of a multivariate data set. While three of them have been already investigated in the statistical literature, the L_1-mean appears to be a new concept. Contrary to the L_1-median, the L_1-mean is proved to be unique in almost all situations.

Key words: Mean, Median, Measure of centrality, Multivariate data, L_1-norm, L_2-norm, Location.

1. Introduction

The median and the mean are two well known criteria used to describe the center of a univariate data set. Let a data set $\mathbf{x} = (x_1, ..., x_n)'$ be a vector of $\mathbf{R}^n$. The median of $\mathbf{x}$ can be defined as the point μ that minimizes the function

$$\sum_{i=1}^{n} |x_i - \mu| \tag{1}$$

and the mean of $\mathbf{x}$ as the point μ that minimizes the function

$$\sum_{i=1}^{n} (x_i - \mu)^2 . \tag{2}$$

The median and the mean of $\mathbf{x}$ may also be defined as the minimizer of respectively the L_1-norm and the L_2-norm of the vector of distances

$$\mathbf{y} = \mathbf{y}(\mu) = \begin{pmatrix} |x_1 - \mu| \\ \cdots \\ |x_n - \mu| \end{pmatrix} .$$

since $\|\mathbf{y}\|_1$ is the same as (1), whereas $\|\mathbf{y}\|_2$ is the square root of (2).

A natural way to extend the definition of the mean from one to many dimensions cases is to define the mean of a multivariate data set as the vector of coordinate means, as we find in all multivariate statistical textbooks. While this is not problematic, similar extension for the median is not straightforward. The vector of coordinate medians is effectively only one of the several possibilities proposed in the statistical literature to define the median of a multivariate data set. Small (1990) surveys several possible definitions of the multivariate median such those proposed by Tukey (1975), Oja (1983) and Liu (1990). These have the common property of producing the usual definition of the median in the univariate case, but give different kinds of medians in multivariate situations.

In this paper, we reconsider the problem of defining the median and the mean for data sets in dimension higher than one. Our point of departure is to define the multivariate median and the multivariate mean as minimizers of respectively the L_1-norm and the L_2-norm of a vector of distances, as done in the univariate case. There are however several kinds of distances in multivariate spaces. In Section 2 we consider a vector of L_1-distances as well as a vector of L_2-distances. In doing so we obtain two types of multivariate medians and two types of multivariate means that are called L_1-median, L_2-median, L_1-mean and L_2-mean. It turns out that the L_1-median is the vector of coordinate medians and the L_2-mean the vector of coordinate means. The L_2-median and the L_1-mean are less well-known. But while the L_2-median has already been investigated in the statistical literature, the L_1-mean appears to be a new concept. Thus the L_1-mean arises to be a surprising alternative to the L_2-mean which is usually presented as the only possibility of generalizing the univariate mean. The question of the uniqueness of the L_1-mean is discussed in Section 3. Contrary to the L_1-median, the L_1-mean is proved to be unique in almost all situations.

2. L_1- and L_2- Medians and Means

Let a data set $\mathbf{X} = (\mathbf{x}_1, ..., \mathbf{x}_n)'$ be a matrix of $\mathbf{R}^{n \times p}$, i.e. each data point $\mathbf{x}_i = (x_{i1}, ..., x_{ip})$ is an element of $\mathbf{R}^p$. In order to be able to define the median or the mean $\boldsymbol{\mu} = (\mu_1, ..., \mu_p)$ of the multivariate data set $\mathbf{X}$ in a similar way as for the univariate case, we have first to choose a vector of distances and then to minimize respectively the L_1- and the L_2-norm of this vector. In this paper we propose to choose either the vector of L_1-distances

$$\mathbf{y}_1 = \mathbf{y}_1(\boldsymbol{\mu}) = \begin{pmatrix} \|\mathbf{x}_1 - \boldsymbol{\mu}\|_1 \\ ... \\ \|\mathbf{x}_n - \boldsymbol{\mu}\|_1 \end{pmatrix}$$

or the vector of L_2-distances

$$\mathbf{y}_2 = \mathbf{y}_2(\boldsymbol{\mu}) = \begin{pmatrix} \|\mathbf{x}_1 - \boldsymbol{\mu}\|_2 \\ \cdots \\ \|\mathbf{x}_n - \boldsymbol{\mu}\|_2 \end{pmatrix}.$$

The L_1-median is then defined as the minimizer of $\|\mathbf{y}_1\|_1$, the L_1-mean as the minimizer of $\|\mathbf{y}_1\|_2$ (or equivalently of $\|\mathbf{y}_1\|_2^2$), the L_2-median as the minimizer of $\|\mathbf{y}_2\|_1$ and the L_2-mean as the minimizer of $\|\mathbf{y}_2\|_2$ (or equivalently of $\|\mathbf{y}_2\|_2^2$).

Thus the L_1-median of $\mathbf{X}$ is the point that minimizes

$$\|\mathbf{y}_1\|_1 = \sum_{i=1}^{n} \left| \sum_{j=1}^{p} |x_{ij} - \mu_j| \right| = \sum_{i=1}^{n}\sum_{j=1}^{p} |x_{ij} - \mu_j| . \tag{3}$$

The L_1-mean of $\mathbf{X}$ is the point that minimizes

$$\|\mathbf{y}_1\|_1^2 = \sum_{i=1}^{n} \left(\sum_{j=1}^{p} |x_{ij} - \mu_j| \right)^2 . \tag{4}$$

The L_2-median of $\mathbf{X}$ is the point that minimizes

$$\|\mathbf{y}_2\|_1 = \sum_{i=1}^{n} \left| \sqrt{\sum_{j=1}^{p} (x_{ij} - \mu_j)^2} \right| = \sum_{i=1}^{n} \sqrt{\sum_{j=1}^{p} (x_{ij} - \mu_j)^2} \tag{5}$$

and the L_2-mean of $\mathbf{X}$ is the point that minimizes

$$\|\mathbf{y}_2\|_2^2 = \sum_{i=1}^{n}\sum_{j=1}^{p} (x_{ij} - \mu_j)^2 . \tag{6}$$

We have therefore two ways to define the median and two ways to define the mean of a multivariate data set.

Let us return to the function (3) minimized by the L_1-median. As

$$\sum_{i=1}^{n}\sum_{j=1}^{p} |x_{ij} - \mu_j| = \sum_{j=1}^{p}\sum_{i=1}^{n} |x_{ij} - \mu_j| ,$$

the jth coordinate of the L_1-median of $\mathbf{X}$ is the point μ_j that minimizes the function $\sum_{i=1}^{n} |x_{ij} - \mu_j|$ and hence is the median of the univariate data set $\mathbf{x}_j = (x_{1j}, ..., x_{nj})'$. As the same argument is valid for each coordinate, it follows that the L_1-median is simply the vector of coordinate medians. This type of median was first introduced by Hayford (1902). It is also studied in Mood (1941) and in Haldane (1948) where it is called the *arithmetic median.* The L_1-median of a data set is not unique when the size n of the data set is even as it is the case for the univariate median.

The identification of the L_2-mean is straightforward too. The function (6) that has to be minimized can be rewritten as follows

$$\sum_{i=1}^{n}\sum_{j=1}^{p}(x_{ij}-\mu_j)^2=\sum_{j=1}^{p}\sum_{i=1}^{n}(x_{ij}-\mu_j)^2 .$$

Thus the jth coordinate of the L_2-mean of $\mathbf{X}$ is the point μ_j that minimizes the function $\sum_{i=1}^{n}(x_{ij}-\mu_j)^2$ and hence is the mean of the univariate data set $\mathbf{x}_j=(x_{1j},...,x_{nj})'$. As the same argument is valid for each coordinate, it follows that the L_2-mean is simply the vector of coordinate means. The L_2-mean is a very old and natural concept. All multivariate books are based on the L_2-mean. The L_2-mean of a data set is always unique.

The L_2-median is also an already known concept. It was first introduced by Gini and Galvani (1929) and by Eells (1930) and was called the *spatial median* by Brown (1983), although Haldane (1948) introduced the term of *geometrical median,* and Gower (1974) the term of *mediancentre.* In his article which treats all kinds of multivariate medians, Small (1990) unfortunately introduces the term of L_1*-median* to denote the spatial median. This last appelation brings some confusion since as mentioned above, the spatial median minimizes a sum of L_2-distances! The vector of coordinate medians which minimizes a sum of L_1-distances should be called L_1-median. The L_2-median of a data set is unique except in the case when the data points are found in a one dimensional subspace, where the L_2-median reduces to the usual median in this subspace (see e.g. Milasevic and Ducharme, 1987).

Let us now differentiate the function (4) minimized by the L_1-mean. We obtain as j^{th} partial derivative (so long as $x_{ij}\neq\mu_j$ for each $i=1,...,n$)

$$\frac{\partial\left\|\mathbf{y}_1\right\|_2^2}{\partial\mu_j}=-2\sum_{i=1}^{n}sgn\left(x_{ij}-\mu_j\right)\cdot\left\|\mathbf{x}_i-\boldsymbol{\mu}\right\|_1$$

where the function $sgn(x)$ is defined to take the value $+1$ and -1 if $x>0$, respectively if $x<0$. Thus if $\boldsymbol{\mu}$ is an L_1-mean of $\mathbf{X}$, we have (for each $j=1,...,p$) that the sum of the L_1-norms of the vectors $\mathbf{x}_i-\boldsymbol{\mu}$ for which $x_{ij}<\mu_j$ is equal to the sum of the L_1-norms of the vectors $\mathbf{x}_i-\boldsymbol{\mu}$ for which $x_{ij}>\mu_j$, that is

$$\sum_{\{i:x_{ij}<\mu_j\}}\left\|\mathbf{x}_i-\boldsymbol{\mu}\right\|_1=\sum_{\{i:x_{ij}>\mu_j\}}\left\|\mathbf{x}_i-\boldsymbol{\mu}\right\|_1 . \tag{7}$$

To our knowledge the L_1-mean has never been studied.

3. Uniqueness of the L_1-mean

The function (4) minimized by the L_1-mean is a continuous real valued convex function defined in $\mathbf{R}^p$, which tends to infinity in all directions of $\mathbf{R}^p$. It follows from the theory of convexity (see e.g. Rockafellar, 1972, p.265) that the minimum of (4) is attained on an non empty closed bounded convex set. This result ensures the existence of the L_1-mean but not its uniqueness.

To see the convexity of (4), let us consider two points $\boldsymbol{\mu} = (\mu_1, ..., \mu_p)$ and $\boldsymbol{\nu} = (\nu_1, ..., \nu_p)$ and a real number λ such that $0 < \lambda < 1$. We have

$$\begin{aligned} \|\mathbf{y}_1((1-\lambda)\boldsymbol{\mu}+\lambda\boldsymbol{\nu})\|_2^2 &= \sum_{i=1}^{n}\left(\sum_{j=1}^{p}|x_{ij}-(1-\lambda)\mu_j-\lambda\nu_j|\right)^2 \\ &= \sum_{i=1}^{n}\left(\sum_{j=1}^{p}|(1-\lambda)(x_{ij}-\mu_j)+\lambda(x_{ij}-\nu_j)|\right)^2 \\ &\leq \sum_{i=1}^{n}\left((1-\lambda)\sum_{j=1}^{p}|x_{ij}-\mu_j|+\lambda\sum_{j=1}^{p}|x_{ij}-\nu_j|\right)^2 \end{aligned}$$

with the equality if and only if

$$\forall i=1,...,n \quad \forall j=1,...,p \quad (x_{ij}\leq\mu_j \text{ and } x_{ij}\leq\nu_j) \text{ or } (x_{ij}\geq\mu_j \text{ and } x_{ij}\geq\nu_j). \tag{8}$$

As we have (a and b being two positive real numbers)

$$(1-\lambda)a^2+\lambda b^2-[(1-\lambda)a+\lambda b]^2=\lambda(1-\lambda)(a-b)^2\geq 0$$

with the equality if and only if $a = b$, we conclude that

$$\begin{aligned} &\sum_{i=1}^{n}\left((1-\lambda)\sum_{j=1}^{p}|x_{ij}-\mu_j|+\lambda\sum_{j=1}^{p}|x_{ij}-\nu_j|\right)^2 \\ \leq\; &(1-\lambda)\sum_{i=1}^{n}\left(\sum_{j=1}^{p}|x_{ij}-\mu_j|\right)^2+\lambda\sum_{i=1}^{n}\left(\sum_{j=1}^{p}|x_{ij}-\nu_j|\right)^2 \end{aligned}$$

with the equality if and only if

$$\forall i=1,...,n \quad \sum_{j=1}^{p}|x_{ij}-\mu_j|=\sum_{j=1}^{p}|x_{ij}-\nu_j|. \tag{9}$$

Putting together these two inequalities we obtain the convexity of (4) that is

$$\left\| \mathbf{y}_1 \left((1-\lambda)\,\boldsymbol{\mu} + \lambda \boldsymbol{\nu} \right) \right\|_2^2 \leq (1-\lambda) \cdot \left\| \mathbf{y}_1 (\boldsymbol{\mu}) \right\|_2^2 + \lambda \cdot \left\| \mathbf{y}_1 (\boldsymbol{\nu}) \right\|_2^2 . \qquad (10)$$

with the equality if and only if (8) and (9) hold.

Now if $\boldsymbol{\mu}$ and $\boldsymbol{\nu}$ are two distinct minima of (4), that is are two distinct L_1-means of the data set $\mathbf{X}$, it is clear that we have the equality in (10). Thus conditions for the non uniqueness of the L_1-mean depend precisely on (8) and (9). The following theorem precises these conditons for bivariate data sets.

Theorem 3.1. *Let $\boldsymbol{\mu} = (\mu_1, \mu_2)$ be an L_1-mean of a data set $\mathbf{X} = (\mathbf{x}_1, ..., \mathbf{x}_n)'$ for which each data point $\mathbf{x}_i = (x_{i1}, x_{i2})$ is an element of $\mathbf{R}^2$.*

(a) If for each $i = 1, .., n$ we have

$$(x_{i1} \geq \mu_1 \text{ and } x_{i2} \geq \mu_2) \text{ or } (x_{i1} \leq \mu_1 \text{ and } x_{i2} \leq \mu_2)$$

that is if one part of the data are on the top right of $\boldsymbol{\mu}$ and the other part of the data are on the bottom left of $\boldsymbol{\mu}$, then the L_1-means of $\mathbf{X}$ are points of the form $(\mu_1 + \varepsilon, \mu_2 - \varepsilon)$ with $\varepsilon \in \mathbf{R}$. The L_1-means of $\mathbf{X}$ are hence found on a segment of line with slope -1.

(b) If for each $i = 1, .., n$ we have

$$(x_{i1} \leq \mu_1 \text{ and } x_{i2} \geq \mu_2) \text{ or } (x_{i1} \geq \mu_1 \text{ and } x_{i2} \leq \mu_2)$$

that is if one part of the data are on the top left of $\boldsymbol{\mu}$ and the other part of the data are on the bottom right of $\boldsymbol{\mu}$, then the L_1-means of $\mathbf{X}$ are points of the form $(\mu_1 + \varepsilon, \mu_2 + \varepsilon)$ with $\varepsilon \in \mathbf{R}$. The L_1-means of $\mathbf{X}$ are hence found on a segment of line with slope $+1$.

(c) If we are not in one of the situations (a) or (b) then the L_1-mean of $\mathbf{X}$ is unique.

Proof. Let $\boldsymbol{\nu}$ be another L_1-mean of $\mathbf{X}$ such that $\boldsymbol{\nu} = (\mu_1 + \varepsilon, \mu_2 + \delta)$ with $\varepsilon, \delta \in \mathbf{R}$ and let γ_i be the difference $\left\| \mathbf{x}_i - \boldsymbol{\nu} \right\|_1 - \left\| \mathbf{x}_i - \boldsymbol{\mu} \right\|_1$ for $i = 1, .., n$. We have

$$\begin{aligned} \left\| \mathbf{y}_1 (\boldsymbol{\nu}) \right\|_2^2 &= \sum_{i=1}^{n} \left\| \mathbf{x}_i - \boldsymbol{\nu} \right\|_1^2 = \sum_{i=1}^{n} \left(\left\| \mathbf{x}_i - \boldsymbol{\mu} \right\|_1 + \gamma_i \right)^2 \\ &= \left\| \mathbf{y}_1 (\boldsymbol{\mu}) \right\|_2^2 + 2 \sum_{i=1}^{n} \gamma_i \cdot \left\| \mathbf{x}_i - \boldsymbol{\mu} \right\|_1 + \sum_{i=1}^{n} \gamma_i^2 . \end{aligned} \qquad (11)$$

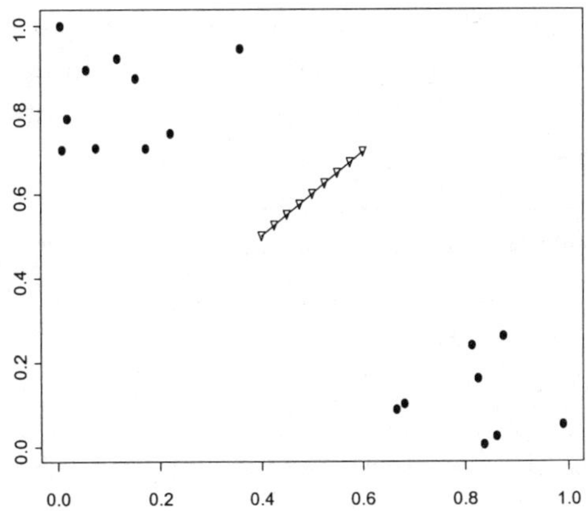

Figure 3.1: The L_1-means of these 18 data points are on the segment of line marked with triangles.

Then partition the data points $\mathbf{x}_i$ as follows: let A_1, A_2, A_3 and A_4 be the sets of data points which are respectively on the top right, on the top left, on the bottom left and on the bottom right of $\boldsymbol{\mu}$. Note that by (8) A_1, A_2, A_3 and A_4 are also the sets of data points which are respectively on the top right, on the top left, on the bottom left and on the bottom right of $\boldsymbol{\nu}$. We have therefore $\gamma_i = -\varepsilon - \delta$ for the n_1 points in A_1, $\gamma_i = \varepsilon - \delta$ for the n_2 points in A_2, $\gamma_i = \varepsilon + \delta$ for the n_3 points in A_3 and $\gamma_i = -\varepsilon + \delta$ for the n_4 points in A_4. It follows that the second term of the right part of (11) is

$$2\varepsilon \cdot \left[\sum_{\{i:\mathbf{x}_i \in A_2 \cup A_3\}} \|\mathbf{x}_i - \boldsymbol{\mu}\|_1 - \sum_{\{i:\mathbf{x}_i \in A_1 \cup A_4\}} \|\mathbf{x}_i - \boldsymbol{\mu}\|_1 \right] +$$
$$2\delta \cdot \left[\sum_{\{i:\mathbf{x}_i \in A_3 \cup A_4\}} \|\mathbf{x}_i - \boldsymbol{\mu}\|_1 - \sum_{\{i:\mathbf{x}_i \in A_1 \cup A_2\}} \|\mathbf{x}_i - \boldsymbol{\mu}\|_1 \right]$$

where each [...] is equal to zero by the property (7) of the L_1-mean given in Section 3. Thus, if $\boldsymbol{\nu}$ is an L_1-mean of $\mathbf{X}$, the third term of the right part of (11)

$$\sum_{i=1}^{n} \gamma_i^2 = (n_1 + n_3) \cdot (\varepsilon + \delta)^2 + (n_2 + n_4) \cdot (\varepsilon - \delta)^2$$

must be equal to zero in order to have $\|\mathbf{y}_1(\boldsymbol{\nu})\|_2^2 = \|\mathbf{y}_1(\boldsymbol{\mu})\|_2^2$. If $n_2 = n_4 = 0$, we must then have $\delta = -\varepsilon$ that is $\boldsymbol{\nu} = (\mu_1 + \varepsilon, \mu_2 - \varepsilon)$ which is the case

(a) of the theorem. If $n_1 = n_3 = 0$, we must then have $\delta = \varepsilon$ that is $\boldsymbol{\nu} = (\mu_1 + \varepsilon, \mu_2 + \varepsilon)$ which is the case (b) of the theorem. In any other case we must have simultaneously $\delta = -\varepsilon$ and $\delta = \varepsilon$ that is $\delta = \varepsilon = 0$ and $\boldsymbol{\nu} = \boldsymbol{\mu}$ which is the case (c) of the theorem. □

The consequence of Theorem 3.1 is that the L_1-mean of a bivariate data set is unique except for very particuliar cases. Figure 3.1 illustrates situation (b) of Theorem 3.1.

References

[1] Brown, B.M. (1983). Statistical uses of the spatal median. *J.Roy. Statist. Soc.* B **45** 25-30.

[2] Eells, W.C. (1930). A mistaken conception of the center of the population. *J. Amer. Statist. Assoc.* **25** 33-40.

[3] Gini, C. and Galvani, L. (1929). Di talune estensioni dei concetti di media ai caratteri qualitativi. *Metron* **8** 3-209.

[4] Gower, J.C. (1974). The mediancentre. *Appl. Statist.* **23** 466-470.

[5] Haldane, J.B.S. (1948). Note on the median of a multivariate distribution. *Biometrika* **35** 414-415.

[6] Hayford, J.F. (1902). What is the center of an area, or the center of the population?. *J. Amer. Statist. Assoc.* **8** 47-58.

[7] Liu, R.Y. (1990). On a notion of data depth based upon random simplices. *Ann. Statist.* **18** 405-414.

[8] Milasevic, P. and Ducharme, G.R. (1987). Uniqueness of the spatial median. *Ann. Statist.* **15** 1332-1333.

[9] Mood, A.M. (1941). On the joint distribution of the median in samples from a multivariate population. *Ann. Math. Statist.* **12** 268-278.

[10] Oja, H. (1983). Descriptive statistics for multivariate distributions. *Statist. Prob. Letters* **1** 327-332.

[11] Rockafellar, R.T. (1972). *Convex Analysis.* Princeton University Press.

[12] Small, C.G. (1990). A survey of multidimensional medians. *Int. Statist. Review* **58** 263-277.

[13] Tukey, J.W. (1975). Mathematics and the picturing of data. In *Proc. International Congress of Mathematicians*, Vol. 2, pp. 523-531. Vancouver.

Simple and Multiple Regression Splines with Bounded Optimal Knots

Jean-François Durand*, Nicolas Molinari†, Robert Sabatier*

⋆Laboratoire de Probabilités et Statistique, Université Montpellier II, Place Eugène Bataillon, 34095 Montpellier, France.

†Laboratoire de Biostatistique, Institut Universitaire de Recherche Clinique, 75 rue de la Cardonille, 34093 Montpellier, France.

∗Laboratoire de Physique Moléculaire et Structurale, 15 av. Ch. Flahaut, 34060 Montpellier, France.

Abstract: Using a B-spline representation for splines, knots seen as free variables, the approximation to data by splines improve greatly. The damage due to the presence of a lot of local optima is very important in the univariate regression context, and things are getting worse in multivariate additive modeling. We present a new algorithm to select knots submitted to constraints for computing least squares spline approximations.

Key words: Additive models; Bound constrained optimization; Free knots selection; Regression splines.

1. Introduction

The approximation to functions by splines has long been known to improve dramatically if the knots are free parameters. A serious problem is the existence of many stationary points for the least squares objective function, and the apparent impossibility of deciding when the global optimum has been found. Free knot splines have not been as popular as might be expected from these results. One problem has been that analytic expressions for optimal knot locations, or even for general characteristics of optimal knot distributions, are not easy to derive. Computationally, things are a little better: there exist several algorithms to find knot locations. Gallant and Fuller (1973) use an iterative algorithm based on the Gauss-Newton method to solve the problem. Jupp (1978) introduces a transformation of the knots to avoid the "lethargy" phenomenon: this transformation puts to infinity the knot set boundaries that makes impossible the free knots coalescence. If we can choose and increase the number of samples, we can use the Agarwal and Studden (1980) algorithm. Guertin (1992) penalizes the distance from equidistant knots. In the following paper, we recall the free knots problem, and the consequences of the "lethargy" theorem. A new algorithm (Molinari 1997) is proposed to compute knot locations in simple and multiple regression by least squares splines. We imagine that a knot can be seen as a convex combination of the data points. It leads us to oblige the knots to lie within disjoint intervals. Efficiency and fastness which characterize the box constrained minimizing algorithm, allow us to propose a method that explores the local minima (knot locations) with an attractive amount of computational time and that definitely avoid the coalescence of free knots.

2. B-spline functions, fixed and free knots for least-squares splines

2.1 B-spline functions

Let $(\xi_0 =)a < \xi_1 < \xi_2 < \cdots < \xi_K < b(= \xi_{K+1})$ be a subdivision of the interval $[a, b]$ by K distinct points. These points are the "knots" of the spline function.

A spline function, $s(t)$, of order $d + 1$ or degree d is a function which
(i) is a polynomial of degree d in each open interval $]\xi_{i-1}, \xi_i[$ for $i = 1, K + 1$,
(ii) has $d - 1$ continuous derivatives in the open interval $]a, b[$.
For a fixed set of knots, $\boldsymbol{\xi} = (\xi_1, \xi_2, \cdots, \xi_K)$, the set of such splines is a linear space of functions with $K + d + 1$ free parameters (de Boor 1978).

A basis for this linear space is given by the Schoenberg's B-splines, or *Basic*-splines. This basis provides a stable method (de Boor 1978) for computing spline functions written as

$$s(x, \boldsymbol{\beta}, \boldsymbol{\xi}) = \sum_{l=1}^{K+d+1} \beta_l B_l(x, \boldsymbol{\xi}),$$

where the $\boldsymbol{\beta} = (\beta_1, \cdots, \beta_{K+d+1})'$ is the vector of the spline coefficients.

2.2 Simple regression and lethargy problem

Let $\{x_i, y_i\}_{i=1,\cdots,n}$ be a set of n observations ranging over $[a, b] \times I\!R$. Denote $\boldsymbol{B}(\boldsymbol{\xi}) = (B_l(x_i, \boldsymbol{\xi}))_{i=1,\cdots,n}^{l=1,\cdots,K+d+1}$ the $n \times (K + d + 1)$ matrix of sampled basis functions, and $\boldsymbol{y} = (y_1, \cdots, y_n)'$ the vector of sampled response.

When knots are fixed, it is a straightforward linear least squares problem to fit the data by splines (Stone 1985)

$$\hat{\boldsymbol{\beta}}(\boldsymbol{\xi}) = \arg\min_{\boldsymbol{\beta}} \|\boldsymbol{y} - \boldsymbol{B}(\boldsymbol{\xi})\boldsymbol{\beta}\|^2 = (\boldsymbol{B}(\boldsymbol{\xi})'\boldsymbol{B}(\boldsymbol{\xi}))^+\boldsymbol{B}(\boldsymbol{\xi})'\boldsymbol{y}, \tag{1}$$

where $\|.\|$ is the usual Euclidian norm and $\boldsymbol{B}^+$ is the Moore-Penrose inverse of $\boldsymbol{B}$. Then, $s(x, \boldsymbol{\beta}, \boldsymbol{\xi})$ is estimated by $\hat{s}(x, \boldsymbol{\xi}) = \sum_{l=1}^{K+d} \hat{\beta}_l B_l(x, \boldsymbol{\xi})$.

When the locations of knots are free variables, the class of splines is no longer linear but form a mixture of linear and nonlinear parameters. Then, the least squares problem is to find

$$\arg \min_{\boldsymbol{\beta},\ \boldsymbol{\xi} \in [a,b]^K} \|\boldsymbol{y} - \boldsymbol{B}(\boldsymbol{\xi})\boldsymbol{\beta}\|^2. \tag{2}$$

When $\boldsymbol{\xi}$ is fixed, the problem (2) reduces to (1). In fact, (Golub and Pereyra 1973) shows that the solutions to (2) are those to

$$\arg \min_{\boldsymbol{\xi} \in [a,b]^K} \|\boldsymbol{y} - \boldsymbol{B}(\boldsymbol{\xi})\hat{\boldsymbol{\beta}}(\boldsymbol{\xi})\|^2. \tag{3}$$

As a consequence, we from now on denote the objective function $F(\boldsymbol{\xi}) = \|\boldsymbol{y} - \boldsymbol{B}(\boldsymbol{\xi})\hat{\boldsymbol{\beta}}(\boldsymbol{\xi})\|^2$.

The following "lethargy" theorem (Jupp 1975) is intrinsic to the free knots problem, and affect the stable and effective computation of optimal knots.

The lethargy theorem: *Denote* $S_K[a,b] = \{\boldsymbol{\xi} \in I\!R^K; a < \xi_1 < \xi_2 < \cdots < \xi_K < b\}$ *the open simplex of knots, and* $S_K^{(p)}$ *the pth (open) main face of* $S_K[a,b]$. $S_K^{(p)}$ *is defined by the system* $(\xi_j - \xi_{j-1}) > 0$, *for* $j \neq p$ *and* $\xi_p = \xi_{p-1}$. *On the pth main face,* $S_K^{(p)}$, $\mathbf{n}_p^{'}\nabla F(\boldsymbol{\xi}) = 0$ *for* $p = 2, \cdots, K$, *where* $\mathbf{n}_p$ *is the unit outward normal to* $S_K^{(p)}$, *and* $\nabla F(\boldsymbol{\xi})$ *the gradient of* $F(\boldsymbol{\xi})$.

The first consequence is the existence of many stationary points of $F(\boldsymbol{\xi})$ on the faces $S_K^{(p)}$, and the second is the poor convergence, or "lethargy" properties, of algorithms that attempt to solve the free knots problem when they are near the boundaries $S_K^{(p)}$. Figure 1 shows $S_2[a,b]$ which is a triangle.

Figure 1. *The simplex* $S_2[a,b]$ *as the parameter space for splines with two variable knots.*

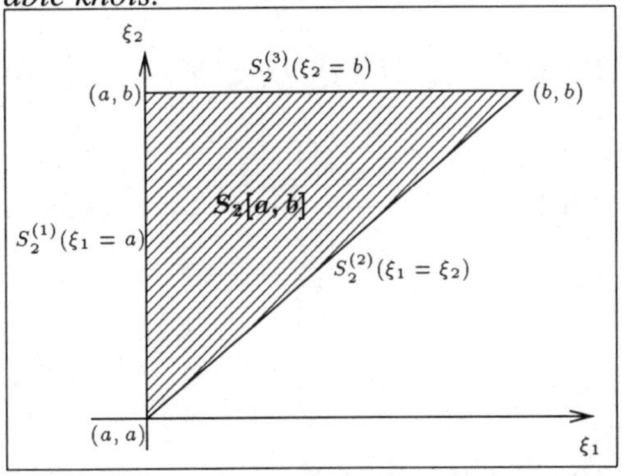

The "lethargy" property is illustrated on the titanium heat data (de Boor and Rice 1968) which is a set of 49 data values that express a thermal property of titanium. As the writers report, this data set is hard to approximate, and has significant experimental error. Figure 2 shows the approximation by cubic spline corresponding to a local minimum with two coalescent knots.

Figure 2. *Titanium heat data approximated by a 5 non optimal knots cubic spline.*

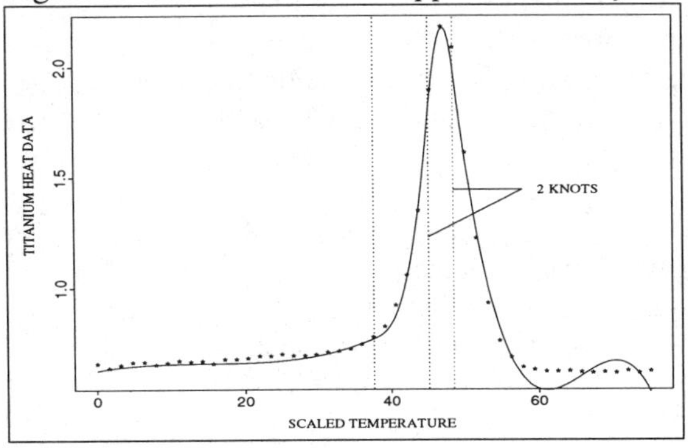

2.3 Additive multiple regression

Let us consider p centered covariates $(X_1, \cdots, X_p)$, and a random centered response Y, all mesured on n observations gathered in the matrices $\boldsymbol{X}$ and $\boldsymbol{Y}$. The problem we consider initially is the estimation of the conditional expectation or regression function $f(x_1, \cdots, x_p) = E(Y/X_1 = x_1, \cdots, X_p = x_p)$.

For constructing an additive estimator $\hat{f}$ of f defined by $\hat{f}(x_1, \cdots, x_p) = \sum_{j=1}^{p} \hat{f}_j(x_j)$, where the $\hat{f}_j$'s are centered, we use the *least squares spline* (LSS) approximation suggested by Stone (1985).

When $\boldsymbol{\xi}^j$ knots are fixed for any predictor j, i.e. given a fixed spline basis $\{B_l^j(., \boldsymbol{\xi}^j)\}_{l=1,\cdots,K_j+d_j}$, the (LSS) estimator of f, is defined by

$$\hat{f}(x_1, \cdots, x_p) = \sum_{j=1}^{p} \sum_{l=1}^{K_j+d_j+1} \hat{\beta}_l^j(\boldsymbol{\xi}) B_l^j(x_j, \boldsymbol{\xi}),$$

where $\hat{\boldsymbol{\beta}}(\boldsymbol{\xi}) = (\hat{\beta}_1^1(\boldsymbol{\xi}), \cdots, \hat{\beta}_{K_1+d_1+1}^1(\boldsymbol{\xi}), \cdots, \hat{\beta}_1^p(\boldsymbol{\xi}), \cdots, \hat{\beta}_{K_p+d_p+1}^p(\boldsymbol{\xi}))'$ is solution to the least squares problem

$$\hat{\boldsymbol{\beta}}(\boldsymbol{\xi}) = \arg\min_{\boldsymbol{\beta}} \|\boldsymbol{Y} - \boldsymbol{B}(\boldsymbol{\xi})\boldsymbol{\beta}\|^2 = (\boldsymbol{B}'(\boldsymbol{\xi})\boldsymbol{B}(\boldsymbol{\xi}))^+ \boldsymbol{B}'(\boldsymbol{\xi})\boldsymbol{Y}. \tag{4}$$

The super vector of spline coefficients is denoted $\boldsymbol{\xi} = (\xi_1^1, \cdots, \xi_{K_1}^1, \cdots, \xi_1^p, \cdots, \xi_{K_p}^p)$ and $\boldsymbol{B}(\boldsymbol{\xi}) = [\boldsymbol{B}^1(\boldsymbol{\xi})| \cdots |\boldsymbol{B}^p(\boldsymbol{\xi})]$ is the $n \times \sum_j (K_j + d_j + 1)$ column centered super coding matrix.

Stone (1985) showed that

$$\hat{f}_j(x_j) = \sum_{l=1}^{K_j+d_j+1} \hat{\beta}_l^j(\boldsymbol{\xi}) B_l^j(x_j, \boldsymbol{\xi})$$

are consistent estimators of the coordinate functions $f_j(x_j)$.

As in the univariate case, when knot locations are free variables, Golub and Pereyra's (1973) result also holds, and we only have to minimize the objective function $F(\boldsymbol{\xi}) = \|\boldsymbol{Y} - \boldsymbol{B}(\boldsymbol{\xi})\hat{\boldsymbol{\beta}}(\boldsymbol{\xi})\|^2$ with respect to $\boldsymbol{\xi}$.

3. Bounded Optimal knots

3.1 Description in the univariate case

Because the knots belong to the interval $[a, b]$, we impose that one knot, ξ_i, lies within some window $[l_i, u_i]$, and, to avoid the lethargy problem, that the windows are disconnected.

Let $\boldsymbol{l} = (l_1, \cdots, l_K)$, $\boldsymbol{u} = (u_1, \cdots, u_K)$ be the vectors of lower and upper bounds. To avoid overlapping windows, we take $l_{i+1} - u_i = \varepsilon > 0$ for $i = 1, \cdots, K-1$, and $l_1 - a = b - u_K = \varepsilon$. Note that $\lim_{\varepsilon \to 0} \bigcup_{i=1,K} [l_i, u_i] = [a, b]$. When the windows are fixed, the *bounded optimal knot* problem is to find

$$\hat{\boldsymbol{\xi}}(\boldsymbol{l}, \boldsymbol{u}) = \arg\min_{\boldsymbol{l} \le \boldsymbol{\xi} \le \boldsymbol{u}} F(\boldsymbol{\xi}). \tag{5}$$

Figure 3. *The $S_2[a,b]$ simplex and a set of bound constrained knots.*

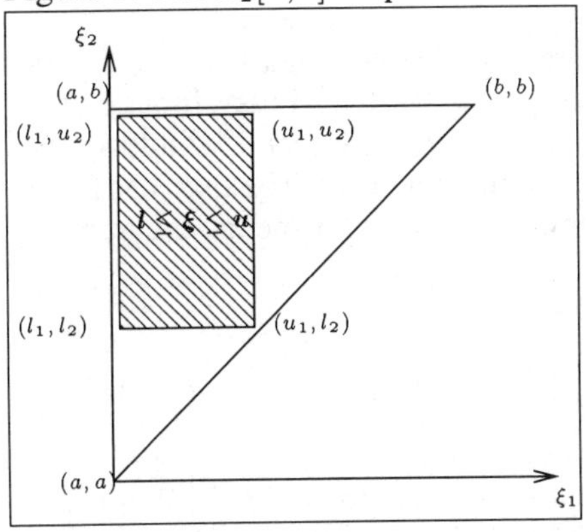

Clearly $\bigcup_{i=1,K}[l_i,u_i] \subset [a,b]$ for $i = 1,\cdots,K$, and (5) doesn't necessarily provides the global minimum to (3) because

$$\min_{\boldsymbol{l}\leq\boldsymbol{\xi}\leq\boldsymbol{u}} F(\boldsymbol{\xi}) \geq \min_{\boldsymbol{\xi}\in([a,b])^K} F(\boldsymbol{\xi}).$$

However, problem (5) is very easy to solve by using classical rapid algorithms. The visual examination of the spline approximation allows the user to experiment different choices for selecting the windows.

To explore most of the local optima, the strategy presented below provides an automatic selection of the windows. We construct a sequence $(\hat{\boldsymbol{\xi}}(\boldsymbol{l}^{(i)},\boldsymbol{u}^{(i)}))_i$ of, say, N solutions to (5) based on N independent uniformly distributed partitions $(l_1^{(i)},u_1^{(i)},\cdots,l_K^{(i)},u_K^{(i)})_{i=1,N}$. In fact, because upper and lower bounds are explicitly linked ($u_j = l_{j+1} - \varepsilon$) we only need to generate a sequence $(l_2^{(i)},\cdots,l_K^{(i)})_{i=1,N}$ of $K-1$ uniformly drawn lower bounds. For sufficiently large N, we expect that

$$\hat{\boldsymbol{\xi}} = \arg\min_{i=1,N} F(\hat{\boldsymbol{\xi}}(\boldsymbol{l}^{(i)},\boldsymbol{u}^{(i)}))$$

provides a good approximation to the optimal knot locations. Clearly, the number N of experiments is unknown, and we heuristically use $N = 100 \times K$.

3.2 The multivariate additive case

Using the notations in 2.3, the natural extension of the precedent procedure leads to solve the next optimisation problem wich is analogous to (5)

$$\hat{\boldsymbol{\xi}}(\boldsymbol{l},\boldsymbol{u}) = \arg\min_{\substack{\boldsymbol{l}^1\leq\boldsymbol{\xi}^1\leq\boldsymbol{u}^1\\ \boldsymbol{l}^2\leq\boldsymbol{\xi}^2\leq\boldsymbol{u}^2\\ \vdots\\ \boldsymbol{l}^p\leq\boldsymbol{\xi}^p\leq\boldsymbol{u}^p}} F(\boldsymbol{\xi}), \tag{6}$$

where the super vector of knots $\boldsymbol{\xi} = (\boldsymbol{\xi}^1,\cdots,\boldsymbol{\xi}^p)$ is constrained to lie within the super vectors of lower and upper bounds.

4. Applications

The examples and simulations we made were implemented on a Ultra-Sparc station, throwgh the so called *bok* (for *bounded optimal knots*) S-Plus© function that uses the native function $nlminb$ applied successively on N sets of windows. By default, initial knots are located at the center of the windows. The user has to decide on 3 types of input parameters: the number, N, of experiments, the number, K, of knots, and the degree, d, of the spline polynomials.

$$bok(x, y, N, K, d).$$

4.1 The titanium heat data

Using $(N, K, d) = (500, 5, 3)$, the global optimum of Jupp (1978) is obtained at the knot location $\xi = (37.62, 43.97, 47.37, 50.20, 59.21)$ after 8400 seconds of cpu time. Of course, the number of simulations (500) is an upper bound which should be improved in the future. Whatever that may be, selecting the initial knots actually located at the center of the windows did not affect the convergence of the procedure in any of the examples we tried the convergence of the procedure.

Figure 4. *Titanium heat data approximated by a 5 knots cubic spline. The location of the optimal knots is indicated by the vertical lines.*

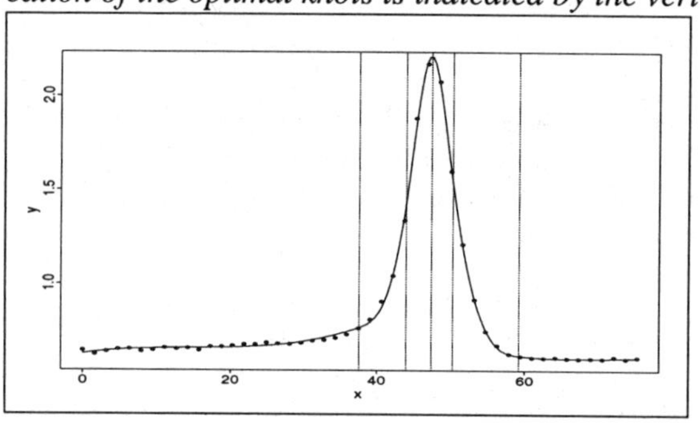

4.2 Simulations for multivariate additive modeling

To illustrate the additive case, the data (Durand 1997) consist of a sample ($n = 30$) of $p = 5$ predictors wich are strongly correlated and generated as follows: X_1 is uniform on $[-1, 1]$, $X_2 = 0.9X_1 + \varepsilon_1$, $X_3 = -1.1X_2 + \varepsilon_2$, $X_4 = 0.9X_3 + \varepsilon_3$ and $X_5 = -1.1X_4 + \varepsilon_4$ where ε_1 and ε_2 are normal $(0, 0.2)$ while ε_3 and ε_4 are normal $(0, 0.1)$. The response is generated by $Y_i = f(\boldsymbol{X}_i) + \varepsilon_i$ for $i = 1, \cdots, 30$, with the ε_i independently drawn from a normal $(0, 0.1)$ distribution. The function f is taken to be

$$f(x_1, \cdots, x_5) = 2\sin(\pi x_1) - 6x_2^3 + 3x_3 - 2x_2 + x_5.$$

B-splines used for the first 2 predictors are of degree 2 with 3 knots, of degree 1 with 3 knots for the 3 last variables.

The author shows that when few samples are used for strongly correlated covariates, least squares spline models obtained with fixed knots, present a large variability of the estimators (see Figure 5).

Figure 5. *A comparison between 10 least squares splines with equidistant fixed knots when the predictors are correlated and the samples are small (n=30).*

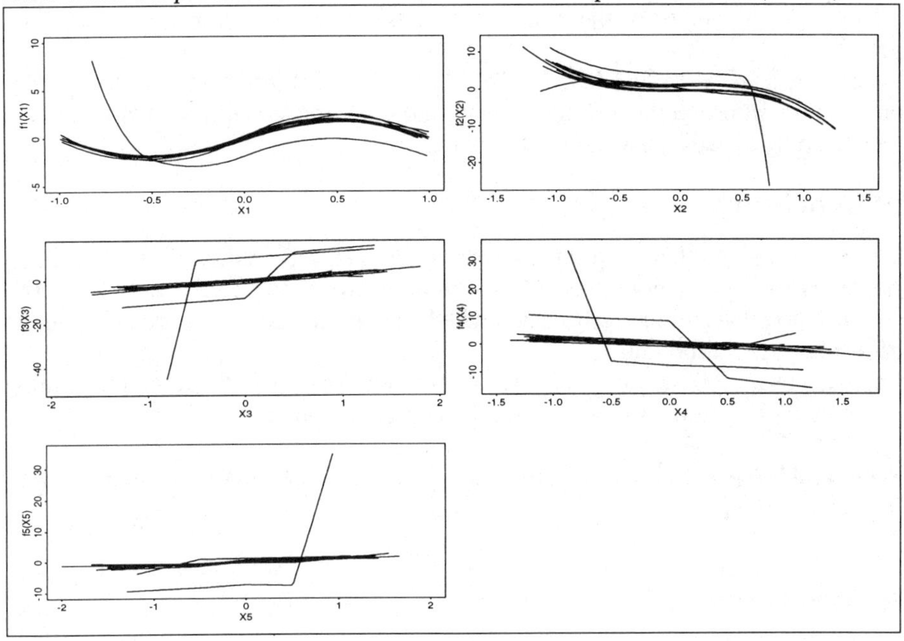

Figure 6. *A comparison between the 10 least squares splines with variable knots.*

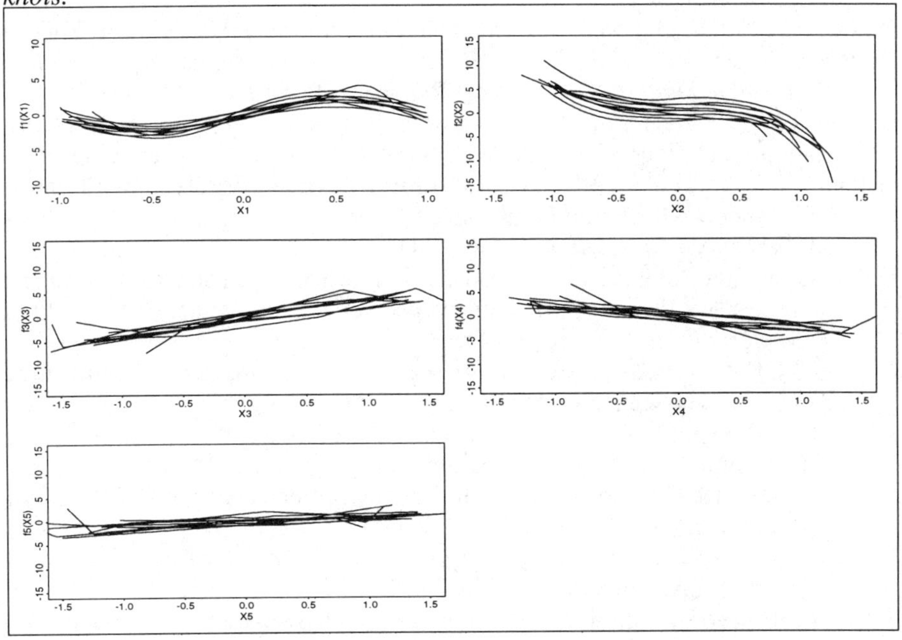

We use the program *mbok* (for *multivariate bounded optimal knots*) which is the analogous of *bok* for the additive case:

$$mbok(\mathbf{X}, y, N, K, d),$$

where K is the vector of the number of knots and d the vector of the spline degrees.
Using $(N, K, d) = (100, (3, 3, 3, 3, 3), (2, 2, 1, 1, 1))$, figure 6 shows that well located knots improve the prediction, and that the bounded optimal knots method can be apply successfully on additive data.

5. Conclusion

The presented method applied in both univariate and multivariate contexts is based on the use of bounded optimal knots to avoid coalescent knots. It constructs a non deterministic algorithm which is not sensitive to the crucial problem of locating the initial knots.
Finally, a cross-validation procedure can be used to estimate the remaining tuning parameters: the number of knots and the degree of spline functions.

Acknowledgments. The authors are grateful to the referee for helpful comments and suggestions.

References

Agarwal, G. G. and Studden, W. J. (1980), Asymptotic Integrated Mean Square Error using Least-Squares and Bias Minimizing Splines, *Annals of Statistics*, 8, 1307-1325.

De Boor, C. (1978), *A Practical Guide to Splines*, Springer-Verlag, New-York.

De Boor, C. and Rice, J. R. (1968), *Least squares cubic spline approximation. II: Variable knots*, CSD Technical Report 21, Purdue University, IN.

Durand, J. F. (1997), *Additive modeling of multivariate data by spline functions*, Rapport technique 97-04, Unité de Biométrie, Université Montpellier II.

Gallant, A. R. and Fuller, W. A. (1973), Fitting Segmented Polynomial Regression Models whose Join Points have to be estimated, *Journal of the American Statistical Association*, vol. 68, 341, 144-147.

Golub, G. H. and Pereyra, V. (1973), The differentiation of pseudo-inverses and nonlinear least squares problems whose variables separate, *SIAM Journal of Numerical Analysis*, 10, 33-45.

Guertin, M. C. (1992), *Sur les splines de régression à noeuds variables*, mémoire de Maitrise es Sciences, Université de Montréal.

Jupp, D. L. B. (1975), The Lethargy Theorem, a Property of Approximation by γ-Polynomials, *Journal of Approximation Theory*, 14, 204-217.

Jupp, D. L. B. (1978), Approximation to Data by Splines with Free Knots, *SIAM Journal of Numerical Analysis*, 15, 204-217.

Molinari, N. (1997), *Splines de Régression à Noeuds Variables Contraints*, mémoire de DEA, Université Montpellier II.

Stone, C. J. (1985), Additive regression and other nonparametric models, *Annals of Statistics*, 13, 689-705.

Interpretation of a Cross-Classification: a New Method and an Application

Sergio Camiz
Università La Sapienza, Roma; Université des Sciences et Technologies, Lille.
e-mail:camiz@mat.uniroma1.it
Jean-Jacques Denimal
Université des Sciences et Technologies, Lille.
e-mail:denimal@alea.univ-lille1.fr

Abstract: For the interpretation of a two-way contingency table cross-classification, based on two hierarchies, new indexes are developed, able at evaluating relative contribution of either hierarchy nodes to the other's nodes or partitions. Indexes are based on groups' inertia projection onto the orthogonal basis, associated to the explanatory hierarchy, in a vector space with χ^2 metrics. An application based on Lombardy municipalities 1991 Census data is given.

Keywords: Contingency tables, Hierarchical classification, Cross-classification.

1. Introduction

The study of cross-classification of two-way contingency tables aims at finding the best one and at explaining either of the two one-way partitions through the other. Orlóci (1978) and Feoli and Orlóci (1979) propose the *analysis of concentration* for the evaluation of a vegetation table *sharpness*: it is a kind of correspondence analysis based on a normalised table, but its use is limited only to presence-absence data. Camiz (1993, 1994) uses it for the search of the *best* cross-classification in a procedure for structuring vegetation tables, but even in this best application the association between groups is detected only through the usual correspondence analysis practice, namely the inspection of scatter diagrams. Govaert (1984) developed techniques based on Diday's (1970) *dynamical clouds*, whose results depend on the *a priori* choice of the suitable number of groups of each partition. Lebart *et al.*(1979) propose statistical methods to identify each group's *typical variables*. For frequency variables, typical are considered those whose occurrence in the group is significantly higher or lower than their expected value, on the basis of hypergeometrical law. This does not allow a direct comparison of two classifications, since they may not be arranged according to a hierarchy: in fact, they may be the same for two different groups of the same partition, but at different frequency levels. Nevertheless, it is a useful method for a deep insight of a classification structure. Benzécri *et al.* (1980) propose a decomposition of nodes inertia on canonical bases, with the same drawbacks as Lebart *et al.*(1979).
The proposed technique, that may be used as a stand-alone cross-classification

procedure, as developed in Denimal (1997), considers two vector bases associated to ascending hierarchical classifications, performed on both rows and columns of the two-way contingency table. The associated indices are used to evaluate the relative contribution of one hierarchy nodes to both other hierarchy nodes and partitions. In this way a relation may be found between hierarchies and partitions, helping in both interpretation and graphical representation of results.

2. The method

Given a two-way contingency table K, for two sets of characters I and J, we consider two hierarchies H^I, H^J built on I and J, respectively, using Ward's (1963; see also Greenacre, 1988) criterion. Two orthogonal bases are associated on vector spaces $\mathbb{R}^{|I|}$ and $\mathbb{R}^{|J|}$ to these hierarchies, where each vector (but one) represents a node in the corresponding hierarchy. Since these bases are orthogonal, they allow us to decompose the squared distance between each group centroid and the grand centroid. Thus, *indices* may be established, to interpret hierarchy and partitions on either set in terms of the hierarchy on the other.

Let be k_{ij}, $i \in I, j \in J$ a generic element of K. As usual, marginal and grand totals may be written as $k_{i\cdot} = \sum_{j\in J} k_{ij}$, $k_{\cdot j} = \sum_{i\in I} k_{ij}$, $k_{\cdot\cdot} = \sum_{(i,j)\in I\times J} k_{ij}$. Given two subsets $p \subseteq I$, $q \subseteq J$, we denote by k_p, k_q, and k_{pq} the partial sums $k_p = \sum_{i\in p} k_{i\cdot}$, $k_q = \sum_{j\in q} k_{\cdot j}$, $k_{pq} = \sum_{(i,j)\in p\times q} k_{ij}$. If we set these in the correspondence analysis frame, where each element $i \in I$ is represented by a point $(k_{ij} / k_{i\cdot})_{j\in J} \in \mathbb{R}^{|J|}$ with mass $k_{i\cdot} / k_{\cdot\cdot}$ (and analogous formulae hold for j), all subsets p and q centroids have representations given by $c^I_p = (k_{pj} / k_p)_{j\in J} \in \mathbb{R}^{|J|}$ and $c^J_q = (k_{iq} / k_q)_{i\in I} \in \mathbb{R}^{|I|}$ respectively, where c^I and c^J are the grand centroids, i.e. the marginals. If we provide the vector spaces $\mathbb{R}^{|I|}$ and $\mathbb{R}^{|J|}$ with the χ^2 metrics, a basis *associated to hierarchy* H^J $(e^0, e^1, \ldots, e^{|J|-1})$ for $\mathbb{R}^{|J|}$ may be given by:

$$\forall j \in J, \quad e^0_j = k_{\cdot j}/k_{\cdot\cdot}$$

$$\forall m = (q_1, q_2) \in H^J, \quad e^m_j = \begin{cases} k_{\cdot j}/k_{q_1} & \text{if } j \in q_1 \\ -k_{\cdot j}/k_{q_2} & \text{if } j \in q_2 \\ 0 & \text{otherwise} \end{cases} \tag{1}$$

with q_1 and q_2 the two subsets that merge at the H^J m-th node. This basis is orthogonal (Benzécri *et coll.*, 1973-82; Weiss 1978; Cazes, 1984) and an analogous one may be given for $\mathbb{R}^{|I|}$.

To interpret H^I hierarchy in terms of H^J, the squared distance $\| c^I_p - c^I \|^2$ of the H^I branch or group p from the centroid c^I is partitioned according to the hierarchy H^J associated basis elements, and we have

$$\delta^I_p = \| c^I_p - c^I \|^2 = \sum_{m=(q_1,q_2)\in H^J} \Delta_{p,(q_1,q_2)} \tag{2}$$

where

$$\Delta_{p,(q_1,q_2)} = \frac{k_{q_1} k_{q_2}}{k_{q_1} + k_{q_2}} k_{..} \left(\frac{k_{pq_1}}{k_p k_{q_1}} - \frac{k_{pq_2}}{k_p k_{q_2}} \right)^2 \tag{3}$$

As a consequence, for each group p of H^I hierarchy, indices $i^I_{p,(q_1,q_2)} = \frac{\Delta^I_{p,(q_1,q_2)}}{\delta^I_{p,(q_1,q_2)}}$ may be calculated for the most significant nodes $m = (q_1, q_2) \in H^J$. In particular, indices i, summing to 1, measure the relative contribution of the hierarchy nodes in the explanation of group's distance from the centroid. In addition, if an H^J node is composed by branches q_1 and q_2, both quantities $f_{pq_1} = \frac{k_{pq_1}}{k_p k_{q_1}}$ and $f_{pq_2} = \frac{k_{pq_2}}{k_p k_{q_2}}$

or their logarithm $lf_{pq1} = \log f_{pq1}$ and $lf_{pq2} = \log f_{pq2}$ may be used to check their association to group p of H^I hierarchy. If $f_{pqi} > 1.$ (i.e. $lf_{pqi} > 0.$) association between groups p and q_i is positive, and negative if $f_{pqi} < 1.$ (i.e. $lf_{pqi} < 0.$). In this way one may outline the additive contribution of the groups forming the hierarchy nodes to the other hierarchy's nodes or partition groups. Analogous indices i^J, Δ^J, and δ^J may be calculated for groups q of H^J hierarchy.

3. Application to Lombardia 1991 Census data

The data are the 575 indicators of 1991 Italian population Census, taken for the 1546 municipalities of Lombardia region, concerning population and dwellings. The aim of the exploratory analysis (Camiz and Tulli, 1996) was to classify the municipalities, based on the similarity of both population and dwellings characters. Correspondence analysis (Benzécri *et coll.*, 1973-82; Lebart *et al.*, 1984) gave three main axes summarizing 83% of total data table information (inertia). This results in a very strong synthesis, that we summarize here: on the first axis, rural lowland areas of Mantova and Brescia regions are opposed to Milan and other main cities; the second axis puts in evidence all holiday towns and villages, with high density of holiday houses; the third opposes Milan and small villages to the other main cities of Lombardia, i.e. it opposes the *houses surface* to all items tied to both *population and houses abundance*. Based on these three axes coordinates, a hierarchical ascending classification was performed, using Ward's (1963) method, on both municipalities and indicators. In both cases, considering the variance within groups, the partitions in 6 and 9 groups were the most interesting. In must be noted that the total information given by these reduced tables was nearly 77% and 80% of the original table, respectively: an outstanding result, since, in particular, the latter corresponds nearly to the amount of inertia explained by the three chosen axes.

The problem was now to choose the best cross-classification and to explain the municipalities groups, based on the values of Census indicators: a task very hard to perform, due to the very high number of items involved. For this reason, the said technique of Lebart *et al.* (1979) was very difficult to be used, at least at a first level; in addition, it was impossible to cross this characterisation with the indicators clustering. Feoli and Orlóci (1987) *concentration analysis* was not applicable, limited as it is to only presence/absence data. In Table 1 indices lf_{pqi} are reported, showing the influence of the hierarchy built on Census indicators on the municipalities hierarchy's nodes and the derived partition into 6 groups. Different cells shadowing indicates the intensity of influence, the lighter background meaning lowest frequencies, thus negative contribution, and the darker meaning highest frequencies, thus positive contribution. Through the inspection of this table and the reciprocal one, based on municipalities hierarchy, we noticed that there was a very strong influence of items of each hierarchy to the other: this suggests a very good agreement among them, nearly a block-diagonal structure.

Table 1: *Indices lf_{pqi} corresponding to the first five nodes of hierarchies and 6×6 cross-classification of Lombardia Census data. Rows represent municipalities nodes (CA...CE) and groups (C1...C6) and columns indicators nodes (RA...RE) and groups (R1...R5, Milan). Within parentheses is given each group frequency.*

CA

RA

	C1 (25)	CB	C2 (89)	C3 (17)	CC	C4 (106)	CE	C5 (215)	CD	C6 (123)
R1 (53)	2.91	.69	-.18	-.69	-.69	-.69	-.69	-.69	-.69	-1.10
RD	1.98	.47	.17	-.15	-.27	-.17	-.29	-.34	-.34	-.64
R2 (125)	2.21	.50	.11	-.19	-.35	-.27	-.35	-.44	-.44	-.75
RE	2.08	.44	.21	-.04	-.26	-.14	-.28	-.31	-.31	-.64
R3 (195)	1.53	.41	.27	-.06	-.18	-.09	-.22	-.26	-.26	-.56
RC	.07	.26	.27	-.00	-.13	-.05	-.16	-.19	-.19	-.44
R4 (1143)	-.59	.24	.28	.01	-.12	-.04	-.15	-.17	-.18	-.42
R5 (29)	-.51	-3.23	- ∞	.45	.34	16	.29	.31	.30	.44
RB	-.27	-2.89	- ∞	.00	.35	15	.40	.45	.45	.77
Milan	.07	-2.64	- ∞	- ∞	.34	14	.57	.65	.66	1.17

In particular, node *CA* highly influences the node *RA*. In fact, the branch *CB*, composed by items defining country sites and tourist areas, has a highest frequency on municipalities in branch *RC*, composed by small centers in country and mountain sites, and lowest on those in branch *RB*; on the opposite, indicators of branch *CC*, related to urban population characters, have high frequencies in municipalities in branch *RB*, namely Milan and other large cities. The influence of second level node was also highly significant, since it splits both the country site municipality and items into purely rural (*C2, R4*) and holiday site characters (*C1, RD*). The further nodes took Milan apart from the other cities (*R5*), due mainly to the items related to densely populated areas, with large houses (*C3*), absent in Milan; on the opposite, in Milan have high frequencies those items tied to rich residential areas and tertiary production. Indices corresponding to further nodes were useful to identify lower level nodes relationships, that were not as evident as the previous. They tell apart rural areas, up to the 6×6 level, that was thus suggested as a best cross-classification. This could mean that from that level on the interpretation of the cross-classification was no longer possible, and that differences among further partition groups of one hierarchy may not be understood only on the basis of the nodes of the other one.

4. Conclusion

Specifically suited for the interpretation of a cross-classification, this method is proved to be useful even when the interpretation of only one classification is important, provided that it may be explained by the other hierarchy. In addition, when cross-classification is an important issue, the method is helpful in detecting the lowest levels where each partition may be explained through the other. In the case of Census data, it was effective in detecting its most interesting features, that were not easily detected through alternative techniques. The fact that the interpretation was limited to the highest hierarchies levels means that deeper partitions, in particular those of municipalities, do not correspond to a cross-classification. Deeper partition may be explained only considering different mean frequencies of the same items in the different groups, something that this technique may not detect. Although very difficult to be used as a first step, when many variables are involved, Lebart *et al.* (1979) technique may be more suited for this purpose as a second step, involving quantitative aspects of only a selected number of items of interest.

Some development directions are now under investigation: the indices distribution and significance, the graphical representation of results, including an improved cells shadowing and the representation of block-structure, as Benzécri *et al.* (1973-82: I, 82 and 365) suggest, the use of the procedure in order to improve the vegetation data tables structuring, very close to Bertin (1977) graphical representation of data, and the generalisation to other cases, as three-way contingency data tables, individual $\times$ characters, and individual $\times$ variables tables.

Acknowledgements

This work was developed during the first author stay at Lille University as a visiting professor in 1997, the Census investigation being performed in the frame of Progetto Finalizzato Trasporti 2 n. PFT2 96.00121.PF74, granted by Italian Consiglio Nazionale delle Ricerche.

References

Benzécri, J.P. *et coll.* (1973-82). *L'Analyse des Données.* 2 voll., Paris, Dunod.

Benzécri, J.P., M.O. Lebeaux, and M. Jambu (1980). "Aides à l'interprétation en classification automatique". *Cahiers de l'Analyse des Données*, 5(1): 101-123.

Bertin, J. (1977). *La graphique et le traitement graphique de l'information.* Paris, Flammarion.

Camiz, S. (1993). "Computer Assisted Procedures for Structuring Community Data". *Coenoses*, 8(2): 97-104.

Camiz, S. (1994). "A Procedure for Structuring Vegetation Tables". *Abstracta Botanica*, 18(2): 57-70.

Camiz, S. and V. Tulli (1996). *I censimenti in Lombardia: una sintesi forte dei risultati.* Communication. Meeting on «Realtà locali e regionali. Metodi quantitativi di analisi». Dipartimento Metodi Quantitativi, Università di Brescia.

Cazes, P. (1984). "Les correspondances hiérarchiques". CEREMADE, Université Paris-Dauphine, *Les Cahiers de Mathématiques de Décision.*

Diday, E. (1970). *La méthode des nuées dynamiques et la reconnaissance des formes.* Paris, IRIA.

Denimal, J.J. (1997). "Aides à l'interpretation mutuelle de deux hiérarchies construites sur les lignes et les colonnes d'un tableau de contingence". *Revue de Statistique Appliquée*, 4.

Feoli, E. and L. Orlóci (1979). "Analysis of concentration and detection of underlying factors in structured tables". *Vegetatio*, 40: pp. 49-54.

Govaert, G. (1984). *Classification croisée.* Université de Paris VI, Thèse d'état.

Greenacre, M. (1988). "Clustering the Rows and Columns of a Contingency Table". *Journal of Classification*, 5: pp. 39-51.

Lebart, L., A. Morineau, and J.P. Fénelon (1979). *Traitement des données statistiques.* Paris, Dunod.

Lebart, L., A. Morineau, and K. Warwick (1984). *Multivariate Descriptive Statistical Analysis.* New York, John Wiley and Sons.

Orlóci, L. (1978). *Multivariate Analysis in Vegetation Research.* The Hague, Junk.

Ward, J.H. (1963). "Hierarchical Grouping to Optimize an Objective Function". *Journal of the American Statistical Association*, 58: 236-244.

Weiss, M.C. (1978). "Décomposition hiérarchique du chi-deux associée à un tableau de contingence à plusieurs entrées". *Revue de Statistique Appliquée*, 26(1): 23-33.

Bidimensional Kalman filter and Iterated Conditional Modes

Luigi Ippoliti - Luca Romagnoli
Department of Quantitative Methods and Economic Theory
University "G. d'Annunzio" - Viale Pindaro, 42 - 65127 Pescara- Italy
e_mail: romagnol@dmqte.unich.it

Abstract: The aim of this paper is to propose a comparison between two different methods for the restoration of degraded images, whose generating model is assumed to be a Gauss-Markov Random Field (GMRF): the former is a Bayesian method, the so-called Iterated Conditional Modes (ICM ; Besag,1986), and the latter is based on a Kalman-type filtering and smoothing algorithm (Moura, Balram,1993). After briefly presenting the fundamental concepts of each of these two approaches, we will give the results of an application on a benchmark image, together with some comments and conclusions.

Keywords: Gauss-Markov Random Fields, Bayesian Image Analysis, Kalman Filter.

1. Gauss-Markov Random Fields in image analysis

The basic problem we are engaged with is the restoration of an image $\boldsymbol{x}$, given the observation of a degraded image $\boldsymbol{y}=\gamma(\boldsymbol{H}(\boldsymbol{x}),\varepsilon)$, where $\boldsymbol{H}$ is a blurring matrix and ε is a noise independent of $\boldsymbol{x}$. In order to establish the background for statistical image modelling, it will be assumed that any image of interest can be represented on a $N \times N(=n)$ square regular lattice S whose pixels co-ordinates are indicated through the pair of integers (h,j), $h=1,\dots,N$; $j=1,\dots,N$. A conventional rastertype scan will be assumed as follows: left to right scan, advance one line, repeat, so that we indicate each pixel variable through one only index, i.e. x_i will denote the grey level assumed by the variable X_i at pixel i $(=1,2,\dots\dots,n)$. In order to assign a probability measure over the space of all the possible configurations $\boldsymbol{x}\in\Omega(=\{0,1,\dots,L\text{-}1\}^n)$ we define a *Markov Random Field* (Mrf) as follows (Besag, 1974):

$$f(\boldsymbol{x})=exp\{-U(\boldsymbol{x})\}/\,Z \qquad (1)$$

where

$$U(\boldsymbol{x})= \sum_{1\le i\le n} x_i G_i(x_i) + \sum_{1\le i\le n}\sum x_i x_j G_{ij}(x_i, x_j) + \dots\dots$$

is the *energy function* of statistical mechanics, the $G(.)$ functions represent the *potentials* of the fields - that is, the local interaction parameters - and Z is the normalising constant which permits to the function to integrate to one. Equation (1) is of a general form and includes higher order interactions, but we will deal only with *pairwise interaction Mrfs*, that is those fields for which the only *cliques* are those consisting of one or two pixels, which can be considered *neighbours* on the basis of some distance criterion. In fact, the key to the definition of a Mrf is the specification of the *neighbourhood system* δ_i on which the Markov property holds:

$$P(x_i|x_j, j\neq i)=P(x_i|x_j, j\in\delta_i) \quad \forall i\in S$$

Then, the Hammersley-Clifford theorem states that the joint density (probability) function of $\boldsymbol{X}$ has the general form (1), and that the $G(.)$ functions are non zero if and only if its arguments are defined over a clique.
In the remainder of the work we will consider the true scene as a realization of a zero-mean Gauss-Markov Random Field (GMRF), that is, the distribution of X_i, conditional upon its neighbours, is Normal, with parameters:

$$E(x_i|\, x_j, j\in \delta_i) = \sum_{j\neq i}\beta_{ij}x_j \quad ; \quad Var(x_i|\, x_j, j\in \delta_i) = \tau^2$$

It follows (Besag, 1974) that the joint distribution of $\boldsymbol{X}$ is

$$f(\mathbf{x})\propto exp\{-\frac{1}{2\tau^2}\mathbf{x}'\boldsymbol{A}\mathbf{x}\} \tag{2}$$

where $\boldsymbol{A}$ is the *(n×n) potential matrix*, with entries equal to 1 along the principal diagonal, to $-\beta_{ij}$ if the zones i and j are neighbours (according to some fixed distance criterion), and zero elsewhere.

2. Bayesian approach to image restoration.

The Bayesian ideas concerning image analysis may be briefly reassumed as follows: given a model for the true scene, $f(\mathbf{x})$, and a suitable choice for the noise term, $f(\mathbf{y}|\mathbf{x})$, we get, by applying Bayes' theorem:

$$f(\mathbf{x}|\mathbf{y})=\frac{f(\mathbf{y}|\mathbf{x})f(\mathbf{x})}{f(\mathbf{y})} \tag{3}$$

so that the best reconstruction, $\hat{\mathbf{x}}$, is the one which maximises $f(\mathbf{x}|\mathbf{y})\propto f(\mathbf{y}|\mathbf{x})f(\mathbf{x})$.
This is the principle of the so called *Maximum a Posteriori* procedures, but it is clear that the search of the maximum of the very complex posterior functions coming from (3) implies great computational problems. Instead, we focus our attention on a particular restoration method known as *Iterated Conditional Modes* (ICM). This is a deterministic, fast algorithm which derives from the following expression:

$$f(x_i|y_i, \hat{x}_{S/i}) \propto f(y_i|x_i)\, f(x_i|\hat{x}_{\delta i}) \tag{4}$$

For the Gaussian case, we can see in (4) that $f(y_i|x_i)$, that is the observation density conditional on the model realisation, is normal, with zero mean and variance σ^2, and that $f(x_i|\hat{x}_{\delta i})$ is the model (Gaussian) conditional density. In order to maximise the posterior distribution, it is very easy to demonstrate that the best estimate of the true scene is obtained by minimising:

$$\sigma^{-2}(y-x)'(y-x) + x'Ax$$

so that the updating formula at pixel i, during a single ICM cycle, is:

$$\hat{x}_i = \frac{(\tau^2 y_i + \sigma^2 \sum_{j \neq i} \beta_{ij} \hat{x}_j)}{(\tau^2 + \sigma^2)}$$

The algorithm is very easy to implement, and is based on an iterative image restoration together with parametric estimation, and has been demonstrated that the posterior $f(\hat{x}|y)$ converges rapidly to a (local) maximum.

3. A “classic” approach

In this case, in order to apply recursive algorithms such as Kalman Filter, it is necessary to put the model, i.e. a GMRF, in a state-space form. The starting point is the equivalence between our conditional model (2) and the simultaneous representation given by Woods (1972): $\boldsymbol{Ax}=\varepsilon$, with ε~MVN$(\boldsymbol{0}, \tau^2\boldsymbol{A})$ and $\boldsymbol{A}$ the potential matrix. Then, by using the Cholesky factorization and the particularly regular structure of the matrix $\boldsymbol{A}$, it is possible (Moura, Balram,1993) to implement a recursive Kalman-type filtering and smoothing algorithm (Rauch, Tung & Striebel, 1965) for the reconstruction of a degraded image $\boldsymbol{y}$ modelled as follows: $\boldsymbol{y}=\boldsymbol{x}+\boldsymbol{w}$, i.e. a GMRF plus an additive zero-mean independent Gaussian noise with $E[\boldsymbol{ww}']=\sigma^2\boldsymbol{I}$.
The principal problem to overcome is that of the estimation of the parameter vector $\theta=[\beta' \; \tau^2 \; \sigma^2]'$, where β is the vector collecting the potentials of the field. As regards σ^2, it has to be said that it is impossible to obtain its estimate with standard statistical methods, so that it will be necessary to “guess” it through a preprocessing of a uniform portion of the image. The maximum likelihood (ML) estimation of the parameters of the field implies the minimisation of the following negative log-likelihood function:

$$L(\boldsymbol{x};\beta,\tau^2) = 0.5ln(\tau^2) - (0.5/N^2)ln\left|\boldsymbol{A}_p(\beta)\right| + [0.5/(\tau^2 N^2)]\, \boldsymbol{x}'\boldsymbol{A}_p(\beta)\boldsymbol{x} \tag{5}$$

The greatest difficulty in minimising (5) is given by the presence of the determinant of the potential matrix, which is a very complex function of the parameters. In any case, it has been demonstrated (Balram & Moura, 1992) that, for first- and a particular class of second-order fields (those with equal diagonal interaction parameters), it is possible to derive an explicit expression for the eigenvalues of $A_p(\beta)$, and in this way the likelihood function (5) may be obtained as a simple function of the parameters.

4. Experimental results and conclusions

We have applied the two methodologies (bidimensional Kalman filter and ICM) to a part of a demo image (a mandrill face) contained in our programming tool, MATLAB v.5; the lattice is of dimensions 100×100 pixels, and the grey levels vary in the range [0,220]. We added a zero-mean white Gaussian noise with variance equal to 2500, so producing a Signal-to-Noise (SNR) ratio of about 1.8. Figs.1(a) and 1(b) show, respectively, the original and the noisy image.

Figure 1: (a) *original image (mandrill's eye);* (b) *noisy image (σ^2=2500)*

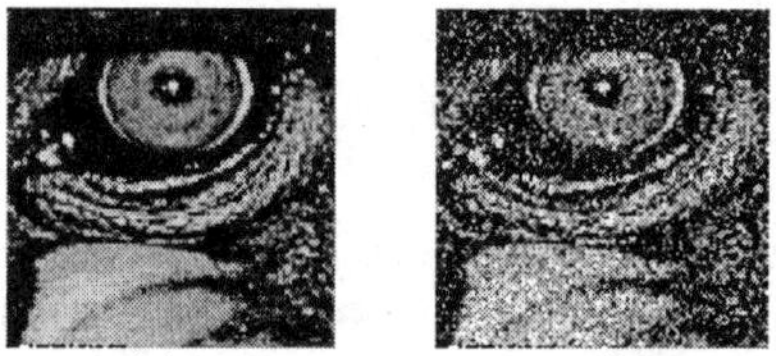

(a) (b)

The following Figs.2(a) and 2(b) are the visual results obtained, respectively, by means of the application of the filtering and smoothing phase of the Kalman-type algorithm for a first-order homogeneous GMRF. The ML parameters estimates are: for β_h (horizontal interaction parameter), 0.34; for β_v (vertical interaction parameter), 0.15; and for τ^2 (conditional variance), 885.42.

Figure 2: *application of bidimensional Kalman algorithm; (a) restoration after filtering phase, and (b) after smoothing phase*

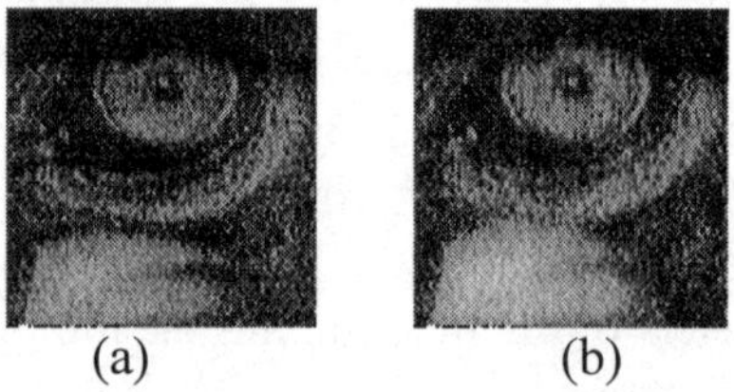

(a) (b)

The goodness of reconstruction has been tested with the calculation of the Mean Squared Error (MSE) between Fig.1(a) and each of the other ones; they are, respectively, equal to 2503.6 (Fig.1(b)), to 2065.4 (Fig 2(a)), and to 1676.1 (Fig.2(b)). As it can be seen, the enhancement between filtering and smoothing

phase is in the order of the 15.55% of the original MSE.
The Bayesian approach followed by ICM has led, in successive iterations, to the images presented in Figs3(a)-(d); the parameters have been estimated by means of the maximum pseudo-likelihood (MPL) method, which is based upon the maximisation of the product of the conditional density functions of the current reconstructed image:

$$PL(\theta) = \prod_{i=1}^{n} [Pr(x_i \mid x_j : j \neq i;\ \theta)]$$

Figure 3: *application of ICM algorithm; from (a) to (d): iterations from 1 to 4*

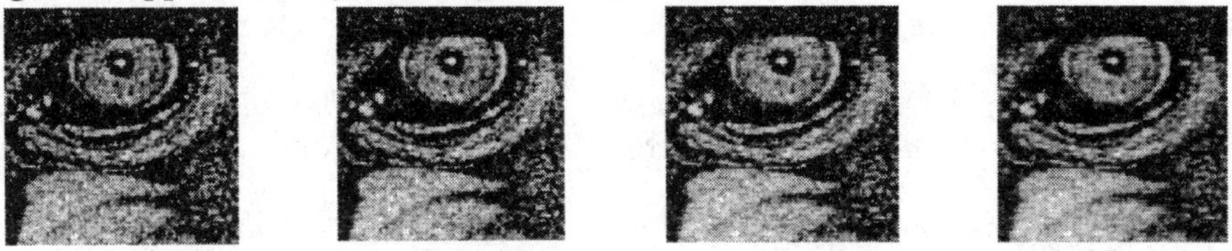

(a) (b) (c) (d)

Numerical results, including parametric estimation and MSE between the original image and the reconstructions, are presented in Table 1.

Table 1: *results of the application of ICM to Fig.1(b).*

Iteration	β_h	β_v	τ^2	σ^2	MSE
1	0.36	0.17	665.69	848.44	1007.50
2	0.39	0.12	629.31	908.65	932.39
3	0.42	0.09	465.30	1058.90	890.20
4	0.48	0.04	235.48	1407.10	836.13
5	0.56	0.02	45.91	2032.00	857.09

As it can be noted, we did non present the resulting image for the 5th iteration, because of the augmented MSE value.

The results of our application would seem to outline an evident better performance of the ICM method, when compared to that of the Kalman filtering and smoothing algorithm. Notwithstanding a very good image restoration, the Bayesian approach we have considered presents at least to problems to be remarked: a) the first one is related to pseudo-likelihood estimation which furnishes, as it can be seen in Tab.1, parameter values always outside of the admissible region [we remember that a sufficient condition, in the case of first-order GMRFs, is $|\beta_h|+|\beta_v|<0.5$]; moreover, it can be seen that MPL is not a well-suited method for parametric estimation in presence of corrupted data: we have to point out that there is a need for some sort of correction to avoid the great bias in the estimated parameters; b) it is not clear when the iteration has to be stopped; we have adopted, at this purpose, the statistical MSE measure, but it seems to be obvious that the algorithm does not converge towards a fixed reconstruction.

References

Balram N. & Moura J.M.F. (1993). *Noncausal Gauss-Markov random fields: parameter structure and estimation*, IEEE Trans. Inform. Theory, **39**, 1333-1355.

Besag J. (1974). *Spatial interaction and the statistical analysis of lattice Systems*, J.R.S.S., B, **36**, 192-236.

Besag J. (1986). *On the statistical analysis of dirty pictures*, J.R.S.S., B, **48**, 259-302.

Geman S. & Geman D. (1984). *Stochastic relaxation, Gibbs distributions, and Bayesian restoration of images*, IEEE Trans. Pattern Anal. Machine Intell., **PAMI-6**, 721-741.

Moura J.M.F. & Balram N. (1993). *Statistical algorithms for noncausal Gauss-Markov fields*, Handbook of Statistics, Vol. 10, Elsevier Science Publishers, 623-691.

Rauch H.E., Tung F., Striebel C.T. (1965). *Maximum likelihood estimates of linear dynamic systems*, J. AIAA, **3**, 1445-1450.

An Iterative Reweighting Procedure to Robustify Projection Pursuit Regression

Annalisa Laghi
Dipartimento di Scienze Statistiche, Università di Bologna, Italy,
e-mail: laghi@stat.unibo.it

Abstract: Projection pursuit regression is generally greatly influenced by outlying points. Aiming at robustness of this non-parametric regression method, we propose to resort to regression M-estimates. Robust estimates of the projection pursuit regression model parameters are obtained, as in the robust linear regression context, by an iterative reweighting procedure, with observation weights determined by the bisquare weight function.

Key words: Projection pursuit regression, robust regression, M-estimators, bisquare weight function.

1. Projection pursuit regression

Projection pursuit regression (PPR), introduced by Friedman and Stuetzle (1981) and refined by Friedman (1984), is a non parametric regression method for modelling a random vector Y, the response variable, as a function of a p-dimensional random vector X, of predictors, on the basis of a sample of n observations $\{(y_i, x_{il});\ \ i=1,2,...,n,\ \ l=1,2,...,p\}$ of (Y,X). The response variable is modelled as a linear combination of smooth functions of linear combinations of the predictor variables. The model takes the form:

$$E[Y|x] = g(x) = E[Y] + \sum_{m=1}^{M} \beta_m f_m(\alpha_m' x) \tag{1}$$

where α_m are p-dimensional unit projection directions, and f_m are univariate smooth functions, with zero mean and unit variance, of the projections $\alpha_m' x$, $m=1,2,..,M$. The coefficients α_m and β_m, and the functions f_m are parameters of the model and are estimated so as to minimise the criterion:

$$L_2 = E\left[Y - E[Y] - \sum_{m=1}^{M} \beta_m f_m(\alpha_m' x)\right]^2. \tag{2}$$

PPR algorithm estimates expected values by weighted means, where the observation weights, w_i, $i = 1,2,\dots,n$, are specified by the user. The default is all weights equal unity. In this paper we evaluate the use of observation weights to implement an iterative reweighting scheme for robustification.

2. Robustification of PPR via regression M-estimates

M-estimation, introduced by Huber (1973), is probably the best known procedure to reduce the influence of outlying points for the linear regression model. This works by replacing the usual sum of squared residuals by a sum of less rapidly increasing functions of the residuals whose general form is:

$$\sum_{i=1}^{n} \rho((y_i - \beta' x_i)/s) \qquad (3)$$

where s is a robust estimate of scale and β is the vector of the regression coefficients. If the parameter estimates are not to be greatly influenced by a few data points which are far from the regression plane, then, in general, $\rho(.)$ must be bounded. The M-estimate of β can be found by minimising (3) or, after taking derivatives, by solving the system of $(p+1)$ equations:

$$\sum_{i=1}^{n} \psi((y_i - \beta' x_i)/s) x_{il} = 0, \quad l = 0,1,\dots,p \qquad (4)$$

where $\psi = \rho'$ and $x_{i0} = 1$, $i=1,2,\dots,n$. The most widely used procedure to solve (4) is iteratively reweighted least squares, with weights $w_i = \psi((y_i - \beta' x_i)/s)/((y_i - \beta' x_i)/s)$. The iterative reweighting procedure is only guaranteed to converge for convex ρ. ρ-Functions associated with redescending ψ functions (such as the biweight one described below) are non convex, so, it is advisable to choose a sufficiently good starting point and perform a small and fixed number of iterations.

PPR algorithm proves to be greatly influenced by outlying points when the observation weights are not purposely specified as robustness weights. This general non-robustness is essentially due to the adopted criterion of fit and smoothing procedure. The algorithm estimates the model (1) parameters by least square (2) and approximates the smooth functions f_m, $m=1,2,..,M$, for each selected projection direction, using a non-robust variable span smoother, called the supersmoother. This smoother is based on a locally linear fit and adopts least squares as criterion of fit.

Roosen and Hastie (1994) have proposed an alternative PPR algorithm in which the smooth functions are estimated by smoothing splines. As the supersmoother, smoothing splines are not robust to outlying points.
We propose to obtain robust estimates of the PPR model parameters by minimising the tapered criterion $\sum_{i=1}^{n} \rho\left(\left(y_i - E[Y] - \sum_{m=1}^{M} \beta_m f_m(\alpha_m' x_i)\right) \Big/ s\right)$.
The minimisation is carried out, as in the robust linear regression context, by an iterative reweighting scheme (PPR-1) which consists in the following steps:

0. Fit a PPR model with no robustness weights: starting values for the residuals $\left(r_i^{(0)}\right)$ and the scale estimate $\left(s^{(0)}\right)$ are computed. Robustness weights $\left(w_i^{(0)}\right)$ are then obtained.
1. Step j: fit a PPR model with robustness weights $\left(w_i^{(j-1)}\right)$.
2. If $\left(\left|\sum_{i=1}^{n} w_i^{(j-2)}\left(r_i^{(j-1)}\right)^2 - \sum_{i=1}^{n} w_i^{(j-1)}\left(r_i^{(j)}\right)^2\right| \Big/ \sum_{i=1}^{n} w_i^{(j-2)}\left(r_i^{(j-1)}\right)^2\right) < \zeta$ where ζ is a user-specified threshold (usually 0.005), or if $j \geq max\,it$, where *maxit* is a user-specified maximum number of iterations (3-5), stop; otherwise update the residuals and the scale estimate, obtain robustness weights $\left(w_i^{(j)}\right)$ and loop back to 1.

Among the different proposals developed for the weight function, we consider Tukey's biweight one, defined by:

$$\begin{aligned} W(z) &= \left(1-(z/k)^2\right)^2, & |z| < k \\ &= 0, & |z| \geq k \end{aligned} \tag{5}$$

where k is a "tuning" constant. The robustness weights are then defined by $w_i = W(r_i/s)$, where s is the median absolute residual.
The bisquare weight function (5) has been shown to perform well for robust linear regression (Gross, 1977) and has been employed by Cleveland (1979) to obtain a robust smoothing technique based on robust locally weighted regression. Gross suggests the value $k=6$ to protect against outliers. The same value is considered in Cleveland (1979).
In the context of non-parametric regression, M-estimate approaches have been also suggested to give rise to robust versions of kernel and spline smoothing (see Simonoff, 1996, for update references) and of additive models (Hastie and Tibshirani, 1990).

3. Example and conclusions

The non-robustness of PPR is shown by a simple example described by Friedman and Stuetzle (1981) to stress the capability of PPR to model interactions between predictors. A sample of *400* observations is generated according to the model:

$$Y = g(X) + \varepsilon = g(X_1, X_2) + \varepsilon = X_1 X_2 + \varepsilon , \tag{6}$$

with (X_1, X_2) uniformly distributed in $(-1,1)^2$.

In case *A* we assume $\varepsilon \sim N(0,0.2)$. In this situation model (6) is expressed as a PPR model with two terms (*M*=2), with estimated coefficients $\hat{\alpha}_1 = (0.7339 \quad -0.6792)'$, $\hat{\alpha}_2 = (0.7159 \quad 0.6982)'$, $\hat{\beta}_1 = 0.1997$, $\hat{\beta}_2 = 0.2207$, and estimated smooth functions $\hat{f}_m$, $m = 1,2$ showing purely quadratic shapes. The considered model is then contaminated by taking $\varepsilon \sim (1-\delta)N(0,0.2) + \delta N(0,2)$, with $\delta = 1$ with probability 0.25 and zero otherwise (case *B*). The two projection directions selected by the standard PPR algorithm are no more correctly determined ($\hat{\alpha}_1 = (0.4009 \quad -0.9161)'$, $\hat{\alpha}_2 = (0.5772 \quad 0.8166)'$). The corresponding estimates for β_m, $m = 1,2$, are $\hat{\beta}_1 = 0.2307$ and $\hat{\beta}_2 = 0.1878$.

The estimated functions are not as smooth as in the no outlier case resulting attracted by outlying points (Figure 1).

Figure 1: $\hat{f}_m$, $m = 1,2$, *obtained in case A (denoted by 1) and in case B (denoted by 2) by the standard PPR algorithm.*

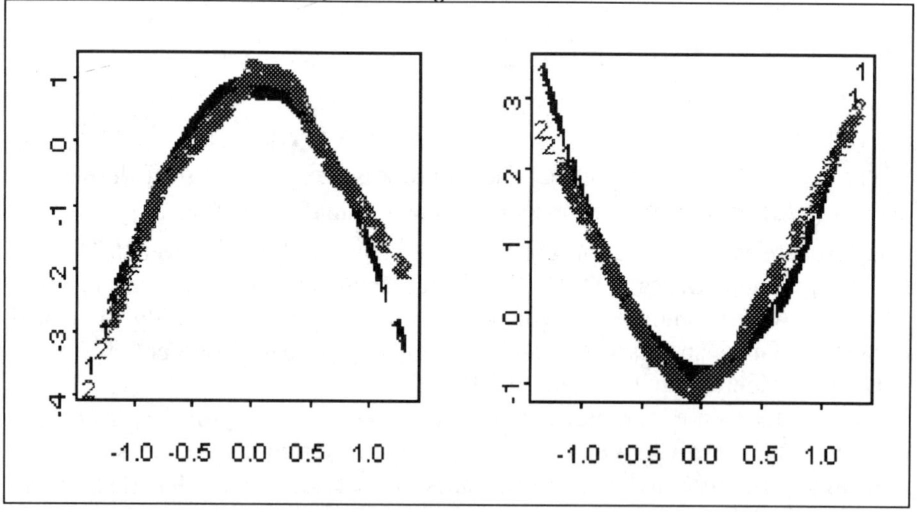

The iterative reweighting procedure, carried out with the bisquare weights (5) with tuning constant *k=6*, shows satisfactory results identifying, after *j=5* iterations, the following parameter estimates: $\hat{\alpha}_1 = (0.7476 \quad -0.6641)'$, $\hat{\alpha}_2 = (0.7534 \quad 0.6576)'$, $\hat{\beta}_1 = 0.1950$ and $\hat{\beta}_2 = 0.2045$. The estimated smooth functions essentially reproduce the behaviour described for the standard non-robust PPR algorithm in the no outlier case (Figure 2). Furthermore the described results obtained by the robust version of PPR in the outlier case, are very similar to those obtained in absence of contamination.

Figure 2: $\hat{f}_m$, $m = 1,2$, *obtained in case A (denoted by 1) by the standard PPR algorithm and in case B (denoted by 2) by PPR-1.*

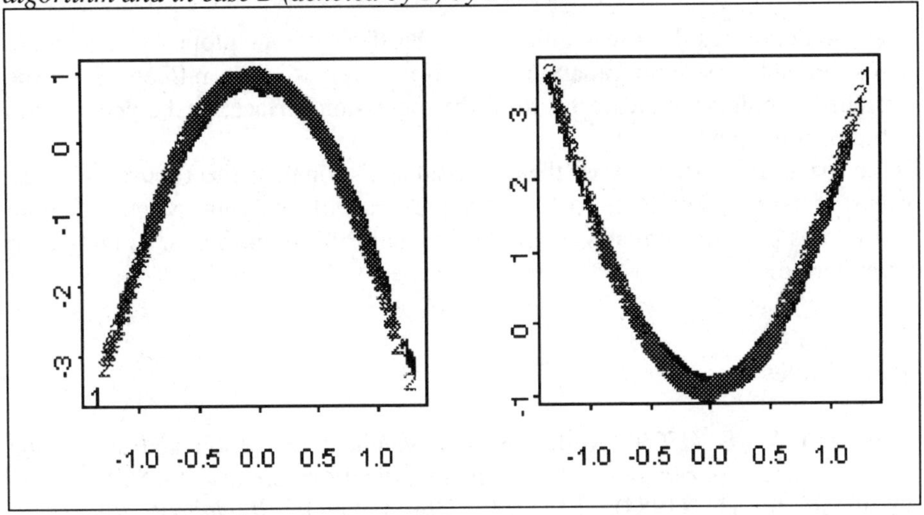

To evaluate the accuracy of our PPR iterative algorithm in approximating the true function, $g(X)$, *100* samples, each of *400* observations, are generated according to model (6) either in absence (case *A*) or in presence (case *B*) of outlying points, with *X* values generated as described above and kept fixed throughout the simulation study.
The mean squared error (*MSE*) is used as a measure of goodness of approximation

$$MSE = \frac{1}{400} \sum_{i=1}^{400} [g(x_i) - \hat{g}(x_i)]^2 \tag{7}$$

with $\hat{g}(x_i)$ denoting the estimate of $g(x_i)$, $i=1,2,\ldots,400$.
In both cases (*A* and *B*) the performances of PPR and PPR-1 are compared on the basis of the mean of the *MSE*'s (*MMSE*), averaged over the *100* replications, and of its standard error.

As shown by the simulation results, reported in Table 1, our iteratively reweighted algorithm outperforms the standard one in the contaminated situation, while both procedure perform very similar, in terms of *MMSE*, in absence of outlying points. In both cases PPR-1 demonstrates a better stability than PPR.

Table 1: *Simulation results on 100 replications*

	Case A		*Case B*	
	PPR	PPR-1	PPR	PPR-1
MMSE	0.0027	0.0024	0.0261	0.0040
se(*MMSE*)	0.0003	0.0001	0.0014	0.0002

The simulation results stress either the robustness of our proposed algorithm, whose satisfactory approximation capability is very slightly influenced by the presence of points which are far from the regression surface, or the general non robustness of standard PPR.

In an exploratory stage, it is, then, advisable to compare the robust and non-robust version of PPR to detect the presence of outlying points. If both procedures perform similarly, then, the original PPR algorithm is, of course, to be preferred.

References

Cleveland, W. S. (1979). Robust locally weighted regression and smoothing scatterplots, *Journal of the American Statistical Association*, 74, 829-836.

Friedman, J. H. (1984). SMART: User's Guide, *Technical Report 1*, Department of Statistics and Stanford Linear Accelerator Center, Stanford University.

Friedman, J. H. & Stuetzle, W. (1981). Projection pursuit regression, *Journal of the American Statistical Association*, 76, 817-823.

Gross, A. M. (1977). Confidence intervals for bisquare regression estimates, *Journal of the American Statistical Association*, 72, 341-354.

Hastie, T. J. & Tibshirani, R. J. (1990). *Generalized Additive Models*, Chapman and Hall, New York.

Huber, P. J. (1973). Robust regression: asymptotics, conjectures, and Monte Carlo, *Annals of Statistics,* 1, 799-821.

Huber, P. J. (1981). *Robust Statistics*, Wiley, New York.

Roosen, C. B. & Hastie, T. J. (1994). Automatic smoothing spline projection pursuit, *Journal of Computational and Graphical Statistics*, 3, 235-248.

Simonoff, J. S. (1996). *Smoothing Methods in Statistics*, Springer-Verlag, New York.

Statistical Multivariate Techniques for the Stock Location Assignment Problem

Angelo M. Mineo*, Antonella Plaia**[•]
* Dep. of Mechanical Technology and Production - University of Palermo
Viale delle Scienze - 90128 - Palermo, Italy, e-mail: amineo@mbox.unipa.it
** Institute of Statistics - Faculty of Economics - University of Palermo
Viale delle Scienze - 90128 - Palermo, Italy, e-mail: plaia@mbox.unipa.it

Abstract: In previous papers we proposed to apply multivariate statistical methodologies, like Multidimensional Scaling (MDS) and Seriation to the stock location assignment problem of a warehouse, often solved by considering the Cube per Order Index (COI). In this paper we compare the results by MDS, Seriation, a COI based method and the Maximum Path criterion, considering the data of a whole year of a Sicilian supermarket chain warehouse. The comparison is based on the simulated times to satisfy a sample of real orders.

Key words: stock location assignment, MDS, seriation, minimum spanning tree, maximum path criterion, COI, simulations.

1. Introduction

The problem of the best stock location assignment in a warehouse has a fundamental role while optimising picking activities. Among the different components of times constituting this activity (Mineo, Plaia, 1997 a), the time to reach the first picking position from the Input/Output (I/O) position (and return) and to reach all the picking positions, represents the most important component: an improvement in stock location assignment can reduce significatively this component of time, reducing in this way warehouse costs.

In Mineo & Plaia (1997 a, c), multivariate statistical methodologies have been applied for the first time in order to optimise stock location assignment; the results by applying MDS and Seriation show the usefulness of such methodologies, even if applied to the data of a warehouse of a Sicilian super and hypermarket chain where a sub-optimal solution has been gained thanks to human experience and know-how. In Mineo & Plaia (1997 b) it has been shown how the MDS approach is very competitive and often better than the methods usually used to cope with such a problem (generally Operational Research techniques as for the travel salesman problem): in particular, the MDS approach has been compared with a method based on the Cube per Order Index, COI,

---

[•] Authors' names are listed in alphabetic order.

(Caron, Marchet, 1994), that is a very popular method.
In this paper we introduce the use of another multivariate statistical technique to solve the considered problem, i.e. an extension of the Maximum Path criterion (Scippacercola, 1997), based on the determination of the Minimum Spanning Tree (Gower, Ross, 1969), and we compare the results obtained by applying this criterion with the MDS, the Seriation and the COI approaches. An important feature of this paper, differently from our quoted papers (Mineo, Plaia, 1997 a, b and c), is that the data of the considered warehouse are relative to a whole year. In this way we can introduce a new variable to deal, in an appropriate way, with seasonal goods and, by comparing the obtained results by means of simulations based on real picking order lists, we can select the best methods, for our considered warehouse, among the proposed ones.

2. The compared methods

In order to apply the MDS, seriation and maximum path criterion a dissimilarity matrix has been considered: this matrix has been computed by considering the Euclidean distances δ_{ij} among the classes and the variables considered, as will be discussed in the next paragraph. In this paper we use the non-metric MDS algorithm (Schiffman, Reynolds, Young, 1981) implemented in the STATISTICA® package and consisting in the minimisation of the raw-stress function:

$$\text{raw-stress} = \sum_{i,j}^{n}\left[d_{ij} - f(\delta_{ij})\right]^2 \tag{1}$$

where d_{ij} are the reproduced distances, given the respective number of dimensions, and δ_{ij} are the input data, i.e. the observed distances. The expression $f(\delta_{ij})$ indicates a non-metric, monotone transformation of the observed distances. Thus, the program will attempt to reproduce the general rank-ordering of distances between the objects in the analysis: in this way we can find an arrangement for placing the goods in the warehouse. For the reason to choose non-metric MDS see Mineo &Plaia (1997 a).
To solve the considered problem we can also use a seriation algorithm, such as that proposed by Wright (1985): in fact, by considering the way the warehouse aisles are crossed, i.e. by following a traversal policy (Caron, Marchet, 1994), we can think to our problem as placing objects along a continuum; so we have to minimise the loss function:

$$LF(\mathbf{x}) = \sum_{i<j}\left(\delta_{ij} - \left|x_i - x_j\right|\right)^2 \tag{2}$$

over $\mathbf{x} = (x_1, x_2, ..., x_n)$, where x_i is the coordinate of object i on the axis and δ_{ij} has the usual meaning.

Finally, by considering the maximum path criterion we can get another layout for the k-dimensional points. This criterion consists in finding the path of maximum length M in the Minimum Spanning Tree (MST) (Ross 1969). In this paper we follow the approach proposed by Scippacercola (1997) consisting in putting the points p_h lying on aside edges of M inside the path at a distance from the extreme vertices proportional to the ratio in the MST, that is by using the relation:

$$g(p_0, p_h) = \frac{d(p_0,p_z)d(p_0,p_h)}{d(p_0,p_h)+d(p_h,p_z)}, \tag{3}$$

where p_0 and p_z represent the extreme points of the Maximum Path in the MST.

Besides these statistical methods, we considered also a method based on COI, defined, for each good, as the rate between the assigned volume on the shelf and the picking rate: if the assigned volumes are equal, increasing COI values correspond to decreasing picking rates.

3. The data set

In order to deepen better the problem described in the precedent section, the data of the warehouse of a Sicilian chain of super- and hypermarkets have been considered. In the warehouse about 1500 different goods are stocked; these can be clustered in respect of good affinity in about 25 classes, according to what suggested by the Organised Distribution Network whom the Sicilian chain belongs to. In our quoted papers, considering just the number of item[1]-per-good picked in the period of time 1.7.1996 to 31.12.1996, we had already perceived a lack of homogeneity inside some classes, so 44 smaller classes were obtained, by cutting off some goods whose behaviour, according to the considered variable, was quite different from the other goods in the class.

In the present paper we consider these classes and, besides the variables already considered, that are:

1. the number of picked item-per-good
2. average volume of the item in the class
3. item fragility

4.-6. average number of items per good per order, for each of the tree different kind of supermarket in the network

7. number of goods in the class;

a new variable, coded by the authors, has been added. In fact, as the period of observation is now a whole year (1.7.1996 to 30.6.1997), a new variable can be

[1] Item: smallest pickable quantity of good.

considered:

8. the seasonal component.

In order to compare the effectiveness of the proposed methodologies for the stock location problem, a C language simulator has been developed.

In particular the four proposed solutions are compared with the current allocation, by drawing a sample of real orders, received by the warehouse from each super/hypermarket in the chain. The time to satisfy each order has been computed.

It is important to highlight that the current allocation is not a starting solution, being actually the results of years of experience and improvement.

By computing the travelled distances to satisfy the sample of real orders, we can choose the best approach to this problem for our specific warehouse.

4. Results

The simulator has been used to compute the distances travelled in order to satisfy a sample of 15 orders for each of the 37 stores of the supermarket chain. The 37 stores are the only retail stores in the chain.

The sample of orders covers the same period of time of the described dataset.

In figg. 1a and 1b the following solutions are compared:

1. the current stock location assignment
2. the COI-based solution
3. the non-metric MDS solution
4. the Maximum Path in the Minimum Spanning Tree solution
5. the Seriation solution.

It is evident how the current solution is not competitive, being worse than the four ones. The COI based solution seems also worse than the others, even if the difference is not so evident in any case.

Figure 1a: *Average distances travelled (m) by the operator to satisfy a sample*

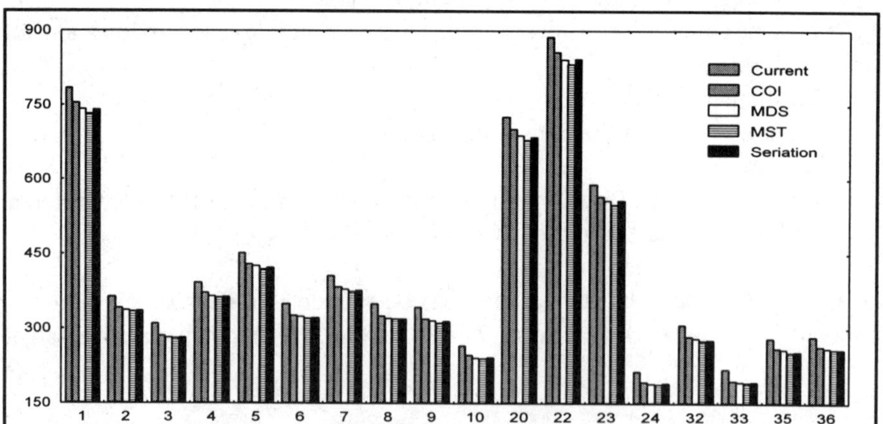

of orders for each of the first 18 stores of the chain with the five techniques
Figure 1b: *Average distances travelled (m) by the operator to satisfy a sample of orders for each of the second 19 stores of the chain with the five techniques*

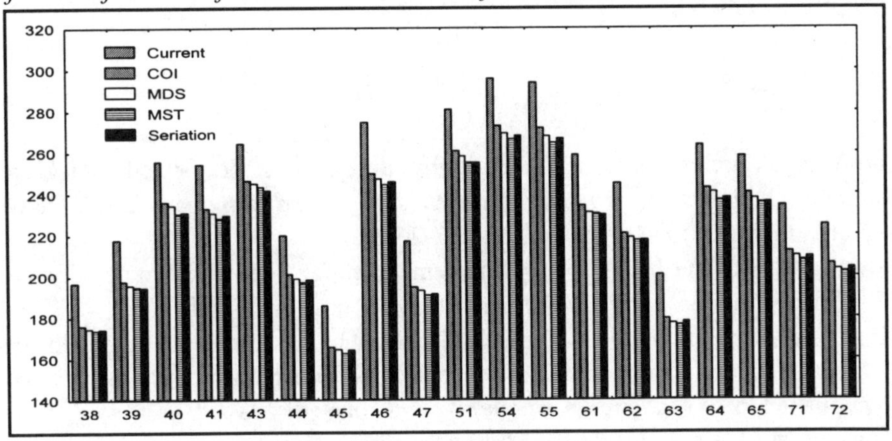

The other three solutions seem to be comparable, but the MST probably produces the best one, even if no tests have been carried out.

5. Conclusion

Different statistical methodologies have been compared in order to improve the current stock location assignment. The results show that all the solutions perform better than the current one, and, moreover, the statistical methodologies seem better than the most popular method to deal with this problem (the COI index).
The new variable (seasonal component) does not change the behaviour of the MDS and Seriation solutions each other (see Mineo & Plaia 1997c).
The new tested methodology, based on the modification of the Maximum Path in the Minimum Spanning Tree proposed by Scippacercola (1997), results to be very interesting and easily adaptable to treat the stock location assignment problem.
In conclusion, all the proposed statistical methodologies seem to solve the stock location assignment problem in a proper way, giving even better results than COI.
Complications could arise if we would consider each good, instead of classes of goods, as a unit (in this case the COI method can be easily applied). The distance matrix would be difficult to handle, and also for this reason we preferred to deal with classes of goods. Anyway, while for small warehouses, the distance matrix can be computed for the single goods, so the methods can be used to get the order of the goods and not of the classes, in large warehouses the management prefers to keep aside goods belonging to the same "class", therefore the application of the proposed methods to the distance matrix of the

classes is both justified and competitive.
Concluding the tested multivariate techniques can be easily and effectively applied in any case to cope with the stock location assignment problem.

References

Caron F., Marchet G. (1994). Politiche di gestione dei sistemi picking, *Advanced topics on production system design and management*, PSD&M Varenna (Lecco) 1-4 giugno 1994, 237-265.

Gower J.C., Ross G.J.S. (1969). Minimum Spanning Tree and Single Linkage Cluster Analysis, *Applied Statistics*, vol. 18, n. 1, pp. 54-64.

Mineo A.M., Plaia A. (1997 a). L'allocazione fisica dei materiali in un magazzino: un approccio basato sul multidimensional scaling, to be published in *La Statistica per le imprese - Convegno SIS 1997*.

Mineo A.M., Plaia A. (1997 b). Multidimensional Scaling and Stock Location Assignment in a Warehouse: An application, *Applied Stochastic Models and Data Analysis - The Ins and Outs of Solving Real Problems, Contributed Papers*, Anacapri 1997, pp. 297-302.

Mineo A.M., Plaia A. (1997 c). A new approach to the stock location assignment problem by multidimensional scaling and seriation, to be published in *Studies in Classification, Data Analysis and Knowledge Organization,* Springer-Verlag.

Ross G.J.S. (1969). Minimum Spanning Tree, *Applied Statistics*, vol. 18, n. 1, pp. 103-104.

Scippacercola S. (1997). Classification by the Maximum Path Criterion, *Applied Stochastic Models and Data Analysis - The Ins and Outs of Solving Real Problems*, *Contributed Papers*, Anacapri 1997, pp. 401-406.

Schiffman S.S., Reynolds M.L., Young F.W. (1981). *Introduction to Multidimensional Scaling - Theory, methods and applications*, Academic Press.

StatSoft, Inc. (1995). *STATISTICA for Windows [Computer program manual for rel. 5.0].* Tulsa, OK, USA.

Wright, R.V.S. (1985). Detecting pattern in tabled archaeological data by principal components and correspondence analysis: programs in BASIC for portable microcomputers, *Science and Archaeology*, 27, 35-38.

Acknowledgements
We would like to thank U. La Commare, Professor of Mechanical Technology at the Dep. of Technology & Mechanical Production of Palermo, and A. Solofra, Informative System Manager at MAR S.p.A. (Palermo), for their precious cooperation, especially for giving us the data for the application in the paper.

A Bayesian Approach to the Analysis of Spatial Time Series [1]

Stefano F. Tonellato
Dipartimento di Statistica – Università Ca' Foscari di Venezia (Italy)

Abstract: In this paper a class of models for conditionally Gaussian space-time processes is proposed in a state-space setup. These models have two main purposes: describing and forecasting the temporal evolution of the observed phenomena. Observed processes are assumed to be the sum of unobservable components, such as a time varying level, a periodic component, a stationary autoregressive component and a measurement error. We give sufficient conditions for model identifiability and show how Bayesian analysis can be performed via Gibbs sampling. We present the results of an application to some simulated spatial time series.

Key words: state–space models; spatial correlation; signal to noise ratio; Gibbs sampler.

1. Introduction

Environmental data are usually collected over space and time. A spatio–temporal analysis can provide benefits not possible either from a spatial–only approach or from the modelling of site specific time–series. Analyzing individual time series separately, for instance, information about spatial interaction or about common features among different sites is lost. On the other hand, a merely spatial analysis, ignoring the underlying dynamics of the observed phenomena, might give misleading results. This is why it is important to analyze both spatial and temporal features of data simultaneously as accurately as possible.

In Tonellato (1997a) we defined a new class of models for spatial time series, which extends the well–known Bayesian dynamic linear models (West and Harrison 1989) to the domain of the analysis of space–time stochastic processes. We believe that such models can be a very useful tool to be used in environmental monitoring. In fact, they can give both a statistical description of the evolution of the observed processes and reliable temporal forecasts, taking into account the spatial interaction among time–series collected over a monitoring network. In this paper we focus our attention on the estimation of unknown parameters using the Gibbs sampler. In section 2 we give a brief description of the models we are dealing with and in section 3 we present the results obtained in some simulations.

[1] This research has been partially supported by M.U.R.S.T. (Ministero dell'Università e della Ricerca Scientifica e Tecnologica).

2. The model

Suppose that observations come from an incomplete sample collected on a fixed region, $\mathcal{S} = \{s_i : s_i \in \mathcal{D} \subset \mathbb{R}^2,\ i = 1, \ldots, n\}$, independent of t. In such a context, a spatial time series is defined as an n–variate time series $\mathbf{y}_t$, $t \in \mathbb{Z}$. For any fixed t, the n–dimensional vector $\mathbf{y}_t$ is a partial realisation of a spatial stochastic process defined on $\mathcal{D}$ and observed on $\mathcal{S}$. The class of models proposed here assumes that $\mathbf{y}_t$ can be represented in terms of unobservable components, each describing salient features of the dynamics of the process. In their most general form, such models give the following representation of the observed process:

$$\mathbf{y}_t = \boldsymbol{\mu}_t + \boldsymbol{\gamma}_t + \mathbf{x}_t + \boldsymbol{\varepsilon}_t, \qquad \forall t \in \mathbb{Z}, \tag{1}$$

where $\boldsymbol{\mu}_t$ and $\boldsymbol{\gamma}_t$ denote respectively a temporal trend and a temporally non stationary periodic component, such as a seasonal or daily periodicity. For any fixed t, $\boldsymbol{\mu}_t$ and $\boldsymbol{\gamma}_t$ might be modelled as spatial processes, or as scalar valued components common to the n sites, or even as n–dimensional vectors of i.i.d. random variables; $\mathbf{x}_t$ is a temporally stationary autoregressive spatio–temporal component; $\boldsymbol{\varepsilon}_t$ denotes an erratic term, distributed as a Gaussian white noise $GWN(\mathbf{0}, \sigma_\varepsilon^2 \mathbf{I}_n)$ ($\mathbf{I}_n$ denotes the identity matrix of order n). A state–space representation of these models is possible and is defined as:

$$\begin{aligned} \mathbf{y}_t &= \mathbf{F}\boldsymbol{\zeta}_t + \boldsymbol{\varepsilon}_t, \qquad \boldsymbol{\varepsilon}_t \sim GWN\big(\mathbf{0}, \sigma_\varepsilon^2 \mathbf{I}_n\big) \qquad (2)\\ \boldsymbol{\zeta}_t &= \mathbf{G}\boldsymbol{\zeta}_{t-1} + \mathbf{B}\mathbf{u}_t \qquad \mathbf{u}_t \sim GWN(\mathbf{0}, \boldsymbol{\Sigma}_\mathbf{u}), \qquad (3) \end{aligned}$$

where $\mathbf{F} \in \mathbb{R}^{n \times k}$, $\boldsymbol{\zeta}_t \in \mathbb{R}^k$, $\mathbf{G} \in \mathbb{R}^{k \times k}$, $\mathbf{u}_t \in \mathbb{R}^{k'}$ $(k' \le k)$, $\mathbf{B} \in \mathbb{R}^{k \times k'}$. Moreover, we shall assume that $\boldsymbol{\varepsilon}_t \perp \mathbf{u}_{t'}\ \forall t, t'$, $\mathbf{u}_t \perp \boldsymbol{\zeta}_{t'}\ \forall t > t'$. Generally, σ_ε^2 and $\boldsymbol{\Sigma}_\mathbf{u}$ are unknown, whereas $\mathbf{G}$ is only partially known, so estimation problems arise. In a previous work (Tonellato 1997a) we gave sufficient conditions for model identifiability (O'Hagan 1994) in order to guarantee likelihood informativeness, avoid overparameterisation problems and reduce the dependency of the predictive distribution on parameter prior. In the same work we considered the inclusion of explanatory variables.

3. Results of a simulation study

We simulated some spatial time series from the following model with $n = 5$ and $1 \le t \le 500$:

$$\mathbf{y}_t = \mu_t \mathbf{1}_5 + \mathbf{x}_t + \boldsymbol{\varepsilon}_t, \quad \boldsymbol{\varepsilon}_t \sim GWN(\mathbf{0}, \sigma_\varepsilon^2 \mathbf{I}_n) \tag{4}$$

(the scalar time varying level μ_t is common to the n sites; $\mathbf{1}_n$ indicates the n–dimensional unit vector), with

$$\begin{aligned} \mu_t &= \mu_{t-1} + \eta_t, \quad \eta_t \sim GWN(\mathbf{0}, \sigma_\eta^2 \mathbf{I}_n) \qquad (5)\\ \mathbf{x}_t &= \boldsymbol{\Phi}\mathbf{x}_{t-1} + \boldsymbol{\omega}_t, \quad \boldsymbol{\omega}_t \sim GWN(\mathbf{0}, \boldsymbol{\Sigma}_\omega \mathbf{R}(\boldsymbol{\theta})\boldsymbol{\Sigma}_\omega) \qquad (6) \end{aligned}$$

Table 1: *Parameter values and estimated marginal posterior percentiles.*

		Parameter value	$q_{0.05}$	$q_{0.5}$	$q_{0.95}$
M1	ϕ_1	0.5	0.337	0.510	0.656
	ϕ_2	-0.7	-0.786	-0.685	-0.570
	ϕ_3	0.2	-0.210	-0.029	0.134
	ϕ_4	-0.4	-0.613	-0.433	-0.367
	ϕ_5	0.7	0.561	0.780	0.932
	σ_η^2	0.01	0.007	0.010	0.014
	σ_ε^2	0.02	0.008	0.023	0.031
	$\sigma_{\omega,1}$	1.41	1.392	1.477	1.565
	$\sigma_{\omega,2}$	1.41	1.304	1.416	1.509
	$\sigma_{\omega,3}$	1.41	1.261	1.354	1.435
	$\sigma_{\omega,4}$	2.00	2.116	2.258	2.445
	$\sigma_{\omega,5}$	2.00	1.938	2.118	2.260
	α	0.95	0.872	0.918	0.979
	κ	0.1	0.084	0.106	0.129
M2	ϕ_1	0.5	0.238	0.408	0.571
	ϕ_2	-0.7	-0.876	-0.778	-0.680
	ϕ_3	0.2	-0.099	0.017	0.154
	ϕ_4	-0.4	-0.568	-0.398	-0.228
	ϕ_5	0.7	0.558	0.650	0.760
	σ_η^2	0.01	0.008	0.011	0.015
	σ_ε^2	0.04	0.014	0.032	0.046
	$\sigma_{\omega,1}$	1.41	1.205	1.302	1.380
	$\sigma_{\omega,2}$	1.41	1.370	1.481	1.582
	$\sigma_{\omega,3}$	1.41	1.268	1.344	1.443
	$\sigma_{\omega,4}$	2.00	1.742	1.904	2.092
	$\sigma_{\omega,5}$	2.00	1.565	1.709	1.852
	α	0.95	0.836	0.893	0.936
	κ	0.1	0.073	0.087	0.104
M3	ϕ_1	0.5	0.151	0.380	0.586
	ϕ_2	-0.7	-0.859	-0.753	-0.636
	ϕ_3	0.2	-0.094	0.016	0.200
	ϕ_4	-0.4	-0.612	-0.418	-0.153
	ϕ_5	0.7	0.485	0.655	0.784
	σ_η^2	0.01	0.006	0.008	0.012
	σ_ε^2	0.08	0.055	0.074	0.088
	$\sigma_{\omega,1}$	1.41	1.360	1.492	1.612
	$\sigma_{\omega,2}$	1.41	1.344	1.450	1.595
	$\sigma_{\omega,3}$	1.41	1.204	1.294	1.368
	$\sigma_{\omega,4}$	2.00	1.837	2.038	2.277
	$\sigma_{\omega,5}$	2.00	1.841	2.046	2.210
	α	0.95	0.753	0.829	0.914
	κ	0.1	0.084	0.092	0.109

Figure 1: *Plots of the 5–th, 50–th and 95–th percentiles of simulated values of α and κ versus iterations of the Gibbs sampler (dotted lines) and true parameter values (bold lines).*

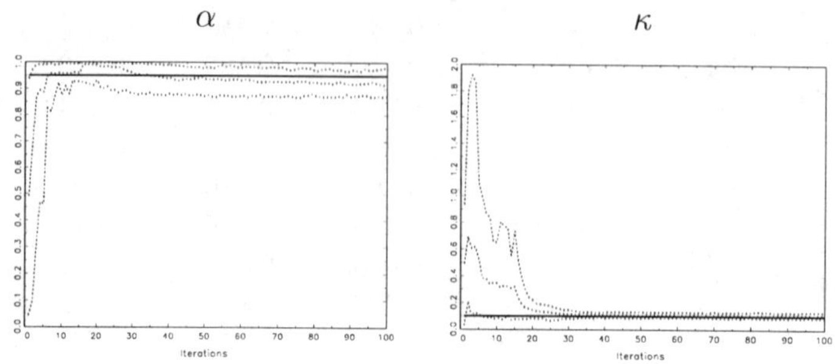

where $\mathbf{\Phi} = \text{diag}[\phi_1, \phi_2, \phi_3, \phi_4, \phi_5]$ is a diagonal matrix containing the unknown autoregressive coefficients. For any fixed t, $\boldsymbol{\omega}_t$ has been defined as a Gaussian spatial process. The matrices $\mathbf{\Sigma}_\omega = \text{diag}[\sigma_{\omega,1}, \sigma_{\omega,2}, \sigma_{\omega,3}, \sigma_{\omega,4}, \sigma_{\omega,5}]$ and $\mathbf{R}(\boldsymbol{\theta})$ contain the site specific standard deviations and the spatial correlation coefficients of $\boldsymbol{\omega}_t$ respectively. The spatial correlation function of $\boldsymbol{\omega}_t$ has been specified as an isotropic exponential one (Cressie 1993), i.e.

$$Corr(\omega_t(s_i), \omega_t(s_j)) = \alpha \exp\{-\kappa d_{ij}\}, \ 0 \le \alpha \le 1, \kappa > 0, \ \boldsymbol{\theta} = [\alpha \ \kappa]'$$

where d_{ij} denotes the Euclidean distance between sites s_i and s_j, $i, j = 1, 2, \ldots, 5$. It is easy to verify that the model defined by eqs. (4)–(6) has a state–space representation as the one introduced in eqs. (2)–(3) with

$$\mathbf{F} = [\mathbf{1}_5 \ \mathbf{I}_5]', \ \boldsymbol{\zeta}_t = \begin{bmatrix} \mu_t \\ \mathbf{x}_t \end{bmatrix}, \ G = \begin{bmatrix} 1 & \mathbf{0} \\ \mathbf{0} & \mathbf{\Phi} \end{bmatrix}, \ \mathbf{B} = \mathbf{I}_{n+1}, \text{ and } \mathbf{u}_t = \begin{bmatrix} \eta_t \\ \boldsymbol{\omega}_t \end{bmatrix}.$$

The unknown parameters are ϕ_i, $\sigma_{\omega,i}$, $i = 1, \ldots, 5$, σ_ε^2, σ_η^2, α and κ.

It is well known in the dynamical systems literature that measurement error variability negatively affects both filtering (i.e. the estimation of the unobservable state variable $\boldsymbol{\zeta}_t$) and parameter estimation. In our simulation study we want to assess how much measurement error variance can affect our inference. Therefore, we generated three independent series, denoted by M1, M2 and M3, from the model defined by eqs. (4)–(6) and characterised by different values of σ_ε^2, which has been fixed equal to 0.02, 0.04 and 0.08 respectively. The values of the remaining parameters have been held constant and their values are reported in Table 1. An indicator of the weight of measurement error variability in the models we are dealing with is given by the ratio between σ_ω^2 and the variance of the stationary signal (we call this quantity signal to noise ratio and denote it by S_{M1}, S_{M2} and S_{M3}). Clearly, due to the spatial heteroschedasticity of $\mathbf{x}_t$, this ratio is not constant over the different sites. It is easy to verify that in the three series we generated, we have: $0.02 \le S_{M1} \le 0.0672$, $0.040 \le S_{M2} \le 0.134$ and $0.089 \le S_{M3} \le 0.269$.

The unknown parameters are assumed to be independent a priori. The prior distribution elicited on σ_ε^2 is an $IG(0.5, 100)$ for M1 and M2 (IG indicates an Inverted Gamma distribution) and an $IG(1.2, 12)$ for M3. The priors elicited on the remaining parameters are the following:

$$\begin{array}{lll} \boldsymbol{\zeta}_0 \sim N\left(\mathbf{0},\, 100\mathbf{I}_6\right), & \sigma_\eta^2 \sim IG(1,\, 100), & \\ \alpha \sim U(0,\, 1), & \kappa \sim U\left(0,\, 1\right), & \\ \phi_i \sim N\left(0,\, 0.5\right) \mathbf{1}_{(-1,1)}, & \sigma_{\omega,i} \sim IG(2.5,\, 1), & i = 1,\, \ldots,\, 5. \end{array}$$

The normal prior for ϕ_i is truncated on the interval $(-1, 1)$ in order to guarantee the stationarity of $\mathbf{x}_t$; the prior distributions of σ_η^2 and σ_ε^2 seem quite concentrated on small values. Actually, the former is appropriate for a smoothly time varying level, which is usually the case for environmental time series, whereas the latter allows for high values of the signal to noise ratio. Suppose our interest centers on the estimation of all the unknown parameters appearing in the model with the exception of state variables. The whole parameter vector has dimension 3020. So we have to cope with a very complex parameterisation and marginalise the joint posterior with respect to a 3006–dimensional nuisance parameter. In spite of the high number of parameters, it can be shown that the model is identifiable (Tonellato 1997a) if $\mathbf{\Phi}$ is nonsingular π–almost surely (π denotes the prior joint distribution of the parameters). Clearly, the complexity of the model does not allow us to compute analytically the posterior distribution of the parameters of interest, so we used an approximate version of the Gibbs sampler (Gelfand and Smith 1990). For the definition of the full conditional distributions required by the Gibbs sampler refer to Tonellato (1997a). Here we want to point out that, in order to sample values of α and κ from their full conditionals, we used an approximated algorithm, the *griddy* Gibbs sampler (Ritter and Tanner 1992). It is worth to note that the same computational tools would work with different specifications of the spatial correlation function of $\boldsymbol{\omega}_t$. For each series we generated 100 independent chains with starting values sampled from the prior distribution of the parameters. The estimated posterior marginal 5–th, 50–th and 95–th percentiles are shown in Table 1. To give an idea of the behaviour of the Gibbs Sampler, in Figure 1 we show the plots of the percentiles of the simulated values of α and κ versus iterations of the Gibbs sampler for M1. We consider these parameters explicitly because they are probably the most difficult to be estimated, since they appear in the spatial correlation function of the noise of the unobservable space–time autoregressive process $\mathbf{x}_t$. The marginal posterior 95% confidence intervals of the autoregressive coefficients contain the true value of each parameter, with the exception of ϕ_3. In fact, ϕ_3 is quite close to the upper bound of the confidence interval. The behaviour of the marginal posterior percentiles of $\sigma_{\omega,i}$, $i = 1, \ldots, 5$, is particularly interesting. The posterior 95% confidence intervals are sensitive to the spatial heteroschedasticity of $\boldsymbol{\omega}_t$, although they do not always cover the true values of the parameters. In any case, such distortions are quite small and they probably would disappear for higher sample sizes. As far as σ_η^2 and σ_ε^2 are concerned, these parameters are always covered with high precision by the posterior confidence intervals. Finally, it is worth noting that the analysis is quite robust to changes of the signal to noise ratio, which means that these models can capture important aspects of the spatio–temporal dynamics of the process even when measurement errors are not negligible.

4. Discussion

We have illustrated some aspects of a new class of models for spatial time series with particular emphasis on parameter estimation. It can be shown that such models are identifiable in spite of the high number of parameters. The results obtained in some simulations support the idea that reliable information about the unknown parameters can be achieved through the posterior analysis. In particular, the dynamics of an unobservable space–time autoregressive process can be analysed at a good level of precision.

The interested reader can find an application to real data in Tonellato (1997b). Here, we want to recall some interesting aspects of that work. We analysed atmospheric carbon monoxide concentrations by defining a nonstationary space–time model and assuming a lognormal distribution of the data conditionally on some unknown parameters. The site specific temporal forecasts resulted quite reliable within a forecast horizon of 24 hours. Moreover, a retrospective analysis of the unobservable state gave a statistical explanation of the temporal heteroschedasticity of the observed process at a particular site.

Possible extensions to discrete data and to spatial prediction problems are under way.

References

Cressie, N. (1993). *Statistics for Spatial Data*. Wiley.

Gelfand, A. E. and A. F. M. Smith (1990). Sampling based approaches to calculating marginal densities. *Journal of the American Statistical Association 85*, 398–409.

O'Hagan, A. (1994). *Kendall's Advanced Theory of Statistics. Bayesian Inference*. London: Edward Arnold.

Ritter, C. and M. A. Tanner (1992). The Gibbs stopper and the griddy Gibbs sampler. *Journal of the American Statistical Association 87*, 861–868.

Tonellato, S. F. (1997a). Bayesian dynamic linear models for spatial time series. Submitted to the *Journal of the Royal Statistical Society*.

Tonellato, S. F. (1997b). A space–time analysis of carbon monoxide atmospheric concentrations. Submitted to the *Journal of the Royal Statistical Society*.

West, M. and J. Harrison (1989). *Bayesian Forecasting and Dynamic Linear Models*. Springer–Verlag.

Dissimilarities Between Trajectories of a Three-Way Longitudinal Data Set

Pierpaolo D'Urso*, Maurizio Vichi**
*Department of Statistica, Probabilità e Statistiche Applicate, University of Roma "La Sapienza", P.le A. Moro, 5, 00185 Roma, Italy
**Department of Metodi Quantitativi e Teoria Economica, University of Chieti "G. d'Annunzio", V.le Pindaro, 42, 65127, Pescara, Italy, e-mail: vichi@dmqte.unich.it

Abstract: This paper deals with the problem of evaluating dissimilarities between trajectories in a three-way longitudinal data set (a set of multiple time series). The dissimilarity between trajectories is defined as a conic combination of the dissimilarities between trends, velocities and accelerations of the pair of trajectories. The coefficients of the linear combination are estimated maximizing its variance. The proposed methodology is applied on a real data set to classify trajectories of Italian regions by their employment dynamic.

Key words: Trajectories, Three-way data, Trend, Velocity, Acceleration, Dissimilarities.

1. Position of the problem

Let $X_1, X_2, \dots, X_k$ be k quantitative variables observed on n units (objects) at u consecutive time points. The observed data can be arranged into a three-way longitudinal data set $\mathbf{X} \equiv \left[\mathbf{x}_{i\cdot t} = (x_{i1t}, x_{i2t}, \dots, x_{ikt}) \colon i \in I,\ t \in T\right]^{(1)} \equiv \left[x_{ijt} \colon i \in I, j \in J, t \in T\right]$, where x_{ijt} is the value of the j-th variable collected on the i-th object at time t; $I = \{1,\dots,n\}$, $J = \{1,\dots,k\}$, and $T = \{1,\dots,u\}$ are the set of indices pertaining to objects, variables and time points, respectively. The observed objects can be represented as points of a vectorial space equipped with a distance, i.e. a real function δ on the set $\mathbf{Y} \equiv \left[\mathbf{y}_{i\cdot t} = (x_{i1t}, x_{i2t}, \dots, x_{ikt}, t) \colon i \in I, t \in T\right]$ from: $\mathbf{Y} \times \mathbf{Y}$ to $\Re^+$, such that: $\delta_t(i,i) = 0$, $\delta_t(i,l) = \delta_t(l,i)$, $\delta_t(i,l) \geq 0$, $\delta_t(i,l) + \delta_t(h,l) \geq \delta_t(i,h)$, $\forall i, l, h \in I$, where $\delta_t(i,l)$ indicates the distance between object i and l of I at time t. Let $M^{k+1} \equiv (\mathbf{Y}, d)$ be the metric space spanning the k variables and time.

For each object i, $Y(i) \equiv \{\mathbf{y}_{i.t} \colon t \in T\}$ describes a *time trajectory* of the i-th object according to the k examined variables. The trajectory $Y(i)$ is geometrically represented by $u-1$ segments connecting u points $\mathbf{y}_{i\cdot t}$ of M^{k+1} (Figure 1).
This paper deals with the problem to measure the dissimilarity between trajectories in a three-way longitudinal data set.

---

(1) The point subscript means that the vector $\mathbf{x}_{i\cdot t}$ has elements taken according to the relative index substituted by the point.

Figure 1: Two time trajectories in M^3

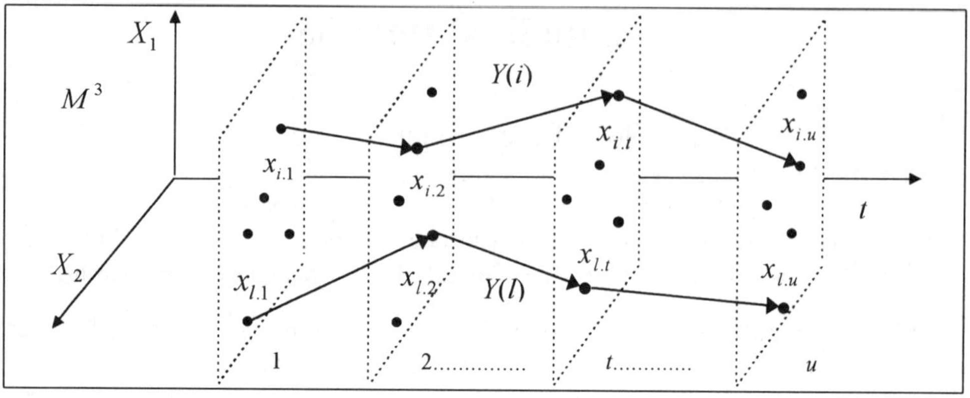

The problem to evaluate a dissimilarity between $Y(i)$ and $Y(l)$ was first introduced by Carlier (1986), who considered as measure the result of two weighted components combined using subjective weights. In this paper the dissimilarity between time trajectories is defined as a linear combination of distances between trends, velocities and accelerations of a pair of trajectories. The coefficients of the combination are obtained maximizing its variance.

The $(n \times n)$ matrix of dissimilarities between time trajectories can be used for successive multivariate analyses. For example, it has been used to classify trajectories into homogeneous sets with a clustering algorithm (Carlier, 1986), or to represent trajectories as points of an Euclidean space of reduced dimensions, with a MDS technique.

Before formalizing the notion of dissimilarity between time trajectories, it is useful to discuss properties that such measures should hold.

Remark 1: a dissimilarity between time trajectories $Y(i)$ and $Y(l)$ takes into account differences: between trends of objects i and l, between velocities and accelerations of $Y(i)$ and $Y(l)$. Velocity and acceleration are two trajectories' characteristics strongly describing changes of objects along time. For example in M^2, velocity of each segment of the trajectory is the slope of the straight line passing through it: if velocity is negative (positive) the slope will be negative (positive) and the angle made by each segment of the trajectory with the positive direction of the t-axis will be obtuse (acute). Geometrically, acceleration of each pair of segments of trajectory represents their convexity or concavity. If acceleration is positive (negative) the trajectory of the two segments is convex (concave).

Remark 2: a dissimilarity between time trajectories degenerates into a classical dissimilarity between two objects when trajectories are reduced to two points (i.e., $u = 1$).

Remark 3: a dissimilarity between time trajectories is a function of distances between multivariate objects, $\delta_t(i, l)$, $t \in T$.

Let us now formalize velocity and acceleration of a trajectory.

Velocity of $Y(i)$ is defined as the rate of change of i-th object position in a fixed time interval and indicating the direction and versus of each segment of the trajectory $Y(i)$ for a given variable.
Therefore, for each time trajectory $Y(i)$, the velocity of evolution of an object i in the interval from t to $t+1$, denoted $s_{t,t+1}$, is, for the j-th variable:

$$v_{ijt,t+1} = \frac{x_{ijt+1} - x_{ijt}}{s_{t,t+1}}.$$

In particular: $v_{ijt,t+1} > 0$ ($v_{ijt,t+1} < 0$) if object i, for the j-th variable, presents an increasing (decreasing) rate of change of its position in the time interval from t to $t+1$; $v_{ijt,t+1} = 0$ if the object i for the j-th variable, does not change position from t to $t+1$.
Acceleration measures the variation of velocity of $Y(i)$ in a fixed time interval.
For each time trajectory $Y(i)$, the acceleration of an object i in the interval from t to $t+2$[(2)], denoted $s_{t,t+2}$, is, for the j-th variable:

$$a_{ijt,t+2} = \frac{v_{ijt+1,t+2} - v_{ijt,t+1}}{s_{t,t+2}}.$$

In particular: $a_{ijt,t+2} > 0$ ($a_{ijt,t+2} < 0$) if the object i, for the j-th variable, presents an increasing (decreasing) variation of velocity in the time interval from t to $t+2$; $a_{ijt,t+2} = 0$ if object i, for j-th variable, does not change velocity from t to $t+2$.
Defined velocity and acceleration, we are now in position to evaluate differences between trends in a time point t, velocities and accelerations in a time interval. Let us first consider:

Remark 4: (a) differences between trend intensities, in a time point t, of objects i and l are evaluated according to a measure of distance between $\mathbf{x}_{i.t}$ and $\mathbf{x}_{l.t}$, $t \in T$; (b) differences between velocities of objects i and l, in a time interval, are evaluated according a measure of distance between $\mathbf{v}_{i.t,t+1} = (v_{i1t,t+1}, \ldots, v_{ikt,t+1})'$ and $\mathbf{v}_{l.t,t+1}$, $t = 1,\ldots,u-1$; (c) differences between accelerations of objects i and l, in a time interval, are evaluated according to a measure of distance between $\mathbf{a}_{i.t,t+2} = (a_{i1t,t+2}, \ldots, a_{ikt,t+2})'$ and $\mathbf{a}_{l.t,t+2}$, $t = 1,\ldots,u-2$.

Considering *remark* 4 and the normalized Minkowski metric of order p, the distances between trends in a time point t, velocities and accelerations in a time interval for object i and l are respectively:

$${}_1\delta_t(i,l) = \frac{1}{\pi_1}\left[\sum_{j=1}^{k}\left|x_{ijt} - x_{ljt}\right|^p\right]^{1/p}, \quad {}_2\delta_{t,t+1}(i,l) = \frac{1}{\pi_2}\left[\sum_{j=1}^{k}\left|v_{ijt,t+1} - v_{ljt,t+1}\right|^p\right]^{1/p}, \quad {}_3\delta_{t,t+2}(i,l) = \frac{1}{\pi_3}\left[\sum_{j=1}^{k}\left|a_{ijt,t+2} - a_{ljt,t+2}\right|^p\right]^{1/p}$$

(trends distance) (velocities distance) (accelerations distance)

where p is an integer ≥ 1 and π_1, π_2, π_3 are suitable weights to normalize distances.

---

(2) Acceleration must be computed on two time intervals $[t, t+1]$, $[t+1, t+2]$.

Remark 5: note that, before computing Minkowski metric, variables have to be standardized to take into account different units of measurements and/or variability (Milligan, Cooper, 1988).

Remark 6: weights π_1, π_2, and π_3 may be defined as the range of respective distances or their standard deviation.

Remark 7: (a) a dissimilarity between trends of $Y(i)$ and $Y(l)$ is a mapping from $\left\{ {}_1\delta_t(i,l),\ t \in T \right\}$ to $\Re^+$; (b) a dissimilarity between velocities of $Y(i)$ and $Y(l)$ is a mapping from $\left\{ {}_2\delta_{t,\ t+1}(i,l),\ t \in T \right\}$ to $\Re^+$; (c) a dissimilarity between accelerations of $Y(i)$ and $Y(l)$ is a mapping from $\left\{ {}_3\delta_{t,\ t+2}(i,l),\ t \in T \right\}$ to $\Re^+$.

2. Dissimilarity between trajectories

Following *remark* 7 for each time trajectory characteristic (trend, velocity, acceleration), three mappings are considered: minimum, maximum and sum:

$$ {}_m d_r(i,l) = \varphi_m\left({}_r\delta_{t,t+r-1}(i,l) \right)^{(3)} \tag{1}$$

where: $m,\ r = 1,2,3$; $t = 1,\dots,u$,, $\varphi_m(\cdot)$ is a function of distance ${}_r\delta_{t,\ t+r-1}(i,l)$.

In particular $\varphi_1(\cdot) = \min_t \left\{ {}_r\delta_{t,\ t+r-1}(i,l) \right\}$, $\varphi_2(\cdot) = \max_t \left\{ {}_r\delta_{t,\ t+r-1}(i,l) \right\}$, $\varphi_3(\cdot) = \sum_{t=1}^{u-2} {}_r\delta_{t,\ t+r-1}(i,l)$.

The total dissimilarity between $Y(i)$ and $Y(l)$ according to the three mappings is:

$$ {}_m d^*(i,l) = \sum_{r=1}^{3} {}_m\gamma_r\ {}_m d_r(i,l),\ m = 1,2,3 \tag{2}$$

where ${}_m\gamma_r$ $(r, m = 1,2,3)$ are suitable positive weights, that indicate the contribution of each trajectory characteristic to determine the total dissimilarity ${}_m d^*(i,l)$.
The weights may be chosen by assigning arbitrary values according to a subjective criterion. A more objective approach is to evaluate weights so that the variance of (2) is maximized as show below.

Let ${}_m\mathbf{D}^* = \left[{}_m d^*(i,l) \right] (m = 1,2,3)$ be the $(n \times n)$ matrices of total dissimilarities between trajectories and let ${}_m\mathbf{D}_r = \left[{}_m d_r(i,l) \right] (m,r = 1,2,3)$ be the $(n \times n)$ matrices obtained considering the dissimilarities between $Y(i)$ and $Y(l)$ for the three $(r = 1,2,3)$ features of each trajectory. Matrices ${}_m\mathbf{D}^*$ and ${}_m\mathbf{D}_r$ may be represented as vectors defined by elements of a triangle below (above) the diagonal of the matrices $\overline{vec}({}_m\mathbf{D}^*)$[4]

---

[3] If $r = 1$: ${}_r\delta_{t,\ t+r-1}(i,l) = {}_1\delta_{t,\ t}(i,l) = {}_1\delta_t(i,l),\ t = 1,\dots,u.$

[4] E.g., $\overline{vec}({}_m\mathbf{D}^*) = \left({}_m d^*(1,2),\dots, {}_m d^*(1,n), {}_m d^*(2,3),\dots, {}_m d^*(n-1,n) \right)'$.

and $\overline{vec}({}_m\mathbf{D}_r)$ $(m,r=1,2,3)$. Therefore, ${}_m\mathbf{D}=\left[\overline{vec}({}_m\mathbf{D}_r)\right]=\left[\overline{vec}({}_m\mathbf{D}_1),\overline{vec}({}_m\mathbf{D}_2),\overline{vec}({}_m\mathbf{D}_3)\right]$ $(m,r=1,2,3)$ is a $(n(n\text{-}1)/2\times 3)$ matrix; while $\overline{vec}({}_m\mathbf{D}^*)=\sum_{r=1}^{3}{}_m\gamma_r\overline{vec}({}_m\mathbf{D}_r)$.

In order to compute the total dissimilarity between each pair of time trajectories, $\overline{vec}({}_m\mathbf{D}^*)$, the follow problem with respect to variables ${}_m\boldsymbol{\gamma}=\left({}_m\gamma_1,{}_m\gamma_2,{}_m\gamma_3\right)'$ has to be solved:

$$\begin{cases} max\ var(\overline{vec}({}_m\mathbf{D}^*))=var(\sum_{r=1}^{3}{}_m\gamma_r\overline{vec}({}_m\mathbf{D}_r)) \\ {}_m\boldsymbol{\gamma}'{}_m\boldsymbol{\gamma}=1 \end{cases} \tag{3}$$

The solution is easily obtained differentiating the Lagrangian function of (3) and equating to zero:

$$\frac{\partial}{\partial_m\boldsymbol{\gamma}}\left[var(\overline{vec}({}_m\mathbf{D}^*))+\lambda(1\text{-}{}_m\boldsymbol{\gamma}'{}_m\boldsymbol{\gamma})\right]=0, \tag{4}$$

where λ is the Lagrange multiplier. Since $var(\overline{vec}({}_m\mathbf{D}^*))={}_m\boldsymbol{\gamma}'\Sigma\,{}_m\boldsymbol{\gamma}$, where Σ is the variance and covariance matrix of ${}_m\mathbf{D}$; thus equation (4) is: $(\Sigma-\lambda\mathbf{I})_m\gamma=0$. Then multiplying for ${}_m\boldsymbol{\gamma}'$: ${}_m\boldsymbol{\gamma}'\Sigma\,{}_m\boldsymbol{\gamma}=\lambda$ corresponding to the maximum eigenvalue of Σ and ${}_m\boldsymbol{\gamma}$ is the associated eigenvector.

3. Clustering Italian Regions by their employment dynamic

For every Italian Region (Table 1), the employment level for each year (1951, 1961, 1971, 1981, 1991) (data were collected from Census of Agriculture and Census of Industry and Services, ISTAT) and for different types of industries were considered (Table 1), defining the three-way dimensions $(19\times 18\times 5)$.

Table 1: Italian Regions and Industry

Italian Regions	Industry
Piemonte (1), Valle d'Aosta (2), Lombardia (3), Trentino (4), Veneto (5), Friuli (6), Liguria (7), Emilia Romagna (8), Toscana (9), Umbria (10), Marche (11), Lazio (12), Abruzzo-Molise (13), Campania (14), Puglia (15), Basilicata (16), Calabria (17), Sicilia (18), Sardegna (19).	Agriculture, Mining, Manufacturing, Construction, Electric, gas, and sanitary services, Wholesale trade, Business services, Retail trade, hotel, and other lodging places, Educational, health, and legal services, Personal services, Auto, trade and services, Transportation, Communication, Finance, Insurance, Real estate, Social services.

Firstly, the data were analyzed using the Factorial Dynamic Analysis (FDA) (Coppi, Zannella, 1978). This strategy of analysis gives information about the structure and the evolution of the employment. The temporal evolution, characterizing the employment level of each Italian Region, was synthesized by the values of the first two factors of FDA, explaining the 86% of the total variance. The first factor is highly correlated with the employment levels of Industry and Tertiary. The second is more correlated with employment of Agriculture, Mining, Construction and Social Service. The time trajectories can be drawn into the factorial plane considering the first two factors. The dissimilarity between trajectories, fixed p=2, in the Minkowski metric were computed, and the matrices $\pi_1\,{}_3\mathbf{D}_1, \pi_2\,{}_3\mathbf{D}_2, \pi_3\,{}_3\mathbf{D}_3$ regarding trend, velocity and acceleration were reported in Table 2. To normalize dissimilarities, $\pi_1 = var({}_3\delta_t(i,l))$, $\pi_2 = var({}_3\delta_{t,t+1}(i,l))$, $\pi_3 = var({}_3\delta_{t,t+2}(i,l))$ were used as indicated in *remark* 6.

Table 2: Dissimilarities matrices $\pi_1\,{}_3\mathbf{D}_1, \pi_2\,{}_3\mathbf{D}_2$ and $\pi_3\,{}_3\mathbf{D}_3$

	1	2	3	4	5	6	7	8	9	10	11	12	13	14	15	16	17	18	19
1		33.96	41.81	27.20	3.33	23.95	18.95	2.19	7.15	29.37	24.73	6.67	25.55	9.96	16.64	31.96	25.46	12.82	24.30
2			75.46	6.82	32.65	10.04	15.07	33.01	30.42	4.77	9.62	40.40	9.19	27.71	20.41	2.33	9.60	27.92	12.14
3				68.81	43.93	65.57	60.51	43.21	47.12	71.01	66.46	35.90	67.33	50.13	57.91	73.59	67.26	51.68	65.98
4					25.87	3.34	8.35	26.23	23.65	2.45	3.04	33.61	2.90	20.96	13.88	4.81	3.66	21.30	6.20
5						22.64	17.86	1.91	4.17	27.95	23.17	8.95	23.88	7.56	14.27	30.56	23.75	9.80	22.13
6							5.28	22.97	20.51	5.44	1.54	30.38	2.92	17.93	11.20	8.04	3.67	18.45	5.38
7								18.15	15.94	10.60	6.33	25.39	7.38	13.43	8.02	13.12	7.49	14.39	7.95
8									5.64	28.33	23.58	7.79	24.33	8.33	14.95	30.94	24.21	11.29	22.78
9										25.67	20.83	12.18	21.42	4.13	11.10	28.25	21.21	5.80	19.30
10											4.85	35.77	4.49	23.00	15.71	2.65	5.01	23.21	7.62
11												31.08	1.53	18.24	11.03	7.44	2.42	18.49	4.41
12													31.82	14.58	22.16	38.35	31.65	17.05	30.32
13														18.68	11.26	6.92	1.26	18.77	3.36
14															8.00	25.49	18.35	3.04	16.55
15																18.12	10.83	7.59	8.74
16																	7.30	25.64	9.83
17																		18.36	2.72
18																			16.17
19																			

	1	2	3	4	5	6	7	8	9	10	11	12	13	14	15	16	17	18	19
1		7.72	10.81	6.34	1.68	5.88	4.99	0.95	1.56	6.60	5.30	2.71	5.77	2.58	3.17	7.18	5.82	2.43	5.56
2			18.40	1.46	8.37	2.18	3.00	8.45	7.12	1.35	2.63	9.89	2.16	5.61	4.96	0.56	1.99	5.99	2.77
3				16.96	10.75	16.62	15.73	10.09	11.70	17.32	16.04	8.94	16.47	13.11	13.75	17.87	16.56	12.94	16.28
4					7.09	1.22	1.89	7.04	5.78	0.82	1.50	8.49	1.03	4.27	3.62	0.98	0.88	4.71	1.95
5						6.30	6.02	1.21	2.06	7.07	5.77	3.15	6.23	3.84	4.19	7.84	6.62	3.26	5.80
6							1.64	6.58	5.12	0.89	0.60	8.28	0.48	3.89	3.31	1.66	0.83	4.11	1.08
7								5.83	4.63	2.05	1.37	7.20	1.68	2.90	2.50	2.49	1.23	3.36	2.06
8									2.12	7.30	5.99	2.27	6.45	3.47	3.99	7.91	6.59	3.34	6.23
9										5.90	4.57	4.07	5.02	2.29	2.62	6.57	5.23	2.01	4.70
10											1.34	8.91	0.92	4.63	4.05	0.84	1.09	4.87	1.58
11												7.70	0.57	3.39	2.86	2.09	1.03	3.63	0.96
12													8.15	4.78	5.42	9.37	8.08	4.91	8.09
13														3.80	3.23	1.63	0.90	4.04	1.07
14															0.79	5.06	3.70	1.22	3.69
15																4.41	3.10	1.63	3.19
16																	1.47	5.43	2.27
17																		4.08	1.33
18																			3.92
19																			

	1	2	3	4	5	6	7	8	9	10	11	12	13	14	15	16	17	18	19
1		5.44	6.82	4.70	1.35	3.98	2.87	0.65	1.07	4.70	3.77	1.81	4.11	1.40	1.73	4.96	4.01	0.97	4.32
2			12.18	0.83	5.61	1.50	2.68	5.23	4.94	0.94	1.72	5.85	1.38	4.93	4.41	0.51	1.47	5.65	1.77
3				11.40	6.78	10.72	9.67	6.99	7.40	11.45	10.53	6.59	10.85	7.50	7.87	11.69	10.75	6.76	11.00
4					5.01	1.05	2.00	4.46	4.23	0.72	0.86	0.86	0.86	4.14	3.63	0.38	0.77	4.92	1.53
5						4.13	3.39	1.09	1.94	4.78	3.91	2.19	4.28	2.33	2.73	5.16	4.40	1.39	4.23
6							1.42	3.78	3.50	0.78	0.34	4.41	0.56	3.50	2.97	1.13	0.58	4.24	1.16
7								2.71	2.37	1.98	1.17	3.38	1.42	2.39	1.98	2.20	1.25	3.09	1.88
8									1.30	4.53	3.61	1.96	3.91	1.28	1.71	4.73	3.79	0.79	4.19
9										4.18	3.27	2.20	3.57	1.00	1.10	4.45	3.52	0.95	3.68
10											0.93	5.04	0.67	4.23	3.70	0.62	0.86	4.92	1.07
11												4.15	0.53	3.33	2.80	1.30	0.62	4.03	1.05
12													4.47	2.29	2.47	5.37	4.50	1.84	4.51
13														3.61	3.08	0.90	0.67	4.29	0.81
14															0.72	4.42	3.48	1.23	3.90
15																3.91	2.97	1.61	3.35
16																	0.99	5.15	1.40
17																		4.23	1.22
18																			4.42
19																			

To obtain weights ${}_m\gamma_1, {}_m\gamma_2, {}_m\gamma_3$ and determine the total dissimilarity the largest eigenvalue and the associated eigenvector of Σ were computed: $\lambda = 2.9412$ and ${}_3\gamma = (0.5802\ \ 0.5783\ \ 0.5735)'$. The total dissimilarity matrix computed through (2) is shown in Table 3.

Table 3: Total dissimilarity matrix ${}_3\mathbf{D}^*$

	1	2	3	4	5	6	7	8	9	10	11	12	13	14	15	16	17	18	19
1		3.49	4.51	2.89	0.65	2.55	2.01	0.36	0.71	3.01	2.45	1.02	2.62	1.02	1.42	3.24	2.60	1.01	2.60
2			7.95	0.63	3.57	0.99	1.55	3.51	3.17	0.57	1.09	4.11	0.93	2.86	2.39	0.27	0.94	3.08	1.21
3				7.34	4.56	7.03	6.49	4.49	4.95	7.48	6.93	3.98	7.10	5.28	5.72	7.70	7.09	5.14	7.06
4					3.02	0.52	1.00	2.91	2.58	0.36	0.51	2.57	0.44	2.26	1.80	0.39	0.42	2.50	0.83
5						2.60	2.23	0.48	0.87	3.04	2.50	1.25	2.67	1.33	1.70	3.33	2.75	1.11	2.53
6							0.73	2.58	2.22	0.49	0.21	3.22	0.29	1.95	1.52	0.77	0.37	2.17	0.60
7								2.06	1.74	1.10	0.68	2.66	0.81	1.41	1.07	1.30	0.71	1.66	0.99
8									0.79	3.03	2.48	1.03	2.63	1.07	1.47	3.25	2.62	1.04	2.61
9										2.66	2.10	1.49	2.25	0.69	1.00	2.91	2.26	0.70	2.16
10											0.57	3.63	0.44	2.40	1.94	0.35	0.52	2.59	0.73
11												3.10	0.25	1.86	1.41	0.85	0.37	2.05	0.52
12													3.26	1.70	2.09	3.86	3.25	1.70	3.21
13														1.99	1.53	0.67	0.32	2.18	0.45
14															0.55	2.59	1.94	0.55	1.97
15																2.13	1.48	0.85	1.50
16																	0.68	2.81	0.98
17																		2.16	0.55
18																			2.10
19																			

Finally, the method of average linkage was applied to the matrix ${}_3\mathbf{D}^*$. The dendrogram is shown in Figure 2 with three classes of similar trajectories according to dissimilarities in Table 3. The number of classes was fixed according to the test proposed by Calinski and Harabasz (1974).

Figure 2: Dendrogram of the time trajectories regarding the employment evolution of the Italian Regions

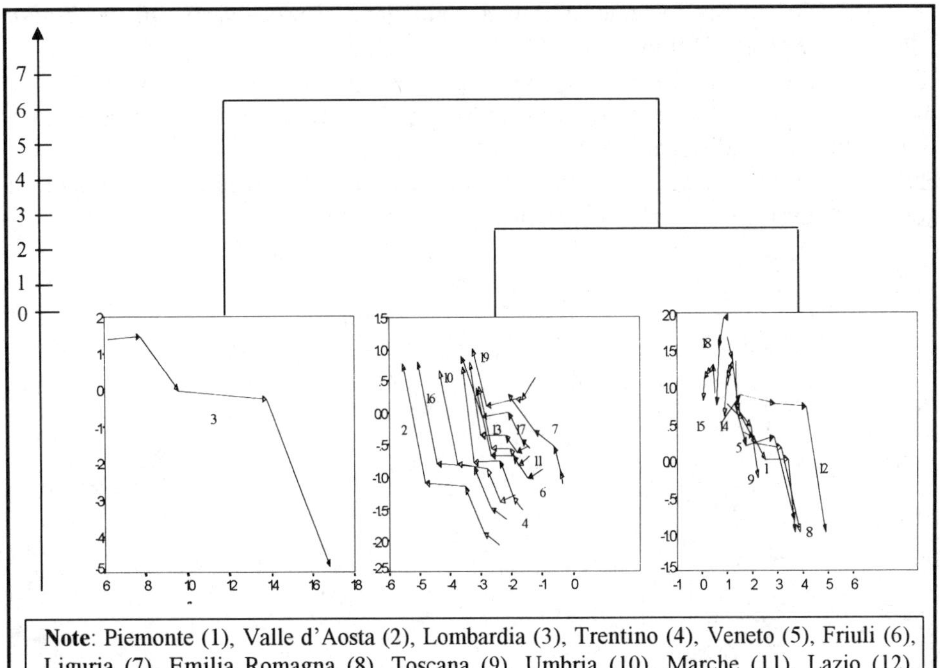

Note: Piemonte (1), Valle d'Aosta (2), Lombardia (3), Trentino (4), Veneto (5), Friuli (6), Liguria (7), Emilia Romagna (8), Toscana (9), Umbria (10), Marche (11), Lazio (12), Abruzzo-Molise (13), Campania (14), Puglia (15), Basilicata (16), Calabria (17), Sicilia (18), Sardegna (19).

4. Discussion

In this paper new dissimilarities between time trajectories $Y(i)$ and $Y(l)$ have been discussed. These are defined as a combination of distances between trends, velocities and accelerations of $Y(i)$ and $Y(l)$, features strongly characterizing trajectories.

The concept of velocity and acceleration was already introduced by Kosmelj (1984) in the study of evolution of phenomena, using these characteristics to measure the dissimilarities between objects. Here, velocity and acceleration, are considered features of each single trajectory (object).

The dissimilarity between trajectories proposed by Carlier (1986) can be considered as a particular case of our dissimilarity when p=2, $\varphi_3(\cdot) = \sum_{t=1}^{u-2} {}_r\delta_{t,\,t+r-1}(i,l)$, acceleration is zero (i.e., ${}_m\gamma_3 = 0$) and velocity regards constant time intervals.

References

Carlier, A. (1986). Factor Analysis of Evolution and Cluster Methods on Trajectories, *Proceedings of COMPSTAT'86*, Physica-Verlag, Heidelberg Wien, 140-145.

Calinski, R.B., Harabasz, J. (1974). A Dendrite Method for Cluster Analysis, *Commun. Stat.*, 3, 1-27.

Coppi R., Zannella F. (1978). L'Analisi Fattoriale di una Serie Temporale Multipla relativa allo stesso Insieme di Unità Statistiche, *Proceedings of the Italian Statistical Society*, Bologna, 61-79.

ISTAT. *Census of Agriculture* and *Census of Industry and Services* (1951-1991).

Kosmelj K. (1984). Means for Analysis of Space-Time Relationships, *Proceedings of COMPSTAT'84*, Physica-Verlag, Wien, 511-516.

Milligan, G.W., Cooper, M.C. (1988). "A study of standardization of variable in cluster analysis", *Journal of Classification*, 5, 181-204.

An Overview of Three-way Analysis and Some Recent Developments

Henk A.L. Kiers
Heymans Institute, University of Groningen, Grote Kruisstraat 2/1,
9712 TS Groningen, The Netherlands, email: h.a.l.kiers@ppsw.rug.nl

Abstract: A brief overview is presented of techniques for the analysis of three-way data. Techniques will be distinguished in terms of the type of data they are meant to analyze, on the basis of the type of analysis pursued, and the three-way models are distinguished with respect to their uniqueness. Next, recent approaches are discussed for eliminating nonuniqueness by imposing constraints and rotation to simple structure.

Keywords: three-way data, three-way models, multi-way data.

1. Introduction

Three-way data are data associated with three sets of units. As examples of three-way data one may, in behavioral research, have scores of a set of individuals on a set of variables at different occasions, or, with fluorescence spectroscopy, emission intensity of light of various excitation wavelengths by various chemical mixtures at various emission wavelengths. In these examples the three sets of units defining the three "ways" of the data are different. However, one also commonly encounters three-way data where two sets of units are equal, for instance in the case of data consisting of the similarities between a number of stimuli as judged by a number of judges, or, data consisting of several covariance matrices for a set of variables. Carroll and Arabie (1980) denoted three-way data pertaining to only two different sets of units (called "modes") as *two-mode three-way data*, and those with three different modes as *three-mode three-way data*. The difference in data type leads to a first difference in techniques for analyzing three-way data.

Techniques for the analysis of three-way data have their origin in techniques for the multivariate analysis of two-way data, notably principal components analysis (PCA) and multidimensional scaling (MDS). They aim at representations for the units of all modes in the data. A number of three-way techniques is based on considering three-way data as two-way data by combining two ways into one. For instance, one may consider J variables measured at K different occasions as a set of JK variables, and analyze the resulting I by JK data set (with I denoting the number of observation units) by PCA or any other two-way analysis technique, thus obtaining a representation for the individuals, and for the combinations of variables and occasions. By combining this analysis with analyses of other two-way data sets constructed from the three-way data, separate representations for the variables and for the occasions are obtained as well. Such techniques are called *combined two-way techniques* here. On the other hand, techniques have been proposed that are based on models specifically designed to represent all modes in the three-way data, and which are consequently fitted to the data. These are called *three-way model fitting techniques*. In Section 2 an overview will be given of both types of techniques.

2. Analysis of Three-way Data

In the present section a brief overview will be given of techniques for the analysis of three-way data. More detailed and comparative information on the techniques can be found, for instance, in the books by Law, Snyder, Hattie and McDonald (1984) and Coppi and Bolasco (1989), a special issue of Computational Statistics and Data Analysis (1994, Volume 18, number 1), a paper by Kiers (1991) and a book chapter by Rizzi and Vichi (1995). Below, first the *combined two-way techniques* will be described, then the *three-way model fitting techniques*. In the following, we assume for simplicity that the three-mode three-way data pertain to scores of I objects on J variables at K occasions, and that two-mode three-way data consist of a set of I by I (dis)similarity matrices obtained from K different judgements or occasions.

Among the first *combined two-way techniques* for the analysis of three-way data is the points-of-view analysis proposed by Tucker and Messick (1963). This technique for the analysis of an $(I \times I \times K)$ two-mode three-way data set consists of collecting the $\frac{1}{2}I(I-1)$ dissimilarities of each set in a vector, and analyzing the $\frac{1}{2}I(I-1) \times K$ matrix consisting of these vectors by means of PCA. The desired number of principal components is rotated to simple structure, and the scores on the resulting components are reordered into dissimilarity matrices again. Finally, each such summarizing dissimilarity matrix is analyzed by (two-way) MDS. Thus, by summarizing the K judgements (first step), and the I objects (second step), both modes of the three-way array have been taken into account.

A similar method, but meant for both three-mode and two-mode three-way data, is STATIS (L'Hermier des Plantes, 1976, also see Carlier et al., 1989). The method is based on first analyzing the "correlations" between a set of data matrices by means of PCA, thus describing the "*interstructure*" for the K units in the third mode. Secondly, the data or cross-products matrix associated with the first principal component, called "*compromise*" is analyzed by PCA, now yielding a representation of the I objects and, if present, the J variables. As with points-of-view analysis, there is no need to restrict oneself to using only one compromise matrix: D'Alessio (1989) proposed to compute and analyze all compromises that play an important role in the interstructure analysis. This idea has also been pursued in the FAMA approach (see Rizzi & Vichi, 1995), which, furthermore, allows to replace the ordinary matrix correlation coefficients (used in the first step) by other correlation measures. A similar approach has been proposed by Vichi (1997) for hierarchical cluster analysis of two-mode three-way data.

Many *three-way model fitting techniques* have been proposed. We focus on exploratory models, thus ignoring models involving external information.

First we describe models for the analysis of *three-mode* three-way data. One of the first such models is Tucker's (1966) *three-mode factor analysis model* (3MFA). In 3MFA, components are found for each mode, and these are linked by a three-way array called "core". The data x_{ijk} are modeled by

$$\hat{x}_{ijk}=\sum_{p=1}^{P}\sum_{q=1}^{Q}\sum_{r=1}^{R} a_{ip}b_{jq}c_{kr}g_{pqr}, \tag{1}$$

$i=1,...,I$, $j=1,...,J$, $k=1,...,K$, where a_{ip}, b_{jq}, c_{kr} are elements of the component matrices **A** $(I\times P)$, **B** $(J\times Q)$, and **C** $(K\times R)$, and the additional parameters g_{pqr} denote the elements of the $P\times Q\times R$ socalled "core array", which indicate how the components from the different modes interact.

Another early model is the PARAFAC model proposed by Harshman (1970, see also Harshman & Lundy, 1984). Like PCA, this model uses only one set of components for all three modes, but it differs from PCA in that the relative contributions of the different components are allowed to differ over occasions. The name *PARAlel FACtor analysis* refers to the components being proportional over occasions. In PARAFAC the data are modeled by

$$\hat{x}_{ijk}=\sum_{r=1}^{R} a_{ir}b_{jr}c_{kr}, \tag{2}$$

$i=1,...,I$, $j=1,...,J$, $k=1,...,K$, where a_{ir}, b_{jr}, c_{kr} again denote elements of matrices **A**, **B**, and **C**. The relation of PARAFAC to ordinary PCA may be more apparent from the matrix description of the model, which, for each frontal plane of the array, denoted as $\mathbf{X}_k$, $k=1,...,K$, reads:

$$\hat{X}_k=AD_kB', \tag{3}$$

where $\mathbf{D}_k$, is a diagonal matrix with the elements of the k-th row of **C**. The PARAFAC model is equal to Carroll and Chang's (1970) CANDECOMP model. They also showed (p.312) that it can be considered as a version of 3MFA in which the core is constrained to be "superdiagonal" (which implies that g_{pqr} is unconstrained if $p=q=r$ and g_{pqr} is constrained to 0 otherwise).

The 3MFA model cannot conveniently be described in terms of matrix products. This is no problem for the interesting special case of 3MFA where no reduction of the third mode is attempted (hence with **C**=**I**), dubbed the *Tucker2 model* by Kroonenberg and De Leeuw (1980):

$$\hat{X}_k = AG_kB', \quad k=1,...,K. \tag{4}$$

Here the matrices $\mathbf{G}_k$ express the relations between components for mode A and mode B, as they differ over occasions. Recently, Harshman and Lundy (1994) proposed a constrained version of this model by the name *PARATUCK2*, with features of both the PARAFAC and the Tucker2 model:

$$\hat{X}_k = AD_k^L G D_k^R B', \quad k=1,...,K, \tag{5}$$

where the single matrix **G** models the relations between components from the two modes, but, as in PARAFAC, differential component weights (in the diagonal matrices $\mathbf{D}_k^{\mathrm{L}}$ and $\mathbf{D}_k^{\mathrm{R}}$) are employed as well.
The final three-mode model mentioned here is meant for three-mode data consisting of (dis)similarities between row- and column-objects, observed at K instances. Such data can be described by "unfolding models", positioning the objects in a Euclidean space such that their distance reflects the (dis)similarity between the objects. Thus, in DeSarbo and Carroll's (1985) *three-way metric unfolding model*, the dissimilarity δ_{ijk} is modeled as

$$\hat{\delta}_{ijk} = \sum_{r=1}^{R} c_{kr}(a_{ir} - b_{jr})^2 + \alpha_k, \tag{6}$$

where **A** and **B** contain the coordinates for the row- and column-objects, **C** contains dimension weights, and α_k is an additive constant.
The best-known model for *two-mode* three-way data is probably the INDSCAL model (Carroll & Chang, 1970), which in fact is equivalent to the three-way unfolding model, (6), with **A**=**B**; actually, the three-way unfolding model was proposed as a generalization of the INDSCAL model. The "scalar products" version of the INDSCAL model reveals its relation to PARAFAC; the k-th transformed (dis)similarity matrix $\mathbf{S}_k$ is modeled as:

$$\hat{S}_k = AD_kA'. \tag{7}$$

Likewise, symmetric variants of 3MFA (*Three-mode scaling*), of the Tucker2 model (*IDIOSCAL*), and of PARATUCK2 (*PARAFAC2*) have been proposed by Tucker (1972), Carroll and Chang (1972), and Harshman (1972), respectively:

$$\hat{S}_k = A\sum_{r=1}^{R} c_{kl}G_lA', \quad \hat{S}_k = AG_kA', \quad \hat{S}_k = AD_kGD_kA', \quad k=1,...,K. \tag{8}$$

In all these models, the intermediate matrices $\mathbf{G}_k$ and $\mathbf{G}$ are constrained to be symmetric, or even positive semi-definite. The INDSCAL model itself has been extended by "specificity" terms for the objects, thus incorporating one unique dimension for each object (see Winsberg & Carroll, 1989):

$$\hat{\delta}_{ijk}=\sum_{r=1}^{R} c_{kr}(a_{ir}-b_{jr})^2+u_k(s_i+s_j). \qquad \textbf{(9)}$$

Finally, some models have been proposed for sets of *asymmetric* two-mode three-way data, such as sets of (asymmetric) friendship data, or brand switching data. Harshman, Green, Wind and Lundy (1982) proposed the three-way DEDICOM model simply as the PARAFAC2 model, now with $\mathbf{G}$ asymmetric. A more specialized model for the analysis of such data has been proposed by Rocci and Bove (1994): Here the objects are represented as vectors in a Euclidean space, and their asymmetric relation can be read from the plot, while differences between occasions are modeled as different weights attached to the lengths of these vectors.

3. Methods for Fitting Three-way Models to Three-way Data

The above models are most often fitted to the data by minimizing the sum of squared residuals. *Least squares* fitting algorithms have been proposed by the authors of the models, except for 3MFA (algorithm by Kroonenberg & De Leeuw, 1980) and PARAFAC2 (algorithm by Kiers, 1993).
Recently, much attention has been paid to the computational efficiency of some least squares fitting procedures. First of all, procedures have been proposed for efficiently handling large data arrays. These are based on compressing the data array by finding good projections onto subspaces for these modes (so as to incur little or no loss of information), and fitting the model to the compressed array; the solution to the compressed array is next expanded so as to obtain estimates for the full data set (see Kiers & Harshman, 1997 for an overview, and Bro & Andersson, 1997, and Kiers, 1998 for the most recent results). Furthermore, procedures for avoiding slow convergence (notably of PARAFAC) have been proposed recently (e.g., Paatero, 1997, Bro, 1998). When one of the modes is considered as random, the possibility arises to formulate factor analytic models for a super-covariance matrix, to be fitted by *maximum likelihood* procedures. Bloxom (1968) thus proposed a factor analytic version of the 3MFA model, and procedures for maximum likelihood fitting of the 3MFA model have been proposed by Bentler and Lee (1978). Mayekawa (1987) likewise proposed a maximum likelihood procedure for fitting the PARAFAC model. Also, a maximum likelihood procedure is available for fitting the INDSCAL model (as part of Ramsay's Multiscale program).

Finally, in chemometrics, closed-form procedures have been proposed that are very efficient, fit noise-free data perfectly, and supposedly work well in cases of low amounts of noise (see Bro, 1998).
In addition to these fitting procedures, variants of the least squares fitting procedures have been proposed that employ special metrics (for examples, see Coppi & Bolasco, 1989), or optimal scaling to handle nonnumerical data (for three-mode three-way data, see Sands & Young, 1980; for two-mode three-way data, see Takane, Young & De Leeuw, 1977, proposing ALSCAL, a nonmetric variant of INDSCAL).

4. Indeterminacy and Constraints

The 3MFA and Tucker2 model are indeterminate: equally well fitting solutions can be obtained after arbitrary nonsingular transformations of all component matrices, provided that these are compensated in the core. On the other hand, their constrained versions PARAFAC and PARATUCK2 do (usually) give unique solutions (e.g., see Harshman & Lundy, 1994). PARAFAC, however, is often too restrictive for describing one's data; PARATUCK2 is an interesting compromise, being less restrictive than PARAFAC, but nevertheless unique. Analogously, for two-mode models, the IDIOSCAL model is not unique, whereas its constrained version INDSCAL and PARAFAC2 (usually) are. Recently, it has been suggested to employ other constrained variants of 3MFA, notably with arbitrary sets of core elements constrained to zero (Kiers, 1992, Rocci, 1992), which, in certain cases leads to unique models as well (Kiers, Ten Berge & Rocci, 1997), or, in combination with some constraints on the component matrices may lead to partial uniqueness (Kiers & Smilde, 1998). Moreover, it has been shown that large numbers of such zero core constraints can be imposed without loss of fit (Murakami, Ten Berge & Kiers, 1998; Ten Berge & Kiers, 1997).
Alternatively, constraints can be used to incorporate external information on the data, and thus, hopefully find better estimates of the parameters underlying the model (for an overview, see Bro, 1998). For instance, when the underlying entities are known to be nonnegative, it is useful to *constrain* the parameters to nonnegativity. Also, in case one of the modes refers to consecutive points in time, one may expect a trend in the data, and model this, or only its smoothness. For descriptive purposes it can also be desirable to impose the above or other constraints, like equality of components, nonnegative correlations of components with variables, or simple structure on the components (e.g., see Krijnen, 1993; Kroonenberg & Heiser, 1998). An extreme case of simplicity is where the component matrices are constrained to be binary. This leads to overlapping cluster models (Carroll & Chaturvedi, 1995), the best-known of which is the INDCLUS model.

5. Solving the 3MFA Indeterminacy by Simple Structure Rotation

In PCA, a common procedure for solving the indeterminacy of the solution is to rotate the loading matrix to "simple structure" (with many elements close to zero). A similar approach can be followed for 3MFA, and would have as an advantage over constraining the solution that by rotation no loss of fit is incurred. However, in 3MFA it is not so clear what aspect of the solution should be simplified. Tucker (1966) mentioned simplifying either the component matrices (by varimax rotation) or the core. As discussed by Kroonenberg (1983, Chapter 5), the first core rotation procedures aimed at diagonalizing the frontal core planes. Later, Kiers (1992) proposed to rotate the core to superdiagonality. The first procedure for rotation of the core, over all three modes, to an *arbitrary* simple structure was the oblique "tri-quartimax" rotation procedure by Kruskal (1988). This paper served as a starting point for the development of Kiers' (1997b) "three-mode orthomax" core rotation procedure, where "orthomax" represents a family of orthogonal simple structure rotation methods. Recently, Kiers (1997a) proposed an extension of the three-mode orthomax procedure in which the core *and* the component matrices are rotated to simplicity jointly. By varying the importance attached to simplicity of the core and each of the component matrices, a very flexible approach arises, of which the core rotation and optimal simplicity rotation of the component matrices are special cases.
Except for Kruskal's procedure, the above techniques pertain to orthogonal rotations. Kiers (1997b) mentioned how *oblique* rotation can be obtained from three-mode orthomax. A procedure aimed explicitly at oblique simple structure is Kiers' (in press) three-mode generalization of SIMPLIMAX for oblique rotation of the core. So far, no procedures have been developed for joint oblique rotation of the core and component matrices.

6. Discussion

A brief overview has been given of techniques for the analysis of three-way data, with special attention to recent developments. By concentrating on techniques that provide representations for all modes of the data, many related techniques are ignored (e.g., all techniques meant for the joint analysis of a number of data sets involving the same variables but different observation units, or vice versa), whereas such techniques may very well be applied successfully to three-way data. For further reading and for keeping up to date with respect to three-way analysis, I refer to two internet home pages: Rasmus Bro's home page (http://newton.foodsci.kvl.dk/rasmus.html) with emphasis on chemometrics, and Pieter Kroonenberg's home page (http://www.fsw.leidenuniv.nl/~kroonenb/) of general three-way interest.

References

Bentler, P. M. & Lee, S.-Y. (1978). Statistical aspects of a three-mode factor analysis model, *Psychometrika, 43*, 343-352.

Bloxom, B. (1968). A note on invariance in three-mode factor analysis, *Psychometrika, 33*, 347-350.

Bro, R. (1998). *Multi-way analysis in the food industry. Models, Algorithms & Applications,* Doctoral Dissertation, University of Amsterdam.

Bro, R. & Andersson, C. A. (1997). *Improving the speed of multi-way algorithms. Part II: Compression*, Manuscript submitted for publication.

Carlier, A., Lavit, Ch., Pagès, M., Pernin, M. O. & Turlot, J.C. (1989). Analysis of data tables indexed by time: a comparative review, in: *Multiway data analysis*, Coppi, R. & Bolasco, S. (Eds.), Amsterdam, Elsevier Science Publishers, 85-101.

Carroll, J. D. & Arabie, P. (1980). Multidimensional scaling, *Annual Review of Psychology, 31*, 438-457.

Carroll, J. D. & Chang, J.-J. (1970). Analysis of individual differences in multidimensional scaling via an N-way generalization of "Eckart-Young" decomposition, *Psychometrika*, 35, 283-319.

Carroll, J. D. & Chang, J.-J. (1972). *IDIOSCAL: A generalization of INDSCAL allowing IDIOsyncratic reference systems as well as an analytic approximation to INDSCAL*, Unpublished manuscript.

Carroll, J. D. & Chaturvedi, A. (1995). A general approach to clustering and multidimensional scaling two-way, three-way and higher-way data, in: *Geometrical representations of perceptual phenomena*, R.D. Luce et al. (Eds.), 295-318.

Coppi, R. & Bolasco, S. (Eds.) (1989). *Multiway data analysis*, Amsterdam, Elsevier Science Publishers.

D'Alessio, G. (1989). Multistep principal component analysis in the study of panel data, in: *Multiway data analysis*, Coppi, R. & Bolasco, S. (Eds.), Amsterdam, Elsevier Science Publishers, 375-381.

DeSarbo, W. S. & Carroll, J. D. (1985). Three-way metric unfolding via alternating least squares, *Psychometrika*, *50*, 275-300.

Harshman, R. A. (1970). Foundations of the PARAFAC procedure: models and conditions for an "explanatory" multi-mode factor analysis, *UCLA Working Papers in Phonetics*, 16, 1-84.

Harshman, R. A. (1972). PARAFAC2: Mathematical and technical notes, *UCLA Working Papers in Phonetics, 22*, 31-44.

Harshman, R. A., Green, P .E., Wind, Y. & Lundy, M. E. (1982). A model for the analysis of asymmetric data in marketing research, *Marketing Science, 1*, 205-242.

Harshman, R. A. & Lundy, M. E. (1984). The PARAFAC model for three-way factor analysis and multidimensional scaling, in: *Research methods for multimode data analysis*, Law, H.G., Snyder, C.W., Hattie, J.A. &

McDonald, R.P. (Eds.), 122-215, New York, Praeger.

Harshman, R. A. & Lundy, M. E. (1994). PARAFAC: Parallel factor analysis, *Computational Statistics and Data Analysis, 18,* 39-72.

Kiers, H. A. L. (1991). Hierarchical relations among three-way methods, *Psychometrika*, 56, 449-470.

Kiers, H. A. L. (1992). TUCKALS core rotations and constrained TUCKALS modelling, *Statistica Applicata*, 4, 659-667.

Kiers, H. A. L. (1993). An alternating least squares algorithm for PARAFAC2 and DEDICOM3, *Computational Statistics and Data Analysis, 16,* 103-118.

Kiers, H. A. L. (1997a). *Joint orthomax rotation of the core and component matrices resulting from three-mode principal components analysis*, Manuscript submitted for publication.

Kiers, H. A. L. (1997b). Three-mode Orthomax rotation, *Psychometrika, 62*, xxx-xxx.

Kiers, H. A. L. (1998). A three-step algorithm for Candecomp/Parafac analysis of large data sets with multicollinearity, *Journal of Chemometrics,* in press.

Kiers, H. A. L. (in press). Three-way SIMPLIMAX for oblique rotation of the three-mode factor analysis core to simple structure, *Computational Statistics and Data Analysis*.

Kiers, H. A. L. & Harshman, R. A. (1997). Relating two proposed methods for speedup of algorithms for fitting two- and three-way principal component and related multilinear models, *Chemometrics and Intelligent Laboratory Systems, 36*, 31-40.

Kiers, H. A. L. & Smilde, A. K. (1998). Constrained three-mode factor analysis as a tool for parameter estimation with applications in chemistry, *Journal of Chemometrics, 12*, 125-147.

Kiers, H. A. L., Ten Berge, J. M. F. and Rocci, R. (1997). Uniqueness of three-mode factor models with sparse cores: The 3×3×3 case, *Psychometrika, 62*, 349-374.

Krijnen, W. P. (1993). *The analysis of three-way arrays by constrained PARAFAC methods*, Leiden, DSWO Press.

Kroonenberg, P. M. (1983). *Three-mode principal component analysis: Theory and applications*, Leiden, DSWO press.

Kroonenberg, P. M. & De Leeuw, J. (1980). Principal component analysis of three-mode data by means of alternating least squares algorithms, *Psychometrika*, 45, 69-97.

Kroonenberg, P. M. & Heiser, W. J. (1998). Parallel factor analysis with constraints on the configurations: An overview, in: C. Hayashi et al. (Eds.) *Data Science, Classification and Related Methods*, Tokyo, Springer-Verlag, 587-597.

Kruskal, J. B. (1988). *Simple structure for three-way data: A new method intermediate between 3-mode factor analysis and PARAFAC-*

CANDECOMP, Paper presented at the 53rd Annual Meeting of the Psychometric Society, Los Angeles, June 27-29.

Law, H. G., Snyder, C. W., Hattie, J. A. & McDonald, R. P. (Eds.)(1984). *Research methods for multimode data analysis*, New York, Praeger.

L'Hermier des Plantes, H. (1976). *Structuration des tableaux à trois indices de la statistique*, Thèse de troisième cycle, Université de Montpellier.

Mayekawa, S. (1987). Maximum likelihood solution to the PARAFAC model, *Behaviormetrika, 21*, 45-63.

Murakami, T., Ten Berge, J. M. F., & Kiers, H. A. L. (1998). A case of extreme simplicity of the core matrix in three-mode principal components analysis, *Psychometrika, 63*, in press.

Paatero, P. (1997). A weighted non-negative least squares algorithm for three-way "PARAFAC" factor analysis, *Chemometrics and Intelligent Laboratory Systems, 38*, 223-242.

Rizzi, A. & Vichi, M. (1995). Three-way data set analysis, in: *Some relations between matrices and structures of multidimensional data analysis*, Rizzi, A. (Ed.), Pisa, Giardini, 93-166.

Rocci, R. (1992). Three-mode factor analysis with binary core and orthonormality constraints, *Journal of the Italian Statistical Society*, *3*, 413-422.

Rocci, R. & Bove, G. (1994). Un metodo per l'analisi simultanea di più matrici di dati quadrate asimmetriche, in: *Atti della XXXVII Riunione Scientifica Società Italiana di Statistica*, San Remo, 263-260.

Sands, R. & Young, F. (1980). Component models for three-way data: An alternating least squares algorithm with optimal scaling features. *Psychometrika, 45*, 39-67

Takane, Y., Young, F. W. & De Leeuw, J. (1977). Nonmetric individual differences scaling: An alternating least-squares method with optimal scaling features, *Psychometrika, 42*, 7-67.

Ten Berge, J. M. F., & Kiers, H. A. L. (1997). *Simplicity of core arrays in three-way component analysis, with applications to the typical rank of* $P \times Q \times 2$ *arrays*, Manuscript submitted for publication.

Tucker, L. R (1966). Some mathematical notes on three-mode factor analysis, *Psychometrika, 31*, 279-311.

Tucker, L. R (1972). Relations between multidimensional scaling and three-mode factor analysis, *Psychometrika*, 37, 3-27.

Tucker, L. R & Messick, S. (1963). An individual differences model for multidimensional scaling. *Psychometrika*, 28, 333-367.

Vichi, M. (1997). *Principal classifications analysis of a three-way data set*, Manuscript submitted for publication.

Winsberg, S. & Carroll, J. D. (1989). A quasi-nonmetric method for multidimensional scaling of multiway data via a restricted case of an extended INDSCAL model, in: *Multiway data analysis*, Coppi, R. & Bolasco, S. (Eds.), Amsterdam, Elsevier Science Publishers, 405-414.

Studying the Diffusion of Three-Mode Analysis in Chemistry: Design Considerations

Pieter M. Kroonenberg
Department of Educational Sciences, Leiden University
Wassenaarseweg 52, 2333 AK Leiden, The Netherlands
kroonenb@rulfsw.fsw.leidenuniv.nl; http://www.fsw.leidenuniv.nl/~kroonenberg

Abstract: The paper presents an example of studying the introduction of methodological innovation in a science, in particular that of three-mode (component) analysis into chemistry, especially in such areas as chromatography, fluorescence, spectrometry, and analytical chemistry in general.

Key words: Bibliometrics, Chemometrics, Multilinear, Multi-Way, PARAFAC, Three-Mode, Three-Way, Trilinear, Tucker Model

1. Introduction

In his 1983 annotated bibliography of three-mode analysis, Kroonenberg (1983a) claimed that all theoretical publications pertaining to three-mode component and factor analysis were included, as well as all published applications of these methods. The claim might be a bit overstated, but in this paper the key problem behind the paper is taken up again: "How does one go about collecting such a set of references, especially the applications?" Keyword searches in titles, keyword and abstract sections, as listed in large data banks such as that of the Institute of Scientific Information, will generally come up with all relevant technical papers, but not necessarily with all applications of the techniques, because authors often use statistical techniques without mentioning this in the abstract. However, techniques are seldom used without a reference to one or more key publications providing background to the techniques. The only exceptions to this are references to techniques which are so included in the lore of science that researchers familiarise themselves with the techniques via general textbooks or refer to the technique without citing anyone, the *t*-test is an example. The techniques for three-mode analysis are mentioned in several general textbooks, but virtually nowhere in sufficient detail to be useful to a practitioner.

In this paper we will describe a first attempt to map the introduction and development of three-mode analysis in (analytical) chemistry starting with a set of key publications from within the discipline and a set of earlier key publications from the field of psychometrics where three-mode analysis originally was developed. Primarily the methodological opening moves will be reported here, rather than the real substantive results mapping three-mode analysis in chemistry itself, nor will the diffusion and synthesis of techniques developed within chemometrics and psychometrics be part of the discussion. Partly such information can already be found in Bro, Workman,Mobley and Kowalski (1997), but in later publications we will go into these issues in greater depth starting from a bibliometric perspective.

All literature searches reported in this paper were carried out via the Internet facility, the *Web of Science®* provided by the Institute for Scientific Information (*http://www.isihost.com*) on a trial basis to the universities of The Netherlands. The trial version unfortunately only contains papers from 1987 on, so that the earliest history could not be traced in this way and will thus not be considered. The reference date for the search carried out in the Science Citation Index Expanded® is (approximately) 31 December 1997.

2. Data Base

2.1 Key Publications

The basic idea of investigating the introduction of a new technique, such as three-mode analysis, via key publications is that most authors develop a technique building upon the work of others or apply the technique explained in certain key publications. Three-mode analysis is here defined as the set of techniques for the analysis of data which can be arranged in a block or a set of matrices, such as chemical compounds which are dissolved in different strength solutions on which certain characteristics are measured. In three-mode analysis one may build upon earlier results obtained in two-mode analysis or upon earlier results in three-mode analysis itself. For the present investigation both these sources are important. Certain chemical problems were such that chemometricians initially invented three-mode analysis for themselves only to discover that psychometricians had been there before them. Notwithstanding, at present, chemistry is one of the most vigorous areas of new developments in the field, often in close co-operation with psychometricians as is evident from joint conferences (TRIC, see Geladi, 1993; TRICAP, see Bro & Kiers, 1997) and publications. In order to discover all or most of the chemical publications in three-mode analysis, one has to recognise the two sources mentioned. In addition, the key publications should be sufficiently general that one may expect that a paper seriously discussing or applying the techniques should quote at least one of them. On the other hand, the set should be small enough to be manageable, and the publications should not be cited for substantive reasons. It is of course impossible to avoid this completely, so that a certain amount of manual checking is unavoidable. Finally, of some authors we have designated more than one paper as a key publication, even if not all of them are highly cited, because it often seemed that authors made a choice between them, in our eyes not necessarily the proper choice. When tracing the dissemination of ideas, authors' awareness of certain work was considered more important than the exact citation count.

On the basis of exposure to the literature, the two sets of key publications listed in Table 1 were chosen for the present project. From the list of psychometric papers, no papers by Kiers (University of Groningen) are included notwithstanding his wealth of papers in this area. This is primarily because he has written very few overview papers to which people naturally refer, and authors citing him generally also cite at least one of the key publications listed in Table 1. Thus even though his work is vital for the further development of the field, in general his publications seem to be too technical to be used as standard references.

Conspicuously absent in the set of chemistry key publications are the two papers based on the unpublished doctoral thesis by Spanjer, i.e. De Ligny, Spanjer, Van Houwelingen and Weesie (1983) and Spanjer, De Ligny, Van Houwelingen and Weesie (1984). These two papers are among the first applications of three-mode analysis to chemical data and they also include a new algorithm for the three-mode principal component model, which has never been published elsewhere. Unfortunately, these papers have not caught the attention of the chemistry community, be it that they are now referred to in recent survey papers. Three other papers have the potential to become key publications (Henrion, 1994; Bro, 1997; Bro, Workman, Mobley, and Kowalski, 1997), but they are too recent to have collected citations by papers which do not cite one of the key publications already present. At the same time, it is expected that the papers by Tucker will gradually fade into the background, especially his 1963 and 1964 ones, because for newcomers to the field, later papers will effectively act as a screen for his seminal work. Some of this is already under way, and one especially expects to see this in applied papers.

To obtain confirmation that the chemical papers mentioned are indeed key publications in this area, their citation count was compared with the 120 entries in the complete list of publications retrieved from all key publications listed in Table 1. On this list there were several papers with higher citation counts than the six key publications. However, two were review papers discussing both three-mode and other techniques and they are thus not eligible. Another two papers from the group around MacGregor and Nomikos at McMasters University, Canada, were especially quoted for their applications. High citation counts were also found for papers by the prolific group around Romà Tauler at the University of Barcelona, but inspection showed that they contained ½ to ¾ self-citations, and could thus not be considered key publications in the area. Within the psychometric field, no separate check was carried out for the psychometric key publications, as they are unquestionably at the core of the field. One of the most cited papers in three-mode analysis is Carroll and Chang (1970), who present, in addition to their main model, a model (and algorithm) for three-mode analysis which is essentially the same as that proposed by Harshman, whose work was only formally and fully published in 1984. Carroll and Chang's paper was, however, not included in the present list, because all papers referring to Carroll and Chang's work also referred to one of the other psychometric publications, but not vice versa. If the search would go further back in time, then their paper should be included because of its early publication data. For instance, Appellof and Davidson (1981, 1983) referred the paper, showing that they were aware of at least some of the psychometric work in three-mode analysis.

2.2 Completeness of the Referring Papers

Unfortunately, it cannot be claimed that the list of citing papers is complete. To do so would be to ignore the unreliability of the ISI data base. First of all the coverage of the ISI is not complete, but its claim that it covers the world's most important science journals is probably true, even though there is a strong English language bias in its coverage. There exists for instance a wealth of Japanese scientific journals which is not included. On the other hand, the impact of such journals is naturally limited due to their Japanese language orientation.

A more serious problem is however the inaccuracy due to author errors in citation. For instance, all key publications suffer from inaccurate citations, either by incorrect first page numbers, incorrect years, etc. Unfortunately, the problem in chemistry might be more serious than for instance in the social sciences, because

Table 1: Chemometric and Psychometric Key Publications in Three-Mode Analysis

Author(s)	**Year**	**Title**	**Source**
Chemometric Key Publications			
Appellof, C.J. & Davidson, E.R.	1981	Strategies for analyzing data from video fluorometric monitoring of liquid chromatographic effluents.	*Analytical Chemistry*
Appellof, C.J. & Davidson, E.R.	1983	Three-dimensional rank annihilation for multi-component determinations	*Analytica Chimica Acta*
Geladi, P.	1989	Analysis of multi-way (multi-mode) data.	*Chemometrics and Intelligent Laboratory Systems*
Sanchez, E. & Kowalski, B.R.	1990	Tensorial resolution: A direct trilinear decomposition	*Journal of Chemometrics*
Smilde, A.K. & Doornbos, D.A.	1991	3-way methods for the calibration of chromatographic systems - comparing PARAFAC and 3-way PLS.	*Journal of Chemometrics*
Smilde, A.K.	1992	Three-way analyses: Problems and prospects.	*Chemometrics and Intelligent Laboratory Systems*
Psychometric Key Publications			
Tucker, L.R.	1963	Implications of factor analysis of three-way matrices for measurement of change.	In Harris, C.W. (Ed).
Tucker, L.R.	1964	The extension of factor analysis to three-dimensional matrices.	In Gullikson, H. & Frederiksen, N. (Eds.)
Tucker, L.R.	1966	Some mathematical notes on three-mode factor analysis.	*Psychometrika*
Harshman, R.A.	1970	Foundations of the PARAFAC procedure: Models and conditions for an "explanatory" multi-mode factor analysis.	*UCLA Working Papers in Phonetics*
Kroonenberg, P.M. & De Leeuw, J.	1980	Principal component analysis of three-mode data by means of alternating least squares algorithms.	*Psychometrika*
Kroonenberg, P.M.	1983b	*Three-Mode Principal Component Analysis: Theory and Applications.*	DSWO Press: Leiden.
Harshman, R.A. & Lundy, M.E.	1984a.	The PARAFAC model for three-way factor analysis and multidimensional scaling.	In Law et al. (Eds.)
Harshman, R.A. & Lundy, M.E.	1984b.	Data preprocessing and the extended PARAFAC model.	In Law et al. (Eds.)

the incomplete referencing system in chemistry journals. Often, titles of articles are not mentioned, and authors are required to use abbreviations for journal titles which is asking for errors to be made. For instance, if a paper appearing in the *Journal of Chemometrics* is referred to as *JChem* or *JChemometrics* rather than *JChemom* the citation will not necessarily be properly recognised by the search program that comes with the *Web of Science*. The problem is even more serious for the most frequently cited journal, *Chemometrics and Intelligent Laboratory Systems*, which should be abbreviated (according to the ISI) as *Chemometr. Intell. Lab.*, but variants are *Chemom. Intell. Lab. Syst.*, *Chemometrics Intell*, *Chemometrics Intelli.*, *Chemo. Lab.* (an often used nickname in the corridors) and all variants exist. To illustrate the seriousness of these problems created by authors, the key publication by Appellof and Davidson (1981) had 1 miss out of 33; Geladi (1989) 4 out of 30, Sanchez and Kowalski (1990) 1 out of 54, and Smilde and Doornbos (1991) 1 out of 26; Smilde (1992) 3 out 28. An even more damaging and curious situation exists in the ISI data base: for some (earlier?) papers in the data base it is stated that they have no references, which means that they cannot be included in this survey because, according to the data base, they do not cite the key references. We have not checked this systematically but at least for the two chemometric journals mentioned above, papers published 1987 through 1991 have been encountered with 0 references, including one of the key publications (Smilde & Doornbos, 1991), which in fact contains 35 references. More specifically, in our list of citing papers, there are no papers during the years mentioned which cite the key references, an unlikely, and demonstrably incorrect, state of affairs.

3. Results

3.1 Introduction

As mentioned above, the key publications pointed to 120 publications (key publications included) in chemistry which refer in one way or another to three-mode analysis. Due to space limitations the complete list of these papers including which papers they cite etc. will not be reproduced here, but may be requested from the author. Different types of papers may quote three-mode analysis for different reasons. They may build on the theoretical developments presented in such a paper, they may be reviews of the state of the art, they may present an application of a three-mode analysis, or they may merely indicate the existence of three-mode methods without explicitly using them. The latter type of papers do not really make a contribution to the field, but they do signify an awareness of the techniques.

3.2 Some basic statistics

During the 11 years under consideration (1987-1997), the yearly numbers of paper in our data base are: 2, 2, 2, 4, 4, 14, 13, 19, 22, 13, 23, but the number of papers in the first five years are too low (see above). Thus, the number of papers in three-mode analysis has been steadily increasing over the years. Of course, one can hardly say that three-mode analysis is one of the basic tools for analysis, but then again one also uses a spark plug spanner only once in a while for a specific kind of job.

The major chemistry journals publishing new developments about three-mode analysis are *Chemometrics and Intelligent Laboratory Systems* (27), *Journal of Chemometrics* (26) and *Analytical Chemistry* (14), taking care of 56% of the publications.

3.3 Research groups

A sign of a relatively healthy state of affairs is that there a large number of research groups from both sides of the Atlantic working in the area with hotbeds at the University of Washington around Kowalski, University of Amsterdam around Smilde, Royal Veterinary and Agricultural University, Frederiksberg, Denmark around Bro and several of his colleagues in other Scandinavian countries such as Geladi, University of Barcelona around Tauler, McMasters University in Canada around MacGregor and Nomikos, while other authors like Booksh (now at University of South Carolina), Burdick (Duke University), Henrion (Humboldt Universität Berlin), Ross (formerly Ohio State University) have produced many initiatives in the field. A special mention should be made of Bruce Kowalski who although never a first author of a three-mode paper is co-author with many of the leading players in the field. The attraction of his Center for Process Analytic Chemistry (CPAC) has been such that several, often young, researchers such as Faber, Smilde and Tauler have been a guest at his institute, and also because of that he has been a major factor in the propagation, integration, and application of three-mode analysis in chemistry. In this connection also Paul Geladi should be singled out as one of the pioneers of three-mode analysis in chemistry, even though this might not be so evident from the data in this study.

3.3 Role of the key publications

The oldest three chemical key publications contain some references to each other, and all contain references to psychometric publications. The first paper to bring together all key publications was the review by Henrion (1994), while thereafter the only other paper to refer to all of them is Bro et al. (1997). One could therefore say that from roughly 1992 on, the integration of the three-mode field was nearing completion and all psychometric information was being used in chemistry. From then on more papers appear which use whatever references were necessary for their purposes.

This does not mean that all research groups seemed to be equally aware of the contributions from psychometrics. For instance, the group around MacGregor and Nomikos do not refer at all to their papers, the main reason is probably that their technique, multiway principal component analysis, does not really deal with three-mode analysis if this is defined by requiring that there are separate parameters for each of the three modes. Interesting is too, that apart from the paper by Sanchez and Kowalski (1990) it took until 1994 for the first paper from the CPAC group to mention work outside the chemometric tradition, and they also mentioned work from other chemometricians very sparingly, which is in a way strange given the number of visitors that passed their centre.

That the initial flow of ideas was from psychometrics to chemometrics is supported by the fact that, searching through the Social Science Citation Index®, only two publications (Harshman, 1994; Kroonenberg, 1994) outside the chemistry sphere refer to the chemometric key publications. So far Kiers and Harshman are the only psychometricians with three-mode publications in chemometric journals (Kiers & Smilde, 1995; Kiers & Harshman, 1997)

3.4 Further work

The above results have of course only a limited breadth and only describe the referencing structure without going into the subject matter at hand. The next step in this research is to read the abstracts and the papers themselves to link explicitly the specific methods developed by the various groups and to assess how the papers that are cited are used in the citing papers. This would also result in setting aside those papers which only mention three-mode analysis in passing. Moreover, it will be possible to pinpoint the specific innovative contributions which were generated within chemometrics, apart from borrowing from psychometrics, and it will be possible to show the (in part) autonomous development in chemometrics.

Furthermore, an analysis of the subject areas that use three-mode methods should be made to establish which fields especially make use of three-mode analysis in their research, and to what extent there is a transfer from one part of chemistry to another.

To produce a more encompassing view of the relations between research groups and papers, a co-citation analysis would be very helpful, i.e. the complete reference lists have to be checked to assess which papers are cited together in the same paper. Thus comparisons should not only be made between the citations to the key publications, but also between all publications among themselves. Such an analysis might also turn up a more articles which do not cite the key publications, but only less prominent ones.

In the future, we hope to report on such issues. Moreover, comparison between different disciplines are planned to answer questions to what extent three-mode analysis is incorporated in different disciplines and to uncover (hopefully) reasons for these differences.

References

Appellof, C.J. & Davidson, E.R. (1981). Strategies for analyzing data from video fluorometric monitoring of liquid chromatographic effluents. *Analytical Chemistry*, 53, 2053-2056.

Appellof, C.J. & Davidson, E.R. (1983). Three-dimensional rank annihilation for multi-component determinations. *Analytica Chimica Acta,* 146, 9-14

Bro, R. (1997). PARAFAC. Tutorial and applications, *Chemometrics and Intelligent Laboratory Systems,* 38, 149-171.

Bro, R. & Kiers, H.A.L. (1997). Meeting Report. TRICAP: Three-way methods in chemistry and psychology, Lake Chelan, WA, USA, 4-9 May 1997. *Journal of Chemometrics, 11*, 463-466.

Bro, R., Workman, J.J., Mobley, P.R., & Kowalski, B.R. (1997). Review of chemometrics applied to spectroscopy: 1985-1995, Part 3 - Multi-way analysis, *Applied Spectroscopy Review*, 32, 237-261.

Carroll, J.D. & Chang, J.J. (1970). Analysis of individual differences in multidimensional scaling via an *N*-way generalization of "Eckart-Young" decomposition. *Psychometrika, 35*, 283-319.

De Ligny, C.L., Spanjer, M.C., Van Houwelingen, J.C. & Weesie, J.M. (1984). Three-mode factor-analysis of data on retention in normal-phase high-performance liquid chromatography, *Journal of Chromatography,* 301, 311-324.

Geladi, P. (1989). Analysis of multi-way (multi-mode) data. *Chemometrics and Intelligent Laboratory Systems,* 7, 11-30.

Geladi, P. (1993). Meeting Report. TRIC, The first conference on three-way methods in chemistry, Epe, Netherlands, 22-25 August, 1993. *Journal of Chemometrics, 8*, 97-98.

Harshman, R.A. (1970). Foundations of the PARAFAC procedure: Models and conditions for an "explanatory" multi-mode factor analysis. *UCLA Working Papers in Phonetics*, 16, 1-84.

Harshman, R.A. & Lundy, M.E. (1984a). The PARAFAC model for three-way factor analysis and multidimensional scaling. In *Research methods for multimode data analysis,* Law, H.G., Snyder Jr, C.W., Hattie, J.A. & McDonald, R.P. (Eds.),. Praeger, New York, 122-215.

Harshman, R.A. & Lundy, M.E. (1984b). Data preprocessing and the extended PARAFAC model. In *Research methods for multimode data analysis,* Law, H.G., Snyder Jr, C.W., Hattie, J.A. & McDonald, R.P. (Eds.),. Praeger, New York, 216-284.

Harshman, R.A. & Lundy, M.E. (1994). PARAFAC - Parallel factor-analysis, *Computational Statistics & Data Analysis,* 18, 39-72.

Henrion, R. (1994). *N*-way principal component analysis: Theory, algorithms and applications, *Chemometrics and Intelligent Laboratory Systems,* 25, 1-23.

Kiers, H.A.L. & Harshman, R.A. (1997). Relating two proposed methods for speedup of algorithms for fitting two- and three-way principal component and related multilinear models, *Chemometrics Intelligent Laboratory Systems*, 36, 31-40.

Kiers, H.A.L. & Smilde, A.K. (1995). Some theoretical results on second-order calibration methods for data with and without rank overlap, *Journal of Chemometrics,* 9, 179-195.

Kroonenberg, P.M. (1983a). Annotated bibliography of three-mode factor analysis, *British Journal of Mathematical and Statistical Psychology,* 36, 81-113.

Kroonenberg, P.M. (1983b). *Three-Mode Principal Component Analysis: Theory and Applications.* DSWO Press: Leiden.

Kroonenberg, P.M. (1994). The TUCKALS line - A suite of programs for three-way data-analysis, *Computational Statistics & Data Analysis,* 18, 73-96.

Kroonenberg, P.M. & De Leeuw, J. (1980). Principal component analysis of three-mode data by means of alternating least squares algorithms. *Psychometrika*, 45, 69-97.

Sanchez, E. & Kowalski, B.R. (1990). Tensorial resolution: A direct trilinear decomposition. *Journal of Chemometrics*, 4, 29-45

Smilde, A.K. & Doornbos, D.A. (1991). 3-way methods for the calibration of chromatographic systems - Comparing PARAFAC and 3-way PLS. *Journal of Chemometrics*, 5, 345-360.

Smilde, A.K. (1992) Three-way analyses: Problems and prospects. *Chemometrics and Intelligent Laboratory Systems*, 15, 143-157.

Spanjer, M.C., De Ligny, C.L., Van Houwelingen, H.C. & Weesie, J.M. (1985). Simultaneous description of the influence of solvent, reaction type, and substituent on equilibrium constants by means of three-mode factor analysis, *Journal of the Chemical Society, Perkin Transactions II*, 1401-1411.

Tucker, L.R. (1963). Implications of factor analysis of three-way matrices for measurement of change. In *Problems in Measuring Change,* Harris, C.W. (Ed.), University of Wisconsin Press, Madison, 122-137.

Tucker, L.R. (1964) The extension of factor analysis to three-dimensional matrices. In *Contributions to Mathematical Psychology,* Gullikson, H. & Frederiksen, N. (Eds.), Holt, Rinehart and Winston, New York, 110-119.

Tucker, L.R. (1966). Some mathematical notes on three-mode factor analysis. *Psychometrika,* 31, 279-311.

Selection of Classification Models Using Data Envelopment Analysis

G. Nakhaeizadeh[(1)] & C.C. Taylor[(2)]
(1) Daimler-Benz Research and Technology, 89013 Ulm, Germany,
email: nakhaeizadeh@dbag.ulm.DaimlerBenz.com
(2) Department of Statistics, University of Leeds, Leeds LS2 9JT, UK,
email: charles@amsta.leeds.ac.uk

Abstract: Model selection is often judged on only one criteria. We investigate the use of the Data Envelopment Analysis to this problem, and consider ways in which this aspect can utilize, as well as inform, existing statistical methodology.

Key words: Evaluation; Model Uncertainty; Selection Criteria

1. Introduction

Various aspects of model selection have been studied in the statistical literature. Classical methods using information criteria (for example, AIC and BIC of Akaike (1974) and Schwartz's (1978) Bayesian criterion SBC) have been used for model selection in time series analysis for some time. The issue of model uncertainty is closely related to model selection, and this has recently been tackled using Bayesian methods (Draper, 1995 and Madigan, Raftery, Volinsky & Hoeting, 1996). In this case, it may be that no one model is selected and some sort of *model average* is used. Some of these methods rely on the models being **nested**, as does a simple likelihood ratio test. More general examples of model selection can be found in the review of Chatfield (1995); see also the recent work of Sclove (1993, 1994).

Apart from the classical classification approaches like linear and quadratic discriminant there are alternative approaches developed by Computer Science researchers. Decision trees (Quinlan, 1993), Neural Networks and rule based methods (for example, CN2 developed by Clark and Niblett (1989)) are well-known examples. Although there have been developments in new classification methods, research into the selection of metrics that can consider all aspects of the classification models has been neglected. As a criterion for selecting between supervised classification methods, the accuracy rate is often the only arbiter; see Michie, Spiegelhalter & Taylor (MST, 1994)). The main aim of this paper is to contribute to this debate by generalizing the model definition to cover all alternative approaches mentioned above and suggest a **multi-criteria metric** for selecting amongst the alternative classification models and take into account various aspects (positive and negative properties) of the models. For this purpose we have developed multi-criteria metrics based on *Data Envelopment Analysis* (DEA).

2. Definition of the "Classification Model"

In this paper we define the classification model as the result of a process which uses the objects in a certain domain (for example, Diagnostics, Risk Management, Image Processing). The objects are characterized by some attributes and are assigned to one of the known classes. The output of this process is a classifier (model) that can be used to find, among others, the class-value of unseen objects. For example, in the case of discriminant analysis this classifier is characterized by a discriminant function. In the case of neural networks it is characterized by the weights of the learned network and in the case of the decision trees by different nodes and leaves of the tree involving the attributes subsets. Rule-based classifiers (models) are characterized by "if-then rules" that also involve various attributes subsets.

The classification models that we are going to evaluate and select amongst belong to each of the above categories. They can also be developed by combining two or more of these approaches, which is known as Multi-Strategy model building. We will discuss various positive and negative properties of such models to construct multi-criteria based evaluation metrics. These will be presented in the next section.

3. Various Properties of the Classification Models and the DEA-approach

A survey of the literature on model selection shows that, until now, in most cases the selection has been based only on one criterion, namely the accuracy rate. This means that most of the available model selection methods in the literature are mono-criterion approaches. In our approach, however, we regard a classification model as a product (for example, a car) or a production unit (for example, a car factory) that can have different properties (Nakhaeizadeh and Schnabl, 1997). Some of these properties are positive and some negative. To the positive properties of the classification models belong the validity, understandability and usability of the results. Complexity, extensive computing time and misclassification costs can be considered as negative properties of the classification model.

In evaluating products (for example, cars) a buyer considers all their positive and negative properties. If we regard classification models as products as well, we should evaluate these models in the same manner. This means that we should use multi-criteria metrics for model selection that are able to take into account all positive and negative properties of the models. This allows us to make a more fair evaluation and select between the alternative classification models.

DEA, which was developed in Operations Research and is used to select between alternative products (Charnes, Cooper & Rhodes, 1978; Charnes, Cooper, Lewin & Seiford, 1996), is an appropriate methodology to compare the alternative classification models using their positive and negative properties (Nakhaeizadeh and Schnabl, 1997). It also allows consideration of the background knowledge and preferences of the user. For example, it can be that for one user the accuracy rate of the classification model is more important than the understandability of the results. In DEA, it is possible to consider such preferences. Originally, DEA tries to develop a ranking system for *Decision Making Units* (DMUs). In our case each DMU is a classification model. In DEA terminology, positive properties are called output components and negative properties input components. In our terminology, a typical example for an output component is the accuracy rate of the classification model. A typical input component is the computation time that the classification model needs for training. Generally, output components are all components where higher values are better and input components are those where lower values are better. Using these components, we can now define the *efficiency* of a classification model as follows:

$$\text{efficiency} = \frac{\text{sum of weighted output components}}{\text{sum of weighted input components}}$$

As the above relation shows, the definition of efficiency covers all positive and negative characteristics of a classification model and efficiency in this form can be regarded as a multi-criteria based metric that can be used for the evaluation and selection between the alternative classification models. In DEA the weights are determined for each classification model individually during the computation by maximizing the efficiency. DEA chooses the weights so that the efficiency of the classification model being evaluated is as close to 100% as possible. At the same time, the other classification models should not have an efficiency more than 100% using the same set of weights.

Obviously this is an optimization problem that can be transformed to a linear program. If the classification model which is under evaluation does not achieve the given threshold (100%), then it is not efficient and there is at least one model amongst the others which dominates it. There are different ways for setting the weights in DEA. The most commonly used ones are input-oriented and output-oriented optimization (Ali and Seiford, 1993). The goal in the input-oriented optimization approach is to reduce radially the input-levels as much as possible. Radially means that all component-values are changed by the same proportion. Conversely in output-oriented optimization approach the main goal is to enlarge radially the output-levels as much as possible. The ranking of the alternative classification models can be done by using their efficiency.

4. Empirical Results and Conclusions

Michie *et al.* (1994) describe the performance of 23 classification algorithms on 22 different domains. To rank different algorithms, when applied to a certain domain, they use only the accuracy rate for the test dataset although they have data about maximum computer storage, training and testing time, training and testing error rates or training and testing misclassification costs (where costs are applicable). To obtain a DEA-based ranking for the classification algorithms applied by MST we have used the input and output-oriented versions of DEA. We illustrate the methodology for the Credit Management dataset reported by MST (section 9.2.1). The dataset was donated by a major British engineering company, and the two classes can be interpreted as the method by which the debts will be retrieved; the class values were determined by an expert on the basis of the given attributes. There were 7 attributes (all of which were numerical) and a total of 20 000 observations of which 15 000 were used as the training data, and 5 000 were used as the test data.

The Andersen and Petersen (1993) model was used with three input components (max. storage, training time and testing time) and one output component (error rate for the test data). As an alternative, we have also used a variant with an additional output component (accuracy rate for training data). Input oriented versions are denoted by 4I (one output and three input components) and by 5I (two output and three input components). Similarly for the output oriented versions denoted by 4O and 5O. The DEA-ranking results, as well as the ranked performance for the individual components, are given in Table 1.

For comparative purposes we carried out a cluster analysis of the components. Since there are three units of measurement (storage, time and error rate), we ensured that each component was treated equally by using the rank performance which are shown in the first five columns of Table 1. This differs from the DEA-models, in which the raw units were used and in which the weighting for each component was chosen optimally. The tree structure and the first two principal components are shown in Figure 1, and the interpretation clearly differs from the DEA-models. However, it can be noted that the efficient algorithms (as defined by the DEA-models) have (ranked) observations which lie closer to the origin (in a suitable metric) than any other algorithm. This can be partially seen in the PC plot of Figure 1 (which is only two-dimensional), but more clearly visualized in a brush-and-spin plot.

We conclude that DEA and perhaps the other multi-criteria based metrics that can take into account all positive and negative criteria of the classification model can lead to a more fair selection between the alternative algorithms. Such approaches open a new perspective to the statistical community in revising the general problem of model selection.

Table 1: *Ranking of algorithms for the Credit Management data set using various criteria and different DEA-models. The column labelled "te.err" is the ranking of MST, based on the error rate for the test data. The figures in italics mean that the algorithm is efficient for that DEA-model.*

Algorithm	Store	tr.tim	te.tim	tr.err	te.err	5I	4I	5O	4O
Discrim	2	2	6	13.5	13	*3*	*3*	*9*	*7*
Quadisc	3	4	10	18.5	21	14	12	18	17
Logdisc	14	5	11	13.5	8	16	13	15	12
SMART	11	18	7	10.0	1	*2*	*2*	*5*	*5*
k-NN	5	19	18	12.0	22	15	14	19	19
CASTLE	1	7	15	18.5	19	*6*	*9*	*9*	*7*
IndCART	17	8	17	5.0	6	17	17	14	13
NewId	4	14	2	2.0	13	*1*	*8*	*9*	*7*
AC²	19	16	19	2.0	8	*9*	19	*7*	16
Baytree	16	3	5	4.0	7	*9*	*5*	*1*	*1*
NaiveBay	15	1	3	16.5	16	*7*	*10*	*9*	*7*
CN2	18	13	9	2.0	12	*9*	16	*8*	15
C4.5	13	6	16	6.0	3	*9*	*4*	*4*	*4*
ITrule	10	15	1	16.5	18	*8*	*11*	*9*	*7*
Cal5	7	10	8	7.0	4	*4*	*6*	*2*	*2*
DIPOL92	8	12	14	8.5	1	*9*	*1*	*6*	*6*
Backprop	6	17	4	8.5	4	*5*	*7*	*3*	*3*
RBF	9	9	12	15.0	10	18	15	16	14
LVQ	12	11	13	11.0	15	19	18	17	18

Figure 1: *Dendogram and Principal Component plot of the ranked performance for the credit management dataset (using the first 5 columns of Table 1).*

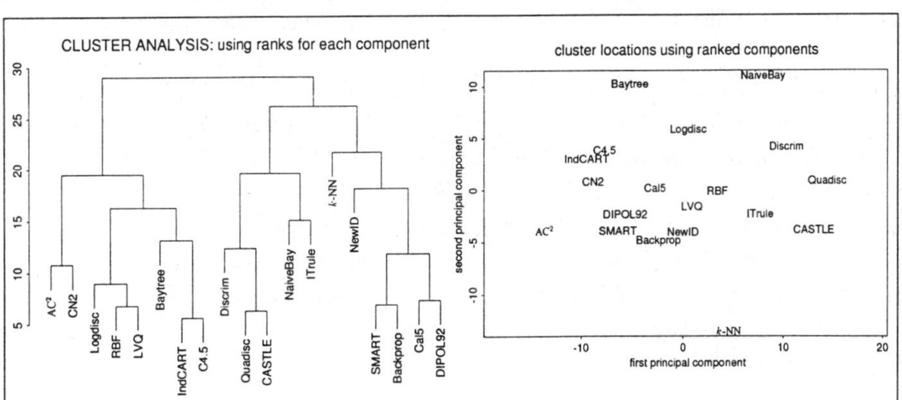

References

Akaike, H. (1974). A new look at the statistical model identification. *IEEE Transactions on Automatic Control*, **AC-19**, 716–723.

Ali, A.I. & Seiford L.M. (1993). The Mathematical Programming Approach to Efficiency Analysis. The Measurement of Productive Efficiency: In Fried, H.O., Lovell, C.A.K. and Schmidt, S.S. (Eds). *Techniques and Applications* pp. 120–159, Oxford University Press.

Andersen, P. & Petersen, N.C. (1993). A procedure for ranking efficient units in Data Envelopment Analysis. *Management Science*, **39**, 1261–1264.

Charnes, A., Cooper, W.W. & Rhodes, E. (1978). Measuring the Efficiency of Decision Making Units. *European Journal of Operational Research* **2**, 429–444

Charnes, A., Cooper, W.W., Lewin A.Y. & Seiford, L.M. (1996). *Data Envelopment Analysis: Theory, Methodology and Applications*. Kluwer Academic Publishers.

Chatfield, C. (1995). Model uncertainty, data mining, and statistical inference (with discussion) *J. Royal Statistical Society A*, **158**, 419–466.

Clark, P. & Niblett, T. (1989). The CN2 induction algorithms. *Machine Learning*, **3**, 261–285.

Draper, D. (1995) Assessment and propagation of model uncertainty (with discussion). *J. Royal Statistical Society B*, **57**, 45–97.

Madigan, D., Raftery, A., Volinsky, C., & Hoeting, J. (1996). Bayesian model averaging. In *Proceedings of the AAI Workshop on Integrating Multiple Learned Models*, Portland, Oregon.

Michie, D., Spiegelhalter, D.J. & Taylor, C.C. (Eds.) (1994). *Machine Learning, Neural and Statistical Classification*, Ellis Horwood, Chichester.

Nakhaeizadeh, G. & Schnabl A. (1997). Development of Multi-Criteria Metrics for Evaluation of Data Mining Algorithms. In proceedings of KDD-97.

Quinlan, R. (1993). *C4.5: Programs for Machine Learning.* Morgan Kaufmann, San Mateo, CA.

Schwartz, G. (1978). Estimating the dimension of a model, *Annals of Statistics*, **6**, 461–464.

Sclove. S.L. (1993). Some aspects of model-selection criteria. In: Bozdogan, H. (Ed.). *Proceedings of the 1st U. S./Japan Conference on the Frontiers of Statistical Modelling : An Informational Approach, 2 (Multivariate Statistical Modelling)*, pp. 37–67. Kluwer Publishers, Netherlands.

Sclove S.L. (1994). Small- and large-sample statistical model-selection criteria. In: Cheesman P. and Oldford, R. W. (Eds.). *Selection Models from Data: Artificial Intelligence and Statistics IV (Lecture Notes in Statistics, No. 89)*, pp. 31–39. Springer Verlag.

Tests for the Rank of Matrices of Hilbert-Schmidt Products

(1) Manuela M. Oliveira and (2) João T. Mexia

(1) Departamento de Matemática, Universidade de Évora, Colegio Luis Antonio Verney, Rua Romão Ramalho 59, 7000 Évora, Portugal manuelaoliveira@mail.telepac.pt.

(2) Departamento de Matemática, Universidade Nova de Lisboa, Faculdade de Ciências e Tecnologia, Quinta da Torre, 2825 Monte da Caparica, Portugal.

Abstract: While applying Statis[1], matrices S of HILBERT-SCHMIDT products are derived. These products express the relations between the studies that are considered. It is shown that, when there is a common structure between the studies, S may be represented as the sum of a rank one matrix with an error matrix. Tests for this hypothesis are presented.

Keywords: STATIS, Matrix Rank, F Tests, Hilbert-Schmidt Products, Hidden Periods.

1. INTRODUCTION

While using STATIS method, either in the direct or in the dual mode, at the end of the first phase (inter-structure), we have a matrix S of HILBERT-SCHMIDT products between the operators representing the studies. Orthogonal projections, of the points representing the operators on the subspace generated by the first two eigenvectors of S, are carried out. When there is a common structure between the studies these projections appear grouped near the first axis. This indicates that a matrix with rank one can then approximate the matrix S. We will show to use this observation to test the existence of a common structure.

2. INFERENCE

2.1 MODELS

Let S is the matrix of HILBERT-SCHMIDT products. We now follow FISHER'S lead in writing a model for these matrices:

---

[1] Structuration des tableaux a trois Indices de la Statistique

$$S = \sum_{j=1}^{r} \lambda_j \vec{\alpha}_j \vec{\alpha}_j^t + E \qquad (1)$$

where the $\lambda_1, \cdots, \lambda_j$, $[\vec{\alpha}_1^k \cdots \vec{\alpha}_j^k]$ are fixed [and mutually orthogonal with norm one], the elements of E being iid, normally distributed with null mean value and variance σ^2.

2.2 F TESTS

When there is inter-structure and S will be the sum of a rank one matrix and an error matrix we then have r=1, and

$$S = \lambda_1 \vec{\alpha}_1 \vec{\alpha}_1^t + E \qquad (2)$$

If the fixed component in S is sufficiently dominant, the first eigenvector of S is a good approximation for α_1. If the remaining eigenvectors of S are the line vectors of matrix K, we then will have

$$KS = KE \qquad (3)$$

(at least approximately). Let us now establish

Proposition 1

The elements of E'=KE are iid normally distributed with null mean value and variance σ^2.

P.

If E=$[e_1^k \cdots e_k^k]$ we have E′=$[e_1^{k-1} \cdots e_k^{k-1}]$ with $e_j^{'} = Ke_j$, j=1,...,k. Since the elements of E are independent the vectors $\vec{e}_j$, j=1,...,k, are independent between themselves as well as the e_j^{k-1}, j=1,...,k. Besides this, see SEBER (1980, pg. 5), the vectors e_j will be normal with mean vector $0^{k-1} = K0^k$ and variance-covariance matrix $\sigma^2 I_{k-1} = K(\sigma^2 I_k)K^t$. Thus the elements of E', which are the components of these vectors will be independent even if they are in the same column.
We also showed that they are normally distributed with null mean values and variance σ^2 and so the proof is complete.

We now use this result to derive a test for

$$H_0 : rank(S) \leq 1 \qquad (4)$$

As we saw this hypothesis leads to

$$Z = KS = E' \tag{5}$$

So, when H_0 holds, the elements in KS will be iid $N(0,\sigma^2)$. We can test this consequence. With $Z = [Z_{ij}]$, g=(k-1)k,

$$D = \sum_{i=1}^{k-1}\sum_{j=1}^{k} Z_{ij}^2 \quad \text{and} \quad T = \sum_{i=1}^{k-1}\sum_{j=1}^{k} Z_{ij}, \tag{6}$$

when the Z_{ij}, i=1,...k-1, j=1,...,k, are iid $N(0,\sigma^2)$, the statistic

$$\mathfrak{I} = \frac{(g-1)T^2}{D - T^2/g} \tag{7}$$

has, see MEXIA (1989), a central F distribution with 1 and g-1 degrees of freedom. Using the acceptance region $[f_{\alpha/2,g,g-1}; f_{1-\alpha/2,1,g-1}]$ we get, see MEXIA (1989), a quite powerful α level test for H_0.

After testing rank $(S) \leq 1$ we may want to test

$$H_0' : \lambda_1 = 0 \tag{8}$$

To do this, we obtain the sum S' of the squares of the components of $\vec{\alpha}_1^{\prime} S$, so that we can compute

$$\mathfrak{I}' = (k-1)\frac{S'^2}{D} \sim F(\ |\ k, g) \tag{9}$$

We now establish

Proposition 2

When H_0' holds $\mathfrak{I}'$ has the central F distribution with k and g degrees of freedom.

P.

When $\lambda_1 = 0$ we have $S=E$, and with P the orthogonal matrix with line vectors $\vec{\alpha}_1^k \cdots \vec{\alpha}_k^k$, we can reason as for the proof of proposition 1 to show that the elements of $E^+ = PS = PE$ are iid $N(0,\sigma^2)$. Thus the sum S' $[D]$ of chi-squares of the elements in the first line [the other lines] of E^{+} will be the product by σ^{2} of a central chi-square with *k [g]* degrees of freedom. Since the elements in E are independent, both chi-square are independent and the thesis follows.

We can now use $\mathfrak{J}'$ to test H_0'.

2.3 TIME-SERIES

When, after applying both tests, we are led to assume that *rank (S) =1* we implicitly are accepting that one parameter is enough to "locate" the studies. The values of that parameter will be the components of $\sqrt{\lambda_1}\alpha_1$ since then we have

$$S \approx (\sqrt{\lambda_1}\alpha_1)(\sqrt{\lambda_1}\alpha_1)' \tag{10}$$

If the different studies are taken at successive moments, these components constitute a time-series. It may be worthwhile to analyse it in order to study the global evolution of the studies.

3. AN APPLICATION

In this application we used the data on AIDS in Portugal from 1984 to 1997. This data was organised by regions, which were the objects. The variables that were considered were disease and transmission classes, sex, age group, and virus type.

Using this data we obtained

Table 1: *Hilbert-Schmidt products*

220.20													
163.58	246.88												
217.62	294.99	408.04											
158.88	219.29	280.50	240.89										
103.28	125.83	158.70	139.12	151.68									
125.78	138.40	170.04	140.78	113.75	142.35								
99.56	102.89	114.66	98.14	91.23	89.73	127.30							
73.77	96.57	99.05	92.09	80.60	82.11	89.27	116.39						
123.75	139.35	173.25	138.60	96.82	110.28	82.37	73.09	139.93					
54.78	54.04	62.39	72.19	74.93	69.43	62.09	64.69	84.24	177.62				
154.37	180.85	231.72	176.47	111.16	123.45	89.45	78.12	120.88	53.91	182.63			
152.15	161.68	203.83	160.64	115.93	120.40	98.38	83.38	120.44	70.09	140.87	171.53		
170.09	229.81	303.99	221.18	126.64	137.48	101.21	89.33	138.53	58.13	188.21	166.89	271.95	
62.02	61.32	54.50	57.79	55.68	51.27	53.66	84.79	45.00	42.86	50.67	58.73	55.37	245.96

The corresponding eigenvalues and first eight eigenvectors (table 2 and table 3)

Table 2 : *Eigenvalues the Hilbert-Schmidt products*

λ_1	λ_2	λ_3	λ_4	λ_5	λ_6	λ_7	λ_8	λ_9	λ_{10}	λ_{11}	λ_{12}	λ_{13}	λ_{14}
1904.73	289.11	182.51	104.00	85.34	57.79	39.65	33.33	32.96	28.95	26.31	23.09	20.73	14.84

Table 3 : *Eigenvectors the Hilbert-Schmidt products*

α_1	0.278	0.337	0.430	0.332	0.220	0.233	0.180	0.162	0.230	0.126	0.283	0.267	0.346	0.116
α_2	0.006	-0.146	-0.330	-0.101	0.158	0.105	0.224	0.322	0.078	0.384	-0.077	0.056	-0.196	0.685
α_3	0.022	0.136	0.165	0.014	-0.191	-0.157	-0.163	-0.028	-0.205	-0.607	0.031	-0.081	0.130	0.656
α_4	0.676	-0.168	-0.219	-0.245	-0.020	0.156	0.302	-0.002	0.025	-0.415	0.101	0.225	-0.199	-0.135
α_5	0.454	-0.111	0.086	-0.074	-0.381	-0.139	-0.423	-0.403	0.186	0.434	0.059	0.059	0.017	0.182
α_6	-0.006	-0.177	-0.024	0.201	0.652	0.294	-0.408	-0.392	-0.021	-0.134	0.026	0.067	-0.250	0.106
α_7	0.129	-0.228	0.103	-0.053	0.315	-0.430	0.245	-0.199	-0.539	0.170	-0.185	0.251	0.343	0.021
α_8	-0.226	-0.284	-0.341	-0.237	-0.024	0.189	-0.108	-0.045	0.005	-0.032	0.576	0.068	0.555	0.010

Using this date we obtained $\mathfrak{J} = 0.563$ which is non-significant as well as $\mathfrak{J}' = 131.028$ which is highly significant. We thus accept that *r=1*.

Since the studies were taken for successive years and may be located by a single parameter we carry out the analysis of the corresponding time-series. The values of this parameter were:

Coord. first axis	12.15	14.71	18.76	14.47	9.60	10.17	7.84	7.06	10.03	5.48	12.35	11.67	15.09	5.08

And the chronogram was

Figure 1: *Chronogram*

This chronogram suggested using the hidden periodicity model, see GRENANDER & ROSENBALTT (1975, pg. 91 to 94). We thus used the last 13 years to adjust the model

$$Y_t = 10.94571 - 3.34127\cos(\frac{2\pi t}{13}) - 2.366508\sin(\frac{2\pi t}{13})$$
$$- 3.340611\cos(\frac{6\pi t}{13}) + 0.322584\sin(\frac{6\pi t}{13})$$
$$+ 3.735936\cos(\frac{12\pi t}{13}) + 1.585178\sin(\frac{12\pi t}{13})$$

In the following graph we placed the adjusted values as abscises and the observed as ordinates:

Figure 2: *Adjusted and observed values*

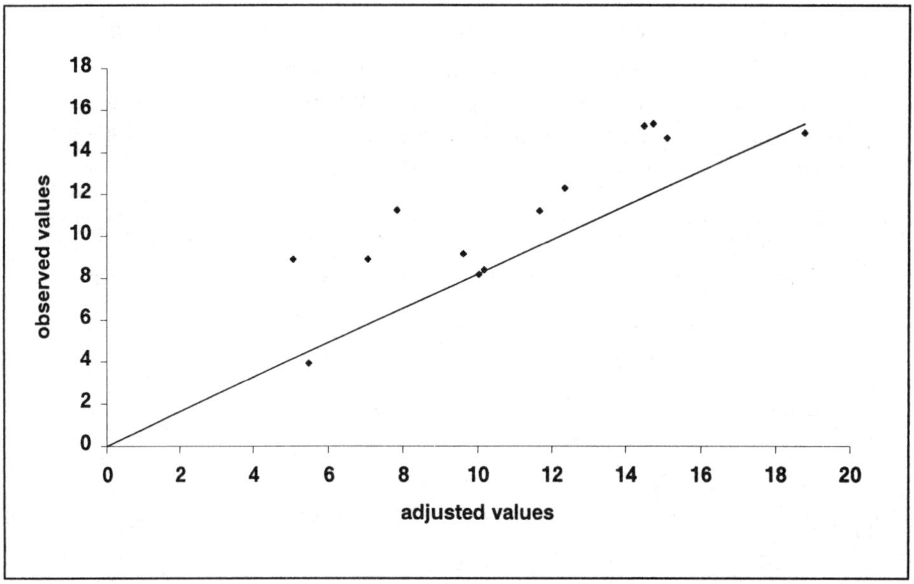

This graph shows an acceptable adjustment, we are thus led to assume the existence of cyclic components with periods 13, 4.33 and 2.16.
As a parting comment we point out that the chronogran we obtained is much easier to analyse than the usual two-dimensional projections.

REFERENCES

Escoufier, Y. (1973). *Le traitement des variables vectorielles*. Biometrics, 29, nº4, 751-760.

Grenander & Rosenblatt (1975). *Statistical Analysis of Stationary Time Seris*, John Wiley.

Lavit C. (1988). *Analyse Conjointe de Tableaux Quantitatifs*, Collection Méthodes + Programmes, Masson, Paris.

Mexia, J. T. (1989). *Controlled Heteroscedasticity, Quotiente Vector Spaces ans F Tests for Hypothesis on Mean Vectors*- Trabalhos de Investigação, nº 1, Dep. Matemática- FCT/UNL.

Seber, G. A. F. (1980). *The Linear Hypothesis: a general theory* 2nd ed, Charles & Co, Londres.

Application of Three-Way Data Clustering to Analysis of Lymphocyte Subset Numbers in Japanese Hemophiliacs Infected with HIV-1

Shinobu Tatsunami[a], Rie Kuwabara[a], Nagasumi Yago[a], Jun-ichi Mimaya[b], Kaneo Yamada[c], Mika Sato[d], and Yosiharu Sato[e]
[a]Radioisotope Research Institute and [c]Institute of Medical Science, St.Marianna University School of Medicine, Kawasaki, Japan 216-8511
[b]Division of Hematology and Oncology, Children's Hospital of Shizuoka Prefecture, Shizuoka, Japan 420-8660
[d]Institute of Policy and Planning Science, University of Tsukuba, Tsukuba, Japan 305-8573
[e]Division of Information and Systems, Hokkaido University, Sapporo, Japan 060-8628

Abstract: The 3-way data clustering was applied to the analysis of a number of CD4+ and CD8+ cells obtained from Japanese hemophiliacs infected with HIV-1 through non-heat treated clotting factor concentrates. A total of 131 hemophiliacs were classified into four clusters, termed Cluster 1, 2, 3 and 4. The members in Cluster 1 and 2 showed almost continuous decline in the number of CD4+ cells, while members in Cluster 3 and 4 showed unclear declining tendency compared to Cluster 1 and 2. The number of CD8+ cells was declining in Cluster 1, 2 and 3, while it was rising in Cluster 4. Nevertheless, the cumulative onset rates of AIDS in the four clusters cannot be divided into the four groups accordingly. The rate was the highest in Cluster 1 while no eminent differences were found among the other three clusters. Therefore, we could identify that there is a large variety in the time courses of CD4+ and CD8+ cell numbers during AIDS incubation period, even if the onset rates were almost identical. These findings may be helpful to find the clue in preventing the onset of AIDS and selecting appropriate therapy for HIV-1 infected patients.

Key words: 3-Way Data, Fuzzy Clustering, AIDS, Hemophilia.

1. Introduction

The 3-way data clustering can analyze data with a 3-way structure composed of objects, attributes and situations (Sato & Sato, 1994). This means that a proper application of 3-way data clustering to time-series clinical data will result in a good data mining, because such data are composed of patients (objects), measured values (attributes) and date of measurements (situation). We applied the 3-way data clustering to the analysis of time-series subset numbers of lymphocytes in a cohort of Japanese hemophiliacs infected with human immunodeficiency virus type 1 (HIV-1). Subsequently we evaluated the results of the classification by referring to the cumulative onset rates of acquired immunodeficiency syndrome (AIDS) in this cohort.

2. Data Description and Classification Method

The Research Committee on Prevention of Developing Illnesses and Therapy for HIV Infected Patients has conducted pathological status surveys of HIV-1 infected people in Japan (Yamada, 1992) from 1987 to 1996. Especially, its Natural History Committee (NHC) has collected survey data from various areas of Japan and constructed a database. The numbers of CD4+ and CD8+ cells after 1985 have been reported every half year: one from the first term (end of June) and another from the second (end of December). Among those data, sets of 13 point values of CD4+ and CD8+ cell numbers collected during the six-year period from the first term of 1986 to the first term of 1992 were forwarded to the program for 3-way data clustering developed by Sato and Sato (1994).
We dealt with the data of CD4+ and CD8+ cells as the three-way data with three suffixes as follows:

$$x_{ia}^{(t)} \quad \left(\begin{matrix} i=1,2,....,131\,;\ a=1,2 \\ t=1,2,.....,13 \end{matrix} \right), \tag{1}$$

where i is a suffix for patients, and a for CD4+ and CD8+ cells. The suffix t with parentheses in Eqn.1 indicates that it is the t-th measurement.
The values of the membership functions of K fuzzy subsets that are named as the degree of belongings of the i-th patient to the k-th cluster, u_{ik}, was obtained as the optimum that minimizes the sum of extended within-class dispersion as follow:

$$J(U,\boldsymbol{v}) = \sum_{t=1}^{T} w^{(t)} J^{(t)}(U,\boldsymbol{v}^{(t)}), \tag{2}$$

where $w^{(t)}$ are weights and $w^{(t)} > 0$.

$$J^{(t)}(U,\boldsymbol{v}^{(t)}) = \sum_{k=1}^{K}\sum_{i=1}^{n} (u_{ik})^m \sum_{a=1}^{p} (x_{ia}^{(t)} - v_{ka}^{(t)})^2 . \tag{3}$$

The solutions U and $\boldsymbol{v}$ for minimising $J(U,\boldsymbol{v})$ are given by

$$u_{ik} = \left[\sum_{l=1}^{K} \left\{ \frac{\sum_{t=1}^{T} w^{(t)} d(\boldsymbol{x}_i^{(t)}, \boldsymbol{v}_k^{(t)})}{\sum_{t=1}^{T} w^{(t)} d(\boldsymbol{x}_i^{(t)}, \boldsymbol{v}_l^{(t)})} \right\}^{\frac{1}{m-1}} \right]^{-1}, \tag{4}$$

$$v_{ka}^{(t)} = \frac{\sum_i (u_{ik})^m x_{ia}^{(t)}}{\sum_i (u_{ik})^m}, \tag{5}$$

where m is a parameter that determines the degree of fuzziness of the cluster (Sato & Sato, 1994) and

$$d(\boldsymbol{x}_i^{(t)}, \boldsymbol{v}_k^{(t)}) = \sum_{a=1}^{p} (x_{ia}^{(t)} - v_{ka}^{(t)})^2. \tag{6}$$

3. Four Clusters

Time-series data of CD4+ and CD8+ cell counts were available from a total of 131 hemophiliacs among 953 registered patients in the database of NHC. We could classify 131 patients into four clusters, termed as Cluster 1, 2, 3 and 4. The number of clusters is determined based on the goodness of fit between the model and observations. The number of patients included, n, and the degree of belongings averaged in each cluster are shown in Table 1. The largest cluster was Cluster 2 (n=49) and the smallest was Cluster 4 (n=13).

Table 1: *Number of patients in each cluster, n, and degree of belonging (Average±SD) to own cluster*

Cluster	n (%)	Degree of belonging
1	36 (27.5%)	0.923 ± 0.123
2	49 (37.4%)	0.850 ± 0.143
3	33 (25.2%)	0.839 ± 0.163
4	13 (9.9%)	0.785 ± 0.213

The motion of the centroids of four clusters is shown in Fig.1. The tendency of steady decline in CD4+ cell numbers was eminent in Cluster 1. Gradual declining was observed also in Cluster 2, although it declined more slowly than Cluster 1. On the contrary, the declining was unclear in Cluster 3 and Cluster 4.

Fig.1: *Motion of centroids of four clusters*

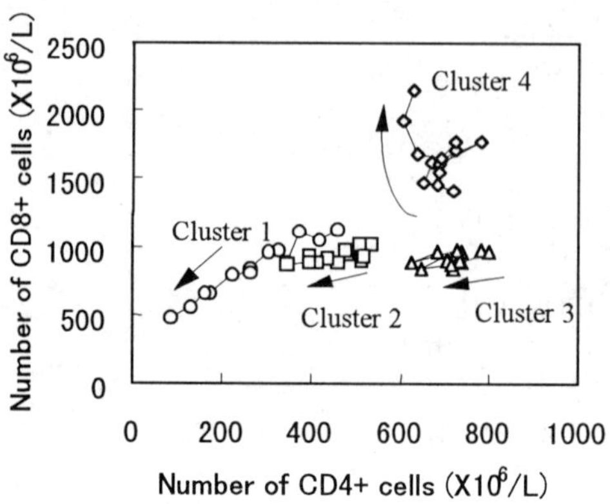

Moreover, most of the reported numbers of CD4+ cell in Cluster 3 and 4 were higher than 500×10^6 /L. The number of CD8+ cell was steadily declining in Cluster 1. Slower declining was shown in Cluster 2 and 3. On the other hand, the number of CD8+ cell showed rising with a gradient of 90.9 (95% CI, -5.1 ~ 186.8)$\times 10^6$/L per year in Cluster 4.

4. Aspects of Each Cluster

Cumulative onset rates of AIDS since the beginning of 1986 in each cluster are shown in Fig.2. The Kaplan-Meier estimates (Kaplan & Meier, 1958) for the fraction of patients with AIDS in Cluster 1 has been the highest among the four clusters. In other clusters, only part of the patients (less than 11%) developed AIDS by the end of March 1994.

Fig. 2: *Cumulative onset rates of AIDS in four clusters*

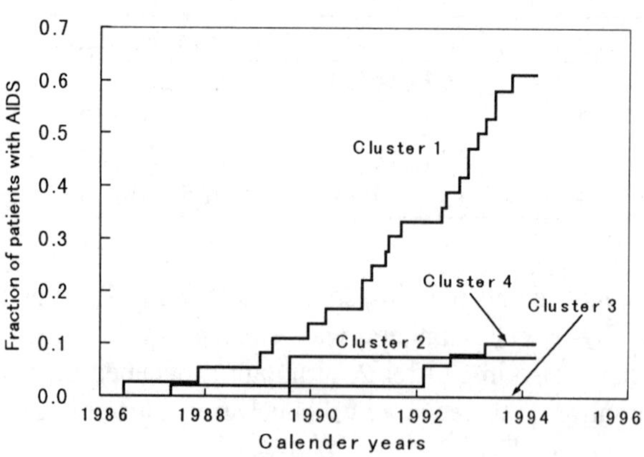

Numbers of cases with clinical disorders in the four clusters are shown in Table 2. Pneumocystis carinii pneumonia that has been one of the AIDS defining clinical diseases was observed only among patients in Cluster 1. Although the onset of clinical AIDS was not reported, herpes zoster and other symptoms have been observed among patients in Cluster 3.
The shapes of distribution of age at the first term of 1986 in Cluster 2, 3 and 4 were almost same. Although the broader range and lower median were observed in Cluster 1, no statistically significant difference was found in age distribution among the four clusters.

Table 2: *Reported number of cases with clinical disorders in the four clusters*

Disorders	Cluster 1	Cluster 2	Cluster 3	Cluster 4
Herpes zoster	9 (25.0%)	6 (12.2%)	3 (9.1%)	2 (15.4%)
Mycosis	7 (19.4%)	3 (6.1%)	0	0
Pneumocystis carinii pneumonia	6 (16.7%)	0	0	0
Intractable diarrhea	5 (13.9%)	3 (6.1%)	0	1 (7.7%)
Others	14 (38.9%)	8 (16.3%)	6 (18.2%)	2 (15.4%)

5. Discussion

The number of CD4+ cell has been considered to be the most informative factor to indicate the disease status (Young et al., 1997). Therefore, a proper classification of time courses of the CD4+ cell numbers during AIDS incubation is valuable to evaluate the natural history of HIV. Using the slope of changes in CD4+ cell count from linear regression is one of the practical methods adopted in some of the definitions of long-term nonprogressors (Strathdee et al., 1996). However, a more appropriate method would be needed to figure out principal characteristics of time courses of CD4+ cell numbers. The present application of 3-way data clustering will be one of the solutions of this problem.
The 131 hemophiliacs can be classified into four clusters by the characteristics shown in Fig.1. Nevertheless, the cumulative onset rates of AIDS in four clusters cannot be divided into four groups (Fig. 2). The rate is the highest in Cluster 1 while no eminent differences are found among the other three clusters. Moreover, the clinical symptoms summarized in Table 2 also indicate the worst conditions of patients in Cluster 1. No remarkable difference is shown among other clusters. Therefore, the present results of clustering seems to be quite meaningful because we could identify that there is a large variety in the time courses of CD4+ and CD8+ cell numbers during AIDS incubation period even if the onset rates were almost identical. For example, the number of CD8+ cell has been rising in Cluster 4 in addition to its highest average among the clusters at the initial point. Thus the patients in Cluster 4 may manifest a more dynamic

response to cytotoxic T-lymphocytes, which requires further investigation to confirm compared to those in other clusters. Identification of such patients may be helpful to find the clue in preventing the onset of AIDS and selecting appropriate therapy for HIV-1 infected patients.

Older age at infection with HIV-1 has been reported to be tightly associated with more rapid AIDS progression in hemophiliac cohorts (Rosenberg et al., 1994; Muñoz & Xu,1996). The individual age at the time of infection has not revealed by NHC. However, a tentative mean date of seroconversion has been proposed by the Committee as April 1983 (Mimaya et al., 1992). Therefore the age at the first term of 1986 can be considered to be approximately 3 years after seroconversion. In the present study, no statistical difference in age at the first term of 1986 was found among the four clusters. Thus the age at infection will not explain the differences of CD4+ cell time courses and AIDS onset rates among clusters. Therefore, further investigation is necessary to detect other background information that resulted in the present classification. The most important of all would probably be the viral loads of patients in each cluster.

References

Ferligoj, A. & Batagelj, V. (1994). Direct multicriteria clustering, *Journal of Classification*, 9, 43-61.

Kaplan, E.L. & Meier, P. (1958). Nonparametric estimation from incomplete observation,. *Journal of American Statistical Association*, 53, 457-481.

Mimaya, J., Meguro, T., Tatsunami, S., et al. (1992). Natural history committee report, in: *Annual Report of Research Committee on Prevention of Developing Illnesses and Therapy for HIV Infected Patients 1991*, Yamada, K.(Ed.), 9-16.

Muñoz, A. & Xu, J. (1996). Models for the incubation of AIDS and variations according to age and period, *Statistics in Medicine*, 15, 2459-2473.

Rosenberg, P.S., Goedert, J.J. & Bigger, R.J. (1994). Effect of age at seroconversion on the natural AIDS incubation distribution, *AIDS*, 8, 803-810.

Sato, M. & Sato, Y. (1994). On a multicriteria fuzzy clustering method for 3-way data, *International Journal of Uncertainty, Fuzziness and Knowledge-Based Systems*, 2, 127-142.

Strathdee, S.A., Veugelers, P.J., Page-Shafer, K.A., et al. (1996). Lack of consistency between five definitions of nonprogression in cohorts of HIV-infected seroconverters, *AIDS* , 10, 959-965.

Yamada, K. (1992). Pathological status and therapy of HIV-infected hemophiliacs in Japan, *Southeast Asian Journal of Tropical Medicine and Public Health*, 23 (Suppl 2), 127-130.

Young, F.H.L., Taylor, J.M.G., Bryant, J.L., et al. (1997). Dependence of the hazard of AIDS on markers, *AIDS*, 11, 217-228.

PART VI

Case Studies

- **Applied Classification and Data Analysis**

On an Error Detecting Algorithm on Textual Data Files

Antonio Bellacicco* and Camillo De Vecchis**
* University of Teramo ,Dept. of Systems Theory and Organizations
** Resfirma s.r.l., Torino

Abstract: The paper presents a new algorithm ,TEFIC, based on the Bayes formula for automatic textual data detection and correction.The algorithm is able to clean any textual file in a data base without a priori knowledge of the semantic interpretation of each word and of the linguistic rules characterizing a given language.The algorithm is able to give a good performance on data bases including names and numbers.The kernel of the algorithm is the Bayes formula and the evaluation of the a priori probability,based on clusters of words.Clustering of words based on the Hamming distance becomes the key for detecting and cleaning successfully errors in a textual data file.

Key Words
Bayes, correction, detection,textual data.

1.Introduction

The aim of the paper is to present a new algorithm, TEFIC, able to clean the errors in a textual data file without any knowledge of the semantic interpretation of the words and of the rules of the language.The area of interest of the paper is somewhere between classical statistics , informatics and linguistics. The main use of the algorithm TEFIC is the automatic cleaning of large textual data files containing thousands of records regarding the features of specific units,like the workers of a given firm as well as the tax payers of a given country.

The main strategies for detecting and correcting errors in large data base,with billions of textual data, is to exploit a thesaurus and to compare each textual data with the most similar elements of the thesaurus..The similarity can be evaluated considering either the simple sequences of the elementary units like letters or numbers,or the meaning and therefore the word with the similar meaning.Finally another approach is to introduce linguistic rules which are constraints on the sequences of elementary units,like letters,commas,points and numbers.All the approaches are based on some a priori knowledge depending on the specific language.Following our point of view the effort of introducing the knowledge can be avoided provided we can build up an algorithm able to recognize the actual rules belonging to a given language from a random sample of records and to increase its knowledge as long as it is running on the data base ameliorating its performance on job.The strategy can be compared with the usual techniques for signal filtering where we deal with the symbols of the signals and the problem is to clean the random noise that is able to modify the symbols,Peirce (1961),Tong (1990). In the coming paragraph we will describe the main features of the algorithm and we will give some information on its performance.

2.The foundations of the TEFIC algorithm

The TEFIC algorithm is based on the probability of occurrence of a given word and on the probability of occurrence of a given character,like a letter,a comma and some others which can be identified. in the text.Moreover we suppose that the source text is randomly modified so that the actual text is a transformation from a source text obtained by a random disturbance.This point of view is quite similar to the usual assumptions on a randomly disturbed channel of a digital transmission.Some other assumptions can be stated as follows:

H_1 .a finite list Θ of N words is considered whose members are both the words before the disturbance and the words after the disturbance;

H_2 . a word ϑ_i can be a random mutation of a subset L_i of other words in Θ;

H_3 . an equiprobability distribution is assigned to the list Θ;

H_4 . the occurrence of a given word in Θ is statistically indipendent from the occurrence of another word.

The consequence of the previous assumptions is that we can identify the list of the words Θ and that their occurrence is submitted to the sampling rules.The full knowledge of the list is obtained as far as the size of the samples increases following a learning process.

Assumption H_2 implies that we can assign to each element of Θ,ϑ_i,a subset L_i of Θ.The subset L_i can be identified either by rules or can be identified by a systematic substitution of each character with all the other characters of the alphabet and finally by permutation of all the characters of the word.Another way can be considered either deleting a character or adding a new character.It is easy to see that the size of the subset L_i can be extremely large and the final purpose is to identify the subset to which the right word belongs.As a final comment ,the identification of a high probable subset to join to each word can be a matter of discussion but its role can be appreciated as far as the exploration of the full set Θ is pratically impossible.As a consequence we have to build up a set of clusters which are a covering of the set Θ.

Actually,each word is the kernel of a cluster and a textual data file can be considered as a set of kernels.The set of all clusters related to the file is a population of overlapping clusters.

The problem we deal is to detect the wrong words and to correct them by a search strategy in the population of clusters.In the next paragraph we will introduce the search algorithm based on the Bayes formula and on the clusters generation.

3.The algorithm TEFIC

In this paragraph we introduce the TEFIC algorithm which is able to detect the wrong words and to correct them.

Let us introduce some notations in order to write down the TEFIC algorithm:

$n(\theta_i) = n_i$: the absolute frequency of the occurrences of θ_i ;
$l(\theta_i) = l_i$: the length of the word θ_i ;
$\theta^i_j \equiv \theta_i \rightarrow \theta_j$: the word θ_i becomes the word θ_j belonging to the same cluster C;
$H(\theta_i, \theta_j)$: the Hamming distance between θ_i and θ_j ;
$\theta_i \equiv \theta^i$;
n^i_j : the frequency of mutations of the word θ_i into the word θ_j .
The Hamming distance is the number of different characters of the two words,in the same position.We can now introduce the following probabilities.Let

i. $P(\theta_i) = n_i / N$ where $N = \Sigma\, n_i$.

$$\text{ii.. } P(\theta_i \cap \theta^i_j) = \frac{n_i}{N} P(\theta^i_j \rightarrow \theta_j). \tag{2}$$

$$\text{iii. } P(\theta_i \mid \theta^i_j) = \frac{P(\theta_i \cap \theta^i_j)}{P(\theta^i_j)} = P(\theta^i_j \rightarrow \theta_i). \tag{3}$$

$$\text{iv. } P(\theta^i_j \mid \theta^i) = \frac{P(\theta^i_j \rightarrow \theta_i)\, n^i_j}{\Sigma_k P(\theta^i_k \rightarrow \theta_i)\, n^i_k} \tag{4}$$

The frequency n^i_j is the number of occurrences of mutations of the word θ^i into the word θ_j belonging to the same cluster of words,C, whose cardinality is $n_i(C)$.
The last equation is the main result of the paper because by (4) it is possible to evaluate the probability of observing a mutation θ^i_j ,where the source word is θ^i.
The previous equatione is the probability of an error and therefore any word in the text can be a transformation of some other word.In other terms the message can be different from the source.
It is possible to set up some hypothesis on the probability of an event :$\theta^i \rightarrow \theta_j$ .The first hypothesis is : $P(\theta^i_j) = 1, \forall j \in C$.The consequence of the previous statement is the following:

$$P(\theta^i_j \mid \theta_i) = \frac{n^i_j}{N} \tag{5}$$

The probability that a word is the transformation of some other word is equal to the relative frequency of the word in the set of all the possible words.
The previous hypothesis is the simplest one and it is based on the actual occurrences in a given text file and was effectively considered in order to test the algorithm.In this case all the informations come from the data and the estimation of the probability is reduced to the estimation of the frequency of the data.

The empirical approach can be justified from the frequentist point of view and does not implies theoretic assumptions.The experimental results will be considered later on in the paper.
Some other hypothesis can be introduced.The simplest one is the equiprobability and the final formula is the following one:

$$P(\theta^{i}_{j} | \theta_{i}) = \frac{(n^{i}_{j})^{2}}{\Sigma_{k}(n^{i}_{j})} \qquad (6)$$

The previous assumption is the simplest one without any support of the data.It is easy to see that (6) is actually a probability so that :
The simplest assumption is the simplest one without any support of the data .It is easy to see that (6) is actually a probability so that :

$$0 \leq P(\theta^{i}_{j}) = 1$$

$$\Sigma_{j} P(\theta^{i}_{j} | \theta_{i}) = 1$$

$$\Sigma_{j} P(\theta^{i} \rightarrow \theta_{j}) = 1$$

A final hypothesis can be considered based on the errors associated to the characters of the alphabet.In this case we are strictly constrained to the choice of the alphabet.Some assumptions should be introduced like the following ones :

i.the errors are transformation of a set of characters into a set of characters;

ii.the characters should be identified in advance;

iii.the probability space of the characters is equiprobable;

iv.the probability of a transformation of a given character into another character is a known number.

We can easily evaluate the following probability:

$$P(\theta^{i} \rightarrow \theta_{j}) = 26^{-H(\theta i;\theta j)}\, p_{c}^{H(\theta i;\theta j)}\, (1 - p_{c})^{n_{i} - H(\theta i;\theta j)} \qquad (7)$$

It is possible to recognize the binomial distribution,where the number 26 comes from the numbers of characters in latin alphabet and the probability p_c is the probability of each character.
We can easily see that all the previous considerations can be introduced for the estimation of the probability of each character.Moreover we still used the Hamming distance for evaluating the previous probabilities.We can introduce some

further assumptions on the occurrences of the words.In case we consider only words that have $H(\theta^i ; \theta_j) \leq 1$,we can rewrite the previous distribution as follows

$$\text{i.If } \theta^i_j \neq \theta_i \text{ then } P(\theta^i \rightarrow \theta_j) = \frac{1}{26} p_c (1 - p_c)^{n_{ij} - 1} \tag{8}$$

$$\text{ii.If } \theta^i_j = \theta_i \text{ then } (1 - p_c)^{n_i} \tag{9}$$

The consequence of the previous formulas is that we can no longer consider the function $P(\theta^i_j)$ as a probability in spite of :

i. $0 \leq P(\theta^i_j) \leq 1$
ii. $\Sigma_j P(\theta^i_j) = 1$

Finally we can write down the explicit formulas for the conditional probabilities:

i.For $\theta^i_j \neq \theta_i$ we have :

$$P(\theta^i_j \mid \theta_i) = \frac{p_c}{26 (1 - p_c) n_i + p_c \Sigma_k n^i_k} \tag{10}$$

ii.For $\theta^i_j = \theta_i$

$$P(\theta^i_j \mid \theta_i) = \frac{26 (1 - p_c) n_i}{26 (1 - p_c) n_i + p_c \Sigma_k n^i_k} \tag{11}$$

The last probability is greater than the previous one depending on the ratios between the frequencies n^i_j and n_i .The equality like (10) = (11) can be combined by the relation :

$$\frac{n^i_j}{n_j} = 26 \left(\frac{1}{p_c} - 1 \right) \tag{12}$$

It is easy to see that the probability p_c plays a basic role in the evaluation of the conditional probabilities of the words.
The TEFIC algorithm can be reduced to the following steps :

Step 1:read a text data file;
Step 2:cluster the words by the Hamming distance ;
Step 3:evaluate the probability of the caracter and the conditional probability of the
words;
Step 4: detect the words with the highest probability and decide which words should be substituted;
Step 5 : find the highest probable words able to substitute the detected words in the previous steps and Stop the algorithm.

4.Concluding remarks

The paper presents a statistical approach for detecting and cleaning the errors in the text data files.The main featuce of the proposed algorithm is its ability to avoid the tesaurus and to exploit the probabilities of the occurrences of the possible errors by a suitable set of clusters of words where the adopted metrics is based on the Hamming distance between the words.The problem of the errors in large data bank is well known and a fast and not very expensive approach to the solution of the problem of detecting and cleaning errors can be appreciated as far as the usual approach are very expensive.

References

Haken H.(1983) .*Synergetics.* Springer -Verlag ,Berlin

Peirce J.R. (1961).*Symbols,signals and noise.* Harper Modern Science Publ. New York.

Tong H. (1990). *Non linear time series* . Claredon Press,New York.

Van der Waerden B.L.(1969),*Mathematical statistics* .Springer-Verlag,Berlin

Hierarchy Techniques of Multidimensional Data Analysis (MDA) in Social Medicine Research

Sabina Popescu-Spineni
Institute of Public Health, Bucharest, Leonte str., 1-3, 76256, ROMANIA

Abstract: Multidimensional Data Analysis (MDA), by means of statistical methods, is used in order to select the main sources of variability and to establish a hierarchy of the characteristics under study. This hierarchy is necessary in social medicine for decision making.

Key words: Cluster, Correspondance Analysis, Social Medicine, Decision

1. Introduction

The MDA is applyed since 1985 in social medicine, to elaborate the strategies for an optimal ratio public health needs/medical staff availability. Multidimensional Data Analysis can solve, through statistical methods, the problem of characteristics under study hierarchy. This hierarchy is necessary in social medicine for decision making, considering the constraints and implicating the priorities.

2. Objectives

A synthetic data representation, with a minimum loss of information and a maximum relevance is focused. A synthetic indicator is theoretically unacceptable (Arrow's theorem). In order to combine the continous methods with the discrete techniques (Naouri's theorem), scientists have to establish a hierarchy. The "without previous hypothesis" (Benzécri) methods, that don't impose a normal distribution of data, may be mentioned.

3. Methods

The factors and the priorities assessement and hierarchisation, the MDA regarding the primary health care, including the number of physicians that is needed, are studied by means of the correspondance analysis (J. P. Benzécri). The applications mentioned above were solved, first, by an adapted FORTRAN (1985) and, later, by SPSS (1991) and SYSTAT (1995) soft. The distances or similarity index computation was made by means of EUCL (euclidean distance), SUMSQ (squared euclidean distance), CHISQ (chisquare index), CORR (dissimilarity index specific to the minus correlation), JACC (dissimilarity inddex specific to the negative Jaccard index). The chisquare distance (SPSS) and the euclidean distance (SYSTAT) were used from those mentioned above.

As a discrete method, the cluster analysis is applied. The "Cluster Analysis Program" by Tim E. Aldrich & J. Wanzer Drane, from South Carolina University & Oak Ridge National Laboratory was used recently (1996), particulary Ohno and Grimson methods, with a cartographic representation on the geographic map of Romania. Ohno method compares rates within geographic cells to determine high rate areas cluster. Ohno program uses chisquare because more than two rate categories are possible. Grimson method uses only a high and low rate group. It calculates and approximates normal (z) statistic to test for non-random mixing of high and low rate (areas) counties. Grimson program is the more precise of the two.

Romania is divided in 41 districts ("judets"), numbered in the alphabetical order. For example, Alba (AB) district has the number 1, so the districts around it are 2, 5, 14, 23, 28, 34 and 40. A geographic referencing file, named Rocounty.geo, is created for all districts. This geographic configuration is created into cluster and then specific data can be entered. Each file can be later edited at will. Mention may be made for the district name abbreviations (i.e. Alba-AB, Bucharest - B, s.o.).

4. Data

Frequency contingence tables (absolute or relative data - physicians per 10000 inhabitants, since 1985) were used for selecting the main objects (41 districts) and/or characteristics (15 medical specialities).

5. Results

The two main factorial axes (F1 and F2) were settled by means of the correspondance method (FORTRAN program): F1 represents 65.35% of the whole variability and F2 represents 21.65% of the whole statistic variability, with 87.00% "quality" of the representation. The chosen model made possible a dual analysis, so the object-lines (districts) and the characteristics-rows (medical specialities) are both represented with respect to the same factorial axes (fig. 1).

B, TM, MS, CJ, DJ and IS, the districts with the main academic centers, are important, being positively correlated with the first factorial axis (F1). The other districts are correlated in opposition with those mentioned above. The center of the factorial axes is occupied by PH district.

Fig. 1: *Dual representation of districts/medical specialities (specialists/10 000 inhabitants)*

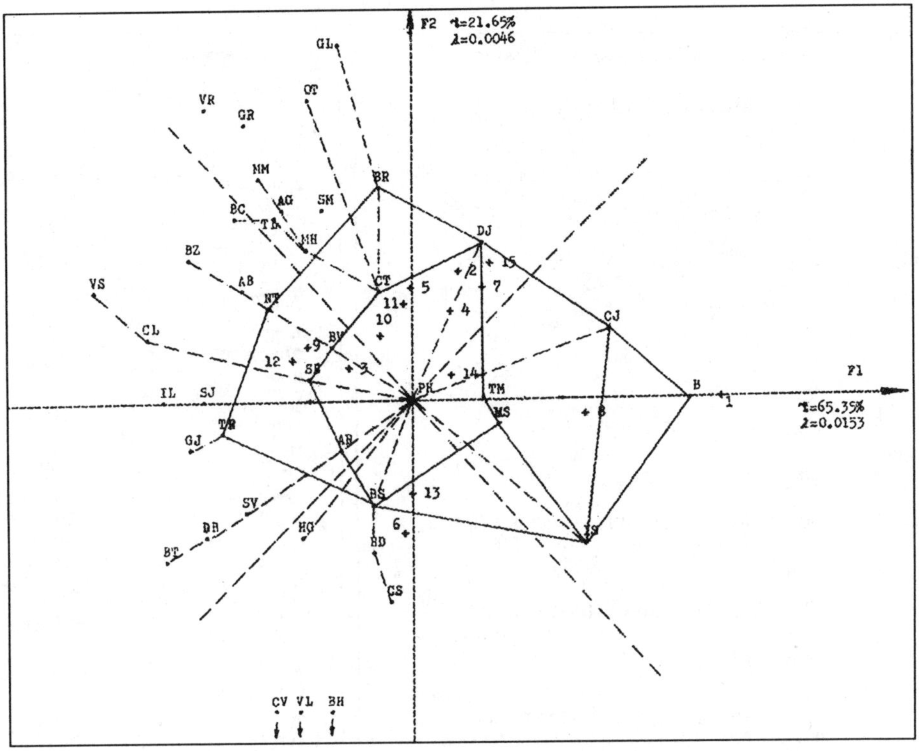

Less variability is recorded with respect to the second factorial axis (F2) which oposes GL and BR districts to CV district, and s.o.

The second representation of the dual analysis makes possible the tight correlation of the medical specialities (physicians per 10 000 inhabitants) with respect to the first factorial axis (F1): pathological anatomy (1) with laboratory (8), epidemiology (3) and radiology (4) being placed in the center of the factorial axes. Other medical specialities, such as internal medicine (6) and psychiatry (13) are more deficitary in certain districts, being opposed to those mentioned above.

SPSS program gives a similar representation for the main objects and/or characteristics (fig.2).

A SYSTAT program (euclidean distance) was used to select the main similarities of the physicians number by medical specialities on districts. At the first level of the diagram, under the reserve of non-standardization of data, the following classification of districts can be seen (fig. 3):

((B, (CJ, PH)); (TM, IS, MS); (CS, HD, (BH, VL, CV));
(IL, (DB, SV, BT), (GJ, SJ), OT, SM); (IL, OT, SM);
((BZ, CL), (TL, VS), (BC, GR), GL, VR, TR, BS);
(DJ, (AG, BR)), CT); (AR, BV, SB, HG, NT); ((AB, MM), MH), s.o.

Fig. 2: *Similar representation of districts/medical specialities*

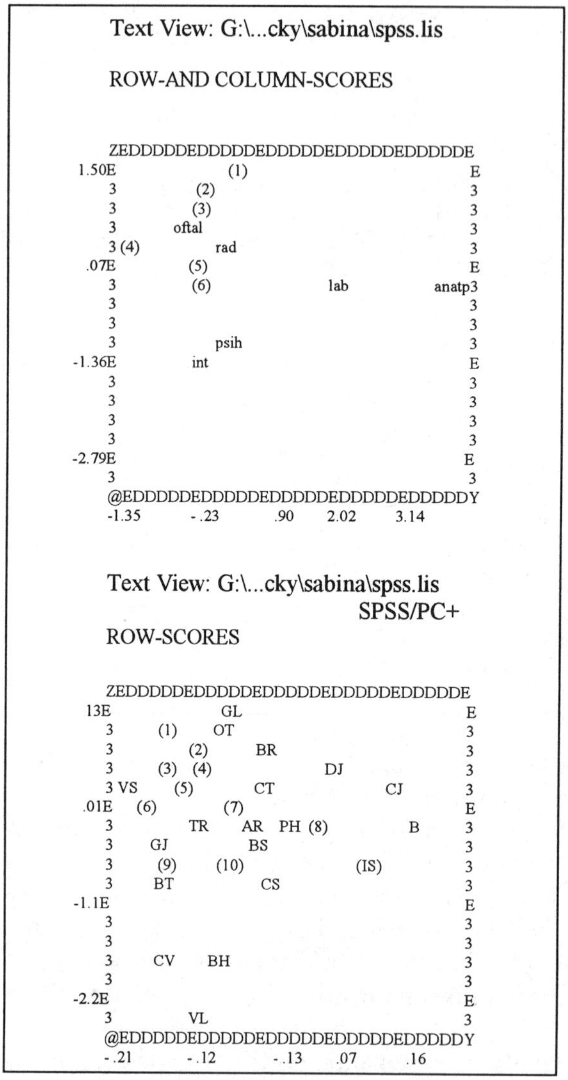

Fig. 3: *Classification of districts/medical specialities*

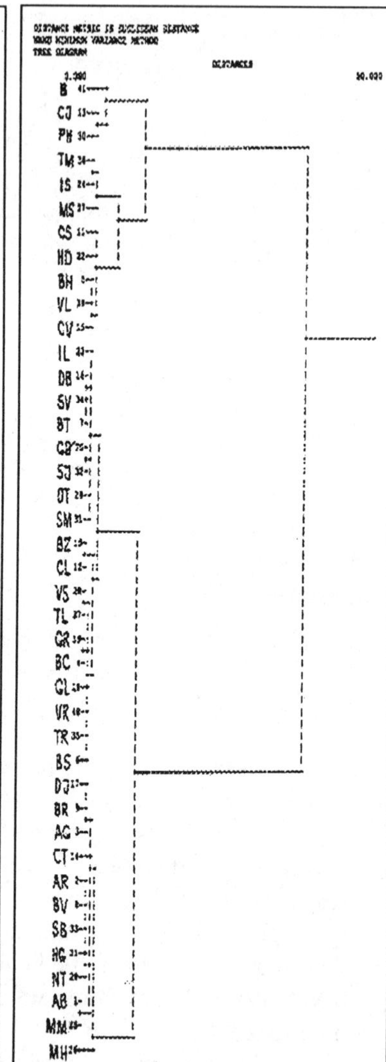

In order to avoid the possible causes of a random aggregation of data, cluster analysis is also motivated. Practically, a public health policy can be based, for example, on a geographic distribution of objects and/or characteristics.

A geographic representation concerning some data was created to be called in a spatial cluster program. A classification of the main areas of interest in decision making became possible, with a cartographic representation on the geographic map of Romania, i.e. regarding the pathology specialists (fig. 4).

The following 12 significant clusters are noticeable (p=0, 10):

(B PH)	(CJ BH)	(CJ MS)	(IS BT)
(IS NT)	(MS HG)	(MS SB)	(TM CS)
(NT HG)	(SB AG)	(CT BR)	(GL BR)

Fig. 4: *Spatial classification of districts ranged by (1)- pathology specialists (1/10 000 inhabitants)*

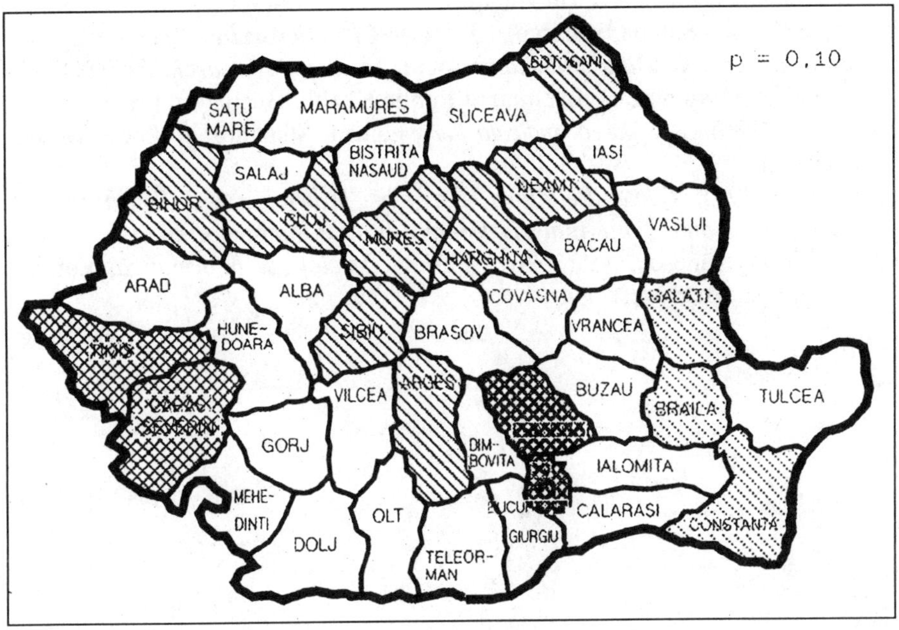

On the map, a special cluster of the west districts TM and CS, of east-districts GL, BR and CT and of the districts B and IL can be mentioned. The other marked districts are located in the north-central side of the country.

6. Conclusions

The development of Multidimensional Data Analysis (MDA) can improve the decision making in the field of social medicine research.

The districts identified by MDA represent areas of priorities in public health for decision making.

References

Benczéri, J. P. (1980) *L'analyse des donnés*, Dunod, Paris.

Dragomirescu, L., Postelnicu, T. (1994) *Specific Numerical Taxonomy Methods in Biological Classification*, Proceedings of the 17 th Course of International School of Mathematics G. Stampachia, Erice, Italy, september 1993, World Scientific, Singapore, 1994.

Popescu-Spineni Sabina, Dragomirescu, L. (1987). *Factorial Analysis and Software in Social Medicine Research*, Scientific Session of the Faculty of Mathematics, Bucharest, (in volume).

Popescu-Spineni Sabina, Spircu, T. (1996) *Hierarchy Techniques of Multidimensional Data Utilised in Social Medicine Research*, the XXXI-th Scientific Session of the Institute of Public Health, Bucharest, (in volume).

Paun, Gh. (1985). *Agregarea indicatorilor*, Editura Stiintificã si Enciclopedicã, Bucharest.

Rizzi, A. (1995). *Some Relations Between Matrices and Structures of Multidimensional Data Analysis*, Giani Editori e Stampatori in Pisa.

Saporta, G., Stefãnescu, V. (1996). *Analiza datelor & informaticã*, Editura Economicã, Bucharest.

Classification and Decision Making in Forensic Sciences: the Speaker Identification Problem

Carla Rossi
Dipartimento di Matematica
University of Rome Tor Vergata, Italy

Abstarct: The decision process in forensic sciences is quite a complex phenomenon, conditioned by incomplete information and wide uncertainty. Setting up a proper model implies modelling an inductive process based upon inferential steps controlled by the evaluation of a set of conditional probabilities and likelihood ratios.
The model proposed in the present contribution is based on an inferential bayesian paradigm and is outlined using a real example concerning the identification of a defendant by means of vocal 'prints', in more technical terms: 'the speaker identification problem'.
A comparison with other methods based on the classical statistical approach allows to appreciate the main features of the present model in computing the proper conditional probabilities of the scientific evidence under the two alternative and exclusive hypotheses: the prosecution hypothesis and the defence hypothesis.
The likelihood estimation problem is also considered.

Key Words: speaker identification, classification errors, likelihoods, kernel density estimation, bayes factor.

1. Introduction

It is clear from everyday experience that the speech signal carries information about its producer. There are specific circumstances in which it is crucial to be able to determine the identity of a speaker from speech alone. For instance, a witness to a crime may have heard the masked perpetrators speaking, or tape recordings of telephone conversations may have to be compared with the voice of a criminal suspect.
Speaker recognition might be defined as any activity whereby a speech sample is attributed to a person on the basis of its phonetic, acoustic or perceptual properties. A distinction can be drawn between naive speaker recognition and technical speaker recognition. In the first case the recognition is performed by untrained observers. The decision is based on what is heard and no special techniques are involved. In the second case much work on the comparison of recordings is required and to establish identity involves acoustic analysis and probabilistic/statistical methodologies. In speaker recognition we can also distinguish two categories: speaker identification and speaker verification problems. In the present paper we consider just the speaker identification problem which includes the usual forensic situation.
The forensic task is generally to compare two samples, one from a known

speaker and one from an unknown source, and ultimately a criterion has to be applied to reach a decision as to whether the samples are similar enough to be from the same speaker.
The speech of an individual is not a constant, either in terms of those properties which result from the physical mechanism of speech or those which are a function of the linguistic system. If we imagine the many parameters which characterize a speaker as defining a location in a multiple dimensional space, we see that it is not a single point which characterizes the speaker but an area of variability, i.e. a probability distribution.
There exists extensive literature dedicated to the identification problem. In the 1970's, there was a prevalent use of spectrographic analysis in voice signals; in the last few years, statistical analysis of properly defined acoustic parameters was developed, so as to take into account stochastic aspects. The almost complete totality of the statistical methodology used is of a parametric type for what concerns the estimation of the likelihoods, and it is based on classical statistical hypothesis testing for what concerns the classification problem. In the following we analyse a non-parametric estimation procedure that avoids the imposition of strong hypotheses that have already been proved inconsistent, as multinormality of the acoustic measures studied, i.e. the `formants' of the vowels (Paronetto, 1995; Beciani, 1998).
For further details on acoustic analysis and data one can see the specific references reported below.

2. The statistical-decision problem

Usually the technical speaker identification is based on the analysis of the frequency distributions of the formants of the vowels. By the formants we mean the acoustic frequencies corresponding to the local maximum intensity on the Fourier transform of the soundwave; depending on the position of these maximum points we can identify one or the other vowels. Because of the effect of anatomical and socio-linguistic characteristics, the formants of a speaker have a particular probability distribution in a multidimensional space 'characterising' that speaker.
One of the foundamental issues in speaker identification is whether each individual in the population corresponds to a different and identifiable formant distribution or whether there is wide overlap.
It is well known that, unfortunately, this latter is the case, thus it follows the crucial problem of dealing with the classification errors, which are related to whatever decision procedure adopted.
Furtherly, we must take into account that the identification process, comprising the estimation of the individual formant distribution, has to be based on a usually small sample of measured formants, thus implying wide uncertainty. These facts place a limitation, in priciple, on the reliability of any act of speaker identification, a limitation which must always be aknowledged. Indeed, ideally, the acoustic parameters extracted should exhibit a large between speaker variability and a low within speaker variability, but for the actual data this is not the case.

2.1. The likelihood estimation problem

The data on which the actual analysis is based relate to 100 different recordings, each coming from a different male italian adult speaker. For each individual 4 formants are measured for four different vowels: "a", "e", "i", "o". The "u" has been excluded due to its low frequency in the italian language. Each recording comprises a different number of replications for each vowel. So, each speaker is represented by a frequency distribution in a 16 dimensional space that can be modelled by a proper probability distribution.
Such distribution has to be estimated before dealing with the classification problem. An extensive description of the data set can be found in Calvani (1996).
As a matter of fact the 16 parameters result highly correlated, thus a reduction of the complexity of the problem is possible by means of multivariate data analysis methods as, for instance, principal component analysis or factor analysis (Beciani, 1998).
Suppose that the independent `factors' can be represented in a k (<16) dimensional space. To determine the proper probability distributions for the classification problem it is possible to use different approaches. We can estimate the k-dimensional distributions and then decide what kind of summary statistic can be used to make decisions about the speaker identification problem, or we can establish a proper vector dissimilarity index to measure the 'distance' between the characteristic measures coming from the recording of the defendant's voice and the same measures coming from the recording of the unknown speaker's voice and estimate the probability distributions of the statistic under the two alternative hypotheses, that means the prosecution and the defence hypothesis.
For the present application we use the second approach and, after defining a proper vector (or scalar) distance between the observations (or the principal components), we estimate the interesting likelihoods, i.e. the conditional probabilities of the evidence under the two alternative hypotheses, that in the present application are related to the within and between speaker distributions, using the kernel estimation method. Cross validation based on Pseudo Maximum Likelihood method is used to optimize the smoothing parameter.
The estimated likelihoods are used to set up the proper classification algorithm based on the likelihood ratio.

2.2. The classification problem

In the forensic arena the commonest speaker recognition task is to decide how likely it is that two recordings are from the same person. This implies that there is the possibility for two kinds of errors. An innocent individual can be convicted, as a consequence of the similarity between the two recordings taken into account, or a guitly person can be released due to the dissimilarity.
The most important information for the decision makers is thus related to the measures of the risks related to the two kinds of errors.
There are judicial cases in Italy where many controversies arose between persons involved in the decision process about the proper approach to the decision problem. In particular, some cases dealt with it by means of significance tests and were highly criticized, due to the fact that the probability of false attribution was not taken into account. Some discussions derived from these cases are

reported in Rossi (1996a and 1996b).
In the present contribution, the decision problem is approached by a bayesian decision paradigm, based on likelihood ratio or Bayes factor of the two hypotheses. The sensitivity analysis with respect to the prior probabilities shows that, for a wide range of values, the maximum likelihood approach is feasible.
The probability of false attribution is easily calculated corresponding to different values of the prior probabilities, thus providing a complete support to decision system for decision makers.
In any case, the evaluation of the posterior probabilities of the two errors is such that any kind of decision seems to be quite critical, due to the uncertainty coming from the incomplete information related to the between speaker distribution as actually its estimation is based on just small samples of voices coming from the reference population.

3. An application to a real data set

To enter in greater details we are going to use the same data set studied by Calvani (1996).

3.1 Exploratory analysis

One of the first issue to be studied is the problem of the 'dimension' of the speaker space generated by the 16 formants. A preliminary exploratory analysis has been conducted using the data about the 100 male adult speakers described in Calvani (1996) by means of principal component analysis using S-plus.
The correlation matrix of the 16 averaged formants has been calculated, showing some rather high correlations.
In particular, we observe correlation coefficients ranging from 0.86 to 0.91 between the f_0 type formants of the four vowels, as it was expected from acoustic models, but also correlation coefficients 0.86 between the f_3 formants of "a" and "e", 0.88 between "a" and "o", 0.85 between "e" and "i" and 0.82 between "e" and "o". The correlation coefficient between the f_2 formants of "e" and "i" results 0.81, to mention only correlations above 0.80.
The results of the principal components analysis has given the following eigenvalues: 5.77 (36%), 3.84 (24%), 1.84 (11.5%), 1.05 (6.6%), 0.93 (5.8%), 0.74 (4.6%), 0.39 (2.4%), 0.34 (2.1%), 0.32 (2%), 0.2 (1.3%), 0.15 (0.9%), 0.12 (0.8%), 0.1 (0.6%), 0.09 (0.6%), 0.08 (0.5%), 0.04 (0.3%). We can observe that the variability related to the first 2 components is 60%, with 7 components we get more than 90%, and using 9 components we obtain 95%, we need 14 components to reach 99%.
The first eigenvector is essentially a weighted mean of the 16 formants, the second is a contrast between the formants (f_0,f_1) and (f_2,f_3), the third is a more complicated contrast and the fourth is a contrast between the formants (f_0,f_2) and (f_1,f_3), the first 4 components are linked to a total variability of 78%.
The scatter plots on the eigenvector planes show sparse (independent) configurations (figure 1).
This suggests using the principal components as independent coordinates of the speaker in the speaker space. In this framework, the easiest scalar distance index is the euclidean distance between the unknown speaker's coordinates and the defendant's coordinates or any other index based on the difference of the

corresponding components Z_i (i=1,2,...16) within and between speakers. Then the proper likelihoods can be estimated. As a first step is thus necessary to study and compare the distributions of such 16 +16 differences.

3.2. The estimation of the two likelihoods

Unfortunately, as shown in Paronetto (1995) and confirmed by new exploratory analyses, using the Bhattacharaya's graphic method (1967), in Beciani (1998), it is not possible to suggest a simple parametric family to model the formants variability, either based on gaussians or gaussian mixtures, and, as a consequence, to model the principal components variability. This means that, as already suggested in Calvani, it is necessary to proceed using non parametric density estimation methods, as the kernel.
First of all a sample of within and between speaker differences is needed. Then the kernel density estimation method can be used, as already done in Calvani for singular original formants.
The sample of the between speaker indices has been generated by sampling 400 couples of different speakers and then calculating the 16 differences between the Z_i. The sample of the within speaker differences has been generated by resampling 400 formant vectors from speaker "66" data, transforming them to obtain the principal components, using the matrix obtained by the principal components analysis of the 100 speakers, and then calculating the 16 differences between the Z_i.
Then the likelihoods have been estimated by kernel method, optimizing the smoothing parameter by pseudolikelihood cross-validation applied to each coordinate separately. Some results are shown in figure 2 and indicate some interesting features. In particular, we can observe that the differences of the corresponding principal components of the within (prosecution hypothesis) and between (defence hypothesis) ditributions show opposite behaviours with respect to their variability. Indeed, the between indices have much higher variability for the first components and lower for the last, while the within indices have lower variability for the first components and higher for the last. This suggest using the coordinates separately, without combining them in a scalar index, to obtain the maximum efficiency for the classification purposes. It must be noted that the overlapping of the distributions of the intermediate indices mirrors the "true dimensionality" of the formants space, already analysed by calculating the correlation coefficients. Higher order differences could allow to increase the different behaviour of the within and between distributions and should be studied more in dept.

3.3. The decision problem

Once the likelihoods of appropriate indices, corresponding to the two alternative hypotheses, have been estimated, the decision (classification) problem can be easily dealt with by means of Bayes factors, as also suggested in Aitken (1995). Thus, in the real cases, using the proper indices and likelihoods, Bayes factors can been calculated for various speakers and several values of the distance indices based on the differences of the principal components. Different tables can then be made available and can be used as a valuable basis to evaluate posterior probabilities and classification errors in forensic cases, once the proper within speaker likelihood has been calculated for the actual defendant. The complete tables and results for the actual case (defendant assumed to be speaker "66") will appear in a future paper (Rossi, in preparation). We must stress now that the present approach is the only one that allows to properly calculate the classification errors and the probative value of any evidence, while, again, we must point out that the approach to speaker identification through tests of significance, used with the aim to accept the null hypothesis and not to reject it, as in Ibba and Paoloni (1993) and by the same authors proposed in some actual forensic cases, is a nonsense, it does not allow to calculate the identification error probability, and may more easily produce the conviction of an innocent person.

4. Conclusions and further developments

Advances in technology will continue to bring gradual progress in the area of technical speaker identification in the forensic context, but the advances will highly depend on a more complete model of 'speaker space' becoming available. That is, if we assume a k-dimensional space defined by acoustic and phonetic parameters which can be analysed, we need to study furtherly what are the relevant statistics for separating speakers and, with respect to the reference population, to what extent the region in the space occupied by one speaker overlaps the regions occupied by other speakers. The research will have to take advantage of databases containing samples from larger numbers of speakers. Some data coming from Italian Scientific Police laboratory are presently available to be studied and the results of the statistical analyses are expected in few months. An important aspects is also connected to efficient dissimilarity indices to be used.

An extensive study of the peculiar discriminant features of different indices is going on presently using the two different data sets: the first is the same data set described in Calvani (1996) and the second is the data set coming from the Italian Scientific Police laboratory that is extracted using a different method with respect to the other one, so that also peculiarities related to the measurement method can be studied.

All these analyses are designed with the aim to reduce the probabilities of the classification errors that may result either in the conviction of an innocent or in the release of a guilty person. Thus, the final result of the studies should be a reduction of the expected social costs deriving from these specific forensic cases, connected to the speaker identification problem.

References

Aitken C.G.G., (1995), *Statistics and the evaluation of evidence in forensic science,* Wiley, Chichester, U.K..

Bhattacharaya C.G., (1967), A simple method of resolution of a distribution into gaussian components, *Biometrics*, 23, 115-135.

Beciani R., (1998), Classificazione e decisione nelle scienze forensi: il problema dell'identificazione del parlatore, *Tesi di Laurea in Matematica*, Università degli Studi di Roma 'Tor Vergata'.

Calvani Francesca, (1996), Il problema dell'errore di assegnazione nel riconoscimento del parlatore, *Tesi di Laurea in Matematica*, Università degli Studi di Roma 'Tor Vergata'.

Cohen A. & Froind I., (1989), On text independent speaker identification using a quadratic classifier with optimal features, *Speech Communication* **8,** 35-44.

Doddington G.R., (1985), Speaker Recognition, Identifying People by their Voices, *Proceedings IEEE* **76** no 11.

Fabi F., (1986), Simulazione ed uso integrato dell'informazione statistica Seminario su *Metodi Statistici per la Ricerca Operativa,* Dipartimento di Statistica, Probabilità e Statistiche Applicate Università degli Studi 'La Sapienza', Roma, 14-22.

Furui S., (1989), *Digital speach processing, synthesis and recognition,* Marcel Dekker inc., New York.

Hand D.J., (1982), *Kernel discriminant analysis,* Wiley, New York.

Ibba G. & Paoloni A., (1993), Analisi delle voci: il parlatore ignoto, *Poste e Telecomunicazioni* **1,** 14-25.

Matsui T. & Furui S., (1993), Concatenated Phoneme Models for Text Variable Speaker Recognition, *Proceedings ICASSP* **2,** 391-394.

Matsui T. & Furui S., (1994) Adaptation of Tied Mixture Based Phoneme Models for Text-Prompted Speaker Verification, *Proceedings ICASSP* **1,** 125-128.

Paoloni A., De Sario N. & Saverione B., (1993), La stima delle frequenze formanti nel riconoscimento del parlatore, *Proceedings XXI AIA Conference*, AIA Padova.

Paronetto Barbara, (1995), Problemi di stima di densità per l'identificazione del parlatore: utilizzo del metodo kernel, *Tesi di Laurea in Matematica*, Università degli Studi di Roma 'La Sapienza'.

Robertson B. & Vignaux G.A., (1995*), Interpreting Evidence: evaluating forensic science in the courtroom,* Wiley, NY.

Rossi C., (1996a), La Statistica per il Riconoscimento del Parlatore: aspetti critici e utilizzo problematico, *Atti delle VI giornate di studio del gruppo di fonetica sperimentale*, Fondazione `Ugo Bordoni', Rome, Italy, 79-86.

Rossi C., (1996b), *Il problema di decisione nell'identificazione del parlatore*, Caratterizzazione del parlatore, F. Fedi & A. Paoloni eds., Fondazione `Ugo Bordoni', Rome, Italy, 173-176.

Rossi C. & Schinaia G., (1993), Classification methods and expert systems from a bayesian viewpoint, *Workshop on Probabilistic Expert Systems,* R. Scozzafava ed. CNR-SIS, Rome, 159-180.

Figure 1: *Scatter plots of some couples of principal components*

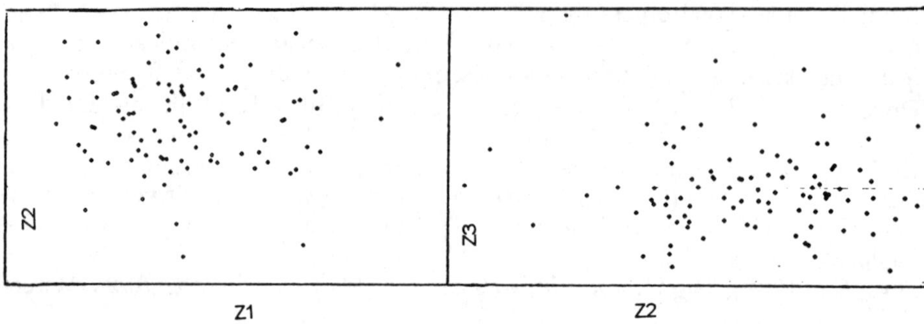

Figure 2: *Estimated densities of within and between differences of some principal components (w=within, b=between)*

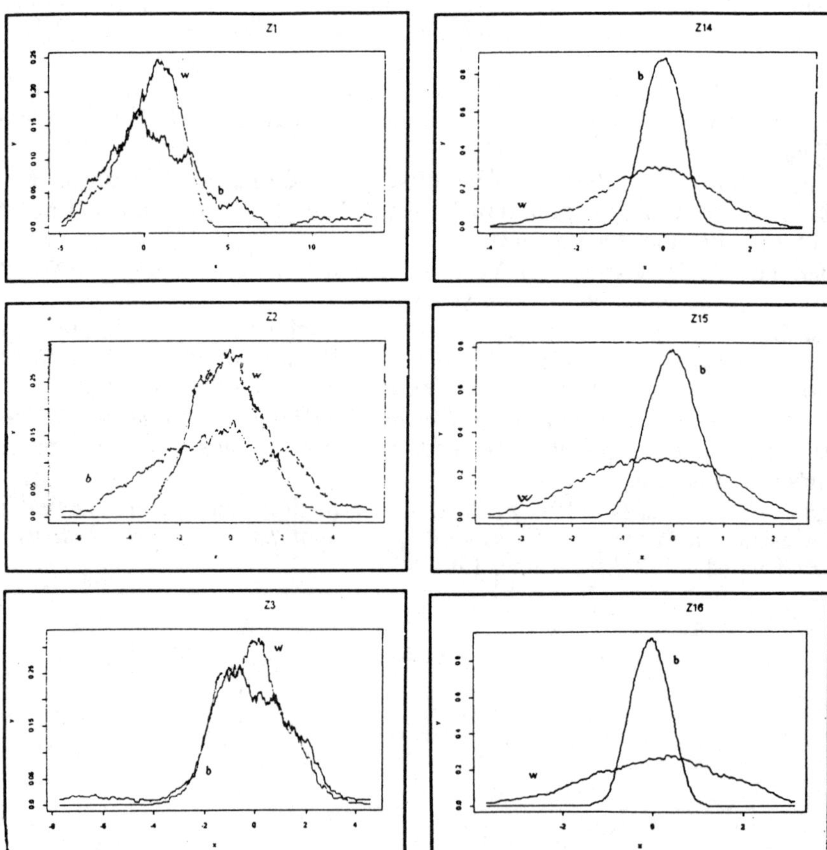

Ordinal Variables in the Segmentation of Advertisement Receivers

Marek Walesiak, Józef Dziechciarz, Andrzej Bak
Wroclaw University of Economics,
Komandorska 118/120, 53-345 Wroclaw, Poland

Abstract: The paper presents segmentation study, which employs methods of classification to single out the segments. The variables measured on the ordinal scale were used as the criteria of market segmentation. Variables used reflected students' attitude towards 20 statements about advertising. Ordinal character of the data required application of specific measure (1) of object's distance. This measure was used in order to evaluate the similarities of objects, which were based on numbers of relations "equal to", "greater than" and "smaller than".

Keywords: Classification, Measurement Scales, Market Segmentation.

1. Market segmentation

Pioneer article of Smith [1956] caused that segmentation appeared as widely researched and applied attitude to marketing. Almost each enterprise observes heterogeneity of its market – that is why the division of consumers into homogeneous sub-groups seems to be promising approach. Due to that fact enterprise may expect increase of profitability of its market operations. Theoretical justification of what had been expected, was published in the work of Frank, Massey and Wind [1972]. They proved that the expected value of obtained profitability level is higher with the diversified approach to the market. This caused quick acceptance of segmentation by both managers and specialists of marketing research.

Marketing strategies are constructed on the results of the undertaken segmentation. Similar tendencies are observed in Poland. Multinationals which operate on Polish market appreciate the role of diversification of marketing strategies according to identified market segments more than the others.

There are two most frequently employed attitudes towards segmentation of practical market researches (see Green [1977]):

- segmentation *a priori*,
- segmentation *post hoc* based on classification procedures.

In a *post hoc* model of segmentation – where classification methods are used differs from the *a priori* model only by the way the base variable was chosen for segmentation. Instead of arbitrary choice of segmentation base (in the *a priori* approach), a variety of consumers characteristics are used in the *post hoc* approach. Additionally, in the next step the classification procedure where neither number of classes nor their descriptive characteristics are known is used. The most frequently used variables for the segmentation study are those identifying needs, attitudes, lifestyle and psychographic characteristics of the consumers. As the result homogeneous consumer groups are isolated and identified. Factor analysis, which aims at reducing the set (number) of diagnostic features often, precedes class isolation process.

The authors of overviewing works by Frank and Green [1968], Saunders [1980], Punj and Stewart [1983], Beane and Ennis [1987] and Arabie and Hubert [1996] pointed out main application fields for classification methods in marketing, mainly in market segmentation.

Empirical segmentation studies, which have been dominated by two mentioned above approaches (*a priori* and *post hoc*), are supplemented by fuzzy and stochastic classification, regression analysis, and conjoint analysis.

Application of classification methods requires formalisation of the "similarity of objects" term. The use of particular construction of similarity measure depends on the scale on which the variables were measured. The choice similarity measures is easy when all the variables describing examined objects are measured on one type of scale. Literature presents plenty of different ways of similarity measurement which can be adopted to variables measured on scale: ratio, interval and (or) ratio, nominal (including binary variables). Wide range of similarity measures has been shown in the works of: Cormack [1971]; Anderberg [1973]; Everitt [1974]; Kaufman and Rousseeuw [1990].

Walesiak [1993], p. 44-45, gives the proposal of the new measure of objects similarity, which can be applied in situation when variables describing those objects are measured only on the ordinal scale. In the construction of the measure under consideration, Kendall's idea of correlation coefficient τ for ordinal variables (see Kendall [1955], p. 19), was used. There is given non-empty set A of objects, each of them is described by m ordinal variables. Because of the fact that on the ordinal scale the only possible empirical operation is counting of events (which means: allotting the number of relations "equal to", "greater than" and "smaller than"), construction of distance measure in the following form is being proposed:

$$d_{ik} = \frac{1}{2} - \frac{\sum_{j=1}^{m} a_{ikj} b_{kij} + \sum_{j=1}^{m} \sum_{\substack{l=1 \\ l \neq i,k}}^{n} a_{ilj} b_{klj}}{2\left[\left(\sum_{j=1}^{m} a_{ikj}^2 + \sum_{j=1}^{m} \sum_{\substack{l=1 \\ l \neq i,k}}^{n} a_{ilj}^2\right)\left(\sum_{j=1}^{m} b_{kij}^2 + \sum_{j=1}^{m} \sum_{\substack{l=1 \\ l \neq i,k}}^{n} b_{klj}^2\right)\right]^{\frac{1}{2}}}, \qquad (1)$$

where:

$$a_{ipj}\left(b_{krj}\right)=\begin{cases}1, & \text{if } x_{ij}>x_{pj}\left(x_{kj}>x_{rj}\right)\\ 0, & \text{if } x_{ij}=x_{pj}\left(x_{kj}=x_{rj}\right),\\ -1, & \text{if } x_{ij}<x_{pj}\left(x_{kj}<x_{rj}\right)\end{cases} \tag{2}$$

$p = k, l;\ r = i, l;\ i, k, l = 1, ..., n$ – number of object,
$j = 1, ..., m$ – number of ordinal variable,
x_{ij} (x_{kj}, x_{lj}) – i–th (k–th, l–th) observation on j–th ordinal variable,

$\sum_{j=1}^{m} a_{ikj}^2 + \sum_{j=1}^{m}\sum_{\substack{l=1\\ l\neq i,k}}^{n} a_{ilj}^2$ – number of relations “greater than” and “smaller than” defined for object i,

$\sum_{j=1}^{m} b_{kij}^2 + \sum_{j=1}^{m}\sum_{\substack{l=1\\ l\neq i,k}}^{n} b_{klj}^2$ – number of relations “greater than” and “smaller than” defined for object k.

A distance measure d_{ik} satisfies conditions: $d_{ik} \geq 0$, $d_{ii} = 0$, $d_{ik} = d_{ki}$ (for all $i, k = 1,..., n$) and does not always satisfy the triangle inequality (based on simulation analysis). Transformation of ordinal data by any strictly increasing function does not change the value of distance (1).

Distance d_{ik} assumes values from the [0; 1] interval. Value 0 indicates that for the compared objects i, k between corresponding observations of ordinal variables, only relations “equal to” take place. Value 1 indicates that for the compared objects i, k between corresponding observations on ordinal variables, relation “greater than” take place or relation “greater than” and relations “equal to” if they are hold for other objects (i.e. objects numbered $l = 1, ..., n$; where $l \neq i, k$).

Discussion about choice of hierarchical classification methods has been shown in the work of Gordon [1987]. In marketing and other behavioural sciences, the most popular hierarchical agglomerative method is complete linkage (see Arabie and Hubert [1994], p. 171). The method of complete linkage has been adopted for the segmentation. This method appears as one of hierarchical agglomerative methods which doesn’t show the tendency to create the classes of chain. Method of complete linkage avoids the disadvantage of other hierarchical methods which show the tendency to incorporate the new objects into the existing groups. New classes are rarely established. As the result, the possibility of presence of dissimilar objects in one class occurs. The distance given by the formula (1) has been adopted as the similarity measure.

2. Segmentation of advertising responders – case study

A group of 244 students of Wroclaw University of Economics were questioned in order to isolate the groups (segments) which similarly reacted to certain statements about advertisements. The experiment has been conducted with the plan to use the classification methods (a *post hoc* approach). The questionnaires were done in the 1996/1997 academic year. Segmentation was based on 20 statements about advertising:

1. Information coming from the advertisement helps me to make a decision.
2. I don't believe such an advertisement which emphasises higher product quality than its competitors.
3. I usually decide to buy advertised product which is accompanied by additional bonuses.
4. Without advertisements no product can be accepted by wide range of consumers.
5. In-store promotion has limited influence on my purchasing decision.
6. Advertising is a good source of information.
7. Advertising tempts people to spend their money foolishly.
8. If advertising were stopped, the customer would be better off.
9. Generally advertisements annoy me.
10. The most effective are TV commercials.
11. On the whole, advertising is believable.
12. I enjoy television commercials.
13. In my opinion advertisements stronger the decision of purchases of elderly people.
14. Information of advertised products saves my precious time.
15. I usually decide to buy newly launched and advertised product.
16. I think, that legal regulations are necessary to limit the advertisement influence on consumers.
17. My brand loyalty could not be changed even by exceptionally good advertisement of competitive brands.
18. Most of the advertisements are based on the credulity of consumers.
19. Before purchasing of newly advertised product I usually seek for my friends' advice.
20. Advertisement is a sign for me that the product should not be purchased.

The questioned students were asked to mark the category on the five-degree ordinal scale, which reflected their attitude towards the given statement: 5 – Strongly agree; 4 – Agree; 3 – Neither agree nor disagree; 2 – Disagree; 1 – Strongly disagree.

For the purpose of this study, the hierarchical agglomerative methods were used to classify students into relatively homogeneous clusters. The complete linkage method proved to be most suitable, because it avoids chain effect. Distance (1) was assumed as the measure of similarity of objects. The optimal number of segments should be between 10 and 5. Due to the fact that the last class has too few objects to consider it a segment, finally seven segments have been established.

In a *post hoc* segmentation, very often the respondents attitudes measured on the Likert scale is used as the segmentation base. In order to establish segments profiles the additional respondents characteristics are analysed (see Green, Tull and Albaum [1988], p. 691-693). The chosen demographic, geographic and economic characteristics are presented in the Table 1.

Table 1: *Characteristics of market segments*

Variables		Segments						
		I	II	III	IV	V	VI	VII
Sex (female / male)		20/25	27/19	22/11	9/14	6/11	29/24	13/12
Average number of household members		3.82	3.67	3.85	3.70	3.94	3.70	3.60
Average time devoted daily for TV watching (in hours)		1.67	1.51	1.87	1.58	1.89	1.15	1.67
Average time devoted daily for radio listening (in hours)		3.20	4.04	4.32	3.02	3.35	3.07	3.58
Median of monthly income being at student's disposal (in Polish Zlotys)		184.4	143.8	129.7	158.3	175.0	182.5	155.0
Place of residence during studies (% of students)	1	51.1	47.8	33.3	30.4	41.2	32.1	36.0
	2	35.6	28.3	48.5	39.1	41.2	43.4	24.0
	3	4.4	6.5	3.0	4.3	5.9	13.2	12.0
	4	8.9	17.4	15.2	26.1	11.8	11.3	28.0
Permanent place of residence (% of students)	A	62.2	39.1	57.6	43.5	35.3	52.8	40.0
	B	6.7	17.4	6.1	4.3	0.0	18.9	24.0
	C	20.0	32.6	27.3	34.8	41.2	20.8	24.0
	D	11.1	10.9	9.1	17.4	23.5	7.5	12.0

Key:
1 – student's hostel
2 – parents' flat
3 – own flat
4 – rented room

A – town with more than 100,000 inhabitants
B – town with 50,000 to 100,000 inhabitants
C – town with less than 50,000 inhabitants
D – village

Segment I treats advertisement as a source of potential benefits. Advertisement for them is an information source of the product and supports their decision of purchasing. It is dominated by women (80%) mainly been accommodated at Students' Hostels or living with parents and coming from the cities of above 100 thousand inhabitants. This cluster has the highest median of monthly income.

Segment II avoids extreme statements when evaluating (they prefer moderate attitudes).

Segment III shows the aversion to risk when buying advertised products they consider friends' advice. They admit that advertisements stronger affect elderly people. TV commercials are considered as the most powerful decision factor. This conclusion may come from radio listening habits. Median of monthly income is the lowest among all segments. Women represent 66.7% of this cluster.

Segment IV shows strong brand loyalty. Men represent 60.9% of this segment. The representatives of this cluster spend relatively less time on radio listening. More than 26% rent a room.

Segment V is the least numerous. It includes the small number of respondents. On one hand they declare reluctance to advertisements but on the other hand they seek for the source of potential benefits in advertisement. Men represent 64.7% of this cluster. This cluster devotes most of its time to TV watching. The highest percentage of them are the flat owners.

Segment VI does not look for the source of future benefits. It shows the reluctance to advertisements. It devotes relatively less time to TV watching and radio listening.

Respondents of **segment VII** do not trust advertisements showing at the same time. Strong brand loyalty the highest percentage wish relation to other segments rents a room.

3. Concluding remarks

The use of segmentation variables measured on ordinal scale is relatively rare in the literature. The specific analytical tools are needed for such information. Important aspect of this study was formulation, in the methodological sense, certain segmentation procedure which implements ordinal variables as a consumer description criteria.

In the empirical study seven segments have been established. The profiles of even segments were analysed. Moreover, this segmentation study may be regarded also as a contribution to evaluation of students' reactions as peculiar social group of advertisement receivers.

Additional result of this study is computer program, which allows computing distances between objects (see Appendix).

Appendix

The computer code in the C++ language computing the value of measure (1) of the distance considered is available at Wroclaw University of Economics in the Dept of Econometrics and Computer Science (e-mail: abak@keii.ae.jgora.pl).

Computational algorithm of distance (1) presented below, omits details of enter and exit operations. The complete computer programme uses files in DBF format, which serve both as the data source and the result files. The Borland® C/C++ v. 4.0 compiler was used for the programming and the CodeBase++™ v. 5.1 library procedures served as a database programming language. As the result of its application the distance matrix is generated. This matrix may be used in the hierarchical agglomerative methods of the classification for division of set of objects into classes. This matrix can be used also for further computations in the *SPSS for Windows* package.

```
procedure dist_mw(X, Y, m, n)
{Computes distance matrix between objects according to the measure (1). X – data matrix (rows
  – objects, columns – variables measured on the ordinal scale); Y – symmetric distance matrix;
  m – number of variables; n – number of objects }
var i, j, k, l, a_1, a_2, b_1, b_2: integer; s_1, s_2, s_3, s_4, s_5, s_6, d: real;
begin
 for i ← 1 to n do begin
  for k ← i+1 to n do begin
   s_1 ← s_3 ← s_4 ← 0;
   for j ← 1 to m do begin
    if X[i,j] > X[k,j] then a_1 ← 1 else if X[i,j] < X[k,j] then a_1 ← −1 else a_1 ← 0;
    if X[k,j] > X[i,j] then b_1 ← 1 else if X[k,j] < X[i,j] then b_1 ← −1 else b_1 ← 0;
    s_1 ← s_1 + a_1 * b_1; s_3 ← s_3 + a_1^2; s_4 ← s_4 + b_1^2;
   end
   s_2 ← s_5 ← s_6 ← 0;
   for l ← 1 to n do begin
    if l ≠ k ∧ l ≠ i then begin
     for j ← 1 to m do begin
      if X[i,j] > X[l,j] then a_2 ← 1 else if X[i,j] < X[l,j] then a_2 ← −1 else a_2 ← 0;
      if X[k,j] > X[l,j] then b_2 ← 1 else if X[k,j] < X[l,j] then b_2 ← −1 else b_2 ← 0;
      s_2 ← s_2 + a_2 * b_2; s_5 ← s_5 + a_2^2; s_6 ← s_6 + b_2^2;
     end
    end
   end
   d ← 0.5 − (s_1 + s_2)/(2 * sqrt((s_3 + s_5)(s_4 + s_6))); Y[k,i] ← d; Y[i,k] ← d;
  end
 end
end
```

References

Anderberg, M. R. (1973). *Cluster Analysis for Applications*, Academic Press, New York, San Francisco, London.

Arabie, P. & Hubert, L. (1994). Cluster Analysis in Marketing Research, in *Advanced Methods of Marketing Research*, Bagozzi R.P. (Ed.), Blackwell, Oxford, 160-189.

Arabie, P. & Hubert, L. (1996). Advances in Cluster Analysis Relevant to Marketing Research, in *From Data to Knowledge*, Gaul, W. & Pfeifer, D. (Eds.), Springer, 3-19.

Beane, T. P. & Ennis, D. M. (1987). Market Segmentation. A Review, *European Journal of Marketing*, 21, (5), 20–42.
Cormack, R. M. (1971). A Review of Classification (with Discussion), *Journal of the Royal Statistical Society*, Ser. A, (3), 321–367.
Everitt, B. S. (1974). *Cluster Analysis*, Heinemann, London.
Frank, R. E. & Green, P. E. (1968). Numerical Taxonomy in Marketing Analysis: a Review Article, *Journal of Marketing Research*, 5, (February), 83–98.
Frank, R. E. Massy, W. F. & Wind, Y. (1972). *Market Segmentation,* Prentice–Hall, Englewood Cliffs.
Gordon, A.D. (1987). A Review of Hierarchical Classification, *Journal of the Royal Statistical Society*, Ser. A, 119-137.
Green, P. E. (1977). A New Approach to Market Segmentation, *Business Horizon*, 20, 61–73.
Green, P. E. & Krieger, A. M. (1995). Alternative Approaches to Cluster-based Market Segmentation, *Journal of the Market Research Society*, 37, (3), 61-73.
Green, P.E., Tull, D.S. & Albaum, G. (1988). *Research for Marketing Decisions*, Prentice–Hall, Englewood Cliffs.
Kaufman, L. & Rousseeuw, P. J. (1990). *Finding Groups in Data: an Introduction to Cluster Analysis*, Wiley, New York.
Kendall, M. G. (1955). *Rank Correlation Methods*, Griffin, London.
Milligan, G. W. & Cooper, M. C. (1985). An Examination of Procedures for Determining the Number of Clusters in a Data Set, *Psychometrika*, 50, (2), 159–179.
Punj, G. & Stewart, D. W. (1983). Cluster Analysis in Marketing Research: Review and Suggestions for Application, *Journal of Marketing Research*, 20, (May), 134–148.
Saunders, J. A. (1980). Cluster Analysis for Market Segmentation, *European Journal of Marketing*, 14, (7), 422–435.
Smith, W. (1956). Product Differentiation and Market Segmentation as Alternative Marketing Strategies, *Journal of Marketing*, 21, 3–8.
SPSS for Windows (1997). *Version 7.51*, SPSS Inc., Chicago.
Walesiak, M. (1993). *Statystyczna analiza wielowymiarowa w badaniach marketingowych [Multivariate Statistical Analysis in Marketing Research]*, Wroclaw University of Economics, Research Papers No. 654.
Walesiak, M. (1996). *Metody analizy danych marketingowych [Methods of Marketing Data Analysis]*, PWN, Warszawa.
Wind, Y. (1978). Issues and Advances in Segmentation Research, *Journal of Marketing*, 15, (August), 317–337.

Some Experimental Trial of Electronic Surveys on the Internet Environments

Osamu Yoshimura 1), Noboru Ohsumi 1),
Yasuyuki Kawaura 2), Naomasa Maruyama 1),
Shin-ichi Yanagimoto 3), Yoriko Anraku 3),
and Mariko Murata 4)

1) The Institute of Statistical Mathematics
4-6-7 Minami-Azabu, Minato-ku, Tokyo 106-8569, Japan
2) Yokohama City University
22-2 Seto, Kanazawa-ku, Yokohama City, Kanagawa 236-0027, Japan
3) Recruit Research Co.,Ltd.
Kachidoki, 1-13-1 Kachidoki, Chuo-ku, Tokyo 104-0054, Japan
4) Sinfonica
6-3-9 Minami-Aoyama, Minato-ku, Tokyo 107-0062, Japan

Abstract: A series of experimental surveys was conducted to compare conventional and electronic surveys from a methodological viewpoint. In this report, some distinguish features of electronic surveys were shown, and the problems, possibility and applicability of the new survey method were discussed.

Keywords: sampling survey, electronic survey, Web-based survey, repeated survey, free-answered data, privacy problems on Internet

1. Introduction

'Electronic surveys' refer to several kinds of surveys, including Internet, online, Web-based and electronic mail (e-mail) surveys, and a lot of these new surveys are being conducted in business and research areas, such as marketing research, consumer behavior, etc. According to the order of historical arrival, and depending on how the questionnaires are represented and how the respondents' answers are collected, these survey methods can be classified into categories such as e-mail, specific systems designed for limited use and Web-based surveys. Especially, in the case of Web-based surveys, the questionnaire sheet is set up as a Web page. The respondents access this electronic page and answer the questions. The method of experimental surveys utilized in this paper is the Web-based survey.

Electronic surveys differ greatly from conventional surveys, and computers and network environments play an important role in designing the survey methods and data acquisition systems and in statistically analyzing the results obtained. However, technological capability in the fields of computer and Internet environments exceeds progress in practical, concrete survey methods, and as a result, despite careful study of such fundamental aspects as representativity, reproducibility and reliability, it seems that further systematic research concerning the actual survey methods are still needed. In this study, we conducted a series of experimental surveys designed to be able to compare conventional and electronic surveys from a methodological viewpoint. The main purpose of the report is to clarify the relationship between the conventional approaches and the more recent survey methods in such aspects as the design of a sampling survey, including sampling methods, how to construct the questionnaire sheet on Web pages, and actual survey procedures.

2. Outline of Experimental Surveys

By receiving the cooperation of the staff of Recruit Research Co., we first established a panel of 2,036 registered samples. Table 1 shows the panel's demographics and the percentages.

Table 1 Panels' Demographics and Their Component Ratio (N=2,036)

Sex	Males	(Married	Unmarried)	Females	(Married	Unmarried)
	60.3	(49.1	50.9)	39.7	(38.0	62.0)
Age Group	Teens	Twenties	Thirties	Forties	Fifties	
	2.4	53.3	37.1	6.0	1.2	
Vocation	Corporate Workers	Civil Service and Teachers	Self-employed	Freelancers and Part-timers	Students	Unemployed or Others
	65.5	4.1	3.4	5.4	13.6	8.0

In Table 1, we see that nearly 80% of the registered samples are office workers and students, and also that their age brackets are concentrated in the twenties and thirties. Table 1 also indicates a tendency often observed in this kind of survey, that is, that there are fewer females in the sample than males. However, in our survey, the ratio of females was higher than in other surveys.

On the panel, twelve Web-based surveys were conducted in all. In each survey, the opinions were collected by following procedures: (1) The panel members were informed by e-mail of the URL of the questionnaire pages and their cooperation in the surveys was requested. (2) Each member of the panel accessed the questionnaire pages and answered the questions shown on the display screen. (3) The tabulated results were obtained by gathering the respondents' answers and merging them together with their demographics. (4) Some incentives were given to the respondents, decided by drawing lots. It must be noted that because of some troubles or other reasons, the number of actually e-mailed members of the panel were less than 2,036 in all surveys. After this, we will call the e-mailed member of the panel 'scheduled sample'.

The themes of the surveys were various, such as 'Drinking and smoking experience,' 'Use of Internet's service sites,' 'Food,' 'Consumer behavior,' etc. We have provided the 3rd, 6th, 7th, 10th and 12th surveys with our original questions to study the construction of the questionnaires. We expressly describe more details of the 6th, the 10th and the 12th surveys below.

Our 6th and 12th surveys are designed to clarify the characteristics of our new method by asking questions with the definite results in conventional sampling surveys available and by asking the same questions repeatedly. The contents of the questions asked were 'Attitudes towards life,' 'Japanese people's nationality,' 'Japanese people's images of their top leaders,' etc., and Free-answer questions about the contents that was asked commonly in almost all the surveys. These surveys were conducted in the period July 13 to July 20, 1997, and November 8 to November 16, 1997, respectively. Reminders were sent once to all members of the scheduled samples during the 12th survey. The number of scheduled samples was 2,029 for the 6th survey and 2,009 for the 12th survey. We received 1,129 complete responses for the 6th survey and 1,157 for the 12th survey.

The 10th survey dealt with privacy problems on Internet. Examples of the questions were: 'Of what matters would you hesitate to be informed about another person?', 'What kind of information do you prefer not to give to others?', and so on. The methods used through all the surveys were more or less the same except for the kinds of incentive. The survey period was from September 18 to September 24, 1997, and the number of complete responses was 954.

3. Main Findings

3.1 Some distinguishing features through the total series of surveys

Fluctuation of completion rate The average of completion rates of all the 12 surveys (54.1%) was higher than we had at first estimated, and there was a general descending tendency in completion rate as the surveys go on. Moreover, there is a little fluctuation in the completion rates through all the surveys.

In the 12th survey, we asked which surveys 'you are interested in,' 'you feel operationally troublesome' and 'you think the contents are difficult to answer.'

Table 2 shows the percentages of the choice and the completion rate of each survey. It is needless to say that the cause of the fluctuation of completion rate cannot be easily determined, but the result shown in Table 2 suggest the possibility that the theme of each survey, complexity of layouts and difficulty of questions were a major cause of the fluctuations, and the opinions stated for the open-ended questions also support this.

Table 2 Completion Rate and Impressions of Each Questionnaire

Survey	1st	2nd	3rd	4th	5th	6th	7th	8th	9th	10th	11th	12th
Completion Rate (%)	70.5	57.2	57.0	59.0	54.1	55.6	50.9	51.7	48.3	47.4	40.2	57.6
Interested in the theme (%)	37.9	57.9	35.2	46.1	31.2	34.7	15.4	18.2	28.8	44.6	14.5	–
Troublesome to answer (%)	18.6	15.8	6.0	6.5	4.2	3.1	12.4	29.1	31.4	5.7	27.9	–
Difficult to answer (%)	9.8	8.1	5.9	21.2	5.8	14.2	42.2	12.2	12.2	21.5	19.4	–

Table 3 The distribution of Participation Frequency

Frequency	0	1	2	3	4	5	6	7	8	9	10	11	12	Total number
Number of participants	288	138	140	100	89	115	120	144	150	206	215	246	85	2,036

Participation frequency Table 3 shows the distribution of access frequency to the Web pages. It is interesting that as many as 288 registered members (14.1%) never answered any questions. This means that quite a few aimed for registration only or only looked at the Web home page without making any response. On the other hand, 85 monitors (about 4.2%) responded to all the surveys. The average was about 6.0 times (including no participated monitors), that means that they made a response to half of the surveys. It should be noted that not a few of the registered members (about 37%) accessed to the Web pages more than 9 times.

The answer rates to the open-ended questions In all of the surveys we set up free-answer questions as the last item. Those who filled in this section were 667 out of 1,129 (59.1%) as for the 6th survey, and 489 out of 954 (about 51.3%) as for the 10th survey. To the questions of the 6th survey: “What is the most important?” and “What is the second most important?”, 1,119 and 1,061 out of 1,129 participants wrote the answers (99.1% and 94.0%, respectively). This seems unusually a lot. We may well say that those interested in these surveys are active enough not to hesitate to fill in.

3.2 Comparison with conventional method and response uncertainty

Comparing with the definite results by conventional method, it was found that our respondents were, in some points, very conservative, but on the other hand they seemed very sensitive about their life. It is important that, in electronic survey, another reliable survey results are needed as comparative indices in interpreting the result.
We used almost the same questions for the 6th and the 12th survey. In the results, generally, we saw a similar tendency as that of the conventional surveys, that is, some inconsistency were observed in between the 6th and the 12th marginal response distributions. On the other hand, there were some questions that showed interesting distributional differences in the cross tabulation of the repetition, despite few differences between the marginal distributions. For example, in the question 'Consensus versus own principle' over 20% of the respondents oppositely changed their opinion between the two surveys (Table 4).

Table 4 The Uncertainty of Response

	Principle	Consensus	N.A.	Total
Principle	29.9	**11.5**	0.0	41.4
Consensus	**12.2**	45.9	0.1	58.2
N.A.	0.2	0.1	0.0	0.3
Total	42.3	57.5	0.1	100.0

Note: The stab is the 6th and the banner is the 12th.
'Consensus versus own principle'
"Which of the two people described on this card would you like best? Just read off the letter.
1: A person who stresses his/her own principles rather than achieving a consensus among other group members.
2: A person who stresses the importance of achieving a consensus among other group member rather than maintaining his/her own principle.

3.3 Privacy on Internet

In the 10th survey, we asked questions about privacy problems on Internet and the distribution of information on the Web. Mainly, the followings are the characteristics we grasped: Many respondents think their income, address, and phone number as their skeleton privacy. They don't think much of the secrecy of sex, age, vocation, e-mail address. The overwhelmingly many reasons for not answering are: 'It's not clear for what my answer will be used.' and 'It's not worth while answering.' When the respondents were asked to input their own information, many of them made no answers. Those who input inaccurate answers intentionally were few. Most of them were active in answering, but some of the options seemed to be difficult for them to answer.

4. Discussion

One of the conclusions is numerical values to compare are needed in interpreting the results of tabulation in electronic surveys. Moreover, it is important to design a panel study for the same respondents and make it possible to repeat the surveys with similar or the same questions in them. This is all natural, but indeed, we can find no electronic survey which fully take account of the interpretability and consistency of responses.

On the other hand, we were also able to get different kinds of information. For example, there were so many highly experienced web users in our surveys. This raises the possibility that the same persons may be participating in different surveys. We may conclude that the population size of the respondents on Internet is limited and the chance that the same persons are frequently captured is very high. Different surveys may being conducted repeatedly on a group with a fixed number of members. There were also a considerable number of duplicated answers and registrations observed. To prevent such applications, we could develop a system getting rid of the answers by the same persons , but such a system might be unable to give opportunities of correcting mistakes. Although people think electronic surveys have a great advantage of real time processing, in fact, there are too many things for us to do: identification and acknowledgments of the respondents' answers, maintenance of data archives, checking of the record contents, and immediate processing. Moreover, although all members of the registered panel had been informed of our purposes at the beginning of the surveys, nevertheless many respondents wrote in the spaces for free answers such questions as 'Why do you conduct such a survey?', and so on. We think it is indispensable to ask the respondents' opinions concerning the release of such information in public, and also give the results back to them as soon as possible. Such actions would help raise awareness of issues related to the administration of surveys.

References

Kehoe, C. M., & Pitkow, J. E. (1996). Surveying the Territory: GVU's Five WWW User Surveys, *The World Wide WebJournal*, Vol. 1, no. 3. [Online]: http://www.cc.gatech.edu/gvu/user_surveys/papers/w3j.html

Smith,C.B. (1997). Casting the Net: Surveying an Internet Population, *Journal of Computer-Mediated Communication*, Vol. 3, no. 1. [Online]: http://www.usc.edu/dept/annenberg/vol3/issue1/smith.html

GUV's WWW User Survey. http://www.cc.gatech.edu/gvu/user_surveys/

CONTRIBUTORS INDEX

—S—

—T—

—U—

—V—

—W—

—Y—

—Z—

KEY WORDS INDEX

—T—

—U—

—V—

—W—

CONFERENCE COMMITTEE

Conference President

Alfredo Rizzi

Scientific Program Committee

Maurizio Vichi *(Co-chairperson)*
Francesco De Antoni *(Co-chairperson)*

International Scientific Program Committee

Phipps Arabie, *U.S.A.*
Helena Bacelar-Nicolau, *Portugal*
Jean-Pierre Barthélemy, *France*
Vladimir Batagelj, *Slovenia*
Hans-Hermann Bock, *Germany*
Peter G. Bryant, *U.S.A.*
J. Douglas Carroll, *U.S.A.*
Renato Coppi, *Italy*
Wolfang Gaul, *Germany*
Allan D. Gordon, *United Kingdom*
John Gower, *United Kingdom*
Chikio Hayashi, *Japan*
Willem J. Heiser, *The Netherlands*
Krysztof Jajuga, *Poland*
Natale Carlo Lauro, *Italy*
Ludovic Lebart, *France*
Jae Chang Lee, *Korea*
Jacqueline J. .Meulman,*The Netherlands*
Noboru Ohsumi, *Japan*
Jean Paul Rasson, *Belgium*
Alfredo Rizzi, *Italy*

Local Scientific Program Committee

Enrica Aureli
Francesco Battaglia
Antonio Bellacicco
Sergio Bolasco
Giuseppe Bove
Luigi Biggeri
Giuseppe Burgio
Carlo Cecchi
Mauro Coli
Annamaria Crescimanni
Luigi D'ambra
Enzo D'Arcangelo
Francesco De Antoni
Agostino Di Ciaccio
Luigi Fabbris
Mary Fraire
Gianfranco Galmacci
Giuseppe Gesano
Giovanni M. Giorgi
Giovanni Girone
Letizia La Tona
Giovanni Latorre
Giuseppe Leti
Martino Lo Cascio
Gianfranco Lovison
Giuseppe Lunetta
Gianni Marliani
Antonino Mineo
Paola Monari
Mario Montinaro
Antonio Mussino
Maria Gabriella Ottaviani
Fortunato Pesarin
Mario Romanazzi
Gabriella Salinetti
Roberta Siciliano
Bruno Simeone
Maurizio Vichi
Sergio Zani
Alberto Zuliani

List of Reviewers

Jaromir Antoch
Phipps Arabie
Enrica Aureli
Daniel Baier
Simona Balbi
Vladimir Batagelj
Francesco Battaglia
Maddalena Barbieri
Mónica Bécue-Bertaut
Antonio Bellacicco
Halima Bensmail
Peter M. Bentler
Hans Hermann Bock
Sergio Bolasco
Simone Borra
Rasmus Bro
Peter Bryant
Giuseppe Burgio
J. Douglas Carroll
Andreas Christmann
Renato Coppi
Annamaria Crescimanni
Marcel Croon
Luigi D'Ambra
Francesco De Antoni
Reinhold Decker
Luc Devroye
Edwin Diday
Yadolah Dodge
Jean François Durand
Phil Thomas Eckes
Hubert Feger
Bernard Fichet
Peter Filzmoser
Mary Fraire
Gianfranco Galmacci
Tommaso Gastaldi
Giovanni Maria Giorgi
Erhard Godehardt
Allan D. Gordon
John Gower
Patrick Groenen
László Gÿorfi
Chikio Hayashi
Willem J. Heiser
Herbert Hoijtink
Lawrence Hubert
Mushtaq Hussain
Peter Ihm
Krysztof Jajuga
Yutaka Kano
Henk A.L. Kiers
Rudiger Klar
Jan Klascka
Christoph Klauer
Reinhard Kohler
Joachim Krauth
Peter M. Kroonenberg
Rudolf Kruse
W.J. Krzanowski
Letizia La Tona
Jean Luc Lambert
François Joseph Lapointe
Natale Carlo Lauro
Ludovic Lebart
Bruno Leclerc
J.C. Lee
Brunero Liseo
Rosaria Lombardo
Gianfranco Lovison
Gabor Lugosi
Gianni Marliani
Rudolf Mathar
Frank Mc Morris
Jacqueline J. Meulman
Antonino Mineo
Sadaaki Miyamoto
Francesco Mola
Paola Monari
Mario Montinaro
Fionn Murtagh
Noboru Ohsumi
M. Gabriella Ottaviani
Roger W. Payne
Emmanuel Périnel
Fortunato Pesarin
Ludovico Piccinato
Greet Pison
Martin Rajman
Jean Paul Rasson
Gunter Ritter
Alfredo Rizzi
Robero Rocci
Mario Romanazzi
Gavin J.S. Ross
Peter J.Rousseeuw
Mika Sato
Martin Schader
Roberta Siciliano
Bruno Simeone
Franz Streit
Jos M.F. ten Berge
M. Terrenoire
Gerarda Tessitore
S. Thiria
Marieke Timmermann
Alfred Ultsch
L. Andries van der Ark
Donatella Vicari
Maurizio Vichi
Cecilia Vitiello
Arndt von Haeseler
Michel Wedel
Sergio Zani